Phytoplankton and Equilibrium Concept: The Ecology of Steady-State Assemblages

Developments in Hydrobiology 172

Series editor
K. Martens

Phytoplankton and Equilibrium Concept: The Ecology of Steady-State Assemblages

Proceedings of the 13th Workshop of the International Association of Phytoplankton Taxonomy and Ecology (IAP), held in Castelbuono, Italy, 1–8 September 2002

Edited by

L. Naselli-Flores[1], J. Padisák[2] & M.T. Dokulil[3]

[1] *University of Palermo, Italy*
[2] *University of Veszprém, Hungary*
[3] *Institute for Limnology, Austria*

Reprinted from Hydrobiologia, volume 502 (2003)

SPRINGER-SCIENCE+BUSINESS MEDIA, B.V.

Library of Congress Cataloging-in-Publication Data

A C.I.P. Catalogue record for this book is available from the Library of Congress.

ISBN 978-90-481-6433-2 ISBN 978-94-017-2666-5 (eBook)
DOI 10.1007/978-94-017-2666-5

Printed on acid-free paper

Originally published by Kluwer Academic Publishers in 2003

TABLE OF CONTENTS

Hydrobiologia **502:** ix–x, 2003.
L. Naselli-Flores, J. Padisák, M. T. Dokulil (eds),
Phytoplankton and Equilibrium Concept: The Ecology of Steady-State Assemblages.

List of participants

Meriç Albay – Istanbul University - Fisheries Faculty, Department of Freswater Biology, Ordu Cad. No: 200, 34470 Laleli, Istanbul, TURKEY

Alla Alster – Kinneret Limnological Laboratory, IOLR, P.O.Box 447, 14950 Migdal, ISRAEL

Rossella Barone – Dipartimento di Scienze Botaniche, Via Archirafi, 38, I-90123 Palermo, ITALY

Silvia Baldon – Università di Padova, Dipartimento di Biologia, Via U. Bassi 58/B, I-35131 Padova, ITALY

Ingrid Chorus – Federal Environmental Agency, P.O. Box 33002, 14191 Berlin, GERMANY

Jean-Pierre Descy – URBO, FUNDP, 61, rue de Bruxelles, B-5000 Namur, BELGIUM

Martin T. Dokulil – Institute for Limnology, Mondseestrasse 9, A-5310 Mondsee, AUSTRIA

Giovanna Flaim – Istituto Agrario, I-38028 S. Michele all'Adige, (TN), ITALY

Véronique Gosselain – URBO, FUNDP, 61, rue de Bruxelles, B-5000 Namur, BELGIUM

Istvan Grigorszky – Botanical Department, University of Debrecen, H-4010 Debrecen, HUNGARY

Vera Huszar – Universidade Federal do Rio de Janeiro, Lab. Ficologia, Depto. Botânica, Museu Nacional-UFRJ, Quinta da Boa Vista, 20940-040 Rio de Janeiro, BRAZIL

Irina Izaguirre – University of Buenos Aires, Depto. Ciencias Biológicas, Facultad de Ciencias Exactas y Naturales, Ciudad Universitaria Pab. II C, 1428 EHA Buenos Aires, ARGENTINA

Pavel Javornický – Bìloveská 1674, 547 01 Náchod, CZECH REPUBLIC

Hedy J. Kling – Algal Taxonomy and Ecology Inc., 31 Laval Dr., R3T 2X8 Winnipeg, CANADA

Jiří Komárek – Faculty of Biological Sciences, University of South Boemia, Branisovska 31, CZ-370 05 České Budějovice, CZECH REPUBLIC

Jarka Komárková – Hydrobiological Institut, Czech Academy of Sciences, Na Sádkách 7, CZ-370 05 České Budějovice, CZECH REPUBLIC

Jørgen Kristiansen – Bot.Inst., Dept.of Phycology, University of Copenhagen, Øster Farimagsgade 2D, 1353 Copenhagen K, DENMARK

Maria Leitão – Bi-Eau, 15, rue Lainé-Laroche, F-49000 Angers – FRANCE

Liisa Lepistö – Finnish Environment Institute, Research Programme for Integrated River Basin Management, P.O. Box 140, FIN-00251 Helsinki, FINLAND

Mariângela Menezes – Museu Nacional/UFRJ, Quinta da Boa Vista s/n, São Cristóvão, 20940-040 Rio de Janeiro, BRAZIL

Ute Mischke – Institute of Freshwater Ecology and Inland Fisheries, Dept. of Shallow Lakes and Lowland Rivers, Müggelseedamm 310, 12587 Berlin, GERMANY

Giuseppe Morabito – CNR Istituto per lo Studio degli Ecosistemi (ex Istituto Italiano di Idrobiologia), L.go Tonolli 50-52, 28922 Verbania-Pallanza, ITALY

Sara M. Morata – Departamento de Ecología y Microbiología, Facultad de Ciencias Biológicas, Universidad de Valencia, Dr. Moliner, s/n, 46100; Burjassot, Valencia, SPAIN

Maria Moustaka-Gouni – Aristotle University of Thessaloniki, School of Biology, Department of Botany, GR-54006 Thessaloniki, GREECE

Luigi Naselli-Flores – Dipartimento di Scienze Botaniche, Via Archirafi, 38, I- 90123 Palermo, ITALY

Verena Niesel – Federal Environmental Agency, P.O. Box 33002, Corrensplatz 1, 14191 Berlin, GERMANY

Brigitte Nixdorf – Brandenburg University of Technology (BTUC), Chair of Water Conservation, Research Station Bad Saarow, Seestraße 45, D-15526 Bad Saarow, GERMANY

Gianfranco Novarino – Department of Zoology, The Natural History Museum, Cromwell Road, London SW7 5BD, UNITED KINGDOM

Inés O'Farrell – University of Buenos Aires, Lab.Limnología, Dep. de Ecología, Genética y Evolución, Facultad de Ciencias Exactas y Naturales, Pab. II, Ciudad Universitaria, C 1428 EHA Buenos Aires, ARGENTINA

Alessandro Oggioni – CNR Istituto per lo Studio degli Ecosistemi (ex Istituto Italiano di Idrobiologia), L.go Tonolli 50-52, 28922 Verbania-Pallanza, ITALY

Judit Padisák – University of Veszprém, Department of Limnology, 10. Egyetem ut, H-8200 Veszprém, HUNGARY

Pierisa Panzani – CNR Istituto per lo Studio degli Ecosistemi (ex Istituto Italiano di Idrobiologia), L.go Tonolli 50-52, 28922 Verbania-Pallanza, ITALY

Sabrina Pozzi – Agenzia Provinciale per la Protezione dell'Ambiente, U.O. Tutela dell'Acqua, P.zza A. Vittoria, 5, I-38100 Trento, ITALY

Carmen Rojo – Instituto "Cavanilles" de Biodiversidad y Biología Evolutiva, University of Valencia, Ap. Oficial 22085, E-46071 Valencia, SPAIN

Eugen Rott – Institute of Botany, University of Innsbruck, Sternwartestrasse 15, A-6020 Innsbruck, AUSTRIA

Nico Salmaso – Dipartimento di Biologia dell'Università, Via U. Bassi 58/B, 35131 Padova, ITALY

Maya Stoyneva – Sofia University "St Kliment Ohridski", Faculty of Biology, Department of Botany, 8, blvd. Dr. Tzankov, 1166 Sofia, BULGARIA

Massimiliano Tardio – Museo Tridentino di Scienze Naturali, Limnology & Phycology Section, Via Calepina 14, I-38100 Trento, ITALY

Maria Temponeras – Bio/consult as, Johs. Ewalds Vej 42/44, 8230 Aabyhøj, DENMARK

Katrin Teubner – Institute of Limnology, Austrian Academy of Sciences, Mondseestrasse 9, A-5310 Mondsee, AUSTRIA

Monica Tolotti – University of Innsbruck, Inst. of Zoology and Limnology, Technikerstrasse 25, A-6020 Innsbruck, AUSTRIA

Renata Trevisan – Università di Padova, Dipartimento di Biologia, Via U. Bassi 58/B, I-35131 Padova, ITALY

Eva Willén – Swedish University of Agricultural Sciences, Department of Environmental Assessment, P.O. Box 7050, SE-750 07 Uppsala, SWEDEN

Santina Zancan – Università di Padova, Dipartimento di Biologia, Via U. Bassi 58/B, I-35131 Padova, ITALY

Tamar Zohary – Kinneret Limnological Laboratory, IOLR, P.O.Box 447, 14950 Migdal, ISRAEL

Hydrobiologia **502:** xi, 2003.
L. Naselli-Flores, J. Padisák, M. T. Dokulil (eds),
Phytoplankton and Equilibrium Concept: The Ecology of Steady-State Assemblages.

Foreword

This special volume of *Hydrobiologia* collects the papers read at the Thirteenth Workshop of the International Association of Phytoplankton Taxonomy and Ecology (IAP), which was held at Hotel Milocca, Castelbuono, near Palermo (Sicily, Italy), from 1–8 September, 2002. The main goals of this Association, as well as its history and a list of the proceedings published up till now, is summarized in the paper by J. Kristiansen opening this volume. Trough the discussion of common problems of phytoplankton ecology, methodology, taxonomy, and identification of the involved organisms, the IAP meetings have become a reference point for many researchers in this field of Science, and especially for its younger practitioners.

The submissions complied well both with the ecological theme "Phytoplankton and equilibrium concept: The ecology of steady state assemblages" and with the taxonomical one "Cryptophytes and filamentous cyanoprokaryotes". In particular, the taxonomical part of the volume contains some useful taxonomical keys (Komárek, 2003; Novarino, 2003; Javornický, 2003), as well as suggestions and recommendations to correctly identify the organisms belonging to these selected groups in temperate (Lepistö & Holopainen, 2003) and in tropical regions (Menezes & Novarino, 2003). New species are also described (Kling & Watson, 2003; Tardio et al., 2003).

With regards to the agreed ecological topic, contributors were invited to explore whether and under what conditions phytoplankton can reach equilibrium in natural environments. The conditions to identify equilibrium were set according to Sommer et al. (1993, *Hydrobiologia* 249: 1–7): a situation where a maximum of three species dominate the assemblage for, at least, three weeks without considerable change in total biomass. In addition, participants were requested to adopt and test the functional classification of freshwater phytoplankton originally proposed by Reynolds (1980, *Holarctic Ecology* 3: 141–159) and updated by Reynolds et al. (2002, *Journal of Plankton Research* 24: 417–428). Participants enthusiastically agreed and several suggestions and improvements are included in most of the papers forming this volume.

The majority of the presented papers concerns the phytoplankton ecology, composition and dynamics of particular lakes or series of lakes, from temperate to tropical and even Antarctic lakes. The wide spectrum of water bodies examined gave us the opportunity to clarify and summarize in an editorial (Naselli-Flores et al., 2003) the emerging environmental features dealing with the existence and occurrence of "equilibrial" species and/or assemblages, to assess the progress that the workshop may have given to our understanding of phytoplankton steady-state assemblages and to summarize the occurrence of functional groups of freshwater phytoplankton among the "equilibrial" assemblages.

We would like to take this opportunity to thank the Director of the "Dipartimento di Scienze Botaniche" of the University of Palermo, Prof. Salvatore Trapani, the Dean of the Faculty of Sciences, Prof. Francesco M. Raimondo, and the Rector of the University of Palermo, Prof. Giuseppe Silvestri, for the financial support they insured to the meeting. Hedy J. Kling, one of the organisers of the 12th IAP meeting, granted the residual money of that workshop to support this 13th. We are also indebted to the Mayor of Castelbuono, Mr. Mario Cicero, who warmly welcomed the participants and supported their strength by offering the social dinner. Special thanks are due to the Director, Mr. Renato Dispenza, and Staff of the Hotel Milocca for hosting our visit and providing such excellent food and accommodation. An earthquake, which scared us a bit but increased the taste of adventure that the IAP meetings always carry on, was kindly offered by Mother Nature, early in the morning (3 am) of September 6.

The "Assessorato Beni Culturali e Ambientali e Pubblica Istruzione" of the Sicilian Region contributed by offering a copy of the proceedings to all the participants.

THE EDITORS

Hydrobiologia **502:** 1–2, 2003.
L. Naselli-Flores, J. Padisák, M. T. Dokulil (eds),
Phytoplankton and Equilibrium Concept: The Ecology of Steady-State Assemblages.

From IAAP to IAP

Jørgen Kristiansen
Botanical Institute, Department of Phycology, University of Copenhagen, Denmark
E-mail: joergenk@bot.ku.dk

Key words: IAP, phytoplankton ecology, phytoplankton taxonomy, associations

Abstract

A short survey is given of the development of the "Internationale Arbeitgruppe Aktiver Phytoplanktologen" into "International Association for Phytoplankton Taxonomy and Ecology" – from informal Central European meetings to a truly international forum. A list of all the thirteen meetings till now is appended.

A brief historical account

This association of phytoplanktologists had its origin at the "Tagung deutschsprachiger Limnologen" in Innsbruck 1976, which was organized by R. Pechlaner. Here it was decided to establish a forum for discussion of current problems in phytoplankton research, especially regarding identification and quantification.

Accordingly, the first workshop of the IAAP, "Internationale Arbeitsgruppe Aktiver Phyto-planktologen", took place the next year, 1977, in Scharfling bei Mondsee, Austria, organized by A. Nauwerck and K. Schwarz. An important function was to establish a bridge between taxonomic specialists and practicians in lake monitoring. The programme included microscopy work, discussions, talks, and excursions. Lectures on selected taxonomic groups ("Schwerpünkte") were given by invited specialists. An important theme was the calibration of identifications and of countings, the so-called Ringtest; the results of the latter were summarized and evaluated by Rott, 1981. The reports of the meetings with the contributions (most of them as abstracts) were published in Schweizerische Zeitschrift für Hydrologie.

From the 3rd Workshop there was a gradual shift from the Central European dominance to a wider, more international participation, also overseas colleagues began to participate. This also resulted in a change of the name to "International Association of Phytoplankton Taxonomy and Ecology", IAP.

An important change took place from the 8th Workshop (Hungary, 1991). At this and the following workshops, a special theme within phytoplankton ecology was agreed upon in beforehand, and the contributions were expected to be more or less related to this theme. The collection of papers from each workshop was then published as independent volumes of Developments in Hydrobiology, which have proved to be of great value. The taxonomic part of the workshop remained along the same lines as previously.

The organization of IAP has always been very loose. At the 3rd Workshop it was decided to have a secretary (E. Rott) for contact and coordination, but this function by degrees was found superfluous. Now each workshop is an independent activity, based on the individual organizers, in accordance with guidelines discussed on the previous meeting.

The international character of the Association has always been important. The original German speaking group soon expanded to include Scandinavians, and by degrees most European countries participated. Overseas members played an increasing role, e.g. since 1984 from Canada and since 1993 from South America. Mediterranean countries such as Israel and Turkey have joined; occasionally also participants from East Asia have been present. The 12th Workshop (1999) in Canada was the first meeting outside Europe, and the 14th is planned for Turkey.

Survey of IAP workshops and proceedings

1. Scharfling, Austria, 1977 (conveners: A. Nauwerck & K. Schwarz). No report.

2. Kastanienbaum, Switzerland, 1978 (convener: H. R. Bürgi). Bürgi, H. R., 1978. 2. Internationaler Workshop aktiver Phytoplanktologen, 21. bis 27. August 1978 in Kastanienbaum. Verh. - Schweizerische Zeitschrift für Hydrol. 40: 306-309, + 310–350.

3. Seeon, Germany, 1980 (convener: W. Tins). Tins, W. (ed.), 1981. 3. Workshop der 'International Association for phytoplankton taxonomy and ecology' (IAP) vom 27. April bis 6. Mai 1980 in Seeon, Chiemgau (BRD). - Schweizerische Zeitschrift für Hydrol. 43: 1–33, + 34–62 (E. Rott: Some results from phytoplankton counting intercalibrations).

4. Salten Skov, Denmark, 1982 (conveners: J. Kristiansen & K. Olrik). Kristiansen, J. & K. Olrik, 1983. Fourth International Phytoplankton Workshop, Salten Skov, Silkeborg, Denmark, May 19–27, 1982. Schweizerische Zeitschrift für Hydrologie 45: 361–363.

5. Stirling. Scotland, 1984 (A. Bailey-Watts). Bailey -Watts, A. E., J. Kristiansen & E. Rott (eds), 1985. Report of the Fifth Workshop of the International Association of Phytoplankton Ecology and Taxonomy. (IAP), Stirling, Scotland, 1–12 September 1984. Schweizerische Zeitschrift für Hydrol. 47: 232–233.

6. Mondsee, Austria, 1986 (convener: M. Dokulil). Dokulil, M. (ed.), 1987. Report of the sixth Workshop of the International Association of Phytoplankton Taxonomy and Ecology (IAP), Mondsee, Austria, 15–25 September 1986. Schweizerische Zeitschrift für Hydrologie 49: 373–392.

7. Konnevesi, Finland, 1989 (convener: P. Eloranta). No report.

8. Baia, Hungary, 1991 (conveners: J. Padisák & A. Schmidt). Padisák, J., C. Reynolds & U. Sommer (eds), 1993. Intermediate Disturbance Hypothesis in Phytoplankton Ecology. Hydrobiologia 249 / Dev. Hydrobiol. 81. 200 pp.

9. Mont Rigi, Belgium, 1993 (conveners: J.-P. Descy & P. Compère). Descy, J.-P., C. S. Reynolds & J. Padisák (eds), 1994. Phytoplankton in turbid environments: Rivers and shallow lakes. Hydrobiologia 289 / Dev. Hydrobiol. 100: I–IX, 1–214.

10. Granada, Spain, 1996 (conveners: M. Álvarez-Cobelas & P. Sanchez-Castillo). Álvarez-Cobelas, M., C. S. Reynolds, P. Sanchez-Castillo & J. Kristiansen (eds), 1998. Phytoplankton and trophic gradients. Hydrobiologia 369/370 / Dev. Hydrobiol. 129: I–XIII. 1–372.

11. Shrewsbury, England, 1998 (convener: C. Reynolds). Reynolds, C. S., M. Dokulil & J. Padisák (eds), 2000. The trophic spectrum revisited. Hydrobiologia 424 / Dev. Hydrobiol. 150: I–XIII, 1–152.

12. Delta Marsh, Canada, 1999 (conveners: H. Kling & P. Hamilton). Hamilton, P. B., H. Kling & M. Dokulil (eds), 2000. Cyanoprokaryotes and chlorophytes across trophic gradients. Hydrobiologia 438: 1–264.

13. Palermo, Italy, 2002 (convener: L. Naselli-Flores) Naselli-Flores, L., J. Padisák & M. Dokulil (eds), Hydrobiologia 502: 1–404.

Hydrobiologia **502:** 3–12, 2003.
L. Naselli-Flores, J. Padisák, M. T. Dokulil (eds),
Phytoplankton and Equilibrium Concept: The Ecology of Steady-State Assemblages.

Are there steady-state phytoplankton assemblages in the field?

Carmen Rojo[1] & Miguel Álvarez-Cobelas[2]
[1]*Institute of Biodiversity and Evolutionary Biology, University of Valencia, Valencia, Spain.*
E-mail: carmen.rojo@uv.es
[2]*Centro de Ciencias Medioambientales, CSIC, Madrid, Spain*

Key words: dynamic equilibrium, phytoplankton competition, assembly mechanisms, controlling stability factors

Abstract

The difficulty in advancing in Ecology is due, in part, to the fact that this science uses a mainly qualitative language instead of a more formal or mathematical one. Therefore, many ecologists' efforts are expended in controversies resulting from the vagueness of ecological concepts, for example: stability, equilibrium, ecosystem, community, and so on. When approaching the study of steady-state phytoplankton assemblage, the different interpretations of these concepts can paralyse fruitful discussion. In the following pages, there is an endeavour to both restrict and precise the meaning of some of the concepts related to this topic and to broaden the range of possibilities of steady-state in the field. It is argued here that, in order to test whether or not there is a steady-state assemblage, first of all a variable or descriptor of such assemblage should be chosen. It is also argued that a steady-state does not necessarily occur as the result of a competition process with a stable equilibrium end. Moreover, we suggest that some other processes and mechanisms could control the assemblage as a steady-state. Examples of steady-state phytoplankton assemblage observed in the field (perturbed and unperturbed situations), but probably not related to a competition system equilibrium, are shown: the alternate dominance of two species (*Cryptomonas erosa* and *Limnothrix redekei*) during seven unperturbed consecutive weeks in *El Porcal* Lake (a gravel pit in Central Spain); the co-dominance of five species (*Planktothrix agardhii, Limnothrix redekei, Dictytosphaerium* sp., *Cyclotella meneghiniana* and *Cryptomonas erosa*) over nine unperturbed weeks in the same lake; the dominance of different species in thirty one fluctuating sites of a wetland (*La Safor*, Mediterranean Spanish coast) and the persistence of some non-dominant species (*Peridinium willei* and *Planktonema lauterbornii*) over more than three weeks in the water column mixing period in *Las Madres* Lake (Central Spain).

Introductory remarks: terms, concepts, definitions

Faced with the topic of the 13th Workshop of the IAP (Castelbouno, Sicily), namely: 'Phytoplankton and equilibrium concept: the ecology of steady-state assemblages' (Naselli-Flores et al., 2003), which is really the goal of many phytoplankton studies (Sommer et al., 1993; Reynolds, 1998), we should be extremely cautious. Unfortunately, this goal combines too many, and too confused, ecological concepts. This is a general problem in Ecological Science and most noticeable in community ecology.

Adhering to Mayr (1982), but in summary, different sciences can be located along a gradient from a hypothetical Type I (paradigm to the classic Physics) to a Type II (as Biology). In this gradient, Type I supposes a formal or mathematical language, to study single or essential objects, laws as mathematical functions and a hypothetical-deductive and experimental method. On the opposite side, science of Type II uses a more conceptual language, its laws are conceptual patterns or historical narration, and it works with the inductive methods, being more observational than experimental. Therefore, using this gradient with respect to Biology, Ecology would tend to Type II, and within Ecology, the community ecology and limnology tend even more to Type II (Pahl-Wostl, 1995; Lampert & Sommer, 1997).

It is for these reasons that many authors publishing on community ecology first describe the sense of the concept they have used, because they know that the same concepts can have many different nuances. Remember for example, the first page of the Reynolds' classic book (1984) 'What is phytoplankton?' In this way, we wish to analyse and describe the concepts implied in this topic, even if this Introduction Section becomes longer than normal, or almost the heart of the manuscript.

What does steady-state phytoplankton assemblage mean? We shall endeavour to analyse the different concepts included in this question. In common language the word assemblage (or assembly) is easy to understand, but in the ecology arena it has many meanings. An assemblage is a group (set or collection) of populations or species of similar organisms (e.g. similar taxa, such as fishes, ciliates or microalgae) that co-occur in a defined area. Therefore, the species of an assemblage do not necessarily interact. When they interact the assemblage becomes a community: 'Community is an interactive assemblage of species occurring together within a particular geographical area, a set of species whose ecological function and dynamics are in some way interdependent (Putman, 1994)'. Some years ago, phytoplankton ecologists used 'phytoplankton community' (e.g. Reynolds, 1984) as it occurred in other ecological areas, for example, 'bird community' was also said. After that, ecologists progressively constrained the use of the community concept to a set of interacting species (trophic relationships, Drake 1990). Thereafter, ecologists rejected the use of 'community' when they talked about a group of similar taxa (within trophic interactions), for example, 'bird assemblage', 'ciliates assemblage', etc can be read. Phytoplankton ecologists gave up the use of the term 'phytoplankton community' and often do not use an alternative, for example, 'diversity of phytoplankton' instead of 'diversity of phytoplankton community'. Moreover, phytoplankton ecologists, in accordance with other ecologists, could describe the set of microalgal species as a phytoplankton assemblage (hereafter the 'first meaning'). In this way it would be 'phytoplankton assemblage in the epilimnion', 'phytoplankton assemblage observed in a lake in October', etc. Unfortunately, because microalgae can be included in the Botany arena, it has been easy to consider the assemblage concept and the association one as similar. However, association, a word used by the phytosociologists, means 'repeatedly encountered assembly of species (species-sets) which were commonly found to co-occur' (Putman, 1994). In this way, Reynolds (1984) wrote 'Though concerned with a relatively small selection of British lakes, Reynolds' (1980) attempt to group together species which collectively made up the assemblages and were, potentially, alternative dominants, provides further evidence of the reproductibility of plankton cycles in given lakes' (quotation marks in the original). That is the question: in one way phytoplankton ecologists can use (as do ecologists of other organisms) assemblage in the sense of community. However, in another way, phytoplanktologists use assemblage as an abstract association of some species, observed together in sufficient similar circumstances to be a referential pattern (labelled with a letter). Therefore, phytoplanktologists can use assemblage in a double sense, for example, in this sentence: the assemblages (first meaning) found in a sample is similar to some assemblage (second meaning = association) described for a particular environment. Now, another attempt to classify the phytoplankton is based on the auto-ecology and functional traits of the species, the new groups are named functional groups (Reynolds et al., 2002). It seems clear that no association-assemblages, nor functional groups, cluster the algae with respect to their temporal stability. Therefore, after this simple attempt to precise the different meanings of the assemblage concept, we conclude that the ecological interest in steady-state phytoplankton assemblage refers to the steady-state of a set of phytoplankton populations sampled together, whether they are similar to a labelled association (Reynolds, 1980), or labelled functional group (Reynolds et al., 2002), or not.

The multiple meanings of the steady-state of an ecosystem can be observed, for example, in the Patten & Jørgensen's book (1995). The steady-state of a variable occurs when there is an increase and decrease (input and output) of the variable, resulting in a variation equal to zero in consecutive measures through time. For example, the dynamics of phosphorus in lakes can be described by a mass balance equation $dP/dt = I-O-(S-R)$ where I is external load, O is the outflow loss, S is the loss to the sediments and R is the return from the sediments or internal load; the steady-state of P (the steady-state equation) may be obtained by setting $dP/dt = 0$ (Imboden & Lerman, 1978). Linked with the steady-state concept is the word 'equilibrium'. When a variable is in steady-state, obviously a sort of equilibrium (invariance) of this variable has been reached. Moreover, it is the result of a dynamic process so the steady-state is a steady dynamic equilibrium. For example, the state of equilibrium in the Island Biogeography Theory by MacArthur & Wilson (1967): some populations of birds get to the island while others disappear, the result can be a steady-state of bird richness in the island (the species can be different, but the number of species is always the same).

Equilibrium is a broad concept in ecology and a discussion of its problems and confused significance can be seen in Pahl-Wostl (1995) for ecosystems, and in Harris (1986) for phytoplankton ecology. This word 'equilibrium' is sometimes related to the succession process in a deterministic way, at other times it is related to the predator and prey densities necessary to be sustained. Other approaches are the models of competition between phytoplankton (see the revision of Tilman's works in Harris, 1986; Keddy, 1989), these models are derived from the equations for nutrient uptake and growth in continuous culture which are explicitly steady-state systems. These examples of steady-state can help us to understand the concept and then to point out what steady-state assemblage means. While growth and loss processes are occurring simultaneously in the populations of an assemblage, it is possible that the result of these processes is the assemblage persistence over time. This means that during some time the populations of this assemblage are the same. It is important to point out that in this case, the constant variable is not, for example, the richness, the diversity index or the total density, but rather the list of species. In other words, it is not a quantitative variable but rather a qualitative description of the assemblage. Therefore, the idea of steady-state assemblage is that by sampling the ecosystem with a convenient frequency (time scale) we observe the same microalgal composition in many samples.

In this paper, steady-state is a dynamic equilibrium because there are losses and growth processes acting through time, the result of which is an invariance of the assemblage. Moreover, the mechanisms or factors controlling the steady-state phytoplankton assemblage would be competition, and/or trophic relationships and/or biogeochemistry processes, etc. Many other assembly mechanisms (selective grazing, allelopathy, bloom from a bank of resistant forms, invasions, parasitism, stochasticity, etc.) in the complex aquatic system could be acting simultaneously, or alternately, possibly resulting in some sets of populations in steady-state.

It is important to emphasise that one of the best-known relationships between species, as much in theoretical as in experimental ecology, is competition. However, it is quite impossible for competitive species to act as an isolated system in the field, so the abiotic and biotic interferences make competitive exclusion or competitive equilibrium (Lampert & Sommer, 1997) extremely difficult (Scheffer et al., 2003). Sommer et al. (1993) describe the assemblage structure resulting from a competitive equilibrium in the field during a sufficiently unperturbed time period. This attempt was made because the goal in that case was the IDH (intermediate disturbance hypothesis) which links competitive exclusion, perturbation and diversity (Connell, 1978). However, our goal is the possible steady-state phytoplankton assemblage in the field and then we should not be limited by this sort of equilibrium concept. For example, Keddy (1989, 2000) or Levins (1979) offered many alternative processes to competition which constrain it (Table 1). In short, these ideas are based on: (a) classic competition mechanisms and results of its perturbation (Connell, 1978; Tilman, 1982; Sommer, 1985); (b) interaction between resources, fluctuations of each resource and different rates of resource usage, implying many more limiting resources (Levins, 1979); (c) the possibility that one species can probably be a consortium of many clones, this would explain the persistence of one species over different circumstances and (d) the importance of the first successful invader, the only survivor of the past, the higher inocula, etc. They can be dominant, but not necessarily the best competitor (Keddy, 2000).

A new question arises: How many consecutive samples should show a similar microalgal composition in order to affirm that we are observing a steady-state of the assemblage? The response, as with other rules in ecology, is a consensus. It is a rule for the phytoplanktologists to sample, at least, weekly. This is because the time generation of microalgae is from 0.3 to 3 days, so a week is a time scale of population response (see, for example, the synopsis of environmental events and phytoplankton responses by Jeffries & Mills, 1990). From a statistical point of view, more than two data are necessary to decide that a similarity between two measures is not accidental. Thus, it is requisite to find a similar assemblage over a period of at least three successive weeks in order to suggest that it is stable.

Finally, the last word of our title: the field. It would appear unquestionable that by sampling in the field, and observing the samples, it would be impossible to conclude which mechanisms trigger the community trajectory in an ecosystem (Harris, 1986, Lampert & Sommer, 1997). However, the repeated observations allow us to achieve a draft, a conceptual or even a mathematical pattern about what is occurring in the

Table 1. Causes for steady-state assemblages in the field. Sources are: 1, Tilman (1982); 2, Keddy (1989, 2000); 3, Levins (1979); 4, Harris (1986); 5, Connell (1978)

	1–3 codominants	No-codominants
Stable environment	Competition = equilibrium coexistence (1) Founder controller (2) Gradual resource changes (2)	Different time to use resources (3) Relax of competition (4) Substitution between competitors (2)
Fluctuating environment	Fluctuation as a new resource (3) Species consortium, sibling species, etc. (3)	Non-competitive: intermediate perturbation (5) Complexity of resources (fluctuations) (3)

study system and thereafter, it is possible to suggest the most probable explanation or underlying mechanisms (Keddy & Weiher, 1999). The mechanisms or factors controlling the assemblage dynamics should then be tested in an experimental way.

In this paper, our aim is to show some steady-state phytoplankton assemblages in the field, and suggest some possibly related processes. We have chosen examples which, furthermore, do not suggest a classic competitive mechanism and, moreover, are from stable and perturbed environments.

Materials and methods

Study sites and sampling methods

The examples of steady-state phytoplankton assemblage used in this work are from three different aquatic systems (Table 2): *El Porcal* lake, *Las Madres* lake and *La Safor* coastal wetland. El Porcal is a warm, monomictic hypertrophic lake sited in a protected area close to Madrid (Central Spain). It was sampled weekly over 60 weeks (Álvarez-Cobelas, 1991, Álvarez-Cobelas et al., 1993). Several perturbation periods were observed in this lake and some equilibrium phases described (in the sense of Sommer et al., 1993), and the phytoplankton assemblages were exhaustively studied (Rojo & Álvarez-Cobelas, 1992, 1993a,b, 1994, 1995a,b). *Las Madres* Lake is a warm, monomictic mesotrophic lake, sampled daily at overturn for 93 days (Rojo et al. 1999, Rojo & Álvarez-Cobelas, 2001). It is situated in a protected area close to Madrid (Central Spain). *La Safor* is a coastal wetland with many different, connected water bodies (shallow ponds, channels, little lakes, etc.). It showed a wide trophic gradient from oligo-mesotrophic to hypertrophic sites (Rodrigo et al., in rev.). To study the spatial heterogeneity 31 sites were sampled at the same time. This sampling effort was repeated a week later and in two different periods: two consecutive weeks in the humid period (spring) and two consecutive weeks in the dry period (end of summer).

Environmental stability (as opposed to disturbed conditions) was always considered with respect to the variability of physical and chemical factors, calculating their average and standard deviation between consecutive samples or a rate of change.

Phytoplankton identification and counting methods. Biovolume calculations

Water samples were preserved with Lugol and additional live samples were retained for *in vivo* identification. Live and fixed materials were identified and counted using an inverted microscope and an Utermöhl counting chamber. At least 400 individuals of the more abundant species were counted from each sample with a 10% error (Lund et al., 1958). Biomass was calculated using measurements of at least 20 individuals of each population and using appropriate geometrical formulae (Rott, 1981). More information about identification of taxa can be found in, for example, Rojo & Álvarez-Cobelas (1992, 1993a,b, 1995a,b).

The functional group (Reynolds et al., 2002) when some species were dominant are indicated.

Results

These results, obtained studying the phytoplankton of different places, highlighted as the assemblage

Table 2. Limnological features of three aquatic systems during the studied periods mentioned in this work. * corresponds to Soluble Reactive Phosphorus data

	El Porcal lake	Las Madres lake	La Safor marsh
Surface area (ha)	45	4	1300
Average depth (m)	3	6	1
Maximal depth (m)	9	19	2
Total Phosphorus (mg l^{-1})	0–2*	0.02–0.06	0.004–20
Chlorophyll a ($\mu g\ l^{-1}$)	20–420	1–5	1–457
Phytoplankton biomass ($mm^3\ l^{-1}$)	5–250	1–5	0–12

structure when it is in a steady-state, shows different patterns and suggest different assembly causes.

The first example is from *El Porcal* lake during an undisturbed period (described in Rojo & Álvarez-Cobelas, 1993c) of 7 weeks (Fig. 1 A). Two species, *Cryptomonas erosa* Ehrenberg and *Limnothrix redekei* (van Goor) Meffert, were co-dominant. These species were, over the given time, more than 80% of the total phytoplankton biomass, this total biomass did not vary significantly during the period (maximum in this year was around 250 $mm^3\ l^{-1}$, Rojo & Álvarez-Cobelas, 1993c). During the first 2 weeks *Limnothrix redekei* achieved a higher percentage of total biomass than *Cryptomonas erosa*, after that, during the following three weeks, the proportion of both algae was fairly similar, while during the last 2 weeks *Cryptomonas erosa* was dominant.

In the same lake, over another stable period (meaning between disturbance events, Rojo & Álvarez-Cobelas, 1993c) five species of phytoplankton, *Planktothrix agardhii* (Gomont) Anagnostidis & Komárek, *Cyclotella meneghiniana* Kützing, *Cryptomonas erosa*, *Dictyosphaerium* sp. and *Limnothrix redekei* persisted for 9 weeks (from October) with a similar percentage of total biomass, while this total phytoplankton biomass slowly decreased, average of total biomass in this period was 21 mm^3/l, standard deviation 9 mm^3/l (Fig. 2).

The last two examples illustrate phytoplankton assemblages during stable environmental periods. The next example shows the phytoplankton composition in Las Madres Lake when the mixing period started. A change was observed in the phytoplankton assemblage (Fig. 3) from few co-dominant species at the end of stratification (*Peridinium willei* Huitfeld-Kass, *Peridiniopsis borgei* Lemm. and *Lagerheimia genevensis* Chod. add up 80% of total biomass). When the mixing period started, five, and later even six, species were co-existing, comprising more than 80% of total biomass. There was a species substitution while the dominance of some of them disappeared. *Peridinium willei* persisted over more than 10 weeks (end stratification, and different states of mixing water column), *Peridinopsis borgei* and *Lagerheimia genevensis* disappeared when mixing period started being replaced by *Planktonema lauterbornii* Schmidle during, *Cryptomonas* and another two or three species during the overturn.

Finally, *La Safor*, a very shallow coastal wetland with great spatial heterogeneity (Rodrigo et al., 2003), is an example of clearly fluctuating conditions (Fig. 4). The total phosphorus concentration changed more than 50% in more than two-thirds of the wetland sites in two consecutive weeks. Under such a circumstance, the great majority of sites (25 of 31 sampled points) showed one dominant species at least each week. Five sites showed an assemblage with dominant species during the first week and the next week few co-dominants; sixteen sites changed one dominant species for another the following week and in four of them the dominance persisted throughout the 2 weeks. Some examples of substitution of dominants were: *Planktothrix agardhii* by *Euglena gracilis* Klebs, *Trachelomonas volvocinopsis* Swirenko by *Synechococcus* sp., *Cryptomonas erosa* by *Nitzschia acicularis* (Kützing) W. Smith or *Cyclotella* cf. *ocellata* Pantocsek by *Cryptomonas erosa* var. *reflexa* Marsson.

Discussion

We would like to discuss these results in a really free way, because the field data, as we commented in the Introduction section, allows us to establish a pattern, but not to make definite conclusions about the underlying mechanisms. Moreover, our data were obtained in the past with the aim of answering other ecolo-

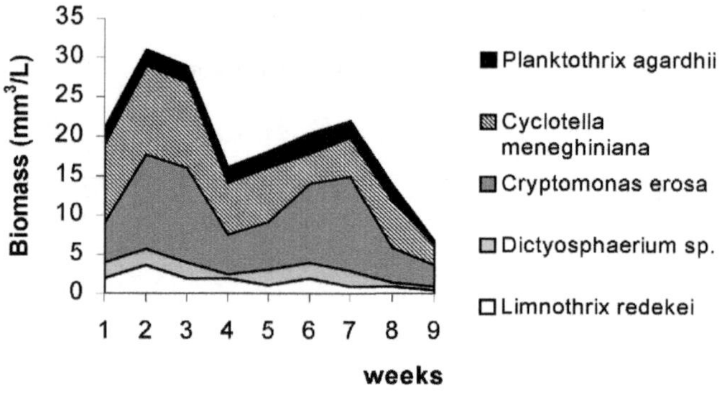

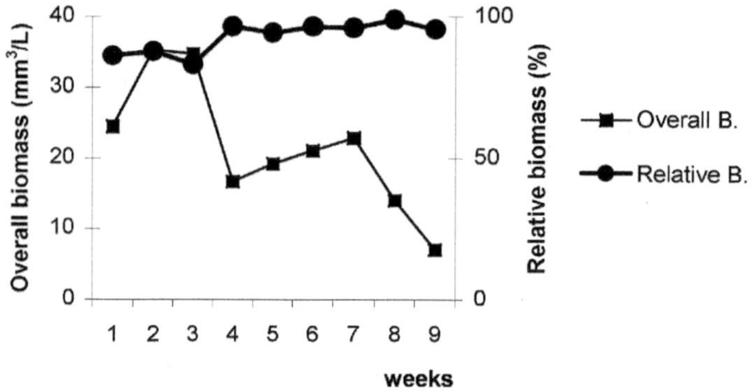

Figure 1. Relative change (log transformed) of total phosphorus concentration between two consecutive weeks in 31 sites of La Safor coastal wetland. Percentage of sites whose phytoplankton assemblage has a dominant species and percentage of sites having the same dominant species over 2 consecutive weeks are shown.

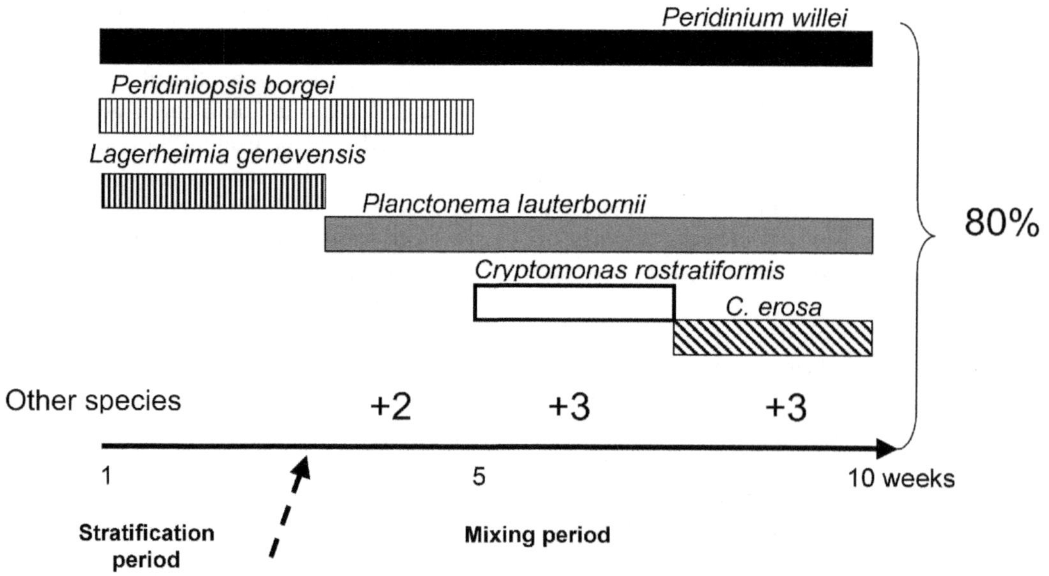

Figure 2. Biomass dynamics of five phytoplankton species over 9 weeks in El Porcal Lake. Total phytoplankton biomass and the relative biomass of the five species are also shown.

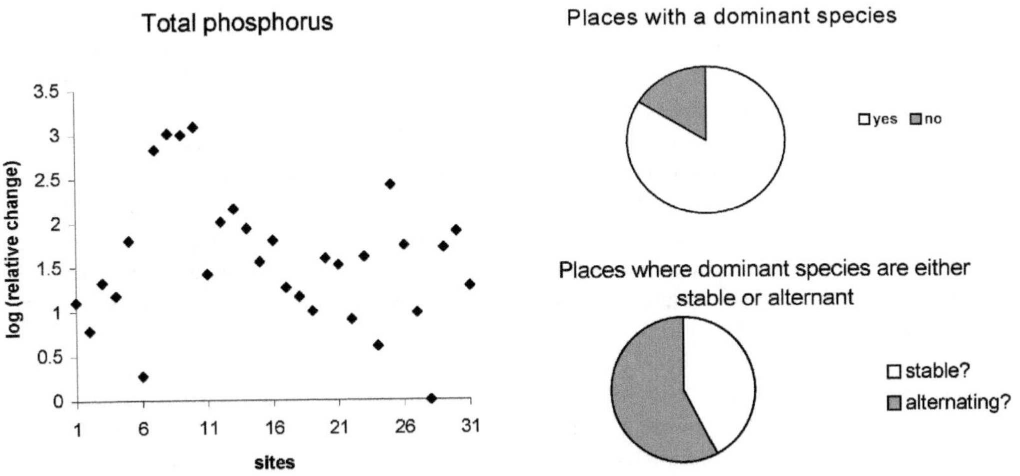

Figure 3. (A) Dynamics of relative biomass of two phytoplankton populations and total biomass over 7 weeks in El Porcal Lake. (B1) pattern of a competitive coexistence during some weeks: *L. redekei* (Lr) is dominant, *L. redekei* and *C. erosa* (Ce) are co-dominants and finally *C. erosa* is the dominant. (B2) *L. redekei* is slowly decreasing being substituted by *C. erosa*, during this process they are together for some weeks.

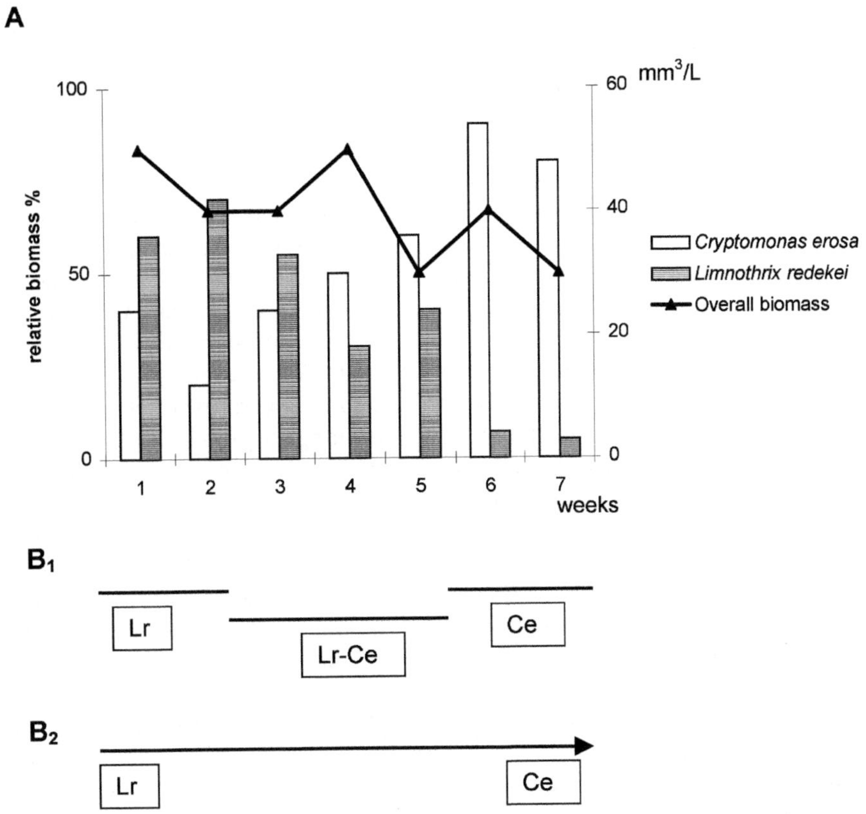

Figure 4. Pattern of substitution of dominant species since the end to the stratification period to start of overturn in Las Madres Lake. The number of species needed to reach 80% of total phytoplankton biomass is also indicated.

gical questions. Now, our goal is to show examples of different steady-states of phytoplankton because they can be useful to illustrate the complexity of processes affecting the dynamic assembly.

The first case shows two co-dominant species in a stable environment over 7 weeks. It is a stable period because the most common physical and chemical variables did not change suddenly by 2 standard deviation from its average value on the preceding dates following the previous disturbance (for details of the dynamics of these variables see Rojo & Álvarez-Cobelas, 1993c). This is a steady-state of phytoplankton assemblage because *Limnothrix redekei* and *Cryptomonas erosa* persisted as a species characteristic of the assemblage over 7 weeks. Over this period a replacement between the 2 species and little change in the total biomass can be observed; the period dominated by *Limnothrix redekei* (functional group S_1) has higher mean value of biomass than the period dominated by *Cryptomonas erosa* (functional group Y). Moreover, during some of these weeks (Fig. 1, central weeks) this assemblage seemed to be in an equilibrium phase (in the context of a competitive system). Equilibrium phase is described as: one, two or three species at the utmost, which together contribute more than 80% of the standing biomass; their coexistence persisted for long enough (more than 1–2 weeks) during which time no significant changes in total biomass occurred (Sommer et al., 1993). In this way, competition seems to be the mechanism that ruled the assemblage and then, according to Lampert & Sommer (1997), two resources were limiting. However, in the hypertrophic El Porcal Lake, nutrients were not limiting (Rojo & Álvarez-Cobelas, 1993c). Moreover, and most importantly, we cannot distinguish a situation of stable equilibrium between two coexisting species and the slow substitution of a dominant 1, for another 2, when the environment changes; obviously both species co-occur during part of the time. Therefore, when we observe Fig. 1A, it is impossible to know if this kind of steady-state (central weeks) is the result of competition (Fig. 1 B_1). For example, they could be the best competitors for that environment (Tilman, 1982; Reynolds, 1997) or the slow substitution between two species over time, whether competitors or not (Fig. 1B_2). Similar dynamic was observed by Stoyneva (2003).

The other case in El Porcal Lake shows a long stable period. Why during this long, undisturbed period, did few species fail to reach dominance? During this period the water column in the lake was mixed but unperturbed (in the same sense as in the above paragraph, Rojo & Álvarez-Cobelas, 1993c). Although total biomass decreases, the period was characterised by a quite constant biomass in comparison to the average biomass in the next period (130 mm^3/l, standard deviation 64 mm^3/l). Therefore, the situation is a steady-state of available resources, quality of light and the possibility of organism distribution in the water column. This seems a case where Levins (1979) explanations can help our interpretation: the assemblage is formed by species with different rates of resources, uptake, mobility and light necessity usages (Reynolds, 1997; Reynolds et al., 2002). Here many different species are together because they do not compete, and have different niches (see Reynolds et al., 2002; Salmaso, 2003). As Dodson et al. (2000) reasoned, from an ecological point of view it is interesting to explain why there is a lack of richness instead the richness, in other words, why some species disappear not why there are there.

The other case of overturn, Las Madres Lake, shows an example of the effect of disturbance. We followed the end of the stratification period and the start of the overturn. Firstly, we would suggest that the mixing interrupted a competitive process (only three species at the end of stratification period). However, some questions arise. Why do two of them, *Peridinium willei* and *Peridiniopsis borgei,* persist? It is difficult to believe that they were the best adapted to the other conditions and still persisted (see a similar case with *Planktothrix* and *Pseudanabaena* described by Morabito et al., 2003). It seems more logical to suppose that some species can inhabit in a broad range of environmental conditions (as *Cryptomonas erosa*, for example) or even that they can need changes from stable to perturbed periods as some Dinoflagellates (Pollingher, 1988). In Las Madres lake *Peridinium willei* (functional group L_o) did not grow (population density did not change) during the overturn. However, in this period the average body size became the double (Rojo & Álvarez-Cobelas, 2001). The result was a biomass increase which was cross-related to the mixing deep (Z_m) with 16 days lag. Anyway, in this example, algae constitute a steady-state assemblage based on this 'generalist' character. To understand this character we need to know, for example, if the species: (a) is really a consortium of populations (many different genotypes with similar morphotypes), (b) it has continuous input of inocula from their resistant forms or/and (c) it requires different conditions linked with its different states of life cycle.

The last example, from the La Safor wetland, shows the dominance of species in a wide trophic spectrum ranging from oligotrophic to highly disturbed hypertrophic sites (Rodrigo et al., in rev). Why is there not more richness here? Are the dominant species found in each site the best competitors? Probably not, other processes can be acting, such as selective grazing, because there is a high zooplankton richness, and a high density of it as well (Rodrigo et al., in rev.), producing monocultures of big body-sized species, such as *Planktothrix* or *Euglena*. Additionally, the polluted conditions and the hydric stress in the wetland provide a strong environmental filter. The situation may be a little pool of species, capable of surviving in this landscape under a remarkable grazing pressure. The dominant species, therefore, are not the best, but rather the remainder.

We hope that these examples and daring explanations challenge you to consider: How many times have you observed persistent assemblages, with few or many species coexisting, in stable or fluctuating environments? Our conclusions are that: (i) there are many different types of steady-state phytoplankton assemblages, (ii) competition does not seem the main assembly mechanism in the field, (iii) there are many different processes which result in steady-state assemblages and (iv) there are steady-state phytoplankton assemblages in stable and fluctuating environments.

Acknowledgements

The authors would like to express their gratitude to their colleagues at the 13th Workshop of the IAP (Sicily, 2002) for the entertaining discussion on this topic. The Spanish Ministry of Science and Technology supports this work with a grant (BOS2002-02333).

References

Álvarez-Cobelas, M. 1991. Optical limnology of a hypertrophic, gravel-pit lake. Int. Rev. ges. Hydrobiol. 76: 213–223.

Álvarez-Cobelas, M., A. Baltanás, J. L. Velasco, M. Valladolid, M. Izquierdo and E. Martín. 1993. Slow overturn in a gravel-pit lake. Int. Ver. theor. angewan. Limnol.: Verh. 25: 83–87.

Connell, J., 1978. Diversity in tropical rain forest and coral reefs. Science 199: 1304–1310.

Dodson, S. I., S. E. Arnott & K. L. Cottingham 2000. The relationship in lake communities between primary productivity and species richness. Ecology 81: 2662–2679.

Drake, J., 1990. Communities as assembled structures: Do rules govern pattern? Trends in Ecol. Evol. 5: 159–164.

Harris, G. P., 1986. Phytoplankton Ecology. Chapman & Hall, London. 384 pp.

Imboden, D. M. & A. Lerman, 1978. Chemical models of Lakes. In Lerman, A. (eds.), Lakes: Chemestry, Geology, Physics. Springer, New York: 341–356.

Jeffries, M. & D. Mills, 1990. Freshwater ecology. Belhaven Press, London. 285.

Keddy, P. A., 1989. Competition. Chapman & Hall, London. 202 pp.

Keddy, P. A., 2000. Wetland Ecology. Cambridge University Press, Cambridge. 614 pp.

Keddy, P. & E. Weiher, 1999. The scope and goals of research on assembly rules. In: Weiher, E. & P. Keddy (eds), Ecological Assembly rules. Cambridge University Press, Cambridge. 418 pp.

Lampert W. & U. Sommer, 1997. Limnoecology: the ecology of lakes and streams. Oxford University Press, New York.

Levins, R., 1979. Coexistence in a variable environment. Am. Nat. 114: 765–783.

Lund, J. W. G., C. Kipling & E. D. Le Cren, 1958. The inverted microscope method of estimating algal numbers and the statistical basis of estimations by counting. Hydrobiologia 11: 143–170.

Morabito, G., A. Oggioni & P. Panzani, 2003. Phytoplankton assemblage at equilibrium in large and deep subalpine lakes: a case study from Lago Maggiore (N. Italy). Hydrobiologia 502 (Dev. Hydrobiol. 172): 37–48.

MacArthur, R. H. & E. O. Wilson, 1967. The theory of island biogeography. Princeton University Press, Pricenton.

Mayr, E., 1982. The growth of biological thought. Harvard University Press, Cambridge. 382 pp.

Naselli-Flores, L., J. Padisák, M. T. Dokulil & I. Chorus, 2003. Equilibrium/steady-state concept in phytoplankton ecology. Hydrobiologia 502 (Dev. Hydrobiol. 172): 395–403.

Pahl-Wostl, C., 1995. The dynamic nature of ecosystems. John Wiley & Sons, Chichester. 267 pp.

Patten, B. C. & S. E. Jørgensen, 1995. Complex Ecology. Prentice Hall PTR, New Jersey. 705 pp.

Pollingher, U., 1988. Freshwater armored dinoflagellates: growth, reproduction strategies, and population dynamics. In Sandgren C. D. (ed.), Growth and Reproductive Strategies of Freshwater Phytoplankton. Cambridge Uni. Press, Cambridge: 134–174.

Putman, R. J. Community Ecology. Chapman & Hall, London. 178 pp.

Reynolds, C., 1980. Phytoplankton assemblages and their periodicity in stratifying lake system. Hol. Ecol. 3: 141–159

Reynolds, C., 1984. The ecology of freshwater phytoplankton. Cambridge University Press, Cambridge. 384 pp.

Reynolds, C., 1998. The state of freshwater ecology. Freshwat. Biol. 39: 741–753.

Reynolds, C. S., V. Huszar, C. Kruk, L. Naselli-Flores & S. Melo, 2002. Towards a functional classification of the freshwater phytoplankton. J. Plankton Res. 24: 417–428.

Rodrigo, M. A., C. Rojo & X. Armengol, 2003. Plankton biodiversity in a landscape of shallow water bodies (Mediterranean coast, Spain). Hydrobiologia 506–509 (in press).

Rojo, C. & M. Alvarez, 1992. Taxonomy and ecology of phytoplankton in a hypertrophic, gravel-pit lake. I. Cyanophyceae. Archiv für Protistenkunde 142: 77–90.

Rojo, C. & M. Alvarez, 1993a. Taxonomy and ecology of phytoplankton in a hypertrophic, gravel-pit lake. II. Cryptophyceae, Euglenophyceae. Nova Hedwigia 57: 47–63.

Rojo, C. & M. Alvarez, 1993b. Population dynamics of Planktothrix agardhii, Limnothrix redekei, Oscillatoria lanceaeformis and

Pseudoanabaena limnetica (Cyanophyceae) in a hypertrophic shallow lake. Hydrobiologia 275/276: 165–171.

Rojo, C. & M. Alvarez, 1993c. Hypertrophic phytoplankton and the intermediate disturbance hypothesis. Hydrobiologia 249: 43–57

Rojo, C. & M. Alvarez, 1994. Taxonomy and ecology of phytoplankton in a shallow, hypertrophic lake. III. Diatomophyceae. Archiv für Hydrobiol. Algol. Stud. 72: 53–70.

Rojo, C. & M. Alvarez, 1995a. Taxonomy and ecology of phytoplankton in a shallow, hypertrophic lake. IV. Chlorophyceae, Chlorococcales. Archiv für Hydrobiol. Algol. Stud. 77: 7–35.

Rojo, C. & M. Alvarez, 1995b. Taxonomy and ecology of phytoplankton in a shallow, hypertrophic lake. V. Chlorophyceae, Volvocales. Archiv für Hydrobiol. Algol. Stud. 79: 19–37.

Rojo, C., K. Kiss, M. Álvarez-Cobelas & M. A. Rodrigo, 1999. Population dynamics of Cyclotella ocellata (Bacillariophyceae): endogenous and exogenous factors. Archiv für Hydrobiol. 145: 479–495.

Rojo, C. & M. Álvarez-Cobelas, 2001. Phytoplankton structure and dynamics at a daily temporal scale: response to the thermal overturn. Archiv für Hydrobiol. 151: 549–569.

Rott, E., 1981. Some results from phytoplankton counting intercalibrations. Schweizerische Zeitschrift für Hydrol. 43: 34–62.

Salmaso, N., 2003. Life strategies, dominance patterns and mechanisms promoting species coexistence in phytoplankton communities along complex environmental gradients. Hydrobiologia 502 (Dev. Hydrobiol. 172): 13–36.

Scheffer, M., S. Reinaldi, J. Huisman & F. J. Weissing, 2003. Why plankton communities have no equilibrium: solutions to the paradox. Hydrobiologia 491: 9–18.

Sommer, U., 1985. Comparison between steady-state and non-steady state competition: experiments with natural phytoplankton Limnol. Oceanogr. 30: 335–346

Sommer, U., J. Padisák, C. S. Reynolds & P. Juhász-Nagy, 1993. Hutchinson's heritage: The diversity-disturbance relationship in phytoplankton. Hydrobiologia 249: 1–7.

Stoyneva, M. P., 2003. Steady-state phytoplankton assemblages in shallow Bulgarian wetlands. Hydrobiologia 502 (Dev. Hydrobiol. 172): 169–176.

Tilman, D., 1982. Resource competition and community structure. Princeton University Press, Princeton. 296 pp.

Hydrobiologia **502:** 13–36, 2003.
L. Naselli-Flores, J. Padisák, M. T. Dokulil (eds),
Phytoplankton and Equilibrium Concept: The Ecology of Steady-State Assemblages.

Life strategies, dominance patterns and mechanisms promoting species coexistence in phytoplankton communities along complex environmental gradients

Nico Salmaso
Dipartimento di Biologia, Università di Padova, Via U. Bassi 58/B, I-35131 Padova, Italy
E-mail: nico.salmaso@unipd.it

Key words: phytoplankton, ecological niches, community change rate, complex environmental gradients, multivariate analyses, deep lakes

Abstract

This paper analyses the life strategies, the dominance patterns and the diversity in phytoplankton communities in large and deep lakes. The study was carried out on the largest Italian Lake (Lake Garda) from 1995 to 2000. Different statistical analyses were applied. For phytoplankton the time variable represents a complex environmental gradient driving annual succession; this gradient was made explicit by the application of PCA analyses to the environmental data. The use of Non Metric Multi Dimensional Scaling applied to Bray-Curtis dissimilarity matrices revealed an ordered and cyclic development of phytoplankton every year; the Bray-Curtis index, calculated between pairs of chronologically contiguous samples, was also used as a measure of the community change rate (β_t) over the temporal succession. A significant relationship between β_t and the complex environmental gradient was assessed. Finally, for every phytoplankton species, the optimum conditions for growth and the realised niches were determined. The positioning of the species on the complex environmental gradient, and the contemporaneous application of cluster analysis based on the different specific environmental optima, highlighted primarily the existence of two groups at the extreme of the complex environmental gradient. The first group included the large late winter/spring diatoms, which developed during high water turbulence and strong physical control, high nutrient concentrations, low light conditions and reduced competition. The second group was composed by many heterogeneous summer species characterised by the ability to contrast losses by grazing and sinking in stratified and stable conditions, and the ability of tolerating nutrient deficiency. A third group of species developed during environmental conditions in the middle of the two previous extremes. These included the three master species *Mougeotia* sp., *Fragilaria crotonensis* and *Planktothrix rubescens/agardhii*. The endogenous and exogenous mechanisms promoting species coexistence are discussed, along with the applicability of competitive and equilibrium/non-equilibrium theories to phytoplankton dynamics.

Introduction

Phytoplankton communities in lakes are composed of many different species. Sometimes dominants and subdominants may be evidenced, along with several rare species coexisting with the more abundant ones; a similar experience is provided by many areas covered by herbaceous or forest vegetation (Hutchinson, 1967).

Most phytoplankton species are potential competitors for the same limited resources of their environment (principally nutrients and light). Under such conditions, the high number of coexisting species contrasts largely with the results that can be predicted from the application of competition theories. The Competitive Exclusion Principle (Gause's Principle) states that two or more species cannot coexist if they make use of the same limiting resources, namely 'complete competitors cannot coexist' (Hardin, 1960) and each community should be dominated by very few species occupying different niches. Niche is intended in the sense of Hutchinson (1967), designating the requirements of an organism abstracted from the spatially extended habitat. However, competitive dominance of the best adapted species is a process requiring a homogeneous habitat, environmental stability and equilibrium conditions; if communities develop in non-equilibrium conditions, then competitive exclusion is prevented. Hutchinson (1961) proposed that the coexistence of many species of phytoplankton could be due to environmental instability preventing equilibrium. Therefore, phytoplankton may be viewed as a non-equilibrium community of competing species and thus are not an exception to the principle of competitive exclusion (Krebs, 2001).

Considering the high potential reproductive rates of phytoplankton organisms (with doubling times spanning from fractions of days to few days), environmental instability comprises factors characterised by very different temporal scales. Some factors act on short-medium time (i.e. days–weeks; e.g., meteorological and hydrological events, grazing, vertical and horizontal chemical gradients), while others evolve regularly on seasonal basis (e.g., changes in solar radiation, development and breaking of thermal stratification, replenishment and depletion of nutrients, seasonality of zooplankton grazers). The first group of factors – depending on the intensity and on the frequency of disturbances – is instrumental for supporting non-equilibrium dynamics and diversity (Connell, 1978; Padisák, 1994) or fast community reorganisation events (reversions or catastrophes). As for the second group of factors, the annual evolution of solar elevation in the medium and high latitude regions is crucial to determine the seasonal replacement of phytoplankton assemblages. In fact, for phytoplankton – and other organisms responding in a similar way to the same temporal scales – the time variable represents a complex environmental gradient driving annual successions. Different life history traits determine the success of different species at a particular region of any environmental gradient. When these traits (e.g., growth rates, nutrient requirements, shade tolerance etc.) are inversely correlated, successional replacement will result (McCook, 1994). On annual basis, the two temporally scaled groups of factors causing instability and environmental seasonal change are essential in supporting the diversity of phytoplankton (the 'species richness' of a waterbody). In a given habitat it is convenient to consider different components in species diversity (e.g., Whittaker, 1972). These include the richness in species of a given sample (α diversity) and the biological diversity along habitat gradients (e.g., elevation or soil moisture: β diversity). Excluding particular cases (e.g., the world's largest freshwater ecosystems, Bondarenko et al., 1996), the main differences in the composition of phytoplankton in the pelagic zone of the lakes evolve along temporal gradients. In this work, the extent of differentiation of the community along the temporal gradient will be indicated as β_t.

The relative importance of the two groups of factors determining environmental instability and change, as well as total phytoplankton diversity in a given waterbody, is mediated by the peculiar morphometric and hydrological characteristics of the different types of lakes. Large and deep lakes have the tendency to operate as large inertial systems, for they minimise the effect of external disturbances. Previous investigations carried out during the 1990s in the largest Italian lake (Lake Garda) revealed an ordered and coherent temporal succession of phytoplankton assemblages in the whole basin (Salmaso, 1996, 2002). On the contrary, the seasonal sequence of phytoplankton species in lakes located at the opposite morphometric and hydrological gradient (e.g., small and shallow mountain lakes) may be strongly influenced by meteorological (rains and storms) and hydrological events (snow melting, high water flushing or drought in the warm period), with different species being dominant in various years (Salmaso & Decet, 1997). The variations concerning phytoplankton tend to be more predictable in large inertial systems, because they are more dependent on the annual evolution of the environmental climatic forcing variables, and less dependent on stochastic events. Disturbances classified as 'strong' for small and shallow lakes may have little consequences for largest basins.

In this work, I will compare the specific environmental requirements and life strategies of different algae, in order to explain their seasonal (successional) adaptation and dominance along temporal gradients in deep and large lakes. The specific objectives of this paper are: (i) to define the average, typical annual development and apparent optimum environmental conditions for growth of the most abundant phytoplankton species in the deep and large Lake Garda, based on 6 years of studying; (ii) to investigate the correlations between the temporal evolution of the main environmental factors and species turnover (β_t) and (iii) to discuss the seasonal changes in the dominant assemblages along complex environmental gradients, taking into account the specific competitive abilities defined for the most abundant taxa, and evidencing the exogenous and endogenous factors promoting phytoplankton diversity and species turnover. The relevance of the above topics will be emphasised in relation to the applicability of equilibrium concepts to phytoplankton dynamics. Hereafter the term 'community' will be used to indicate the whole pool of species present in an annual cycle, i.e. the potential competitors and winners along the temporal gradient, whereas 'assemblage' will be used to indicate a generic seasonal phytoplankton group.

Study site

Lake Garda is the largest Italian lake. Along with the lakes Iseo, Como, Lugano and Maggiore, it is one of the deep lakes (Insubrian lakes) located south of the Alps. Lake Garda has a volume of 49 km^3, a surface of 368 km^2 and a maximum depth of 350 m. The long theoretical water renewal time (27 years) is due to its low catchment (lake included)/lake surface ratio (around 6) and to low annual rainfall (790–1150 mm, IRSA, 1974). The main inflow is River Sarca, at the northern edge of the lake. The outflow, with an average discharge of 58 m^3 s^{-1}, is River Mincio, at the southern edge of the lake. Details of the catchment and the lake are reported in IRSA (1974).

Materials and methods

Methods in the field and laboratory

The data refer to samples collected normally every four weeks between January 1995 and December 2000 in the deepest zone of the lake (west basin, Brenzone). The average values of the chemical variables and phytoplankton abundance in the upper 20 m were estimated from samples collected at discrete depths.

From 1996 to 2000, water samples for phytoplankton analyses were collected at the integrated depths of 0–2 m, 9.5–11.5 m and 19–21 m with a 5 l, 0.5 m long Niskin bottle (for a total final volume of 20 l); in 1995 samplings were not carried out in the middle layer (9.5–11.5 m). Chlorophyll *a* was determined by spectrophotometry following the methods proposed by Lorenzen (1967). Subsamples of 200 ml were fixed with Lugol's solution and stored in bottles of glass kept in the dark at 4 °C for subsequent phytoplankton analyses. Algal cells were counted using inverted microscopes, following the criteria reported by Lund et al. (1958). The counts include, besides the identified fraction, ultraplankton (naked or flagellate cells around 4 μm) and undetermined nanoflagellates (around 5–10 μm). A detailed description of the procedures used in the laboratory is reported by Salmaso (2002).

Water samples for chemical analyses were collected at the surface and at 20 m depths, with the exception of December 2000. Soluble reactive (RP) and total phosphorus (TP), nitrate (NO_3-N) and ammonium nitrogen (NH_4-N), reactive silica and pH have been measured by the Veneto Region Environment Protection Agency (ARPAV, District of Belluno) following standard methods (APHA et al., 1989). The analytical procedures are described by Decet & Salmaso (1997).

On each sampling, profiles of temperature were carried out with an underwater multiparameter probe. The differences of water density between 0–20 m ($\Delta\delta_{0-20}$) and 0–150 m ($\Delta\delta_{0-150}$) were taken as measures of the water column stability in the layer sampled for phytoplankton analyses and in the mixolimnion, respectively. Water density was computed from temperature measurements (Chen & Millero, 1986).

Secchi disk transparency (z_s) was estimated using a bathiscope to reduce uncertainties in the measurements due to light reflections and wave motions. The euphotic depth (z_{eu}) was estimated from Secchi disc readings using the relationship:

$$Z_{eu} = 4.8 \times Z_S^{0.68}$$

(Salmaso, 2002).

The measurements of total solar radiation were obtained at the Agricultural Institute of S. Michele all'Adige (Section of Agrometeorology, publically available information). The measurement station was located at Arco, at the northern border of the lake, approximately 25 km away from the sampling station. As a measure of light availability for phytoplankton growth, the average values of solar radiation during the 3 days prior to sampling dates (I_{3d}) were considered (cf. Bleiker & Schanz, 1997). In only few cases, missing observations were supplied with the corresponding measurements made at S. Michele all'Adige (35 km from Arco). The 3-day averages were compared with the average annual solar radiation trend at Arco, obtained on the basis of fortnightly averages computed for the period 1983–2000.

Data analysis

Individual samples were ordered by principal components analysis (PCA) calculated from the correlation matrix of the physical and chemical untransformed variables.

Biovolume based Shannon (H') diversity (α diversity) was estimated using natural logarithms (Magurran, 1988); unidentified ultraplankton and nanoflagellates were not considered in the calculation. Phytoplankton data were analysed by Nonmetric Multidimensional Scaling (NMDS) (Kruskal & Wish, 1978; Salmaso, 1996); the ordinations were applied to Bray & Curtis' dissimilarity matrices (Bray & Curtis, 1957) computed on biovolume values. Unidentified phytoplankton and rare species found in one occasion were neglected. Double square root transformation of the original data was applied to reduce the weight of the most abundant species (Field et al., 1982). The Bray-Curtis dissimilarity calculated between pairs of chronologically contiguous samples, was also used as an estimate of the differentiation of the community along the complex environmental gradient (community change rate, β_t) (cf. Whittaker, 1972; Salmaso, 1996).

The single species were placed on the complex environmental gradient defined by the first two PCA axes (cf. Fabbro & Duivenvoorden, 2000) taking into account their biovolume variations (cf. Bray & Curtis, 1957) (Fig. 7). Each taxon k has been located computing the average values – weighted by the corresponding biovolume values – of the coordinates of the samples where it was determined. For the coordinates along the horizontal axis, x_k is:

$$x_k = \sum_{i=1}^{n} b_{ik} X_i \Big/ \sum_{i=1}^{n} b_{ik}$$

where x_k is the value of the coordinate of species k on the first PCA axis, b_{ik} the abundance of taxon k in sample i ($b_{ik} \geq 0$), x_i the value of the coordinate of sample i on the first PCA axis (cf. Fig. 2) and n the number of samples (n = 75).

For every species, the observed optimum environmental conditions for growth were computed – analogously to the previous formula – by weighted averaging estimates (ter Braak & van Dam, 1989). A species k with a particular optimum for a variable, will be most abundant in samples where this variable is close to its optimum. The weighted averaging estimate of the optimum u_k is:

$$u_k = \sum_{i=1}^{n} b_{ik} v_i \Big/ \sum_{i=1}^{n} b_{ik}$$

and the species' tolerance, t_k, or weighted standard deviation, is:

$$t_k = \left[\sum_{i=1}^{n} b_{ik} (v_i - u_k)^2 \Big/ \sum_{i=1}^{n} b_{ik} \right]^{1/2}$$

where b_{ik} is the biovolume of taxon k in sample i ($b_{ik} \geq 0$) and v_i the value of the variable of interest in sample i; the computations were carried out on the whole set of samples (n = 76 or, in the case of algal nutrients, n = 75).

Finally, the optimum values (u_k) were utilised to identify the principal life strategies, classifying the phytoplankton species by cluster analysis (Pearson distance, Ward's method).

Phytoplankton diversity and dissimilarity indices were computed with SIMDISS 2.0 (http://www.bio.unipd.it/limno/simdiss/), whereas multivariate analyses were carried out with SYSTAT packages.

Results

Physical and chemical variables

The principal environmental abiotic variables affecting the seasonal evolution of the phytoplankton community are reported in Figure 1.

Solar radiation (I_{3d}) showed an annual oscillation, with values ranging from 2.5–5 MJ m^{-2} d^{-1} to 20–25 MJ m^{-2} d^{-1} (Fig. 1a). Some differences emerged from the comparison of the I_{3d} values with the average annual solar radiation trend (1983–2000). However, the 3-day averages were strongly correlated with the corresponding averages computed during the week before the sampling (7-day averages; $r = 0.94$; $P < 0.01$), so they seem to represent quite well the available income radiation for the growth of phytoplankton in a determined period of the annual cycle.

From late spring to early autumn the lake displayed a marked thermal stratification. Maximum temperatures in the first metre reached 22–25 °C, whereas the metalimnetic layer deepened down to 30–40 m (Fig. 1b). The maximum winter euphotic depths ranged around 30–45 m (Fig. 1b). During summer, the lower limit of the euphotic layer was located between

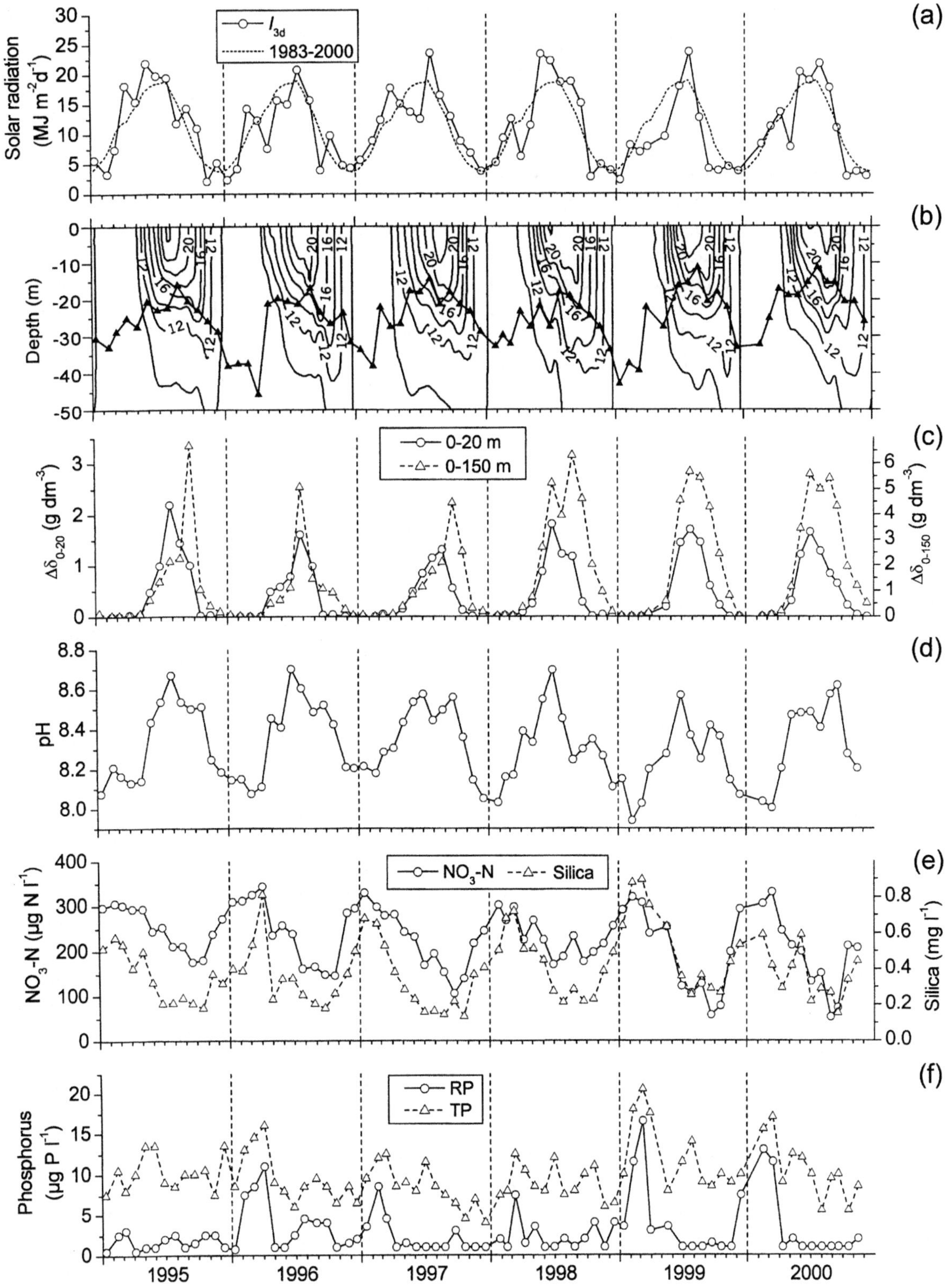

Figure 1. (a) Temporal variations of the average values of solar radiation during the three days prior to sampling dates (I_{3d}), and average annual solar radiation trend estimated for the period 1983–2000. (b) Isopleths of temperatures (°C) with superimposed seasonal variations of the euphotic depth (m) and (c) temporal variations of the differences of water density between 0–20 m ($\Delta\delta_{0-20}$) and 0–150 m ($\Delta\delta_{0-150}$). (d) Temporal variations of pH, (e) nitrate nitrogen and reactive silica, (f) reactive (RP) and total (TP) phosphorus in the layer 0–20 m.

15 and 25 m from 1995 to 1998, and between 11 and 20 m in 1999 and 2000. The euphotic depth showed similar or greater values than the mixing depth ($z_{eu}/z_{mix} \geq 1$) from May-June to September.

In the layer 0–20 m complete mixing ($\Delta\delta_{0-20} = 0$ g dm^{-3}) was generally observed in the period between October–November and March–April (Fig. 1c); the higher stability of the water column ($\Delta\delta_{0-20} >$ 1 g dm^{-3}) was observed between June-July and August. The isopycnic layer extended down to 150 m ($\Delta\delta_{0-150} \cong 0$ g dm^{-3}) between December and March–April. This layer may be roughly considered the mixolimnion of the lake, because it undergoes thermal cooling and mixing during the late winter and early spring months. During the 1990s, complete vertical cooling (down to 350 m) and circulation was documented only in 1999 and 2000 (Salmaso et al., 2001).

The seasonality of pH and nutrients is related to higher phytoplankton growth during the summer months and also to the vertical mixing of the water column occurring from late autumn to early spring. In the layer 0–20 m the pH values showed a regular seasonal evolution, with values ranging from 7.9 to 8.7 (Fig. 1d).

Nitrate nitrogen and silica decreased together during the warmest months (Fig. 1e; $r = 0.68$; $P < 0.01$). The lower concentrations of NO_3-N in the 20-m layer were around 120–175 μg N l^{-1} from 1995 to 1998, and 60–70 μg N l^{-1} in 1999 and 2000. The minimum Si concentrations fluctuated from 0.15 to 0.25 mg Si l^{-1} in the whole case study. After the minimum summer values, the epilimnetic increase of NO_3-N and Si concentrations took place with the deepening of the mixing layer (Fig. 1c). NH_4-N concentrations were generally below 20 μg N l^{-1} and contributed only secondarily to the total amount of nitrogen.

In 1999 and 2000, reactive and total phosphorus showed higher concentrations (up to 20 μg P l^{-1} of TP in 1999) in comparison to those measured in 1995–1998 (Fig. 1f). These differences were caused by the different extent of the spring vertical mixing, which determined a major recycling of phosphorus from the deepest layers to the surface during the 2 years (1999–2000) of complete overturn (Salmaso et al., 2001). A clear influence of the extent of the vertical water mixing was not evident for Si and NO_3-N.

Table 1. (a) Principal components analyses computed on the original physical and chemical variables and (b) on the same set of variables averaged over two consecutive samples. The two panels show the percentage of explained variance and the correlations between the first two components and the input variables. Significant correlations are reported in bold (P<0.01) and italic (P<0.05)

	(a)		(b)	
PCA Axis	I	II	I	II
Variance explained	64.8%	13.7%	68.7%	12.8%
T_{0-20}	**0.93**	0.00	**0.93**	–0.06
pH	**0.87**	0.09	**0.89**	0.13
$\Delta\delta_{0-20}$	**0.81**	**0.46**	**0.84**	**0.43**
I_{3d}	**0.61**	**0.68**	**0.66**	**0.68**
z_{eu}	**–0.87**	0.13	**–0.91**	0.09
NO_3-N	**–0.85**	*0.23*	**–0.85**	*0.26*
Si	**–0.83**	*0.29*	**–0.84**	*0.29*
RP	**–0.62**	**0.51**	**–0.67**	**0.47**

PCA

The ordination of the individual samples by principal components analysis is reported in Figure 2; arabic numbers indicate the month of sampling. The input environmental variables were chosen to represent the most important abiotic factors influencing the development of phytoplankton. These include water column stability ($\Delta\delta_{0-20}$) and average temperature (T_{0-20}) in the layer sampled for phytoplankton analyses, the extent of the euphotic depth (z_{eu}), algal nutrients (NO_3-N, Si and RP), pH (which is dependent on CO_2 concentrations) and solar radiation (I_{3d}). Compared with the present ordination, initially a PCA with TP was also computed, giving a two-axis ordination and component loadings less interpretable.

The first two principal components explain 64.8% and 13.7% of the total variance (Table 1a). The first component is highly correlated with all the considered variables, showing a positive correlation with temperature, pH, water column stability and solar radiation, and a negative correlation with the algal nutrients and the euphotic depth. The first component separates the late winter and early spring samples from those of late summer. This component reflects the temporal evolution of the climatic variables and the phytoplankton abundance (nutrient consumption and decrease of z_{eu}; *see next section*). The second component is highly correlated with solar radiation, RP and $\Delta\delta_{0-20}$, and separates the samples placed in the middle of the first component (i.e. the late autumn and early winter

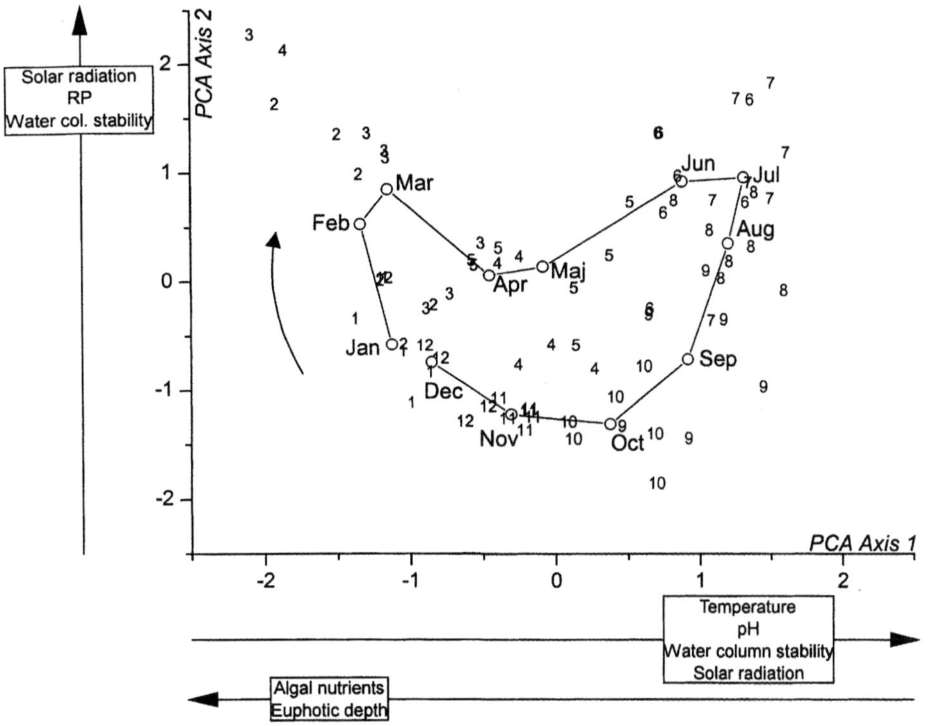

Figure 2. Ordination of samples by principal components analysis; arabic numbers refer to the sampling months. The correlations of the original variables with the two principal components (Table 1a) are highlighted. The circles corresponding to consecutive centroids of the different months have been joined with a continuous line.

samples from those of late spring and early summer); this component shows only a slight correlation with Si and NO_3-N (Table 1a). The second axis reflects the effects of the annual climatic evolution, along with the 'residual' high availability, and low concentrations of RP after the spring overturn and the autumn months, respectively. Overall, the plane defined by the first two components describes a complex environmental gradient, expressed by the combination of the considered variables.

The temporal path of the individual samples follows an annual cycle. This becomes clearer taking into account the seasonal evolution of the centroids of the different months, whose position in Figure 2 was determined by computing the monthly averages of the coordinates x_i and y_i for the complete set of samples.

Chlorophyll a, *phytoplankton biovolume and diversity*

The highest epilimnetic chlorophyll *a* concentrations (over 7 μg l^{-1}) were recorded in 1999 and 2000, i.e. during the years of complete overturn (Fig. 3a). Annual average values in 1999 and 2000 were, respectively, 3.7 and 4.1 μg l^{-1}. From 1995 to 1998, the corresponding annual average values ranged between 3.0 and 3.2 μg l^{-1}. The seasonal evolution of the biovolume values (Fig. 3b) followed closely those of chlorophyll *a* ($r = 0.78$; $P < 0.01$). Chlorophyll *a* and biovolume showed a positive correlation ($P < 0.01$) with pH (*r* equal to 0.41 and 0.56, respectively), and negative correlations with z_{eu}, NO_3-N, Si and RP ($P < 0.01$; $-0.64 < r < -0.33$).

Biovolume peaks were mainly caused by the development of single or few dominant species belonging to the Conjugatophyceae, Bacillariophyceae and Cyanobacteria; owing to their modest size, the contribution of ultraplankton and nanoflagellates was negligible (Fig. 3b). At least three species characterised the time course of the phytoplankton biovolume: *Mougeotia* sp., *Fragilaria crotonensis* and filaments ascribed to the *Planktothrix rubescens/agardhii* group (hereafter this taxon is referred to as *P. rubescens/ag.*). Owing to their higher biovolumes in comparison to the remaining phytoplankton, these three taxa have been defined as 'master species' (Salmaso, 2002). The three master species influenced significantly

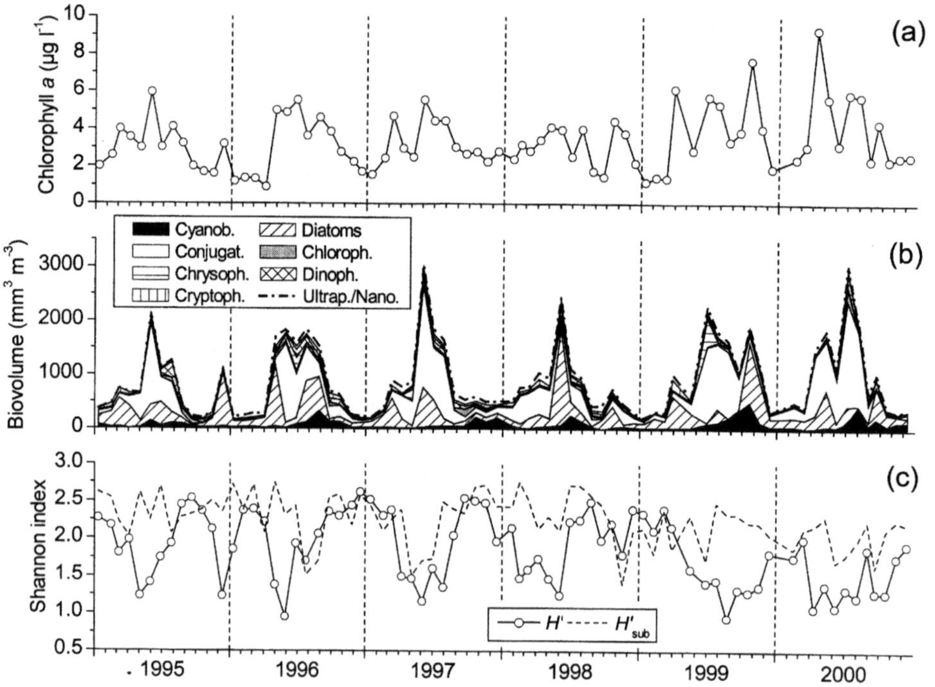

Figure 3. (a) Temporal variations of chlorophyll *a*. (b) Biovolume variations of phytoplankton subdivided by algal groups and ultraplankton/nanoflagellates in the layer 0–21 m. (c) Temporal variations of Shannon diversity computed for the whole phytoplankton community (H') and disregarding the contribution of the three master species *Mougeotia* sp., *F. crotonensis* and *P. rubescens/ag.* (H'_{sub}).

($P < 0.01$) the temporal evolution of the biovolumes of their corresponding algal classes, with correlation coefficients r equal to 0.997 (*Mougeotia* sp. vs. conjugatophytes), 0.91 (*F. crotonensis* vs. diatoms), and 0.89 (*P. rubescens/ag.* vs. cyanobacteria). Together, these three taxa strongly determined the time course of the total biovolume ($r = 0.91$; $P < 0.01$). *Mougeotia* sp. developed regularly in the period from spring to early autumn; *F. crotonensis* was mainly present during the spring or autumn months, but with irregular, significant biovolumes also in the remaining periods; *P. rubescens/ag.* developed mostly in summer and autumn, with populations persisting in winter and spring.

Phytoplankton diversity showed a regular temporal evolution (Fig. 3c). With few exceptions (e.g., December 1995), maximum values were found from summer to winter. The time course of diversity was strongly affected by the development of the most abundant taxa. In particular, the elevated growth of *Mougeotia* sp. in spring and early summer was important for the decrease of diversity. On the other side, diversity was sustained by the seasonal development of many other dominant or subdominant species. The Shannon index computed disregarding the contribution of *Mougeotia* sp., *F. crotonensis* and *P. rubescens/ag.* (H'_{sub}) showed greater values in comparison with those calculated on the whole data matrix.

Seasonal evolution of the phytoplankton community

Besides the three master species, many other taxa became important along the annual temporal gradient. A group of species, with biovolume peaks > 50 mm^3 m^{-3}, was distinguished from the remaining taxa. These species belong to the Bacillariophyceae (*Asterionella formosa*, *Aulacoseira granulata*, *A. islandica*, *Cyclotella* spp., *Stephanodiscus* sp. and *Tabellaria fenestrata*), Conjugatophyceae (*Closterium aciculare* and *C. pronum*), Chlorophyceae (*Coelastrum* spp., *Ulothrix* sp. and Ulothrichales), Cyanobacteria (Limnotrichoideae (*Limnothrix* sp. ?) *sensu* Anagnostidis & Komárek (1988), *Microcystis aeruginosa* and *Planktolyngbya limnetica*), Chrysophyceae (*Dinobryon divergens*, *D. sociale* and Ochromonadaceae), Cryptophyceae (*Plagioselmis nannoplanctica* (Novarino et al., 1994) (which, in Lake Garda, was previously reported as *Chroomo-*

nas acuta), *Cryptomonas* cf. *ovata* and *Rhodomonas minuta*) and Dinophyceae (*Ceratium hirundinella*). These taxa developed with minor biovolumes in comparison to the three master species (<400 mm^3 m^{-3}), with the exclusion of two peaks of *Cyclotella* spp. during summer 1996 (522 and 724 mm^3 m^{-3}).

Other three important algal components developed significantly. The first is *Anabaena lemmermannii*; during the 1990s this cyanobacterium developed episodic blooms strictly localised in the first centimetres of the water column and with low overall biovolumes in the first 20 m (Salmaso, 2000). The other components include two groups belonging to the Chlorococcales/coccal greens (mainly *Ankyra judayi*, *Monoraphidium* spp., *Oocystis* sp., *Pediastrum* spp. and undetermined coccal greens) and to the Chroococcales, mainly represented by *Aphanothece/Aphanocapsa*-type colonies and *Snowella* cf. *arachnoidea* (previously reported entirely as *Coelosphaerium kuetzingianum*).

To highlight the major seasonal patterns, for every taxon or group the biovolume data recorded during the 6 study years were plotted against the respective calendar days of sampling, smoothing the points by distance weighted least squares (DWLS; Wilkinson, 1990) (Fig. 4). The analysis does not include the filaments ascribed to the Ulotrichales, which were found with an isolated peak (203 mm^3 m^{-3}) on July 2000.

Taxa in Figure 4 have been roughly ordered considering the beginning of their seasonal increase over the 6-year period. The annual development followed an unimodal seasonal pattern for most of the considered taxa, with a single annual oscillation. Only few species presented high relative biovolumes in different parts of the year, with bimodal annual patterns (*F. crotonensis* and *Cryptomonas* cf. *ovata*). A summarising table, with a classification of phytoplankton taxa based on their seasonal maximum development, is reported in Table 2. A delimited group of diatoms was present with high biovolumes during the late winter and/or spring months (Table 2: seasonal group I). This included *Stephanodiscus* sp., *A. islandica*, *A. granulata*, *A. formosa* and *T. fenestrata*; these two last species, however, began their increase earlier, during the autumn months, with a broader seasonal development. Another important species during the spring months, which showed isolated peaks during late summer and autumn, was *R. minuta*. *Cryptomonas* cf. *ovata* and *F. crotonensis* (IIa and IIb) were found with high biovolumes during early spring and summer/autumn, and during late spring and autumn, respectively. *Mougeotia* sp. and *Ulothrix* sp. (IIc) began their increase during the late winter and spring months, maintaining or (in the case of *Mougeotia* sp.) increasing their abundance during the warmest months. *D. divergens* and *Ochromonadaceae* (IId) represented typical late spring taxa. An ample group of species developed during the summer months. *C. hirundinella*, Chlorococcales/coccal greens and Chroococcales (IIIa) showed an ample presence during the warmest months, whereas *D. sociale*, Limnotrichoideae, *Cyclotella* spp., *A. lemmermannii*, *P. limnetica* and *Coelastrum* spp. (IIIb) stood out for a marked delimitation of their growth period. Though represented also in other seasons, *P. rubescens/ag.* and *C. aciculare* (IVa) had their maximum development in summer and autumn, whereas *M. aeruginosa* (IVb) appeared more limited to the autumn months. The remaining two species had a wide or not wholly defined distribution (*P. nannoplanctica* and *C. pronum*).

Many phytoplankton species developed within delimited temporal periods, contributing to the ordered temporal successions of phytoplankton assemblages. Figure 5 reports the ordination of phytoplankton samples by NMDS. The cyclic character in the phytoplankton development was explicit. In every configuration the first axis separates the winter and early spring samples from those of summer. The second axis distinguishes the late spring and/or early summer samples from those of autumn; this distinction is less evident in 1999.

The distance between two successive dates in the NMDS configurations is proportional to the differences in phytoplankton composition. As evidenced in previous applications of this ordination technique (Salmaso, 1996, 2002), these distances take different values over the year. The high degree of aggregation of the winter samples in comparison to the more "dispersed" spring and summer samples, indicates a greater level of stability and homogeneous composition during the coldest months.

Community change rate (β_t)

Figure 6a reports an estimate of the community change rate; the X-axis reports the second date of the two chronologically adjacent samples involved in the comparisons. To compare the results, only the computations between successive samples separated by equivalent time intervals were considered. The reported values of β_t are strongly dependent from the time in-

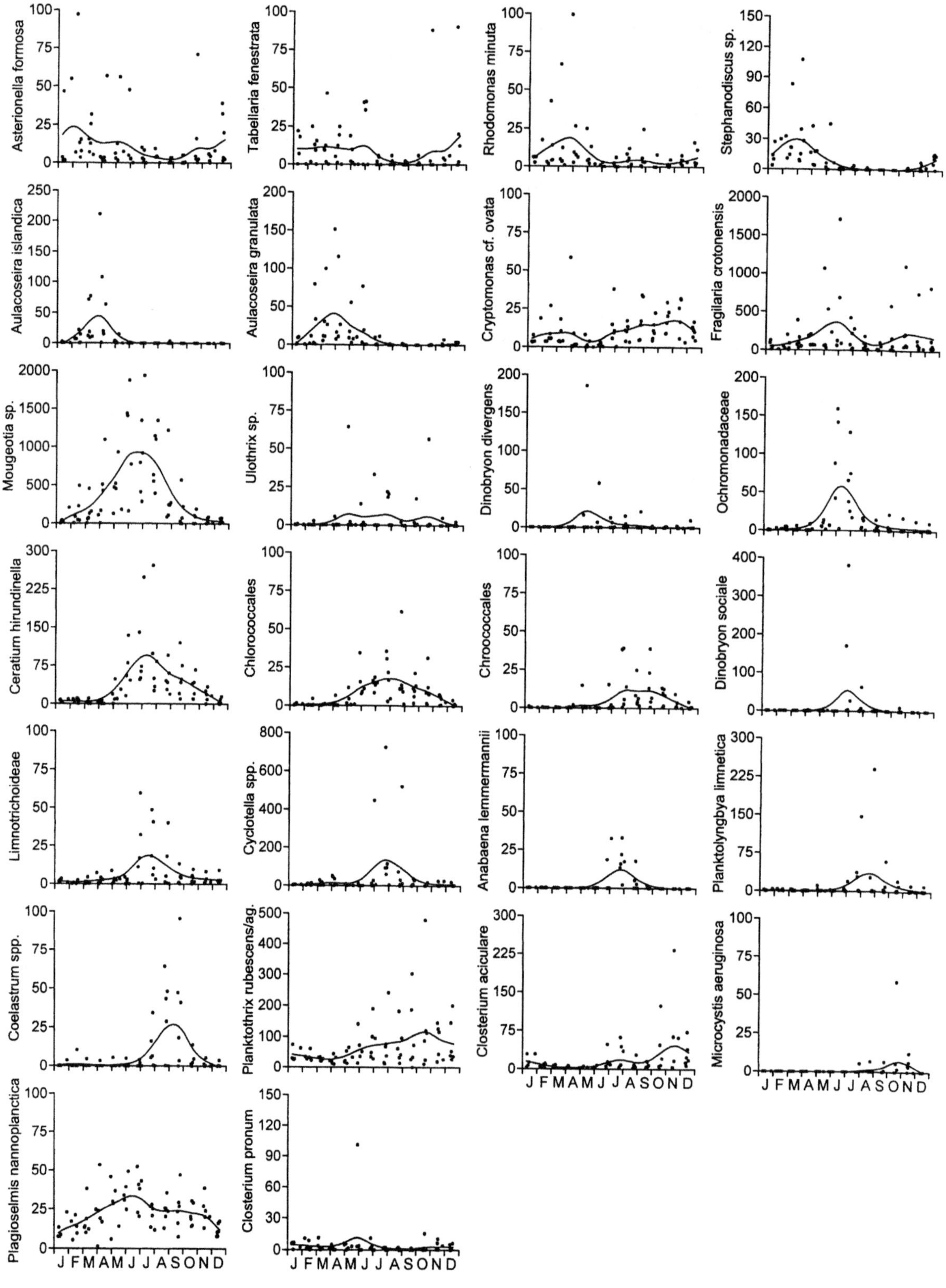

Figure 4. Annual distribution of biovolume values (mm^3 m^{-3}) for the most abundant taxa. Data refer to the period 1995–2000. The points were smoothed by distance weighted least squares (DWLS).

Table 2. Classification of phytoplankton taxa based on their seasonal maximum development. Abbreviations of taxa: Aste fo, *Asterionella formosa* Hassal; Tabe fe, *Tabellaria fenestrata* (Lyngbye) Kützing; Rhod mi, *Rhodomonas minuta* Skuja; Step sp, *Stephanodiscus* sp.; Aula is, *Aulacoseira islandica* (O. Müller) Simonsen; Aula gr, *A. granulata* (Ehrenberg) Simonsen; Cryp ov, *Cryptomonas* cf. *ovata* Ehrenberg; Frag cr, *Fragilaria crotonensis* Kitton; Moug sp, *Mougeotia* sp.; Ulot sp, *Ulothrix* sp.; Dino di, *Dinobryon divergens* Imhof; Ochromo, Ochromonadaceae; Cera hi, *Ceratium hirundinella* (O. Müller) Dujardin; Chloroc, Chlorococcales and coccal greens; Chrooco, Chroococcales; Dino so, *Dinobryon sociale* Ehrenberg; Limnotr, Limnotrichoideae (*Limnothrix*-like filaments); Cycl sp, *Cyclotella* spp.; Anab le, *Anabaena lemmermannii* P. Richter; Plan li, *Planktolyngbya limnetica* (Lemm.) Kom.-Legn. et Cronb.; Coel sp, *Coelastrum reticulatum* (Dang.) Senn and *C. polychordum* (Korš.) Hind.; Plan ru, *Planktothrix rubescens/agardhii* group; Clos ac, *Closterium aciculare* T. West; Micr ae, *Microcystis aeruginosa* (Kütz.) Kütz.; Plag na, *Plagioselmis nannoplanctica* Novarino, Lucas et Morral; Clos pr, *Closterium pronum* Bréb.

	Winter and spring species (I)	
Ia	Wide seasonal development, with autumn or late summer peaks	Aste fo, Tabe fe, Rhod mi
Ib	Exclusively late winter and spring species	Step sp, Aula is, Aula gr
	Species increasing during the spring months, with or without successive biovolume maintenance or increase (II)	
IIa	2nd increase during summer and autumn (bimodal)	Cryp ov
IIb	2nd autumn increase (bimodal)	Frag cr
IIc	Increase in spring and early summer, and autumn decline	Moug sp, Ulot sp
IId	Late spring species	Dino di, Ochromo
	Summer species (III)	
IIIa	Wide distribution in the warmest months	Cera hi, Chloroc, Chrooco
IIIb	Delimited distribution in the warmest months	Dino so, Limnotr, Cycl sp, Anab le, Plan li, Coel sp
	Late summer/Autumn (IV)	
IVa	Increase from late spring to the autumn maxima	Plan ru, Clos ac
IVb	Autumn	Micr ae
	Species with wide or not defined distribution	Plag na, Clos pr

terval used, i.e. they tend to low values for samples collected during short time intervals, and to their maximum values for samples taken in different seasons (cf. the corresponding distances in Fig. 5). For example, the maximum values of the Bray-Curtis index in the input matrices used in the NMDS analyses of Figure 5 were equal to 0.56 (March vs. September in 1995, and February vs. July in 2000). However, the regular evolution of the phytoplankton community and the cyclic and predictable patterns reported in the NMDS configurations (Figs 3 and 5) indicate that, in this large lake, the sampling frequency adopted (four weeks) was sufficient to grasp the large portion of the seasonal phytoplankton variations.

With some differences among the six years, the summer and the autumn months were characterised by a high community change rate. The lower values were found from December to March. The turnover rate computed disregarding the contribution of the three master species (β_{t-sub}) showed a high correlation (r = 0.94, $P < 0.01$) and values comparable with those computed taking into account the whole set of species (β_t).

To evaluate the relative weight of the most important environmental variables in the control of the community change rate, the β_t values were related to a complex environmental factor, obtained by PCA analysis. In order to obtain a set of coordinates corresponding to the β_t values, the PCA was carried out on a set of variables averaged over two consecutive samples (Table 1b). The results do not differ from those obtained with the original data (Table 1a), and will not be further examined. Figure 6b reports the relationship between the community change rate and the coordinates of the first component obtained from the PCA. The relationship is highly significant (r^2 = 0.31; $P < 0.01$). The first axis was the most representative in explaining the phytoplankton variability; the second axis was not correlated with the β_t values ($P > 0.05$).

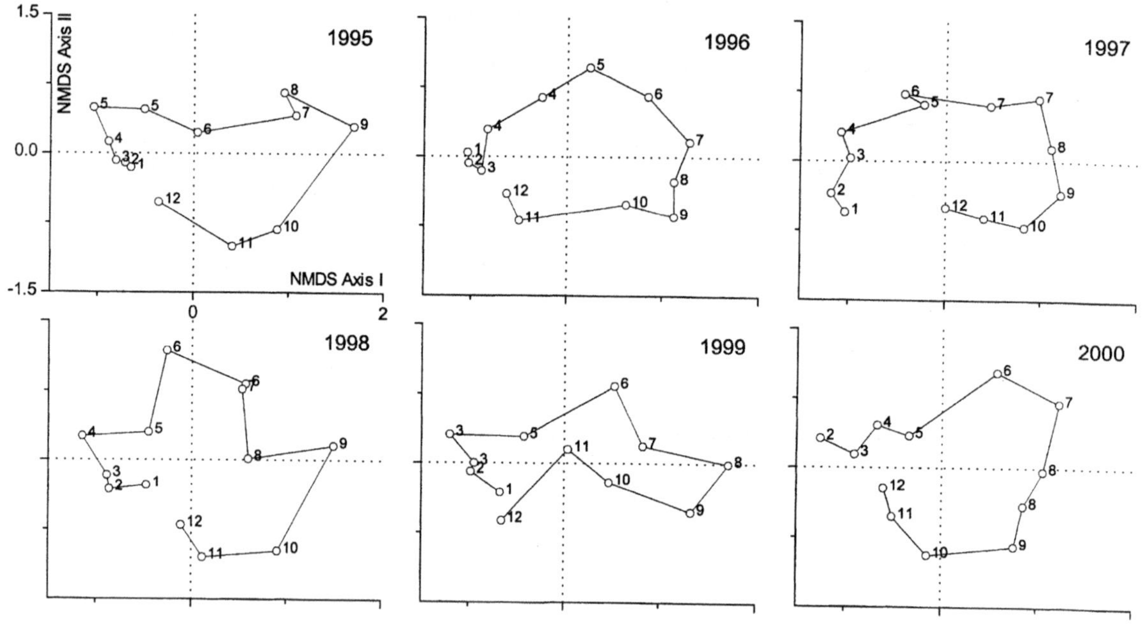

Figure 5. Ordination of phytoplankton samples in the two-dimensional NMDS configurations; the arabic numbers indicate the month of sampling. The stress values (Kruskal & Wish, 1978) of the six configurations are comprised between 0.05 and 0.11.

Apparent optimum conditions for growth

In this section, the adjective 'apparent' will be used in the meaning of 'observed', and will be clarified further on (*see 'Discussion'*). In Figure 7, the plane defined by the first two components of Figure 2 has been redrawn, evidencing the annual cycle of the centroids of the different months. Inside this outline, a 'species environmental positioning' has been carried out. This technique may be applied to various types of environmental gradients, as long as it exists as an interpretable structure for a chosen number of components (≥ 1) and an unambiguous seasonal or spatial pattern in the distribution of the species. In Figure 7 the species have been associated by highlighting the main four seasonal groups defined in Table 2. The position of the different taxa reflects their apparent optimum conditions for growth in a specific zone of the complex environmental gradient. In particular, the opposite life strategies of the winter/spring groups (I, left quadrants), compared with the summer groups (III, right quadrants), are well evidenced. The winter/spring taxa develop almost exclusively with high nutrient concentrations and high z_{eu}, low stability of the water column and low values of temperature, solar radiation and pH (cf. Table 1a). The opposite conditions characterise the development of the summer species. At the middle of the plane, phytoplankton is represented by 'transition species' from the two opposite seasonal poles.

An estimate of the apparent optimum conditions of the species growth is reported in Figure 8. The variables are those used in the PCA analysis (Figs 2 and 7), with the addition of TP. In every graph, species were ordered based on their coordinates x_k along the first axis of Figure 7. Similarly, the different variables were ordered taking into account their correlations with the first PCA axis (Table 1a). An ordered increase (T_{0-20}, pH, $\Delta\delta_{0-20}$, I_{3d}) or decrease (z_{eu}, nutrients) of the values was observed from the spring species to the summer ones (i.e. from *Stephanodiscus* sp. to *A. lemmermannii*). The main irregularities were found in solar radiation and in phosphorus concentrations. In the first case, the quantity I_{3d} is subject to transient perturbations (cloud cover), which poorly affects the other variables measured in the water column. In the second case, the gradients are poorly accentuated, due to the general low availability of P in this oligo-mesotrophic lake. A further difference is due to *M. aeruginosa* that, for some variables (pH, $\Delta\delta_{0-20}$, I_{3d}, nutrients), does not take the values expected from the corresponding gradients. This is due to its rank in the different graphs of Figure 8, which was determined considering only the variability along the first axis, whereas this species stood out for its

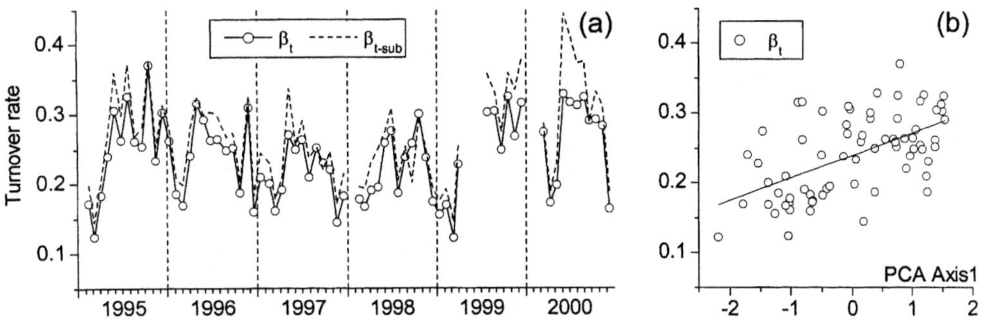

Figure 6. (a) Community change rate (β_t) estimated by the Bray-Curtis index calculated, from biovolume data, on the couples of chronologically contiguous phytoplankton samples; the dashed line reports the temporal variations of the community change rate (β_{t-sub}) calculated disregarding the contribution of the three master species; the X-axis reports the second date of the two samples involved in the comparisons. (b) Relationship between the community change rate (β_t) and the coordinates of the first component obtained from the PCA analysis summarised in Table 1b.

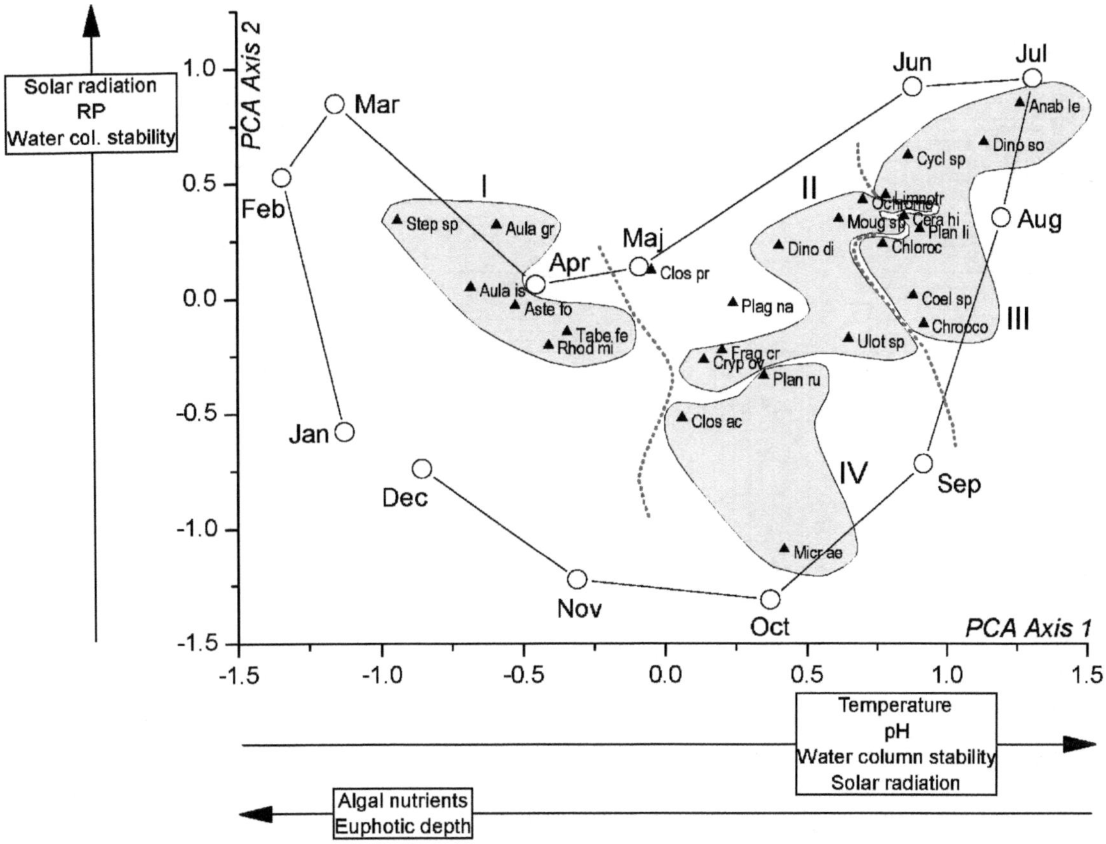

Figure 7. Positioning of the dominant taxa on the complex environmental gradient defined by the two PCA axes of Fig. 2. The taxa have been associated by highlighting their membership to one of the four main seasonal groups defined in Table 2. The dotted lines indicate the principal heterogeneities within the taxa based on the differences in the environmental optima reported in Fig. 8. Abbreviations of taxa as in Table 2 (see explanation in the text).

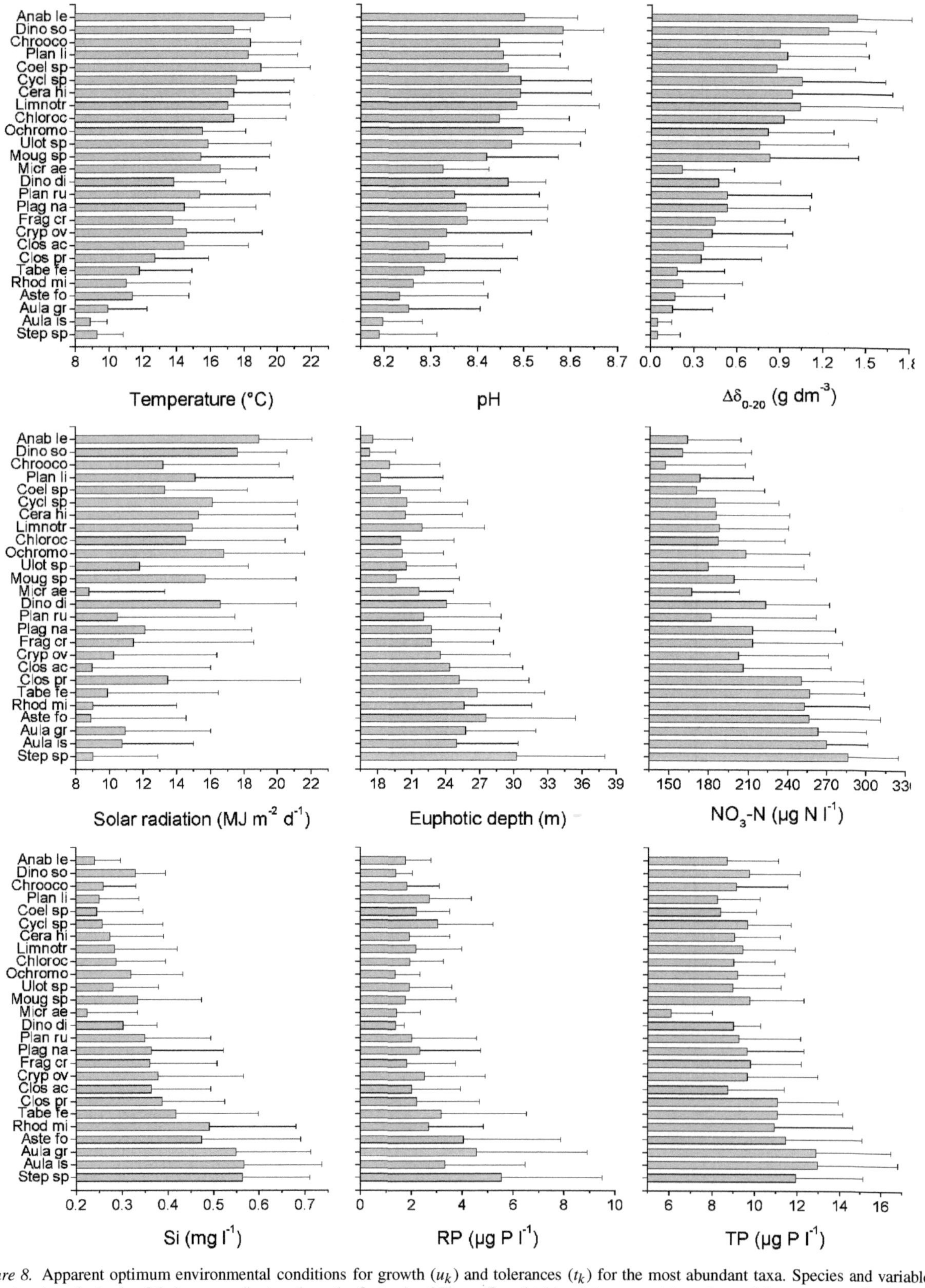

Figure 8. Apparent optimum environmental conditions for growth (u_k) and tolerances (t_k) for the most abundant taxa. Species and variables are ordered on the basis of their x_k coordinates along the first axis of Figure 7, and taking into account the values of the component loadings in Table 1a, respectively. Abbreviations of taxa as in Table 2.

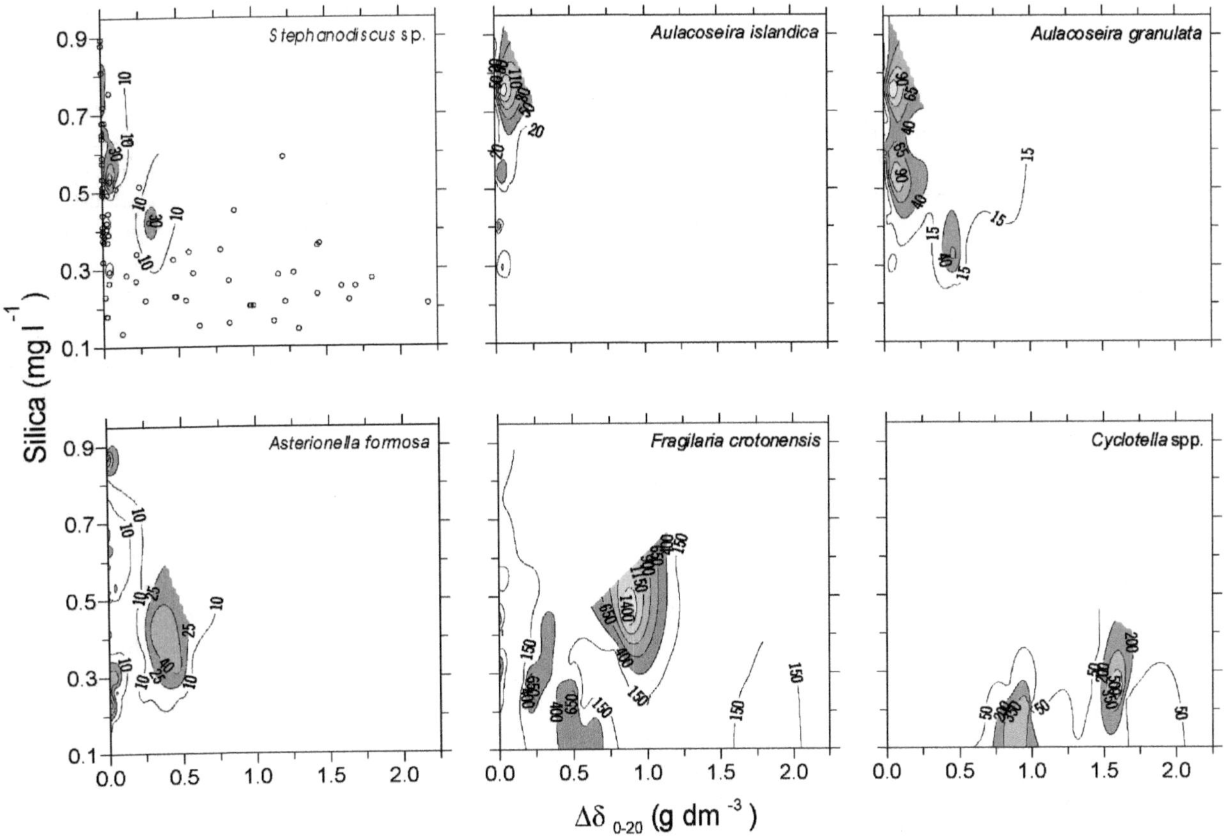

Figure 9. Niche response surfaces for the diatoms. Contours represent biovolume values at a range of differences of water density between 0 and 20 m ($\Delta\delta_{0-20}$) and Si concentrations; for every taxon, the minimum values of the contour lines represent around 10% of the respective maximum biovolumes. In the first panel the points used to construct the surfaces are indicated with hollow circles.

marked positioning in the lower quadrants of Figure 7. The tolerance values indicate a high degree of adaptability of many species at different environmental conditions. However, the differences are notable when the comparison is made between the winter/spring and summer species (I, III).

The environmental optima computed in Figure 8 are compatible with the position of the taxa in Figure 7. A same ordination of the species, fully compatible with that of Figure 7, was obtained by PCA calculated from the correlation matrix of the untransformed apparent optimum values of Figure 8 (the percent of total variance explained by the first two components was, in this case, equal to 91%; figure not shown).

To aid the identification of the principal heterogeneities within the species, the single taxa in Figure 7 were subjected to cluster analysis (Pearson distance, Ward's method) based on their environmental optima. The major group separations are reported by dashed lines. The winter/spring taxa form a compact cluster (I). A second large cluster includes the summer taxa (III). The last group includes many transition species (II and IV in Table 2).

Discussion

The ultimate aim of ecological research is to identify the factors and to study the interactions that control the distribution and abundance of the organisms in the environment. This task may be carried out using two approaches (Sommer, 1988). The ecophysiological approach seeks differences in distributional patterns by differences in the specific experimentally established growth requirements, competitive abilities and mortality rates (e.g., Tilman, 1977; Tilman et al., 1986). The comparative descriptive approach tries to find distributional patterns from field studies and recorded abundances. This approach may follow different strategies (Gauch, 1982), whose choice depends on the temporal and spatial scales involved and on the life cycles of the studied organisms. In terrestrial vegetational studies, differences in the contemporary composition and abundance may be followed along spatially extended habitats (e.g., altitudinal and moisture gradients; Whittaker, 1956) or along temporal gradients, during primary successions (e.g., Wood & del Moral, 1987). With the comparative approach, phytoplankton studies try to extrapolate specific life traits from samplings carried out generally with weekly or monthly frequency; eventually, the species that dominate the annual cycles from different lakes ordinated along environmental gradients may be compared, to find their optimum conditions for growth for a specific factor (e.g., trophic level or acidification; Reynolds, 1998; Lepistö, 1999). The study of the temporal changes in phytoplankton composition and abundance (cf. Fig. 4) may be conceived as a direct gradient analysis performed along a complex of seasonally changing environmental variables. These variables may present a high degree of correlation (e.g., Table 1a), acting jointly as a complex forcing factor that selects seasonal groups of species sharing similar requirements. At the same time, it is not always easy to extract the correct set of environmental variates influencing the wax and wane of phytoplankton from complex gradients. The intensity of competition and the action of other biotic factors (e.g., grazing) occurring within the potential competitors at a particular region of any complex environmental gradient determine the ultimate structure of the phytoplankton community, its diversity and seasonal dominance patterns.

The complex environmental gradient

The whole set of environmental variables considered in Figure 1 were characterised by a strong temporal component. $\Delta\delta_{0-20}$ values were highly correlated with the corresponding I_{3d} values ($r = 0.73$; $P < 0.01$). Maximum average water temperatures in the layer 0–20 m were shifted of 1–2 months in comparison to the I_{3d} maxima, with correlation coefficients equal to 0.73 and 0.79 for time lags of 1 and 2 months, respectively, against an r value of 0.48 computed on the original data. The high degree of correlation of I_{3d} and $\Delta\delta_{0-20}$ is reflected by their high loadings on the first two components (Table 1). The remaining variables (excluding RP), were strongly correlated only with the first component, reflecting the existence of temporal lags also in the response of these variates against the increase of solar radiation.

The seasonal increase of pH values was strongly associated to phytoplankton activity. Salmaso & Decet (1998) showed that, in Lake Garda, from 1995 to 1996, $H_2CO_3^*$ ($CO_{2(aq)}$ + H_2CO_3) concentrations in the layer 0–20 m exceeded the equilibrium values 1.5–2.5 fold from November to April. During the warmest months (particularly June–August), a slight undersaturation (by 0.8–0.5 fold) was observed only at the surface, whereas at 20 m $H_2CO_3^*$ concentrations remained around or a little above the air equilibrium values, suggesting the presence of only a partial limitation of the dissolved carbon on the phytoplankton growth in the layer 0–20 m for pH values ranging from 8.0 to 8.8. A computation of CO_2 saturations on the data from 1997 to 2000 essentially confirms these conclusions.

The decrease of nutrients in the first 20 m began after the maximum spring replenishment, and in coincidence with the stabilisation of the water column. Water flux from rivers is low, and mainly limited to the northern shore of the lake, so that the spring supply from the deep waters constitutes a significant input of new nutrients to the productive layers every year. With the progressive formation of a stable and isolated epilimnion (Fig. 1b,c), nutrients originating from the hypolimnetic waters decrease their importance, therefore the phytoplankton growth from May to September appears to rely also on nutrients regenerated by microbial mineralisation and zooplankton grazing (Salmaso & Decet, 1998).

The high regularity (Fig. 1) and high temporal synchronicity (cf. Table 1a) that characterise the environmental variables is due to a mix of causes like the annual climatic evolution, the vertical mixing of the water column from late autumn to early spring (a 'homogenisation factor'), and the high inertial characteristics of this large system, which is able to minimise local perturbations (e.g., meteorological events), so that the lake clearly bears the signs of the modifications brought about by phytoplankton during summer.

Phytoplankton niches

The arrangement of the seasonal groups in Table 2 aims to identify the most probable temporal location of each species, leaving the differences in absolute abundance with the other taxa out of consideration. In fact, if absolute biovolumes are used (Salmaso, 2002), the classification scheme in Table 2 should be strongly affected by the presence of the most abundant species, in particular by the three master species (*Mougeotia* sp., *F. crotonensis* and *P. rubescens/ag.*).

Species that have a common seasonal development tend to share similar resource requirements and life strategies. Resource is any environmental factor necessary to maintain a positive net growth balance (e.g., nutrients, light, water turbulence). However, only few species are able to develop with high biomasses, typifying strongly the algal assemblages in different periods of the year. These considerations raise two kinds of problems, i.e. the definition of the optimum environmental conditions needed for the growth of different groups of phytoplankton species (Figs 7 and 8), and the causes of the observed different abilities to obtain high dominance along the complex environmental gradient (Figure 7; *next section*).

A key problem is represented by the interpretation and reliability of the estimated environmental apparent optima (Fig. 8). In fact, these estimates, which are derived from field observations, do not necessarily coincide with the physiological optima. The concept of niche given by Hutchinson (1967) may help to clarify the problem. For each of the *n* variables that control the development of a particular species, a set of corresponding *n* axes defining a hyperspace may be conceived. For each axis a range of values exists over which the organism can develop successfully. The 'fundamental niche' is the *n*-dimensional hypervolume defined by the range of values where all the environmental conditions are favourable for growth in the absence of competition and other interactions. The 'realised niche' is the reduced hypervolume in which a species exists as consequence of competition and limitation by other most critical factors, i.e. the observed resource use of a species. Some species may not occupy their best part of the fundamental niche, while others do. In Figure 8 the ranges of the coordinates defining favourable conditions for growth have been substituted by the estimates of the observed optimum values and tolerances; the limits of the hypervolume defined in this way delimit the relative success of the species population within the realised niche.

The variables considered in Figure 8 have different critical importance for the selection of taxa along the temporal gradient. For example, silica is essential for diatoms and pH can be important for species with particular acidity tolerances and CO_2 requirements. An important task for ecophysiologists is to seek the most relevant resource axes for any population of species. An important conceptual tool of physiological ecologists is the Shelford's law of tolerance, which states that the distribution of a species will be controlled by that environmental factor for which the organism has the narrowest range of tolerance; these tolerance limits define the fundamental niche of the species (Krebs, 2001).

In the real world, the attainment of limiting values for some critical environmental variables narrows the ranges of the other variables, with a sort of 'switch effect'. The growth of a species may be confined to a well-delimited temporal period, defined by the tolerance to a critical variable. In turn, this controls the observed ranges of the other variables. Along with competition and other biotic interactions, these processes act in modelling the reduced hypervolumes of the realised niches. The elucidation of the processes controlling the wax and wane of populations requires the identification of the most valuable variables controlling the development of different phytoplankton groups along the temporal gradient. In the next section, for example, the development of diatoms will be analysed more in detail as a function of water column stability and silica concentrations.

A further problem is represented by the difficulty to generalise the information about the properties of the realised niches for various algae. The deep southern subalpine lakes have a comparable seasonal phytoplankton development (Buzzi, 2002; Morabito et al., 2002; Salmaso, 2002). However, many discrepancies may be found considering different typologies of lakes. Nygaard (1996), studying the temporal and spatial development of individual species of phytoplankton from European lakes, found that, in several cases, the species were characterised by one individual behaviour in one group of lakes, while the same species behaves differently in another group of lakes. Nygaard suggested that these differences would be the result of the existence, within several species, of two or more types with different tolerances and environmental requirements. Perez Martinez & Sanchez Castillo (2001, 2002) analysed the temporal occurrence of *Ceratium hirundinella* in 100 Spanish reservoirs. They found that this species was very commonly present in winter time, during mixing conditions, low temperatures and low light availability, in contrast with the traditional observations (including this work), which report the occurrence of *C. hirundinella* in warm and stratified waters. Perez Martinez & Sanchez Castillo (2001) concluded that the temporal occurrence of this dinophyte in the southern north-temperate systems could be regulated by different factors than those operating in the northern north-temperate zone. The contrasting behaviour of many species in different type of lakes may be explained by postulating the existence of different ecotypes ('types', Nygaard, 1996), or by assuming different occupations of the fundamental niche of the same species in different lakes with diverse climatic, morphometric, hydrological and trophic characteristics.

Selection of functional groups along the complex environmental gradient

Without experimental work, identifying the intensity of competition for resources among phytoplankton species is only a matter of speculation. To solve these problems, an experimental approach would involve the removal of selected species within an assemblage and the comparison of the growth of the remaining species with untreated controls.

In disturbed or fluctuating environments the strength of the interactions among species is strongly reduced and the direction of competition can be inverted, preventing extinction of the worst competitors. In aquatic environments, the stability of the water column is considered a key factor in the control of the structure of phytoplankton assemblages (Harris, 1983; Dokulil & Teubner, 2003; Morabito et al., 2003). On a temporal scale of months, the passage from non stratified, turbulent water column, to conditions of high stratification and water column stability is instrumental for the selection of species with different life strategies. In this section, the phytoplankton changes will be analysed taking into account the specific competitive abilities of the most important groups. The temporal dynamics of the single species and seasonal assemblages have been analysed in detail by Salmaso (2000, 2002).

During the winter and early spring months, the high extention of water mixing (down to or over 150 m, Fig. 1) determines high nutrient accessibility and, also, a decreasing light availability due to low solar radiation and, despite high z_{eu} values (Fig. 1b), low z_{eu}/z_{mix} ratios. Competition for nutrients among species is reduced. In Lake Garda a typical assemblage of the winter and early spring months (group I) is composed by the large diatoms *A. islandica, A. granulata, Stephanodiscus* sp., *A. formosa* and *T. fenestrata* (Figs 4 and 7; Table 2). The seasonal development of diatoms is strongly controlled by the thermal stability of the water column and by the availability of silica (Kilham, 1971; Reynolds, 1984). In Figure 9 the biovolume values of the winter and early spring diatoms – along with *F. crotonensis* (group II) and *Cyclotella* spp. (group III) – have been superimposed on the plane defined by the concentrations of Si and by the estimate of water column stability ($\Delta\delta_{0-20}$). In this way, a measure of the population response of the species at each point of the two-dimensional niche is given (cf. Begon et al., 1996). The contrasting strategy between the winter–early spring diatoms (*Stephanodiscus* sp., *Aulacoseira* spp. and, partly, *A. formosa*) and the small summer *Cyclotella* spp. is quite evident. *F. crotonensis* appears to have an intermediate response between these two extremes (see also Fig. 7, group II). *T. fenestrata* (figure not shown) has intermediate characteristics between the winter–early spring diatoms and *F. crotonensis*. Species of the cold months develop almost exclusively during high Si concentrations and high water turbulence. The relative importance of these two variables is difficult to ascertain, for high losses by sinking and mortality may override elevated growth rates, even when Si is not limiting (Reynolds, 1984; Sommer, 1987). However, taking into account that half saturation constants for Si-limited growth are generally between <0.1–0.2 mg l^{-1}, a negative net growth balance for the winter and early spring diatoms is more probably linked to increasing thermal stability and increasing sinking losses. Besides water turbulence and silica concentrations, many diatoms are favoured by their ability to become light saturated at low PAR, so that they compete successfully with other phytoplankters in mixed waters where z_{mix} exceeds z_{eu} (Reynolds, 1984; Sommer, 1987). Moreover, owing to the good competition for phosphorus under sufficient Si supply, high concentrations of this element are a positive factor for the development of diatoms (Tilman et al., 1986; Sommer, 1987). In terms of adaptive strategies (Grime, 1977; Reynolds, 1988) the diatoms belonging to the group I (Fig. 7) consist of ruderal species (R-strategists). The cryptophyte *R. minuta* (a competitive, C-strategist) attains its maximum biovolume during spring time. During the warmest months the growth of this species – and of the related *P. nannoplanctica* – may be strongly limited by zooplankton grazing. The winter and early spring assemblages are strongly controlled by allogenic factors (high mixing and disturbance, low biotic coupling, low nutrient limitation and competition).

At the other extreme of the environmental gradient (Fig. 7), the abundance of the small and light *Cyc-*

lotella spp. during summer is due to their ability to remain in suspension and to compete at low Si:P ratios with the larger pennates (cf. Fig. 9) (Tilman & Kilham, 1976; Tilman, 1977; Sommer, 1987). Besides the small centric diatoms, the summer group (III) is represented by many species belonging to different algal orders and sharing very similar apparent optimum conditions for growth (Figs 7 and 8). Overall, the summer assemblages develop during strong thermal stability of the water column and low availability of nutrients (high stress), i.e. under conditions favouring stress-tolerant algae (S-strategists). However, within this group many fine adaptations promote phytoplankton coexistence and diversity (*see below*). In this period grazing is high, with an elevated development of copepods (*Copipodiaptomus steueri*) and cladocerans (*Diaphanosoma brachyurum, Bosmina (Eub.) longicornis kessleri* and *Daphnia hyalina-galeata* complex) (Salmaso & Naselli-Flores, 1999). The control by the winter and early spring environmental constraints (mixing and very low z_{eu}/z_{mix} ratios) is terminated, whereas it increases the importance of species interactions and potential competition. However, a final outcome of competition processes, with a definitive winner, is never attained, and recurrent mixed and highly diversified assemblages may be identified every year (*see* Salmaso, 2000, 2002). The high number of subdominant taxa and higher turnover rates (β_t) in the warmest months strongly contrast with the Principle of Competitive Exclusion. Under the assumption of strong competition, environmental stability and low disturbances, a community in equilibrium should be composed of very few species. The minimisation of competition by disturbances (e.g., size-differential grazing) is essential in supporting species coexistence. However, the heterogeneity of phytoplankton composition during the stratified months could be simply promoted also by different life strategies.

Usually, the Chlorococcales (here represented by *Coelastrum* spp. and a heterogeneous group of Chlorococcal greens), Chroococcales (*Aphanothece/Aphanocapsa*-type colonies and *Snowella*) and Nostocales (*Anabaena lemmermannii*) are typical orders that respond positively to the increasing thermal stability in temperate lakes (cf. Reynolds, 1996, 1997). These orders are morphologically adapted to contrast large sinking losses in stratified lakes by means of adaptations including small size, mucilage formation and buoyancy control (gas vacuolated algae). Moreover, colony size of cyanobacteria and/or the presence of gelatinous envelopes are a protection against zooplankton grazing. In the small chlorococcal algae, the gelatinous envelopes and the thick cellulose cell wall may render the coccoid greens difficult to digest by the zooplankton (Porter, 1973). Some of these adaptation traits (low sinking rates, moderate or low susceptibility to grazing) are partly shared also by the two thinnest Oscillatoriales (*Pseudanabaena limnetica* and filaments ascribed to the Limnotrichoideae).

The delimited temporal development of *Anabaena lemmermannii* during periods when nitrate concentrations are low conforms to the general ability of Nostocales to fix atmospheric nitrogen when other N sources are depleted. In the first two metres, NO_3-N during summer may occasionally reach very low concentrations (below 10 μg N l^{-1}), whereas, on average, the nitrogen pool in the first 20 m never falls below 50 μg N l^{-1} (Fig. 1). On the other hand, extended blooms of *A. lemmermannii* in Lake Iseo developed during high surface concentrations of nitrate (L. Garibaldi, pers. comm.); this is an observation that agrees with the dominance of Nostocales documented in other basins during high nitrate concentrations (e.g., Jensen et al., 1994), thus confirming that the success of heterocystous cyanobacteria is determined, besides the ability to fix nitrogen, by other factors (cf. Blomqvist et al., 1994; Hyenstrand et al., 1998). Anyhow, the surface blooms of *Anabaena* in Lake Garda were observed exclusively during high thermal stability, high solar radiation (cf. Fig. 8) and low wind velocities (Salmaso, 2000).

Some species developing during late spring and during the summer months are mixotrophs (e.g., *Dinobryon* spp. and *Ceratium hirundinella*). This modality of nutrition, in environments where inorganic nutrients have been consumed, provides a further means of growth (Nygaard & Tobiesen, 1993; Isaksson, 1998, Jones, 2000). When phagotrophy predominates, mixotrophs may release soluble nutrients facilitating phototrophs (Rothhaupt, 1996). Other mixotrophs are present also during spring and relative high nutrient concentrations (Ochromonadaceae and *Cryptomonas* cf. *ovata*), suggesting that this alternative mode of nutrition may not represent exclusively a competitive adaptation against nutrient depletion. *Cryptomonas* is considered a genus whose primary mode of nutrition is phototrophy, ingesting particles at low rates. However, recent detailed studies supported the idea that this genus utilises efficiently nitrogen and phosphorus from bacteria under conditions of nutrient depletion (Urabe et al., 2000).

An additional strategy to supplement nutrients in phytoplankton may involve luxury uptake of N- and P-compounds (Reynolds, 1984). Nutrients may be accumulated within cells as storage products deposited when their concentrations become high, e.g., in micropatches from zooplankton excretion. P and N limitations are considered episodic in time and usually weak. However, Hudson et al. (2000), using new radio-bioassay techniques, found that phosphate levels in many lakes were two to three orders of magnitude lower than previously thought; this suggests that phosphorus is the most essential of nutrients and so the staff of life (Karl, 2000).

These different ways to obtain nutrients reduce the competition among species (see also Steinberg & Geller, 1993). A further way to promote coexistence is vertical migration. Phytoplankton during thermal stratification may be irregularly distributed in the euphotic layer. For example, *Anabaena lemmermannii* in Lake Garda forms wide surface patches with diurnal dynamics, whereas Oscillatoriales sometimes develop metalimnetic maxima (Salmaso, 2000). Unhomogeneities in the phytoplankton distribution along the water column guarantee a major niche diversification through the exploitation of different environmental gradients (cf. Levin, 1974). The positioning in different places in a given habitat to reduce competition is a common strategy in nature. For instance, in a classic study, MacArthur (1958) explained the bird diversity in boreal forests by differences in their feeding positions on the tree branches and by differences in food. The development of many fine evolutionary adaptations, which favour niche differentiation and α diversity, is a general characteristic of many plant and animal communities (Wilson, 1992).

The different life strategies and biology of the component species described above contribute to increase the diversity by means of endogenous mechanisms. In general, the organisation of the species through niche and habitat partitioning tends to shape the communities around their local equilibrium conditions.

Besides the above endogenous mechanisms, exogenous factors can also promote diversity. Disturbance effects are considered as important promoters of community diversity (Connell, 1978). A disturbance is any episode that changes the status of resources, disrupting community structure and equilibrium dynamics. Several studies about the relationships between phytoplankton diversity and disturbance have been made (Sommer et al., 1993; Reynolds et al., 1993). In large and deep lakes characterised by low water renewal and with algal production developing in a large euphotic layer, physical disturbances are reduced in comparison to small systems (i.e. large and deep lakes have higher stability and resilience, *sensu* Pimm, 1984). This finds support in the high regularities that characterise the development of the variables measured in-lake in comparison to the higher variability of solar radiation. However, the existence of both endogenous and exogenous mechanisms promoting diversity is consistent with the view that real communities are spread along a continuum from equilibrium to non-equilibrium (Krebs, 2001). These mechanisms are temporally scaled, providing a variety of ways in which the probability of coexistence is enhanced and diversity increased (cf. Begon et al., 1996). From a general point of view, near-equilibrium communities are not necessarily destined to become dominated by very few species.

The late spring/early summer species (group II) and the late summer/autumn species (group IV) have apparent optimum conditions for growth at the middle of the environmental gradient (Table 2; Figs 7 and 8). These two groups include several ruderal species different from the filamentous centric diatoms. A few of the species belonging to these intermediate groups are adapted to develop within a wider range of environmental conditions; in this case, however, the maximum growth is attained in different periods (Table 2; Fig. 4). This is evident for the three master species *Mougeotia* sp., *Fragilaria crotonensis* and *Planktothrix rubescens/agardhii*. High biovolume levels of these three taxa were not reached during periods characterised by high physical disturbance (deep mixing and low light) or during high thermal stratification, when equilibrium dynamics and competition are at their seasonal maxima. The extreme physical control and the increase of competition for resources in oligotrophic or mesotrophic lakes appear both incompatible with the attainment of high biovolume levels.

The factors causing the seasonal wax and wane of the three master species have been recently discussed (Salmaso, 2000, 2002). The dominance of *Mougeotia* sp. is promoted by its good competition for phosphorus and its resistance against sinking (Padisák et al., 2003) and grazing. Laboratory experiments carried out in Lake Constance showed that *M. thylespora* was the most successful competitor for P at low Si:P ratios among non diatoms; moreover, in-lake experiments carried out in the field demonstrated that *Mougeotia* presented annual average sinking velocities in the layer 0–20 m between 0.06–0.15 m d^{-1}, against corresponding average velocities of 1.3–1.5 m d^{-1} found for the colonies of *Fragilaria crotonensis* (Sommer, 1987). The large filaments of *Mougeotia* are poorly edible by the zooplankton. A recent study carried out in Lake Garda on the susceptibility of phytoplankton to grazing showed that the inedible unicellular fraction and filamentous algae became abundant during the maximum development of zooplankton, which is coincident with the development of herbivorous cladocerans (Salmaso, 2002). The collapse of this species during the summer months takes place during the deepening of mixing, decreasing z_{eu}/z_{mix} ratios and maximum nutrient depletion.

In contrast with the large centric diatoms, the higher development of *F. crotonensis* is limited to moderate mixed conditions (Fig. 9). Its absence during the warmest months is due to high mortality by enhanced sinking and low Si availability.

Filaments of *Planktothrix rubescens/agardhii* are particularly adapted to conditions of low irradiance (Reynolds, 1984), showing a vertical zonation during summer, with metalimnetic maxima around the limit of the euphotic zone and vertical homogenisation and dilution in autumn and winter (Salmaso, 2000). Walsby et al. (1998) and Micheletti et al. (1998) showed that this seasonal evolution is linked to the perennation strategy of *Planktothrix* in the deep lakes, with the autumn populations that survive in the mixing column during the successive coldest months. Other two typical autumn species in Lake Garda are *Closterium aciculare* and *Microcystis aeruginosa*. Apparently, these two taxa appear to respond positively to moderate water mixing and illumination.

Conclusions

The most abundant species living in the deep and large Lake Garda are characterised by regular annual developments, which result from different selection of specific life traits along the complex environmental gradient. The development of the different species during their favourable growth period may be characterised by biovolume differences from year to year. However, at the community level, the temporal regularities that distinguish the annual development of phytoplankton are favoured by the high stability and resilience that characterise large and deep lakes with low water renewal times.

The application of multivariate analyses based on the different specific environmental optima highlighted the existence of three large algal groups. The first was represented by the winter and early spring diatoms (*Aulacoseira* spp., *Stephanodiscus* spp., *Asterionella formosa* and *Tabellaria fenestrata*). This group develops during high water turbulence, high nutrient concentrations, low light conditions and low z_{eu}/z_{mix} ratios. The competition is reduced and the community is under strong physical control. At the opposite extreme of the complex environmental gradient (summer months), the second group was composed by many species belonging to cyanobacteria (thin Oscillatoriales, Nostocales and Chroococcales), Chlorococcales and coccal greens, *Dinobryon sociale* and *Ceratium hirundinella*. Common life history traits in this group include the ability to contrast losses by grazing and sinking in stratified and stable conditions, and the ability to tolerate nutrient deficiency. The remaining group developed in the 'intermediate' seasons (mainly late spring/early summer and/or autumn), i.e. during environmental conditions (and, possibly, stress competition) in the middle of the two previous extremes. This group is strongly typified by the three master species *Mougeotia* sp., *F. crotonensis* and *P. rubescens/ag.*. These taxa are characterised by higher biovolumes and extended periods of dominance in comparison to the remaining subdominants. Their success appears favoured by different abilities to contrast grazing and sinking. High dominance of the three master species is prevented during high physical disturbance (mixing and low light) or during high stability of the water column, when equilibrium dynamics and competition are at their seasonal maximum.

During the summer months, the increasing thermal stability of the water column coincides with the maximum nutrient depletion; compared with the groups developing during mixing and higher nutrient concentrations, the interaction and competition among species increase, but without the selection of exclusive winners. In contrast with the Principle of Competitive Exclusion, the elevated number of coexisting species does not find support with equilibrium dynamics controlled by competition. Many factors, regulated by phytoplankton or under external control, can minimise competition, allowing the coexistence of many phytoplankton species in stratified and stable environments. Exogenous factors are represented by disturbance events which tend to disrupt equilibrium dynamics. Endogenous factors consist of many fine adaptations, which include vertical positioning, supplemented assimilation of nutrients by mixotrophy, N-fixation by Nostocales and luxury consumption; these factors tend to shape the community around its local equilibrium, by means of niche and habitat partitioning. The overall balance is consistent with the existence of a phytoplankton community oscillating along a continuum, from equilibrium to non-equilibrium. Anyhow, the enhancement of competition among species with similar requirements and the coexistence of different life strategies, has the property to increase the community change rate, which, during the stratified period, is at its maximum.

The disturbances acting on short to medium time and the existence of fine adaptations reducing competition increase α diversity. These factors, which act with different intensity over the seasons, may also contribute to the temporal differentiation of algal assemblages. On the other side, besides the above factors, β_t diversity in medium and high latitude regions is strongly promoted by slow and regular annual evolution of the climatic variables. Under these assumptions, a hypothetical attainment of equilibrium conditions in phytoplankton communities could be only transient (locally stable equilibrium). Equilibrium communities tend to be stable (*sensu* Pimm, 1984), recovering quickly from disturbances and with high persistence over time. However, time changes the rules of the game continually, interfering with the competitive exclusion dynamics. Over a temporal scale of years, phytoplankton undergo cyclic compositional changes which are comparable, as magnitude, to those that may be documented in terrestrial vegetation over the millennia during the glacial-interglacial cycles (e.g., Miyoshi et al., 1999). In these two cases, the identification of equilibrium conditions strictly depends on the choice of the temporal window. If scaled over many generations, vegetation communities never approach stable and continuous equilibrium conditions.

For every phytoplankton species, the estimates of the optimum conditions for growth and the realised niches must be considered mostly specific for this peculiar type of lake. The potential colonising capacity, survival and high reproductive growth of a particular species in response to the most relevant environmental axes (defined by its fundamental niche) is strongly reduced both by environmental constraints typical of the characteristics of the colonised lakes, and by competition dynamics. More abiotic and biotic factors must be considered in the study of species distribution in different types of lakes. Many discrepancies and the lack of unifying concepts in the studies regarding the distribution of phytoplankton along trophic gradients originated from the implicit reductive assumption of the existence of selective factors acting almost exclusively along a unique environmental axis. In particular, the significance of the morphometric, hydrological and climatic characteristics of lakes as selective factors have been largely overlooked.

Acknowledgements

I am grateful to Prof. Ireneo Ferrari and Prof. Pierluigi Viaroli (Dipartimento di Scienze Ambientali, Università di Parma) for their hepful comments on an earlier version of this work. The manuscript received helpful suggestions from two anonymous referees. I am grateful to Dr Rosario Mosello (CNR-Istituto per lo Studio degli Ecosistemi, Pallanza) for motivating discussions on the limnology of the deep southern subalpine lakes. I wish to thank Prof. Paolo Cordella (Dipartimento di Biologia, Università di Padova), Dr Fabio Decet (ARPAV-Belluno) and Dr Giorgio Franzini (ARPAV-Verona) for critical discussions and logistic support. This work was partially funded by the Veneto Region and ARPAV (Veneto Region Environment Protection Agency).

References

Anagnostidis, K. & J. Komárek, 1988. Modern approach to the classification system of Cyanophytes. 3-Oscillatoriales. Archiv für Hydrobiol./Algol. Stud. 50–53: 327–472.

APHA, AWWA & WEF., 1989. Standard methods for the examination of water and wastewater. 17th edn, American Public Health Association, Washington.

Begon, M., J. L. Harper, C. R. Townsend, 1996. Ecology. Individuals, populations and communities. 3rd edn, Blackwell Science Ltd, Oxford. 1068 pp.

Bleiker, W. & F. Schanz, 1987. Influence of environmental factors on the phytoplankton spring bloom in Lake Zürich. Aquat. Sci. 51: 47–58.

Blomqvist, P., A. Petterson & P. Hyenstrand, 1994. Ammonium-nitrogen: a key regulatory factor causing dominance of non-nitrogen-fixing cyanobacteria in aquatic systems. Archiv für Hydrobiol. 132: 141–164.

Bondarenko, N. A., N. E. Guselnikova, N. F. Logacheva & G. V. Pomazkina, 1996. Spatial distribution of phytoplankton in Lake Baikal, spring 1991. Freshwat. Biol. 35: 517–523.

Bray, J. R. & J. T. Curtis, 1957. An ordination of the upland forest communities of Southern Wisconsin. Ecol. Monogr. 27: 325–349.

Buzzi, F., 2002. Phytoplankton assemblages in two sub-basins of Lake Como. J. Limnol. 61: 117–128.

Chen, C. T. & F. J. Millero, 1986. Precise thermodynamic properties for natural waters covering only the limnological range. Limnol. Oceanogr. 31: 657–662.

Connell, J. H., 1978. Diversity in tropical rain forests and coral reefs. Science 199: 1302–1310.

Decet, F. & N. Salmaso, 1997. Indagini preliminari sulle caratteristiche chimiche dei principali affluenti e dell'emissario del Lago di Garda. Acqua Aria 7: 91–97.

Dokulil, M. T. & K. Teubner, 2003. Steady state phytoplankton assemblages during thermal stratification in deep alpine lakes. Do they occur? Hydrobiologia 502 (Dev. Hydrobiol. 172): 65–72.

Fabbro, L. D. & L. J. Duivenvoorden, 2000. A two-part model linking multidimensional environmental gradients and seasonal succession of phytoplankton assemblages. Hydrobiologia 438: 13–24.

Field, J. G., K. R. Clarke & R. M. Warwick, 1982. A practical strategy for analysing multispecies distribution patterns. Mar. Ecol. Prog. Ser. 8: 37–52.

Gauch, H. G., 1982. Multivariate analysis in community ecology. Cambridge University Press, Cambridge. 298 pp.

Grime, J. P., 1977. Evidence for the existence of three primary strategies in plants and its relevance to ecological and evolutionary theory. The American Naturalist 111: 1169–1194.

Hardin, G., 1960. The competitive exclusion principle. Science 131: 1292–1297.

Harris, G. P., 1983. Mixed layer physics and phytoplankton populations: studies in equilibrium and non-equilibrium ecology. Prog. Phycol. Res. 2: 1–52.

Hudson, J. J., W. D. Taylor & D. W. Schindler, 2000. Phosphate concentrations in lakes. Nature 406: 54–56.

Hutchinson, G. E., 1961. The paradox of the plankton. Am. Nat. 95: 137–145.

Hutchinson, G. E., 1967. A treatise on limnology, 2. Wiley & Sons, New York. 1115 pp.

Hyenstrand, P., P. Blomqvist & A. Petterson, 1998. Factors determining cyanobacterial success in aquatic systems – a literature review. Archiv für Hydrobiol./Adv. Limnol. 51: 41–62.

IRSA, 1974. Indagini sul Lago di Garda. IRSA Quaderni 18, Roma. 540 pp.

Isaksson, A. 1998. Phagotrophic phytoflagellates in lakes – a literature review. Archiv für Hydrobiol./Adv. Limnol. 51: 63–90.

Jensen, J. P., E. Jeppesen, K. Olrik & P. Kristensen, 1994. Impact of nutrients and physical factors on the shift from cyanobacterial to chlorophyte dominance in shallow Danish lakes. Can. J. Fish. aquat. Sci. 51: 1692–1699.

Jones, R. I., 2000. Mixotrophy in planktonic protists: an overview. Freshwat. Biol. 45: 219–226.

Karl, D. M., 2000. Phosphorus, the staff of life. Nature 406: 31–33.

Kilham, P., 1971. A hypothesis concerning silica and the freshwater plankton diatoms. Limnol. Oceanogr. 16: 10–18.

Krebs, C. J., 2001. Ecology. 5th edn, Addison-Wesley, San Francisco. 695 pp.

Kruskal, J. B. & M. Wish, 1978. Multidimensional scaling. Sage Publications, Beverly Hills and London. 93 pp.

Lepistö, L., 1999. Phytoplankton assemblages reflecting the ecological status of lakes in Finland. Monogr. Boreal Environ. Res. 16. 43 pp.

Levin, S. A., 1974. Dispersion and population interactions. Am. Nat. 108: 207–228.

Lorenzen, C. J., 1967. Determination of Chlorophyll and Pheo-Pigments: spectrophotometric equations. Limnol. Oceanogr. 12: 343–346.

Lund, J. W. G., C. Kipling, & E. D. Le Cren, 1958. The inverted microscope method of estimating algal numbers and the statistical basis of estimations by counting. Hydrobiologia 11: 143–170.

MacArthur, R. H., 1958. Population ecology of some warblers of northeastern coniferous forests. Ecology 39: 599–619.

Magurran, A. E., 1988. Ecological diversity and its measurement. Croom Helm, London. 179 pp.

McCook, L. J., 1994. Understanding ecological community succession: causal models and theories, a review. Vegetatio 110: 115–147.

Micheletti, S., F. Schanz & A. E. Walsby, 1998. The daily integral of photosynthesis by *Planktothrix rubescens* during summer stratification and autumnal mixing in Lake Zurich. New Phytol. 139: 233–246.

Miyoshi, N., T. Fujiki & Y. Morita, 1999. Palynology of a 250-m core from Lake Biwa: a 430 000-year record of glacial-interglacial vegetation change in Japan. Rev. Palaeobot. Palynol. 104: 267–283.

Morabito, G., A. Oggioni & P. Panzani, 2003. Phytoplankton assemblage at equilibrium in large and deep subalpine lakes: a case study from Lago Maggiore (N. Italy). Hydrobiologia 502 (Dev. Hydrobiol. 172): 37–48.

Morabito, G., D. Ruggiu & P. Panzani, 2002. Recent dynamics (1995-1999) of the phytoplankton assemblages in Lago Maggiore as a basic tool for defining association patterns in the Italian deep lakes. J. Limnol. 61: 129–145.

Novarino, G., I. A. N. Lucas & S. Morrall, 1994. Observations on the genus *Plagioselmis* (Cryptophyceae). Cryptogamie, Algologie 15: 87–107.

Nygaard, G., 1996. Temporal and spatial development of individual species of plankton algae from European lakes. Hydrobiologia 332: 71–91.

Nygaard, K. & A. Tobiesen, 1993. Bacterivory in algae: a survival strategy during nutrient limitation. Limnol. Oceanogr. 38: 273–279.

Padisák, J., 1994. Identification of relevant time-scales in non-equilibrium community dynamics: conclusions from phytoplankton surveys. New Zealand J. Ecol. 18: 169–176.

Padisák. J., É. Soróczki-Pintér & Zs. Rezner, 2003. Sinking properties of some phytoplankton shapes and the relation of form resistance to morphological diversity of plankton: an experimental study. Hydrobiologia 500: 243–257.

Perez Martinez, C. & P. Sanchez Castillo, 2001. Temporal occurrence of *Ceratium hirundinella* in Spanish reservoirs. Hydrobiologia 452: 101–107.

Perez Martinez, C. & P. Sanchez Castillo, 2002. Winter dominance of *Ceratium hirundinella* in a southern north-temperate reservoir. J. Plankton Res. 24: 89–96.

Pimm, S. L., 1984. The complexity and stability of ecosystems. Nature 307: 321–326.

Porter, K. G., 1973. Selective grazing and differential digestion of algae by zooplankton. Nature 244: 179–180.

Reynolds, C. S., 1984. The ecology of freshwater phytoplankton. Cambridge University Press, Cambridge. 384 pp.

Reynolds, C. S., 1988. Functional morphology and the adaptive strategies of freshwater phytoplankton. In Sandgren C. D. (ed.), Growth and Reproductive Strategies of Freshwater Phytoplankton. Cambridge University Press, Cambridge: 388–433.

Reynolds, C. S., 1996. The plant life of the pelagic. Int. Ver. theor. angewan. Limnol.: Verh. 26: 97–113.

Reynolds, C. S., 1997. Vegetation processes in the pelagic: a model for ecosystem theory. Ecology Institute, Oldendorf. 371 pp.

Reynolds, C. S., 1998. What factors influence the species composition of phytoplankton in lakes of different trophic status? Hydrobiologia 369/370: 11–26.

Reynolds, C. S., J. Padisák & U. Sommer, 1993. Intermediate disturbance in the ecology of phytoplankton and the maintenance of species diversity: a synthesis. Hydrobiologia 249: 183–188.

Rothhaupt, K. O., 1996. Laboratory experiments with a mixotrophic chrysophyte and obligately phagotrophic and phototrophic competitors. Ecology 77: 716–724.

Salmaso, N., 1996. Seasonal variation in the composition and rate of change of the phytoplankton community in a deep subalpine lake (Lake Garda, Northern Italy). An application of nonmetric multidimensional scaling and cluster analysis. Hydrobiologia 337: 49–68.

Salmaso, N., 2000. Factors affecting the seasonality and distribution of cyanobacteria and chlorophytes: a case study from the large lakes south of the Alps, with special reference to Lake Garda. Hydrobiologia 438: 43–63.

Salmaso, N., 2002. Ecological patterns of phytoplankton assemblages in Lake Garda: seasonal, spatial and historical features. J. Limnol. 61: 95–115.

Salmaso, N. & F. Decet, 1997. Seasonal and interannual changes of chemical characteristics and phytoplankton in a mountain lake of the eastern Italian Alps (Lake Calaita, Trentino). Int. Rev. ges. Hydrobiol. 82: 15–31.

Salmaso, N. & F. Decet, 1998. Interactions of physical, chemical and biological processes affecting the seasonality of mineral composition and nutrient cycling in the water column of a deep subalpine lake (Lake Garda, Northern Italy). Archiv für Hydrobiol. 142: 385–414.

Salmaso, N. & L. Naselli-Flores, 1999. Stud. on the zooplankton of the deep subalpine Lake Garda. J. Limnol. 58: 66–76.

Salmaso, N., F. Decet, G. Franzini & P. Cordella. 2001. Evoluzione trofica del Lago di Garda in relazione alla completa circolazione delle acque del 1999 e 2000. Atti Ass. Ital. Oceanol. Limnol. 14: 109–123.

Sommer, U., 1987. Factors controlling the seasonal variation in phytoplankton species composition. A case study for a deep, nutrient rich lake. Prog. Phycol. Res. 5: 123–178.

Sommer, U., 1988. Growth and survival strategies of planktonic diatoms. In Sandgren, C. D. (ed.), Growth and Reproductive Strategies of Freshwater Phytoplankton. Cambridge University Press, Cambridge: 227–260.

Sommer, U., J. Padisák, C. S. Reynolds & P. Juhász-Nagy, 1993. Hutchinson's heritage: the diversity-disturbance relationship in phytoplankton. Hydrobiologia 249: 1–7.

Steinberg, C. E. W. & W. Geller, 1993. Biodiversity and interactions within pelagic nutrient cycling and productivity. In Schulze, E.-D. & H. A. Mooney (eds), Biodiversity and Ecosystem Function. Ecological Studies 99: 43–64.

Ter Braak, C. J. F. & H. van Dam, 1989. Inferring pH from diatoms: a comparison of old and new calibration methods. Hydrobiologia 178: 209–223.

Tilman, D., 1977. Resource competition between planktonic algae: an experimental and theoretical approach. Ecology 58: 338–348.

Tilman, D. & S. S. Kilham, 1976. Phosphate and silicate uptake and growth kinetics of the diatoms *Asterionella formosa* and *Cyclotella meneghiniana* in batch and semicontinuous culture. J. Phycol. 12: 375–383.

Tilman, D., R. Kiesling, R. Sterner, S. S. Kilham & F. Johnson, 1986. Green, bluegreen and diatom algae: taxonomic differences in competitive ability for phosphorus, silicon and nitrogen. Archiv für Hydrobiol. 106: 473–485.

Urabe, J., T. B. Gurung, T. Yoshida, T. Sekino, M. Nakanishi, M. Maruo & E. Nakayama, 2000. Diel changes in phagotrophy by *Cryptomonas* in Lake Biwa. Limnol. Oceanogr. 45: 1558–1563.

Walsby, A. E., A. Avery & F. Schanz, 1998. The critical pressures of gas vesicles in *Planktothrix rubescens* in relation to the depth of winter mixing in Lake Zurich, Switzerland. J. Plankton Res. 20: 1357–1375.

Whittaker, R. H., 1956. Vegetation of the Great Smoky Mountains. Ecol. Monogr. 26: 1–80.

Whittaker, R. H., 1972. Evolution and measurement of species diversity. Taxon 21: 213–251.

Wilkinson, L., 1990. SYGRAPH: The system for graphics. Evanston, IL, Systat, Inc.. 547 pp.

Wilson, E. O., 1992. The diversity of life. Penguin Books, London. 406 pp.

Wood, D. M. & R. del Moral, 1987. Mechanisms of early primary succession in subaline habitats on Mount St. Helens. Ecology 68: 780–790.

Hydrobiologia **502:** 37–48, 2003.
L. Naselli-Flores, J. Padisák, M. T. Dokulil (eds),
Phytoplankton and Equilibrium Concept: The Ecology of Steady-State Assemblages.

Phytoplankton assemblage at equilibrium in large and deep subalpine lakes: a case study from Lago Maggiore (N. Italy)

Giuseppe Morabito, Alessandro Oggioni & Pierisa Panzani
CNR Istituto per lo Studio degli Ecosistemi, Pallanza, Verbania, Italy.
E-mail: g.morabito@ise.cnr.it

Key words: deep lakes, phytoplankton assemblages, seasonal succession, equilibrium, metalimnion

Abstract

In 1996, we studied the phytoplankton seasonal succession in Lago Maggiore (N. Italy) through weekly sampling. Such a frequency enabled us to evaluate the changes of the phytoplankton assemblage in the light of the equilibrium and non-equilibrium theories. The distinct phases of changing of the species composition were identified separating the samples by means of cluster analysis and non-metric multidimensional scaling (NMDS). We recognised well distinct phytoplankton associations, whose seasonal succession followed a clear cyclic path throughout the year, with spring and summer phases respectively characterised by a rapid turn-over of the assemblages and by a relative stability. Moreover, we observed an increase of species number and Shannon-Wiener diversity during the spring, followed by a summer decline of the diversity in spite of an unchanged species number. Because of the dominance of the same few species for about two months during summer, coupled with small fluctuations of the total biomass, we could identify the summer assemblage as a steady state assemblage. The aim of the present contribution, although describing the whole seasonal succession, is to draw the attention towards the species composition at the steady state, taking into account the functional properties of the species involved. The possible role of the metalimnetic niche in selecting a particular summer assemblage in deep and large lakes will be discussed.

Introduction

When analysing a phytoplankton succession trying to recognise different assemblages, and, in particular, 'steady-state' or 'equilibrium' assemblages, the phytoplanktologist must face some fundamental concepts, such as disturbance, diversity, species competition: in a few words, he has to deal, once again, with the renowned Hutchinson's (1961) "paradox of the plankton". The idea that phytoplankton species are, for most of their time, in a non-equilibrium condition, due to the variability of the pelagic environment, has often been claimed to solve for the paradox. This hypothesis is, in fact, strongly supported, because, in nature, non steady-state conditions are the rule rather than the exception: as Reynolds (1988) pointed out, few cases can be mentioned where equilibrium conditions are met in natural phytoplankton assemblages, most of them occurring in water bodies characterised by a long-living stability of the water column, because of peculiar climatic conditions (permanently stratified tropical lakes, ice-covered Antarctic lakes, lakes exposed to a constant wind mixing).

Otherwise, the phytoplankton assemblages are maintained far from equilibrium. The succession developments observed in natural environment are, then, mostly allogenic (Reynolds, 1980), the changes in species composition being brought about by externally imposed physical perturbations. As a consequence, it is often difficult to determine whether a given succession step can be identified as an equilibrium phase: in controlled chemostat conditions, comparable to a steady-state situation, the competitive exclusion produces a decrease of diversity, according to the equilibrium theory, as clearly demonstrated by the well known experiments by Tilman (1977, 1981); Tilman et al. (1982), Tilman & Sterner (1984) and Kilham (1984). Although in nature the competitive exclusion principle seems to be rarely effective and it is certainly difficult to demonstrate if it is acting or not, the pattern of diversity changes can help in recognising a state of low disturbance (probably the closer to an equilibrium

Table 1. Main morphometric and hydrological features of Lago Maggiore

Altitude (m a.s.l.):	194
Drainage basin area (km^2):	6599
Volume ($m^3 * 10^6$):	37500
Area (km^2):	212
Max depth (m):	370
Mean depth (m):	177.5
Turn over time (yrs):	4.1

in natural environment), according to the relationship between diversity and disturbance postulated by the Intermediate Disturbance Hypothesis (Connell, 1978). The Connell's IDH has been successfully applied to the study of the phytoplanktonic successions, both in controlled (Sommer, 1993, 1995) and in natural conditions (Hambright & Zohary, 2000). Starting from the IDH postulates, Sommer et al. (1993) provide a set of conditions that, when satisfied, can help to identify an equilibrium phase: (1) 1, 2 or 3 species of algae contribute to more than 80% of total biomass; (2) their existence or coexistence persists for long enough; (3) during that period the total biomass does not increase significantly. These three conditions can be met if the system is sufficiently persistent: Sommer's (1985, 1989) experimental investigations demonstrated that some 35–60 days was required to achieve an equilibrium. Such a long period of stability could easily take place during summer in deep and large lakes, because such a kind of lacustrine ecosystems usually show a more or less high resistance to disturbance, either on a short or long time scale: in this paper we analyse the seasonal succession observed during a single year (1996) in Lago Maggiore, one of the deep southern subalpine lakes, trying to characterise the phytoplankton assemblage at equilibrium. The equilibrium association will be compared with those found in comparable seasonal situations in other Italian and European subalpine lakes.

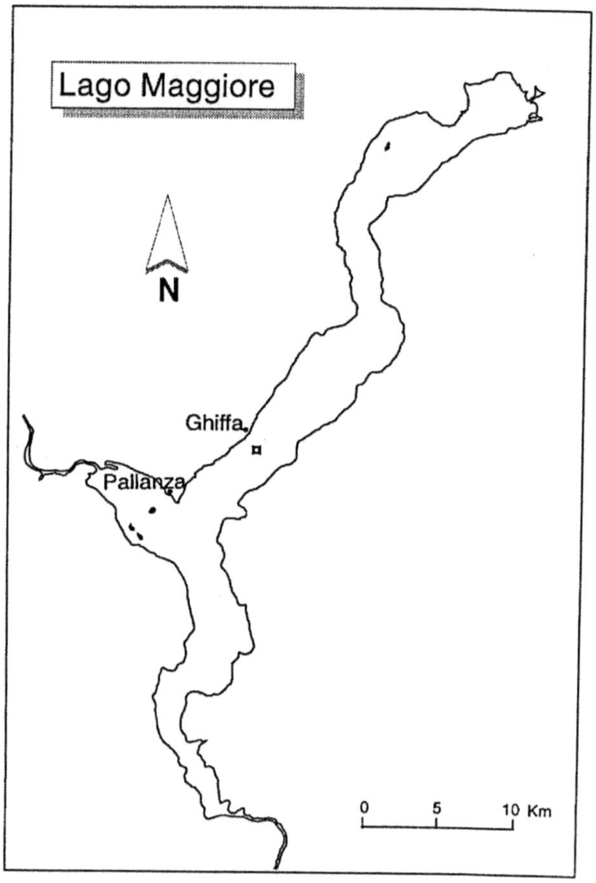

Figure 1. Map of Lago Maggiore: the sampling station is indicated.

Study site and earlier work

Lago Maggiore, the second largest Italian subalpine lake (Fig. 1; Table 1), is oligotrophic by nature, as testified by early limnological studies (Monti, 1929; Baldi, 1949; Vollenweider, 1965) and by the analysis of the sedimentary pigments (Guilizzoni et al., 1983; Marchetto et al., 2000). The eutrophication process started in the sixties: the algal nutrient concentration in the lake water started to increase and was soon followed by an increase of the phytoplankton abundance, biovolume and primary productivity (Ravera & Vollenweider, 1968; Ruggiu & Saraceni, 1977; Morabito & Pugnetti, 2000). The lake reached a trophic state close to eutrophy in the late seventies, when the P loads peaked and the maximum in-lake TP concentration at winter mixing was recorded (30 $\mu g\ l^{-1}$; Mosello & Ruggiu, 1985). Since that time, the P loads have been gradually reduced by various means, among which the adoption of treatment plants and the reduction of total phosphorus in detergents were the most important. As a result, the values of TP at winter mixing gradually decreased, to values around 10 $\mu g\ l^{-1}$ in the most recent years (Calderoni et al., 1997). The slow reversal of the trophic state of Lago Maggiore is documented by many papers: from a biological point of view, in the eighties strong emphasis was put on the apparent resilience of the plankton communities against falling phosphorus (de Bernardi et al., 1988). However, starting from 1987 to 1988, major biological changes were at last manifest, especially in the phytoplankton (Manca et al., 1992; Ruggiu, 1993). Notable changes were also recorded in the structure of the phytoplankton assemblages with oligotrophication (Ruggiu et al., 1998): among the most important of these, we must mention a remarkable decrease of the average cell size due to an increased importance of the smaller sized phytoplankters, a pattern which could involve deep changes in the whole ecosystem functioning (Harris, 1994).

Methods and data treatment

The data reported here refer to a research carried out during 1996. Water samples for phytoplankton analysis were collected weekly from April to September, about fortnightly in March and October, monthly in February and November, at the station of Ghiffa, corresponding to the deepest point of the lake. A bottle designed to take an integrated sample in the 0–20 m water layer was used. Phytoplankton determinations were carried out on subsamples preserved in acetic Lugol's solution; algal cells (including ultraplankton cells of about 1–2 μm diameter) were counted on a Zeiss Axiovert microscope, following Lund et al. (1958), until 400 cells for the most important species were counted. Phytoplankton biomass was estimated from the density data and the original measurements of the cell volume (Smayda, 1978). Species were identified using the series *Süßwasserflora von Mitteleuropa*, founded by Pascher (Vol. I–XXIV; 1978–1999) and *Das Phytoplankton des Süßwassers*, founded by Hüber-Pestalozzi (Vol. I–XVI; 1938–1983). Oscillatoriales were determined following also Anagnostidis & Komárek (1988).

Water for chemical analysis was collected monthly from March to November and the chemical parameters have been determined at the chemical laboratory of the Istituto per lo Studio degli Ecosistemi (Pallanza), following the methods reported in Mosello & Ruggiu (1985).

Temperature profiles (1 measure every meter) were obtained by means of a thermistor probe.

The thermal stability of the water column was estimated from the change of density according to temperature, calculating the Brünt-Väisälä frequency, by means of the equation reported in Mortimer (1974). We calculated the dissimilarity between pairs of samples using the Bray and Curtis index computed on biovolumes of identifiable algae, after a double square root transformation of the raw data, as suggested by Salmaso (1996). The computation of the Bray and Curtis index was carried out with SIMDISS 2.0 (http://www.bio.unipd.it/limno/simdiss/), whereas the samples ordination (cluster analysis and Non-metric Multidimensional Scaling) was performed with Statistica package by StatSoft Inc. (1997).

Results

Chemical and physical environment

The monthly values of the main chemical parameters recorded during 1996 in the euphotic zone are shown in Figure 2.

The seasonal pattern of pH in the surface waters was typical, paralleling the fluctuations of the phytoplankton photosynthetic activity, i.e. with the minimum values in autumn and winter and the maxima in summer. The summer pH, however, rarely exceeds 8 units in Lago Maggiore, mirroring the oligotrophic nature of the waters and the absence of epilimnetic blooms.

The trend of nitrate nitrogen in the surface waters is probably related to the algal consumption: after the replenishment of the epilimnetic waters with mixing, when nitrate reaches about 0.85 mg l^{-1}, its concentration decreased from spring to the end of summer, when the annual minimum values of around 0.600 mg l^{-1} are usually recorded. With these values, a nitrogen limitation can be excluded in Lago Maggiore.

Epilimnetic reactive silica fluctuates in the range following an usual pattern, dependent on the consumption by diatoms and on the supply by winter mixing. The maximum concentration (around 1.5 mg l^{-1}) was recorded in March, then the epilimnetic concentration decreased to 0.5 mg l^{-1}, mirroring the pattern of diatom growth. The annual minimum of 0.4 mg l^{-1} was reached during summer and in autumn a new supply to epilimnetic waters was observed, when the metalimnetic temperature gradient weakened and the mixing of the upper layers took place.

Soluble reactive phosphorus remained always close to detection limit of the analytical method, fluctuating around 4 μg l^{-1}. Total phosphorus concentration also showed small variations around 10 μg l^{-1} during the whole year.

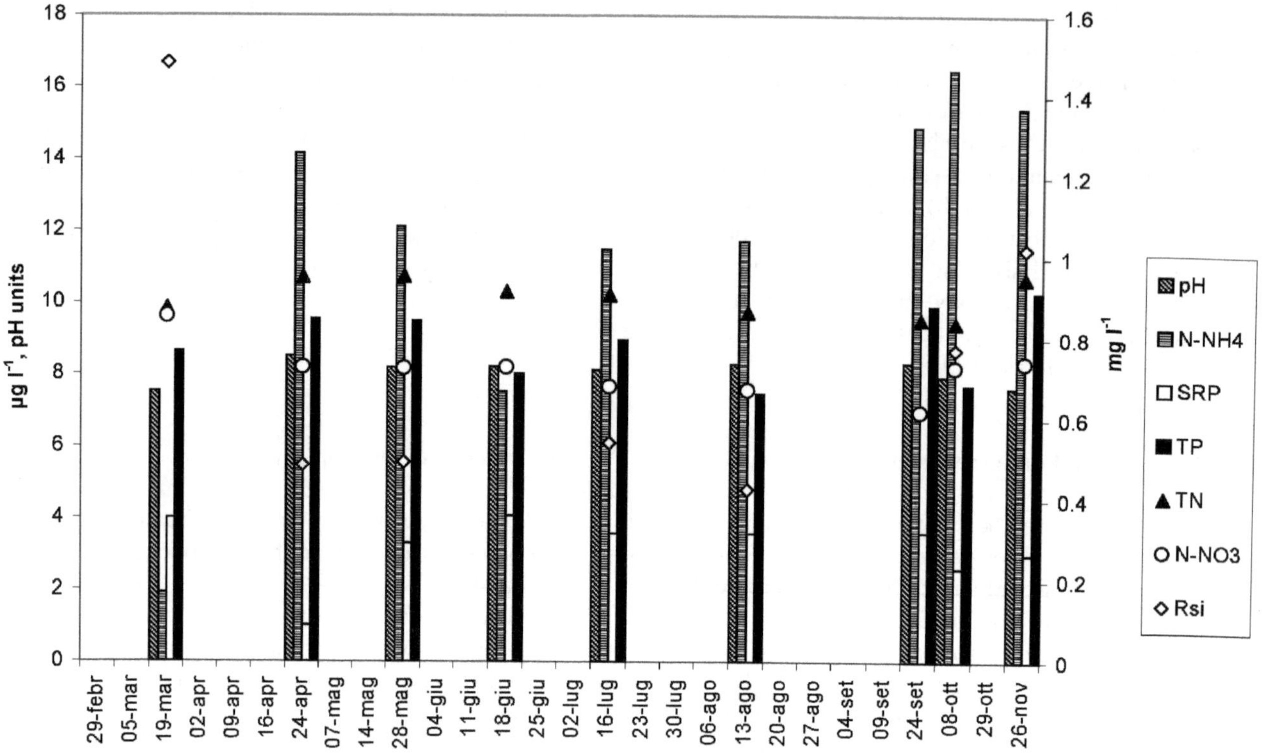

Figure 2. Seasonal trends of the basic chemical parameters as weighted average in the 0–20 m water layer. On the left scale, all as bars: pH, total phosphorus (TP), soluble reactive phosphorus (SRP) and ammonium nitrogen (N-NH4). On the right scale: nitrate nitrogen (N-NO3), total nitrogen (TN) and reactive silica (Rsi).

The thermal cycle described here covers the 0–20 m water layer, corresponding to the euphotic zone. The minimum yearly temperatures were 6.1 °C, characteristic of the whole column of water in February and March, whereas the maximum (22.1 °C) was recorded in August. During spring the temperature gradually increased and during May the thermal stratification started to appear.

The lake became clearly stratified since June, when a thermocline was evident at around 5 m depth, deepening to 8 m in July, to 10 m August and, finally, to 15 m in September.

During October, the metalimnion disappeared and the temperature in the euphotic zone became more homogeneous, with values declining from 14 to 10 °C. In November and December the euphotic zone was mixed, with temperature values around 10 °C.

The seasonal variations of the Brünt-Väisälä Frequency (Fig. 3) indicate a period of low water column stability from the beginning of the year to the end of April, including the maximum diatom development. Starting from May the stability gradually increases, reaching the peak values during summer, when the maxima located around 10 meters depth indicate the existing of a stable thermocline. In late summer, this metalimnetic layer became less stable and deeper, as a consequence of the cooling of epilimnetic waters. The metalimnetic waters were still recognisable between 15 and 20 m at the end of September, but in October the stratification was no more evident.

Phytoplankton succession

The recognition of at least three well distinct phases in the course of the phytoplankton succession appeared, first of all, by comparing the seasonal trends of species number and Shannon-Wiener diversity, calculated on biovolume values (Fig. 4). A spring phase (March – beginning of June), characterised by the parallel increase of both parameters, is followed by a summer phase, during which the species number does not change significantly, while the diversity gradually drops towards the annual minimum. Finally, during autumn, after a small increase of diversity, both para-

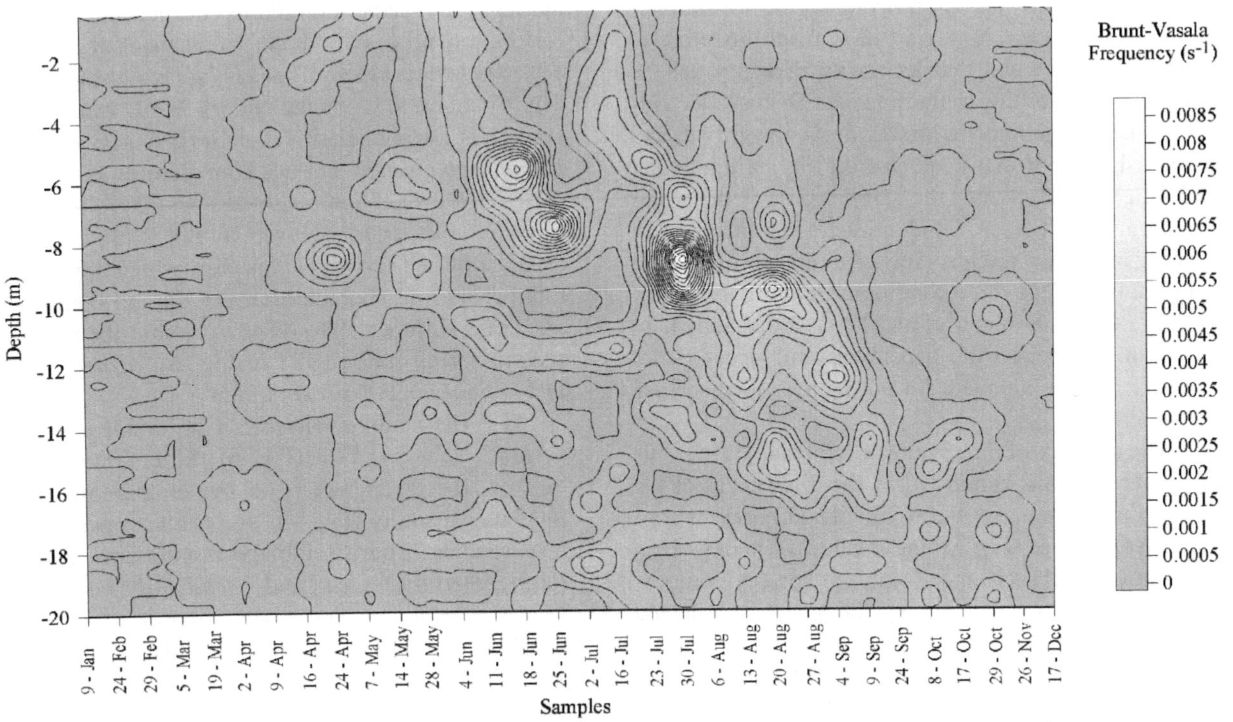

Figure 3. Seasonal changes of the Brünt-Väisälä frequency in the 0–20 m water layer.

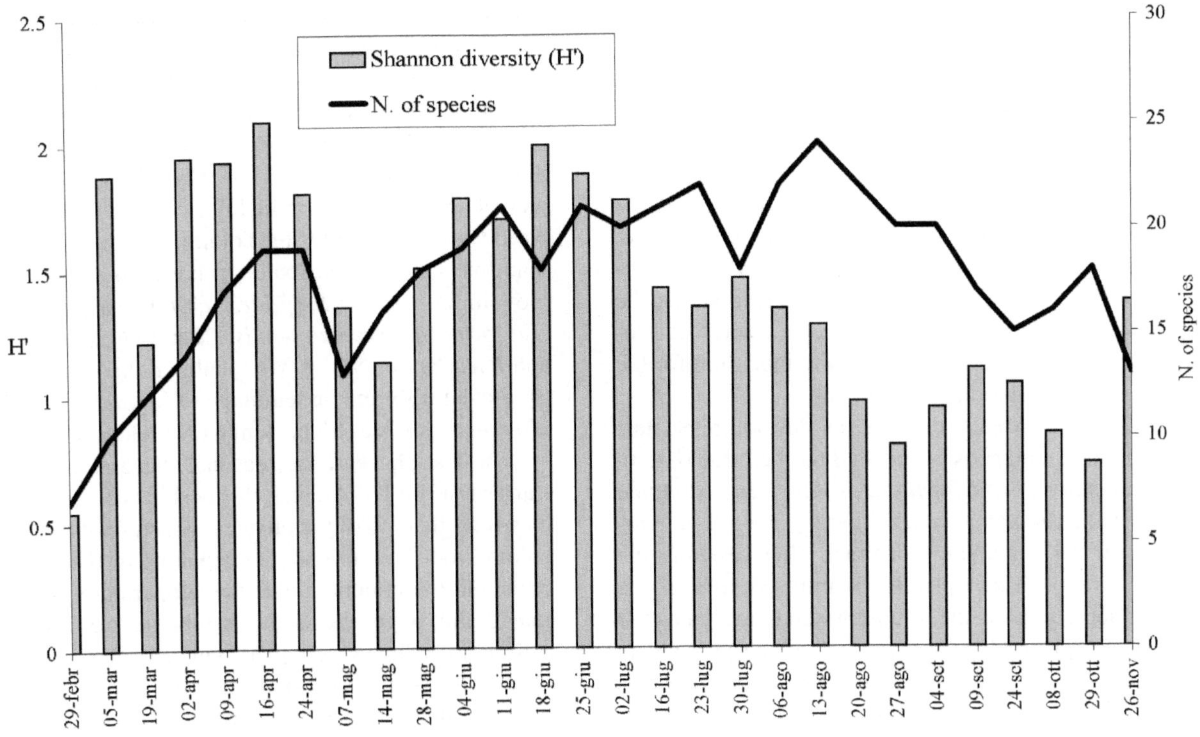

Figure 4. Seasonal variation of the Shannon-Wiener diversity and of the total number of species.

meters declined. The shift from spring to summer phases can be seen as a transition from moderate to low disturbance and the decline of diversity can be interpreted according to Intermediate Disturbance Hypothesis: the epilimnetic mixing drove the growth and establishment of many species during spring, many of which declined after the stabilisation of the water column. In particular, we identified a summer period, lasting about two months (July–August) during which at least two of the three conditions defining the 'equilibrium' *sensu* Sommer et al. (1993) are satisfied. In fact, in this period the species assemblage did not modify and the biomass did not increase or decrease significantly.

The stability of the species assemblage during this summer period is confirmed by the differences of the Bray-Curtis similarity index calculated between consecutive samples (Fig. 5): these values, which give an indication of the change inside the species composition, were very small among the samples collected from July until August.

Cluster analysis (Fig. 6) and NMDS ordination (Fig. 7) allow to get a deeper inside into the species succession, separating the seasonal assemblages according to the similarity of the species composition. Four main seasonal groups, described in the following paragraphs, were identified.

The winter season includes three samples: February 29, March 5 and 19, quite different from each other, as the Bray-Curtis distance and their respective position in NMDS ordination reveal. All these samples were characterised by a very low biomass (Fig. 8) and showed the dominance of *Fragilaria crotonensis* (Ehr.) Grun., *Planktothrix rubescens* Anagnostidis & Komàrek and *Asterionella formosa* Hass. Few cryptophytes are also present. The winter period should be considered as a phase of 'dormancy', where the species recorded probably represent the inoculum for the successive population development.

After mid March the diatom development started: the assemblage is dominated by the large diatom species *Asterionella formosa*, *Aulacoseira islandica* O. Muell., *Diatoma tenuis* Agardh, *Fragilaria crotonensis*. They gradually declined after reaching the biovolume peak of April 16, mainly because of the silica depletion. In the southern subalpine lake district the spring development of the large pennate diatoms (*Asterionella formosa*, *Fragilaria crotonensis* and *Aulacoseira* spp.) usually characterises the deep oligo-mesotrophic lakes (Salmaso, 2000, 2003; Mosello et al., 2001). The same species reappearead among the spring dominants during the years of restoration process in Lake Constance (Kümmerlin, 1998; Gaedke, 1998). This first spring phase (SP1 in Fig. 6) ended towards the end of April, when the filamentous cyanobacteria (*Planktothrix rubescens* and *Pseudanabaena* sp.) replaced the diatoms as dominant *taxa*. However, *Diatoma tenuis* and *Fragilaria crotonensis* still persist in lower numbers, together with an unidentified *Cyclotella* species, which appeared in this period and characterised the diatom community until mid summer. Two other *taxa* contributed to the second spring phase (SP2 in Fig. 6): *Mougeotia* sp. and *Dinobryon bavaricum* Imhof.

Also the summer phase can be divided into two periods, SU1 and SU2 (Fig. 6). SU1, from May 28 to July 16, is characterised by an increase of the Shannon diversity (Fig. 4), due to the appearance of new species. The assemblage is still dominated by *Planktothrix rubescens* and *Pseudanabaena* sp., but we observed a new increase of *Fragilaria crotonensis* and *Diatoma tenuis*, mainly at the end of the period. *Mougeotia* sp. continues its growth, reaching the maximum development at mid June. Moreover, many other species gave a significant contribution to the total biovolume: the most important were the diatoms *Aulacoseira islandica*, *Cyclotella comensis* Grum. and *Cyclotella* sp., the chrysophytes *Dinobryon bavaricum* and *Uroglena americana* Calkins, the cryptophytes *Rhodomonas minuta* Skuja and *Plagioselmis nannoplanctica* (Skuja) emend. Novarino, Lucas et Morrall (formerly *Rhodomonas minuta* var. *nannoplanctica* Skuja, cf. Novarino et al., 1994). The second summer phase, starting at mid July, corresponded to the period of maximum water column stability: no more than five species contributed to build up 80% of total biovolume, namely *Fragilaria crotonensis*, *Asterionella formosa*, *Diatoma tenuis*, *Planktothrix rubescens* and *Pseudanabaena* sp. We identified this assemblage as the equilibrium assemblage. *Cyclotella* sp. and *Mougeotia* sp. are still present with a relevant biomass.

The last phase of the seasonal succession (Fig. 6) starts at the end of August: this period is characterised by a modification of the physical structure of the water column, due to the gradual deepening of the thermocline and the attenuation of the temperature gradient. Inside the phytoplankton assemblage we observed a dramatic change, coupled with a strong decrease of the diversity (Fig. 4). Only two among the species dominating during summer remained still dominant across the first autumn phase (Fig. 8), namely *Pseudanabaena* sp. and *Planktothrix rubescens*: their

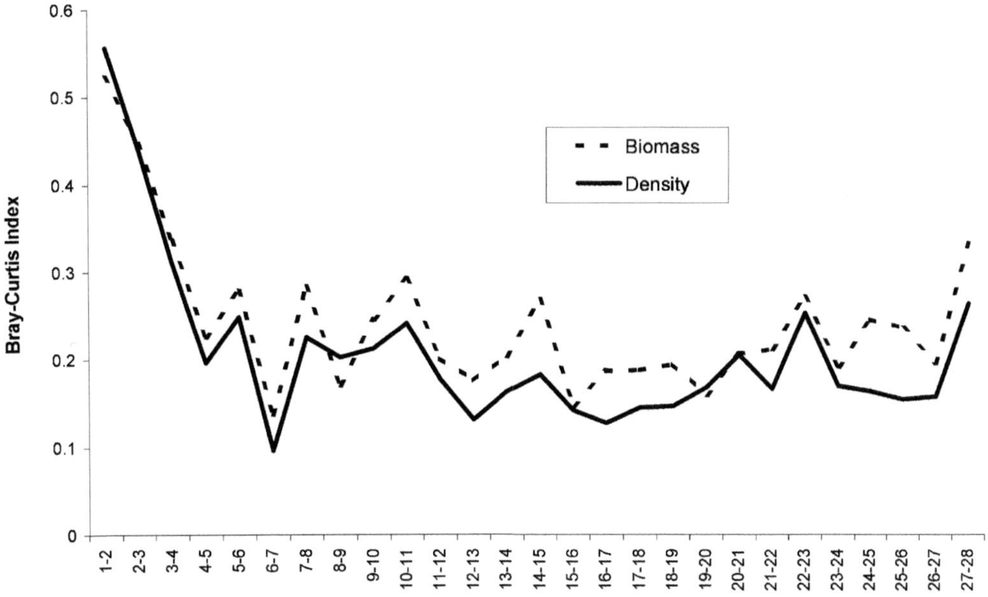

Figure 5. Changes of the difference of the Bray & Curtis index between pairs of consecutive samples.

high efficiency of light harvesting makes them the best competitors under the new environmental conditions of mild light, due to both the decrease of average solar incident radiation and the deepening of the metalimnetic layer where these species are probably mostly located. The last autumn phase is characterised by a reappearance of *Diatoma tenuis* and *Fragilaria crotonensis*, probably stimulated by a new silica supply and by the decrease of the water column stability.

Discussion

In this paper we presented an analysis of the seasonal succession observed during 1996 in Lago Maggiore: however, because the ecology of steady-state assemblages is the topic of the 13th IAP Workshop (Naselli-Flores et al., 2003), the following discussion will be focused on the summer assemblage, that we recognised as equilibrium assemblage. The species composition and the ecology of the assemblages described for the remaining seasons resemble those described and analysed in a recent paper (Morabito et al., 2002) dealing with the phytoplankton dynamics in Lago Maggiore in the period 1995–1999.

The onset and breakdown of thermal stratification represent the most important disturbances affecting the phytoplankton species composition in a lake. The thermal cycle of a deep and large subalpine European lake is typically monomictic, characterised by a period of mixing (from late autumn to late winter), followed by the stratification phase (early summer to early autumn): these lacustrine ecosystems can then be classified as environments affected by infrequent disturbances (2–8 yr^{-1}; Reynolds, 1993), where the environment can be sufficiently persistent to allow the establishment of a low-diversity, near-climatic assemblage (*sensu* Reynolds, 1993).

The summer of 1996 was characterised by a long period (about 60 days) of water column stability: according to the equilibrium theory, this time interval should be long enough to complete the process of competitive exclusion, reducing the species diversity: our data do not allow to evaluate if a competitive exclusion took place, however the conditions listed by Sommer et al. (1993) defining the equilibrium phase were satisfied. A decrease of diversity was recorded, few species dominated and their dominance did not change significantly from the beginning of July to the end of August. Accordingly, this summer period can be identified as the equilibrium condition and the species inhabiting the euphotic zone represent the equilibrium assemblage.

Following Tilman (1994), the thermal stratification during summer, should create environmental heterogeneity, giving rise to a niche diversification and to an increase of diversity (see also Salmoso, 2003). The pattern we observed, however, is opposite, similarly to

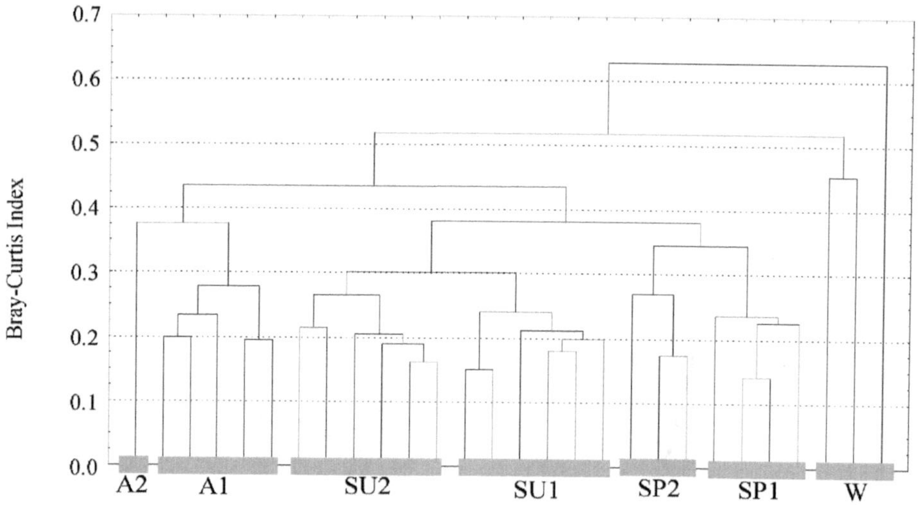

Figure 6. Ordination of the samples in seasonal clusters, according the Bray & Curtis similarity index. W = Winter; SP 1 & 2 = Early Spring & Late Spring; SU 1 & 2 = Early Summer and Late Summer; A1 & 2 = Early Autumn & Late Autumn. The samples are clustered according to the dates reported in Figure 8.

the finding by Elliott et al. (2000), who also recorded a diversity decrease in stable conditions: the authors attributed this result to the vertical segregation of the water column, which makes the environment more selective with reduced mixing depth. In Lago Maggiore, a lake with a clear and nutrient poor summer epilimnion, two growth strategies are probably favoured: invasive or C-strategy, typical of the small flagellates, such as *Plagioselmis nannoplanctica*, able to migrate across the epilimnion down to the metalimnion, exploiting the nutrient patchiness and the best light condition of the upper layers and attuning or R-strategy, typical of either *r*- or K-selected species, both adapted to the low light climate of the metalimnion. During the summer the epilimnetic waters became probably gradually nutrient depleted and the setting up of a steep thermocline prevents any replenishment from the lower layers: this new set of physico-chemical conditions probably determines the decline of the migrating C-strategists, giving an advantage to the metalimetic stratifying R-strategists in acquiring the nutrients, by virtue of their high growth rates (pennate diatoms) or their superior storage ability (*Planktothrix rubescens* and *Pseudanabaena* sp.). The result is a decline of the species diversity following the increase of the column stability.

The dominant and sub-dominant species recorded at the equilibrium in Lago Maggiore are characteristic of five among the functional groups identified by Reynolds et al. (2002): **C**, **P**, **T**, $\mathbf{S_1}$ and **R** (Table 2). The common trait shared by the species included in these groups is their high tolerance to mild or low light conditions. As reported in Reynolds (1997), the elongated shape of the diatoms *Asterionella formosa*, *Diatoma tenuis* and *Fragilaria crotonensis*, of the cyanoprokaryota *Planktothrix rubescens* and *Pseudanabaena* sp. and of the green alga *Mougeotia* sp. make these algae very efficient in utilising the available underwater PAR, giving them a competitive advantage at low light intensities. In the stratified water column of Lago Maggiore they probably find the optimum light climate at the depth of the metalimnetic layer: as regards the Oscillatoriales of the **R** and $\mathbf{S_1}$ groups, their deep summer stratification is a well know phenomenon, observed in many lakes, where these species take advantage of their peculiar chromatic adaptation (see i.e. Lindholm & Meriluoto, 1991; Walsby et al., 1998; Walsby, 2001; Dokulil & Teubner, 2003). On the other side, the important presence of the pennate diatoms of the groups **C** and **P** into this summer association could appear quite surprising, although typical of the phytoplankton summer composition in Lago Maggiore (Morabito et al., 2002). The large diatoms are considered by Reynolds et al. (2002) very sensitive to water column stratification, then they should decline when the stability increases. However, some opposite evidences exist: Agbeti et al. (1997) described the development of lightly silicified and spindle-shaped diatoms (*Rhizosolenia* and single celled *Fragilaria*) during a period of high wa-

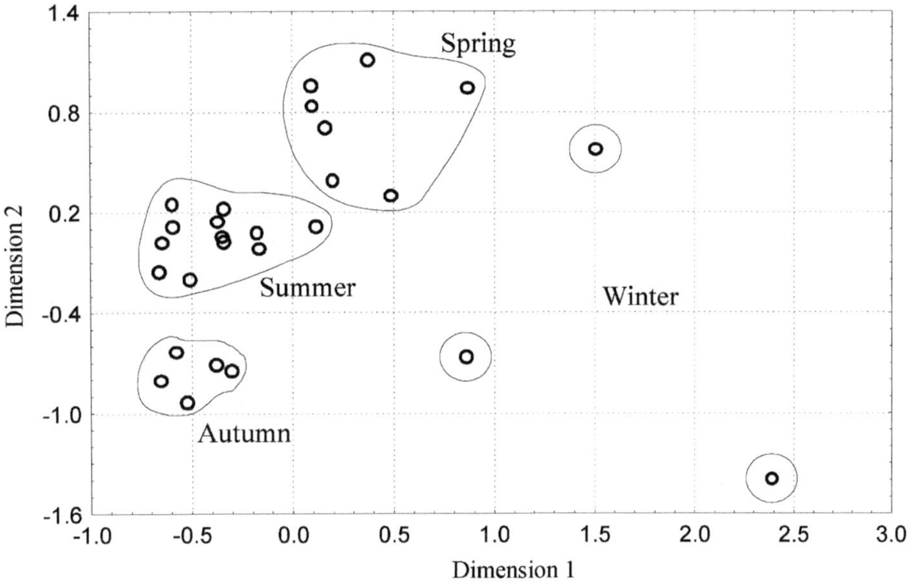

Figure 7. Non-metric Multidimensional Scaling ordination of the samples, clustered in four main seasonal groups, according the Bray & Curtis similarity index.

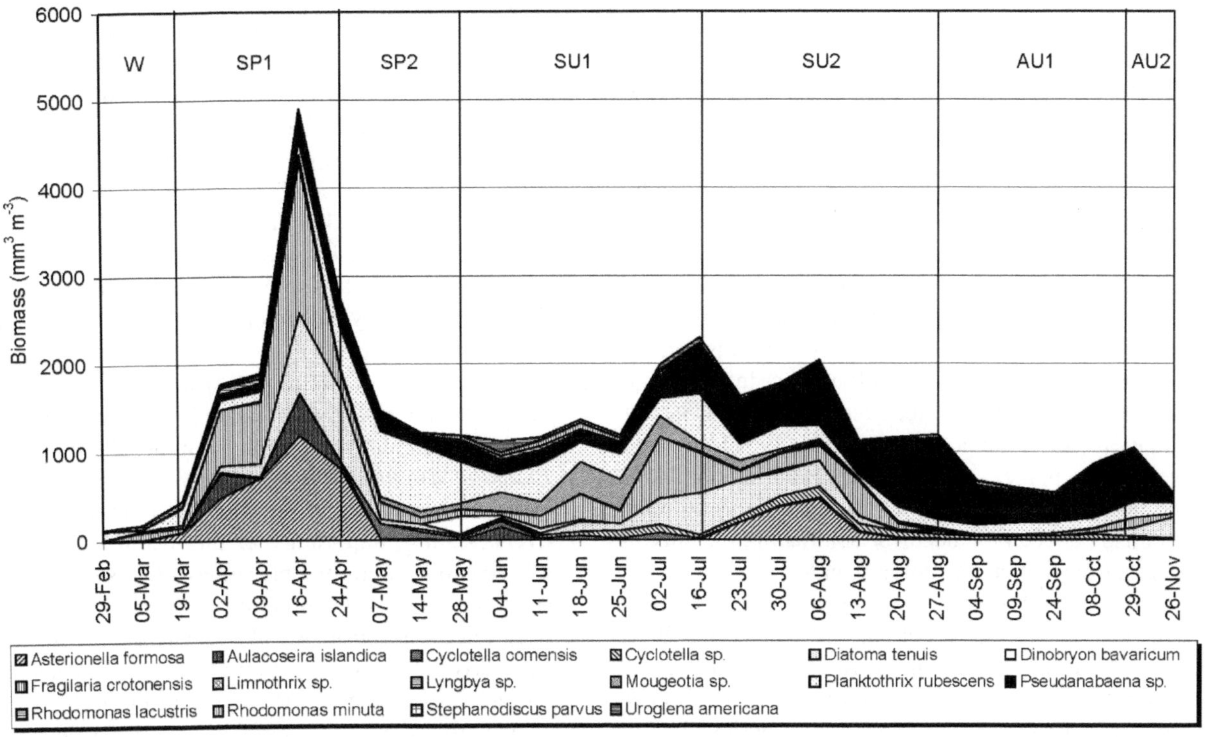

Figure 8. Seasonal changes of the biomass of the dominant species (the species contributing to build up 80% of the total biomass at least once during the year). The groups of samples are divided according to the cluster analysis.

Table 2. Functional groups found in Lago Maggiore at equilibrium, following the classification in Reynolds et al. (2002): in bold type, the species present in summer 1996; underlined, among sensitivities and habitat, the traits not correspondent to the euphotic zone of Lago Maggiore during stratification

Codon	Habitat	Typical representatives	Tolerances	Sensitivities
C	Mixed, eutrophic small-<u>medium lakes</u>	*Asterionella formosa* *Aulacoseira ambigua* *Stephanodiscus rotula*	Light, C deficiencies	Si exhaustion, <u>stratification</u>
P	Eutrophic epilimnia	*Fragilaria crotonensis* *Aulacoseira granulata* *Closterium aciculare*	Mild light and C deficiencies	<u>Stratification</u> Si depletion
T	Deep, well mixed epilimnia	*Geminella*, **Mougeotia**, *Tribonema*	Light deficiency	Nutrient deficiency
S1	Turbid mixed layers	*Planktothrix agardhii*, *Limnothrix redekei*, **Pseudanabaena**	Highly light deficient conditions	Flushing
R	Metalimnia of mesotrophic stratified lakes	**Planktothrix rubescens**, *P. mougeotii*	Low light, strong segregation	Instability

ter column stability in Lake Upper Rock, a dimictic lake in Ontario. Willén (1991), in her review of the ecology of planktonic diatoms, reports that the sinking could be halted when the cells reach the thermocline: some past (1994–1995) observations on phytoplankton vertical distribution in Lago Maggiore (Morabito, unpublished data) seem to confirm this finding, showing an increase of diatom biovolume close to the metalimnetic layer.

Moreover, during 1996, the number of coenobia and the average number of cell per colony of the large diatoms were counted: both *Diatoma tenuis* and *Fragilaria crotonensis* showed, between July and August a decrease of the average number of cells per colony, from 8 to 3 and from 21 to 5 respectively: it is unquestionable that the increase of cell number into a colony decreases the sinking rate, according to the Stokes' equation (see examples in Reynolds, 1997), however we could hypothesise that, in a stratified water column, the decrease of cell number would be a good compromise between the need for contrasting the sinking and that for making the colony lighter.

A further discrepancy between our observations and the traits of the functional groups **C** and **P** regards the trophic preferences: Reynolds et al. (2002) ascribe *Asterionella formosa* and *Fragilaria crotonensis* to eutrophic habitats, but Lago Maggiore is now close to oligotrophy. However, we have to consider that metalimnia are often quite rich in nutrients (i.e. Amand & Carpenter, 1993), due to the accumulation and recycling of organic matter, so this layer can have a trophic level higher than the upper water column, especially in oligo-mesotrophic waters. In Lago Maggiore the metalimnion, because of its peculiar chemical and physical features, could provide a sort of refuge for most of the species which colonise the water column during summer, playing a key role in the establishment of the equilibrium assemblage. This 'metalimnetic niche', characterised by mild light, nutrient richness and high thermal stability, represent, in the oligotrophic Lago Maggiore, a kind of 'mesotrophic microhabitat', where the pennate diatoms and the filamentous cyanobacteria find their ideal growth conditions.

An analysis of the recent dynamics of the phytoplankton assemblages in Lago Maggiore (Morabito et al., 2002), showed that the summer dominance of *Planktothrix rubescens* and *Fragilaria crotonensis* is a common feature, reinforcing the hypothesis of the overwhelming importance of the metalimnetic niche as a selective environment. Moreover, this summer association typifies the summer euphotic zone of other deep oligo-mesotrophic lakes, where the metalimnetic layer is sufficiently illuminated (surface shading blooms are extremely rare) and nutrient rich: for instance, this kind of assemblage has been described also in lakes Como (Bettinetti et al., 2000) and Garda (Salmaso, 2000). In Lake Garda this association has been recorded at 20 m depth, thus confirming the adaptation of this two species to the metalimnetic niche.

The studies carried out in many deep and large European lakes seem to indicate that the selection of a peculiar equilibrium assemblage, forced by the existence of the metalimnetic niche, became stronger moving from eutrophy to meso-oligotrophy: for instance, during the oligotrophication of Lac Leman (Anneville et al., 2002) a deepening of the nutrient rich layer was observed, coupled with the summer dominance of mild light tolerant species, such as *Diatoma tenuis*, *Planktothrix rubescens* and *Mougeotia gracillima* (Hassall) Wittrock, stratifying close to the thermocline.

Conclusions

Our results seem to indicate that in the deep and large lakes, characterised by a low frequency of disturbance, the summer phase of seasonal succession can be viewed as an equilibrium phase: the decrease of diversity, the stability of the species composition and the small changes in the phytoplankton biomass support this view.

The studies carried out in the deep European subalpine lakes, most of which are in oligo-mesotrophic conditions, show that the composition of this equilibrium assemblage is forced, by the physico-chemical structure of the water column, towards the dominance of species physiologically adapted to grow under low light conditions: the vertical gradient of nutrients and temperature of the transparent water column represent a further advantage to their establishment. Then, the existence, in this lacustrine ecosystems, of a metalimnetic niche represents a major steering factor towards the selection of a peculiar equilibrium assemblage (see also Teubner et al., 2003).

As a final suggestion, we would stress the need for deeper investigations of the vertical distribution of phytoplankton in deep lakes, in order to achieve a better understanding of the species selection according to the changes in the vertical chemical and physical structure of the water column.

References

Agbeti, M. D., J. C. Kingstom, J. P. Smol & C. Watters, 1997. Comparison of phytoplankton succession in two lakes of different mixing regimes. Archiv für Hydrobiol. 140: 37–69.

Amand, A. St. & S. R. Carpenter, 1993. Metalimnetic phytoplankton dynamics. In Carpenter S. R. & J. F. Kitchell (eds), The Trophic Cascade in Lakes. University Press, Cambridge: 210–224.

Anagnostidis, K. & J. Komárek, 1988. Modern approach to the classification system of Cyanophytes. 3-Oscillatoriales. Algological Stud. 50–53: 327–472.

Anneville, O., S. Souissi, F. Ibanez, V. Ginot, J. C. Druart & N. Angeli, 2002. Temporal mapping of phytoplankton assemblages in Lake Geneva: Annual and interannual changes in their patterns of succession. Limnol.Oceanogr. 47: 1355–1366.

Baldi, E., 1949. La situation actuelle de la recherche limnologique après le Congrès de Zurich. Rev. suisse d' Hydrol. 11: 637–649.

Bettinetti, R., G. Morabito & A. Provini, 2000. Phytoplankton assemblage structure and dynamics as indicator of the recent trophic and biological evolution of the western basin of Lake Como (N. Italy). Hydrobiologia 435: 177–190.

Calderoni, A., R. Mosello & R. de Bernardi, 1997. Le indagini sulla evoluzione dei laghi profondi sudalpini svolte dal C.N.R. Istituto Italiano di Idrobiologia. Documenta Istituto italiano di Idrobiol. 61: 19–32.

Connell, J., 1978. Diversity in tropical rain forests and coral reefs. Science 199: 1304–1310.

de Bernardi, R., G. Giussani, M. Manca & D. Ruggiu, 1988. Long-term dynamics of plankton communities in Lago Maggiore (N. Italy). Int. Ver. theor. angewan. Limnol.: Verh. 23: 729–733.

Dokulil, M. T. & K. Teubner, 2003. Steady state phytoplankton assemblages during thermal stratification in deep alpine lakes. Do they occur? Hydrobiologia 502 (Dev. Hydrobiol. 172): 65–72.

Elliott, J. A., C. S. Reynolds & T. E. Irish, 2000. The diversity and succession of phytoplankton communities in disturbance-free environments, using the model PROTECH. Archiv für Hydrobiol. 149: 241–258.

Gaedke, U., 1998. Functional and taxonomical properties of the phytoplankton community of large and deep Lake Constance: Interannual variability and response to re-oligotrophication (1979-1993). Archiv für Hydrobiol./Adv. Limnol. 53: 119–141.

Guilizzoni, P., G. Bonomi, G. Galanti & D. Ruggiu, 1983. Relationship between sedimentary pigments and primary production: evidence from core analyses of 12 Italian lakes. Hydrobiologia 103: 103–106.

Hambright, K. D. & T. Zohary, 2000. Phytoplankton species diversity control through competitive exclusion and physical disturbance. Limnol. Oceanogr. 45: 110–122.

Harris, G. P., 1994. Pattern, process and prediction in aquatic ecology. A limnological view of some general ecological problems. Freshwat. Biol. 32: 143–160.

Hüber-Pestalozzi, G. (ed), 1938, 1941, 1942, 1950, 1955, 1961, 1972, 1983. Das Phytoplankton des Süsswassers. Die Binnengewasser. Vols. I–XVI. Schweizerbart'sche Verlagsbuchhandlung, Stuttgart.

Hutchinson, G. E., 1961. The paradox of plankton. Am. Nat. 95: 137–147.

Kilham, S. S., 1984. Silicon and phosphorus growth kinetics and competitive interactions between *Stephanodiscus minutus* and *Synedra* sp. Int. Ver. theor. angewan. Limnol.: Verh. 22: 435–439.

Kümmerlin, R. E., 1998. Taxonomical response of the phytoplankton community of Upper Lake Constance (Bodensee-Obersee) to eutrophication and re-oligotrophication. Archiv für Hydrobiol./Adv. Limnol. 53: 109–117.

Lindholm, T. & J. A. O. Meriluoto, 1991. Recurrent depth maxima of the hepatotoxic cyanobacterium *Oscillatoria agardhii*. Can. J. Fish. aquat. Sci. 48: 1629–1634.

Lund, J. W. G., C. Kipling & E. D. Le Cren, 1958. The inverted microscope method of estimating algal numbers and the statistical basis of estimations by counting. Hydrobiologia 11: 143–170.

Manca, M., A. Calderoni & R. Mosello, 1992. Limnological research in Lago Maggiore: studies on hydrochemistry and plankton. Mem. Ist. ital. Idrobiol. 50: 171–200.

Marchetto, A., A. Lami, S. Musazzi, L. Guzzella, L. Langone, J. Massaferro, R. Bettinetti, G. Morabito & P. Guilizzoni, 2000. Reconstruction of the recent eutrophication and oligotrophication history of a large sub-alpine lake (Lake Maggiore, N. Italy). 8th International Symposium on Paleolimnology. Kingston, Ontario, Canada (poster).

Monti, R., 1929. Limnologia comparata dei laghi insubrici. Int. Ver. theor. angewan. Limnol.: Verh. 4: 462–497.

Morabito, G. & A. Pugnetti, 2000. Primary productivity and related variables in the course of the trophic evolution of Lake Maggiore. Int. Ver. theor. angewan. Limnol.: Verh. 27: 2934–2937.

Morabito, G., D. Ruggiu & P. Panzani, 2002. Recent dynamics (1995–1999) of the phytoplankton assemblages in Lago Maggiore as a basic tool for defining association patterns in the Italian deep lakes. J. Limnol. 61: 129–145.

Mortimer, C. H., 1974. Lake hydrodynamics. Mitt. int. Ver. theor. angewan. Limnol. 20: 124–197.

Mosello, R. & D. Ruggiu, 1985. Nutrient load, trophic conditions and restoration prospects of Lake Maggiore. Int. Rev. ges. Hydrobiol. 70: 63–75.

Mosello, R., M. C. Brizzio, F. Buzzi, L. Colzani, E. Pizzotti & D. Mocellin, 2001. Qualità delle acque dei tributari e del Lario nel biennio 1998–1999. Atti dell'Ass. ital. Oceanol. Limnol. 14: 137–146.

Naselli-Flores, L., J. Padisák, M. T. Dokulil & I. Chorus, 2003. Equilibrium/steady-state concept in phytoplankton ecology. Hydrobiologia 502 (Dev. Hydrobiol. 172): 395–403.

Novarino, G., I. A. N. Lucas & S. Morrall, 1994. Observations on the genus *Plagioselmis* (Cryptophyceae). Cryptogamie, Algologie 15: 87–107.

Pascher, A. (ed), 1978–1999. Süsswasserflora von Mitteleuropa. Vols. I–XXIV. Gustav Fischer Verlag, Jena, Stuttgart.

Ravera, O. & R. A. Vollenweider, 1968. *Oscillatoria rubescens* D.C. as an indicator of Lago Maggiore eutrophication. Schweizerische Zeitschrift für Hydrol. 30: 374–380.

Reynolds, C. S., 1980. Phytoplankton assemblages and their periodicity in stratifying lake systems. Hol. Ecol. 3: 141–159.

Reynolds, C. S., 1988. The concept of ecological succession applied to seasonal periodicity of freshwater phytoplankton. Int. Ver. theor. angewan. Limnol.: Verh. 23: 683–691.

Reynolds, C. S., 1993. Scales of disturbance and their role in plankton ecology. Hydrobiologia 249: 157–171.

Reynolds, C. S., 1997. Vegetation processes in the pelagic: a model for ecosystem theory. Ecology Institute, Oldendorf/Luhe, Germany. 371 pp.

Reynolds, C. S., V. Huszar, C. Kruk, L. Naselli-Flores & S. Melo, 2002. Towards a functional classification of the freshwater phytoplankton. J. Plankton Res. 24: 417–428.

Ruggiu, D., 1993. Osservazioni conclusive. Considerazioni generali sull'evoluzione a lungo termine dei popolamenti planctonici. Fitoplancton. In C.N.R. Istituto Italiano di Idrobiologia, Ricerche sull'evoluzione del Lago Maggiore. Aspetti limnologici. Campagna 1992 e Rapporto quinquennale 1988–1992. Ed. Commissione Internazionale per la protezione delle acque italo-svizzere: 115–116.

Ruggiu, D. & C. Saraceni, 1977. Fitoplancton, clorofilla e produzione primaria nel Lago Maggiore durante gli anni 1972–1973. Mem. Ist. ital. Idrobiol. 34: 57–78.

Ruggiu, D., G. Morabito, P. Panzani & A. Pugnetti, 1998. Trends and relations among basic phytoplankton characteristics in the course of the long-term oligotrophication of Lake Maggiore. Hydrobiologia 369/370: 243–257.

Salmaso, N., 1996. Seasonal variation in the composition and rate of change of the phytoplankton community in a deep subalpine lake (Lake Garda, Northern Italy). An application of nonmetric multidimensional scaling and cluster analysis. Hydrobiologia 337: 49–68.

Salmaso, N., 2000. Factors affecting the seasonality and distribution of cyanobacteria and chlorophytes: a case study from the large lakes south of the Alps, with special reference to Lake Garda. Hydrobiologia 438: 43–63.

Salmaso, N., 2003. Life strategies, dominance patterns and mechanisms promoting species coexistence in phytoplankton communities along complex environmental gradients. Hydrobiologia 502 (Dev. Hydrobiol. 172): 13–36.

Smayda, T. J., 1978. From phytoplankters to biomass. In Sournia A. (ed.), Phytoplankton Manual. UNESCO 23: 273–279.

Sommer, U., 1985. Comparison between steady state and non-steady state competition: Experiments with natural phytoplankton. Limnol. Oceanogr. 30: 335–346.

Sommer, U., 1989. The role of competition for resources in phytoplankton species succession. In Sommer, U. (ed.), Plankton Ecology – Succession in Plankton Communities. Springer, Berlin: 57–106.

Sommer, U., 1993. Disturbance-diversity relationships in two lakes of similar nutrient chemistry but contrasting disturbance regimes. Hydrobiologia 249: 59–65.

Sommer, U., 1995. An experimental test of the intermediate disturbance hypothesis using cultures of marine phytoplankton. Limnol. Oceanogr. 40: 1271–1277.

Sommer, U., J. Padisák, C. S. Reynolds & P. Juhász-Nagy, 1993. Hutchinson's heritage: the diversity-disturbance relationship in phytoplankton. Hydrobiologia 249: 1–7.

StatSoft Inc., 1997. Statistica for Windows (Computer program manual). StatSoft, Inc., Tulsa, OK. 958 pp.

Teubner K., M. Tolotti, S. Greisberger, H. Morscheid, M. T. Dokulil & H. Morscheid, 2003. Steady state phytoplankton in a deep pre-alpine lake: species and pigments of epilimnetic *versus* metalimnetic assemblages. Hydrobiologia 502 (Dev. Hydrobiol. 172): 49–64.

Tilman, D., 1977. Resource competition between planktonic algae: An experimental and theoretical approach. Ecology 58: 338–348.

Tilman, D., 1981. Test of resource competition theory using four species of Lake Michigan algae. Ecology 62: 802–815.

Tilman, D., 1994. Competition and biodiversity in spatially structured habitats. Ecology 75: 2–16.

Tilman, D. & R. W. Sterner, 1984. Invasions of equilibria: Tests of resource competition using two species of algae. Oecologia 61: 197–200.

Tilman, D., S. S. Kilham & P. Kilham, 1982. Phytoplankton community ecology: The role of limiting nutrients. Ann. Rev. Ecol. Syst. 13: 349–372.

Vollenweider, R. A., 1965. Materiali ed idee per una idrochimica delle acque insubriche. Mem. Ist. ital. Idrobiol. 19: 213–286.

Walsby, A. E., 2001. Determining the photosynthetic productivity of a stratified phytoplankton population. Aquat. Sci. 63: 18–43.

Walsby, A. E., A. Avery & F. Schanz, 1998. The critical pressures of gas vesicles in *Planktothrix rubescens* in relation to the depth of winter mixing in Lake Zürich, Switzerland. J. Plankton Res. 20: 1357–1375.

Willén, E., 1991. Planktonic diatoms – an ecological review. Algol. Stud. 62: 69–106.

Hydrobiologia **502:** 49–64, 2003.
L. Naselli-Flores, J. Padisák, M. T. Dokulil (eds),
Phytoplankton and Equilibrium Concept: The Ecology of Steady-State Assemblages.

Steady state phytoplankton in a deep pre-alpine lake: species and pigments of epilimnetic *versus* metalimnetic assemblages

Katrin Teubner[1], Monica Tolotti[2], Sonja Greisberger[1], Heike Morscheid[3], Martin T. Dokulil[1] & Harald Morscheid[3]
[1]*Institute of Limnology, Mondseestrasse 9, A-5310 Mondsee, Austria. E-mail: katrin.teubner@oezw.ac.at*
[2]*Institut für Zoologie, Abt. Limnologie, Univ. Innsbruck, Technikerstrasse 25, A-6020 Innsbruck, Austria*
[3]*Bayerisches Landesamt für Wasserwirtschaft, Demollstrasse 31, D-82407 Wielenbach, Germany*

Key words: DCM, stratification, *Planktothrix*, nutrients, carotenoids

Abstract

The vertical distribution pattern of algal species, chlorophylls and specific carotenoids present in the dimictic pre-alpine Ammersee (Bavaria, Germany) are given for the year 2001. A detailed taxonomic list of the phytoplankton species is recorded, along with light micrographs and detailed descriptions of the flagellates. A deep chorophyll maximum, mainly built by *Planktothrix rubescens*, was common in this deep mesotrophic lake. The three most dominant species among 83 identified taxa alternated seasonally and reached significant biovolumes in both the epi- and the metalimnion (*Planktothrix rubescens* > *Ceratium hirundinella* > unicellular centric diatoms > *Asterionella formosa* > *Fragilaria* spp. > *Anabaena lemmermannii* > *Phacotus lenticularis* and less frequent dominant was *Rhodomonas minuta*). We define a steady state phytoplankton assemblage in Ammersee as a stable community in terms of species composition and standing crop. The stability of species composition was measured by Bray-Curtis similarity between monthly samples and indicate the change of individual biovolumes of species from month to month. The stability of standing crop was evaluated by the net change of total biovolume for the same time intervals. Focussing on steady state phytoplankton assemblages we compared three spatially heterogeneous environments for vertical niche separation within the top 12 m: the euphotic epilimnion (2 and 5 m), the euphotic metalimnion (7 m) and the metalimnion below the euphotic zone with dim-light less than 1% (10 and 12 m). The definition of a steady state assemblage on both ends hold true only for metalimnetic layers at dim-light levels below 1% in Ammersee. At this metalimnetic layer more than 80% similarity in phytoplankton composition between successive monthly samples was reached, associated with almost zero net changes of total biovolume only. The greater the contribution of the three most dominant taxa to biovolume, the higher were the Bray-Curtis similarities at metalimnetic depths below the euphotic zone. Zooplankton biomass had very little effect on species assemblages in the metalimnion, while parameters related to stratification (Schmidt stability) as well as those of trophy (TP, Chl) correlated with species changes. The similarity values between successive monthly samples from all the euphotic layers never reached more than 60% and were usually significantly lower, even if biovolume net changes were around zero. Both the high fluctuations of the ratio of photosynthetic *versus* photo-protective carotenoids (PSC:PPC) and the statistical significance of correlations between the change of species and environmental-biotic parameters separate the euphotic layer of the top 7 m as a homogenous community from deeper strata. At all sampling depths within this euphotic zone the increase of sunshine duration was associated with an increase of the carotenoid ratio PSC:PPC, but no relationship was found for the deeper layers. The change of species in the euphotic layer was not significantly related to thermal stability, TP or the dominance structure of phytoplankton, but linked with the zooplankton biomass and therefore seemed to be top-down controlled. From our observations, we can conclude that only during stratification and only in the metalimnion below the euphotic zone steady state assemblages can be expected in the deep mesotrophic Ammersee.

Introduction

The thermodynamic term 'steady state' originally coined by studies on chemical reaction kinetics is also used to define growth in continuous culture systems, to describe the persistence of biota, and relatively stable conditions as a whole in an ecosystem like the rain forest or parts of the ocean (e.g. Bannister, 1974; Jannasch, 1974; Sommer, 1985). Here we define a steady state phytoplankton assemblage as a stable community in terms of species composition and standing crop.

Steady state assemblages in the microbiological sense could exist only if production and losses were perfectly balanced independent of time. Intermediate disturbance hypothesis and rescource ratio hypothesis focus on the dynamic of abiotic factors, mainly nutrients, and emphasise the bottom up control on the structure of algal community (e.g. Tilman et al., 1982; Reynolds et al., 1993). Species diversity shaped by nutrients is seen as the response to the number of limiting nutrients (e.g. Interlandi & Kilham, 2001) and to the dynamic of nutrients defined by both quantities – the magnitude and the frequency of pulses (e.g. Sommer, 1995; Polishchuk, 1999). The top-down control on diversity is complex as well as reviewed in Hixon (2002). Approach on chaos theory stresses that the feedback in plankton system itself generates internal dynamics preventing the system from coming to equilibrium (Scheffer et al., 2003).

Results from multispecies chemostat experiments by Sommer (1983) suggest that an inoculum-independent equilibrium in phytoplankton would need at least 3–6 weeks of constant environmental conditions in a lake. We can not expect such a continuous culture situation of constant environmental conditions in nature because of (i) a great variety of factors which may have an influence on phytoplankton succession, and (ii) the seasonality in temperate lakes. Steady state conditions may, however, at least be found in stratified lakes developing simultaneous spatially separated phytoplankton associations in different strata (Haffner & McNeely, 1989; Reynolds, 1992; Lindholm, 1992). We expect that most of the abiotic factors which could have a major influence on phytoplankton dynamics are strongly fluctuating in the surface layer, favouring higher dynamics in species change in the surface than in metalimnetic layers. We performed this study on a mesotrophic, dimictic alpine lake with a deep chlorophyll maximum, mainly built by *Planktothrix rubescens*. During summer stratification from 1997 to 2001, the metalimnetic phytoplankton below the euphotic zone reached biovolumes on average 30% higher than those in the epilimnion (Teubner, unpubl.). The main emphasis in this paper is on the vertical distribution pattern of algal species and their specific pigments. For practical purposes three criteria must be met according to Sommer et al. (1993) to consider an algal community to be in steady state: (i) no more than three species contribute more than 80% to the total biomass, (ii) the persistence of these species for more than 1–2 weeks and (iii) during this period the total biomass does not significantly change. In this phytoplankton study in Ammersee, we searched for such time periods and depth layers in which steady states are most likely.

We hypothesise that significant differences between the epilimnetic and metalimnetic phytoplankton become evident by both (i) the composition and (ii) the change of composition. In particular, we used a measure for the similarity between successive phytoplankton samples in the respective depth layers to evaluate the monthly change of species composition at different depth layers. We evaluate the applicability of the definition of steady state assemblages given above.

Method

Site description

Ammersee is a mesotrophic dimictic pre-alpine lake near Munich (Bavaria in Germany) with a maximum depth of 81 m and a water volume of 1750 10^6 m^3. Further details on morphometry and trophy were described in Morscheid & Mayr (2002).

Sampling and data treatment

Monthly samples were taken between 10 and 11 a.m. at the deepest point in Ammersee from February to November 2001. Biovolume of phytoplankton of the depth samples 2, 5, 7, 12 and 15 m was estimated from microscopically counted abundance and size measurements of single units of cells or colonies. Samples for pigments and chemical parameters were taken from 0, 2, 5, 7, 10, 12, 15, 20 and 30 m. Pigments were analysed by high performance liquid chromatography (HPLC) in accordance with Wright et al. (1991). One litre of lake water was filtered (Whatman GF/F). The filter was frozen and extracted by 90% acetone. To maintain the calibration settings, a mixed pigment standard was used containing caroten-

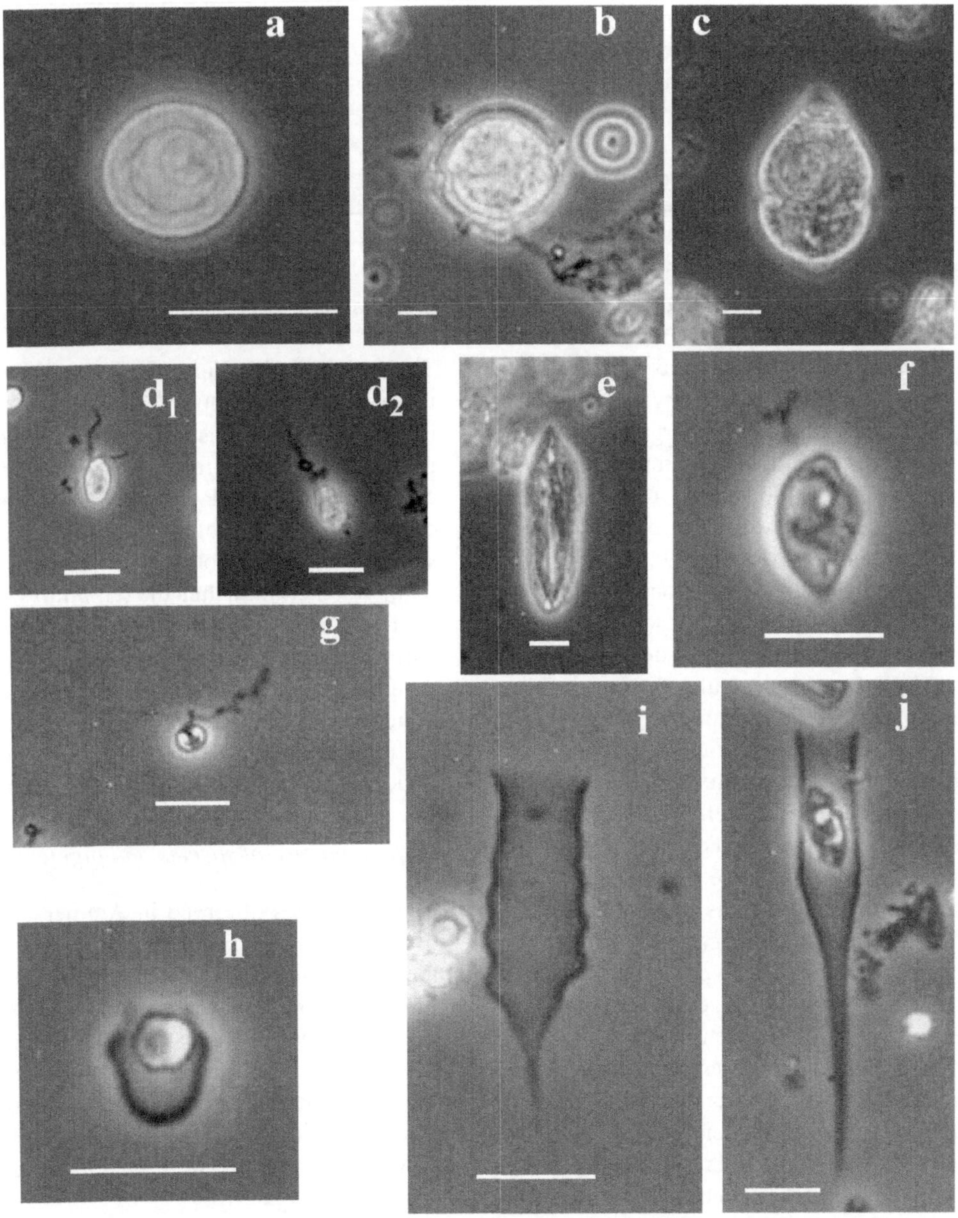

Plate 1. Flagellates from Ammersee. (A) *Phacotus lenticularis* (B) *Peridiniopsis* spec. (C) *Gymnodinium helveticum*, (D 1–2) *Katablepharis* cf. *ovalis*, (E) *Cryptomonas* cf. *erosa*, (F) *Rhodomonas lens*, (G) *Chrysochromulina parva*, (H) *Kephyrion* cf. *rubri-claustri*, (I) *Dinobryon acuminatum*, (J) *Dinobryon bavaricum* (bars = 10 μm).

oids in quantitative proportion to chlorophyll-*a* as measured in algal cultures and described in Wilhelm et al. (1991), respectively. Carotenoids in Figure 4 were categorised according to their distinct function as 'photo-synthetic carotenoids (PSC)', such as fucoxanthin and fucoxanthin-like pigments, peridinin, prasinoxanthin and α-carotene involved in light harvesting, and photo-protective carotenoids (PPC), which are the remaining carotenoids protecting the cells against photo-oxidation (Rowan, 1989; Bricaud et al., 1995; Stuart et al., 1998). The ratio PSC:PPC was related to the sunshine duration averaged over the 21 days before sampling, because this time interval was best related to the pigment ratio when compared with stepwise intervals from 1 to 33 days.

For phytoplankton biovolume ($mm^3\ l^{-1}$) the mean values of single species of a normal or a log-normal distribution were calculated depending on the best fit. The method for estimation of the zooplankton biomass and the zooplankton species composition in Ammersee is described in Morscheid & Mayer (2002). The zooplankton biomass in Figure 6 includes rotifers, cladocerans and copepods.

Temperature, turbidity (NTU) and conductivity in the water column were measured *in situ* by a multi-parameter profiler (YSI 6920). The pattern of vertical distribution of algae was detected by in situ chlorophyll-fluorescence using a Backscat-fluorometer (Haardt). Dissolved inorganic nitrogen calculated as the sum of nitrate and ammonia (DIN), total phosphorus (TP), and soluble reactive silica (SRSi) were estimated from discrete depth samples using standard techniques (APHA, 1992).

Photosynthetically active radiation (PAR) was measured with a 4 π quantum sensor (LI-COR). Underwater-light profiles were used to define the layer of 1% surface light intensity calculated from the vertical attenuation coefficient. The mixing depth (z_{mix}) was calculated from the depth with maximum relative thermal resistance against mixing by individual temperature profiles (RTR; Wetzel, 2001), the stability of stratification in the water column by Schmidt stability according to Livingstone & Schanz (1994). The metalimnetic layer defined as the water stratum of steep thermal gradient demarcated by an almost homoiothermal epilimnion and hypolimnion (Wetzel, 2001) was calculated by RTR-values and was in average of about 5–8 m for Ammersee in 2001.

Notched box-whisker plots were created in SYSTAT 8.0 for Windows (Figs 1–3 and 5 D). Boxes are notched at the median, with the length of the notches indicating the 95% confidence intervals. The length of the box shows the range of the central 50% of the values. The phytoplankton dissimilarity measure between each pair of successive monthly samples was based on continuous data (biovolumes of individual species). These dissimilarity values were calculated by Bray-Curtis ranging from 0–1 using SYSTAT. For easier interpretation this dissimilarity (d) was converted to similarity (s) by s = 1-d. This similarity considers the biovolume of all individual species with more than 3% contribution to total biovolume at a single depth ($n = 80$) including also non-obligate autotrophic taxa as e.g. *Gymnodinium helveticum.* The variables used for Pearson correlation in Figure 5 have passed the normality test (Kolmogorov-Smirnov). Net change of phytoplankton biomass

$$k_p = \frac{\ln p_2 - \ln p_1}{t_2 - t_1}$$

as d^{-1} was calculated by monthly data.

Results

The variability of abiotic factors in the vertical profile is summarised in Figure 1. Temperature, light, turbidity and concentrations of SRSi were highly fluctuating throughout the year in the surface layer compared to deeper layers. For TP and DIN a slightly higher range of variation was recognised at 5–7 m and at 2 m, respectively, even if both were more evenly distributed in the water column than the remaining variables. Conductivity is the only shown variable which was most variable at layers deeper than 10 m. Most of the abiotic parameters in Figure 1 were therefore strongly fluctuating in the surface layer supporting the argument for higher dynamics in species change in the surface than in deeper layers advanced in the introduction.

Vertical pattern of species and pigments

The 83 species observed in Ammersee in the course of the study are listed in Table 1. About 15% of these taxa contributed 5% or more to total phytoplankton biovolume in the epilimnetic layer at least once between May and August and almost the same number of species (16%) add to that volume in the metalimnion. For each individual species, the preferred layer is indicated in Table 1. More than 40% of the taxa observed in Ammersee were flagellates, of which some are shown in Plate 1.

The annual variation of the biovolume of algal species and concentrations of pigments in the water column is shown in Figures 2 and 3. Species of the orders haptophytes, chlorophytes, cryptophytes, dinoflagellates and diatoms such as *Chrysochromulina parva*, *Pseudokephyrion* cf. *entzii*, *Phacotus lenticularis*, *Ceratium hirundinella*, *Cryptomonas erosa*, *Rhodomonas minuta* and unicellular centric diatoms had their centre of gravity only in the epilimnetic layer as indicated by high median values in the 2–7 m depth-range (Fig. 2A–F, H, Plate 1 a, e, g, h). *Tabellaria flocculosa* was a diatom species with highly fluctuating biovolumes in the epilimnion. The flagellates

Table 1. The 83 phytoplankton taxa identified for Ammersee in 2001. The preference of taxa for the epilimnion (EPI) or metalimnion (META) during thermal stagnation (May-October) is marked by: + = present; ++ = frequent; — = absent (further detailed description of species see Lenhart, 1987, 2000; Morscheid, 2002)

Division and Taxa	EPI	META
CYANOPHYCEAE (12 taxa)		
Chroococcales		
Chroococcus limneticus LEMM.	+	—
Chroococcus minutulus KÜTZ. (NÄG.)	+	+
Chroococcus sp.	+	+
Snowella lacustris (CHOD) KOM. et. HIND.	+	+
Merismopedia sp.	+	—
Microcystis flos-aquae KÜTZ.	—	+
Microcystis sp.	+	+
Radiocystis sp.	+	+
Oscillatoriales (2 taxa)		
Planktolyngbya sp.	+	+
Planktothrix rubescens (DC ex. GOM) ANAGN. et KOM.	+	++
Nostocales (2 taxa)		
Anabaena cf. *spiroides* KLEBAHN	++	+
Anabaena lemmermannii BORY	+	+
BACILLARIOPHYCEAE (12 taxa)		
Centrales		
Aulacoseira islandica (MÜLLER) SIMONSEN	—	+
Aulacoseira granulata (EHRENB.) SIMONSEN	+	++
Cyclotella spp. (KÜTZ.) BRÉBISSON	++	+
Stephanodiscus spp.	+	+
Pennales		
Asterionella formosa HASSALL	+	+
Cymbella sinuata GREGORY	—	+
Fragilaria crotonensis KITTON	++	+
Fragilaria ulna (NITZSCH) LANGE-BERTALOT	+	+
Synedra acus var. *angustissima* GRUNOW	++	+
Navicula radiosa KÜTZ.	—	+
Nitzschia sp.	—	+
Tabellaria flocculosa (ROTH) KÜTZ.	++	+
CRYPTOPHYCEAE (7 taxa) *Chroomonas* sp.	—	+
Cryptomonas erosa EHRENB.	++	+
Cryptomonas marssonii SKUJA	+	+
Cryptomonas rostratiformis SKUJA	+	+
Katablepharis cf. *ovalis* SKUJA	+	+
Rhodomonas lens PASCHER et RUTTNER	+	+
Rhodomonas minuta var.*nannoplanctica* SKUJA	++	+
EUGLENOPHYCEAE (1 taxon) *Trachelomonas volvocina* EHRENB.	—	+
CHLOROPHYCEAE (23 taxa) **Volvcales (3 taxa)**		
Carteria sp.	+	—
Phacotus lenticularis STEIN	++	+
Tetraselmis sp.	+	+
Chlorophyceae sensu lato		
Coelastrum microporum NÄGELI	—	—

Table 1. Continued

Crucigenia tetrapedia (KIRCHNER) W. et G. S. WEST	+	—
Elakatothrix genevensis HIND.	+	+
Eutetramorus fottii (*Coenococcus fottii*) (HIND.) KOM.	+	+
Monoraphidium arcuatum (KORŠIKOV) HIND. —	—	
Monoraphidium sp.	+	—
Oocystis cf. *borgei* SNOW	+	—
Oocystis lacustris CHODAT	+	+
Oocystis parva WEST	+	—
Pediastrum boryanum (TURPIN) MENEGHINI	+	+
Pediastrum tetras (EHRENB.) RALFS	+	—
Pseudodyctiosphaerium minusculum HIND.	+	+
Radiococcus sp.	—	+
Scenedesmus armatus (CHOD.) CHOD.	—	+
Scenedesmus quadricauda(TURPIN) BRÉBISSON sensu CHODAT	—	+
Scenedesmus sp.	—	+
Tetrachlorella incerta HIND.	+	+
Tetraedron minimum (A.BRAUN) HANSGIRG	+	+
Tetrastrum komarekii HIND.	+	—
Desmidiales		
Cosmarium depressum (NÄGELI) LUND	+	+
HAPTOPHYCEAE (1 taxon) *Chrysochromulina parva* LACKEY	++	+
CHRYSOPHYCEAE (18 taxa) *Bitrichia chodati* (REVERDIN) HOLLANDE	+	+
Chromulina spp.	++	+
Chrysococcus rufescens KLEBS	+	+
Chrysolykos planktonicum MACK	++	+
Dinobryon bavaricum IMHOF	++	+
Dinobryon acuminatum RUTTNER	+	—
Dinobryon divergens IMHOF	+	+
Dinobryon sociale EHRENB.	++	+
cf. *Gloeobotritis* sp. PASCHER	—	+
Kephyrion cf. *mastigophorum* SCHMID	+	—
Kephyrion rubi-claustrii CONRAD	+	—
Mallomonas caudata IVANOV emend. KRIEGER	+	+
Mallomonas cf. *tonsurata* TEILING emend. KRIEGER	++	+
Ochroomonas cf. *margaritata* SKUJA	+	+
Pseudokephyrion cf. *entzii* CONRAD	++	+
cf. *Pseudopedinella erkensis* SKUJA	++	+
Stichogloea doerderleinii (SCHMIDLE) WILLE	+	—
Uroglena sp.	++	+
DINOPHYCEAE		
Amphidinium sp. CLAPARÈDE & LACHMANN	—	+
Ceratium hirundinella (O. F. MÜLLER) DUJARDIN	++	+
Gymnodinium helveticum PENARD	+	+
Gymnodinium spp.	+	+
Gymnodinium uberrimum (ALLMAN) KOFOID et SWEZY	+	+
Peridiniopsis penardiformis	+	+
Peridinium inconspicuum LEMM.	+	+
Peridinium willei HUITFELD-KAAS	+	+
Woloszinskia / *Peridiniopsis* sp.	+	++

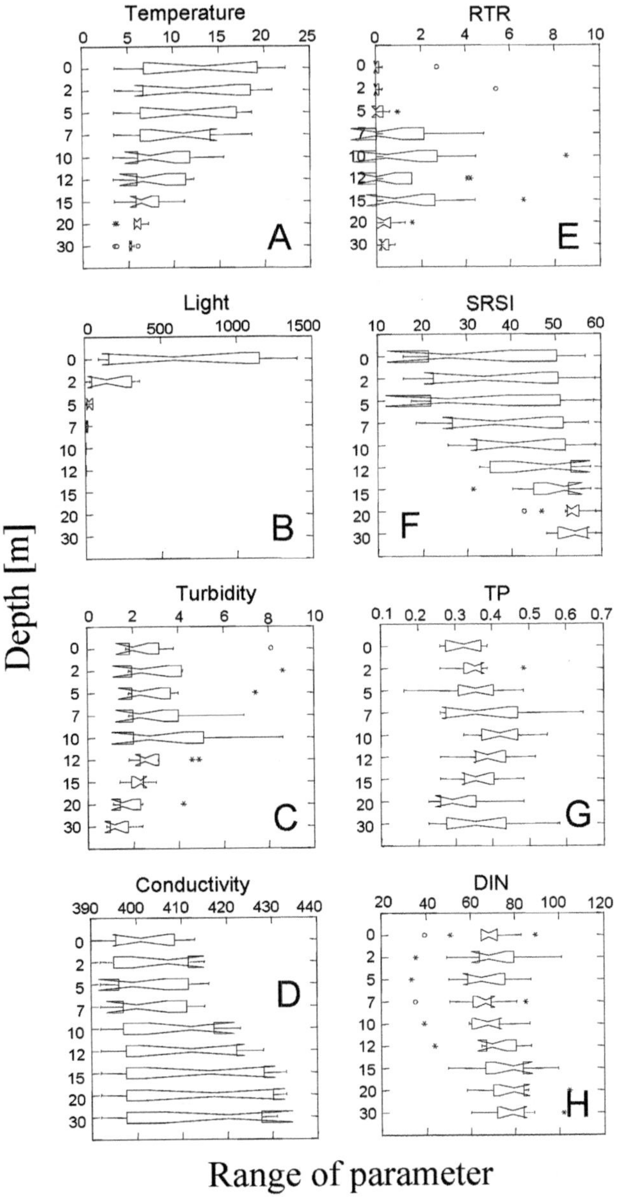

Figure 1. Annual profiles of temperature (A), light (B), turbidity (C), conductivity (D), relative thermal resistance (RTR; E), soluble reactive silica (F), total phosphorus (G) and dissolved inorganic nitrogen (H) as notched box-whisker plots (see 'Method]).

Rhodomonas lens, *Gymnodinium helveticum*, *Woloszinskia/Peridiniopsis* spec. (Fig. 2G, J–K, Plate 1 c, f) and *Cryptomonas rostratiformis* had biovolume maxima both in the epilimnion and in the metalimnion. With a median biovolume of 1 $mm^3 \, l^{-1}$ *Planktothrix rubescens* was the absolute dominant species. This cyanobacterium contributed up to 95% of the total biovolume at 10–12 m during stratification in August–September but also up to 75% biovolume at 2–5 m during mixing in April and October. *Katablepharis* cf. *ovalis* was common only at 12 m depth (Fig. 2 L, Plate 1 d) in late summer and early autumn. *Aulacoseira granulata* and *A. islandica* contributed up to 7% of the total biovolume at 15–20 m in June, *Uroglena* spec. 7–10% at 7–10 m in August. Other diatoms such as *Fragilaria crotonensis*, *Asterionella formosa*, and the chrysophytes *Dinobryon divergens* (Fig. 2 N–P), *D. acuminatum* and *D. bavaricum* (Plate 1 i–j) were

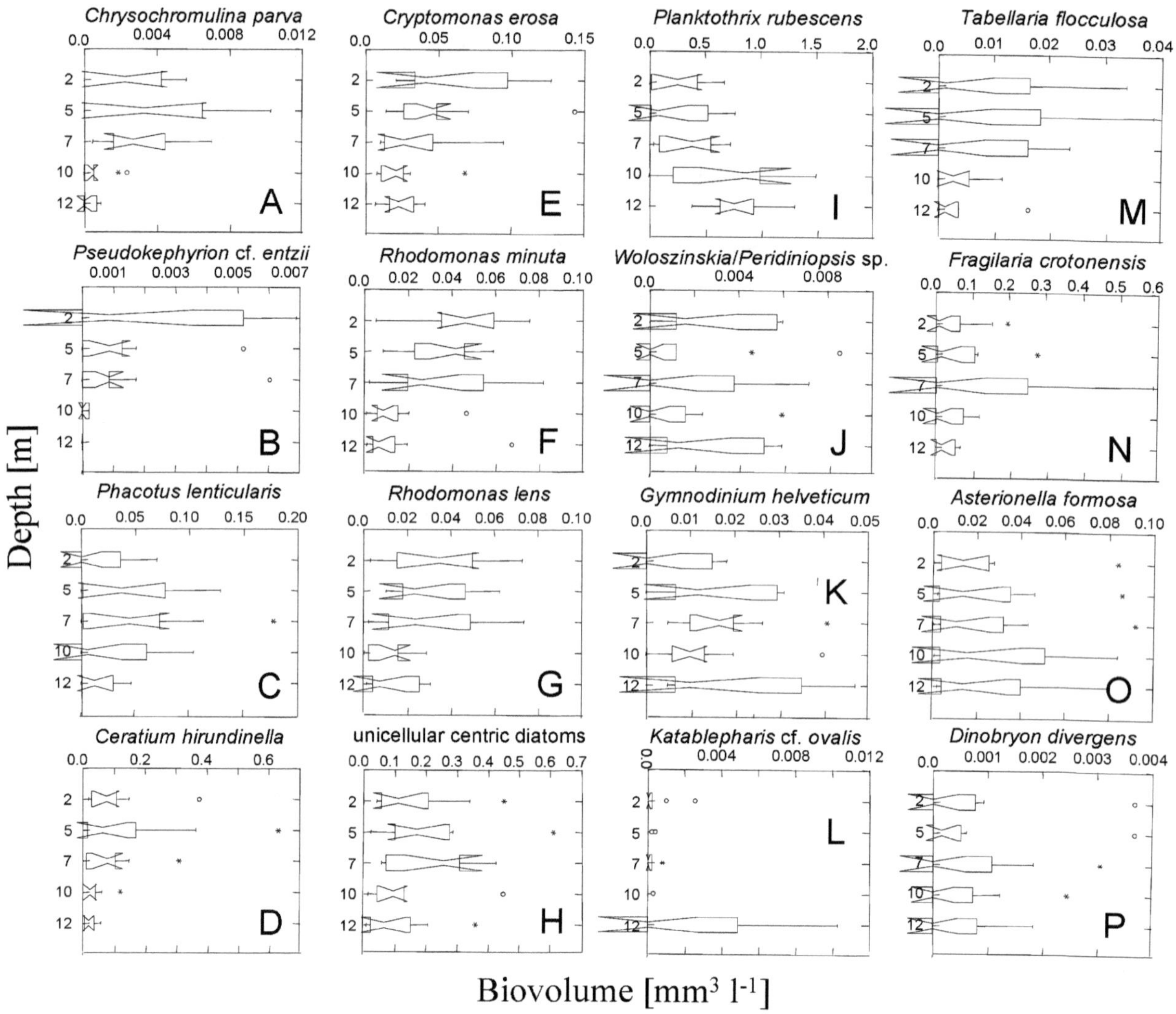

Figure 2. Vertical distribution of the biovolume in $mm^3\ l^{-1}$ of haptophytes (A – *Chrysochromulina parva*) chrysophytes (B – *Pseudokephyrion* cf. *entzii*), chlorophytes (C – *Phacotus lenticularis*), dinoflagellates (D – *Ceratium hirundinella*, J – *Woloszinskia/Peridiniopsis*, K – *Gymnodinium helveticum*), cryptophytes (E – *Cryptomonas erosa*, F – *Rhodomonas minuta*, G – *R. lens*, L – *Katablepharis* cf. *ovalis*), diatoms (H – unicellular centric diatoms, M – *Tabellaria flocculosa*, N – *Fragilaria crotonensis*, O – *Asterionella formosa*) and cyanobacteria (J – *Planktothrix rubescens*). Notched box-whisker plots as in Figure 1.

more evenly distributed in the water column. Significant differences in the vertical distribution of individual species are indicated by non-overlapping notches in the box-plots of Figure 2. In particular, *Ceratium hirundinella*, *Cryptomonas erosa* and *Rhodomonas minuta* had significant higher biovolumes at 2 or 5 m, and *Chrysochromulina parva* at 7 m when compared with deep layers. In contrast, *Planktothrix rubescens* developed significant higher biovolumes only at 12 m.

The general vertical distribution pattern for the different algal groups is shown by their respective pigments in Figure 3. Both chlorophyll-*a* and β-carotene, which are ubiquitous or virtually present in all algae showed the same vertical distribution with a maximum at 10 m indicating the highest phytoplankton biovolume in the deep layer (Fig. 3A, D). Both were significantly related to total biovolume of phytoplankton (chlorophyll-*a*, $r = 0.86$; β-carotene: $r = 0.65$, $n = 70$). Major and minor carotenoids for cyanobacteria shown in Figure 3 E–H and echinenone (not shown) had the similar vertical pattern as chlorophyll-*a* and β-carotene. The close relationship

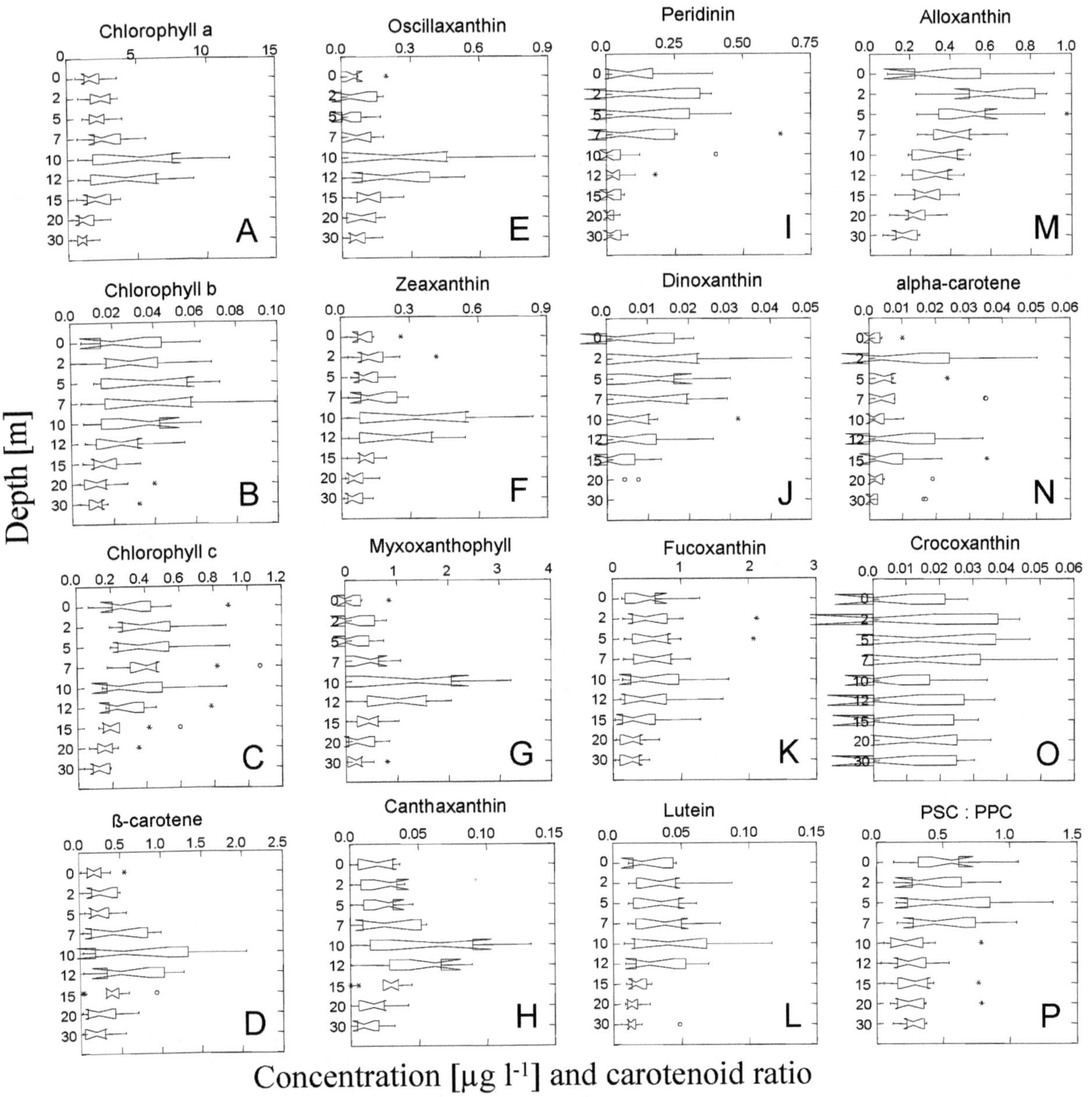

Figure 3. Vertical distribution of the concentration of pigments in μg l^{-1} (A–C chlorophylls, D–O carotenoids) and the carotenoid ratio PSC:PPC (P, photo-synthetic carotenoids: photo-protective carotenoids).

between these pigments and both the total biovolume of the cyanobacteria and of *Planktothrix rubescens* indicates the dominance of this single species among the photoautotrophic procaryotes (oscillaxanthin: $r = 0.80$ and 0.81, zeaxanthin: $r = 0.87$ and 0.87, myxoxanthophyll: $r = 0.79$ and 0.80, canthaxanthin: $r = 0.79$ and 0.79, echinenone: $r = 0.88$ and 0.88). Peridinin and dinoxanthin, characteristic pigments in photosynthetic dinoflagellates, correlate with the total biovolume of dinoflagellates as well as with that of *Ceratium hirundinella* (Fig. 2D and 3I–J; peridinin: $r = 0.65$ and 0.64; dinoxanthin $r = 0.37$ and 0.33, respectively). The predominantly epilimnetic occurrence of *Ceratium hirundinella* was highlighted by both the biovolume of this species and the total amount of the respective carotenoids. The major carotenoid alloxanthin in cryptophytes was closely related to the biovolume of this algal group, the sum of *Cryptomo-*

nas species and, in particular, to *C. erosa* ($r = 0.68$, 0.69 and 0.73, respectively), while crocoxanthin was correlated to all *Rhodomonas species* but especially to *R. lens* ($r = 0.54$ and 0.58, respectively). The epilimnetic dominance of the cryptophytes in general is illustrated by alloxanthin, while a more even distribution from the surface down to the deepest layers was emphasised by crocoxanthin. The major pigments of the chlorophytes and of the amalgamated group of diatoms and chrysophytes were closely related to the respective algae (lutein: $r = 0.65$, fucoxanthin: $r = 0.84$). Both pigments are again examples for evenly distributed pigments in the vertical 0–12 m.

Vertical pattern of the ratio of photo-synthetic versus photo-protective carotenoids

The ratio of lipophilic accessory photo-synthetic versus photo-protective carotenoids (PSC:PPC) was higher and more fluctuating in the surface layer 0–7 m than at deeper depth (Fig. 3 P). Different behaviour of phytoplankton in the two layers was indicated by the relationship between this carotenoid ratio and the sunshine duration throughout the year (Fig. 4). At all sampling depths within the euphotic zone ($\leq$7 m, Fig. 5 B) the phytoplankton was related to increased sunshine duration by the relative increase of photosynthetic versus photo-protective carotenoids (Fig. 4 A), but no relationship was found for the deeper layers (Fig. 4 B). The greatest response to sunshine duration was found at 5 m as indicated by a higher slope when compared with other surface layers. The mean relative light intensities in the euphotic zone at 2 m, 5 m and 7 m were 33%, 6.5% and 2.3%, respectively, in the metalimnion below the euphotic depth at 10 m and 12 m still 0.51% and 0.19%.

Seasonal dynamics of phytoplankton and controlling factors

The seasonal dynamics of chlorophyll-*a* concentration, the stability of stratification in the water column (Schmidt stability), the depth of the euphotic zone (z_{eu}) and the mixing layer (z_{mix}) are shown in Figure 5. During thermal stratification the euphotic layer was deeper than the mixing depth. The sampling depths 2 and 5 m were representative for the epilimnetic layer during thermal stratification. Samples from 7 m represent the bottom layer of the euphotic zone and were already in the metalimnion. 10 and 12 m were always the dim-light layers below the euphotic zone

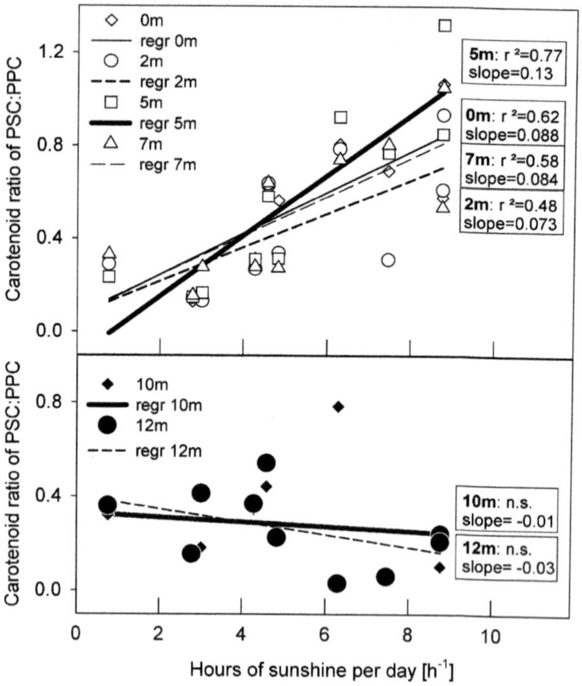

Figure 4. The hours of sunshine per day versus the carotenoid ratio PSC:PPC as in Figure 3 P for epilimnetic (A) and metalimnetic layers (B).

in the metalimnion. With the onset of stratification the chlorophyll concentration in the metalimnion became much higher than in the epilimnion (Fig. 5 A). From May to September, the three most dominant species reached on average 77–75% of total biovolume at 2 and 5 m respectively, 82% and 88% at 10 and 12 m. Most of these taxa were dominant in both the epi- and the metalimnion (*Planktothrix rubescens* > *Ceratium hirundinella* > unicellular centric diatoms > *Asterionella formosa* > *Fragilaria* spp. > *Anabaena lemmermannii* > *Phacotus lenticularis* and less frequent dominant was *Rhodomonas minuta*). In contrast, *Cryptomonas erosa* and *Gymnodinium helveticum* were dominant only in the epi- or the metalimnion respectively. In terms of the number of species, up to 9 taxa, on average 3–4 taxa contributed to at least 80% of total biovolume in the epilimnion, but only up to 6 taxa, on average 2–3 to that in the metalimnion.

The change of species in phytoplankton was evaluated by the similarity between two successive monthly samples (Bray-Curtis, see method). A low value was calculated between the samples February and March (2–3) for each depth (Fig. 5 A). Almost the same low value of similarity was calculated for the sampling

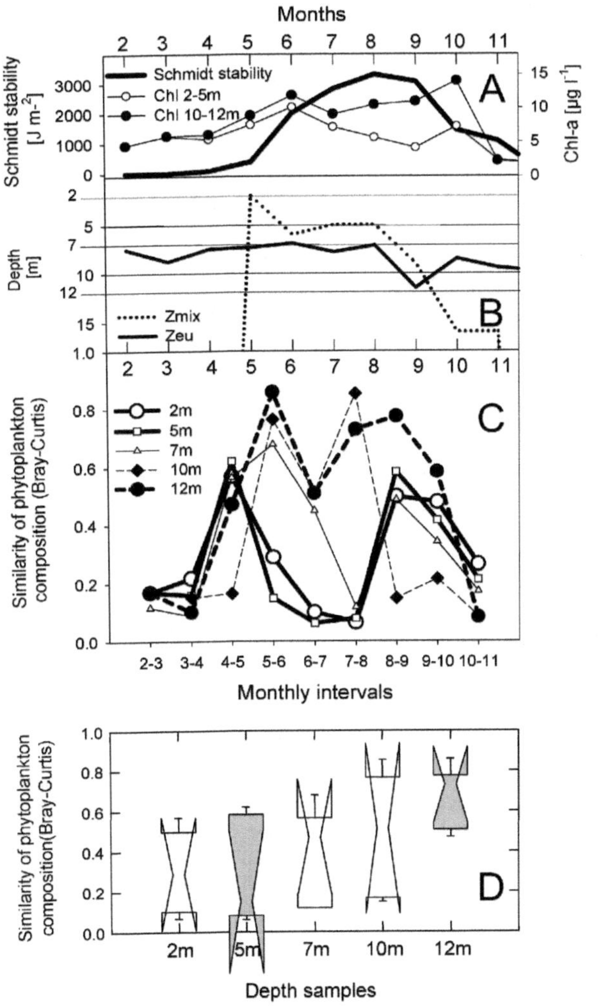

Figure 5. (A) Stability of stratification in the water column (Schmidt stability), concentration of chlorophyll-*a* in the upper epilimnion (2–5 m) and in the metalimnion (10–12 m). (B) Euphotic (z_{eu}) and mixing depth (z_{mix}). (C–D) Change of individual species biovolumes in phytoplankton as similarity by Bray-Curtis between successive monthly samples (C) and between layers during thermal stratification from May to September as box-whisker plot (D). Significant differences between depth layers 5 m and 12 m in D are indicated by non-overlapping confidence (see 'Method').

pair March–April (3–4) for all depths. In other words, the share of biovolume of individual species in March and April was quite different from that of the previous month. Significant higher similarity values were reached from April to May before the onset of stratification indicated by low values of Schmidt stability and the narrow mixing depth (Fig. 5 A–C). A further increase in the similarity at 7, 10 and 12 m was concomitant with the onset of thermal stratification between May and June. Particularly high similarities, indicating a stable species composition, were measured for the phytoplankton at 12 m from May to September. The similarity at 10 m follows the same trend but dropped off between August and September. At the same time we measured a decrease of stability of the thermal stratification, a mixing down to about 11 m (z_{mix}) and an increase in euphotic depth (z_{eu}).

The opposite trend was observed for the epilimnetic layer. With the onset of stratification, the similarity decreased again (5–6). Stable phytoplankton species composition could not be observed at any time between May and August. With decreasing stability of the thermal stratification and deeper mixing between August and September the similarity of phytoplankton composition decreased even at 2 and 5m.

The similarity of phytoplankton between October and November (10–11) was as low as between March and April (3–4). Therefore for both mixing periods, the same unstable phytoplankton composition was observed for all the investigated layers.

All similarity values for the stratified period May to September are compared in the box-plots of Figure 5D for all individual depths. High median values indicate the general trend of a stable phytoplankton composition in the metalimnion (7–12 m). At the same time, the epilimnetic phytoplankton changed significantly from month to month, as shown by low similarity values. In particular, a significant difference in the change of phytoplankton composition at 5 m and 12 m is shown by non-overlapping notches.

The correlation between Bray-Curtis similarity and different variables is given in Figure 6. At all sampling depths within the euphotic zone ($\leq$7 m, Fig. 5 B) the Bray-Curtis similarity was significantly related only to the biovolume of zooplankton (rotifers and metazoan zooplankton). Bray-Curtis values for the metalimnetic zone below the euphotic zone, 10 and 12 m, were significantly related to all parameters except zooplankton. The close correlations to z_{eu}/z_{mix} and Schmidt stability (SCHMIDT) were in accordance with the general trend of low similarity values during mixing in spring and autumn, but high values during stratification in summer as shown in Figure 5. The higher the concentrations of chlorophyll-*a* and TP, the more similar was the phytoplankton composition between successive months, especially at 10 m. The more the biovolume amount contributed by the three most dominant taxa, the higher were the Bray-Curtis similarities at both metalimnetic depths below the euphotic zone.

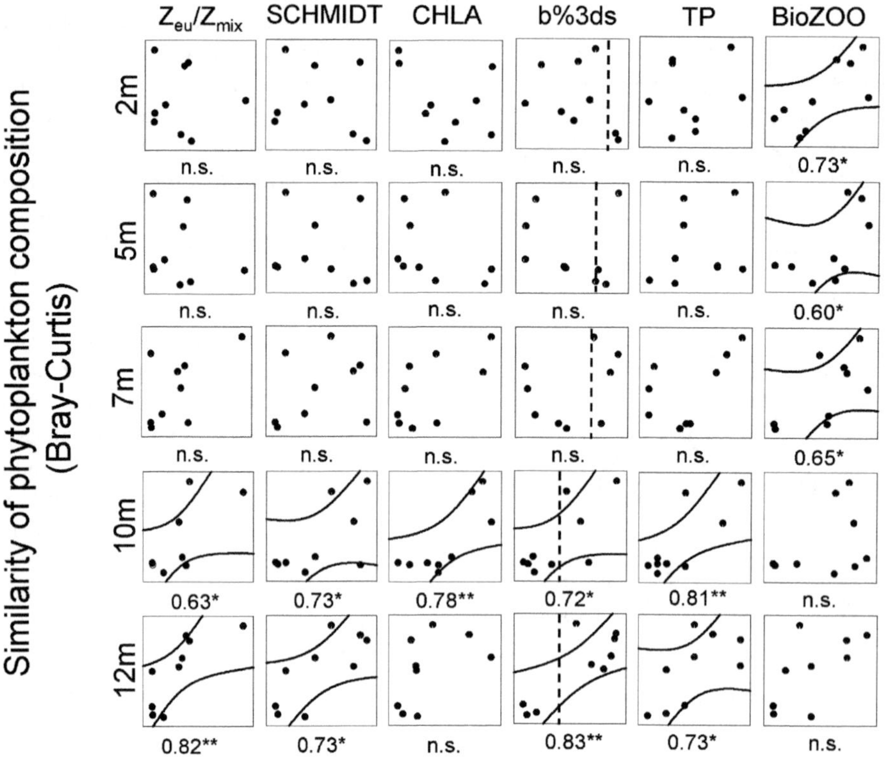

Figure 6. Pearson correlation between the similarity of phytoplankton between successive samples (Bray-Curtis) and variables: the ratio between euphotic to mixing zone (z_{eu}/z_{mix}), the stability of stratification in the water column (Schmidt stability, Schmidt), Chlorophyll-*a* concentration (Chl-*a*), the percentage contribution of the three most dominant species to phytoplankton biovolume at each sampling date (b%3ds , 80% marked dashed line), total phosphorus concentration (TP) and the biovolume of zooplankton (BioZoo). Except z_{eu}/z_{mix} and Schmidt stability the correlations were calculated between both values of the respective depths.Data cover all sampling intervals ($n = 9$, February–November). Coefficients and 0.95 confidence intervals of regression lines are shown only for significant correlations. Significance is marked as: *-$P < 0.05$, **-$P < 0.01$, n.s.- not significant.

Discussion

A metalimnetic deep chlorophyll maximum below the euphotic zone mainly built by *Planktothrix rubescens* as observed in Ammersee is common in deep meso-eutrophic lakes and can be assigned to one of the four main types of deep-living algal communities suggested by Adler et al. (2000). A number of species were common at deep layers in Ammersee even if less dominant than *Planktothrix rubescens*. This occurrence of species of various phytoplankton groups in the region of the deep chlorophyll maximum has also been described by other studies (e.g. Haffner & McNeely, 1989; Lindholm, 1992; Gervais, 1998; Flaim et al., 2003). The distribution of species ranging from the surface down to layers below the euphotic zone is discussed as temporary or stationary vertical niche separation, to avoid washing out and grazing, overcoming nutrient limitation by obtaining access to enhanced or other nutrient reserves than at surface (e.g. Sommer, 1982; Jones, 1991, 1993; Arvola et al., 1991; Gervais, 1998). Oligotrophic alpine lakes, however, rarely have a deep chlorophyll maximum (e.g. Teubner, 2003).

Considering thermal stratification and light as relevant for vertical co-colonisation three spatially distinct environments for phytoplankton were allocated within the top 12 m in Ammersee: the euphotic epilimnion, the euphotic metalimnion and the metalimnion below the euphotic zone with dim-light less than 1%. The phytoplankton assemblages in these three spatial layers were confirmed by both the results of phytoplankton counting and the analysis of specific pigments (Schmid et al., 1998; Descy et al., 2000). Photosynthetic dinoflagellates as identified by peridinin, and particularly *Ceratium hirundinella*, were frequent in the euphotic epilimnion, which is in accordance with the hypothesis of vertical migration in dinoflagellates being mainly controlled by the underwater light (Heaney & Talling, 1980; Moore, 1981). *Chrysochromulina parva, Dinobryon divergens, Rhodomonas species, Gymnodinium helveticum* and some diatoms were common in the epilimnion but had also pronounced biovolume peaks in the euphotic metalimnion. Most of them are flagellated species known for specific patterns of diel vertical migration in response to daily fluctuating conditions in the environment (e.g. Sommer, 1982; Rott, 1983; Sommer, 1988; Arvola, 1991; Gervais, 1997 a). Many species do not avoid migration even through steeper gradients near the metalimnion which all could support a highly diverse phytoplankton structure along deep verticals in principle. The carotenoid ratio PSC:PPC suggests however, that the euphotic layer ($<=7$ m) can be seen as a homogenous community of photosynthetic organisms functionally related to the duration of sunshine throughout the year. The carotenoid ratio of phytoplankton layers below the euphotic zone was not related to sunshine duration nor were other ratios of lipophilic pigments such as, e.g., all photosynthetic pigments including chlorophylls versus photo-protective carotenoids (not shown). Phycobilins became increasingly important for cyanobacteria and cryptophytes under dim-light conditions and narrow spectral range. Gervais (1997b) concluded from experiments on light-dependent growth and dark survival that acclimatisation to low-light environment is an important pre-adaptation for the dominance of *Cryptomonas* species near the chemocline. Light acclimatisation below the compensation point to dim-light stimulates uptake of organic compounds in *Planktothrix rubescens* as indicated by in situ experiments on photo-heterotrophy by Zotina, Köster & Jüttner (pers. comm.). Both *Cryptomonas* and *Planktothrix* contribute to the deep chlorophyll maximum by active growth rather than by migration or sinking of cells, or depth-adjustment of filaments respectively (Gervais, 1997a; Bright & Walsby, 2000). The distinct carotenoid ratio PSC:PPC between 2–7 m and 10–12 m referred therefore mainly to the distinct phytoplankton composition between euphotic layer and the metalimnetic layer below 1% light intensity and did not show in detail the acclimatisation to underwater light climate (e.g. Descy et al., 2000; Teubner et al., 2001).

Mixing in both spring and autumn led to comparable low species similarities, indicating the same rapid change of species/rapid shift of individual biovolumes during rapid nutrient turnover. During stratification, the only two layers which were significantly different in terms of change in species composition were 5 m and 12 m. These layers were representative for the two extreme conditions in the vertical, the epilimnetic euphotic zone with significant response to sunshine duration and the metalimnion below the euphotic zone with dim-light less than 1%, respectively.

Reynolds (1993) and Padisák (1994) supposed that a near-equilibrium state at environmental constancy requires 12–16 generations spanning 35–60 days in summer. Calculating the Bray-Curtis similarity measure from monthly intervals implies integration over several generations. Doubling times estimated from maximum growth rates under laboratory conditions range from 3.3 days in slow growing species such as *Ceratium hirundinella* to 1/2 day in fast growing unicellular diatoms. The doubling time of 2.8–5.3 days for a mixture of *Cryptomonas* measured by in situ growth rates in a deep living population (Gervais, 1998) was much longer than the 0.85 days expected from culture experiments (Morgan & Kalff, 1979). Similarly, a natural metalimnetic population of *Planktothrix rubescens* in Lake Zürich had doubling times of 7 days (Zotina, Köster & Jüttner, pers. comm.) compared to 1.7 days in culture (Meffert, 1971). We can therefore expect that this species has at least 4 generations during a monthly interval, while for most other species 30–75 generations can be expected (Reynolds, 1984). The dominance of individuals with long generation time, as we can expect for species in the metalimnion below the euphotic zone, implies a low change between the individual biovolumes of species, hence high Bray-Curtis similarities, if no other factors disturb the constancy of phytoplankton composition.

Significant relationships between zooplankton and phytoplankton in the epilimnetic layers substantiate this zone as homogeneous. Although only three species made up 80% of biomass at times, fluctuations within the community remained large. Judging from the small vertical migration amplitude of zooplankton in Ammersee (Morscheid & Mayr, 2002), separate impacts can be ascertained for epi- and metalimnion (Lampert, 1992; Adrian et al., 2001). Zooplankton had very little effect on phytoplankton species assemblages in the metalimnion while parameters related to stratification such as z_{eu}/z_{mix} and the Schmidt stability as well as those of trophy (TP, Chl) correlated with species changes.

To summarise, we defined steady state conditions in lake Ammersee as a stable community in terms of species composition and standing crop. Both aspects of stability in phytoplankton are shown in Figure 7.

The stability of species composition was measured by Bray-Curtis similarity between monthly samples. The 'perfect' equilibrium state measured by Bray-Curtis similarity is theoretical given by the value 1 (given as percentage it is equal to 100%), and would indicate no change of individual biovolumes of species from month to month. The stability of standing crop was evaluated by the net change of total biovolume from month to month. The 'perfect' equilibrium state in terms of standing crop is given by net change values of zero, means no increase or decrease of biomass, a perfect balance between production and losses. The definition of a steady state assemblage on both ends holds true only for metalimnetic layers at light levels below 1% in Ammersee. High values for similarity of phytoplankton composition of more than 80% are reached at almost zero net changes of phytoplankton. In contrast, similarity indices from all the euphotic layers never reached 60% and were usually significantly lower even if biovolume net changes were around zero. This result confirmed non-steady state assemblages at the epilimetic layer in other deep lakes (e.g. Dokulil & Teubner, 2003; Salmaso, 2003) and the hypothesis that steady state conditions occur less regularly in deep lakes and are less predictable compared to surface layyers in shallow polymictic lakes (Mischke & Nixdorf, 2003; Nixdorf et al., 2003).

The arched-shaped pattern in Fig. 7 implies that the lowest changes of species composition can be assumed at zero net change rates of total biovolume. The similarity of species composition between consecutive monthly samples at both layers, the epilimnion and the metalimnion, however, did not exceed 60 and 80% respectively. Hence the internal structure of plankton community altered even if no changes on higher level of aggregation were measured, a sign for compensatory changes in species populations within a community. It supports the idea by Scheffer et al. (2003), that plankton dynamics at higher level of aggregation, such as e.g. the total biovolume, are more predictable than on species level.

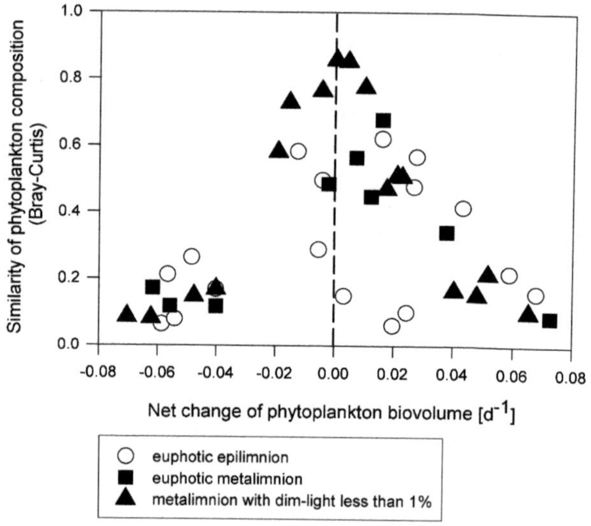

Figure 7. Relationship between the similarity of species composition (Bray-Curtis as measure for the change of individual biovolumes of species from month to month) and the net change of biovolume for the phytoplankton in the euphotic epilimnion (2 and 5 m), the euphotic metalimnion (7 m) and the metalimnion below the euphotic zone with dim-light less than 1% (10 and 12 m).

Conclusions

Steady state assemblage defined as a stable community hold true on both ends, a stable composition by low changes of individual biovolumes of species (Bray-Curtis) and stable standing crop by no net changes of total phytoplankton biovolume for metalimnetic layers at light levels below 1% in Ammersee. No steady state assemblages, however, were found for the euphotic layer. Focussing on phytoplankton composition these findings were further substantiated by relationships between Bray-Curtis similarities and environmental-biotic parameters. Stable phytoplankton composition in the metalimnion below 1% light intensity was associated with the increase of thermal water stability, high TP and with times when the three dominant species contributed more than 80% of total phytoplankton. No statistically significant influence of zooplankton biomass on species changes was observed in the metalimnion. Therefore the metalimnetic plankton community below 1% light can be seen as bottom-up controlled. The ratio of photosynthetic versus photoprotective carotenoids functionally related to the duration of sunshine throughout the year suggests, that the euphotic layer (<=7 m) can be seen as a homogenous community of photosynthetic organisms. The change of species in this euphotic layer was not significantly related to thermal stability, TP or the dominance structure of phytoplankton, but linked with the zooplankton biomass and therefore seemed to be top-down controlled. From our observations, we can conclude that only during stratification, and only in the metalimnion below the euphotic zone, can steady state assemblages be expected in the deep mesotrophic Ammersee.

Acknowledgements

We thank Christian Fesl and Christian Kamenik for discussion on statistics of species changes, Veronika Kucklentz for supporting this research study and the Bavarian ministry for financial support. We are very grateful to David Livingstone for stability calculations and Frank Gervais, Stephen Francis Lowry and two anonymous reviewers for improving an earlier draft of the manuscript. We thank Luigi Naselli-Flores for convening the '13th Workshop Of The International Association Of Phytoplankton Taxonomy And Ecology (IAP) on Phytoplankton and Equilibrium Concept: The Ecology of Steady State Assemblages' in Castelbuono in summer 2002.

References

Adler, M., F. Gervais & U. Siedel, 2000. Phytoplankton species composition in the chemocline of mesotrophic lakes. Archiv für Hydrobiolog./Adv. Limnol. 55: 513–530.

Adrian, R., A. Wickham & M. N. Butler, 2001. Trophic interactions between zooplankton and the microbial community in contrasting food webs: the epilimnion and deep chlorophyll maximum of a mesotrophic lake. Aquat. Microbial Ecol. 24: 83–97.

American Public Health Association (APHA), 1992. Standard methods for examination of water and wastewater. 18th edition, American Public Health Association, Washington, DC.

Arvola, L., A. Ojala, F. Barbosa & S. I. Heaney, 1991. Migration behaviour of three cryptophytes in relation to environmental gradients: an experimental approach. British Phycol. J. 26: 361–373.

Bricaud, A., M. Babin, A. Morel and H. Claustre, 1995. Variability in the chlorophyll-specific absorption coefficients of natural phytoplankton: analysis and parameterization. J. Geophys. Res. 100: 13321–13332.

Bright, D. I. and A. E. Walsby, 2000. The daily integral of growth by Planktothrix rubescens calculated from growth rate in culture and irradiance in Lake Zürich. New Phytol. 146: 301–316.

Descy, J.-P., H. W. Higgins, D. J. Mackey, J. P. Hurley & T. M. Frost, 2000. Pigment ratios and phytoplankton assessment in Northern Wisconsin Lakes. J. Phycol. 36: 274–286.

Dokulil, M. T. & K. Teubner, 2003. Steady state phytoplankton assemblages during thermal stratification in deep alpine lakes. Do they occur? Hydrobiologia 502 (Dev. Hydrobiol. 172): 65–72.

Flaim, G., E. Rott, F. Corradini, G. Toller & B. Borghi, 2003. Long-term trends in species composition and diurnal migration of dinoflagellates in Lake Tovel (Trentino, Italy). Hydrobiologia 502 (Dev. Hydrobiol. 172): 357–366.

Gervais, F., 1997a. Diel vertical migration of Cryptomonas and *Chromatium* in the deep chlorophyll maximum of a eutrophic lake. J. Plankton Res. 19: 533–550.

Gervais, F., 1997b. Light-dependent growth, dark survival & glucose uptake by cryptophytes isolated from a freshwater chemocline. J. Phycol. 33: 18–25.

Gervais, F., 1998. Ecology of cryptophytes coexisting near freshwater chemocline. Freshwat. Biol. 39: 61–78.

Haffner, G. D. & R. Mc. Neely, 1989. Community structure in epilimnetic and metalimnetic phytoplankton assemblages. Hydrobiologia 182: 59–71.

Heaney, S. I. & J. F. Talling, 1980. Dynamic aspects of dinoflagellate distribution patterns in a small productive lake. J. Ecol. 68: 75–94.

Hixon, M. A., S. W. Pacala & S. A. Sandin, 2002. Population regulation: historical context and contemporary challenges of open vs. closed systems. Ecology 83: 1490–1508.

Interlandi, S. J. & S. S. Kilham, 2001. Limiting recourses and the regulation of diversity in phytoplankton communities. Ecology 82: 1270–1282.

Jones, R. I., 1991. Advantages of diurnal vertical migrations to phytoplankton in sharply stratified, humic forest lakes. Archiv für Hydrobiolog. 120: 257–266.

Jones, R. I., 1993. Phytoplankton migration: patterns, processes and profits. Archiv für Hydrobiol. Beih. Ergebn. Limnol. 39: 67–77.

Lampert, W., 1992. Zooplankton vertical migration: implications for phytoplankton–zooplankton interactions. Arch. Hydrobiol. Beih. Ergebn. Limnol. 35: 69–78.

Lenhart, B., 1987. Limnologische Studien am Ammersee. Informationsberichte Bayerisches Landesamt für Wasserwirtschaft 2: 1–105.

Lenhart, B., 2000. Langfristige Entwicklungen -Eutrophierung und Reoligotrophierung-am Ammersee. Münchner Beiträge zur zur Abwasser,- Fischrei- und Flussbiologie 47: 185–205.

Lindholm, T., 1992. Ecological role of depth maxima of phytoplankton. Archiv für Hydrobiol. Beih. Ergebn. Limnol. 35: 33–45.

Livingstone, D. M. & F. Schanz, 1994. The effects of deep-water siphoning on a small, shallow lake: a long-term case study. Archiv für Hydrobiolog. 123: 15–44.

Meffert, M.-E., 1971. Cultivation and growth of two planktonic *Oscillatoria* species. Mitt. int. Ver. theor. angewand. Limnol. 19: 189–205.

Mischke, U. & B. Nixdorf, 2003. Equilibrium phase conditions in shallow German lakes: How Cyanoprokaryota species establish a steady state phase in late summer. Hydrobiologia 502 (Dev. Hydrobiol. 172): 123–132.

Moore, J. W., 1981. Seasonal abundance of Ceratium hirundinella (O. F. MÜLLER) SCHRANK in lakes of different trophy. Archiv für Hydrobiol. 92: 535–548.

Morgan, K. C. & J. Kalff, 1979. Effect of light and temperature interactions on the growth of *Cryptomonas erosa* (Cryptphyceae). J. Phycol. 15: 127–134.

Morscheid, H., 2002. Langzeitentwicklung des Phytoplanktons im Ammersee während der Oligotrophierung. PhThesis, Wien.

Morscheid, H. & C. Mayr, 2002. Supply and demand – The interplay of vertical distribution of whitefish (*Coregonus lavaretus* L.) and zooplankton and its consequence for seasonal succession and development of zooplankton. Archiv für Hydrobiol. / Adv. Limnol., Biol. Manage. of Coregonid Fishes 57: 411–421.

Nixdorf, B. U. Mischke & J. Rücker, 2003. Phytoplankton assemblages and steady state in deep and shallow eutrophic lakes – an approach to differentiate the habitat properties of Oscillatoriales. Hydrobiologia 502 (Dev. Hydrobiol. 172): 111–121.

Padisák, J., 1994. Identification of relevant time-scales in non-equilibrium community dynamics: conclusions from phytoplankton surveys. New Zealand J. Ecol. Forum 18: 169–176.

Polishchuk, L. V., 1999. Contribution analysis of disturbance-caused changes in phytoplankton diversity. Ecology 80: 721–725.

Reynolds, C. S., 1984. The ecology of freshwater phytoplankton., Cambridge University Press, Cambridge. 384 pp.

Reynolds, C., 1992. Dynamics, selection and composition of phytoplankton in relation to vertical structure in lakes. Archiv für Hydrobiol. Beih. Ergebn. Limnol. 35: 13–31.

Reynolds, C. S., 1993. Scales of disturbance and their role in plankton ecology. Hydrobiologia 249: 157–171.

Reynolds, C. S., J. Padisák & U. Sommer, 1993. Intermediate disturbance in the ecology of phytoplankton and the maintenance of species diversity: a synthesis. Hydrobiologia 249: 183–188.

Rott, E., 1983. Sind die Veränderungen im Phytoplanktonbild des Piburger Sees Auswirkungen der Tiefenwasserableitung? Archiv für Hydrobiologie / Algological Studies 67: 29–80.

Rowan, K. S., 1989. Photosynthetic pigments of algae. Cambridge University Press. 334 pp.

Salmaso, N., 2003. Life strategies, dominance patterns and mechanisms promoting species coexistence in phytoplankton communities along complex environmental gradients. Hydrobiologia 502 (Dev. Hydrobiol. 172): 13–36.

Scheffer, M., S. Reinaldi, J. Huisman & F. J. Weissing, 2003. Why plankton communities have no equilibrium: solutions to the paradox. Hydrobiologia 491: 9–18.

Schmid, H., F. Bauer & H.B. Stich, 1998. Determination of algal biomass with HPLC pigment analysis from lakes of different trophic state in comparison to microscopically measured biomass. J. Plankton Res. 20: 1651–1661.

Sommer, U., 1982. Vertical niche separation between two closely related planktonic flagellate species (Rhodomonas lens and Rhodomonas minuta v. nannoplanctica). J. Plankton Res. 4: 137–143.

Sommer, U., 1983. Nutrient competition between phytoplankton species in multispecies chemostat experiments. Archiv für Hydrobiol. 96: 399–416.

Sommer, U., 1985. Comparison between steady state and nonsteady state competition: experiments with natural phytoplankton. Limnol. Oceanogr. 30: 335–346.

Sommer, U., 1988. Some relationships in phytoflagellate motility. Hydrobiologia 161: 125–131.

Sommer, U., J. Padisák, C. S. Reynolds & P. Juhász-Nagy, 1993. Hutchinson's heritage: the diversity-disturbance relationship in phytoplankton. Hydrobiologia 249: 1–7.

Sommer, U., 1995. An experimental test of the intermediate disturbance hypothesis using cultures of marine phytoplankton. Limnol. Oceanogr. 40: 1271–1277.

Stuart, V., S. Sathyendranath, T. Platt, H. Maass & B. D. Irwin, 1998. Pigments and species composition of natural phytoplankton populations: effect on the absorption spectra. J. Plankton Res. 20: 187–217.

Teubner, K., A. Sarobe, M. R. Vadrucci & M. T. Dokulil, 2001. ^{14}C photosynthesis and pigment pattern of phytoplankton as size related adaptation strategies in alpine lakes. Aquat. Sci. 63: 310–325.

Teubner, K., 2003. Phytoplankton, pelagic community and nutrients in a deep oligotrophic alpine lake: Ratios as sensitive indicators of the use of P-resources (DRP:DOP:PP and TN:TP:SRSi). Water Res. 37: 1583–1592.

Tilman, D., 1982. Resource competition and community structure. Monographs in Population Biology 17., Princeton University Press. 296 pp.

Wetzel, R. G., 2001. Limnology: lake and river ecosystems., Academic Press, San Diego, San Francisco, New York, Boston, London, Sydney, Tokyo. 1006 pp.

Wilhelm, C., I. Rudolph & W. Renner, 1991. A quantitative method based on HPLC-aided pigment analysis to monitor structure and dynamics of the phytoplankton assemblage – A study from Lake Meerfelder Maar (Eifel, Germany). Archiv für Hydrobiol. 123: 21–35.

Wright, S. W., S. W. Jeffrey, R. F. C. Mantoura, C. A Llewellyn, T. Bjornland, D. Repela & N. Welschmeyer, 1991. Improved method for the analysis of chlorophylls and carotenoids from marine phytoplankton. Mar. Ecol. Progr. Ser. 77: 183–196.

Hydrobiologia **502:** 65–72, 2003.
L. Naselli-Flores, J. Padisák, M. T. Dokulil (eds),
Phytoplankton and Equilibrium Concept: The Ecology of Steady-State Assemblages.

Steady state phytoplankton assemblages during thermal stratification in deep alpine lakes. Do they occur?

Martin T. Dokulil & Katrin Teubner
Institute of Limnology, Austrian Academy of Sciences, Mondseestrasse 9, A-5310 Mondsee, Austria
E-mail: martin.dokulil@oeaw.ac.at

Key words: phytoplankton, steady state assemblages, equilibrium phase, alpine lake, Mondsee

Abstract

Phytoplankton seasonal and long-term succession can be described and functionally classified by associations similar as in terrestrial vegetation studies. Such a concept has to include 'climax' into pelagic succession which in turn leads to the question to what extent steady state assemblages occur and if during periods of dynamic equilibrium can be identified. Here we explore the situation with respect to the above question for deep, nutrient poor, alpine lakes in Austria. We first track the long-term development of phytoplankton biomass, their taxonomic structure and their relation to total phosphorus and chlorophyll-*a* as predictors of trophic state over the past 35 years. We then analyse this data set for coherent algal associations which can be ascribed to trait separated functional groups according to Reynolds et al. (2002). A three year period of stable environmental conditions has then be extracted from the progression of trophic state indices, having similar dominating species each year. These years were finally analysed for steady state conditions according to definitions given in Sommer et al. (1993). During thermal stratification, achievement of an equilibrium could be ruled out although coexistence of several dominating species lasted for several weeks. Habitat templates were constructed from environmental variables prior to biomass peaks for two species important in summer assemblages, the dinoflagellate *Ceratium hirundinella* and the diatom *Fragilaria crotonensis*. In summary, functional groups proved to be a valid and useful concept to describe species succession of phytoplankton in deep alpine lakes while pelagic climax is much less clear and steady state conditions were never met.

Introduction

The seasonality of freshwater phytoplankton has attracted considerable attention during the limnological venue of last century summarised in e.g. Munawar & Talling (1986). Much of the periodicity in the wax and wane of the phytoplankton taxa has been gathered through, often long-time, serial investigations in a wide variety of lake types and reservoirs. Although additional, essential information has been generated by controlled laboratory experiments, interpretation of field results has been hampered by the difficulty to test hypotheses properly. Whole lake experiments and the advent of aquatic enclosures offered an alternative and generated a brake-through in the testing of ecological hypotheses in freshwaters (e.g. Reynolds, 1986). Detailed information has been accumulated since then by experimental manipulations and field observations on adaptive strategies of phytoplankton, importance of their functional morphology and species selection across trophic gradients summarised in e.g. Álvarez-Cobelas et al. (1998), Reynolds et al. (2000) and Hamilton et al. (2000). These and other results had influential impacts on the progressive development of a coherent theory of dynamic processes in the pelagic (Reynolds, 1997). This topic was further developed towards functional classification and prediction of freshwater phytoplankton (Reynolds & Irish, 1997; Reynolds at al., 2002). In essence, the theory hypothesizes succession leading to a climax in the pelagic similar to terrestrial vegetation (Grime, 1979).

If this concept has relevance, we must hypothesize that phytoplankton assemblages or associations sensu Reynolds et al. (2002) shall achieve steady state

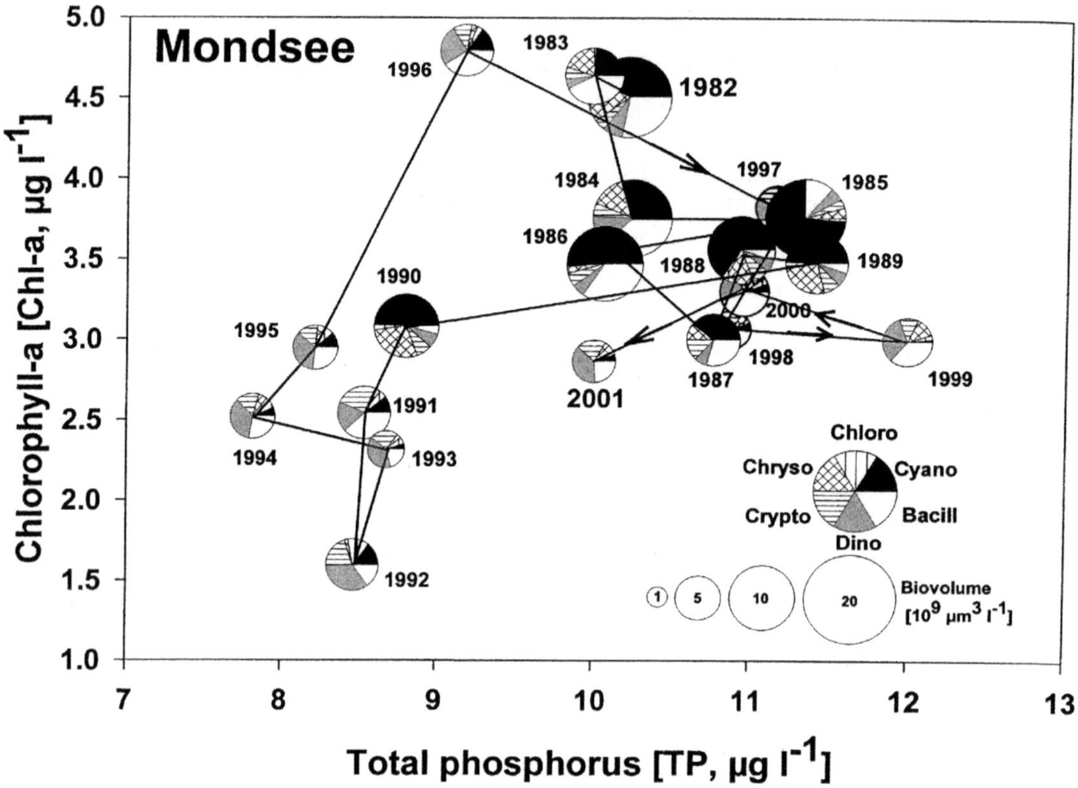

Figure 1. Long-term development of annual average total biovolume and algal group contribution for 1982 – 2001 in Mondsee depicted in the total phosphorus – chlorophyll-*a* space. The diameter of the bubbles is equivalent to total biovolume.

or reach equilibrium at some stage of their seasonal succession. Following Sommer et al. (1993), the equilibrium phase is defined here as a period of more than two weeks of coexistence of one, two or a maximum of three species contributing not less than 80% of total biomass with no significant changes in biovolume.

The questions addressed in this paper are:

1. How are functional groups related to long-term trophic gradients?
2. Do steady state assemblages occur during thermal stratification in deep alpine lakes?
3. What are the ecological conditions which may lead to steady state conditions?
4. Do species show any physiological plasticity?

These problems are explored using a long term data set from an alpine lake in Austria. Considering the spatial and temporal coherence within the lake district (Dokulil & Teubner, 2002), results from this particular lake may be relevant for other water bodies in the region.

Methods

Site description

The deep alpine lake Mondsee (surface area 14.2 km^2, maximum depth 68.3 m, average theoretical retention time 1.7 years) is located in the 'Salzkammergut' lake district of Austria. As many other lakes in the world, Mondsee underwent cultural eutrophication in the 1970s but has considerably improved since then. At present, the lake can be classified as oligo-mesotrophic (OECD, 1982). More detailed information on this and other lakes in the region, their spatial coherence and their relationship to regional climatic conditions can be obtained from Livingstone & Dokulil (2001) and Dokulil & Teubner (2002).

Phytoplankton sampling

Phytoplankton has been regularly monitored since 1982. The sampling interval ranged from biweekly to monthly, but was biweekly most of the time. Periods of particular interest were sampled twice a week in specific years (Dokulil, 2000). Sampling depths varied somewhat during the investigation period. Beginning in 1987, a rigorous protocol was adopted with intervals every three metres from the surface to 20 m and larger spacing further down. In addition, an integrated sample from the top 20 m was taken at all occasions (Schröder, 1969). Samples for phytoplankton enumeration were preserved with 200 μl Lugol's solution added to 100 ml brown glass-bottles prior to filling and stored in the dark at 4 °C until further analysed. Cell numbers were estimated on an inverted microscope using the sedimentation technique developed by Utermöhl (1958) using a computerised counter (Hamilton, 1990). Cell dimensions were measured with an image analysis system (Lucia, V 3.1) and cell volumes calculated using simple geometric approximations (Rott, 1981). Species abundance was then converted to biovolume.

Chlorophyll-*a* (Chl) was measured spektrophometrically after extraction in hot-ethanol following ISO 10260 (1992). Concentrations, corrected for degradation products, were calculated according to Lorenzen (1967).

Water quality sampling

Transparency was estimated on each occasion with a white 20 cm-diameter Secchi disk (z_{SD}). Vertical profiles of water temperature, pH, conductivity, oxygen concentration and turbidity were measured prior to 1997 from thermometer readings and individual hand held instruments while post-1997 data originated from a multi-parameter profiler (YSI6920). Both discrete depth and integrated water samples were analysed for total-, total dissolved and phosphate-phosphorus (TP, TDP and PO_4-P), total nitrogen (TN), nitrate- and nitrite nitrogen (NO_3-N, NO_2-N) and silicate (SiO_3-Si) in the laboratory following standard analytical techniques (Greenberg et al., 1992).

Data treatment

In the present study, only results from the integrated samples are presented. All annual averages were calculated from all individual measurements. The depth of the mixing layer (z_{mix}) was estimated from individual temperature profiles. The ratio z_{mix}/z_{SD} was used to characterise under-water light conditions. Euphotic depth (1% light level) was obtained either directly from vertical profiles of under-water PAR (Li-Cor, 4π) or converted from Secchi-depth using $z_{eu} = 2.5\ z_{SD}$. The conversion factor was substantiated from a large data set systematically combining Secchi disk readings with direct evaluations of z_{eu}. Trophic state indices for TP and Chl have been calculated according to Carlson (1977)

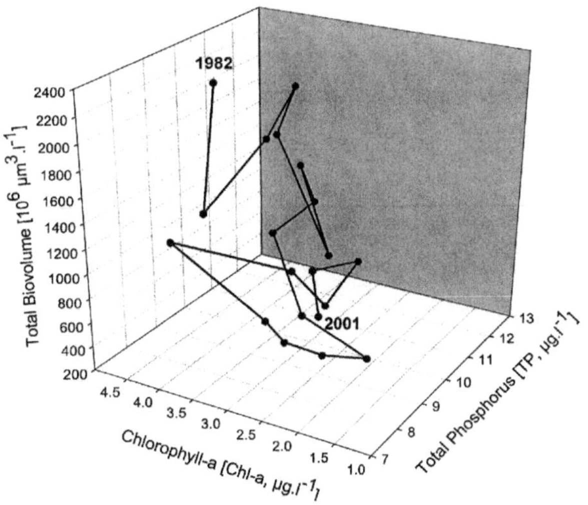

Figure 2. Trajectory plot of total phytoplankton biovolume *versus* total phosphorus and chlorophyll-*a* for 1982 – 2001 in Mondsee.

Results

Long-term changes of phytoplankton assemblages are depicted in Figure 1 as annual average biovolume bubbles in the phosphorus – chlorophyll space for the years 1982–2001. Each biovolume bubble shows the average contribution of taxonomic algal groups. For a clearer picture of the development over time, the trajectory of biovolumes is repeated in 3D-space in Figure 2. Highest measured TP and Chl concentrations (18–25 μg l^{-1} TP, 4-8 μg l^{-1} Chl), not shown here, occurred prior to 1981, a period which was characterised by the dominance of *Planktothrix rubescens.* Since 1982, biovolume was mainly composed of cyanobacteria and diatoms associated with annual TP concentrations of 10 – 11.5 μg l^{-1} and average chlorophyll-*a* levels of 3.5 – 4.5 μg l^{-1} indicating mesotrophic conditions (Fig. 1). From 1989 to 1990, the annual average TP concentration dropped from >11 to less than 9 μg l^{-1} with no significant change in either biovolume, chlorophyll-*a* or algal composition. Thereafter, the first half of the 1990s was characterised by low concentrations of all vari-

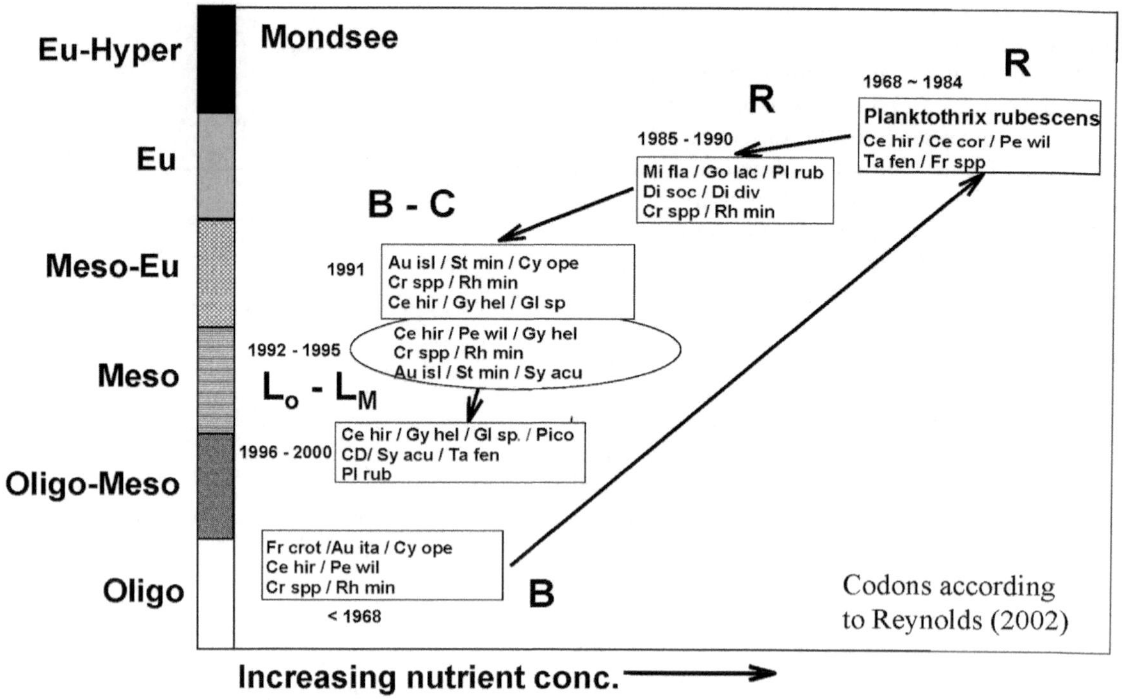

Figure 3. Conceptual diagram for the development of dominant and sub-dominant phytoplankton species in the nutrient/trophic level space including information prior to 1968. Functional group codons according to Reynolds (2002). Further explanation in the text. Abbreviations: Au isl = *Aulacoseira islandica*, Au ita = *Aulacoseira italica*, CD = small centric diatoms, Ce cor = *Ceratium cornutum*, Ce hir = *C. hirundinella*, Cr spp. = *Cryptomonas* spp., Cy ope = *Cyclotella operculata*, Di soc = *Dinobryon sociale*, Di div = *D. divergens*, Fr crot = *Fragilaria crotonensis*, Gl sp = *Glenodinium* sp., Go lac = *Gomphospheria lacustris*, Gy hel = *Gymnodinium helveticum*, Mi fla = *Microcystis flos-aquae*, Pe wil = *Peridinium willei*, Pico = Picoplankton, Pl rub = *Planktothrix rubescens*, Rh min = *Rhodomonas minuta*, St min = *Stephanodiscus minutulus*, Sy acu = *Synedra acus*, Ta fen = *Tabellaria fenestrata*.

ables (Figs 1 and 2) and a shift in species composition. The contribution of cyanobacteria, especially the metalimnetic population of *P. rubescens,* was largely reduced while dinoflagellates gained much in importance (Fig. 1). In the second half of that decade TP levels increased slightly to 12 $\mu g\ l^{-1}$ in 1999 not reflected in chlorophyll or biovolume concentrations (cf. Figs 1 and 2). During this period, dinoflagellates and diatoms contributed between 50 and 75% of total biovolume while the cyanobacterial share substantial decreased. In 1999 *P. rubescens* became undetectable but recovered somewhat afterwards. A closer look into the dominating associations is provided by the conceptual diagram in Figure 3 which includes information on the phytoplankton community prior to the invasion by *P. rubescens* in 1968 (Findenegg, 1965, 1969). Although no direct measurements of P-concentrations are available earlier than 1977, all parameters from the 1950s and 1960s as well as indirect estimates of phosphorus levels estimated from palaeolimnological records (Schmidt, 1991) indicated oligotrophic conditions. From the few records we have prior to 1968 (Findenegg, 1965), the phytoplankton assemblage was dominated by diatoms (*Fragilaria crotonensis* > *Aulacoseira italica* > *Cyclotella operculata*). The dinoflagellates *Ceratium hirundinella* and *Peridinium willei* appeared sub-dominant, followed by several species of the genus *Cryptomonas* and *Rhodomonas minuta*. This association may provisionally be ascribed to the functional grouping '**B**' of Reynolds et al. (2002). A substantial increase in P-content of the lake was associated with the rapidly rising tourism in the lake region in the 1960s (Dokulil & Jagsch, 1989) finally resulting in the appearance of *P. rubescens* in 1968 which overwhelmingly bloomed in the lake then for several years and then retreated into the metalimnion where the species remained until now (group **R**). Sub-dominants remained essentially the same as in the period before but included *Tabellaria flocculosa* var. *fenestrata* (Dokulil & Kofler, 1995). During the 1980s dominance shifted to *Microcystis flos-aquae*, *Gomphosphaeria* and often short lived outbreaks of

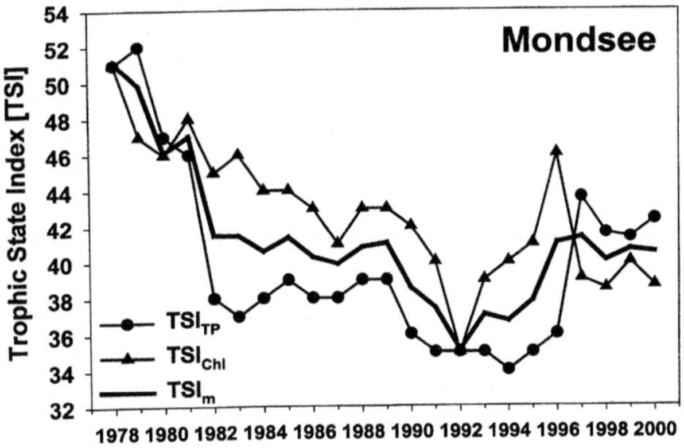

Figure 4. Trophic state indices for total phosphorus (TP), chlorophyll-*a* (Chl-*a*) and the average of both according to Carlson (1977) for the years 1978 – 2000 in Mondsee.

Dinobryon sociale and *D. divergens*. This association is still termed **R** although it might as well be called $\mathbf{L_M}$. The transitional grouping of 1991, dominated by *A. islandica* now, *Stephanodiscus minutulus* and *Cyclotella operculata* can be ascribed to **B–C**. During the 1990s dinoflagellates became increasingly important (Fig. 3) and are best described by the functional groups $\mathbf{L_M}$ and later $\mathbf{L_O}$. The progression and recession of the trophic development in Mondsee across these 50 years may be summarised by $\mathbf{B} \rightarrow \mathbf{R} \rightarrow \mathbf{C} \rightarrow \mathbf{L_M} \rightarrow \mathbf{L_O}$ (Fig. 3).

The re-oligotrophication of the lake since the late 1970s is also reflected in the course of the trophic state indices of TP and Chl-*a* depicted in Figure 4. Since both the associations in Figure 3 and the average trophic state index in Figure 4 indicate more stable situations in recent years (1998–2000), we focused on this period. Seasonal development of total biovolume in relation to environmental variables is exemplified for 1999 in Figure 5. Annual variation of the variables shown here are surprisingly similar in the two other years and so is phytoplankton biovolume. Summers were characterised by nutrient depletion, low Secchi-disk visibility, severe thermal stratification and high surface water-temperatures. Species composition during peaks were almost identical between years and peaks appeared at approximately the same time each year. During the stratified period from early June to late September indicated by the dotted lines in Fig. 5, the dinoflagellates contributed between 40 and 60% of total biovolume to the peak in July every year with a 90 to 96% share of *C. hirundinella* (Table 1). At decreasing biovolume levels during August and September (Fig. 5), dinoflagellates still contributed the

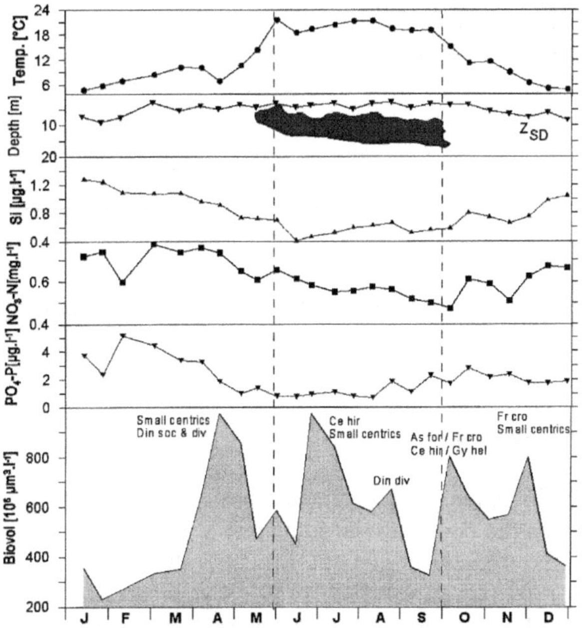

Figure 5. Seasonal fluctuation of total phytoplankton biovolume in Mondsee 1999 (gray shaded area) in relation to changes in surface temperature (°C), the depth of Secchi disk visibility (Z_{SD}), the vertical extent of the metalimnion (black shaded area), and the concentrations of silica (Si), nitrate-nitrogen (NO_3-N) and phosphate-phosphorus (PO_4-P). The genera dominating the phytoplankton biovolume at different seasons are identified by abbreviations: Small centrics = mixture of centric diatoms, Abbreviations: As for = *Asterionella formosa*; all others as in Figure 3.

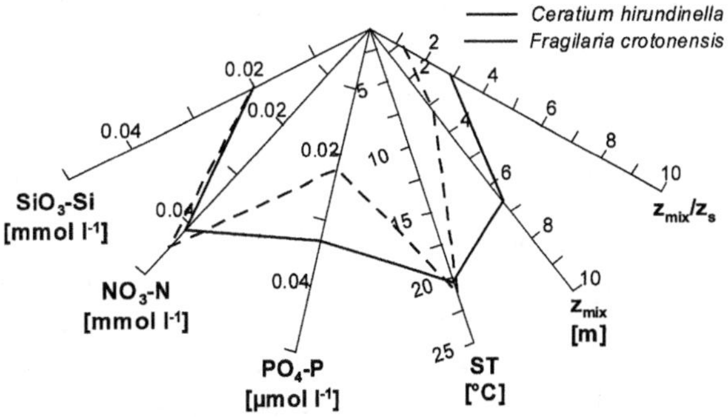

Figure 6. Graphical summaries of algal attributes for *Ceratium hirundinella* and *Fragilaria crotonensis*. Biovolume is represented as a function of mean available light in terms of the ratio mixed depth (Z_{mix}) to Secchi-depth (Z_S), mixed depth (Z_{mix}), surface temperature (ST), and the molar concentrations of three major nutrients, silica (SiO_3-Si), nitrate-nitrogen (NO_3-N) and phosphate-phosphorus (PO_4-P) Axes are arranged as 'hexacles' according to Reynolds (1997). All variables are averaged for two pre-peak weeks over the three years 1998–2000. For more information refer to the text.

most (37–42%B) and the mixotrophic *Gymnodinium helveticum* became more prominent. In June at the end of the clear-water phase, the 35–39% share of the Bacillariophyceae consisted mainly of centric diatoms and *F. crotonensis*. Algal attributes for *Ceratium* and *Fragilaria* were graphically summarised by 'hexacles' (Reynolds, 1997), constructed by averaging variables for the two pre-peak weeks of all three years, and plotted together in Figure 6. Requirements for silica and nitrate are similar for both species during pre-summer surface temperatures in June while *Fragilaria crotonensis* can better cope with phosphorus depletion under highly stable conditions.

Discussion

Phytoplankton biovolumes considerably declined during the post-eutrophication period in accordance with the decreased nutrient loading (Dokulil & Jagsch, 1992; Dokulil et al., 1999). Cyanobacterial contribution however, remained substantially high until TP dropped in 1990 below 10 $\mu g\ l^{-1}$ in the epilimnion, a value much below the threshold discussed by Chorus (1995). The following year 1991 was a transition period dominated by diatoms and cryptophyceans. Thereafter, dinoflagellates began to dominate both summer and annual average assemblages. The delayed response in both biomass (Fig. 1), chlorophyll-*a* (Figs 1 and 4) and compositional change (Fig. 3) to decreased nutrient supply is a phenomenon of hysteretic persistence associated with the homeostatic behaviour commonly observed in aquatic ecosystems (Reynolds, 2002).

Adopting the concept of 'functional groups' (Lavorel et al., 1997; Reynolds et al., 2002), for species frequently co-existing and increasing or decreasing in biomass simultaneously, dominating and sub-dominating clusters of algal taxa can be identified for certain periods or years in Mondsee. Their long-term progression can be best approximated by the following sequence of codons sensu Reynolds et al. (2002):

$$\mathbf{B} \rightarrow \mathbf{R} \rightarrow \mathbf{C} - \mathbf{B} \rightarrow \mathbf{L_M} \rightarrow \mathbf{L_O(Z)}$$

The pre-eutrophication period, dominated by diatoms (Dokulil, 1991), can best be ascribed to an association (codon B) tolerating low temperatures, moderate light availability and low levels of SRP (see Table 3 in Reynolds et al., 2002). A large contribution of picoplankton in these years can not be ruled out. The peak phase of eutrophication and the early period of nutrient decline following external restoration was dominated for quite some time by *P. rubescens* recognised in the functional group **R.** This species has a remarkable capacity for buoyancy regulation, chromatic acclimation and is highly grazing resistant. It therefore strives well in the dim light, cold metalimnetic layers of deep, clear alpine lakes, but also tolerates holomictic entrainment when nutrients are abundant. Species of the functional group **R** therefore have a special ecological function by using nutrients accumulated in the thermocline efficiently. Their persistent occurrence in the metalimnion largely influences the overall trophic status of the lake. Although the species assemblage in the epilimnion is already indicating improvement. In other words, these two assemblages shall be treated independently.

The summer epilimnia of these years were inhabited either by species typical for eutrophic lakes (**$\mathbf{L_M}$**), tolerating high temperatures and light intensities at depleted nutrient levels, or opportunistic forms such as species of *Dinobryon* and *Cryptomonas* (Dokulil & Skolaut, 1986) which can augment low levels of essential resources through phagotrophy (Psenner & Sommaruga, 1992). When phosphorus concentrations and cyanobacterial biomass in the lake had finally reached a minimum in 1991 the phytoplankton association changed to diatom dominated (grouping **C** or **B**) similar to what it has been before trophic enhancement (Dokulil, 1991; Schmidt, 1991).

Analysis of steady state conditions during the stratified summer period was restricted to three consecutive years of greatest ecological stability, considered as a manifestation of a mature ecosystem with little fluctuations (Holling, 1973). Steady state conditions however, as defined in the introduction have never been met. In all cases, more than three species comprised 80% of the biovolume never reaching a dynamic equilibrium, although coexistence persisted for more than 1–2 weeks. Inspection of other phytoplankton data sets from deep, stratifying, clear-water lakes (Dokulil et al., 2000; Morabito et al., 2003; Salmoso, 2003), reveals that steady state conditions seem to be highly unlikely at the lower end of the trophic spectrum. In such lakes, the explanation of species domination is still puzzling (Huisman & Weissing, 2001). At higher trophic levels however, equilibrium is easily reached and sometimes maintained for considerable periods of time, mainly because environmental conditions constrain species diversity to one or two dominating species (e.g. Berman et al., 1992; Dokulil & Mayer, 1996; Kruk et al., 2002) unless when rapidly flushed (Huszar et al., 2003) or in the metalimnetic niche (Nixdorf et al., 2003; Teubner et al., 2003). Environmental variables leading to summer peaks were evaluated for two important species. Habitat templates used here are different from those in Reynolds (1997) because diagnostic axes start from zero at the root of the hexacle. Peak biomass was used instead of net-density changes from field data, equivalent to growth rates, because rates derived from fortnightly intervals might be misleading. Results for both species however, are in good agreement with hexacles hypothesized for functional group **L** by Reynolds (1997, Fig. 47, p. 218). Differences between the two habitat templates indicate more physiological plasticity of the diatom, *F. crotonensis*. Considered as a typical energy-efficient R-strategist with low P- and Si requirements in pronounced stable situations, *Fragilaria* may as well appear earlier in succession (Table 1). The large motile dinoflagellate, *C. hirundinella* can offset its poor growth capability (low S/V-ratio) by the capacity of self-regulating its position in the water column and to counter passive sinking (S-strategist). Both species resist serious grazing losses (Reynolds, 1997). This outcome of the seasonal progression can be considered as short-lived climax in temperate lakes where nutrients are depleted during stratification.

Conclusions

The concept of functional groups has successfully been applied to long-term succession of average annual phytoplankton biomass. Groups have undergone significant modifications during long-term progression from low trophic levels through anthropogenic nutrient enrichment (eutrophic phase), and finally reoligotrophication after external restoration measures were applied. The *Planktothrix*-dominated stage (**R**) lasted particularly long and severely affected the rehabilitation phase which is interpreted as resilient behaviour of the system.

A period of relative stable ecological conditions with persistent functional groups was identified (1998 – 2000). Environmental variables and phytoplankton biomass during stratification in summer were searched for possible equilibrium phases. More than three species always dominated the phytoplankton which never reached a stable equilibrium of no significant biomass changes similar to results of long term laboratory experiments by Scheffer et al. (2003). Coexistence of plankton species however, lasted for more than two weeks. The initial hypothesis that steady state assemblages occur during thermal stratification in the pelagic has therefore been falsified for deep alpine lakes in Austria. Habitat templates for two of the dominating species coincided with the provisional trait-separated functional group **L**. Physiological plasticity was observable at least for the diatom species.

Overall, functional groups proved to be a valid and useful concept to describe species succession of phytoplankton in deep alpine lakes while pelagic climax is much less clear.

Acknowledgements

We kindly acknowledge the enthusiastic and enormous help over the many years of Claudia Skolaut during field collections, sample processing and elaboration in the laboratory. Further thanks extend to Lotte Eisl and Hiltraud Gollmann for chemical analysis as well as many others who have helped in one or the other way during the last twenty years. Many thanks also to Albert Jagsch and Kurt Schwarz which allowed us to use their data prior to 1981. Finally, this paper has greatly benefited from discussions during IAP meetings, especially with Judit Padisák and Luigi Naselli-Flores.

References

Álvarez-Cobelas, M., C. S. Reynolds, P. Sanchez-Castillo & J. Kristiansen (eds.), 1998. Phytoplankton and trophic gradients. Dev. Hydrobiol. 129: 1–372.

Berman, T., Y. Z. Yacobi & U. Pollingher, 1992. Lake Kinneret phytoplankton: stability and variability during twenty years (1970–1989). Aquat. Sci. 54: 104–127.

Carlson, R. E., 1977. A trophic state index for lakes. Limnol. Oceanogr. 22: 361–369.

Chorus, I., 1995. Müssen in der Seesanierung Gesamtphosphat-Schwellenwerte unterschritten werden, bevor das Phytoplankton eine Reaktion zeigt? In Jaeger, D. & R. Koschel (eds), Verfahren zur Sanierung und Restaurierung stehender Gewässer. Limnologie Aktuell 8 Fischer, Stuttgart: 21–28.

Dokulil, M. T., 1991. Populationsdynamik der Phytoplankton-Diatomeen im Mondsee seit 1957. Wasser und Abwasser 35: 53–75.

Dokulil, M. T., 2000. Die Bedeutung hydroklimatischer Ereignisse für die Dynamik des Phytoplanktons in einem alpinen Klarwassersee (Mondsee, Österreich). Beiträge zur angewandten Gewässerökologie Norddeutschlands 4: 87–93.

Dokulil, M. T. & A. Jagsch, 1989. Some aspects of the impact of tourism on Mondsee, Austria. Symp. Biol. Hung. 38: 415–428.

Dokulil, M. T. & A. Jagsch, 1992. Dynamics of phosphorus and nitrogen loading and its effects on phytoplankton in Mondsee, Austria. Hydrobiologia 243/244 (DH79): 389–394.

Dokulil, M. T. & S. Kofler, 1995. Ecology and autecology of *Tabellaria flocculosa* var. *asterionelloides* Grunow (Bacillariophyceae) with remarks on the validation of the species name. In Marino, D. & M. Montresor (eds), Proc. 13th International Diatom Symposium 1994. Biopress Limited, Bristol: 23–38.

Dokulil, M. T. & J. Mayer, 1996. Population dynamics and photosynthetic rates of a *Cylindrospermopsis - Limnothrix* association in a highly eutrophic urban lake, Alte Donau, Vienna, Austria. Algol. Stud. 83: 179–195.

Dokulil, M. & C. Skolaut, 1986. Succession of phytoplankton in a deep stratifying lake: Mondsee, Austria. Hydrobiologia 138: 9 – 24.

Dokulil, M. T. & K. Teubner, 2002. The spatial coherence of alpine lakes. Int. Ver. theor. angewan. Limnol. Verh. 28: 1861–1864.

Dokulil, M. T., K. Schwarz & A. Jagsch, 1999. Stand der Reoligotrophierung der großen Alpenseen im Salzkammergut. Deutsche Gesellschaft für Limnologie (DGL) – Tagungsberichte 1998 (Klagenfurt). Tutzing 1999: 154–158.

Dokulil, M. T., K. Schwarz & A. Jagsch, 2000. Die Reoligotrophierung österreichischer Seen: Sanierung, Restaurierung und Nachhaltigkeit – Ein Überblick. Münchener Beiträge Abwasser-, Fischerei- und Flußbiologie 53: 307–321.

Findenegg, I., 1965. Limnologische Unterschiede zwischen den österreichischen und ostschweizerischen Alpenseen und ihre Auswirkung auf das Phytoplankton. Vierteljahresschrift der Naturforschenden Gesellschaft in Zürich 110: 289–300

Findenegg, I., 1969. Die Eutrophierung des Mondsees im Salzkammergut. Wasser und Abwasser Forschung 4: 139–144

Greenberg, A. E., L. S. Clesceri & A. D. Eaton, 1992. Standard methods for the examination of water and wastewater. American Public Health Ass. Washington.

Grime, J. P., 1979. Plant strategies and vegetation processes. Wiley, Chichester.

Hamilton, P. B., 1990. The revised edition of a computerised plankton counter for plankton, periphyton and sediment diatom analyses. Hydrobiologia 194: 23–30.

Hamilton, P. B., H. Kling & M. T. Dokulil (eds), 2000. Cyanoprokaryotes and Chlorophytes Across Trophic Gradients. Hydrobiologia 438: 1–264.

Holling, C. S., 1973. Resilience and stability of ecological systems. Ann. Rev. Ecol. Syst. 4: 1–23.

Huisman, J. & E. J. Weissing, 2001. Fundamental unpredictability in multispecies competition. Am. Nat. 157: 488–494.

Huszar, V., C. Kruk & N. Caraco, 2003. Steady-state assemblages of phytoplankton in four temperate lakes (NE U.S.A.). Hydrobiologia 502 (Dev. Hydrobiol. 172): 97–109.

ISO 10260, 1992. Water quality measurements of biochemical parameters – Spectrophotometric determination of the chlorophyll-*a* concentration. 5 pp.

Kruk, C., N. Mazzeo, G. Lacerot & C. S. Reynolds, 2002. Classification schemes for phytoplankton: a local validation of a functional approach to the analysis of species temporal replacement. J. Plankton Res. 24: 901–912.

Lavorel, S., S. McIntyre, J. Landsberg & T. D. A. Forbes, 1997. Plant functional classification: from general groups to specific groups based on response to disturbance. TREE 12: 474–478.

Livingstone, D. M. & M. T. Dokulil, 2001. Eighty years of spatially coherent Austrian lake surface temperatures and their relationship to regional air temperatures and to the North Atlantic Oscillation. Limnol. Oceanogr. 46: 1220–1227.

Lorenzen, C. J., 1967. Determination of chlorophyll and phaeopigments: spectrophotometric equations. Limnol. Oceanogr.. 12: 343–346.

Morabito, G., A. Oggioni & P. Panzani, 2003. Phytoplankton assemblage at equilibrium in large and deep subalpine lakes: a case study from Lago Maggiore (N. Italy). Hydrobiologia 502 (Dev. Hydrobiol. 172): 37–48.

Munawar, M. & J. F. Talling (eds), 1986. Seasonality of Freshwater Phytoplankton. Hydrobiologia 138: 1–236.

Nixdorf, B. U. Mischke & J. Rücker, 2003. Phytoplankton assemblages and steady state in deep and shallow eutrophic lakes – an approach to differentiate the habitat properties of Oscillatoriales. Hydrobiologia 502 (Dev. Hydrobiol. 172): 111–121.

OECD, 1982. Eutrophication of waters. Monitoring, assessment and control. OECD, Paris.

Psenner, R. & R. Sommaruga, 1992. Are rapid changes in bacterial biomass caused by shifts from top-down to bottom-up control? Limnol. Oceanogr. 17: 1092–1100.

Reynolds, C. S., 1986. Experimental manipulations of the phytoplankton periodicity in large limnetic enclosures in Blelham Tarn, English Lake District. Hydrobiologia 138: 43–64.

Reynolds, C. S., 1997. Vegetation processes in the pelagic: a model for ecosystem theory. In Kinne, O. (ed.), Excellence in Ecology 9. Ecol. Inst., Oldendorf: 1–371.

Reynolds, C. S., 2002. Resilience in aquatic ecosystems-hysteresis, homeostasis and health. Aquat. Ecosyst. Health Management 5: 3–17.

Reynolds, C. S. & A. E. Irish, 1997. Modelling phytoplankton dynamics in lakes and reservoirs: the problem of *in-situ* growth rates. Hydrobiologia 349: 5–17.

Reynolds, C. S., M. T. Dokulil & J. Padisák (eds), 2000. The Trophic Spectrum Revisited. Hydrobiologia 424: 1–152.

Reynolds, C. S., V. Huszar, C. Kruk, L. Naselli-Flores & S. Melo, 2002. Towards a functional classification of the freshwater phytoplankton. J. Plankton Res. 24: 417–428.

Rott, E., 1981. Some results from phytoplankton counting intercalibrations. Schweiz. Zeitsch. Hydrol. 43: 35–62.

Salmaso, N., 2003. Life strategies, dominance patterns and mechanisms promoting species coexistence in phytoplankton communities along complex environmental gradients. Hydrobiologia 502 (Dev. Hydrobiol. 172): 13–36.

Scheffer, M., S. Reinaldi, J. Huisman & F. J. Weissing, 2003. Why plankton communities have no equilibrium: solutions to the paradox. Hydrobiologia 491: 9–18.

Schmidt, R., 1991.Diatomeenanalytische Auswertung laminierter Sedimente für die Beurteilung trophischer Langzeittrends am Beispiel des Mondsees (Oberösterreich). Wasser und Abwasser 35: 109–123.

Schröder, R., 1969. Ein summierender Wasserschöpfer. Archiv für Hydrobiol. 66: 241–243.

Sommer, U., J. Padisák, C. S. Reynolds & P. Juhász-Nagy, 1993. Hutchinson's heritage: the diversity-disturbance relationship in phytoplankton. Hydrobiologia 249: 1–7.

Teubner K., M. Tolotti, S. Greisberger, H. Morscheid, M. T. Dokulil & H. Morscheid, 2003. Steady state phytoplankton in a deep pre-alpine lake: species and pigments of epilimnetic *versus* metalimnetic assemblages. Hydrobiologia 502 (Dev. Hydrobiol. 172): 49–64.

Utermöhl, H., 1958. Zur Vervollkommnung der quantitativen Phytoplankton Methodik. Mitt. int. Ver. theor. angewand. Limnol. 9: 1–38.

Hydrobiologia **502:** 73–83, 2003.
L. Naselli-Flores, J. Padisák, M. T. Dokulil (eds),
Phytoplankton and Equilibrium Concept: The Ecology of Steady-State Assemblages.

The Effect of perturbations on phytoplankton assemblages in a deep reservoir (Vouglans, France)

M. Leitão[1], S.M. Morata[2], S. Rodriguez[3] & J.P. Vergon[4]
[1]*Bi-Eau, 15 rue Lainé-Laroche, 49000 Angers, France. E-mail: bieau@unimedia.fr*
[2]*Dpto. Microbiología y Ecología & Instituto Cavanilles de Biodiversidad y Biología Evolutiva, Universitat de València, E-46100 Burjassot, Spain*
[3]*Bureau d'études Sylvaine Rodriguez, 3 rue Jean d'Abbans, 25220 Thise, France*
[4]*DIREN de France-Comté, 5 rue du général Sarrail, 25014 Besançon cedex, France*

Key words: disturbance, freshwater red tide, functional groups, man-made lake, *Peridiniopsis durandi*, phytoplankton dynamics, reversion, stability

Abstract

Disturbances, such as rainfall and hydrological management of Vouglans reservoir, seem to influence the establishment of the ecological system inside this man-made lake. In this paper, the main effect of these disturbances are changes in the z_{mix}:z_{eu} ratio, with a shift in the composition and phytoplankton dynamics, and consequently in the Shannon's diversity. The seasonal changes found in the surface layers of Vouglans reservoir were typical for mesotrophic and temperate lakes, but some perturbations induced the 'reversion' to an earlier stage in the expected successional pattern during the stratification period. This early-state displayed an intensive proliferation of the filamentous ulotrichal *Planctonema lauterbornii*, which remained dominant in terms of abundance from the middle to the end of the stratification. The beginning of this period was characterized by the highest value of Shannon diversity index, followed by its marked decrease, whilst z_{mix}:z_{eu} remained around 1. Despite the environmental variations which induced the reversion in this expected pattern, we found some sort of equilibrium, that is, stability in the assemblage simply expressed as a 'no-change situation', although not in the sense of equilibrium inside a competitive system. Moreover, peculiar freshwater red tides occurred in this reservoir involving *Rhodomonas lacustris* and *Peridiniopsis durandi*, whose dynamical events were also influenced by external forces.

Introduction

Reservoirs are an important framework inside ecological studies of assemblage organization, mainly due to the abiotic instability generated in its water body (Thornton et al., 1990). The effect of environmental variations can be measured as changes in the phytoplankton dynamics and according to the Intermediate Disturbance Hypothesis (IDH) elaborated by Connell (1978), and adapted by Reynolds (1988) and Padisák et al. (1993), *perturbations in intermediate intensities support high values of diversity in phytoplankton communities*. Nevertheless, Rojo & Álvarez-Cobelas (1993) pointed out that diversity changes are not entirely related to pelagic community changes.

In temperate lakes, an autogenic sequence of different phytoplankton associations or functional groups as a temporal model has been described (Reynolds, 1997, 2001, 2002). However, many aquatic ecosystems are highly perturbed, like in the hydroelectric reservoirs. In these cases, allogenic factors change the expected trajectory, and other functional groups of algae adapted to those perturbations can appear (Reynolds, 1993). In a temporal scale, phytoplankton shows patterns of a strong relationship between physical changes in the environment and biological processes, especially inside the euphotic zone. These interactions usually prevent equilibrium conditions (Sousa, 1984), affecting the growth and persistence of phytoplankton population and setting its communities far away from a steady-state (Reynolds, 1994, 1997).

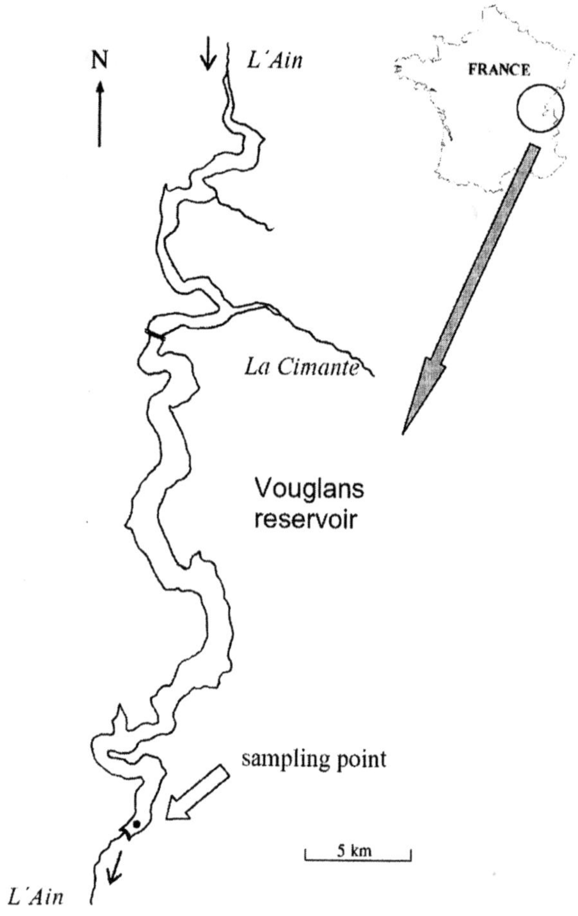

Figure 1. Study area and sampling point in Vouglans reservoir.

The aim of this study is to highlight what environmental factors are perturbations on phytoplankton assemblage in a hydroelectric man-made lake (Vouglans, France). In Vouglans reservoir, physical perturbations such as rainfall and anthropogenic management of the reservoir are considered as the main external disturbances, and the z_{mix}:z_{eu} ratio also alters, as occurs in other reservoirs (Naselli-Flores, 2000; Naselli-Flores & Barone, 2000).

Material and methods

Study site

The study was carried out in Vouglans reservoir (35 km length; 0.4 km large), built across the river Ain (Fig. 1). It is located in the Superior Jurassic calcareous Jura Mountains (NE of France) and was filled in 1968. The capacity of the reservoir is 605 $\times 10^6$ m^3, with a mean retention time of 184 days and a watershed area of 1120 km^2. It is a deep reservoir (over 100 m near the dam), with deficiencies in dissolved oxygen at the deepest layers during stratification (Rodriguez, 2001).

Hydroelectric management controlled by EDF (Electricité de France) induces two events of dewatering each year, displaying a water-level fluctuation which prevents the development of littoral vegetation. Turbines are located at 10 m above the bottom.

Sampling and analytical methods

The survey station was a central point situated near the dam, at the maximum depth (Fig. 1). The physical effect of the river Ain inflow is not considered here, due to the great distance between the course and the dam. The sampling was carried out monthly during 2000, from March to May and from June to October, and every 2 weeks between May and June.

Temperature, pH, conductivity and oxygen profiles were recorded *in situ* with WTW instruments, and water transparency was estimated with a Secchi disk. Water-level fluctuation were also recorded *in situ*. Subsurface water samples (0–1 m) were collected for chemical analyses. Chlorophyll *a* (Chl *a*), ortophosphates ($PO^{3-}{}_4$), total phosphorus (TP), nitrates ($NO^-{}_3$), ammonium ($NH^+{}_4$) and Kjeldhal nitrogen (N_{kj}) were measured by AFNOR standard methods (Brémond & Perrodon, 1979). Climatologic registers were provided by the Meteorological Station of Cernon. Rainfall values were taken daily. Euphotic depth was measured as the depth at which 1% of superficial irradiation was reached. Mixing depth was considered as the upper mixed layer during thermal stratification and as the maximum depth in absence of thermal stratification. Both are indicated in relation to the water level in each sampling.

Phytoplankton samples were collected with a hydrographic bottle from the surface to 10 m, and aliquot volumes were mixed as an integrated sample. The samples were preserved *in situ* with 10% formalin. Algae were counted using the Utermöhl method (1958) according to Lund et al. (1958), and the results expressed as cell density. The algal biomass was estimated using specific biovolumes, obtained by geometrical approximations in the measure of several population individuals (Rott, 1981), using Hagmeier (1961) correction for dinoflagellates. The volume of mucilage in colonial algae was ignored. For species identifica-

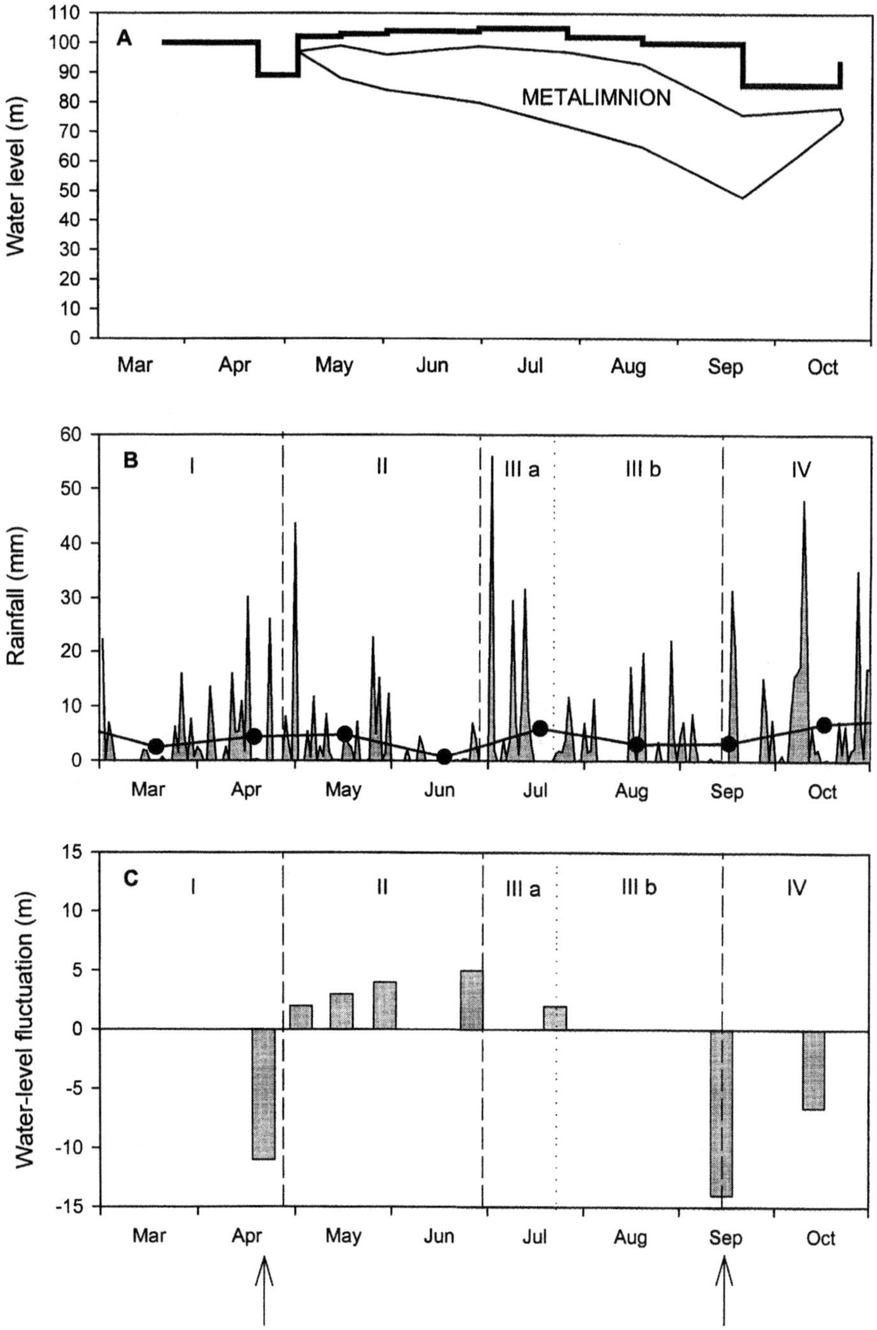

Figure 2. (**A**) Metalimnion (established by temperature profiles) related to the water level of the reservoir (upper marked line). (**B**) Shaded area: Dialy rainfall (mm) in the study area. Line and scatter plot: Mean monthly rainfall (mm) (**C**) Water-level fluctuations (m). Maximum depth indicated as level 0 m. Vertical discontinuous lines and upper numbers indicate the parts in which we divided the study period (see text). Arrows point out the dewatering events in Vouglans during 2000.

tion the following works were consulted: Ramanathan (1964), Fott (1968), Bourrelly (1968, 1970, 1972), Kómárek & Fott (1983), Starmach (1985), Popovskỳ & Pfiester (1990) and Krammer & Lange-Bertalot (1986, 1988, 1991a, b). Diversity index (in terms of individual units) was calculated according to Shannon & Weaver (1963).

Table 1. Minimum, maximum and average values of the main physico-chemical characteristics in the surface layers of Vouglans reservoir during the study period in 2000. chl a = chlorophyll a; TP = Total phosphorus; N_{kj} = Kjeldhal nitrogen

	Surface Temperature (°C)	Conductivity (μS cm^{-1})	Secchi (m)	cha a (μg l^{-1})	TP (μg l^{-1})	PO_4^- (μg l^{-1})	NO_3^- (μg l^{-1})	N_{kj} (μg l^{-1})	NH_4^+ (μg l^{-1})
Min.	7.6	258	0.6	1.3	<10.0	<10	970	100	20
Max.	21.5	343	12.5	16.7	20.0	<10	3240	400	110
Average	17.7	290	3.9	5.6	12.5	<10	1614	188	50

Results

Study of perturbations

The main physico-chemical characteristics of Vouglans surface waters throughout the study period are summarized in Table 1. Applying OECD classification (OECD, 1982), based on some parameters showed in Table 1, during 2000 the trophic level in the reservoir was mesotrophic.

Thermal stratification was established in May until the overturn succeeded in October (Fig. 2A), and thermocline deepened and increased in extension until its breakdown during the autumn mixing.

Figure 2B shows the daily rainfall in the study area (March–October 2000). July was one of the rainiest months in the past century, with a total of approximately 156 mm for the first fifteen days. This period was followed by intensive levels of sunshine (250–270 hours in August) and more than 30 °C in maximum air temperature. In addition, rainfall in October exceeded the values compared to previous years.

Two events of hydrological management in the reservoir occurred in 2000, each time inducing a vertical movement in the water level of about 12–14 m in relation to the maximum depth (Fig. 2C). Between these episodes, the water level increased weakly until August due to the summer storms, which were not reflected in the following water level due to the evaporation losses and the outflow. After the second dewatering the high values in October rainfall recovered the hydric reserve.

The study period can be divided into five intervals that correspond to the main events observed in the physical perturbation patterns detailed above (Fig. 2B, C).

The mixing to euphotic depth ratio (z_{mix}:z_{eu}) reached its highest value (54.91) in April (Fig. 3). The complete water column mixing in that month seems to have been the cause of this significant value. During the second dewatering event, the ratio did not change significantly, as the effect of water stratification was more important in September. The lowest ratio value was obtained in early stratification coinciding with the maximum value of water transparency. During stratification, z_{mix}:z_{eu} ratio maintained around 1, increasing weakly at the end of this period due to autumn mixing.

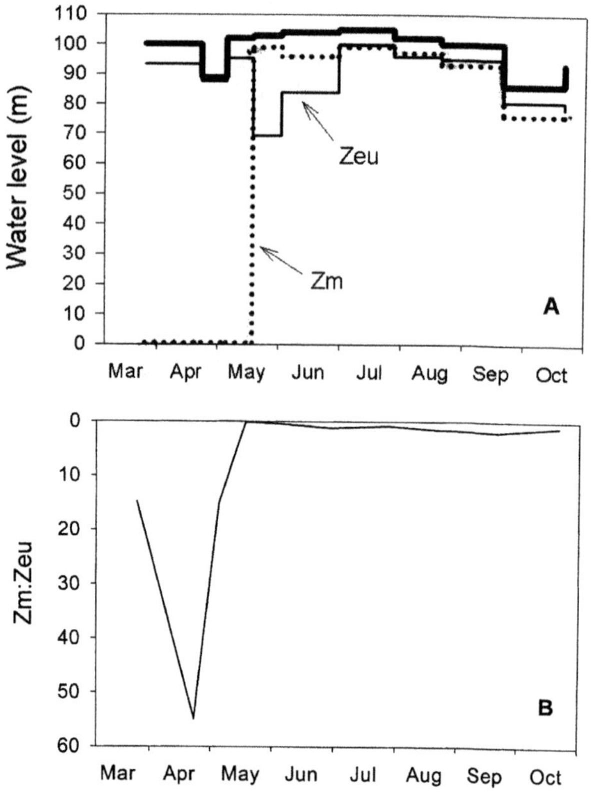

Figure 3. (**A**) Euphotic depth (Zeu) and mixing depth (Zm) related to the water level in the reservoir (upper marked line). (**B**) z_{mix}:z_{eu} ratio during the study period.

Table 2. Phytoplankton taxonomic composition and occurrence in Vouglans reservoir. + = sporadical presence

CHLOROPHYTES	
Chlorophyceae	occurrence
Actinastrum hantzschii Lagerheim	+
Ankyra judayi (G. M. Smith) Fott	+
Ankyra sp.	50
Chlamydomonas sp.	60
Chlorogonium sp.	+
Closteriopsis sp.	+
Coelastrum microporum Nägeli *in* A. Braun	+
Coenocystis subcylindrica Koršikov	+
Coronastrum sp.	+
Crucigenia sp.	+
Crucigeniella sp.	+
Dictyosphaerium subsolitarium Van Goor	+
Didymocystis bicellularis (Chodat) Komárek	30
Didymocystis fina Komárek	+
Didymogenes sp.	+
Eudorina elegans Ehrenberg *sensu* Conrad	+
Golenkinia radiata Chodat	+
Gonium pectorale Müller	+
Korschikoviella sp.	+
Micractinium pusillum Fresenius	+
Monoraphidium arcuatum (Koršikov) Hindák	10
Monoraphidium contortum (Thuret) Komárková-Legnerova	20
Monoraphidium komarkovae Nygaard	10
Nephrochlamys sp.	+
Nephroselmis olivacea Stein	+
Oocystis lacustris Chodat	+
Oocystis rhomboidea Fott	+
Oocystis sp.	40
Pediastrum boryanum var. *longicorne* Reinsch	+
Phacotus lendneri Chodat	+
Pteromonas sp.	10
Pyramimonas sp.	10
Scenedesmus linearis Komárek	10
Scenedesmus quadricauda (Turpin) Brëbisson *sensu* Chodat	10
Sphaerellopsis sp.	+
Sphaerocystis planctonica (Kořikov) Bourrelly	40
Tetraedron incus (Teiling) G. M. Smith	+
Tetraedron minimum (A. Braun) Hansgirg	+
Tetraselmis sp.	+
Tetrastrum sp.	+
Treubaria triappendiculata Bernard	+
Westella botryoides (W. West) De-Wildeman	+
Willea vilhelmii (Fott) Komárek	10
Ulothricophyceae	
Elakatothrix gelatinosa Wille	20
Planctonema lauterbornii Schmidle	30

CHROMOPHYTES (cont.)	
	occurrence
Chrysamoeba sp.	+
Chrysococcus sp.	+
Chrysophyceae	
Dinobryon crenulatum W. & G. S. West	+
Dinobryon cylindricum Imhof	+
Dinobryon divergens Imhof	+
Dinobryon sertularia Ehrenberg	+
Kephyrion litorale Lund	+
Mallomonas akrokomos Ruttner	10
Mallomonas tonsurata Teiling	+
Mallomonas sp.	30
Ochromonas sp.	+
Pseudokephyrion sp.	+
Salpingoeca sp.	+
Synura sp.	+
Bacillariophyceae	
Achnanthes sp.	+
Actinocyclus normanii (Gregory) Hustedt	10
Asterionella formosa Hassal	60
Aulacoseira sp.	10
Cocconeis placentula Ehrenberg	10
Cyclostephanos dubius (Fricke) Round	10
Cyclotella radiosa (Grunow) Lemmermann (ex *C. comta)*	20
Cyclotella cf. cyclopuncta Håkansson & Carter	+
Cyclotella wuethrichiana Druart & Straub	60
Cyclotella sp.	+
Cymbella sp.	+
Fragilaria crotonensis Kitton	20
Fragilaria ulna var. *acus* (Kützing) Lange-Bertalot	+
Fragilaria sp.	20
Gomphonema sp.	+
Navicula sp.	+
Nitzschia spp.	10
Rhizosolenia sp.	+
Stephanodiscus hantzschii Grunow	10
Stephanodiscus hantzschii fo. *tenuis* (Hustedt) Håkansson & Locker	10
Stephanodiscus minutulus (Kützing) Cleve & Möller	30
Stephanodiscus neoastrea Håkansson & Hickel	10
Surirella sp.	10
PYRRHOPHYTES	
Cryptophyceae	
Cryptomonas spp.	100
Cryptomonas erosa Ehrenberg	+
Rhodomonas lacustris Pascher & Ruttner	90
Rhodomonas sp.	+

Table 2. Continued

Zygophyceae	occurrence	Dinophyceae	occurrence
Cosmarium sp.	+	*Amphidium* sp.	+
Staurastrum sp.	+	*Ceratium hirundinella* (O. F. Müller) Schrank	30
		Gymnodinium sp.	20
EUGLENOPHYTES		*Peridiniopsis durandi* Rodriguez et al.	20
Euglenophyceae		*Peridinium* spp.	10
Euglena sp.	+		
Trachelomonas sp.	+	**CYANOPHYTES**	
		Cyanophyceae	
CHROMOPHYTES		*Aphanocapsa elachista* West & West	20
Xanthophyceae		*Aphanothece minutissima* (West) Komárková–Legnerová & Cronberg	+
Centritractus belenophorus Lemmermann	20	*Coelosphaerium* sp.	+
Nephrodiella lunaris Pascher	+	*Microcystis* sp.	10
		Snowella sp.	+

Effect of perturbations on the phytoplankton assemblage

About 100 taxa were identified in the surface layers during the study period, with occurrence values indicating that within the phytoplankton population, Cryptophyceae was the most represented group, followed by Bacillariophyceae and Chlorophyceae (Table 2).

The phytoplankton assemblages found in each period and related to the perturbations considered (Fig. 2B, C) were:

(I) Mixing period, with high values of rainfall and one event of dewatering. Such as in terms of cell density and biomass, this period was characterized by the presence of cryptophytes (Fig. 4A, B), with an intensive bloom due to *Rhodomonas lacustris* Pascher *et* Ruttner in April. The small centric diatoms were also important in terms of cell density, especially *Stephanodiscus minutulus* (Kützing) Cleve & Möller. In this period, both richness and individual diversity decreased until the middle of April (Fig. 5).

(II) Period of early stratification, with values of rainfall decreasing and a weak increase of water level. In terms of cell abundance small diatoms and chlorophytes were dominants, especially *Cyclotella wuethrichiana* Druart & Straub and *Ankyra judayi* (G.M. Smith) Fott. But, with respect to biomass, the main species were *Fragilaria* sp. and *Chlamydomonas* sp., both unicellular. Richness and diversity showed low values.

(III) Period of stratification, which can be divided into two episodes (III a and III b; Fig. 2B, C) separated by the highest values of rainfall recorded during the first days of July, together with the decrease in water level due to the intensive levels of sunshine and the second dewatering event at the end of this period. During the storms (III a) larger chlorophytes and diatoms succeeded, the significance of which was marked by the abundance and biomass of *Sphaerocystis planctonica* (Koršikov) Bourrelly and *Fragilaria crotonensis* Kitton. After these events (III b), in terms of cell density, the cyanophyte *Aphanocapsa elachista* West & West and the small chlorophyte ulotrichal *Planctonema lauterbornii* Schmidle were dominant in August, followed by complete dominance of this chlorophyte. This filamentous species was responsible for the annual cell maximum (32×10^6 cell l^{-1}, in September). Nevertheless, in terms of biomass, August was dominated by the dinoflagellate *Peridiniopsis durandi* Rodriguez et al. (with the highest value of total biomass: 1872 $\mu g\ l^{-1}$), followed by *Planctonema lauterbornii*. Richness and diversity increased (III a) and decreased (III b; Fig. 5), diversity reaching its highest value in July (3.10 bits per individual) and its lowest in the middle of September (0.53 bits per individual).

(IV) End of stratification-beginning of autumn overturn, with high values of rainfall and an increase of water level. *Planctonema lauterbornii* was dominant in cell density, despite the high decrease of total concentration, but in biomass the domin-

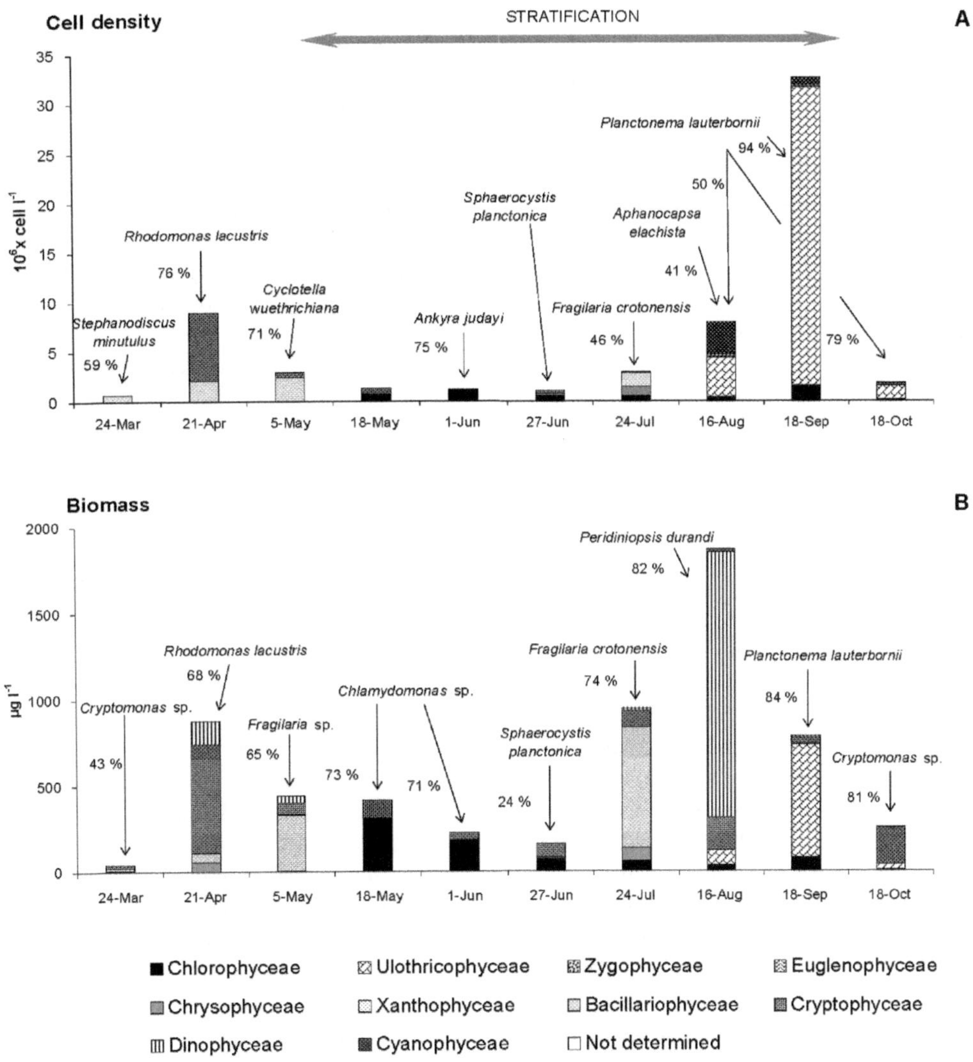

Figure 4. (**A**) Cell density ($10^6 \times$ cell l^{-1}) and (**B**) biomass ($\mu g\ l^{-1}$) of phytoplankton groups, and higher species proportion found in each sample from March 2000 to October 2000. The upper row indicates the stratification period.

ance was due to *Cryptomonas* sp. Richness and diversity turned to higher values.

There were few dominant species in terms of biomass: five species reached over 10% of the total annual biomass (*Planctonema lauterbornii*, *Fragilaria crotonensis*, *Rhodomonas lacustris*, *Cryptomonas* sp. and *Peridiniopsis durandi*), and only *Peridiniopsis durandi* exceeded the 20%. Considering species representing over 20% of each sample biomass, a temporal pattern was found (Fig. 6), showing an intense growth but short persistence in most of the species.

Discussion

According to the alphanumerical code proposed by Reynolds et al. (2002), the phytoplanktonic functional groups found in the surface layers of Vouglans reservoir in each period, determined by allogenical forces, show the following algal assemblage: (**C-X$_2$**)→(**C-X$_1$**)→ (**F-P**)→(**K-T**)→(**T**) in terms of cell density and (**Y-X$_2$**)→(**P-X$_2$**)→ (**F-P**)→(**L$_0$-T**)→(**Y**) in terms of biomass. In both cases these functional groups are typical of mesotrophic environment (Reynolds et al., 2002).

During the spring, factors like high turbulence, low temperature and availability of nutrients favoured the

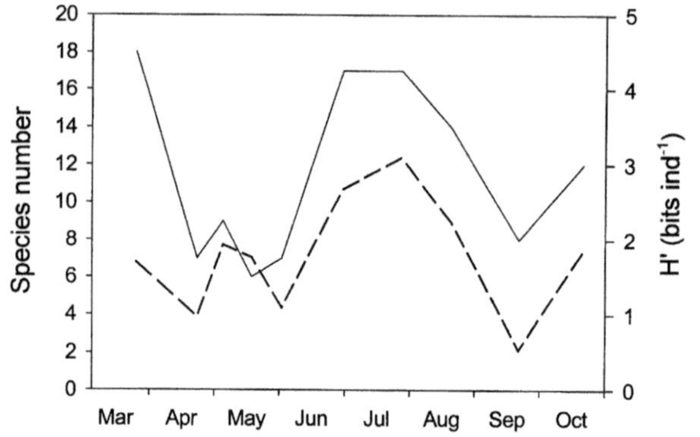

Figure 5. Richness (species number; solid line) and Shannon diversity (bits individual^{-1}; discontinuous line) during the study.

development of small centric diatoms, and also allowed the bloom of *Rhodomonas lacustris*. In early stratification, rainfall seems to have affected the algal proliferation, with diatoms and small chlorophytes present. In spite of strong storms in July, the phytoplankton assemblage followed the typical pattern in temperate lakes during stratification, with colonial chlorophytes and diatoms. In this case, the strong growth of *Fragilaria crotonensis* in summer could be due to an intermittent mixing of the epilimnion (Lindenschmidt & Chorus, 1998). Nevertheless, this perturbation led to a high development of the small ulotrichal chlorophyte *Planctonema lauterbornii*. This event can be considered as a 'reversion' in the expected assemblage pattern (*sensu* Reynolds, 1980, 1993), setting this sequence back to an earlier stage. Just before this development, the dominance was due to *Aphanocapsa elachista* in terms of cell density and *Peridiniopsis durandi* in terms of biomass. The reversion is due to external perturbations which affected the surface layers of the water column, weakly deepening the mixing zone, promoting the growth and persistence of chlorophytes (Barlow, 1984) which is reflected in a significant change in individual diversity. The dominance of this ulotrichal is also maintained in the beginning of the autumn overturn. In general terms, the phytoplankton assemblage found in the surface layers of this reservoir is typical in mesotrophic and temperate lakes (Reynolds, 1980; Sommer et al., 1986).

Although significant variations were not found in z_{mix}:z_{eu} ratio, meteorological perturbations seem to affect the superficial assemblage of phytoplankton (reflected in the diversity changes), rather than the vertical movement of water level due to the dam management (the very deep outflow does not disturb the superficial stratified levels). Moustaka-Gouni et al. (2000) described similar effects of external forces in phytoplankton dynamics in another deep reservoir, although in that case the hydraulic regime created by the reservoir management was significant. These outcomes reflect how the perturbations can induce changes in the composition and autogenic-succession of phytoplankton (Reynolds, 1988; Reynolds et al., 1993), and in the state of community organization and diversity (Reynolds, 1993).

Peculiar freshwater red tides occurred in Vouglans reservoir, involving *Rhodomonas lacustris* and *Peridiniopsis durandi*. In 2000, the bloom was due to *Rhodomonas lacustris*, coinciding with maximum z_{mix}:z_{eu} ratio.

Historical data showed a strong *Peridiniopsis durandi* bloom in 1995 (Rodriguez et al., 1999), but in 2000 this dinoflagellate proliferation was probably prevented by the effect of superficial turbulence during the stratification, phenomena also detailed in the subtropical lake Kinneret (Berman & Shteinman, 1998).

The presence of *Planctonema lauterbornii* in high numbers during the stratification, and the importance of chlorophytes and dinophytes in this period, were observed in others meso-eutrophic man-made lakes. In some reservoirs in eastern Spain the phytoplankton population in summer was dominated by *Planctonema lauterbornii* and *Peridinium cinctum* Penard (Dasí et al., 1998). Furthermore, these results were found in others Spanish reservoirs (Sabater & Nolla, 1991; Riera et al., 1992), specially the high values of *Planctonema lauterbornii*. Rojo (in this volume) detailed that in the oligo-mesotrophic lake Las Madres (Central Spain) this ulotrichal reached high abundance within the mixing period, in addition to a significant presence of dinoflagellates throughout the study. In Vouglans reservoir, as detailed above, the perturbations during the stratification favoured the development of *Planctonema lauterbornii*, due to its good light antenna physiology (Reynolds, 1994), and also its persistence during the subsequent 3 months. Moreover, this chlorophyte was found in the deep and eutrophic Rosamarina reservoir (Sicily, Italy) during thermal stratification, period that showed z_{mix}:z_{eu} < 1.5 and diversity above 2.0 bits per individuals (Naselli-Flores & Barone, 1998), that are similar to the values observed in Vouglans reservoir. These outcomes agree with the idea that external perturbations increase diversity and promote species with small cellular sizes

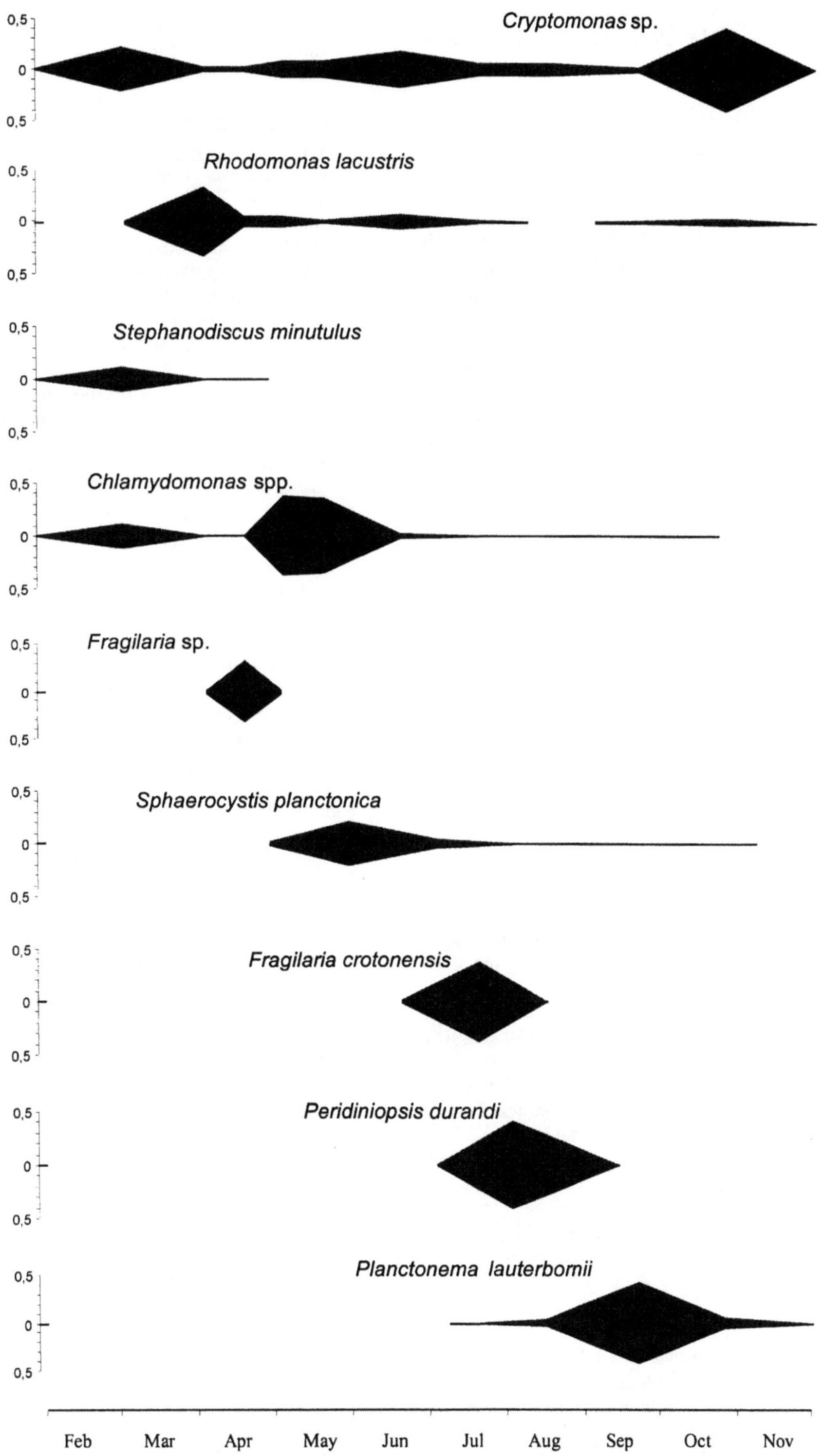

Figure 6. Temporal biomass rates for the main phytoplankton species during the study.

(Padisák et al., 1993). The constancy of z_{mix}:z_{eu} ratios indicates a relative stability which can affect the phytoplankton assemblages (Naselli-Flores, 2000; Naselli-Flores & Barone, 2000).

Conclusions

The study carried out in 2000 in Vouglans reservoir has revealed that, at this temporal scale, environmental factors generate an early-state in the expected phytoplankton assemblage, followed by stability in its progression simply expressed as a 'no-change situation'. Therefore, some sort of equilibrium is achieved although it was not possible to determine the underlying mechanisms.

Acknowledgements

We thank François Straub & Pierre Compère for their help in species identification. We are also very grateful to Carmen Rojo for her helpful comments and suggestions. S.M. Morata was a holder of a scholarship from the Education, Culture and Sport Spanish Ministry. This study was funded by Conseil Géneral du Jura.

References

Barlow, R. G., 1984. Physiological responses of phytoplankton to turbulence and stable environments in an upwelling region. J. Plankton Res. 6: 385–397.

Berman, T. & B. Shteinman, 1998. Phytoplankton development and turbulent mixing in Lake Kinneret (1992–1996). J. Plankton Res. 20: 709–726.

Bourrelly, P., 1968, 1970, 1972. Les algues d'eau douce. Initiation à la systématique. Tome II: Les algues jaunes et brunes. Chrysophycées, Phéophycées, Xanthophycées et Diatomées. 438 pp. Tome III: Les algues bleues et rouges. Les Eugléniens, Peridiniens et Cryptomonadines. 512 pp. Tome I: Les algues vertes. 569 pp. Editions N. Boubée, Paris.

Brémond, R. & C. Perrodon, 1979. Les paramètres de la qualité des eaux. 2° édition. Ministère de l'Environnement et du cadre de vie. 260 pp.

Connell, J. H., 1978. Diversity in tropical rain forests and coral reefs. Science 199: 1302–1310.

Dasí, M. J., M. R. Miracle, A. Camacho, J. M. Soria & E. Vicente, 1998. Summer phytoplankton assemblages across trophic gradients in hard-water reservoirs. Hydrobiologia 369/370: 27–43.

Fott, B., 1968. *Cryptophyceae, Chloromonadophyceae, Dinophyceae*. 3 Teil. E. Schweizerbart'sche Verlag. (Nägele u. Obermiller), Stuttgart. 322 pp.

Hagmeier, E., 1961. Plankton-Aquivalente. Kieler Meeresfosch. 17: 32–47.

Komárek, J. & B. Fott, 1983. Das Phytoplankton des Süsswassers. Systematik und Biologie. *Chlorophyceae* (Grünalgen). Ordung: *Chlorococcales*. Teil 7. In Elster, H. J. & W. Ohle (eds), Die Binnengewässer. Einzeldarstellungen aus der Limnologie und ihren Nachbargebieten. E. Schweizerbart'sche Verlag. Stuttgart. 1044 pp.

Krammer, K. & H. Large-Bertalot, 1986, 1988, 1991a, 1991b. Teil 2: *Bacillariophyceae*. (2/1): Naviculaceae, 876 pp.; (2/2): Bacillariaceae, Epithemiaceae, Surirellaceae. 596 pp.; (2/3): Centrales, Fragilariaceae, Eunotiaceae. 576 pp.; and (2/4): Achnanthaceae. 437 pp. In Ettl, H., J. Gerloff, H. Heynig & D. Mollenhauer (eds), Sübwasserflora von Mitteleuropa, Gustav Fischer Verlag. Stuttgart, Jena.

Lindenschmidt, K. E. & I. Chorus, 1998. The effect of water column mixing on phytoplankton succession, diversity and similarity. J. Plankton Res. 20: 1927–1951.

Lund, J. W. G., C. Kipling & E. D. Le Cren, 1958. The inverted microscope method of estimating algal numbers and the statistical basis of estimation by counting. Hydrobiologia 11: 143–170.

Moustaka-Gouni, M., K. Albanakis, M. Mitrakas & A. Psilovikos, 2000. Planktic autotrophs and environmental conditions in the newly-formed hydroelectric Thesaurus reservoir, Greece. Archiv für Hydrobiol. 149: 507–526.

Naselli-Flores, L., 2000. Phytoplankton assemblages in twenty-one Sicilian reservoirs: relationships between species composition and environmental factors. Hydrobiologia 424: 1–11.

Naselli-Flores, L. & R. Barone, 1998. Phytoplankton dynamics in two reservoirs with different trophic state (Lake Rosamarina and Lake Arancio, Sicily, Italy). Hydrobiologia 369/370: 163–178.

Naselli-Flores, L. & R. Barone, 2000. Phytoplankton dynamics and structure: a comparative analysis in natural and man-made water bodies of different trophic state. Hydrobiologia 438: 65–74.

OECD, 1982. Eutrophication of waters. Monitoring, assessment and control. Paris. 154 pp.

Padisák, J., C. S. Reynolds & U. Sommer (eds), 1993. Intermediate Disturbance Hypothesis in Phytoplankton Ecology. Developments in Hydrobiology 81, Kluwer Academic Publishers, Dordrecht. 200 pp. Reprinted from Hydrobiologia 249.

Popovskỳ, J. & L. A. Pfiester, 1990. Teil 6: *Dinophyceae* (Dinoflagellida). In Ettl, H., J. Gerloff, H. Heynig & D. Mollenhauer (eds), Sübwasserflora von Mitteleuropa. Gustav Fischer Verlag. Stuttgart, Jena. 272 pp.

Ramanathan, K. R., 1964. *Ulothricales*. I.C.A.R., New Delhi. 185 pp.

Reynolds, C. S., 1980. Phytoplankton assemblages and their periodicity in stratifying lake systems. Hol. Ecol. 3: 141–159.

Reynolds, C. S., 1988. The concept of biological succession applied to seasonal periodicity of phytoplankton. Int. Ver. theor. angewan. Limnol.: Verh. 23: 683–691.

Reynolds, C. S., 1993. Scales of disturbance and their role in plankton ecology. Hydrobiologia 249: 157–171.

Reynolds, C. S., 1994. The long, the short and the stalled: on the attributes of phytoplankton selected by physical mixing in lakes and rivers. Hydrobiologia 289: 9–21.

Reynolds, C. S., 1997. Vegetation processes in the pelagic: a model for ecosystem theory. Excellence in Ecology series. In O. Kinne (ed), Ecology Institute, Oldendorf/Luhe. Germany. 371 pp.

Reynolds, C. S., 2001. Emergence in pelagic communities. Sci. Mar. 65 (Suppl. 2): 5–30.

Reynolds, C. S., J. Padisák & U. Sommer, 1993. Intermediate disturbance in the ecology of phytoplankton and the maintenance of species diversity: a synthesis. Hydrobiologia 249: 183–188.

Reynolds, C. S., V. L. Huszar, C. Kruk, L. Naselli-Flores & S. Melo, 2002. Towards a functional classification of the freshwater phytoplankton. J. Plankton Res. 24: 417–428.

Riera, J. L., D. Jaume, J. De Manuel, J. A. Morgui & J. Armengol, 1992. Patterns of variation in the limnology of Spanish reservoirs: a regional study. Limnetica 8: 111–124.

Rodriguez, S., 2001. La retenue de Vouglans: diagnose rapide, données 1999. Conseil général du Jura, Agence de l'eau RMC, CEMAGREF, Diren Franche-Comté et B.E.. 33+XVI pp.

Rodriguez, S., A. Couté, L. Ten-Hage & G. Mascarell, 1999. *Peridiniopsis durandi* sp. Nova (Dinophyta), une nouvelle Dinophycée d'eau douce responsable de marées rouges. Algological Studies 95: 11–29.

Rojo, C. & M. Álvarez-Cobelas, 1993. Hypertrophic phytoplankton and the Intermediate Disturbance Hypothesis. Hydrobiologia 249: 43–57.

Rott, H. E., 1981. Some results from phytoplankton counting intercalibrations. Schweizerische Zeitschrift für Hydrologie 43: 34–62.

Sabater, S. & J. Nolla, 1991. Distribution patterns of phytoplankton in Spanish reservoirs: first results and comparison after fifteen years. Int. Ver. theor. angewan. Limnol.: Verh. 24: 1371–1375.

Shannon, C. E. & W. Weaver, 1963. The mathematical theory of communication. Univ. Illinois Press, Urbana, U.S.A. 117 pp.

Sommer, U., M. Z. Gliwicz, W. Lampert & A. Duncan, 1986. The PEG-model of seasonal succession of planktonic events in fresh waters. Archiv für Hydrobiol. 106: 433–471.

Sousa, W. P., 1984. The role of disturbance in natural communities. Ann. Rev. Ecol. Syst. 15: 353–391.

Starmach, K., 1985. Teil 1: *Chrysophyceae* und*Haptophyceae*. In Ettl, H., J. Gerloff, H. Heynig & D. Mollenhauer (eds), Sübwasserflora von Mitteleuropa, Gustav Fischer Verlag. Stuttgart, Jena. 515 pp.

Thornton, K. W., B. L. Kimmel & F. E. Payne (eds), 1990. Reservoir Limnology: Ecological Perspectives. John Wiley & Sons, New York. 246 pp.

Utermöhl, H., 1958. Zur Vervollkommung der quantitativen Phytoplankton-Methodik. Mitt. int. Ver. theor. angewan. Limnol. 9: 1–38.

Hydrobiologia **502:** 85–95, 2003.
L. Naselli-Flores, J. Padisák, M. T. Dokulil (eds),
Phytoplankton and Equilibrium Concept: The Ecology of Steady-State Assemblages.

Factors influencing the phytoplankton steady state assemblages in a drinking-water reservoir (Ömerli reservoir, Istanbul)

Meriç Albay & Reyhan Akçaalan
Istanbul University, Fisheries Faculty, Ordu Cad. No: 200 34470 Laleli, Istanbul, Turkey
E-mail: merbay@istanbul.edu.tr; merbay08@yahoo.com

Key words: Ömerli reservoir, steady state assemblages, copper sulphate treatment, functional groups

Abstract

In this study, algal assemblages, phytoplankton biomass, chlorophyll-*a* and selected physico-chemical parameters were investigated in a drinking water reservoir in Ömerli, Istanbul from October 1999 to November 2000. Water samples were collected from three sampling stations at the surface, a depth of 1 m and at 5 m depth intervals to the bottom (approximately 20 m). Particular attention was given to the ecology of steady state phytoplankton assemblages. When the euphotic depth (z_{eu}) was less then 1.52 m and the ratio of the euphotic to mixing zone (z_{eu}/z_{mix}) was 0.2 m (during October 1999 and September and October 2000), *Microcystis aeruginosa* Kütz. dominanted, accounting for 85 – 95% of the total phytoplankton biomass. A shift to *Fragilaria crotonensis* Kitton and *Asterionella formosa* Hass. occurred during April and early May 2000 when $z_{eu} > 4.1$ m. Other dominant species were *Coelastrum microporum* Nag. *Pediatrum boryanum* (Turp.) Menegh. and *P. duplex* (May to mid June 2000); *C. microporum, Staurastrum* spp. and *Cryptomonas* spp. (late June to mid July 2000); and *Scenedesmus* spp., *S. gracile* and *Nitzschia holsatica* Hust. (late July to late August 2000). Below 10 meters, a high biomass of *Mougeotia* sp. was recorded at all sites. However, in October and early November 2000, this species appeared in the surface water. During late May to late August 2000, the z_{eu} depth changed between 2.7 and 3.38 m. pH and total organic carbon (TOC) concentrations were found to be relatively high in the surface waters, whereas total phosphorus (TP) and Soluble Reactive Silicate (SRSi) concentrations were high below the 5 meters. The total nitrogen (TN) concentrations were high either in the surface waters or in deep layers. It is suggested that the observed increase in cell biomass of some species was controlled by the nutrient concentrations. The changes in the phytoplankton assemblages resulted in a reduction of the euphotic depth.

Introduction

The growth of phytoplankton in reservoirs is regulated by various physical and chemical factors (Komárková & Hejzlar, 1996; Naselli-Flores & Barone, 2000). Although the growth and biomass of phytoplankton in reservoirs and lakes are controlled by similar parameters, Wetzel (1990) and Komárková & Hejzlar (1996) have reported marked differences between these two types of aquatic ecosystems. In particular, factors including precipitation, evaporation, flow rate, stratification, irregular water use (including summer drawdown), concentrations of suspended solids, and flora and fauna diversity may influence the growth of phytoplankton as well as steady state assemblages.

Ömerli reservoir is the largest reservoir in Istanbul, Turkey and is located in north east of the city. The reservoir provides approximately 48% of the city's drinking water. Morphometrically, the reservoir has a surface area of 23.5 km^2, a drainage area of 621 km^2 and a volume of 2.2×10^6 m^3 (Eruz, 1995).

In the last few decades, Ömerli reservoir has become heavily polluted by sewage, industrial wastewater and soil run-off which have promoted the formation of toxic cyanobacterial (blue–green algae) blooms from late summer to mid autumn. Recently, several thousands fish (mainly *Cyprinus carpio*) died when the cyanotoxin (microcystin) was detected in the reservoir at a concentration of 2.1 $\mu g\ l^{-1}$. After this phenomenon, the Istanbul Water Authority decided to apply algicide (copper sulphate) to control the blue–green algal biomass at specific sites in the reservoir. The results of the effects of the algicide treatment on the blue–green algae were given Albay et al. (in press).

The objective of this study was to describe the steady state assemblages and investigate the mechanisms affecting the equilibrium in natural phytoplankton assemblages of the Ömerli reservoir. Additionally, the effects of algicide addition on Steady State Assemblages (SSA) of treated and untreated sites of the reservoir were investigated and compared.

Materials and methods

The investigation can be divided into two distinctive periods. The first period (period 1) was from November 1999 to March 2000 and the second period (period 2) was from April 2000 to October 2000. In the second period a copper sulphate treatment was applied at station 1 by the Istanbul Water Authority. The effect of the $CuSO_4$ treatment on cyanobacteria and microcystin production has been reported previously by Albay et al. (in press).

Samples were taken from three monitoring stations at regular monthly intervals from October 1999 to April 2000, at weekly intervals from mid April to mid August 2000, and then at 3-week intervals early September to mid November 2000. Water samples were collected from the surface, a depth of 1 m and then at 5 m intervals to the bottom (approximately 20 m) using a Hidrobios vertical water sampler (Nansen bottle). Water temperature and dissolved oxygen were measured *in situ*, using WTW model oxygen meter and pH was measured in the laboratory using WTW model pH meter. Water transparency was measured using a Secchi disc and the euphotic zone (z_{eu}) was calculated as 2.7 times the Secchi disc depth. The mixing zone (z_{mix}) was identified using temperature profiles considering density gradients greater than 0.02 kg m^{-3} m^{-1} (Reynolds, 1992). Dissolved inorganic nutrients, (nitrate+nitrite, Soluble Reactive Phosphorus), Total Phosphorus and Total Nitrogen were determined colorimetrically using Technicon Industrial Methods (1977 a,b) and Strickland & Parsons (1972). Total Organic Carbon was measured according to Crompton (1989) using a Shimadzu TOC – 500 model analyser.

Phytoplankton samples were preserved in Lugol's solution, and counted using the inverted–microscope method (Utermöhl, 1958). Wet weight biomass was calculated from the recorded abundance and specific biovolume estimates, which were based on geometric approximations (Rott, 1981), assuming a specific density of phytoplankton cells of 1 g cm^{-3}. In general, average cell size was based on at least twenty measurements. Chlorophyll-*a* was determined by spectrophotometry (Nush, 1980). Diversity of phytoplankton (H') was calculated according to the Shannon–Wiener (1949) function.

Statistical analysis

Relationship between environmental parameters and phytoplankton groups (including SSA) were evaluated using Spearman rank correlation test. To compare the $CuSO_4$ addition effects where used only station 1 to the other sites, t-tests were used and results were considered significant agreement at $p < 0.01$ and $p < 0.05$.

Results

During the first period of study, environmental parameters in the surface waters at the various monitoring stations did not change very much. The trends in z_{eu} (euphotic depth) increased with time towards the end of February 2000, ranging from 3.98 to 8.2 m (mean 6.32 m). The thermocline formation was observed between a depth of 5 and 10 m during period 2. Average z_{eu} decreased considerably compared to the previous period (mean 3.12, 2.92 and 2.72 m at stations 1, 2 and 3, respectively). z_{eu} showed strong negative correlation with cyanobacteria and a weak negative correlation with chlorophytes (Table 1). The z_{eu}/z_{mix} ratio, which is closely related to phytoplankton biomass and composition, decreased to 0.18 m at station 3 during September 2000. No significant differences were found among the stations ($p > 0.05$).

The dissolved oxygen (DO) concentrations were high at all sites and there was no anoxia during November 1999 to March 2000. Dissolved oxygen concentrations changed vertically, however never decreased below 2 mg l^{-1}. As the water column stratified (from April to October), the distribution of DO concentrations showed significant vertical differences. At station 3, maximum DO concentrations (17.4 mg l^{-1}) occurred at the surface during September 2000, and the minimum concentration (0.2 mg l^{-1}) was recorded near the bottom (about 20 m) during October 2000.

Table 1. Correlation coefficients (r) between selected variables. Only those significant ($p<0.05$) are shown. Abbreviations: Temp= Temperature; Tot. Biom. = Total Biomass; z_{eu}= euphotic zone; z_{mix}= mixing zone; TOC= Total Organic Carbon; TN= Total Nitrogen; TP= Total Phosphorus

	Station 1		Statiion 2		Station 3	
	period 1	period 2	period 1	period 2	period 1	period 2
Temp. vs Tot. Biom.	0.69	0.76	0.56	0.65	0.7	0.75
SRSi vs diatoms	0.26	–	0.26	–	–	–
SRSÍ vs Tot. Biom.	–	0.61	–	–	0.71	–
z_{eu} vs blue-greens	–	–0.86	–	–0.79	–	–0.84
z_{eu} vs diatoms	0.49	–	0.44	0.51	–	0.53
z_{eu} vs greens	–0.95	–0.4	–0.84	–0.43	–0.3	–0.58
Blue-greens vs greens	–0.22	–0.37	0.81	–0.34	0.8	–0.33
pH vs Tot. Biom.	0.84	0.81	0.75	0.74	0.73	0.77
pH vs blue-greens	0.96	0.65	0.86	0.76	0.92	0.66
pH vs greens	0.38	0.78	0.74	0.62	0.63	0.74
NH_4 vs Tot. Biom.	0.42	–	0.61	–	0.42	–
TOC vs Tot. Biom.	–	0.82	–	0.76	–	0.82
TN vs Tot. Biom.	0.34	0.87	0.28	0.42	0.36	0.45
TP vs Tot. Biom.	0.58	–	0.31	–	0.71	–

Table 2. Main characteristics of studied water bodies (average and range values).Abbreviations as for Table 1

		Station 1				Station 2				Station 3			
		Period 1		Period 2		Period 1		Period 2		Period 1		Period 2	
	unit	mean	interval	mean	interval	mean	interval	mean	interval	mean	interval	mean	interval
Temp.	°C	11.2	8–14	21.8	14.4–27.4	11.2	8-14.1	21.8	14.3–27.2	11.3	8.1–14.1	21.9	14.4–27.4
pH		7.4	7.2–7.8	9.7	8.2–10.6	7.4	7.2–7.8	9.8	8.2–10.6	7.4	7.2–7.9	9.9	8.2–10.7
z_{eu}	m	6.32	3.98–8.2	3.12	1.66–6.6	6.3	4–8.2	2.92	1.58–6.5	6.2	4.1–8	2.74	1.52–4.3
z_{eu}/z_{mix}		–	–	0.35	0.2–0.48	–	–	0.37	0.2–0.48	–	–	0.35	0.2–0.48
SRP	$\mu g\, l^{-1}$	96.7	54–142	69.6	21–142	93.2	56–132	83.9	44–142	101	58–146	90.4	48–146
TP	$\mu g\, l^{-1}$	129.7	83–206	104	82–162	122.3	78–160	128.6	96–168	139	82–226	136	100–170
NO_2+NO_3	$\mu g\, l^{-1}$	823	665–1243	665	4–1508	814	665–1124	685	12–1508	764	610–1154	630	24–1406
NH_4	$\mu g\, l^{-1}$	376	180–610	380	18–754	380	174–622	372	36–740	417	250–610	390	54–760
TN	$\mu g\, l^{-}$	1728	1143–2080	1754	1468–2670	1659	920–2124	2157	1450–2750	1880	1260–2340	2420	1640–3200
TN/TP		14.6	8.52–23.1	28.7	11.1–4	13.1	10.2–17.2	20.4	10.7–26.3	14.4	8.1–23.1	18.6	9–26.6
SRSi	$\mu g\, l^{-}$	667	330–1530	729	283–1410	593	130–1330	705	280–1250	689	520–1420	796	550–1510
Chl-*a*	$\mu g\, l^{-}$	3.2	0.67–6.9	35.7	9.2–6.6	2.4	0.9–4.6	52.1	10.4–132.7	3.7	1.2–8.1	54.6	11.8–146
Tot. Biom.	$mg\, l^{-1}$	5.29	1.8–13.89	81.01	19.91–206.1	5.06	1.44–13.56	77.32	21.26–256	5.52	2.25–14.52	95.69	23.4–308.9
TOC	$mg\, l^{-1}$	4.31	3.25–5.44	8.25	4.13–12.6	4.63	3.42–5.44	8.14	4.22–10.84	4.82	3.5–5.8	8.86	4.48–13.6

Average surface water temperatures were relatively low (average 11.2 °C, between November 1999 and early March 2000) at all sites in period 1 (Table 2) compared to period 2. The vertical distribution of temperature was nearly homogeneous at all stations. The average water temperature for period 2 (between late April and early October 2000) was 21.8 °C. During the summer stratification period, temperature differences of 0.1 and 1.4 °C were observed. Maximum and minimum temperatures were recorded at station 3 as 27.4 °C (surface) in late July and as 6.8 °C (near the bottom) in April 2000.

Chlorophyll–*a* (chl–*a*) concentrations ranged from 0.67 to 6.99 $\mu g\ l^{-1}$ at station 1 (mean 3.21 $\mu g\ l^{-1}$), from 0.98 to 4.66 $\mu g\ l^{-1}$ at station 2 (mean 2.43 $\mu g\ l^{-1}$) and from 1.2 to 8.12 $\mu g\ l^{-1}$ at station 3 (mean

3.76 μg l^{-1}) during period 1 (Table 2). Chl–*a* concentrations increased considerably and three peaks were observed during period 2. The first peak was observed in early June 2000, when it reached 43.7 μg l^{-1}, 76.8 μg l^{-1} and 68.2 μg l^{-1} at stations 1, 2 and 3, respectively. During the early and mid summer periods, chl–*a* concentrations fluctuated slightly at station 2 and 3, whereas a marked increase was recorded in early August 2000 at station 1 which coincided with a bloom of *Scenedesmus* spp. The third peak was observed at station 2 and 3 as 132.4 and 146 μg l^{-1}, respectively, in early autumn and occurred in parallel with a cyanobacterial bloom. Similar to phytoplankton biomass, the chl-*a* concentration did not fall significantly by a depth of 5 m. However, a considerable decline (approximately 82%) was observed between a depth of 5 and 10 m.

The pH varied between 7.2 and 7.8 (average 7.4) and significant ($p > 0.05$) differences were not found at the surface waters or throughout the water column among the sites. The highest pH was measured in November 1999, however this continuously decreased until February 2000. This increase and decrease in pH showed strong positive correlation with the phytoplankton fluctuation during period 1 at all sites (Table 1).

The pH values were significantly higher during period 2 than period 1. Between late July and September 2000, pH values never decreased below 10 in the surface waters. However, marked differences were observed below the thermocline. The mean pH value decreased to 7.32 at 20 m. It is important to point out that the mean pH values were slightly lower at station 1, where surface waters were treated repeatedly by $CuSO_4$, compared to the other stations during period 2.

Dissolved Inorganic Nitrogen (DIN), (as NO_2 + NO_3 and NH_4^+) were generally extremely high, whereas total phosphorus (TP) and Soluble Reactive Phosphorus (SRP) concentrations were moderately high at all sites during period 1. The highest value of TN (2.34 mg l^{-1}) occurred during March 2000 at station 3. It ranged between 1.14 and 2.08 mg l^{-1} at station 1, 0.92 and 2.12 mg l^{-1} at station 2, and 1.26 and 2.34 mg l^{-1} at station 3. Mean DIN values as NO_2 + NO_3 and NH_4^+ for period 1 are presented in Table 2. There were only small differences among the stations. The concentration of NO_2 + NO_3 at station 1 and station 2 was about 7% higher than station 3 and the concentrations of NH_4^+ at station 3 was about 7% higher than the other stations (Table 2).

Similar to TN, the highest SRP and TP concentrations were recorded during March 2000 at station 3 as 164 and 226 μg l^{-1}, respectively. The mean TP concentration was 12% higher than the other stations. The mean TN/TP ratios did not change among the stations ($p > 0.05$) and it varied between 13.09 and 14.58 (Table 2). The minimum and maximum TN/TP ratios were measured at station 3 as 8.12 (March 2000) and 23.11 (November 1999), respectively.

The concentration of TN in the surface waters increased approximately 20 and 23% at station 2 and 3, where this part of the lake had been fertilized throughout the year by waters from a nutrient rich stream. TN did not differ at station 1 during the period 2 (Table 2). Nitrite+nitrate and ammonia presented higher values. However, both parameters increased with depth and reached maximum values (1127 $\pm$ 290 and 710 $\pm$ 234 μg l^{-1}, respectively) at a depth of 10 m. A distinctive effect of the $CuSO_4$ treatment was observed in the TP and SRP at station 1 during period 2. Average TP and SRP concentrations were measured as approximately 21% lower than previous sampling periods at this site. In addition, the overall TP concentrations at station 1 were 20 and 24% lower than station 2 and 3, respectively.

Soluble Reactive Silicate (SRSi) concentrations were relatively low (mean 667, 593 and 689 μg l^{-1} at stations 1, 2 and 3, respectively), at the surface during period 1. Vertical distribution of SRSi was homogeneous between November 1999 to March 2000. In October 1999, maximum SRSi values were recorded at a depth of 15 m (3.4 mg l^{-1}) and also near the surface (4.1 mg l^{-1}).

Silicate concentrations presented relatively higher values in the surface waters during period 2 compared to period 1 (Table 2). In addition, vertical distribution of this nutrient differed considerably; whereas the mean concentration was 1152$\pm$321 μg l^{-1} at a depth of 5 m, and reaching 2527 μg l^{-1} at a depth of 10 m.

Total Organic Carbon (TOC) concentrations were relatively high in the surface waters. No significant differences were found among the stations. However, considerable vertical difference was recorded at all sites. Maximum TOC values (5.8 mg l^{-1}) occurred in the surface waters at station 3 during March 2000 and minimum values (2.4 mg l^{-1}) was measured at 10 m at station 1 during February 2000.

There were significant increases in TOC concentrations during period 2 at all sites. The mean concentration was found to be twofold that of the previous period (Table 2). The vertical distribution of TOC concentrations were positively correlated with algal biomass and varied from r = 0.76 to 0.82. It decreased to 5.09±1.57 mg l^{-1} at a depth of 5 m and 3.91±0.93 mg l^{-1} a depth of 10 m.

Phytoplankton seasonal cycles and steady state assemblages

During period 1 and period 2, a total of 78 taxa were identified in the Ömerli reservoir. Of the total phytoplankton species observed Chlorophyta was the dominant taxonomic group (55.13%) followed by Bacillariophyta (23.08%) and Cyanophyta (11.54%). Cryptophyta, Euglenophyta and Pyrrophyta were present but very few species were identified.

The phytoplankton biomass and composition showed great variability during period 1 and 2 (Fig. 1 and Table 2). At the beginning of the investigation (October 1999), an algal bloom was observed which was dominated by *Microcystis aeruginosa* (Albay et al., in press). During periods of *M. aeruginosa* domination, a few other species proliferated (such as some cyanobacteria, *Aphanizomenon* sp. *Gomphosphaeria* sp., *Chroococcus minutus,* few chlorophytes and diatoms). The *M. aeruginosa* grew mainly above a depth of 5 m and accounted for 90.09 – 92.6% of the total biomass in the surface waters. In October 1999, the *M. aeruginosa* accounted for 95.29 – 97.13% of the total phytoplankton biomass at a depth of 1 m and 75.37 – 85.97% at a depth of 5 m. One important point was that 96.8 – 97.2% of the total algal biomass grew above the 5 m during the *M. aeruginosa* dominance.

Although the investigation started from October 1999, this month was not included in period 1. The similarity observed among the stations in terms of the algal biomass was significant ($p > 0.05$; Table 2). The total phytoplankton biomass varied from 1.44 to 14.52 mg l^{-1} at the surface, 1.24 to 7.78 mg l^{--1} at a depth of 1 m, 0.7 to 6.36 mg l^{-1} at a depth of 5 m and 0.32 to 2.32 mg l^{-1} at a depth of 10 m between November 1999 and March 2000. During late autumn and mid winter, the biomass was dominated by Chlorophyta (such as *Pediastrum boryanum, Scenedesmus quadricauda, S. magnus, Staurastrum* spp.). However, during late winter and early spring the biomass at all stations was dominated mainly by diatoms (*Aulacoseira granulata, Fragilaria crotonensis* and *Asterionella formosa*) and *Cryptomonas* spp. The maximum biomass was recorded at the end of period 1 (March 2000). *Cryptomonas* spp. accounted for 96.14 – 98.6% of the total biomass. The dominance of *Cryptomonas* spp. declined with increasing depth.

During period 2, mean phytoplankton biomass at all stations was considerably higher then in period 1 ($p < 0.01$; Table 2). Definite steady state assemblages were observed between late April and early October 2000. For instance, from late April to late May, the steady state assemblage (SSA) was made up by *A. formosa*, and *F. crotonensis* (Fig. 2). During this period z_{eu} was measured as 6.6 m and z_{eu}/z_{mix} was 0.65. *A. formosa*, and *F. crotonensis* accounted for 77.24 – 91.19% of the total biomass. These ranges did not change vertically and biomass contributions of this group were 95.6% at 5 m and 86% at a depth of 10 m. Subsequent declines in these species resulted in chlorococcal chlorophytes, *Coelastrum microporum* (dominant species), *Pediastrum boryanum* and *P. duplex* (codominant species) dominating the community from late May (Fig. 2). These species reached concentrations of 65.1 mg l^{-1}, 74.1 mg l^{-1} and 78.2 mg l^{-1}at station 1, 2 and 3, respectively, and gave rise to an increase in pH (9.1 at the surface).

Marked differences were not observed in SSA until mid June and the total biomass fluctuated between 74.2 and 82.2 mg l^{-1} at station 1, 84.3 and 91.5 mg l^{-1} at station 2 and 89.3 and 101.8 mg l^{-1} at station 3. The dominance of these species did not change vertically, however, their biomass decreased abruptly below 5 m.

Staurastrum cf *gracile, S. planctonicum* and *Cryptomonas* spp. replaced the chlorococcal chlorophytes between late June and mid July. The number of observed species decreased to 11 at the surface and to 4 at depth of 15 m. The second biomass peak of chlorococcalen chlorophyte, *S. quadricauda* (dominant species), *S. planctonicum* and *Nitzschia holsatica* (subdominant species), was observed during the three week period between late July and early August 2000. *S. quadricauda, S. planctonicum* and *Oocystis lacustris* grew extremely well at station 1 after the addition of algicide (Fig. 2) and similar peak densities occurred between mid August 2000 to early September 2000 which gave rise to a marked decline in z_{eu} (2.4 m) and a large increase in pH (10.5) in the surface waters.

Vertical dominance of this algal group did not differ significantly among the stations. The contribution to the phytoplankton biomass of this group was 94.1% at the surface and 86.1% at a depth of 10 m at station 1; 93.2% at the surface and 91.3% at a depth of 10 m at station 2 and 91.9% at the surface and 89.8% at a depth of 10 m.

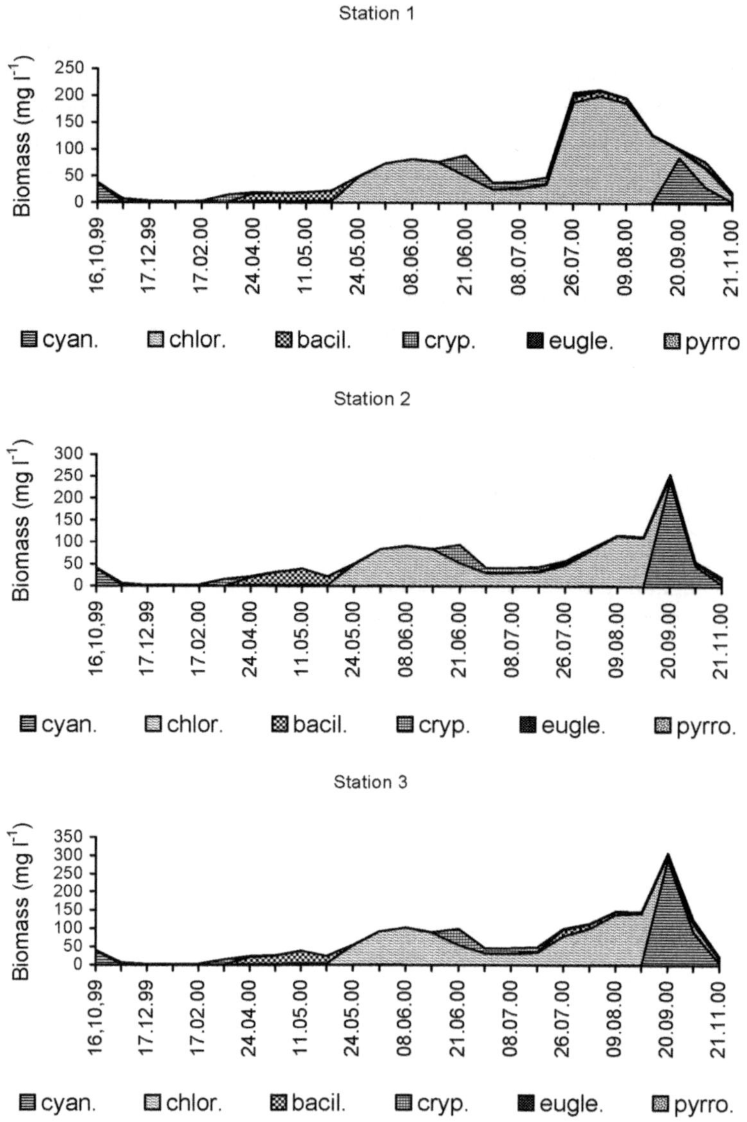

Figure 1. Biomass of phytoplankton groups (fresh weight mg l^{-1}) from October 1999 to November 2000 in surface layer of Ömerli reservoir. Abbreviations: Cyan= Cyanobacteria; Chlor= Chlorophyta; Bacil= Bacillariophyta; Cryp= Cryptophyta; Eug= Euglenophyta, Pyrro= Pyrrophyta.

Between 1 and 19 September 2000 the algicide treatment was stopped. Chlorophytes started to decline in the surface waters and *Microcystis aeruginosa* became pioneer species. The mean biomass of *M. aeruginosa* decreased fifty twofold at a depth of 5 m and almost disappeared at a depth of 10 m. *S. quadricauda* and *Mougeotia* sp. were recorded as dominant and subdominant species.

Community features

Ömerli reservoir presented relatively low numbers of species during period 1 (mean 10, 10 and 11 species at stations 1, 2, and 3) than period 2 (mean 13, 14 and 16 at station 1, 2, and 3). Species richness showed small fluctuation in the surface waters during period 1. However, rather big modifications were observed during period 2 (Fig. 3). Contrary to common belief, high richness values were recorded during the bloom of *M. aeruginosa* (18, 20 and 21 taxa at stations 1, 2 and 3,

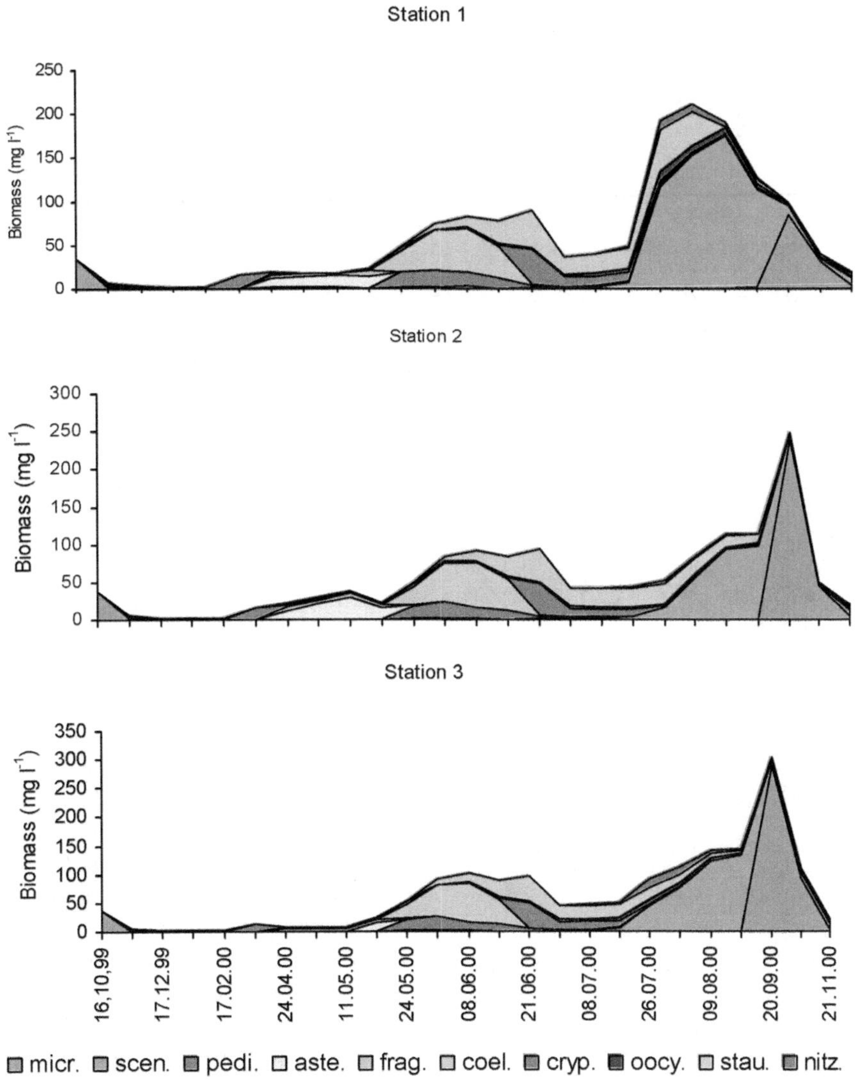

Figure 2. Temporal variations of biomass (freshweight mg l^{-1}) for dominant species which constituted of the steady state assemblages in surface layer of Ömerli reservoir. Abbreviations: micr= *M. aeruginosa*; scen= *Scenedesmus* spp.; aste= *A. formosa*; frag= *F. crotonensis*; coel= *C. microporum*; cryp= *Cryptomonas* spp. oocy= *Oocystis* spp.; stau= *Staurastrum* spp.; nitz= *N. holsatica.*

respectively). Shannon diversity indices for the phytoplankton communities presented higher values during period 1 than period 2 (Fig. 4). The lowest values occurring in late spring, summer and early autumn when the phytoplankton was in equilibrium state, whereas the highest values were detected in late autumn.

Discussion

Our data show that there were substantial differences between periods 1 and 2 in terms of the phytoplankton biomass, number of species and environmental variables. Since water samples were collected at monthly intervals during period 1 for practical reasons, it is not possible to compare the SSA to period 2.

Charpin et al. (1998) revealed that physical factors that lead to seasonal cycles of mixing and stratification may considerably affect the photosynthetic physiology of phytoplankton through their effects on available irradiance in lakes. The extent of these factors varies related to geographical location. Naselli-Flores (2000) argued that a change in the structure of the underwater

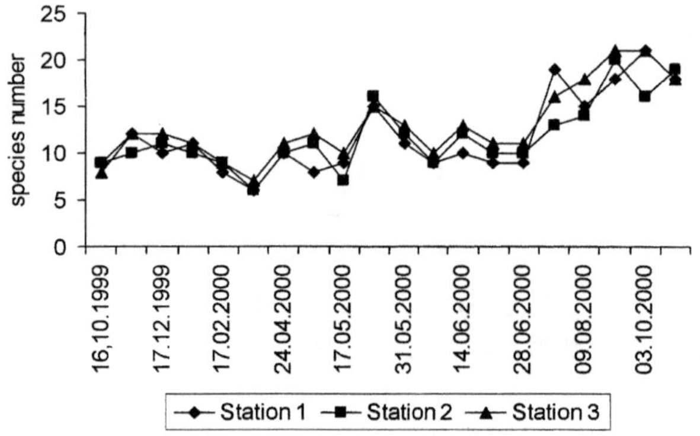

Figure 3. Species number of phytoplankton in Ömerli reservoir to between October 1999 and November 2000.

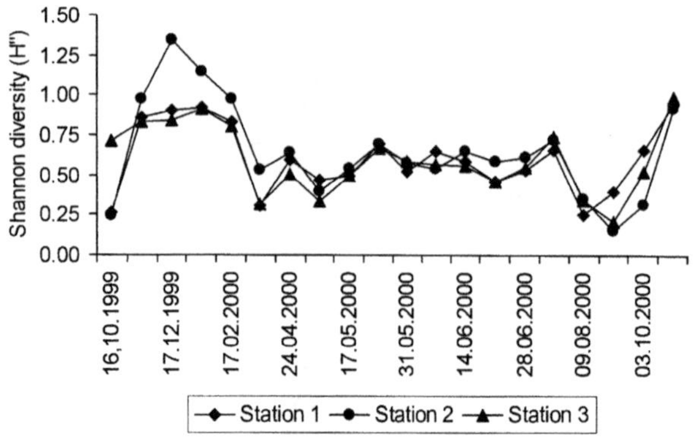

Figure 4. Shannon diversity of phytoplankton in Ömerli reservoir to between October 1999 and November 2000.

light field may lead to a modification in the structure of phytoplankton assemblages. He also pointed out that water-level fluctuations and summer drawdown interfered with the periodicity and stability of stratifications in Sicilian reservoirs (Naselli-Flores, 2003). If approximately 826×10^3 m^3 day^{-1} water drawdown for daily consumption is considered, this may be very important for the Ömerli reservoir.

In addition to light, mixing and water-level fluctuation including water drawdown, the concentration of nutrients (mainly N and P) are other factors which govern the phytoplankton growth and distribution (Naselli-Flores & Barone, 2003). Overall concentrations of nutrients in reservoirs never limit the phytoplankton growth. On the contrary, TP, SRP, TN, NO_2+NO_3 and NH_4^+ concentrations were always above the required levels (Table 2) for phytoplankton growth saturation (Reynolds, 1997). However, vertical distribution of some nutrients such as NO_2+NO_3 and NH_4 showed great variability between the layers when algal bloom was observed. For instance when *S. quadricauda* peak occurred in early August 2000, NH_4 concentration was 754 $\mu g\ l^{-1}$ at a depth of 10 m, whereas it fell to 18 $\mu g\ l^{-1}$ in the surface. Albay & Pokorny, (1999) recorded that at high pH, NH_4^+ converts into NH_3 (gas) and is released into the air which explains the low concentration of $NH_4^+$the surface.

Another finding was that there was a marked decline (sevenfold) in euphotic depth (z_{eu}) compared to Kaplan's (1989) results. For instance, he measured z_{eu} depth as 10.8 m in September 1987, whereas we found only a depth of 1.52 m in September 2000 which coincided with the *M. aeruginosa* bloom.

Dominant phytoplankton association during period 1

There are numerous studies related to typical species associations to environmental conditions (Komárková & Hejzlar, 1996; Temponeras et al., 2000; Dokulil & Teubner, 2000). Reynolds et al. (2002) separated 31 functional groups (associations) which included major abundant species of temperate lakes. We were able to recognise the phytoplankton species from Ömerli reservoir to 3 of the 31 assemblages during period 1 and 8 of the 31 assemblages during period 2. The codon **J**, which was the first dominant group recorded in phytoplankton in November 2000, assemblages formed by chlorococcal chlorophytes *Pediastrum boryanum, Scenedesmus quadricauda* and *Coelastrum microporum* dominate in shallow lakes and ponds (Reynolds et al., 2002). Although this group was identified as a dominant assemblage, it never reached high quantities. The codon **Y** assemblage was formed by *Cryptomonas* spp. which are known for their rapid phosphorus uptake rates and relatively fast growth, are common in moderately enriched systems (Sommer, 1981; Huszar et al., 2000). The *Cryptomonas* spp. peaks were observed in March 2000 (before the establishment of the thermocline) and coincided with the lowest TN/TP ratios at all stations. Cryptophyta dominance was in accordance with the investigation of Temponeras et al. (2000) in a shallow lake.

The codon **B**, assemblage dominated by the centric diatom *Aulacoseira granulata*, showed a higher biomass in the Ömerli reservoir during period 1. This species prefers eutrophic and mesotrophic conditions (Moustaka-Gouni & Tsekos, 1989; Negro et al., 2000).

Steady state assemblages during period 2

The ecological conditions changed considerably (Table 2) and it is possible to conclude that these parameters lead to the steady state assemblages during period 2. For instance, temperature, increased approximately two fold (mean 21.8 °C) and showed strong positive correlation with total biomass (Table 1). pH never decreased below 8.2 (mean 9.74, 9.82 and 9.9 at stations 1, 2 and 3) and TN/TP ratios increased 15 – 40% in the surface waters. Similarly the concentrations of TOC increased 25 – 31% during period 2 and light intensity was considerably higher than in period 1. Finally, stratification was an important factor which governed the vertical distribution of algal biomass. According to the outlined course of the classification of the freshwater phytoplankton, proposed by Reynolds (1997), Padisák & Reynolds (1998) and Reynolds et al. (2002) eight types were recognisable during period 2. The **C** assemblage formed by *Asterionella formosa* (dominant species) and **P** assemblage formed by *Fragilaria crotonensis* (codominant species), which bloomed later, successfully colonized within approximately 4 weeks after the establishment of thethermocline. During this period, TN/TP ratio varied between 10.1 and 15.22; silica varied between 280 and 1125 $\mu g\ l^{-1}$ and total algal biomass varied between 17.6 and 22.7 mg l^{-1}. *A. formosa* and *F. crotonensis* accounted for 77.2 and 91.2% of the total biomass. *A. formosa* has been shown to be resistance to grazers, with relatively low sinking losses (Leitão & Leglize, 2000) and has moderate nutrient uptake (Tilman, 1977) whereas *F. crotonensis* is defined as a common species in large mesotrophic lakes in low latitudes (Huszar et al., 2003). With consideration to the water quality parameters of Ömerli reservoir indicated a eutrophic character in many respects, and numerous studies have shown that these species can reach high biomass in oligo-mesotrophic to eutrophic lakes (Leitão & Leglize, 2000; Salmaso, 2002; Reynolds et al., 2002). It is necessary to emphasize that stratification enabled *A. formosa* to increase in biomass above the thermocline where mean biomass was 21.6 mg l^{-1}, whereas it decreased to 4.5 mg l^{-1} under the thermocline (a depth of 10 – 15 m). As total algal biomass varied between 17.6 – 22.7 mg l^{-1} and these two species accounted for 77.2 to 91.2% of the total biomass and water chemistry parameters did not change considerably, we may say that steady state assemblages occurred, starting from late April 2000 in Ömerli reservoir.

Although the TN/TP ratio changed slightly between late May and mid June 2000, marked differences were observed in steady state assemblages. Standing biomass was formed by **J** assemblages chlorococcal chlorophytes, mainly *Coelastrum microporum* and *Pediastrum boryanum* and **P** assemblage *Staurastrum* cf *gracile* and *S. planktonicum.* During this phase temperature increased approximately 5 °C and algal biomass increased fourfold compared to the previous phase (Fig. 1). This finding might be explained by an increase in temperature and light intensity. Similar to *A. formosa* and *F. crotonensis, C. microporum* and later *Staurasrum cf gracile, S. planktonicum* and *Crytomonas* spp. (**Y** assemblages) reached high biomass above the thermocline. Negro et al. (2000) concluded that the increase of *Staurasrum* was related to organic matter content in Valparaiso reservoir, Spain. It is also worth mentioning that desmid contributions to the biomass is high during warm and sunny season.

S. quadricauda, which became the dominant species between late July and late August 2000, was more resistant to stratification and were able to reach high biomass in large TN/TP ratios. Biomass of this species at a depth of 15 m was only fourfold less than in surface waters. In addition to *S. quadricauda* the **D** and **P** assemblage, formed by mainly *Nitzschia* spp., and *S. planctonicum*, respectively, and **F** assemblages, formed by mainly *Oocystis* species (*O. lacustris* and *O. borgei*) became subdominant groups consecutively in steady state assemblages during this period. **D** assemblage formed by *Nitschia* spp. occur usually in shallow, nutrient enriched, well–ventilated lakes and fish ponds (Şen & Çağlar, 1994; Köhler & Hoeg, 2000) and **F** assemblage grows best in clear water and has strong representation among mesotrophic lakes. However, this group is also sensitive to nutrient enrichment (Reynolds et al., 2002). Mean water temperatures were 18.8 °C and 27.6 °C in July and 19.8 °C and 28 °C in August at night and in the daytime, respectively.

M assemblage formed by *Microcystis aeruginosa* became the dominant group between mid September and early October 2000 in Ömerli reservoir. It is necessary to emphasize that this season is a quite relevant time for *Microcystis* bloom in Turkey (Cirik & Cirik, 1989; Aykulu et al., 1999). DIN decreased considerably; for example, $NO_2 + NO_3$ decreased to 4 $\mu g\ l^{-1}$ at the surface and 3 $\mu g\ l^{-1}$ at a depth of 1 m. Vertical distribution of *M. aeruginosa* showed an inverse relationship with $NO_2 + NO_3$ ($r = -0.77$) during this phase. The vertical position of *M. aeroginosa* can be adjusted by buoyancy alteration and highest biomass values were detected above a depth of 1 m (94.0 and 99.0% at stations 1, 2 and 3, respectively). Van Lieve & Walsby (1982) stated that *M. aeruginosa* can effectively increase its light–dose rate and trap more CO_2 whereas Shapiro (1997) revealed that *M. aeruginosa* has better kinetics for CO_2 uptake than most non-cyanophytes.

The *Microcystis* bloom gave rise to marked differences in environmental variables under 1 m depth in Ömerli reservoir. NO_2+NO_3 reached 535 μg l^{-1} at a depth of 10 m, whereas TOC, TN and TP decreased approximately two fold at the same depth. pH increased to 10.7 at the surface and z_{eu} decreased to 1.52 m. Under these conditions, low values of subsurface species numbers are expected. Surprisingly, species number did not change vertically. Furthermore, a filamentous chlorophyte, *Mougeotia* sp. reached a maximum biomass of 18.7 mg l^{-1} and 6.8 mg l^{-1} at a depth of 5 and 10 m at station 2, respectively.

As we have pointed out in the 'Materials and methods' section, copper sulphate was used at one site to control cyanobacterial bloom. Our data show that whereas copper sulphate had remarkable effects on blue-greens at station 1 compared to the reference stations, it was relatively ineffective in controlling other algal groups. Furthermore, we observed marked increase in copper resistant species of chlorophytes (Fig. 2). It is widely accepted that resistance to metal toxicity is a phenomenon that occurs among the Chlorophyceae members (Takamura et al., 1993; Beyruth, 2000). The common belief is that treatment with lime or copper sulphate additions generally result in a decrease in biomass. Furthermore, recovery of a lake after the treatment is a slow process and re-colonization usually takes some years (Hörnström, 1993). However, there are contradictory results that shortly after the copper sulphate or lime additions phytoplankton communities experienced remarkable changes e.g. biomass increase, re-appearance of many new species and a significant increase in biological diversity (Morabito & Curradi, 1997; Beyruth, 2000).

According to the Sommer et al. (1993) definition, we should say that steady state assemblages were occurred between the mid spring and early autumn 2000 in Ömerli reservoir. However, it is worthwhile the point out that steady state assemblages were constituted by eight groups. According to the outlined course of classification of the freshwater phytoplankton Reynolds et al. (2002), 5 of the 8 assemblages characterized of nutrient enriched lakes. It clearly shows that water quality is prone to deterioration.

Acknowledgements

We remain extremely grateful to Dr Luigi Naselli-Flores for his kind support and invitation to the 13th IAP-meeting. We thank H. Tufekci for his kind help with some chemical analysis of waters and that of D. Moore for improving English. We also thank Istanbul Water Authority (ISKI) allowing us to use data from copper sulphate treatment. This work was supported by the Research Fund of the University of Istanbul. Project number: 993/090597 and UDP-34/18072002.

References

Albay, M., R. Akçaalan, G. Aykulu, H. Tufekci, K. A. Beattie & G. A. Codd, 2003. Occurrence of toxic cyanobacteria before and after copper sulphate treatment in a water reservoir, Istanbul, Turkey. Archiv für Hydrobiol./Algological Studies (in press)

Albay, M. & J. Pokorny, 1999. Evolution of fishponds restoration project in South Bohemia (Czech Republic). J. aquat. Prod. Special Issue: 251–261.

Aykulu, G., K. Doğan & S. Hasirci, 1999. A study on the phytoplankton communities of lakes Taşkisi and Poyrazlar (Adapazam, Turkey). J. Aquat. Prod. Special Issue: 157–184.

Beyruth, Z., 2000. Periodic disturbance, trophic gradient and phytoplankton characteristics related to cyanobacterial growth in Guarapiranga Reservoir, Sao Paolo State, Brazil. Hydrobiologia 424: 51–65.

Charpin, M. F., N. Maurin, C. Amblard & J. Devaux, 1998. Seasonal variations of phytoplankton photosynthate partitioning in two lakes of different trophic level. J. Plankton Res. 20: 901–921.

Cirik, S. & Ş Cirik, 1989. Planktonic algae of Gölcük. J. Aquat. Prod. 3: 131–150.

Crompton, T. R., 1989. Analysis of Seawater. Hartnolls ltd. Bodmin, Cornwall. 423 pp.

Dokulil, M. T. & K. Teubner, 2000. Cyanobacterial dominance in lakes. Hydrobiologia 438: 1–12.

Eruz, E., 1995. Ecological problems of water reservoirs of Istanbul. The symposium on water reservoirs of Istanbul. Proceedings book. ISKI publ.: 145–149.

Hörnström, E., C. Ekström, E. Fröberg & J. Ek, 1993. Plankton and chemical-physical development in six Swedish west coast lakes under acidic and limed conditions. Can. J. Fish. aquat. Sci. 50: 688–702.

Huszar, V., C. Kruk & N. Caraco, 2003. Steady-state assemblages of phytoplankton in four temperate lakes (NE U.S.A.). Hydrobiologia 502 (Dev. Hydrobiol. 172): 97–109.

Huszar, V. L. M., L. H. S. Silva, M. Marinho & C. L. Sant'Anna, 2000. Cyanoprokaryote assemblages in eight productive tropical Brazilian waters. Hydrobiologia 424: 67–77.

Kaplan, H., 1989. Seasonal distribution of zooplankton groups in Ömerli dam lake. M. Sc. thesis. Istanbul University, Institute of Science.

Komárková, J. & J. Hejzlar, 1996. Summer maxima of phytoplankton in the Rimov reservoir in relation to hydrologic parameters and phosphorus loading. Archiv für Hydrobiol. 136: 217—236.

Köhler, J. & S. Hoeg, 2000. Phytoplankton selection in a river lake system during two decades of changing nutrient supply. Hydrobiologia 244: 13–24.

Leitao, M. & L. Leglize, 2000. Long term variations of epilimnetic phytoplankton in an artificial reservoir during a 10- year survey. Hydrobiologia 424: 39–49.

Marinho, M. M. & V. L. M. Huszar, 2002. Nutrient availability and physical conditions as controlling factors of phytoplankton composition and biomass in a tropical reservoir (Southeastern Brazil). Archiv für Hydrobiol. 153: 443–468.

Morabito, G. & M. Curradi, 1997. Phytoplankton community structure of a deep supalpine Italian lake (Lake Orta, N. Italy) four years after the recovery from acidification by liming. Int. Rev. ges. Hydrobiol. 82: 487–506.

Morabito, G., 2001. Six years' (1992 – 1997) evolution of phytoplankton communities after recovery by liming in lake Orta, northern Italy. Lakes & Reservoirs: Res. Manage. 6: 305–312.

Naselli-Flores, 2003. Man-made lakes in Mediterranean semi-arid climate: The strange case of Dr Deep Lake and Mr Shallow Lake. Hydrobiologia 506–509 (in press).

Naselli-Flores, L., 2000. Phytoplankton assemblages in twenty one Sicilian reservoirs; relationship between species composition and environmental factors. Hydrobiologia 424: 1–11.

Naselli-Flores, L. & R. Barone, 2000. Phytoplankton dynamics and structure: a comparative analysis in natural and man-made water bodies of different trophic state. Hydrobiologia 438: 65–74.

Naselli-Flores, L. & R. Barone, 2003. Steady-state assemblages in a Mediterranean hypertrophic reservoir. The role of *Microcystis* ecomorphological variability in maintaining an apparent equilibrium. Hydrobiologia 502 (Dev. Hydrobiol. 172): 133–143.

Negro, A. I., C. De Hoyos & J. C. Vega, 2000. Phytoplankton structure and dynamics in lake Sanabria and Valparaiso reservoir (NW Spain). Hydrobiologia 424: 25–37.

Nush, E. A., 1980. Comparison of different methods for chlorophyll and phaeopigment determination. Archiv für Hydrobiol./Ergebnisse der Limnol. 14: 14–36.

Padisák, J. & C. S. Reynolds, 1998. Selection of phytoplankton association in lake Balaton, Hungary, in response to eutrophication and restoration measures, with special references to the cyanoprokaryotes. Hyrobiologia 384: 41–53.

Reynolds, C. S., 1992. Dynamics, selection and composition of phytoplankton in relation to vertical structure in lakes. Archiv für Hydrobiol./Ergebnisse der Limnol. 35: 13–31.

Reynolds, C. S., 1997. Vegetation process in the pelagic: A Model for Ecosystem Theory. –Ecol. Institute, Oldendorf / Luhe.

Reynolds, C. S., V. L. M. Huszar, C. Kruk, L. Naseli-Flores & S. Melo, 2002. Towards a functional classification of the freshwater phytoplankton. J. Plankton Res. 24: 417–428.

Rott, H. E., 1981. Some results from phytoplankton counting intercalibration. Schweizerische Zeitschrift für Hydrol. 43: 35–62.

Salmaso, N., 2002. Ecological patterns of phytoplankton assemblages in lake Garda: Seasonal spatial and historical features. J. Limnol. 61: 95–115.

Shannon, C. E. & W. Weaver, 1949. The mathematical Theory of Communication. University of Illinois Press, Urbana. 117 pp.

Shapiro, J., 1997. the role of carbodioxide in the initiation and maintenance of blue-green dominance in lakes. Freshwat. Biol. 37: 307–323.

Sommer, U., 1981. The role of r- and K selection in the succession of phytoplankton in lake Constance. Acta Oecologica 2: 327–342.

Sommer, U., J. Padisák, C.S. Reynolds & P. Juhasz-Nagy, 1993. Hutchinson's heritage: the divesity-disturbance relationship in phytoplankton. Hydrobiologia 249: 1-7.

Strickland, J. D. H. & T. R. Parsons, 1972. A practical handbook of seawater analysis, 2nd Edition. Bull. Fish. Res. Bd Can. 167.

Şen, B.& M. Çağlar, 1994. A study on the diatoms (Bacillariophyta) of fish ponds in Cip Fishery Research Center. J. aquat. Prod. 1–2: 67–81.

Takamura, N., F. Kasai & M. M. Watanabe, 1989. Effects of Cu, Cd and Zn on photosynthesis of freshwater benthic algae. J. Appl. Phycol. 1: 39–52.

Technicon Industrial Methods, 1977a. Nitrate and nitrite in water and wastewater. No. 158–71, W/A.

Technicon Industrial Methods, 1977b. Phosphate and silicate analysis in water and seawater. No. 253–280 E. Application note, p. 5 – 7, U.K.

Temponeras, M., J. Kristiansen & M.Moustaka-Gouni, 2000. Seasonal variation in phytoplankton composition and physical-chemical features of the shallow Lake Doirani, Macedonia, Greece. Hydrobiologia 424: 109–122.

Tilman, D., 1977. Resource competition between planktonic algae: an experimental and theoretical approach. Ecology 58: 338–348.

Utermöhl, H., 1958. Zur Vervollkommung der quantitativen Phytoplankton – methodik. Mitt. int. Ver. theor. angewan. Limnol. 9: 1–38.

Van Liere, L. & A. E. Walsby, 1982. Interaction of cyanobacteria with light. In Carr N. G. & B. A. Whitton (eds), The Biology of Cyanobacteria. Blackwell Scientific, Oxford: 9–45.

Wetzel, R. G., 1990. Reservoir Ecosystems: Conclusions and Speculations. In Thornton, K. W., B. L. Kimmel & F. E. Payne (eds), Reservoir Limnology: EcologicalPerspectives. John Wiley & Sons, Inc. New York: 227–238.

Hydrobiologia **502:** 97–109, 2003.
L. Naselli-Flores, J. Padisák, M. T. Dokulil (eds),
Phytoplankton and Equilibrium Concept: The Ecology of Steady-State Assemblages.

Steady-state assemblages of phytoplankton in four temperate lakes (NE U.S.A.)

Vera Huszar[1], Carla Kruk[2] & Nina Caraco[3]
[1]*Lab. de Ficologia, Museu Nacional/UFRJ, São Cristóvão, Rio de Janeiro (RJ) 20940-040, Brasil.*
E-mail: vhuszar@gbl.com.br
[2]*Limnology Section, Universidad de La República, 11400 Montevideo, Uruguay*
[3]*Institute of Ecosystem Studies, Box AB, 12 545, Milbrook, NY, U.S.A.*

Key words: phytoplankton assemblages, steady-states, driving forces, stratified lakes

Abstract

For four temperate lakes (Northeast U.S.A.) we identify periods of persistent phytoplankton assemblages and investigate the ecological conditions that correlate to these persistent assemblages. Periods of persistent assemblages, here considered as steady-state phases, were defined according to equilibrium criteria (two or three coexisting species, contributing to 80% of the standing biomass, for at least 2 weeks) defined by Sommer et al. (1993, Hydrobiologia 249: 1–7). For all four lakes, samples were taken weekly during the ice-free season and phytoplankton attributes (biomass, assemblages, diversity, species richness, change rates) and abiotic variables (temperature, I^* – as light mean in the mixing zone – z_{mix}, and nutrients) were analysed. Chodikee (CH), an eutrophic and rapidly flushed lake, did not show any persistent phase. The remaining three lakes showed single steady-state phases that occurred at varying times during the ice-free season. Steady-state phases occurred during early stratification in late spring in the stably stratified oligotrophic Mohonk Lake (MO), in the late summer stratification in the meso-eutrophic Stissing Lake (ST), and during spring mixing in Wononscopomuc Lake (WO). MO showed a 3-week period with dominance of **F** assemblage (*Botryococcus braunii, Willea wilhelmii* and *Eutetramorus planctonicus*), characteristic for clear epilimnia, tolerant to low nutrient and sensitive to high turbidity. For three weeks, ST had a stable assemblage with dominance of **Lo** (*Woronichinia* sp.), common assemblage in summer epilimnion of mesotrophic lakes and sensitive to prolonged or deep mixing; and **P**, assemblage able to live in eutrophic epilimnia with mild light and sensitive to stratification and silica depletion. In contrast, the mesotrophic Wononscopomuc Lake (WO) showed persistent assemblages during a 4-week period of spring circulation, when a dinoflagellate (**Lo**) was co-dominant with *Nitzschia acicularis* (**C**). The latter species is characteristic for mesotrophic lakes, tolerant to low light and sensitive to stratification and silica depletion. Both **Lo** and **P** assemblages, among seven others, had before been quoted, in literature, as dominant in maturing stages. We could not find consistent statistical differences between the periods classified as steady-state and non-steady-state. However, the data demonstrated that prolonged period of both mixing and stratification can maintain dominant assemblages. Although, historically sensed as opposite mechanisms, both mixing and stratification, if persistent, were observed maintaining dominant assemblages because both scenarios are characterized by environmental constancy.

Introduction

Equilibrium and non-equilibrium hypothesis have often been used to explain community ecology. In the non-equilibrium hypothesis, forces driving the equilibrium are weak and the competitive interactions are minimal. Under such circumstances, environmental disturbances occur frequently enough to disrupt the course of competitive exclusion (Harris, 1986). Higher diversity is reached when environmental and internal disturbances can prevent the establishment of equilibrium conditions (Sommer et al., 1993). In non-equilibrium, each dominant species occupies a different niche which results from and reduces direct competition. Then, species composition is more or less stable according to environmental predictability (Whittaker, 1975). In this sense, both temporal and spatial niche segregation are essential to maintain communities in equilibrium. In aquatic environments, this segregation is uncommon (Harris, 1986) and can be reached only in absence of disturbance, when diversity is reduced to minimal levels by competitive exclusion.

Selection of dominant phytoplankton species in lakes depends upon a complex and mainly unpredictable combination of factors. The physical structure of the system (hydrology, temperature and light), the availability of nutrients (phosphorus, nitrogen, silicate and carbon) and the zooplankton community are the most important phytoplankton conditioning factors (Reynolds, 1980). Phytoplankton grow very fast and most species are cosmopolitan implying that colonisation of new habitats is hardly a limiting factor (Reynolds, 1997). We can assume that everything can grow everywhere and filter the spectrum of possible assemblages using information about their environmental tolerances and sensitivities. The longer certain conditions persist, the poorer is the survival of the sensitive groups and the better the survival of the tolerant ones. In less severe environments, more species will potentially operate successfully, but as their growth depletes nutrients and reduces light availability, conditions become more restrictive and the identity of survivors becomes more predictable. After constant environmental conditions over 12–16 generations, the phytoplankton community reaches a climax or climactic equilibrium (Reynolds, 1993). When disturbances conditions prevail, plagioclimatic assemblages can develop (Reynolds, 1988). In these situations, we expect to find dominant species or particular species assemblages with different levels of stability and diversity according to characteristics of the system. In this sense, Sommer et al. (1993) ensure that a equilibrium phase development is attained when two or three species, contributing to more than 80% of the standing biomass, coexist with no changes for, at least, 2 weeks. Considering that the phytoplankton community is usually far from thermodynamic equilibrium and most pelagic systems are neither well-organised nor successionally well advanced (Reynolds, 1997), hereinafter we will use the word steady-state since it can be held by nutrients, or light, or flushing, or stratification, or spatial transport, or grazers (Rojo & Álvarez-Cobelas, 2003), but always it is dominated by few species for which the conditions are the best.

A number of models have been developed to understand the regulatory influence of environmental factors acting upon phytoplankton assemblages (e.g. Moss et al., 1996). Most of the predictive models of spatial and seasonal variation treat phytoplankton either as a single entity or by major taxonomic divisions. Far from being a uniform group, however, freshwater phytoplankton is composed of organisms drawn from 11 or so algal phyla, with highly diverse characteristics, which can be grouped in assemblages, based on the physiological, morphological and ecological attributes of the species that potentially and alternatively may dominate or co-dominate the system.

Opposing the traditional view, which concentrates on taxonomic dominance at division level, the morphological-functional framework (Reynolds, 1997; Reynolds et al., 2002) has performed better than the approach (Huszar & Caraco, 1998; Fabbro & Duivenvorden, 2000; Kruk et al., 2002) in simulating compositional changes. The last version of the scheme (Reynolds et al., 2002) outlined 31 assemblages (labelled as an alpha-numeric code) with the basic pattern of their distinctive ecologies. According to Reynolds (1997) dominant, near monospecific populations of some assemblages appear strong candidates to qualify and, under specific cases, to achieve the maturing stage *sensu* Odum' rules (Odum, 1969). We hypothesized that, provided steady-state phases in a lake, phytoplankton will present a higher and less variable biomass, lower diversity and change rates, and will be represented by particular assemblages conditioned by the environmental attributes.

In this paper, we apply data from the phytoplankton community of four temperate lakes with different mixing regimes and trophic states, located in northeast U.S.A., in order to identify periods of low variability in biomass and composition of the assemblages (steady-states) and to understand the ecological conditions which lead, maintain (tolerances) and disrupt (sensitivities) such persistence.

Materials and methods

Study sites

The four lakes (Chodikee, Stissing, Wonoskopomuc and Mohonk) are located in northern U.S.A. and have different residence times, mixing patterns and trophic states (Table 1). A detailed discussion about the relationship between phytoplankton composition and

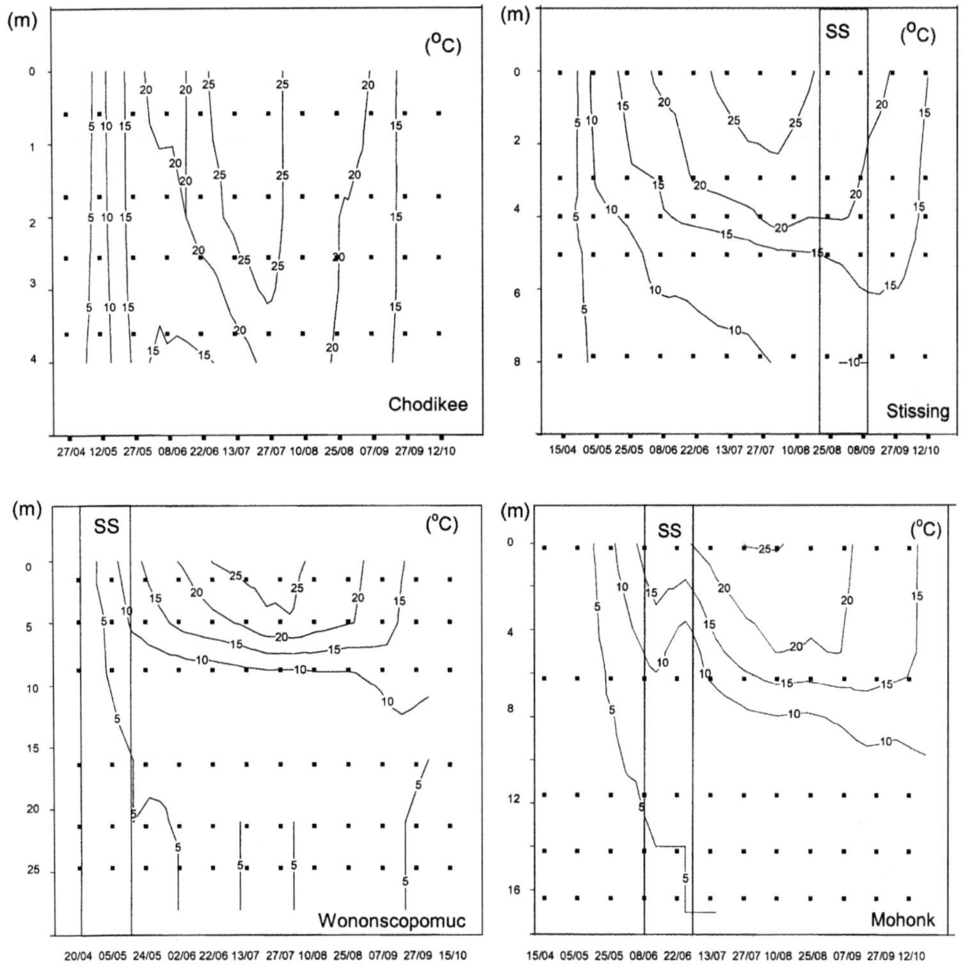

Figure 1. Depth-time diagrams of water temperature (°C) recorded for the four lakes. SS=steady-states.

Table 1. Morphometric data for the studied lakes. Loc = localization; z_{max} = maximum depth; Ret.time = retention time

Lakes	Loc.	Area (km^2)	z_{max} (m)	Ret.time (years)	Mixing	Trophic State[1]
Chodikee	NY	0.24	6	0.3	Discontinuous Cold Polymictic	Eutrophic
Stissing	NY	0.28	8	0.7	Dimictic	Meso- eutrophic
Wononscopomuc	CT	1.43	32	3	Dimictic	Mesotrophic
Mohonk	NY	0.07	19	3	Dimictic	Oligotrophic

[1]Vollenweider & Kerekes (1982); Nürnberg (1996).

physical and chemical variables of these same lakes can be found in Huszar & Caraco (1998).

Field sampling

Samples were taken weekly during the ice-free season (April–October). Limnological samples were taken from the surface (0.5 m) at a fixed point in the deepest area of each system, with exception of temperature which was measured biweekly in profile (four or six depths). Phytoplankton and nutrients were taken in polyethylene bottles; dissolved inorganic carbon (DIC) from gas-tight, 60 ml containers. pH was measured in situ (Fisher accumet 1001-meter) and temperature with YSI temperature oxygen meter. Water transparency by Secchi depth and conductivity with YSI 3000 T-L-C meter.

Before samples were analysed or preserved, they were kept in the dark on ice. Samples for dissolved nutrients were filtered within 8 h using Whatman GF/F filters. Nutrients and DIC samples were preserved with clean H_2SO_4 to pH 2.0 and 1.0, respectively (Caraco et al., 1993). Phytoplankton was preserved in Lugol solution.

Sample analysis

Nutrient concentrations were estimated by colorimetric analysis using a Technicon or ALPKEM autoanalyser (APHA, 1992). DIC was measured on a Shimadzu ASI-5050 analyser. Alkalinity was calculated from DIC, pH, temperature and conductivity (Cole et al., 1994).

Phytoplankton was concentrated prior to counting from 250 ml to 20 ml. Concentration was accomplished by settling and using peristaltic pump to remove overlying water. Following concentration, phytoplankton, except picoplankton, were enumerated in random fields (Uhelinger, 1964), using the settling technique (Utermöhl, 1958). The units (cells, colonies and filaments) were enumerated, at least to 100 specimens of the most frequent species (counting error <5%, Lund et al., 1958). Specific biomass was estimated from the product of the population density and mean unit volume of each species (Edler, 1979), assuming a specific gravity of phytoplankton cells of 1 g cm^{-3}. In general, average cell size was based on measurements of at least 30 organisms.

Data analysis

The mixing zone (z_{mix}) was identified using temperature profiles, and the euphotic zone (z_{euf}) calculated as 2.7 times the Secchi depth (Cole, 1994). CO_2 is a calculated value based on DIC concentrations. A measure of light availability in the mixing zone was calculated according to Reynolds (1997). The seven major taxonomic groups considered were: Cyanobacteria, Cryptophyta, Dinophyta (dinoflagellates) Bacillariophyta (diatoms), Chrysophyta, Euglenophyta (euglenoids) and Chlorophyta (chlorophytes). Species richness was measured as the number of taxa per sample; diversity according to Shannon and Wiener Index (Shannon & Weaver, 1949), based on biomass data and using log 2; and evenness according to Pielou (1966). Change rates were evaluated following the Summed Differences Index, proposed by Lewis (1978) based on Jassby & Goldman (1974). Phytoplankton assemblages were established according to Reynolds (1997) and Reynolds et al. (2002), as groups of species (>5% of relative abundance) with similar morphological and ecological features.

Periods of persistent assemblages, considered as steady-state phases, were defined according to Sommer et al. (1993), where a maximum of three species dominate (70–80%) the community for, at least, three weeks without significant change in total biomass. Non-parametric Kruskall-Wallis H-tests (H statistic) were used to test for differences among different conditions in the phytoplankton community according to total biomass and differences among the steady-state indicators and assemblages between steady and non-steady-state phases. Significant differences were considered with p values lower than 0.05 and reported with the H statistic and p number values. The results about lake chemical, physical regimes and phytoplankton seasonal cycles are reported for each lake and by periods, which were identified based on the variation in phytoplankton biomass and species composition.

Results

Chodikee Lake (CH)

This was stratified only in June and July and was intermittently mixed during the rest of the ice-free season (Fig. 1). Despite the fact that it was the most productive (1.1 μM annual mean of total-P) among the

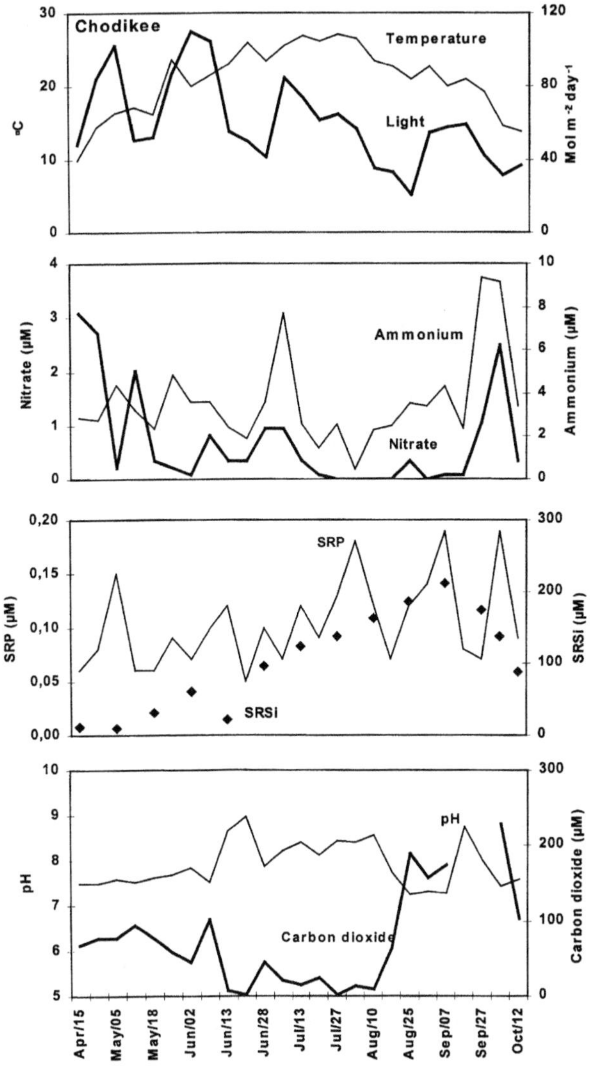

Figure 2. Seasonal variation of environmental variables in Chodikee Lake (SRP = soluble reactive phosphorus; SRSi = soluble reactive silicon; SS = steady-states).

four lakes, this slightly eutrophic system was relatively poor in dissolved inorganic N and P (annual mean SRP = 0,1 μM and DIN 4.5 μM); SRSi occurred in high levels (annual mean = 104 μM), with increasing concentrations from spring to late summer (Fig. 2).

According to the established criteria, the seasonal cycle of phytoplankton in Chodikee lake was divided to five periods (Fig. 3). Total biomass was significantly different among these periods (H = 19,35; $p < 0.001$). Period **I** (April 15–May 27) showed low biomass and co-dominance of N-fixing cyanobacteria, $\mathbf{H}_1$ and $\mathbf{H}_2$ assemblage (*Anabaena spiroides* Klebahn and *A. solitaria* Klebahn, respectively) and **C** and **P** diatom assemblages [*Aulacoseira ambigua* (Grunow) Simonsen and *Fragilaria crotonensis* Kitton, respectively] (Table 2). During period **II** (June 08–28), high biomass and dominance of $\mathbf{H}_1$ assemblage (*Aphanizomenon gracile* Lemmermann) were registered. Period **III** (July 6–27), with low biomass and a diversified community, showed similar contribution of *A. gracile* ($\mathbf{H}_1$) and *Woronochinia* sp. (**Lo**), *Pandorina morum* O. Müller (**G**) and *Strombomonas* sp. ($\mathbf{W}_1$). Period **IV** (August 2–30) had the highest biomass and dominance of cyanobacteria, mainly *Woronichinia* sp. (**Lo**) and *A. gracile* ($\mathbf{H}_1$). Finally, during period **V** (September 7–October 12), low biomass and co-dominance of *A. gracile* ($\mathbf{H}_1$) and *A. ambigua* (C) were observed.

According to the selected criteria we did not identify any steady-state in this rapidly flushed shallow lake.

Stissing Lake (ST)

This was stratified from May to September (Fig. 1). Light in the mixing zone was highest at the beginning of the stratification. This meso-eutrophic lake (0.6 μM annual mean of total-P) showed low annual mean and similar changes of the DIN and SRP concentrations (14.1 and 0.07 μM, respectively), with peak nutrient concentration in the spring, decreasing from April to June (Fig. 4). As the other three lakes, Stissing had relatively high levels of SRSi (75 μM).

Phytoplankton biomass (H = 10.45; $p < 0.01$) and composition were very variable in ST and three periods were identified (Fig. 3). Period **I** (April 20–June 02) had a relatively high biomass, which decreased through the end of the period. Diatoms of **C** and **P** assemblages (*Asterionella* sp. and *F. crotonensis*, respectively), a non-identified Centrales of **A** and *Cryptomonas curvata* Ehrenberg emmend. Penard and *C. marsonii* Skuja of **Y** assemblages were dominant during this phase (Table 2). Period **II** (June 08–August 10) was marked by very low biomass shared among $\mathbf{H}_2$ (*A. solitaria*), **Lm** [*Ceratium hirundinella* (O.F. Müeller) Dujardin], **F** (*Botryococcus braunii* Kützing) and the same **A** and **P** assemblages of period **I.** During period **III** (August 17–October 12), biomass increased again and *Woronichinia sp.* (**Lo**) and *F. crotonensis* (**P**) were co-dominant.

During part of period **III,** in the summer and according to the established criteria, we identified a phase of three weeks classified as steady-state. During this time (August 17–September 8), *F. crotonensis* (**P;** H = 4.176; $p < 0.05$), *Woronichinia* sp. (**Lo;** H

Table 2. Percentages of biomass (means) of phytoplankton assemblages (Ass.) as dominant groups of species, by period in each lake according to Reynolds (1997) and Reynolds et al. (2002)

	Period I	%	Ass.	Period II	%	Ass.	Period III	%	Ass.	Period IV	%	Ass.	Period	%	Ass.
Chodikee	15/04–27/05			08/06–28/06			06/07–27/07			02/08–30/08			07/09–12/10		
	Anabaena solitaria	6	H_2	*Aphanizomenon gracile*	52	H_1	*A. gracile*	14	H_1	*A. solitaria*	5	H_2	*A. solitaria*	8	H_2
	Anabaena spiroides	15	H_1	*C. hirundinella*	17	Lm	*Woronichinia* sp.	19	Lo	*A. spiroides*	9	H_1	*A. gracile*	22	H_1
	Cryptomonas curvata	5	Y	*F. crotonensis*	18	P	*Synechocystis aquatilis*	5	X1	*A. gracile*	20	H_1	*Woronichinia*	8	Lo
	Ceratium hirundinella	5	Lm				*Pandorina morum*	14	G	*Woronichinia* sp.	38	Lo	*A. ambigua*	23	C
	Aulacoseira ambigua	13	C				*Strombomonas* sp.	15	W_1				*P. morum*	10	G
	Fragilaria crotonensis	9	P												
	Chlorophyceae 3	7													
Stissing	20/04–02/06			08/06–10/08			17/08–12/10								
	C. curvata	12	Y	*A. solitaria*	9	H_2	*Woronichinia* sp.	26	Lo						
	C. marsonii	6	Y	*C. hirundinella*	14	Lm	*C. curvata*	10	Y						
	Asterionella sp.	20	C	*F. crotonensis*	11	P	*C. hirundinella*	4	Lm						
	F. crotonensis	19	P	Centrales	10	A	*F. crotonensis*	27	P						
	Centrales	19	A	*B. braunii*	14	F	*B. braunii*	7	F						
Wononscopomuc	20/04–18/05			24/05–22/06			28/06–02/08			10/08–15/10					
	Dinoflagelado 3	53	Lo	*Anabaena flos-aquae*	21	H_1	*A. flos-aquae*	21	H_1	*A. flos-aquae*	6.7	H_1			
	Nitzschia acicularis	27	C	*Aphanothece nidulans*	8	K	*A. solitaria*	27	H_2	*A. nidulans*	8.5	K			
				C. hirundinella	21	Lm	*A. nidulans*	27	K	*Aphanocapsa elachista*	6.9	K			
				C. bodanica	16	A	*F. crotonensis*	8	P	*Oscillatoria* sp.	7.2	S			
				F. crotonensis	12	P	*C. hirundinella*	10	Lm	*C. hirundinella*	19	Lm			
				Dinobryon divergens	5	E	*C. bodanica*	12.4	A						
										Chrysamoeba radians	7.5	X_2			
Mohonk	15/04–18/05			02/06–13/09			27/09–12/10								
	Tabellaria fenestrata	58	N	*Eutetramorus planctonicus*	19	F	*B. braunii*	34	F						
	Dinobryon cylindricum	18	E	*B. braunii*	25	F	*W. wilhelmii*	52	F						
	Botryococcus braunii	4	F	*Willea wilhelmii*	37	F	*Oocystis borgei*	8	F						

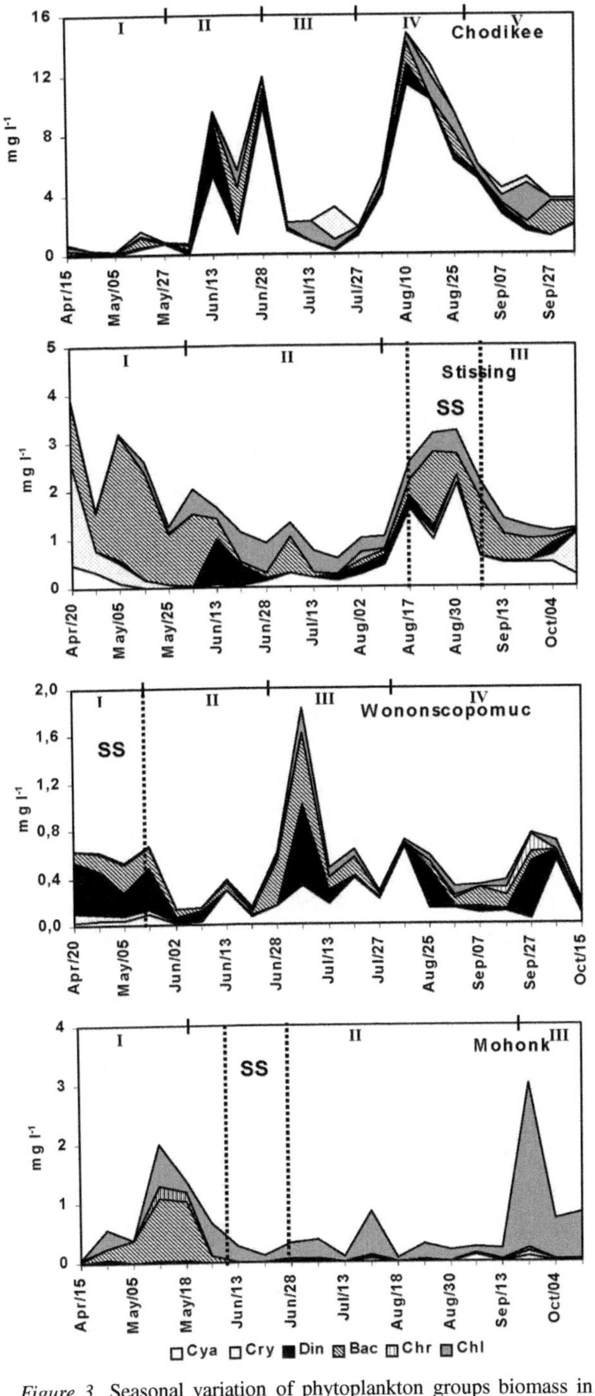

Figure 3. Seasonal variation of phytoplankton groups biomass in the four lakes, by periods (SS = steady-states; I. II. III. IV = periods).

= 8.978; $p < 0.01$) and *B. braunii* (**F**; H = 4.176; $p < 0.05$) were the dominant populations, contributing 63–77% to the total biomass (Table 4). During that period, at the end-stratification, all the DIN was under

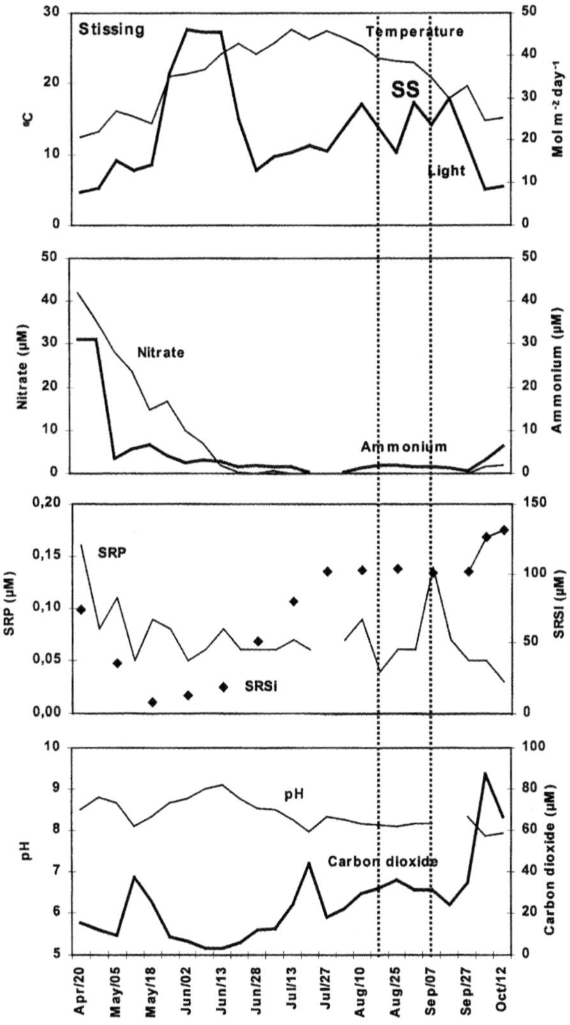

Figure 4. Seasonal variation of environmental variables in Stissing Lake (abbreviations as in Fig. 2).

ammonium form and SRP went to very low and less variable concentrations (Table 3). Biomass and change rates were significantly higher during the steady-state, but nonsignificant differences were found for diversity, evenness and species richness (Table 5), neither for abiotic factors between steady and non-steady-states.

Wononscopomuc Lake (WO)

A mesotrophic system (0.34 μM annual mean of total-P) was consistently stratified during summer with full circulation early spring (Fig. 1). Very low concentrations of dissolved inorganic N and P (annual mean = 1.81 and 0.08 μM, respectively) were registered. Nitrate decreased from the spring toward summer going

Table 3. Water variables weekly, during steady-state phases in Stissing (ST), Wonoscopomuc (WO) and Mohonk (MO) lakes. z_{euf} = euphotic zone; z_{mix} = mixing zone; z_{max} = maximum depth; I* = mean light in the z_{mix}; DIN = dissolved inorganic nitrogen; SRSi = soluble reactive silicon

	ST				WO					MO			
	Aug-17	Aug-25	Aug-30	Sep-7	Apr-20	Apr-27	May-5	May-12	May-18	Jun-8	Jun-13	Jun-22	Jun-28
Temperature (°C)	23.5	23.2	23	20.8	7.4		13.4	16	13.5	19.4	21.1	22.9	21.4
z_{euf} (m)	5.9	6.8	7.6	6.5	7.3	8.6	7.3	7.3	7.3	12.2	10.8	9.7	17.6
z_{mix} (m)	4.0	4.0	5.0	6.0	17.0	12.0	9.0	9.0	8.0	4.0	5.0	7.0	7.0
z_{max} (m)	8.0	8.0	7.9	8.0	31	31	31	31	31	19	19	19	19
I*(mol/m^2/day)	23.2	17.2	29.0	23.8	2.5	8.1	13.6	8.0	9.4	70.3	41.4	26.9	43.9
pH	8.1	8.1	8.2	8.2	8.2	8.4	8.6	8.8	8.8	8.7	7.7	7.4	7.5
Alkalinity (μEq l^{-1})	1928	1988	1945	2009	2183	2258	2241	2390	2599	215	172	216	181
N NO_3^- (μM)	0.0	0.0	0.0	0.0	1.1	1.1	0.0	0.3	0.1	0.8	0.8	1.1	1.7
N NH_4^+ (μM)	1.9	1.9	1.7	1.7	2.5	2.6	2.9	3.5	3.4	1.1	1.2	1.3	1.6
DIN (μM)	1.9	1.9	1.7	1.7	3.5	3.6	2.9	3.9	3.5	1.1	1.3	1.3	1.7
P $PO^{3-}{}_4$ (μM)	0.04	0.06	0.06	0.14	0.07	0.06	0.06	0.07	0.06	0.01	0.03	0.01	0.03
SRSi (μM)		104		101	32		17		14		13		13
CO_2 (μM)	32	36	31	31	43		16	10	9	1	9	19	15

undetectable through its end. SRSi was lower than in CH and ST, but still relatively high considering algal requirements. Light was lower during spring time (Fig. 5).

According to the changes in phytoplankton biomass ($H = 25.85$; $p < 0.001$) and composition, a unimodal pattern of biomass was observed and three periods were recognised (Fig. 3). Period **I** (April 20–May 18) had intermediate biomass levels and a high dominance of a non-identified dinoflagellate (**Lo**) and *Nitzschia acicularis* (Kützing) W. Smith (**C**). Period **II** (May 24–June 22) presented, in general, low biomass and a mixed population of *Anabaena flos-aquae* Brébisson ex. Bornet & Flahault (**H**$_1$) and *C. hirundinella* (**Lm**), among others. Period **III** (June 28–August 2) showed a highly variable biomass with a peak in July 6, dominated mainly *by C. hirundinella* (**Lm**) and *Cyclotella bodanica* Grunow (**A**). This period also had a high contribution of **H**$_1$ and **H**$_2$ (*A. flos-aquae* and *A. solitaria*, respectively) and **K** (*Aphanothece nidulans* Richter) assemblages. During period **IV** (August 10–October 15) a diversified community was registered with similar contribution of cyanobacteria (**H**$_1$ and **K**), dinoflagellates (**Lm**) and diatoms (**A**).

A steady-state phase was found (Table 3) during early mixing in period **I** (April 20–May 18), with dominance (79–84%) of *N. acicularis* (**C** assemblage; $H = 16.60$; $p < 0.001$) and a non-identified dinoflagellate (**Lm** assemblage; = 16.60; $p < 0.001$). The biomass of these assemblages (Table 5) was not significantly different from the rest of the study, and, unlike the results in ST, the dominant species in each assemblage was different (*N. acicularis*: $H = 16.60$; $p < 0.001$; dinoflagellate: $H = 16.60$; $p < 0.001$). Total mixing, low light conditions, relatively high concentrations of DIN and SRP were observed in this phase (Table 3). Different from ST, many abiotic variables were significantly different during the steady-state period. The tendency of fluctuations in abiotic conditions was increasing light ($H = 9.273$; $p < 0.01$) and temperature ($H = 9.273$; $p < 0.01$) and decreasing SRP ($H = 5.631$; $p < 0.05$), N NO_3^- ($H = 9.273$; $p < 0.01$), SRSi (NS) and CO_2 (NS). Lower pH ($H = 7.229$; $p < 0.01$) and water transparency ($H = 8.868$; $p < 0.01$) and higher z_{mix} ($H = 9.935$; $p < 0.01$) were also registered during the steady-state. As expected, diversity, species richness and change rates were significantly lower during the steady-state period than in the rest of samplings (Table 5). Total biomass and evenness did not show significant differences between steady and non-steady states.

Mohonk Lake (MO)

A very clear and oligotrophic lake (0.25 μM annual mean of total-P), was consistently stratified from late-spring to late summer (Fig. 1) and showed decreasing DIN and SRP concentrations from spring

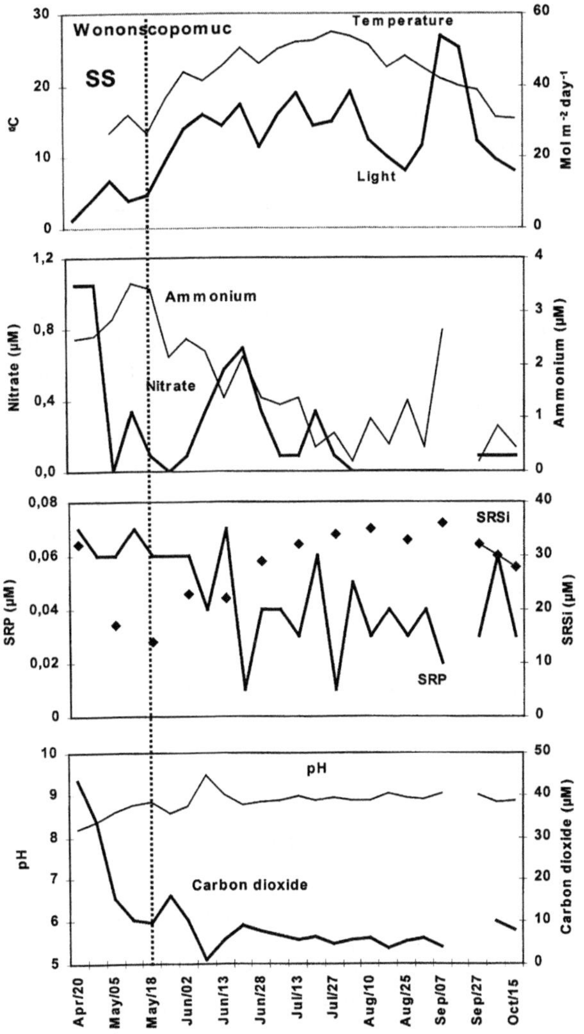

Figure 5. Seasonal variation of environmental variables in Wononscopomuc Lake (abbreviations as in Fig. 2).

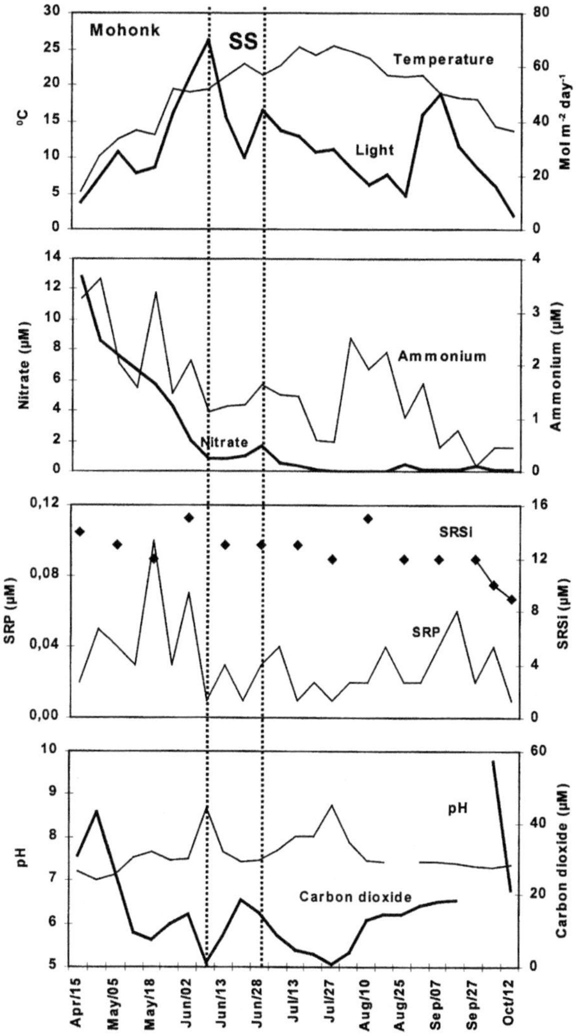

Figure 6. Seasonal variation of environmental variables in Mohonk Lake (abbreviations as in Fig. 2).

to mid-summer. N-NO_3^- went to undetectable from mid-summer to fall (Fig. 6).

A bimodal seasonal cycle (Fig. 3) was reported in MO and three periods were identified (total biomass; H = 6.39; $p < 0.05$). During the spring, period **I** (April 15–May 18) was dominated mainly by *Tabellaria fenestrata* (Lyngbye) Kützing (**N** assemblage). In period **II** (June 2–September 13) a very low biomass was recorded also with low variability and dominance of greens of the **F** assemblage [*Eutetramorus planctonicus (Korsikov)* Bourrelly, *Botryococcus braunii* and *Willea wilhelmii* (Fott) Komárek]. Period **III** (September 27–October 12) was marked by higher biomass and showed a peak of *Willea wilhelmii* during 1 week, followed by co-dominance with other greens also of **F** assemblage (*B. braunii* and *Oocystis borgei*).

A steady-state phase occurred during three weeks in the end-spring (June 8–28) during period **II** (Table 3), when colonial green-algae of **F**-assemblage (*B. braunii, E. planctonicus* and *W. wilhelmii*) were co-dominant (71–94%, Table 4). Nonsignificant differences between steady and non-steady-states were found. At that time the lake was consistently stratified, being z_{mix} the only significantly different abiotic variable between steady (5.0 m) and non-steady-state periods (7.5 m) (H = 4.941; $p < 0.05$). Low DIN and SRP concentrations were registered, always in limiting concentration for phytoplankton growth. The general tendency during that phase was decreasing light in the mixing zone, and increasing temperature and CO_2. Unexpectedly high diversity and evenness were found during this phase (Table 5).

It is worthwhile to mention that, considering all the three lakes together where steady-states were identified, we could not find any statistical difference between steady and non-steady-state, using biotic and abiotic variables.

Discussion

We consider that the functional approach does capture much of the ecology of the phytoplankton and can be used as a verifiable quantitative method of describing community structure and changes (Fabbro & Duivenvorden, 2000; Reynolds et al., 2002; Kruk et al., 2002). In this context, we analysed the identified steady-state periods in the lakes using functional groups, with assemblages as descriptors (Naselli-Flores et al. 2003; Teubner & Dokulil, 2003). Assemblages were useful for describing community conditions during phases with different stability. Important differences were found between steady and non-steady states. In these situations, dominant near monospecific community with different levels of stability and diversity, according to characteristics of the conditions, were recorded.

Lo and **Lm** assemblages (dinoflagellates and cyanobacteria of summer epilimnion of mesotrophic and eutrophic lakes), **M** (colonial cyanobacteria of diel mixed eutrophic waters of low latitudes), **S** (cyanobacteria of turbid and mixed layers of enriched temperate systems), **R** (cyanobacteria forming maximum metalimnetic), **H** (N-fixing cyanobacteria), **N** (diatoms and desmids of mesotrophic epilimnion) and **P** (of eutrophic epilimnion) had been quoted as possible to achieve maturing stages (Reynolds, 1997) *sensu* Odum' rules (Odum, 1969). Among these seven assemblages, two were found in this study (**Lo** and **P**). Now, we add other two (**F** and **C**) as candidates to integrate the list.

According to criteria by Sommer et al. (1993), we found one steady-state in three of the four studied lakes. Chodikee lake (CH), a rapidly flushed and shallow system did not show any steady-state as expected by Sommer's indicators of stability. However, clear periods of constant community structure were identified in the other lakes. Stissing (ST), Wononscopomuc (WO) and Mohonk (MO) lakes, showed a relative constancy of composition during, at least, 3 weeks without significant change in biomass. Steady-state phases occurred in the very-well stratified ST during the summer and in MO during end-spring, but also during mixing times in WO.

The main factor of persistence that caused and maintained the steady-state in ST was stratification. As a consequence, low nutrient levels were registered during this state. The disruption was a consequence of decreasing temperature and mixing, leading to decreasing light and increasing CO_2. During the steady-state period in ST, **P** and **Lo** assemblages were dominant and occurred until the end of summer stratification, when the mixing zone was deepening (4–6 m). **Lo** refers to stratified mesotrophic lakes, which tolerate segregated nutrients through the water column and are sensitive to deep mixing. **P** assemblage includes diatoms, which are depending on physical mixing, requiring a continuous or semi-continuous mixed layer of, at least, 2–3 m in thickness (Reynolds et al., 2002). As expected, decreasing temperature and deepening of mixing zone favoured the disruption of the **Lo** dominance. However, **P**-assemblage, which should be favoured by those conditions, also declined. Among the stability indicators only total biomass and change rates were significantly different between states. The observed total biomass, as expected, was higher during the stable state, but a non-expected higher change rate was found. The coexistence of representatives of both assemblages with 22 other species did not make it possible to find the expected lower diversity, resulting of the decreasing environmental and internal disturbances, which prevents the establishment of equilibrium conditions (Sommer et al., 1993).

The dominant **F** assemblage during the steady-state period in MO was represented by non-motile but near-neutrally buoyant colonial green-algae, which perform better in clear waters and are tolerant of deep mixing and low nutrient concentrations. Indeed, phytoplankton growth in MO was considered, during a whole ice-free season, strongly limited by both phosphorus and nitrogen (Huszar & Caraco, 1998). In addition, during the steady-state phase, the lake had a deep and very clear mixing zone. **F** assemblage, composed by almost the same species, was dominant during the rest of the stratification period. However, the absolute and relative biomass were much more variable during non-steady-state. The main factor of constancy was stratification with low nutrient concentrations. The factors of disruption were associated with

Table 4. Weekly phytoplankton attributes during steady-state phases in Stissing (ST). Wonoscopomuc (WO) and Mohonk (MO) lakes (Div = diversity; Ch. Rates = change-rates)

	ST	ST	ST	ST	WO	WO	WO	WO	WO	MO	MO	MO	MO
	Aug 17	Aug 25	Aug 30	Sep 07	Apr 20	Apr 27	May 05	May 12	May 18	Jun 08	Jun 13	Jun 22	Jun 28
Biomass (mg l^{-1})	2.6	3.2	3.2	2.2	0.6	0.6	0.5		0.7	0.7	0.3	0.1	0.3
Div. (bits mg $^{-1}$)	2.7	2.9	2.4	2.3	1.6	1.9	1.9		1.9	2.5	2.1	1.9	2.3
Evenness (%)	61	63	56	55	49	50	54		60	69	54	53	58
Ch. rates (day^{-1})	0.20	0.32	0.12	0.35	0.05	0.09	0.04		0.12	0.15	0.07	0.28	0.16
Species richness (taxa per sample)	22	25	21	19	10	13	11		8.9	12	16	11	15
Assemblages (%)	64	63	77	76	80	84	82		79	71	89	94	85
	Lo -*Woronichinia* sp.				Lm- *Dinoflagellate*					F-*Botryococcus braunii*			
	P – *Fragilaria crotonensis*				C -*Nitzschia acicularis*					F- *Willea wilhelmii*			
	F– *Botryococcus braunii*									F-*Eutetramorus planctonicus*			

Table 5. Mean of phytoplankton attributes during the total study, and the steady and non-steady states in Chodikee (CH), Stissing (ST), Wononscopomuc (WO) and Mohonk (MO) lakes. Results from Kruskall-Wallis median test are shown for difference between steady and non-steady states for each lake, including H and *p* values. ns. refers to non significant results. Sp. rich = species richness. The significantly different mean values are marked with * and **. * Refers to results concordant with the expected and ** refers to significant results opposed to that expected by Sommer's criteria (Sommer et al., 1993)

Period	Biomass mg l^{-1}	Diversity Bits mg^{-1}	Eveness %	Sp. rich. Taxa/sample	Change day^{-1}
TOTAL					
CH	5.1	2.3	56	18	0.2
ST	1.8	2.6	63	18	0.1
WO	0.5	2.2	56	17	0.1
MO	0.9	1.6	42	15	0.2
STEADY-STATE					
ST	2.8*	2.6	59	22	0.24**
WO	0.6	1.8*	53	11*	0.08*
MO	0.3	2.2**	59**	13	0.17
NON STEADY-STATE					
ST	1.6*	2.6	64	17	0.11**
WO	0.5	2.5*	57	21*	0.16*
MO	0.7	1.6**	43**	15	0.18
KRUSKALL-WALLIS MEDIAN TEST					
ST	5.283	ns	ns	ns	5.689
	0.022				0.017
WO	ns	4.246	ns	8.001	4.827
		0.039		0.005	0.028
MO	ns	4.523	4.331	ns	ns
		0.033	0.037		

deepening of the mixing zone. None of the stability indicators observed showed the expected pattern according to Sommer et al. (1993). Moreover and unexpectedly, diversity and evenness were higher during the steady-state.

According to the literature, a situation not expected during equilibrium conditions, differently from the other, was registered in WO. In this system, the factors of constancy were associated with persistent mixing, low light and high nutrient concentrations. The factors of disruption were increasing stratification and light, as well as decreasing nutrient lakes. In this lake, the steady-state phase occurred during the vernal development, with dominance of a non-expected large dinoflagellate. **Lo** was not expected, because the lake was in early mixing with relatively high nutrient concentrations and **Lo** performed better in stratified columns, with segregated nutrients, and is sensitive to prolonged or deep mixing (Reynolds et al., 2002). The spatial transport within a system can complicate the interpretation, because dinoflagellates can be transported from the benthos or littoral area to the pelagic region during the mixing. A similar situation was reported to lake Kinneret (Israel), where, for many years, the lake supported a periodic spring sequence, reaching a regular dominance of *Peridinium gatunense* (**Lo**), which does not resist the nutrient deficiency through summer time and enters in benthic resting stages (Berman et al., 1992). The dominance was shared with the diatom *Nitzschia acicularis* (**C**), whose habitat template in WO was different to other species of *Nitzschia*, which have been included in **D** assemblage of shallow mixed and enriched lakes (Reynolds et al., 2002). It can be ascribed to a deep mixed zone, but with low light, and to soft, but not enriched waters. These features are close to **C** assemblage, which can develop in mixed conditions and are tolerant to low light and carbon deficiency, but sensitive to stratification and SRSi depletion. The probable spatial transport of the dinoflagellate, during the mixing in WO, made possible the co-existence with the diatom, both with different requirements. During the steady-state, diversity, species richness and change rate were lower and concordant with the expected.

According to the observations presented in this paper, the steady-state periods occurred during strong stratification, but also in mixed water columns. What it can tell us is that long periods of spring mixing can also set up and maintain dominant assemblages. Assemblages persistence can be hold as long as the physical conditions are unchanged and the current crop of organisms can fill (or partially fill) the carrying capacity of the system. Steady-states were ended only by major hydrographic changes, stratification or mixing. In the shallow and rapidly flushed Chodikee Lake, constant hydrological conditions maintained a highly variable environment, preventing the appearance of a differentiable steady-state, concordant at least to some extent to definition by Sommer et al. (1993), but also resembling the plagioclimaxic conditions described by Reynolds (1993).

In synthesis, constancy of hidrological conditions, characterised either by mixing or stratification, during 3 or 4 weeks, lead to dominance of particular assemblages composed by few species. However, the expected equilibrium with consistent reduction of diversity, species richness and change rates were not observed in the community. Considering the habitat template and phytoplankton attributes, not consistent statistical differences between steady-state and non-steady-state phases were found in the phytoplankton stability indicators. Some lakes, for example, did not show the expected lower diversity and species richness during steady-states (ST) and other showed even higher values than in non-steady-states (MO). The comparison of steady-states among systems, as a whole, was not possible, probably because criteria by Sommer et al. (1993) are not useful for defining periods in any situation. Its application should be narrowed to those cases when plankton is not changing rapidly (it is not in ascendancy, collapse or rapid succession), or when physical conditions coincide with what Reynolds (1993) called plagioclimax, or any other condition which can be maintained along with high diversity or low biomass. Steady-states can be promoted and ended by major hydrographic changes (either it stratifies or it mixes). What matters to promote and maintain persistent assemblages is the environmental constancy, independently of its relation with a particular physical condition.

Acknowledgements

The authors would like to thank Dr Luigi Naselli-Flores for financial support to attend the 13th Workshop of the International Association of Phytoplankton Taxonomy and Ecology; David Fisher for technical support; Rich Miller and Peter Raymond for collection of samples and nutrient analysis. This research was financially supported by Conselho Nacional de Desenvolvimento Científico e Tecnológico, Brasil,

Fundação de Amparo à Pesquisa do Estado do Rio de Janeiro, Brasil and National Science Foundation, U.S.A. This is a contribution of Institute of Ecosystem Studies.

References

APHA, 1992. Standard Methods for Examination of Waste and Wastewater. American Public Health Association, New York.

Berman, T. Y. Z., J. Yacobi & U. Pollingher, 1992. Lake Kinneret phytoplankton: stability and variability during 20 years (1970–1989). Aquat. Sci. 54: 104–127.

Caraco, N., J. J. Cole & G. Likens, 1993. Sulfate control of phosphorus availability in lakes. Hydrobiologia 253: 275–280.

Cole, G. A., 1994. Textbook of Limnology. Waveland Press Inc., Illinois.

Cole, J., N. Caraco, N. F. Kling & T. K. Kratz, 1994. Carbon dioxide supersaturation in the surface waters of lakes. Science 265: 1568–1570.

Dokulil, M. T. & K. Teubner, 2003. Steady state phytoplankton assemblages during thermal stratification in deep alpine lakes. Do they occur? Hydrobiologia 502 (Dev. Hydrobiol. 172): 65–72.

Edler, L. (ed.), 1979. Recommendations for Marine Biological Studies in the Baltic Sea. Phytoplankton and Chlorophyll. 38 pp. (UNESCO, Working Group 11, Baltic Marine Biologists).

Fabbro, L. D. & L. J. Duivenvorden, 2000. A two-part model liking multidimensional environmental gradients and seasonal succession of phytoplankton assemblages. Hydrobiologia 438: 13–24.

Harris, G. P., 1986. Phytoplankton ecology: structure, function and fluctuation. Chapman and Hall, London.

Huszar, V. L. M. & N. Caraco, 1998. The relationship between pytoplankton composition and physical-chemical variables: a comparison of taxonomic and morphological approaches in six temperate lakes. Freshwat. Biol. 40: 679–696.

Jassby, A. D. & C. R. Goldman, 1974. Loss rates from a phytoplankton community. Limnol. Oceanogr. 19: 618–627.

Kruk, C., N. Mazzeo, G. Lacerot & C. S. Reynolds, 2002.Classification schemes for phytoplankton: a local validation of a functional approach to the analysis of species temporal replacement. J. Plankton Res. 24: 1191–1216.

Lewis Jr., W. M., 1978. Analysis of succession in a tropical phytoplankton community and a new measure of succession rate. Am. Nat. 112: 401–414.

Lund, J. W. G., C. Kipling & E. D. Le Cren, 1958. The inverted microscope method of estimating algal numbers and statistical basis of estimating by counting. Hydrobiologia 11: 143–170.

Moss, B., P. Johnes & G. Phillips, 1996. The monitoring of ecological quality and the classification of standing waters in temperate regions: a review and proposal based on worked scheme for British waters. Biol. Rev. 71: 301–399.

Naselli-Flores, L., J. Padisák, M. T. Dokulil & I. Chorus, 2003. Equilibrium/steady-state concept in phytoplankton ecology. Hydrobiologia 502 (Dev. Hydrobiol. 172): 395–403.

Nürnberg, G. K., 1996. Trophic state of clear and coloured, soft-and hard-water lakes with special consideration of nutrients, anoxia, phytoplankton and fish. J. Lakes Res. Manage. 12: 432–447.

Odum, E. P., 1969. The strategy of ecosystem development. Science 164: 262–270.

Pielou, E. C., 1966. The measurement of diversity in different types of biological collections. J. theor. Biol. 13: 131–144.

Reynolds, C. S, V. L. M. Huszar, C. Kruk, L. Naselli-Flores & S. Melo, 2002. Towards a functional classification of the freshwater phytoplankton. J. Plankton Res. 24: 417–428.

Reynolds, C. S., 1980. Phytoplankton assemblages and their periodicity in stratifying lake systems. Hol. Ecol. 3: 141–159.

Reynolds, C. S., 1993. Scales of disturbance and their roles in plankton ecology. Hydrobiologia 249: 151–171.

Reynolds, C. S., 1997. Vegetation Processes in the Pelagic: A Model for Ecosystem Theory. Ecology Institute, Oldendorf/Luhe.

Rojo, C. & M. Álvarez-Cobelas, 2003. Are there steady-state phytoplankton assemblages in the field? Hydrobiologia 502 (Dev. Hydrobiol. 172): 3–12.

Shannon C. E. & W. Weaver, 1949. The mathematical theory of communication. Urbana, Univ. Illinois Press.

Sommer, U., J. Padisák, C. S. Reynolds & P. Juhasz-Nagy, 1993. Hutchinson's heritage: the diversity-disturbance relationship in phytoplankton. Hydrobiologia 249: 1–7.

Tilman, D., 1982. Resources competition and community structure. Princeton Univ. Press, Princeton.

Uhelinger, V., 1964. Étude statistique des méthodes de dénobrement planctonique. Arch. Sci. 17: 121–123.

Utermöhl, H., 1958 Zur Vervollkomnung der quantitativen Phytoplankton- methodik. Mitt. int. Ver. theor. angewan. Limnol. 9: 1–38.

Vollenweider, R. A. & J. Kerekes, 1982. The loading concept as basis for controlling eutrophication philosophy and preliminary results of OECD programme on eutrophication. Prog. Water Technol. 12:5–38

Whittaker, R. H., 1975. Communties and Ecosystems. MacMillan, New York.

Hydrobiologia **502:** 111–121, 2003.
L. Naselli-Flores, J. Padisák, M. T. Dokulil (eds),
Phytoplankton and Equilibrium Concept: The Ecology of Steady-State Assemblages.

Phytoplankton assemblages and steady state in deep and shallow eutrophic lakes – an approach to differentiate the habitat properties of Oscillatoriales

Brigitte Nixdorf[1], Ute Mischke[2] & Jacqueline Rücker[3]
[1,3]*Brandenburg Technical University of Cottbus, Seestraße 45, D-15526 Bad Saarow*
E-mail: b.nixdorf@t-online.de
[2]*IGB, Institute of Freshwater ecology and Inland Fisheries, Dept. of Shallow Lakes and Lowland Rivers, Müggelseedamm 310, D-12587 Berlin, Germany*

Key words: steady state assemblages, cyanobacteria, Oscillatoriales, *Planktothrix agardhii,* dimictic lakes, polymictic lakes, under water irradiance, nutrients

Abstract

Ecological conditions and phytoplankton succession in two shallow hypertrophic lakes (Langer See and Melangsee) and a dimictic, eutrophic lake (Scharmützelsee) in a lake chain in Eastern Germany were analyzed from 1999 to 2001 in order to find situations of phytoplankton steady state assemblages and variables controlling the phytoplankton composition according to Reynolds et al. (2002). Long term background data from 1993 to 2001 suggest steady state conditions in shallow lakes, whereas the deep lake exhibited irregular fluctuations between various phytoplankton stages. Since the phytoplankton composition in the shallow lakes was similar in all the 3 years, it was highly predictable. Steady state conditions dominated by different species of Oscillatoriales were detected during the summer period 1999 and 2000 in Langer See and in Melangsee (see Mischke & Nixdorf, this volume). This dominant assemblage found in both lakes (group $\mathbf{S_1}$ acc. to Reynolds et al., 2002): *Planktothrix agardhii* (Gom.) Anagn. et Kom., *Limnothrix redekei* (Van Goor) Meffert, *Pseudanabaena* (Lauterb.) is typical in turbid mixed layers with highly light deficient conditions, but it is also regularly dominant in the dimictic lake Scharmützelsee as observed in 1999 and 2001 (*Pseudanabaena limnetica* (Lemm.) Kom. The Nostocales *Cylindrospermopsis raciborskii* (Wolz.) Seenayya et Subba Raju and *Aphanizomenon gracile* (Lemmerm.) Lemmerm. were important in the shallow lakes as well as in lake Scharmützelsee. Nevertheless, the occurrence of filamentous cyanobacteria in the dimictic lake was not regular and an unpredictable change in phytoplankton development was observed in 2000. It is discussed, whether this phenomenon of regular succession in shallow hypertrophic lakes is caused by adaptation to a resilient and an extreme environment or by the pool of species that can live or survive in that environment. This was checked through comparison of the depth of the mixed layer, the mean daily irradiance within this layer and the nutrient resources. Although the nutrient resources in both types of lake are near threshold levels, indicating growth inhibition by dissolved nutrients (DIP, DIN, TIC, DSi), the under water light supply seems to be the key factor favoring the dominance of filamentous cyanobacteria belonging to the functional group $\mathbf{S_1}$.

Introduction

The phenomenon of plankton succession has been explained by several theories, e.g. as a low diversity equilibrium (Hardin, 1960) due to the competitive exclusion of species. According to the adaptation of the IDH theory (Connell, 1978) to planktonic succession by Reynolds & Lund (1988), plankton algal species are simultaneously limited by only one to three resources (nutrients as phosphorus, nitrogen or carbon and/or light). Most field observations show more than 1 – 3 dominant species at any phase of seasonal development as predicted by the competitive exclusion theory. The reasons are found in the different responses of phytoplankton on the frequency of disturbances or changes in abiotic resources conditions at different time scales (Reynolds, 1984): (a) shorter than one generation time induce physiological responses, (b) frequencies between 200 and 20 h interact with the phytoplankton growth rate and c) disturbances at up to 10 days intervals can initiate a successional sequence in phytoplankton development. Due to our sampling frequency of 1–4 weeks in the 3 years, we confine our analyses to successional sequence in phytoplankton development and influencing abiotic factors. Using a long term data set of nutrients and chlorophyll *a*-concentrations, we can detect regularities in the phytoplankton succession. The phytoplankton composition was already studied in former years by Zippel & Nixdorf (1997), Wiedner (1999) and Wiedner et al. (1996, 2002).

Concerning one topic of the 13th IAP workshop (Naselli-Flores et al., 2003) 'steady state and equilibrium', we test the aspects of equilibrium conditions according to Sommer et al. (1993) using the following criteria:

1. 1, 2 or 3 species of algae contribute more than 80% of total biomass
2. their existence or coexistence persists for more than 1–2 weeks
3. during that period, total biomass does not increase significantly

We try to find and to describe 'equilibrial species' which dominate the assemblages in periods of equilibrium states in deep and shallow lakes in the German lowlands. These results will be the basis for an attempt to draw the template of dominant species and trait-separated functional groups according to the Reynolds scheme (Reynolds et al., 2002).

Study sites and trophic states of the lakes

The investigated lakes are situated in the Scharmützelsee region (57° 20′ N, 54° 35′ E) in Brandenburg (Germany) about 60 km East from Berlin and belong to a lake chain (Fig. 1). They originated from the Weichsel glacier and are located at the border of the southern plateau of the Berlin glacial valley. The catchment area is dominated by forests and agriculture with an increasing importance for tourism. The lakes differ substantially in hydrology, morphometry, size and catchment area (Table 1). Theoretical retention times vary between 22 and 60 days for lake Melangsee (MEL) and lake Langer See (LAN) and 16 years for lake Scharmützelsee (RIE). Lakes MEL and LAN are polymictic and 'very shallow' according to Nixdorf & Deneke (1997), while lake RIE is a dimictic lake with a maximum depth of 29 m. All lakes are in a eutrophic to hypertrophic state (Table 2; OECD, 1982) but differ considerably in trophic parameters. Very shallow lakes are more productive due to a higher efficiency in converting phosphorus into algal biomass, which is reflected by a higher Chl *a*: TP ratio (Nixdorf & Deneke, 1997).

Materials and methods

The lakes in the Scharmützelsee region have been investigated since 1993. For this study data from monthly to biweekly (RIE) and weekly to biweekly (LAN, MEL) sampling was used for the years 1999 – 2001. Samples were taken as mixed samples from the mixed water layer (i.e. the entire water column or the epilimnion only) with a 2.3 L-LIMNOS sampler. Sampling stations were above the deepest points, where Secchi depth was also recorded. Depth profiles of physical and chemical parameters were measured at 0.5 m intervals with a HYDROLAB H20 probe connected to a field computer (HUSKY Hunter). Secchi depth (SD), total phosphorus (TP), dissolved inorganic phosphorus (DIP), total nitrogen (TN) and different inorganic dissolved nitrogen fraction (DIN), dissolved silica (DSi) as well as chlorophyll *a* (Chl *a*) were determined according to standard methods (DEV, 1986–1998). Chl *a* concentrations were corrected for phaeophytin.

Phytoplankton was counted after fixation with Lugol's solution according to Utermöhl (1958). Biovolume was calculated by measuring the corresponding dimensions using the geometric formulas given by Rott (1981) and Willén (1976).

Mean PAR in the water column I^* was calculated according to Riley (1957):

$$I^* = 0.45 \cdot I_0 \cdot \left(\frac{1 - e^{-K_d \cdot Z_{mix}}}{K_d \cdot Z_{mix}} \right)$$

where z_{mix} is the depth of the mixed water column which was assumed to be equal to the mean depth of the lakes during circulation and K_d is the attenuation coefficient (m^{-1}). During stratification periods, the results of the depth profiles measurements were taken to estimate the mixing depth. I_0 is the average daily sum of global radiation. For the share of PAR on total global radiation and the reflection at the water surface, a correction factor of 0.45 is used as proposed by Behrendt & Nixdorf (1993). Global radiation (J cm^{-2}

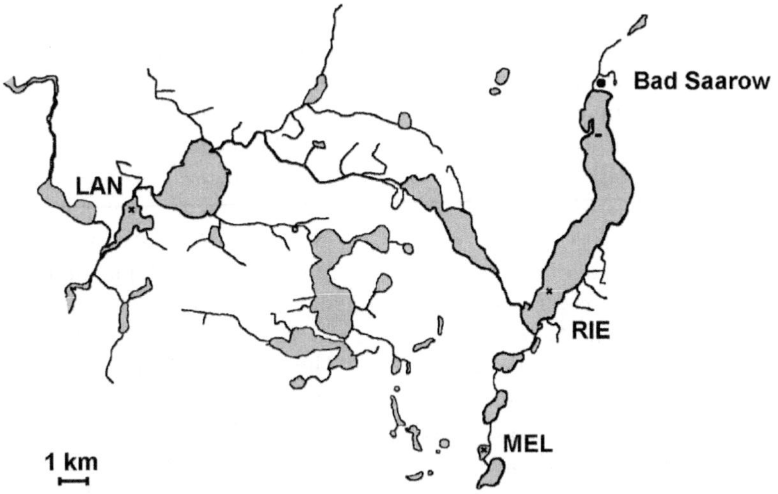

Figure 1. Investigated lakes in the Scharmützelsee region: Melangsee (MEL), Scharmützelsee (RIE) and lake Langer See (LAN).

Table 1. Characteristic data of lake morphometry and hydrology of 3 lakes in the Scharmützelsee region. Abbreviations: Z_{max}: maximum depth, Z_{mean}: mean depth, t: theoretical retention time in days (d) or years (y), CA: catchment area, di: dimictic, poly: polymictic

Lakes	Abbr.	Z_{max} m	Z_{mean} m	Area km^2	Volume 10^6 m^3	t d or y	CA km^2	Mixis
Scharmützelsee	RIE	29.5	9.0	12.09	108.23	16 y	112	di
Melangsee	MEL	3.3	1.6	0.11	0.17	22 d	5	poly
Langer See	LAN	3.8	2.1	1.55	3.27	60 d	395	poly

Table 2. Annual means of total phosphorus (TP), chlorophyll-*a* (Chl *a*) and dissolved inorganic nitrogen (DIN) concentrations and mean Secchi depths (SD) for 1994–97 and 1998–2001, respectively

Lakes	Abbr.	TP [μg l^{-1}]		Chl *a* [μg l^{-1}]		SD [m]		DIN [μg l^{-1}]	
		94–97	98–01 98–00*	94–97	98–01	94–97	98–01	94–97	98–01
Scharmützelsee	RIE	68.7	57.0	26.1	22.1	1.8	2.1	252.8	215.9
Melangsee	MEL	60.8	69.0*	62.4	57.3	1.0	1.0	142.6	148.8
Langer See	LAN	113.0	99.6*	88.1	77.4	0.7	0.8	274.4	175.5

d^{-1}) was measured by the meteorological station in Lindenberg located 15-30 km from the lakes. It was converted to daily photon flux density (E m^{-2} d^{-2}) by a factor of 0.0445.

The approach of Reynolds (1997) is to calculate the mean irradiance in a mixed water column (I^*) by:

$$\ln I^* = (\ln I_0{}' + \ln I_{\mathrm{m}})/2$$

It is not identical with the equation of Riley (1957). The recommended threshold number of 1.5 E m^{-2} d^{-1} given in Reynolds et al. (2002) corresponds with 3 E m^{-2} d^{-1} using the modified Riley (1957) equation. This threshold describes under water light situations and the functional response of different assemblages. Dominance of a codon denotes tolerance or positive response even under conditions with values of under water irradiance in the mixed layer below the threshold. Negative response denotes no positive benefit on the assemblage due to the mean daily irradiance levels experienced in the mixed water column.

Results

The two shallow lakes Melangsee and Langer See show a regularly recurring phytoplankton composition at a high biomass level during the long term investigation since 1993 (for details see Mischke & Nixdorf, this volume, Zippel & Nixdorf, 1997, Rücker et al., in press, Schmitt & Nixdorf, 1999, Wiedner et al., 2002). Steady state conditions were detected in summer 2000 for 7 consecutive weeks in lake Langer See (dominated by *Planktothrix agardhii, Aphanizomenon gracile* and *Pseudanabaena limnetica*). In lake Melangsee, the very thin *Limnothrix* species *L. amphigranulata* (Van Goor) Meffert was the most common species accompanied by *Pseudanabaena limnetica* and *Planktothrix agardhii* (see Mischke & Nixdorf, 2003). Here, the results of all these studies are not shown in detail, but are used to compare with those of the dimictic lake Scharmützelsee.

Long term development of phytoplankton composition and biomass, main resources (nutrients and light) in lake Scharmützelsee

In Figure 2, the long-term development of dissolved and total phosphorus and Chl *a* concentration and Secchi depth for lake Scharmützelsee from 1993 to 2001 is illustrated. We observed a tendency for decreasing nutrient concentrations (TP concentration: annual means from 70 to 57 mg m^{-3}) without corresponding response of phytoplankton biomass (Chl *a*-concentration annual average between 20 and 25 mg m^{-3}). The shallow lakes MEL and LAN have comparable TP-concentrations (between 60 and 113 mg m^{-3}) but much higher phytoplankton biomass (Chl *a* 58–90 mg m^{-3} as annual mean between 1993 and 2001, see Table 2).

All lakes are dominated by filamentous cyanobacteria which comprise high proportion on total biomass. There were only 2 years (1994 and 2000) with a quite different and very low phytoplankton development in Lake Scharmützelsee (Fig. 2, Table 3). In 'normal' years a phytoplankton biovolume between 6.8 and 10.7 mm^3 l^{-1} with a maximum of 15.5 mm^3 l^{-1} was detected in late summer with a portion between 61 and 87% of cyanobacteria (maximum 89%).

Under water light supply and nutrient resources in Lake Scharmützelsee

In general, a typical seasonal pattern can be described for all three lakes during the investigation period: Lake Scharmützelsee is turbid during the vegetation season (Secchi depth about 1 m) and clearer in winter (Secchi depth between 2 and 4 m) due to a drastic reduction in phytoplankton biomass (Fig. 3). In early summer, clear water phases do occasionally occur but were distinct in the untypical years 1994 and 2000. The mean irradiance in the mixed water layer of lake Scharmützelsee varied between 0.08 and 16.97 E m^{-2} d^{-1} (see Fig. 4). It is obvious that the threshold of 3 E m^{-2} d^{-1} was almost fallen below during circulation periods in autumn and spring and during winter. In 1999 and 2001 this value also fell below the threshold during summer, especially in late summer during the mass development of filamentous cyanobacteria. The mean summer irradiance in 2000 was mostly above the mentioned threshold value.

The hypertrophic shallow lakes are very turbid (Secchi depth during summer below 0.5 m, annual average below or about 1 m). The mean underwater light intensity is relatively low, ranging from 4.7 and 8.3 E m^{-2} d^{-1} in lake Melangsee and 2.4 and 6.0 E m^{-2} d^{-1} in lake Langer See (see Mischke & Nixdorf, 2003).

The development of the main nutrient resources inorganic dissolved phosphorous (as DIP), dissolved inorganic nitrogen (DIN) and dissolved inorganic silica (DSi) is shown in Figure 4. Concentration of DIP decreased to limiting levels of detection about or less than 0.1 μmol l^{-1} from late spring to September. After the autumnal overturn and mixing of the P-enriched hypolimnetic water into the whole water column concentration increased up to 1 – 2 μmol l^{-1}. This pattern was regularly observed, also during long term investigations (see Fig. 2). Values below 0.1 μmol l^{-1} were detected regularly during summer 1999, seldom in 2000 and they were never reached in 2001.

The dynamic of the DIN concentration is similar to the course of the SRP concentration: High values (between 20 and 50 μmol l^{-1}) from autumn to early summer, followed by a decrease in summer to 0.4 μmol l^{-1}. Silica (DSI) concentrations declined during the spring bloom periods in Lake Scharmützelsee and will not be considered for the dynamic of the cyanobacteria within this study.

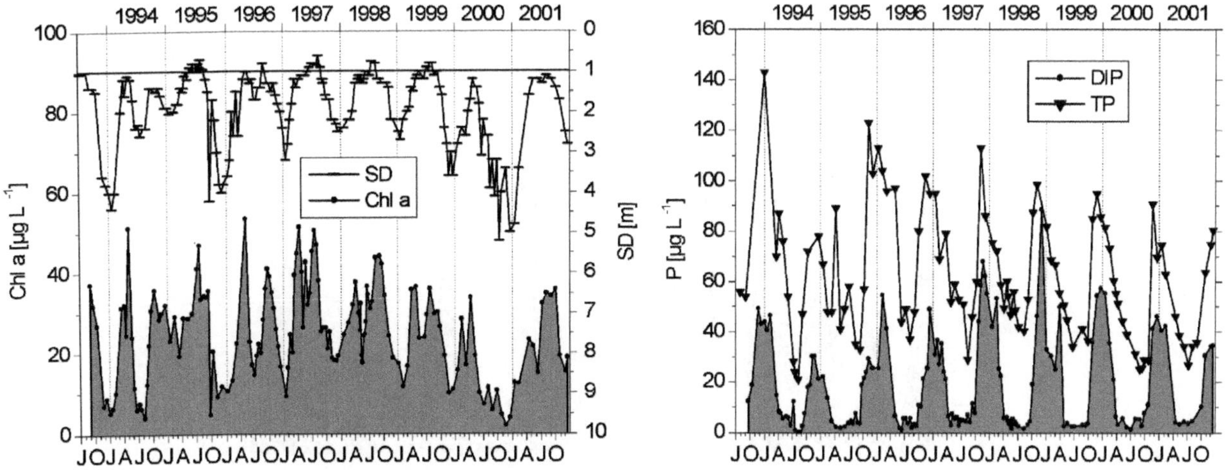

Figure 2. Long term development of TP, SRP and Chl *a* concentration and Secchi depth in lake Scharmützelsee from 1993 to 2001.

Table 3. Total biovolume of phytoplankton during late summer (July – September, average and maximum value) and portion of Cyanobacteria on total biovolume [%] and maximum value from 1993 to 2001 in lake Scharmützelsee (values from 1997/98 were converted from HPLC analyses of different planktonic groups using a relation between Chl *a* – concentration and phytoplankton biovolume of $5*10^{-3}$)

Year	1993	1994	1995	1996	1997	1998	1999	2000	2001
Average Biovolume [$mm^3\ l^{-1}$].	9.7	3.4	8.6	8.1	6.3	6.5	5.3	1.1	6.3
Maximum Biovolume [$mm^3\ l^{-1}$].	15.4	6.5	14.2	13.9	10.1	12.1	9.4	2.7	10.7
Average cyanobacterial biovolume [%]	71	58	53	66	68	66	64	18	80
Maximum cyanobacterial biovolume (%)	86	81	83	89	79	79	84	58	89

Phytoplankton composition and succession in Lake Scharmützelsee from 1999 to 2001

Development of phytoplankton biomass in lake Scharmützelsee was similar in 1999 and 2001 reflecting the common eutrophic situation in this lake. In 2000, phytoplankton biomass was lower and algal composition was quite different.

Phytoplankton composition in lake Scharmützelsee was similar in 1999 and 2001 and characterized by the dominance of cyanobacteria from early summer to autumn. In 2000, phytoplankton biomass was lower and the algal composition was quite different with very low portions of cyanobacteria (maximum 38% of total biovolume in October). In 1999, phytoplankton started from a low biomass level with cyanobacteria and diatoms occupying one third to one half of the total biomass in winter and early spring and reached a maximum in April 1999. All phytoplankton groups were detected during spring with a dominance of cyanobacteria and cryptophytes until the end of the mass development interrupted by a clear water state at the end of May 1999. After that period, cyanobacteria dominated the phytoplankton with more than 80% of the total biomass (maximum = 9.45 $mm^3\ l^{-1}$). This phytoplankton assemblage was stable until October. Autumnal circulation led to a decrease of algal biomass to about 1 $mm^3\ l^{-1}$ during November to December with negligible portions of cyanobacteria. The winter/spring development in 2000 started very early in January/February with peaks over 3 $mm^3\ l^{-1}$. Dominating species were diatoms (*Cyclotella* (Kütz.) Brébisson, *Aulacoseira italica* (Ehrenb.) Simonsen and other diverse Centrales larger 10 μm in diameter). An irregular phytoplankton pattern at a very low total biomass level (below 2 $mm^3\ l^{-1}$) and diverse species composition was observed during the whole year with a notable dominance of *Cryptomonas erosa/ovata* (Ehrenb.) and *Rhodomonas minuta* (Skuja). The initial phytoplankton pattern in 2001 was comparable to that of 2000. A very early diatom peak (3.19 $mm^3\ l^{-1}$) from January to April with the dominance of *Stephan-*

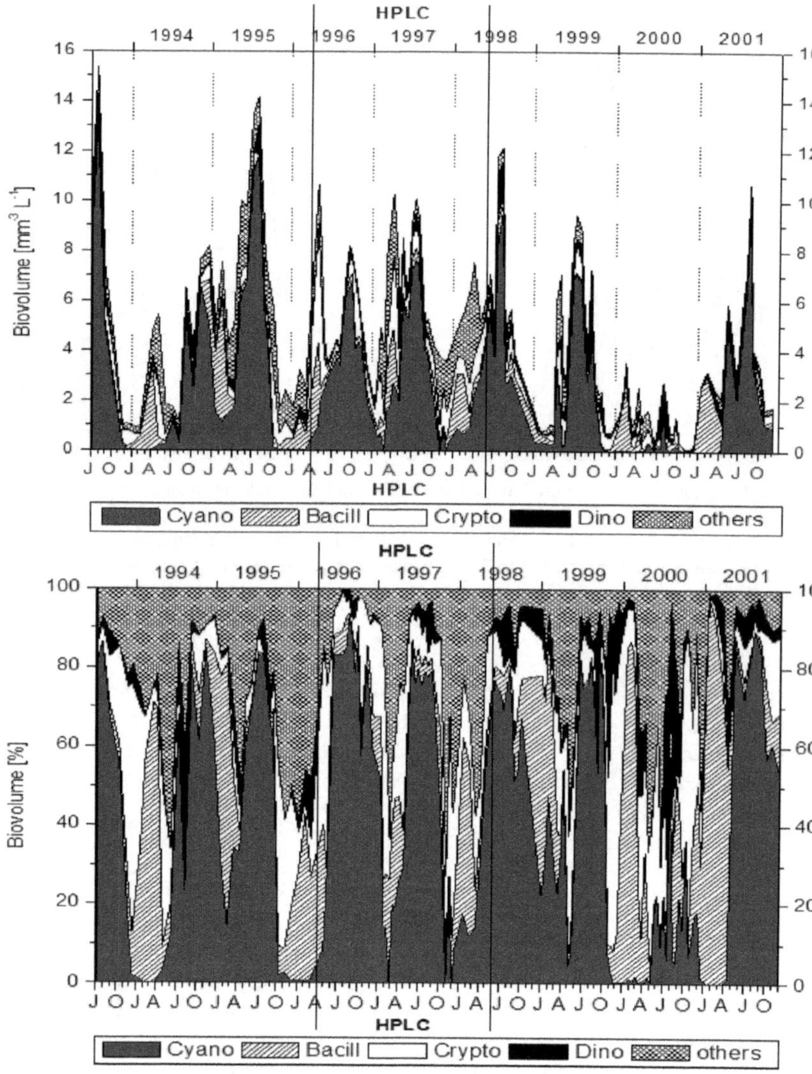

Figure 3. Phytoplankton development in lake Scharmützelsee from 1993 to 2001 (upper part: most important taxa or groups as biovolume, lower part: portion of dominant groups and the remaining phytoplankton on total biovolume).

odiscus spp. (Ehrenb.) was replaced in April by cyanobacteria (*Aphanizomenon gracile, Pseudanabaena limnetica, Planktothrix agardhii*) with a short decrease in total biomass in June 2001 during a clear water phase. From June to September, a dominance of cyanobacteria was observed with very high portions of more than 80% on total biomass, i.e. a steady state for 6 weeks. With autumnal circulation, the biomass decreased without a shift in the dominance structure.

Summarizing the taxonomic analysis, 119 taxa were distinguished and on average 25 taxa per sampling date were counted (17 – 42). Centrales as well as Chroococcales were not determined at species level and no differentiation between *Cryptomonas ovata* and *C. erosa* was done.

In Table 4, the summer algal taxa with more than 10% on total biovolume are listed according to their dominance spectra. Dominant cyanobacteria with highest biomass belong to the codon $\mathbf{S_1}$ (*Pseudanabaena limnetica, Planktothrix agardhii*), $\mathbf{S_N}$ (*Cylindrospermopsis raciborskii*) and **H1** (*Aphanizomenon gracile*).

Table 4. Phytoplankton list of the dominant species (>10% of total biomass) and their biovolume (average and maximum value) in lake Scharmützelsee in July – October (upper table) and November – June (table above) in 1999 – 2001, sorted acc. portion% = Maximum portion on total biomass. Members of the Cyanobacteria are marked (bold). Heterotrophic species were excluded

Date	Biovolume [$mm^3\ m^{-3}$]	Portion %	Taxon or species	Average biovolume [$mm^3\ m^{-3}$]
Taxa dominant from July-October				
23.08.2001	8594	80.3	**Pseudanabaena limnetica**	2745
28.10.1999	1171	59.8	*Cryptomonas erosa/ovata*	480
24.07.2001	3379	55.5	**Aphanizomenon gracile**	1418
28.08.2000	337	43.5	*Fragilaria crotonensis*	300
15.08.2000	303	35.5	*Ceratium hirundinella*	238
09.08.1999	1413	23.9	**Cylindrospermopsis raciborskii**	1413
12.09.2000	153	23.3	*Erkenia subaequiciliata*	153
24.10.2000	212	22.2	*Rhodomonas minuta*	129
10.09.1999	1293	17.8	*Peridinium* spp., small	1293
11.10.2000	44	15.2	**Microcystis flos-aquae**	44
28.08.2000	112	14.4	*Dinobryon sociale* var. *americana*	112
13.10.1999	319	14.0	*Cosmarium punctulatum* var. *subpunctulatum*	319
28.10.1999	214	13.4	*Gymnodinium* cf. *ubberrium*	214
29.10.2001	205	12.3	**Planktothrix agardhii**	205
26.09.2000	155	11.7	**Anabaena flos-aquae**	155
12.09.2000	68	10.3	**Anabaena** cf. **planctonica**	68
10.09.1999	746	10.3	*Peridinium umbonatum*-complex	746
Taxa dominant in winter–early summer (Nov – Jun)				
12.02.2001	2714	87.0	*Stephanodiscus* spp.	1536
12.11.1999	1045	72.8	*Cryptomonas erosa/ovata*	270
14.05.2001	4165	70.9	**Aphanizomenon gracile**	1538
22.06.1999	5921	70.2	**Pseudanabaena limnetica**	1221
02.02.2000	1582	55.8	*Cyclotella* spp.	999
20.06.2000	134	49.8	*Eutetramorus fottii*	134
06.06.2000	1849	43.2	*Rhodomonas minuta*	474
18.12.2000	112	38.1	*Closterium aciculare*	112
04.02.1999	852	36.5	*Centrales* div. 12–30 μm	333
19.05.1999	1036	32.3	*Erkenia subaequiciliata*	391
25.04.2000	402	26.8	*Chlorococcales*	288
04.12.2001	433	25.3	**Planktothrix agardhii**	433
04.12.2000	17	18.4	*Closterium acutum* var. *variabile*	17
18.01.2000	249	16.9	*Aulacoseira italica*	129
24.11.1999	119	16.3	*Gymnodinium* sp.	119
09.12.1999	101	15.0	*Ceratium hirundinella*	101
06.06.2000	146	14.4	**Planktolyngbya limnetica**	146
20.06.2000	39	14.4	*Ankyra* spp.	39
12.04.1999	838	14.0	*Volvocales*	838
25.04.2000	751	13.2	*Monoraphidium contortum*	337
21.12.1999	106	12.4	*Rhodomonas lens*	89
20.11.2000	7	11.7	**Anabaena flos-aquae**	7
25.04.2000	123	11.2	*Centrales* div. <12 μm	123

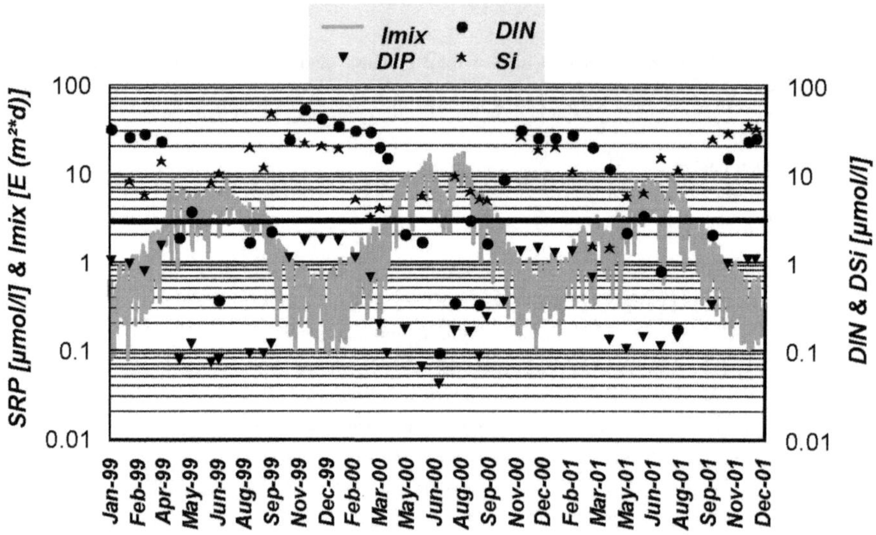

Figure 4. Main nutrients (DIP, DIN and DSi in μmol l^{-1}) and under water light intensity (E m^{-2} d^{-1}) calculated with the approach of Riley (1957). The bold line marks the threshold for under water light intensity.

Discussion

According to the topics of the 13th IAP workshop we discuss in this study: (1) the ecological conditions of steady state (or non steady state) assemblages in eutrophic shallow and deep lakes, (2) the mechanisms affecting the attainment of steady state in natural phytoplankton assemblages in different types of lakes.

Scheffer et al. (2003) contributed remarkable ideas to the equilibrium debate based on Hutchinson's "paradox of plankton". The authors suggest that even in homogenous and constant environment plankton will never never attain equilibrium and its dynamics is therefore unpredictable. Nevertheless, algal biomass may show regular patterns.

Steady state conditions occur more regularly in shallow lakes and are more predictable compared to deep lakes

Concerning the steady state and equilibrium condition according to Sommer et al. (1993), we focus on the summer when cyanobacteria were dominant with the exception of 2000 in the deep and dimictic lake Scharmützelsee, although cryptophytes and diatoms were very abundant during the non summer seasons. Unfortunately, the taxonomic estimation of these algal groups was done only at genus level in most cases of centric diatoms.

Mischke & Nixdorf (2003) showed distinct late summer steady states in the highly eutrophicated shallow lakes Langer See and Melangsee of between 3 – 4 and 7 weeks. The dominant taxa belong to the filamentous cyanobacteria, i.e. *Planktothrix agardhii, Aphanizomenon gracile, Pseudanabaena* spp. in lake Langer See and *Limnothrix amphigranulata, Pseudanabaena* spp., *Planktothrix agardhii* in lake Melangsee.

In the deep and dimictic lake Scharmützelsee, the same species were dominant from summer to autumn in 1999 and 2001 (*Pseudanabaena limnetica, Planktolyngbya* cf. *limnetica/subtilis* (Anagn. et Kom.), *Planktothrix agardhii, Cylindrospermopsis raciborskii, Aphanizomenon gracile*). The succession described for the years 1999 and 2001 is a regular and recurring phenomenon and could be observed in the past 8 years with the exception of 1994 and 2000 (Rücker et al., 2003). Within the studied period 1999 – 2001, steady state was only found in 2001 lasting at least 6 weeks from the end of July to the beginning of September with dominance of *Pseudanabaena limnetica* and *Aphanizomenon gracile* and to some extent *Cylindrospermopsis raciborskii*. In one instance (23.8.2001), one species (*Pseudanabaena limnetica*) contributed to more than 80% to the total biovolume.

A comparable phytoplankton composition was observed in lake Scharmützelsee in 1999 but steady state conditions according to Sommer et al. (1993) were not fulfilled. This confines especially to the first criteria: in no case 1, 2 or 3 species of algae contributed more than 80% of total biomass. Phytoplankton in 2000 was characterized by a high diversity and the absence or low portion of cyanobacteria. No phytoplankton steady states could be found in 2000.

The most successful associations in all three lakes belong to the $\mathbf{S_1}$ assemblages: *Planktothrix agardhii, Pseudanabaena* spp. and the very thin *Limnothrix* species like *L. amphigranulata.* These assemblages (group $\mathbf{S_1}$) are typical in turbid mixed layers with highly light deficient conditions and are often accompanied by *Cylindrospermopsis raciborskii* and *Aphanizomenon gracile* in all three lakes. The permanent dominance of Oscillatoriales during summer and autumn has often been reported for eutrophic lakes in central Europe (Berger & Sweers, 1988). Almost all of the 12 lakes in the lake chain of the Scharmützelsee region are or were dominated by filamentous cyanobacteria independent on morphometry and mixis (Nixdorf & Deneke, 1997, Rücker et al., 2003). This phenomenon was first described by Wundsch (1940) and Czensny (1938) and referred to as the 'Oscillatoriales disease' for the lakes in that region.

In contrast to the common hypothesis which was formulated for the 13th IAP workshop "The repetitive phytoplankton assemblages of deep lakes compared to the unpredictable assemblages of shallow ones: Is it only connected to a more resilient environment or is it someway linked to the pool of species that can live in that environment?" we found predictable assemblages in the shallow lakes but three different stages of phytoplankton succession in the dimictic lake from 1999 to 2001. The phenomenon of a regular steady state in eutropic shallow lakes could be a result of adaptations to a resilient and extreme environment. The pool of species that can live or survive in that environment will be considered by a comparison of habitat properties, such as the depth of the mixed layer, the mean daily irradiance within this layer and the nutrient resources in the different study sites.

Functional assemblages of species in relation to basic habitat properties

The ability of the $\mathbf{S_1}$ species to live on minimal income of light energy has long been recognized (Gibson & Foy, 1983; Foy, 1983; Post et al., 1985) and is confirmed by our investigations. Although the nutrient resources in both types of lakes are near threshold levels, indicating limitation (or better control or positive or negative responses as suggested by Reynolds et al. [2002]) by dissolved nutrients (DIP, DIN, TIC, DSi,), the under water light supply seems to be the key factor favoring the dominance of Oscillatoriales. In Table 5, a comparison of habitat properties and threshold values defined by Reynolds et al. (2002) and values measured in late summer in lake Scharmützelsee (RIE), Melangsee (MEL) and lake Langer See (LAN) is shown. The expected responses of selected functional groups of phytoplankton to these factors are given.

Table 5. Habitat properties and threshold values defined by Reynolds et al. (2002) and values measured in late summer in lake Scharmützelsee (RIE), Melangsee (MEL) and lake Langer See (LAN). Below, the expected responses to them of selected functional groups of phytoplankton defined by Reynolds et al. (2002)

	Mixing depth	Light	Nutrients		
Variable	h_m	I^*	[P]	[N]	[Si]
Unit	m	E m^{-2} d^{-1}	mol l^{-1}	mol l^{-1}	mol l^{-1}
Threshold	< 3	I^*<3	$<10^{-7}$	$<10^{-6}$	$<10^{-5}$
		Scharmützelsee (RIE)			
Average	>	≥	≥	≥	>
Min	4.0	0.8	0.4	0.9	5
Max	6.0	9.6	2.5	2.8	49
		Melangsee (MEL)			
Average	<	>	>	<	>
Min	0.2	4.7	2	0.2	3.5
Max	1.6	8.3	6	1.7	20.3
		Langer See (LAN)			
Average	<	≥	>	>	>
Min	0.5	2.4	1	0.2	2.3
Max	2.1	6.0	4	2.2	18.0
		Functional groups and their response			
S_1	+	+	-	-	+
S_2	+	+	-	-	+
S_N	+	+	-	+	+
H_1, H_2	+	-	-	+	+

It is obvious that the nutrient conditions in the shallow lake Melangsee and in the dimictic lake Scharmützelsee are very similar. In lake Langer See, the concentrations of total and dissolved inorganic N and P are higher. The main difference in habitat properties are the mixing behavior and the under water light regime in the different lakes. The large and wind-exposed shallow lake Langer See was very turbid and during the steady state at or below the threshold of the mean under water light in the mixed column. Using data sets of phytoplankton, light and nutrients measured in different years of investigations, Wiedner et al. (2002) and Mischke (2003) concluded independently that Oscillatoriales are able to outcompete Nostocales under declining photosynthetically active radiation (PAR). This ability was promoted by the growth of cyanobacteria itself and had a positive feed back on the development of Oscillatoriales but a negative one on Nostocales. The authors showed that *P. agardhii* outcompeted *Limnothrix* spp. under higher phosphorus concentrations due to higher resuspension of phosphorus into the water column in lake Langer See. In general, *P. agardhii* is favoured in turbid mixed and highly light deficient layers but it is most successful in shallow turbid mixed layers with intensive interaction between sediment and water. This species is often able to persist throughout the year as a monoculture. Scheffer (1998) described the result of this very successful behavior as the 'third stable state' for shallow lakes which is often observed in the very shallow hypertrophic lakes in the German lowlands. It is therefore recommended that this species should be included in the codon, the association of the shallow mixed layers (Reynolds et al., 2002).

The light climate in the deep lake Scharmützelsee is generally similar to that of lake Langer See although the phytoplankton biomass and therefore the turbidity is lower. This dimictic lake has a 2 – 3 fold greater epilimnetic mixing depth. During the steady state conditions in late summer 2001, the under water light irradiance decreased to less than 1 E m^{-2} d^{-1}. This relative darkness combined with low DIP concentration supported the development of the thin filamentous Oscillatoriales (*Pseudanabaena limnetica*).

The very shallow, small and wind-protected lake Melangsee is the lake with the highest under water light intensity. During steady state conditions, it is never below the threshold given by Reynolds et al. (2002). On the other hand, the TP and DIP concentrations are relatively low and comparable to the conditions in the epilimnion of the dimictic lake Scharmützelsee. The late summer dominance of *Limnothrix amphigranulata* can be explained by the better ability to use inorganic nutrients at low concentrations and without pulsed supply from the sediment.

Our results are confirmed by the "absolute, relative, functional and pragmatic" characterization of shallow lakes in the essay of Padisák & Reynolds (2003). The authors described "functional shallowness" including the structural segregation of the water column and the processes that segregation permits, favours or controls ecological behaviour in shallow and deep lakes. This behaviour as well as nature and distribution of their biota is related to the extent of bottom sediment within the trophogenic part of most shallow lakes as it was shown in our results in Langer see.

Conclusions

The dominance of Oscillatoriales in lakes of the Scharmützelsee region is a result of the anthropogenic induced eutrophication process and has been observed since the 1930s.

Our results concerning steady state conditions within the phytoplankton community according to Sommer et al. (1993) confirmed the general allocation of the Oscillatoriales within the $\mathbf{S_1}$ codon (Reynolds et al., 2002).

Steady state conditions were characterized by the dominance of single solitary filamentous cyanobacteria: *Planktothrix agardhii* in lake Langer See, *Limnothrix amphigranulata* in lake Melangsee and *Pseudanabaena limnetica* in the dimictic lake Scharmützelsee. All lakes are characterized by turbid mixed layers and highly light deficient conditions.

It is supposed that *Planktothrix agardhii* can be more successful in very shallow, wind-exposed lakes with intense sediment water interactions and induced short term nutrient pulses. We did not observe a dominance of this species in turbid, dimictic lakes. It is recommended to allocate this species into the codon (shallow turbid mixed layers).

Limnothrix-like species (this means thin filaments like *Limnothrix amphigranulata, Pseudanabaena limnetica*) are favoured in environments with lower mixing intensity or deeper mixing depth (e.g. in the epilimnion of deep lakes). This confirms the hypotheses of Rücker et al. (1997) on the difference between Planktothrix and Limnothrix dominated lakes.

Acknowledgements

The authors thank all members of the laboratory staff at the Limnological Research Station in Bad Saarow, especially Gudrun Lippert, Tina Hanke, Ute Abel and Kornelia Tahedel for the chemical analyses. Ingo Henschke and Wolfgang Terlinden are greatly thanked for the reliable field work. Phytoplankton analyses were done by LBH Freiburg. This work was partly supported financially by the German Research Foundation (DFG) contract AZ NI347/5-1 and SFB 565.

References

Behrendt, H. & B. Nixdorf, 1993. The carbon balance of phytoplankton production and loss processes based on in situ measurements in a shallow lake. Int. Rev. ges. Hydrobiol. 78: 439–458.

Berger, C. & H. E. Sweers, 1988. The Ijsselmeer and its phytoplankton – with special attention to the suitability of the lake as a habitat for *Ocillatoria agardhii* GOM. J. Plankton Res. 10: 579–599.

Connell, J., 1978. Diversity in tropical rain forests and coral reefs. Science 199: 1304–1310.

Czensny, R., 1938. Die Oscillatorienerkrankung unserer Seen, Biologie und Chemismus einiger märkischer Seen. Vom Wasser 8: 36–57.

DEV (Deutsche Einheitsverfahren zur Wasser-, Abwasser- und Schlammuntersuchung), 1976–1998. Schlammuntersuchung. Verlag Chemie, Weinheim, D11, E5, D9, C9, E1, H7.

Foy, R. H., 1983. Interaction of temperature and light on the growth rates of two planktonic *Oscillatoria species* under a short photoperiod regime. British Phycol. J. 18: 267–273.

Gibson, C. E. and R. H. Foy, 1983. The photosynthesis and growth efficiency of a planktonic blue-green algae, *Oscillatoria redekei*. British Phycol. J. 18: 631–638.

Hardin, G., 1960. The competitive exclusion theory. Science 131: 1292–1297.

Mischke, U. & J. Rücker, 2001. Veränderungen der Zusammensetzung der Algenzönose in Standgewässern des Scharmützelseegebietes. In Krumbeck, H. & U. Mischke (Hrsg.) AR 6/2001: 19–38.

Mischke, U., 2003. Cyanobacteria associations in shallow polytrophic lakes: influence of environmental factors. Acta Oecologica, 24 (Supplement 1): 11–24.

Mischke, U. & B. Nixdorf, 2003. Equilibrium phase conditions in shallow German lakes: how Cyanoprokaryota species establish a steady state phase in late summer. Hydrobiologia 502 (Dev. Hydrobiol. 172): 123–132.

Naselli-Flores, L., J. Padisák, M. T. Dokulil & I. Chorus, 2003. Equilibrium/steady-state concept in phytoplankton ecology. Hydrobiologia 502 (Dev. Hydrobiol. 172): 395–403.

Nixdorf, B. & R. Deneke, 1997. Why 'very shallow' lakes are more successful opposing reduced nutrient loads? Hydrobiologia 342/343: 269–284.

OECD, 1982. Eutrophication of Waters, Monitoring, assessment and control. OECD, Paris. 154 pp.

Post, A. F., R. de Witt & L. Mur, 1985. Interaction between temperature and light intensity on growth and photosynthesis of the Cyanobacterium *Oscillatoria agardhii*. J. Plankton Res. 7: 487–495.

Padisák, J. & C. S. Reynolds, 2003. Shallow lakes: the absolute, the relative, the functional and the pragmatic. Hydrobiologia 506–509 (in press).

Reynolds, C., V. L. M. Huszar, C. Kruk, L. Naselli-Flores & S. Melo, 2002. Towards a functional classification of the freshwater phytoplankton. J. Plankton Res. 24: 417–428.

Reynolds, C. S. & J. W. G. Lund, 1988. The phytoplankton of an enriched, soft-water lake subject to intermittent hydraulic flushing (Grasmere, English Lake District). Freshwat. Biol. 19: 379–404.

Reynolds, C. S., 1984. Phytoplankton periodicity: the interactions of form, function and environmental variability. Freshwat. Biol. 14: 111–142.

Reynolds, C. S., 1997. Vegetation processes in the pelagic: a model for ecosystem theory, Ecology Institute, D-21385 Oldendorf/Luhe, Germany.

Riley, G. A., 1957. Phytoplankton in the north central Sargasso Sea 1950–1952. Limnol. Oceanogr. 2: 252–272.

Rott, E., 1981. Some results from phytoplankton counting intercalibrations. Schweizerische Zeitschrift für Hydrol. 43: 34–62.

Rücker, J., C. Wiedner & P. Zippel, 1997. Factors controlling the dominance of *Planktothrix agardhii* and *Limnothrix redekei* in eutrophic shallow lakes. Hydrobiologia 342/343: 107–115.

Rücker, J., B. Nixdorf, R. Deneke, A. Kleeberg & U. Mischke, 2003. Unterschiedliche Reaktionen von Seen im Scharmützelseegebiet (Brandenburg) auf die Reduzierung der externen Belastung. Wasser & Boden 55/4: 4–10.

Scheffer, M., 1998. Ecology of shallow lakes. Chapman & Hall, London.

Scheffer, M., S. Reinaldi, J. Huisman & F. J. Weissing, 2003. Why plankton communities have no equilibrium: solutions to the paradox. Hydrobiologia 491: 9–18.

Schmitt, M. & B. Nixdorf, 1999. Spring phytoplankton dynamics in a shallow eutro- phic lake. Hydrobiologia 408/409: 269–276.

Sommer, U., J. Padisák, C. S. Reynolds & P. Juhász-Nagy, 1993. Hutchinsons heritage: the diversity-disturbance relationship in phytoplankton. Hydrobiologia 249: 1–7.

Utermöhl, H., 1958. Zur Vervollkommnung der quantitativen Phytoplankton-Methodik. Mitt. int. Ver. theor. angewan. Limnol. 9: 1–38.

Wiedner, C., 1999. Toxische und nicht-toxische Cyanobakterien in Gewässern der Scharmützelseeregion: Ihr Vorkommen in Gewässern unterschiedlicher Trophie und Morphometrie und Steuermechanismen ihrer Dynamik in polymiktischen Flachseen. Dissertation BTU Cottbus.

Wiedner, C., B. Nixdorf, R. Hinze, B. Wirsing, U. Neumann & J. Weckesser, 2002. Regulation of cyanobacteria and microcystin dynamics in polymictic shallow lakes. Archiv für Hydrobiol. 155: 383–400.

Wiedner, C., J. Rücker & P. Zippel, 1996. Besonderheiten des Blaualgenregimes in ausgewählten Gewässern des Scharmützelseegebietes. In Nixdorf, B. & A. Kleeberg (Hrsg.), Gewässerreport Scharmützelseegebiet, Teil II. BTUC-AR 2/96: 50–60.

Willén, E., 1976. A simplified method of phytoplankton counting. British Phycol. J. 11: 265–278.

Wundsch, H. H., 1940. Beiträge zur Fischereibiologie märkischer Seen, VI. die Entwicklung eines besonderen Seentypus (H2S-Oscillatorien-Seen) im Flußgebiet der Spree und Havel, und seine Bedeutung für die Fischereibiologischen Bedingungen in dieser Region. - Z. Fischerei XXXVIII: 443–648.

Zippel, P. & B. Nixdorf, 1997. Die Entwicklung des Phytoplanktons im Scharmützelsee (1993–1996) und Storkower See. In Deneke, R. & B. Nixdorf (Hrsg.), Gewässerreport (Teil III). BTUC-AR 5/97: 60–71.

Hydrobiologia **502:** 123–132, 2003.
L. Naselli-Flores, J. Padisák, M. T. Dokulil (eds),
Phytoplankton and Equilibrium Concept: The Ecology of Steady-State Assemblages.

Equilibrium phase conditions in shallow German lakes: How Cyanoprokaryota species establish a steady state phase in late summer

Ute Mischke[1] & Brigitte Nixdorf[2]
[1]*IGB, Institute of Freshwater Ecology and Inland Fisheries, Dept. of Shallow Lakes and Lowland Rivers, Müggelseedamm 310, D-12587 Berlin, Germany*
E-mail: b.nixdorf@t-online.de
[2]*Brandenburg Technical University of Cottbus, Seestraße 45, D-15526 Bad Saarow*
E-mail: nixdorf@t-online.de

Key words: shallow lakes, steady state assemblage, cyanobacteria, mixing conditions, diversity, phytoplankton

Abstract

In 2000, a field study in two shallow, polytrophic lakes (Langer See and Melangsee) in eastern Germany revealed an equilibrium state assemblage of Cyanoprokaryota in late summer. During 4 successive weeks in Langer See *Planktothrix agardhii* (Gom.) Anagn. et Kom., *Aphanizomenon gracile* (Lemmerm.) Lemmerm. and *Pseudanabaena limnetica* Lemmerm. were more than 80% of the standing biomass of phytoplankton, and their cumulative biovolume was around 33 $mm^3 l^{-1}$ ((±3.2 SD). In Melangsee, the very small *Limnothrix* species *L. amphigranulata* (Van Goor) Meffert was the most common species, accompanied by *Pseudanabaena limnetica* and *Planktothrix agardhii*. For 3 weeks, their cumulative biovolume was about 23 $mm^3 l^{-1}$ ((±3.4 SD), which represented 75 – 82% of total biovolume. The dominant species all belong to the functional group **S1** defined by Reynolds (1997), except for *A. gracile*, which we suggest to be included in group $\mathbf{S_N}$. In both lakes mean light intensities ranged between 2.2 and 8.3 E $m^{-2} d^{-1}$. Overall species spectra were very similar in both lakes, but dominance by *Limnothrix* and by *Planktothrix* in the respective lakes is observed repeatedly. The success of these species is discussed in the context of the habitat properties in August/September. Summer mixing events represented no disturbances in the sense of Connell (1978), since they do not interrupt the species dominance. More frequent mixing events and higher concentrations of dissolved nitrogen occurred in Langer See than in the more shallow, but wind protected Melangsee. In Langer See light deficient conditions were intensified by an increasing biomass of *P. agardhii*, and this species probably benefited from nutrient input by more frequent resuspension. The light deficiency also affected the diversity, expressed as Shannon–Wiener Index (H), which was reduced more in lake Langer See (H = 0.51) than in Melangsee (0.74) during steady state periods. Recognizing the important effects of mixing, we suggest an additional variable to describe habitat properties: the number of full mixing days as a proportion of total days of observation should help to discriminate between shallow habitats with intermittent mixing events, and those with more regularly mixing in summer period.

Introduction

Steady state conditions are expected to be more likely in deep lakes than in shallow mixed ones. Shallow habitats are more responsive to changes in weather conditions. Strong wind events are a frequent summer disturbance and lead to nutrient pulses due to a more intensive sediment resuspension (Wiedner et al., 2002). In these habitats, environmental factors such as nutrients, light and stagnation undergo frequent change between deficiency and superfluity. It is obvious that these irregularly occurring mixing events are the predominant disturbance to the succession of phytoplankton in such shallow lakes. In the sense of the 'Intermediate Disturbance Hypothesis of Connell (1978), the occurrence of frequent disturbances should lead to a niche for pioneers, which can co-exist beneath those species, which are specialized to specific conditions, subsequently the diversity and the species richness should increase.

In late summer, in the two hypertrophic shallow lakes (Langer See and Melangsee) regularly occurring populations of Cyanoprokaryota could be observed (Zippel, 1996; Rücker et al., 1997, Wiedner et al., 2002), which consisted of few species of the functional group **S1**, $\mathbf{S_N}$ and **H1** defined by Reynolds et al. (2002). Different species were successful however, in these geographically closely located lakes (Mischke, 2003). Competition for nutrients and for light has been discussed as responsible for the different development of the genus *Planktothrix* versus *Limnothrix/Pseudanabaena* species (Rücker et al., 1997; Wiedner et al., 2002; Mischke, 2003). Using the criteria of Sommer et al. (1993) the periods of their dominance can be denoted as characteristic equilibrium phases. Mischke (2003) described the seasonal development of phytoplankton, in the context of environmental conditions and the thermal stratification as measured by two continuously-measuring data loggers in 1999 and 2000.

This paper analyses the summer habitat properties of both lakes using the criteria of Reynolds et al. (2002), to give a better insight into the conditions, which favor Cyanoprokaryota species, and how they manage to suppress the development of pioneers and other competitors. Additionally, it is a test if the threshold values given by Reynolds et al. (2002) are suitable predictors of the defined functional groups.

Description of sites studied

Melangsee and Langer See are two shallow lakes in the eastern part of Brandenburg in Germany. Both lakes are located on the border of the south plateau of the Berlin glacial valley. They are parts of two different lake systems with water retention times about respectively 22 and 60 days. Melangsee belongs to the Glubig-lake-chain, originating from the Weichsel glacier. Both sites are polymictic and polytrophic lakes. More frequent, deeper and longer lasting mixing events could be observed in Langer See than in the more shallow, but wind protected Melangsee (Nixdorf & Deneke, 1997; Mischke, 2003). Fishponds in the catchment area of Melangsee can cause flushing events in spring (Schmitt & Nixdorf, 1999). The sampling stations were located near the lake centers, but the lake depth at sampling site was not the deepest point of Lake Langer See (Table 1).

Materials and methods

Water samples were collected in the morning every week in summer 2000 using a Ruttner-sampler. For the rest of the year sampling was at least biweekly. A depth profile was taken in 0.5 m steps and these samples were mixed together giving a single-mixed (depth-integrated) sample for chemical and biological analysis.

Water temperatures, oxygen saturation and pH were measured *in situ* with a multi-parameter probe (Hydrolab Corporation). Global radiation (J cm^{-2} d^{-1}) was kindly provided by the meteorological station in Lindenberg located 15–30 km from the two lakes and was converted into photon flux (E m^{-2} d^{-2}) by a factor of 0.0445 as photosynthetic active radiation (PAR) at water body surface (I_o). Underwater photosynthetic active radiation (PAR) was measured using two scalar LI 193 SB Spherical Quantum Sensors (LiCor) at 0.5m intervals through the water columns. The parameters light attenuation factor (k_d) and depth of euphotic zone (z_{eu} in [m]) were calculated from these variables. Mean light intensities (I^* in E m^{-2} d^{-1}) were calculated after Behrendt & Nixdorf (1993) using surface light intensity (I_o), light attenuation factor (k_d) and mixing depth ($Z_{mix} = Z_{mean}$ see Table 1).

Mixed samples were analyzed for dissolved nutrients (soluble reactive phosphorus [SRP], nitrate and ammonia) according to standard methods. The sum of the two dissolved nitrogen fractions is given as DIN, since nitrite concentrations were negligibly small. Concentrations of all nutrients ($\mu g\ l^{-1}$) were calculated as atomic mol per liter (mol l^{-1}), making them more easily comparable to the thresholds of Reynolds et al. (2002). The detection limits of the methods were 1.4 N 10^{-6} mol l^{-1} for ammonia, 0.1 10^{-6} N mol l^{-1} for nitrate and 0.03 10^{-6} P mol l^{-1} for soluble reactive phosphorus.

Phytoplankton analysis

Phytoplankton species were identified by light microscopic observation (Zeiss Axioskop) of living and Lugol-fixed samples. Commonly used taxonomic monographs about cyanobacteria (Anagnostidis & Komárek, 1985, 1888; Komárek & Anagnostidis, 1989) were supplemented with recently published original literature for species discrimination (Horecká & Komárek, 1979; Meffert et al., 1981; Kohl et al., 1985; Hindák, 1988; Meffert, 1988). Phytoplankton biomass was determined in Lugol fixed subsamples by determination of cell concentration according to Utermöhl (1958) and estimation of cell volumes according to Rott (1981) using an inverted microscope (Zeiss Ax-

Table 1. Morphometric and hydrological parameters of the investigated lakes

	Z_{max} [m]	Z_{mean} [m]	$Z_{sampling\ point}$ [m]	Volume [10^6 m^3]	Area [km^2]	Mixing type
Melangsee	3.3	1.6	1.6	0.17	0.11	Polymictic
Langer See	3.8	2.1	2.7	3.27	1.55	Polymictic

iovert). For cell counts all Lugol-fixed samples from May until October were diluted 1:19 by filtered drinking water and 10 ml volume of that dilution were used for settling in counting chambers to achieve an equal distribution of algal filaments. The given number of taxa per sampling date included categories discriminated on the level of species, genus or in six cases at the class level. Categories on class level included centric diatoms, which were determined as *Stephanodiscus hatzschii* and *Cyclotella* species, but were enumerated only in three size classes during cell counting procedure. The logarithm (log_2) of Pi (Pi = species biovolume / total biovolume) was used for Shannon–Wiener Index ($H = -\sum(Pi\ log_2[Pi])$) and Evenness ($E = H/log_2(S)$; S = total number of species). Both diversity indices were separately calculated for each sampling date.

Equilibrium periods were indicated, when (1) three species contribute more than 80% of the standing biomass and (2) the total biomass of phytoplankton deviated less then 15% from the value at the previous sampling date, so no significant changes in total biomass occurred (see definition given by Sommer et al., 1993).

Results

In summer, the plankton of the two lakes were dominated either by *Planktothrix agardhii* (Gom.) Anagn. et Kom. or by the *Limnothrix* species *L. amphigranulata* (Van Goor) Meffert (see Figures 1 and 2).

Equilibrium phase development

In Langer See, the equilibrium phase development is dominated by a maximum of three species, which together contribute more than 80% of the standing biomass. Their coexistence persisted from August to the mid of November in 2000. During four weeks no significant changes in total biomass occur (33.2 $mm^3\ l^{-1}$ $\pm$ 3.23, see Fig. 1). The equilibrium phase started on 6 September and was ended by a biomass decline after 4 October, when their high portion of total biomass still persisted (see Fig. 1, upper graph). *P. agardhii* constituted the greatest specific population biovolume of 31.4 $mm^3\ l^{-1}$ at the end of August (see Fig. 1). In both lakes *Aphanizomenon gracile* (Lemmerm.) Lemmerm. was the most significant Nostocales and made up more than 20% of the biovolume in early summer. In Langer See, it remained important into late summer. The third dominant species was the very fragile species *Pseudanabaena limnetica* Lemmerm. Its biovolume was included into the counting group *Pseudanabaena* spp. with few filaments of other *Pseudoanabaena* species, which could not be differentiated during counting process.

Not all the criteria of an equilibrium phase defined by Sommer et al. (1993) were fulfilled in Melangsee. The cumulative biovolume of the three most abundant Oscillatoriales surpassed a 80% portion of total biovolume only in late August, but presented more than 75% in the two subsequent weeks and was the most important phytoplankton assemblage throughout the whole summer (see Fig. 2). In late summer, the portion of *L. amphigranulata* decreased by about one third, while in contrast, *Pseudanabaena* became more important in Melangsee.

Diversity

In contrast to the strong dominance of few species, the species richness of phytoplankton was high in both lakes. Up to 100 taxa per sampling date contributed to total biomass and on average 65 in Langer See and 70 taxa in Melangsee per sampling date from May until October were seen (see Fig. 3). A total of 167 taxa were distinguished in the phytoplankton, 89% of them were common to both sites. 43 species had a biovolume of more than 1 $mm^3\ l^{-1}$ at any one sampling date. At the beginning of summer, cyanobacteria already formed 70% of total biovolume with a significant contribution of the 12 – 19 Cyanoprokaryota species, which belong to the functional

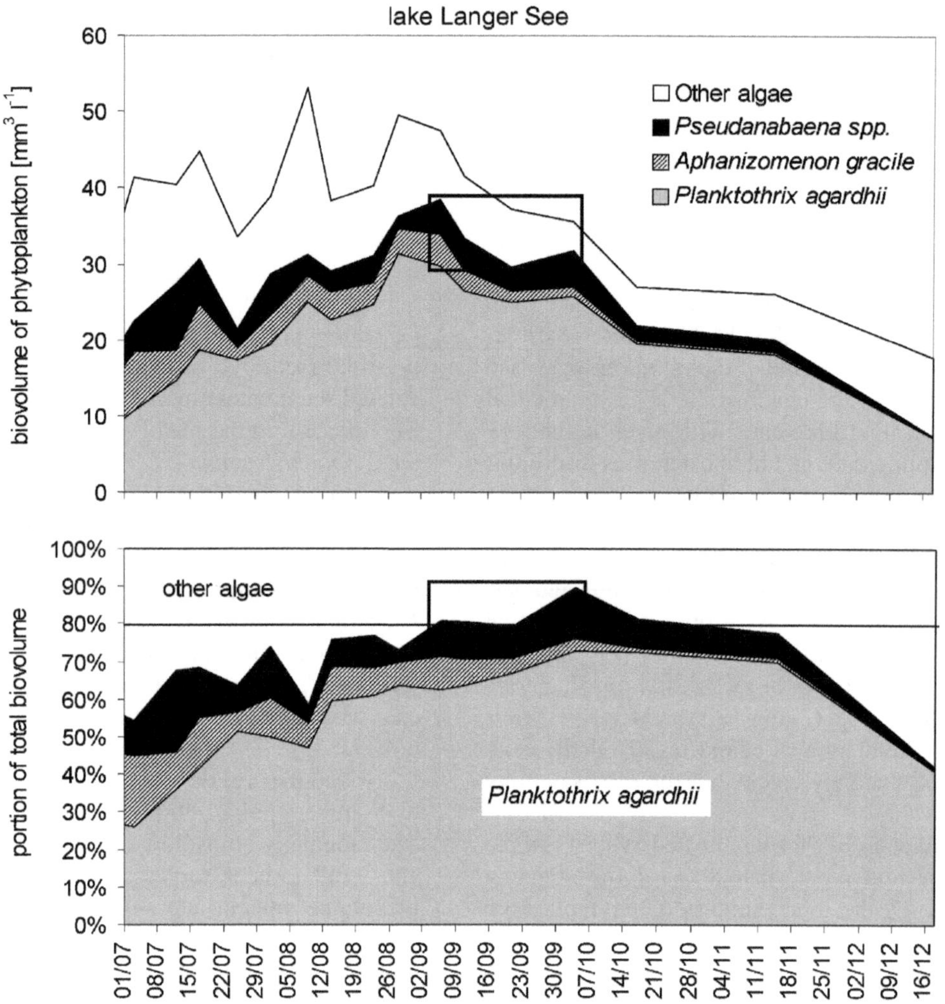

Figure 1. Seasonal development of the three, most abundant species and other algae (white areas) in Langer See from July 1st to the end of the year 2000. Period of equilibrium phase is circled by a quarter field. Definition see text. Upper graph: Cumulated biovolume of three species and the rest of phytoplankton. Below: Contribution of the same species (areas) to total biovolume of phytoplankton.

groups **S1**, **S2**, $\mathbf{S_N}$ and **H1** defined by Reynolds et al. (2002) and by Padisák & Reynolds (1998). The number of significant cyanobacterial species (portion of total biomass >1%) decreased during summer. Especially noteworthy, the biomass of N-fixing Cyanoprokaryota such as *Cylindrospermopsis raciborskii* (Wolz.) Seenayya et Subba Raju and *Aphanizomenon issatschenkoi* (Usac.) Prosk.-Lavr. declined (Mischke, 2003). The calculated Shannon–Wiener Index (H) and the Evenness (E) of all sampling dates are given in Figure 3. During the equilibrium phase defined above, both diversity indices reached their minima of the year ($H(\log_2) = 1.68$ in Langer See and 2.44 in Melangsee). On average from May to October, the phytoplankton of Langer See differed from that of Melangsee having a significantly lower diversity (H = 2.96 e.g. H = 3.54, see also Figure 3).

Habitat properties

Using the criteria and threshold values of habitat properties defined by Reynolds et al. (2002), Table 2 gives the limiting characteristics of the habitats in the lakes Melangsee and Langer See in August and September 2000. (1) The depth of surface mixed layer is mainly restricted to about 2 m by lake morphometry at both study sites (see Table 1), but lake Melangsee is shallower and is better wind protected. This can be inferred from a detailed record of the lake temperatures, measured by two continuously-measuring

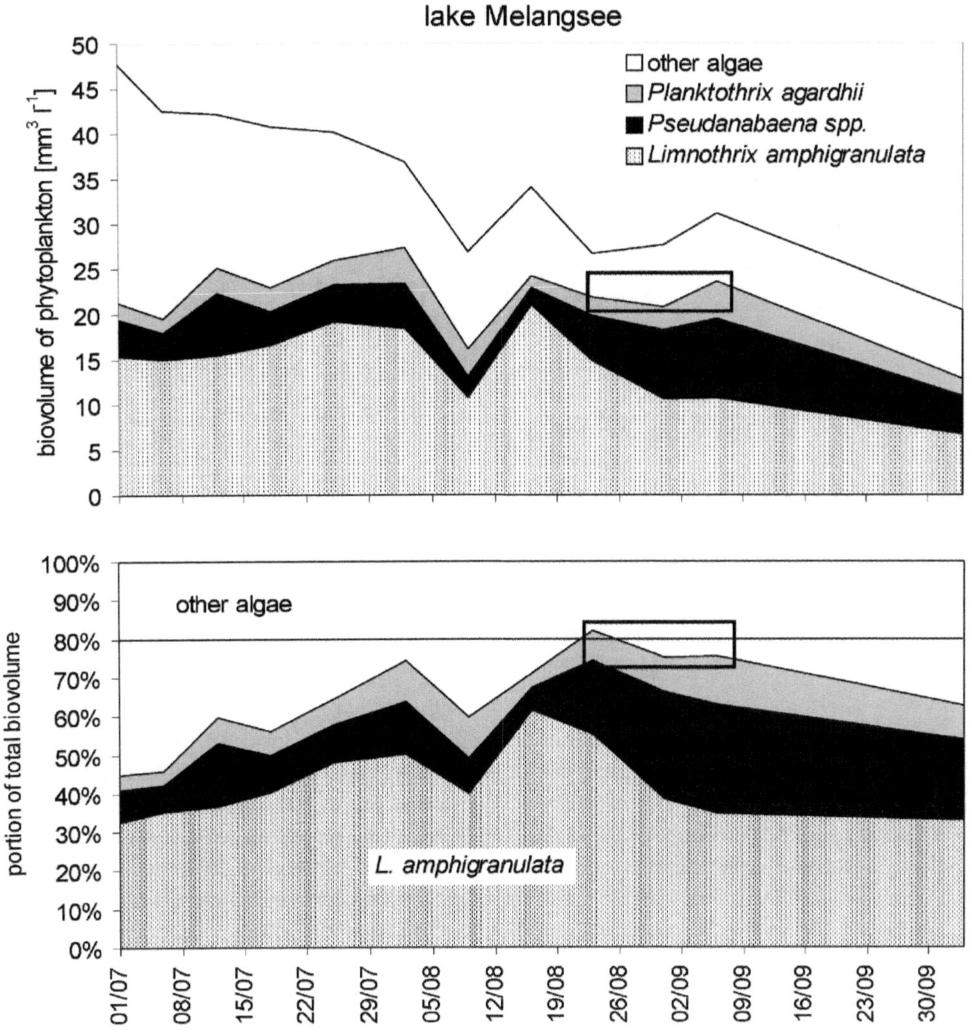

Figure 2. Seasonal development of the three, most abundant species and other algae (white areas) in Melangsee from July 1st to the end of the September in 2000. Period of equilibrium phase is circled by a quarter field. Definition see text. Upper graph: Cumulated biovolume of three species and the rest of phytoplankton. Below: Contribution of the same species (areas) to total biovolume of phytoplankton. Note shorter investigation period in comparison to Figure 1.

data loggers. Mischke (2003) showed that the number of days with complete mixing was different in both lakes: From 25 July until the 23 October 2000 the *in-situ* temperature difference (15 min. values of near bottom and near surface probe) was more than 0.2 °C for 38 complete days in Melangsee and only for 28 days in Langer See. (2) The phytoplankton experienced mean daily irradiance levels (I*) between 2.4 and 10.0 mol photons m^{-2} d^{-1} (Figure 4, lowest graphs). High self-shading effects were indicated by water Secchi depth of less than 0.5 m (Figure 4). The surface light intensity averaged 63 mol photons m^{-2} d^{-1} and was strongly reduced by a light attenuation factor of 2.78 and of 3.92 m^{-1} in the water bodies of Melangsee and Langer See, respectively. (3) The average water temperature was very similar in both lakes (18.4 °C and 18.6 °C) and the surface water warmed up to 22.9 °C. (4) Concentrations of soluble reactive phosphorus remained just above 2 10^{-7} mol l^{-1} (Fig. 4). (5) The concentration of dissolved inorganic nitrogen were below 10^{-6} mol l^{-1} in Melangsee and slightly above 10^{-6} mol l^{-1} in Langer See. Nitrate was always less than 0.36 10^{-6} mol l^{-1}, except for one occasion at the beginning of September in Langer See (Fig. 4). (6) The concentrations of soluble reactive silicon exceeded 10^{-5} mol l^{-1}, and were sufficient to

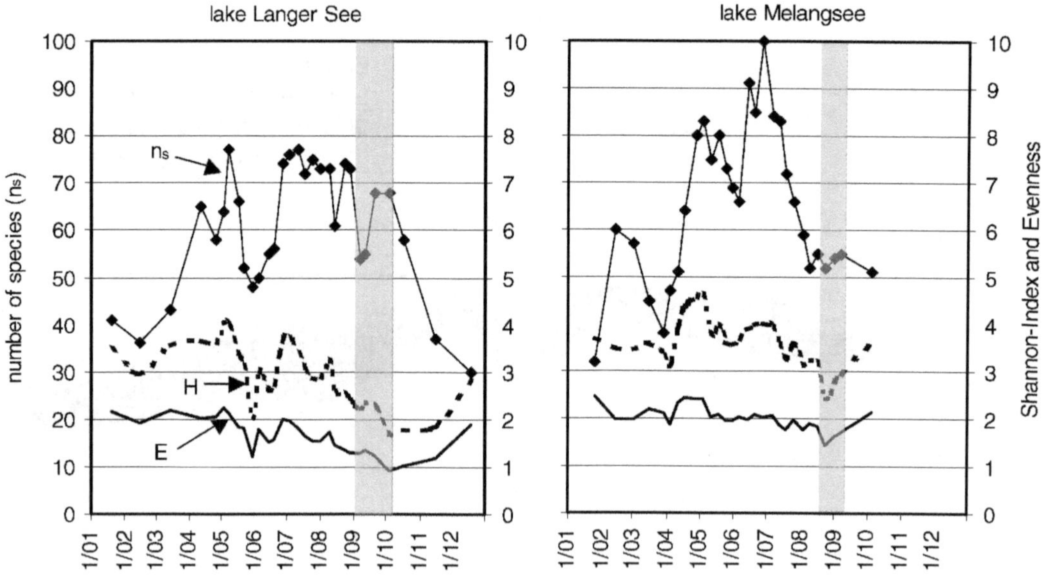

Figure 3. Seasonal development of total number of phytoplankton species (ns), Shannon–Wiener Index (H ($\log_2$)) and Evenness (E) in Langer See and Melangsee in the year 2000. Equilibrium phases are highlighted by gray shadows.

allow diatom growth. (7) Due to frequent mixing the pH values remained below 8.2 during the period of high primary production and thus concentration of dissolved carbon dioxide was not limiting. (8) The daily grazing rate by rotiferan and crustacean zooplankton is unknown.

Discussion

A small number of species of the association **S1** were able to replace the diverse Cyanoprokaryota community present at the beginning of summer, which resulted in a steady state phase in late summer. The succession described for 2000 is a regular and recurring phenomenon, noted over the past 8 years of observation (Rücker et al., 1997; Wiedner et al., 2002; Mischke, 2003). Our conceptual model explaining this pattern of behaviour is summarized schematically in Figure 5 and is discussed below.

We suggest that the Cyanoprokaryota assemblage, which has established in summer, induces light deficient conditions by high shading effects, that other phytoplankton association are suppressed. The shading effect operates even during turnover/mixing events, because the lakes are very shallow and therefore, the total biomass is not diluted by plankton poor, deeper waters, as it happens in deeper lakes during summer mixing events. The growth of other species, such as *Cylindrospermopsis* or *Aphanizomenon issatschenkoi* which are known to tolerate low light deficient conditions and mixed layers, is suppressed by the low water temperatures. *Cylindrospermopsis* is more successful in tropical lakes (Padisák, 1997; Komárková & Tavera, 2003) still a gradual adaptation to lower water temperatures cannot be excluded in Europe (Dokulil & Mayer, 1996; Mischke, 2003; Nixdorf et al., 2003).

Different members of the successful association **S1** found in both lakes showed a differential tolerance for mixing and nutrient deficient conditions, therefore the ecological niches of the two habitats were occupied either by toxic (*Planktothrix agardhii*) or non-toxic species (*Limnothrix/Pseudanabaena*).

Due to the hypertrophic status of the lakes, phosphorus is not a limiting factor. In Figure 5 this fact is indicated by the relatively high level of the horizontal line, presenting the soluble reactive phosphorus concentrations (SRP) availability (Naselli-Flores & Barone, 2003; Stoyneva, 2003) and with blooms of *Microcystis* the development of cyanoprokaryotes may be suppressed by the relatively low SRP concentrations found in our shallow lake systems. The light- and the nitrogen deficient conditions are thus the most important factors, which not only decide the outcome of competition within the functional group **S1**, but also influence the possibility of other phytoplankton communities.

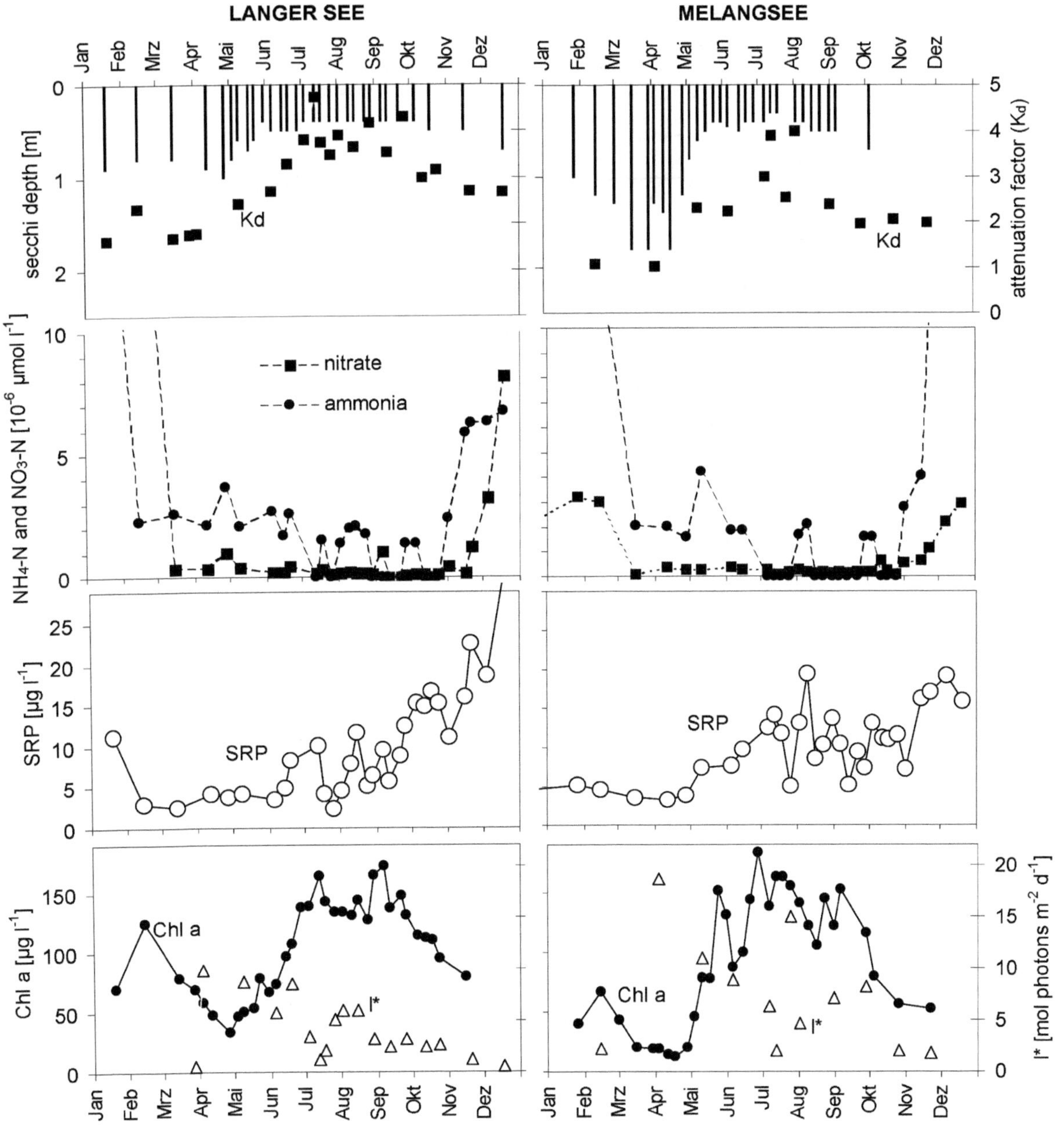

Figure 4. Secchi depths and attenuation factors (Kd; both upper graphs), concentrations of dissolved inorganic nitrogen (nitrate, ammonia, graphs below), of soluble reactive phosphorus (SRP, third graphs) and of chlorophyll a (Chl a) and daily irradiance levels (I*, lowest graphs) in the lakes Langer See (left side) and Melangsee (right side) in 2000.

Nitrogen

The threshold value to indicate nitrogen deficient conditions defined by Reynolds et al. (2002, see Table 2) is very low, and is less than the detection limit of the method for ammonia applied in this study. It cannot be excluded that even in Melangsee the concentrations of DIN were still above this threshold value, but they never exceeded 50 μg l^{-1} in summer (Mischke, 2003), another threshold value in widespread use to detect N-limitation.

However, since the Oscillatoriales and not the N-fixing Nostocales ultimately dominated the summer succession in both lakes over several years (Wiedner et al., 2002; Mischke, 2003), it can be concluded, consistent with Reynolds's threshold value, that the

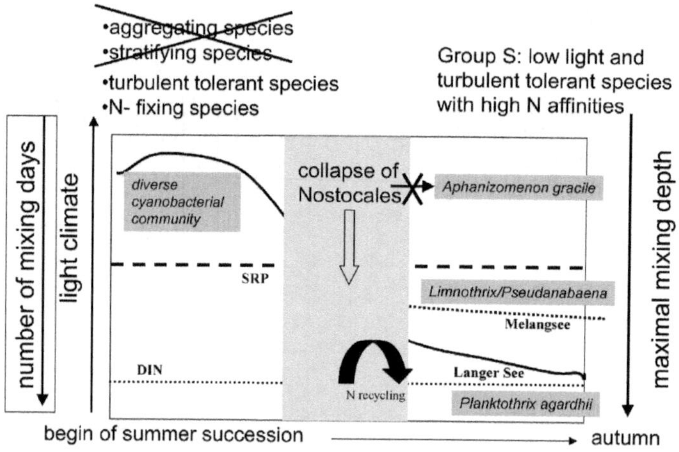

Figure 5. Model of phytoplankton succession in two shallow polytrophic lakes in summer. In late summer the phytoplankton composition changed from a divers one to a composition, which is dominated by *Planktothrix agardhii* in Langer See or by *Limnothrix/Pseudanabaena* in Melangsee. Higher number of mixing days, lower light regime (y-axis on left side) and deeper mixing depth (x-axis on right side) are the characteristic environmental conditions of Lake Langer See, and differ from those of Lake Melangsee. Only phosphorus supply (indicated as the SRP horizontal line) is sufficient in both study sites. Light deficiency caused collapse of Nostocales, subsequently, nitrogen was set free and recycled by Oscillatoriales (black circular arrow).

Table 2. Habitat properties and threshold values defined by Reynolds et al. (2002) in the two study sites in summer (August/September). Additionally, the new habitat property 'mixing' [Mix] is defined as the proportion of complete mixing days relative to total days of the investigation period. Surpassing (>) or limiting (<) of threshold values is given for the summer average (August/September) in the same units and exponential as the threshold values and minima (Min) and maxima (Max) for both lakes during the same period

	Mixing depth	Light	Temperature	Nutrients			Carbon	Grazing	Mixing/ Stagnation
Variable	h_m	I^*	θ	[P]	[N]	[Si]	$[CO_2]$	f	[Mix]
Unit	m	$E\ m^{-2}\ d^{-1}$	°C	$mol\ l^{-1}$	$mol\ l^{-1}$	$mol\ l^{-1}$	$mol\ l^{-1}$	%/100	
Threshold	< 3	I^*<1.5	<8	$<10^{-7}$	$<10^{-6}$	$<10^{-5}$	$<10^{-5}$	<0.4	<33% of d
Melangsee									
Summer average	<	>	>	>	<	>	>	High?	<
Min	0.2	4.7	13.0	2	0.2	3.5			
Max	1.6	8.3	22.9	6	1.7	20.3			
Langer See									
Summer average	<	>	>	>	>	>	>	High?	>
Min	0.5	2.4	12.8	1	0.2	2.3			
Max	2.1	6.0	22.4	4	2.2	18.0			

supply of nitrogen did not determine the composition of Cyanoprokaryota. The Nostacales *Aphanizomenon gracile* is an exception, since this species turned out to be tolerant of light deficient, and mixing conditions and did behave like the members of group **S1** and $\mathbf{S_N}$ found in this study.

All the species of functional group **S1** must have a high nitrogen affinity to manage population growth under the given environmental conditions. In contrast, species of *Limnothrix* and *Planktothrix agardhii* are known to suffer nitrogen limitation when DIN concentration falls below 50 $\mu g\ l^{-1}$. More frequently Nostocales species became dominant under nitrogen deficient conditions (Barica et al., 1980), a situation found also at the beginning of summer in both study sites (Mischke, 2003). It is unclear, how the cyanobacteria meet their nitrogen demand. In mid-summer, nitrogen was probably set free subsequently after the population collapse of Nostocales, which was induced by increasingly lower light conditions. This recycling process presented a potential N source to the Oscillatoriales, whose biomass increased thereafter. This hypothesized N-recycling is represented in Figure 5 by the black arrow at the horizontal line as a symbol for DIN concentrations. Furthermore, it should be kept in mind that the pelagial of shallow lakes receives nutrients pulse-wise due to resuspension of the lake sediment during mixing events (Wiedner, 1999). This is true for the phosphorus release also, and especially *Planktothrix agardhii* is able to use such pulses by rapid incorporation as shown by Wiedner (1999) in enclosure experiments. In the more stagnant Melangsee, as in the nearby dimictic lake Scharmützelsee, *Limnothrix* and *Pseudanabaena* species are more successful (Nixdorf et al., 2003).

Light conditions

In very shallow lakes the mixing depth is strongly limited by lake morphometry. During full circulation periods enough light remains for low light adapted Cyanoprokaryota species, when water depths is restricted to about 2m as is the case for both study sites. In spite of intensive phytoplankton development, mean light intensity does not fall beneath the threshold value defined by Reynolds et al. (2002). In contrast the light attenuation factor is extremely high especially in lake Langer See, which is a certain sign of high turbidity. Thus, the threshold is not applicable to detect low light conditions in such a case.

When summer mixing events are regarded as disturbances in shallow lakes, these disturbances were too weak to replace specialist species and to promote pioneers.

Small differences in the light climate favor either one or other of the Oscillatoriales species of the functional group **S1** in both the shallow habitats. Taking account of morphometric factors such as the deeper water depth, and more frequent mixing days in Langer See, together with the higher light attenuation factors, the under water light climate was reduced much more strongly in Langer See, due to one third higher phytoplankton biomasses than in Melangsee (see Figs 1 and 2). Consistent with this fact, the reduction of the diversity was stronger at this study site. In summary, the shading effect controlled the algal composition. Disturbances by mixing events were not strong enough to replace the low light specialist *Planktothrix agardhii*. In conclusion, the reduced success of *P. agardhii* in Melangsee is accounted for by its inability to build up a comparable high biomass because of the highly nutrient deficient conditions, so effects of shading could not develop as strongly as in Langer See.

Mixing conditions

Mixing conditions alter very frequently in shallow lakes, thus weekly measurements can miss important mixing events occurring in between observations. Therefore, it is difficult to describe mixing conditions correctly. To solve this problem, continuously measuring data loggers should be applied. In the sites studied here, our results show clearly, that a wind protected shallow lake such as Melangsee experience complete mixing much less frequently than deeper, but wind-exposed one such as Langer See (Mischke, 2003). Taking into consideration the overwhelming importance of mixing for the resuspension of nutrient rich sediment matter, we suggest a further variable to describe habitat properties: the portion of full mixing days *vis-à-vis* total days of observation should help to discriminate between shallow habitats with intermittent mixing events, from those with more regular and complete mixing in summer period.

Conclusions

Repetitive phytoplankton assemblages even occur in shallow lakes. Intermittently stagnant conditions replace a high number of functional groups, whose success depends on those conditions, which they are adapted to, continuing for a critical time period. Few species tolerate the combination of intermittent nutrient deficiency and low light conditions produced by the frequent but irregular mixing. The remaining species are able to build up very dense populations, which intensifies turbidity, so most of the Nostocales populations collapse. The steady state period is a self- induced habitat, in which competitors fail because of low-light conditions combined with effective exploitation of nutrient resources by the dominant Cyanoprokaryota. Therefore, we conclude, that members of the group **S1** are able to change the limiting factor actively in shallow lakes in full correspondence to Zevenboom & Mur (1980). The extent of summer mixing determines the nutrient exchange between sediment and pelagial zone and subsequently determines the potential phytoplankton biomass, which can develop. Summer mixing events in very shallow, polytrophic lakes do not represent disturbances in the sense of Connell (1978), since they do not interrupt the species dominance. The habitat of the functional group **S1** is well defined as a 'turbid, mixed layers'. Similar, members of group $\mathbf{S_N}$ as *Cylindrospermopsis* and *Aphanizomenon issatschenkoi* are successful in this habitat. *Aphanizomenon gracile* should be included here. The threshold value of 1.5 E m^{-2} d^{-1} for mean light intensities suggested by Reynolds et al. (2002) seems to be too low and does not predict the dominance of low light adapted species of group **S1**, which occurs above 10 E m^{-2} d^{-1}.

Acknowledgements

The work was carried out at the research station Bad Saarow of the Brandenburg Technical University of Cottbus. The authors thank all members of the laboratory staff, especially Gudrun Lippert, Tina Hankem, Ute Abel and Kornelia Tahedel for the chemical analysis. Ingo Henschke and Wolfgang Terlinden are greatly thanked for the reliable field work. Claudia Wiedner greatly assisted this work by discussions. This work was supported financially by the German Research Foundation (DFG) contract AZ NI347/5-1.

References

Anagnostidis, K. & J. Komárek, 1988. Modern approach to the classification system of cyanophytes. 3. Oscillatoriales. Archiv für Hydrobiol. Supplement 80 1/4: 327–472.

Anagnostidis, K. & J. Komárek, J., 1985. Modern approach to the classification system of cyanophytes. 1. Introduction. Archiv für Hydrobiol. Supplement 71 1/2: 291–302.

Barica, J., H. Kling & J. Gibson, 1980. Experimental manipulation of algal bloom composition by nitrogen addition. Can. J. Fish. aquat. Sci. 37: 1175–1183.

Behrendt, H. & B. Nixdorf, 1993. The carbon balance of phytoplankton production and loss processes based on *in situ* measurements in a shallow lake. Int. Rev. ges. Hydrobiol. 78: 439–458.

Connell, J., 1978. Diversity in tropical rain forests and coral reefs. Science 199: 1304–1310.

Dokulil, M. T. & J. Mayer, 1996. Population dynamics and photosynthetic rates of a *Cylindrospermopsis - Limnothrix* association in a highly eutrophic urban lake, Alte Donau, Vienna, Austria. Algol. Stud. 83: 179–195.

Hindák, F., 1988. Planktonic species of two related genera *Cylindrospermopsis* and *Anabaenopsis* from Western Slovakia. Algological Studies 50–53: 283–302.

Horecká, M. & J. Komárek, 1979. Taxonomic position of three planktonic blue-green algae from the genera *Aphanizomenon* and *Cylindrospermopsis*. Preslia 51: 289–312.

Kohl, J.-G., G. Dudel, M. Schlangstedt & H. Kühl, 1985. Zur morphologischen Abgrenzung von *Aphanizomenon flos-aquae* RALF ex. BORN. et FLAH. and *A. gracile* (LEMM.) LEMM., Archiv für Protistenkunde 130: 119–131.

Komárková, J. & R. Tavera, 2003. Steady state of phytoplankton assemblage in the tropical Lake Catemaco (Mexico). Hydrobiologia 502 (Dev. Hydrobiol. 172): 187–196.

Komárek, J. & K. Anagnostidis, 1989. Modern approach to the classification system of cyanophytes. 4. Nostocales. Archiv für Hydrobiol. Supplement 82: 247–345.

Meffert, M.-E., 1988. *Limnothrix* Meffert nov. gen.. Archiv für Hydrobiol. Supplement 80: 269–276.

Meffert, M.-E., R. Oberhäuser & J. Overbeck, 1981. Morphology and Taxonomy of Oscillatoria redekei (Cyanophyta). British Phycol. J. 16: 107–114.

Mischke, U., 2003. Cyanobacteria associations in shallow polytrophic lakes: Influence of environmental factors. Acta Oecologica, 24: 11–23.

Naselli-Flores, L. & R. Barone, 2003. Steady-state assemblages in a Mediterranean hypertrophic reservoir. The role of *Microcystis* ecomorphological variability in maintaining an apparent equilibrium. Hydrobiologia 502 (Dev. Hydrobiol. 172): 133–143.

Nixdorf, B. & R. Deneke, 1997. Why 'very shallow' lakes are more successful opposing reduced nutrient loads? Hydrobiologia 342/343: 269–284.

Nixdorf, B. U. Mischke & J. Rücker, 2003. Phytoplankton assemblages and steady state in deep and shallow eutrophic lakes – an approach to differentiate the habitat properties of Oscillatoriales. Hydrobiologia 502 (Dev. Hydrobiol. 172): 111–121.

Padisák, J., 1997. *Cylindrospermopsis raciborskii* (Woloszynska) Seenayya et Subba Raju, an expanding, highly adaptive cyanobacterium: worldwide distribution and review of its ecology. Archiv für Hydrobiolie Supplement 107: 563–593.

Padisák, J. & C. S. Reynolds, 1998. Selection of phytoplankton associations in Lake Balaton, Hungary, in response to eutrophication and restoration measures, with special reference to the cyanoprokaryotes. Hydrobiologia 384: 41–53.

Reynolds C., V. L. M. Huszar, C. Kruk, L. Naselli-Flores & S. Melo, 2002, Towards a functional classification of the freshwater phytoplankton. J. Plankton Res. 24: 417–428.

Reynolds, C. S., 1997. Vegetation processes in the pelagic: a model for ecosystem theory, Ecology Institute, d^{-2}1385 Oldendorf/Luhe, Germany.

Rott, E., 1981. Some results from phytoplankton counting intercalibrations. Schweizerische Zeitschrift für Hydrol. 43: 34–62.

Rücker, J., C. Wiedner & P. Zippel, 1997. Factors controlling the dominance of *Planktothrix agardhii* and *Limnothrix redekei* in eutrophic shallow lakes. Hydrobiologia 342/343: 107–115.

Schmitt, M. & B. Nixdorf, 1999. Spring phytoplankton dynamics in a shallow eutrophic lake. Hydrobiologia 408/409: 269–276.

Sommer, U., J. Padisák, C. S. Reynolds & P. Juhász-Nagy, 1993. Hutchinsons heritage: the diversity-disturbance relationship in phytoplankton. Hydrobiologia 249: 1–7.

Stoyneva, M. P., 2003. Steady-state phytoplankton assemblages in shallow Bulgarian wetlands. Hydrobiologia 502 (Dev. Hydrobiol. 172): 169–176.

Utermöhl, H., 1958. Zur Vervollkommnung der quantitativen Phytoplankton-Methodik. Mitt. int. Ver. theor. angewan. Limnol. 9: 1–38.

Wiedner, C., 1999. Toxische und nicht-toxische Cyanobakterien in Gewässern der Scharmützelseeregion: Ihr Vorkommen in Gewässern unterschiedlicher Trophie und Morphometrie und Steuermechanismen ihrer Dynamik in polymiktischen Flachseen. Dr, Thesis. Brandenburgischen Technischen University: 1–131.

Wiedner, C., B. Nixdorf., R. Hinze, B. Wirsing, U. Neumann & J. Weckesser, 2002. Regulation of cyanobacteria and microcystin dynamics in polymictic shallow lakes. Archiv für Hydrobiol. 155: 383–400.

Zevenboom, W. & L. R. Mur, 1980. N2-fixing cyanobacteria: Why they do not become dominant in Dutch hypertrophic ecosystems?. In Barica, J. & L. R. Mur (eds), Hypertrophic Ecosystems, Developments in Hydrobiology 2. Dr W. Junk Publishers, The Hague: 123–130.

Zippel, P., 1996. Phytoplanktonsukzession und -dynamik in den Gewässern des Scharmützelseegebietes. In Nixdorf, B. & A. Kleeberg. Gewässerreport (Teil II). Aktuelle Reihe 2/96: 44–60. Published by University of Cottbus.

Hydrobiologia **502:** 133–143, 2003.
L. Naselli-Flores, J. Padisák, M. T. Dokulil (eds),
Phytoplankton and Equilibrium Concept: The Ecology of Steady-State Assemblages.

Steady-state assemblages in a Mediterranean hypertrophic reservoir. The role of *Microcystis* ecomorphological variability in maintaining an apparent equilibrium

Luigi Naselli-Flores & Rossella Barone
Dipartimento di Scienze Botaniche, Università di Palermo, Via Archirafi, 38, I-90123 Palermo, Italy
E-mail: luigi.naselli@unipa.it

Key words: thermal stability, mixing regime, phytoplankton morphology, C-S-R-strategy

Abstract

Lake Arancio is a hypertrophic Mediterranean man-made lake, located on the southern coast of Sicily. Its artificial origin and the climate make it a very dynamic environment, strongly characterised by very wide water-level fluctuations. These vertical water movements interfere with the thermal stability of the water body often causing the breaking of the thermocline in mid-summer. In addition, the summer level-decrease influences the nutrient dynamics and modifies the z_{mix}/z_{eu} ratio. All these modifications were observed to support a high environmental variability, which was reflected by the richness of its phytoplankton composition and by its dynamics. Nevertheless, an investigation carried out from March 2001 to March 2002 showed that the assemblage was strongly dominated by a few species, one by one. In particular, two different *Microcystis* morphotypes dominated the assemblage from mid-April till the beginning of October. The prolonged dominance of these 'species' should suggest that a steady state condition took place in Lake Arancio during spring and summer 2001. This is in contrast with previous investigations, which showed high diversity values especially occurring in the period of strong environmental instability when the continuous dewatering caused the breaking of the thermocline in the middle of summer. Nevertheless, this dominant species showed a very wide morphological variability and alternated among 'more S', 'S', and 'R' (*sensu* Reynolds) ecotypes. The ever-changing morphological features suggest a different ecological behaviour of the species involved. They seem to confirm that the environmental variability of Mediterranean reservoirs sustains high diversity values, even though this diversity has to be sought in the amplitude of morphological plasticity of one or a few species, rather than in the coexistence of a variety of species.

Introduction

In ecological works, scientists often refer to equilibrium and non-equilibrium hypotheses to try to explain the dynamics of specific communities (Naselli-Flores et al., 2003).

The non-equilibrium hypothesis posits that equilibrium forces at work are weak and that competitive interactions are minimal. A non-equilibrium model requires that environmental disturbances occur with sufficient frequency to disrupt the course of competitive exclusion (Harris, 1986). This view highlights the influence of disturbance on biological diversity. In particular, higher diversity values are attained when environmental and internal disturbances can prevent the establishment of equilibrium conditions (Sommer et al., 1993). In non-equilibrium models, species richness is decreased through random local extinction in a stochastic dynamic system (Scheffer et al., 2003). However, species richness can be maintained when the rates of local extinction are low and rapidly balanced by local immigration (inocula).

If systems operate hierarchically, then spatial and temporal heterogeneity play a central role in understanding population, community and ecosystem processes (Whittaker & Levine, 1977). Hierarchy theory is a formal approach to the complex influences of scale, and suggests that different phenomena (e.g. long term processes and disturbances) influence systems on different scales (Allen & Starr, 1982). In addition, the different scales (phenomena) are nested and interact. Climate changes alter vegetation over long periods of time. However, within a specific climate such as our current climate, disturbance can alter the vegetation at a much smaller temporal scale. While the vegetation reacts to the disturbance, it is still constrained by the current climate in how it can react (i.e., what species are available for colonisation) thus, disturbance is nested within the climatic regime. Disturbance is recognised as an important factor that increases heterogeneity at all scales. The frequency of disturbance events has been a focus of investigations in phytoplankton dynamics in Padisák et al. (1993).

According to Whittaker (1975), in a steady-state community, each species occupies a different niche which results from and reduces direct competition and thus species composition is more or less stable according to environmental predictability. In this respect, both temporal and spatial niche segregation is essential to maintain communities in steady state. In aquatic environments, this segregation can be reached only in the absence of disturbance: an eternal steady state where diversity is reduced to minimal levels by competitive exclusion.

Anyway, this absence of disturbance is unlikely to occur in most of the aquatic environments for long periods. Especially in Mediterranean reservoirs where management procedures applied to the water bodies produce wide water-level fluctuations and contribute to deeply alter hydraulic regimes, thermal stability, nutrient dynamics and, ultimately biological structure (Naselli-Flores, 1999). This peculiar hydrological behaviour coupled with meteorological constraints (e.g. amount of precipitation, wind regime, solar radiation trends) and trophic state of water bodies, contribute to making these aquatic environments very 'unstable' at a yearly, seasonal or daily scale. Different temporal scales in the occurrence of disturbance may require different adaptive strategies to phytoplankton. Thus, it was observed that regularly fluctuating environments showed a higher number of rare species 'ready to develop' when the conditions became suitable; on a shorter scale, morphological variability, commonly and frequently observed within species, may also help to overcome adverse conditions (Naselli-Flores & Barone, 2000).

An investigation carried out from March 2001 to March 2002 on phytoplankton dynamics in the hypertrophic, Mediterranean reservoir, Lake Arancio highlighted the occurrence of a long period of dominance of species, one by one, with constant low values of diversity and a strongly reduced species richness. This event should suggest the establishment of steady state conditions in Lake Arancio during spring and summer 2001 despite the fact that, in this period, the reservoir exhibited a strong environmental variability due to the continuous dewatering to fulfil irrigation purposes.

The aim of this paper is to investigate in details phytoplankton dynamics in Lake Arancio to better understand how steady-state assemblages may develop in very unstable environments and the possible role played by phytoplankton morphological variability in the maintenance of populations under different environmental conditions.

Materials and methods

Lake Arancio is a hypertrophic Mediterranean man-made lake, located on the southern coast of Sicily. A detailed morphological description of the reservoir is given in Barone & Naselli-Flores (1994). The mechanisms governing nutrient dynamics and thermal stability of the reservoir in the studied period were analysed in detail in Naselli-Flores (2003), whereas alkalinity and major ion dynamics were reported in Naselli-Flores et al. (2003b).

Phytoplankton was collected weekly from March 2001 to March 2002. Samples were taken 250 m away from the dam at depths corresponding to 100%, 50% and 1% of the surface irradiance, measured by a Li-Cor quantum sensor. These samples were mixed and a 100 ml subsample was immediately fixed with Lugol's solution for counting. Live samples were also taken for species identification using the most up-to date phycological literature. In particular, *Microcystis* taxa were identified according to Komárek et al. (2002). Cell counting was performed using a Zeiss-Axiovert 100 inverted microscope according to the sedimentation method developed by Utermöhl. In several cases, it was necessary to dilute the sample up to 10 times to perform counting. *Microcystis* density evaluations were simplified by the fact that colonies tended to disintegrate rather easily in Lugol's solution once the sample was squashed. Cell numbers in the few colonies intact after squashing were evaluated by counting their number in a reference area and then extrapolating to the entire colony. Biomass was estimated from biovolume, assuming unit specific gravity, by geometrical approximations according to Hillebrand et al. (1999). At least 200 specimens or colonies of the most abundant phytoplankton taxa were measured in each sample to calculate biovolume. This was calculated for each organism or colony and then the average biovolume was used as the final value. Growth rates were calculated by comparing species cell numbers in two consecutive samples (N_i and N_{i-1}) collected at the time t_i and t_{i-1} (in days) according to the formula:

$$r = \frac{lnN_i - lnN_{i-1}}{t_i - t_{i-1}} [day^{-1}]$$

Mixing depth was estimated from profiles of temperature measured with a multiparameter probe (Idromar IM5052) every 0.50 m, between 09:30 and 10:30 hours, at the same time intervals. In particular, it was considered equal to the epilimnetic zone when the reservoir showed stratification. In case of absence of stratification, mixing depth was taken equal to the average depth of the reservoir.

Euphotic depth was estimated from irradiance profiles as 1% of subsurface value.

The dimensionless parameter Relative Water Column Stability (RWCS) was calculated, according to Padisák et al. (2003), by comparing the density difference between bottom (D_b) and surface (D_s) water to the density difference between 4 °C (D_4) and 5 °C (D_5) of pure water, using the formula:

$$RWCS = \frac{D_b - D_s}{D_4 - D_5}$$

Water density was calculated from temperature values using a Water Density Calculator, available on the Internet, which calculates water density at a given temperature between –8 and 108 °C using 5-point Lagrange interpolation (Senese, 2003).

Results and discussion

The physical scenario

The nature of the lake as a water storage reservoir and the Mediterranean climate of its location make it a dynamic environment, strongly characterised by wide water-level fluctuations. The amplitude of these fluctuations is variable from year to year depending on the amount of water stored in winter and thus, on precipitation. During the studied period the water level decreased by 11 m. The depth at the sampling station varied between a maximum of 15 m at the beginning of July 2001, to a minimum of 4 m at the beginning of October 2001 (Fig. 1A). This extreme waterlevel drawdown disrupted the thermal stability of the water body causing the elimination of the thermocline in mid-summer. As a consequence, the summer level-decrease influenced the nutrient dynamics promoting a huge removal of nutrients from the sediment and a continuous transfer of phosphorus and nitrogen to the upper layers (Naselli-Flores, 2003). The storage of water in winter, and the lack of outflows from the reservoir, approximates the residence time to an infinite value and transforms the reservoir into a nutrient accumulator (Straškraba et al. 1995). A strong increase of the internal loading can be deducted by comparing the studied period to former investigations (Barone & Naselli-Flores, 1994; Naselli-Flores & Barone, 1998). Some data useful to better understand phytoplankton dynamics, and dealing with the physical conditions of the water column during the studied period, are reported in Figure 1A and 1B.

The stored volumes fluctuated between 21.5 and 2.8 Mm^3 in the study period (Fig. 1A). The maximum holding was recorded at the end of April 2001 and until the end of June the level decreased slowly mainly because of evaporative losses and due to the checking procedures to control the regular opening of the outlets. Since July, starting the irrigation period, the water losses increased constantly and an average value of 1.25 Mm^3 per week was utilised until the end of September. The average weekly reduction of the water column depth amounted to 0.86 m. On the 6th of October, to meet agriculture demands in the district, the management board started re-filling the reservoir with water from another reservoir, Lake Garcia. In total, 13 Mm^3 of water were added to Lake Arancio in six weeks, from October 6 to November 10. Of these, about 7 Mm^3 were used to continue irrigating and 6 Mm^3 increased the reservoir volume to 8.9 Mm^3 and the depth at the sampling station passed from 4 to 8 m. This value remained more or less stable throughout all winter since no precipitation occurred during this period of prolonged drought striking the island, as well as other southern Mediterranean zones, in the years 2000–2001.

From a thermal point of view, Lake Arancio is a potentially monomictic water body. The onset of stratification occurred at the end of March when the temperature difference between the surface and the bottom was about 2 °C. This value increased constantly up to 8.5 °C at the end of June. The pycnocline in the reservoir was found between 4.5 and 6 m of depth. Since the beginning of July an increasing tendency to mixing could be observed in the water column due to the decrease of depth and to the deep location of water withdrawal (Kennedy et al., 2002).

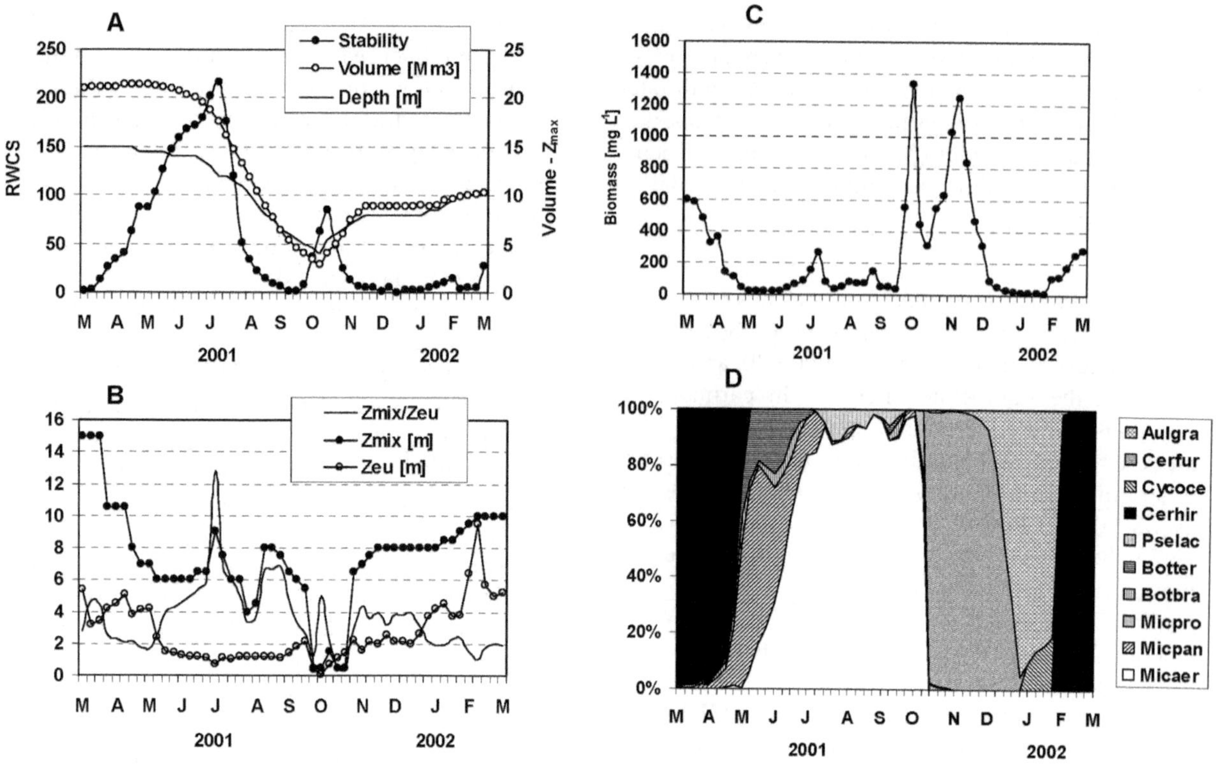

Figure 1. **A**: seasonal variations of relative water column stability (RWCS), depth at the sampling station (Zmax), and volume stored in Lake Arancio in the studied period. **B**: seasonal variations of mixing depth – euphotic depth ratio (Zmix/Zeu), mixing depth (Zmix) and euphotic depth (Zeu) in Lake Arancio in the studied period. **C**: seasonal variation of total phytoplankton biomass in Lake Arancio in the studied period. **D**: relative biomass of dominant species in Lake Arancio in the studied period. Alagra: *Aulacoseira granulata*; Cerfur: *Ceratium furcoides*; Cycoce: *Cyclotella ocellata*; Cerhir: *Ceratium hirundinella*; Pselac: *Pseudosphaerocystis lacustris*; Botter: *Botryococcus terribilis*; Botbra: *B. braunii*; Micpro: *Microcystis protocystis*; Micpan: *M.* cf. *panniformis*; Micaer: *M. aeruginosa*.

Accordingly, relative stability of the water column stratification reached its maximum of 220 at the end of June (Fig. 1A). With the onset of irrigation, relative stability rapidly decreased and the water column of the reservoir experienced a 2-week period of isothermy in mid-September. A second peak in relative stability was observed in mid-October. This occurred just after the starting of water income from Lake Garcia, when Lake Arancio doubled its volume and its water level increased by more than 2 m in a few days. Winter circulation started in mid-November.

The ratio between mixing depth and euphotic depth is generally above 2, reaching a peak above 12 at the beginning of July (Fig. 1B), due to a deepening of the mixing depth probably related to the opening of the outlets to start irrigation. Lowest values were recorded in mid-October due to the lack of wind, which caused the formation of diel thermoclines located in the uppermost 50 cm of the water column and to a contemporary increase of the euphotic depth.

It is rather to be expected that these extreme water level fluctuations, due to the operational procedures applied to the reservoir, along with the strong decrease of winter precipitation, make sufficient environmental variation to get phytoplankton responses that will support or confound present perceptions on natural selection mechanisms.

Phytoplankton dynamics

The lack of nutrient limitation is well depicted by the phytoplankton biomass values (Fig. 1C), generally above 25 mg l^{-1}, which reached numbers above 10^3 mg l^{-1} in late summer and autumn.

Previous investigations on Lake Arancio phytoplankton showed that the assemblages are generally formed by one or a few dominant species and several rare species (see Naselli-Flores & Barone, 1998). Conversely, the study period was characterised by a few species, rather clearly alternating. The reduction in species richness could be related to the progressive increase of eutrophication processes (Naselli-Flores, 2003). The adaptation required to these environmental conditions, especially because of the scarce underwater light availability, reduced the spectrum of variability of possible competitors. Spring was characterised by the dominance of a very large four-horned *Ceratium hirundinella* (O.F. Müller) Dujardin (Fig. 1D). *Microcystis* spp. and *Botryococcus* spp. accompanied this species. At the beginning of May the relative importance of *Microcystis* progressively increased compared to that of *Botryococcus* spp. and it resulted as the absolute dominant of the assemblage till mid-October.

Botryococcus was represented by two species: *B. braunii* Kützing and *B. terribilis* Komárek & Marvan. The first one is able to float because of its accumulation of oil. The second one, described by Komárek & Marvan (1992), is characterised by long, often ramified protuberances ornamented by oil droplets, which likely slow down its sinking velocity. Both these species showed to be able to coexist with *Microcystis* under relatively high z_{mix}/z_{eu} conditions. The colonies of *Botryococcus* spp., especially those of *B. terribilis*, were often observed inside larger colonies of *Microcystis*. It is not clear whether *B. terribilis* is a good species or it is an ecotype of *B. braunii* developing under alkaline conditions (Komárek, pers. com.). In any case, it is reasonable to hypothesise that the intense photosynthetic activity of *Microcystis* alkalinises the microenvironment and allows the development of this 'species'.

In mid-July *Botryococcus terribilis* was replaced by rather large, spherical colonies (200 μm) of *Pseudosphaerocystis lacustris* (Lemmermann) Nováková. Also this coexisted with *Microcystis* reaching its relative maximum value in the same underwater light conditions of *Botryococcus* spp.

Autumn months showed the dominance of the three-horned *Ceratium furcoides* (Levander) Langhans followed in winter months by *Aulacoseira granulata* (Ehrenberg) Simonsen and *Cyclotella ocellata* Pantocsek. The re-appearing of *Ceratium hirundinella* marked the end of winter period.

According to Reynolds et al. (2002), the sequence of associations $\mathbf{L}_M \rightarrow \mathbf{M}$ (with **F** representatives) $\rightarrow \mathbf{L}_M \rightarrow \mathbf{P}$ characterised the phytoplankton dynamics in the studied period. In particular, $\mathbf{L}_M$ association developed in the period of higher clarity of the water column and before the onset of thermal stratification. The large dimensions of the dinoflagellate *Ceratium hirundinella* (147 000 μm^3) allow it to easily cover

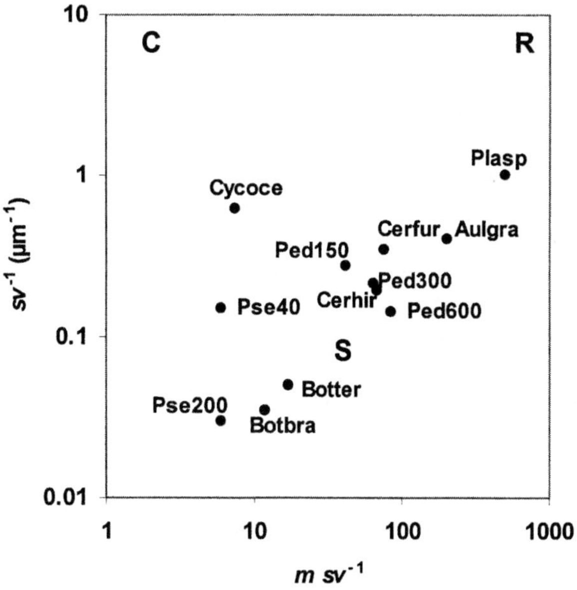

Figure 2. Morphological ordination of some planktonic algae recorded in Lake Arancio in the studied period, plotted against maximal linear dimension (*m*), surface area (*s*) and volume (*v*) of cells or colonies and the ecological C-, S-, and R-tendencies. Alagra: *Aulacoseira granulata*; Cerfur: *Ceratium furcoides*; Cycoce: *Cyclotella ocellata*; Cerhir: *Ceratium hirundinella*; Pse40: colony 40 μm in diameter of *Pseudosphaerocystis lacustris*; Pse 200: colony 200 μm in diameter of *Pseudosphaerocystis lacustris* Botter: *Botryococcus terribilis*; Botbra: *B. braunii*; Ped150: colony 150 μm in diameter of *Pediastrum* spp.; Ped300: colony 300 μm in diameter of *Pediastrum* spp.; Ped600: colony 600 μm in diameter of *Pediastrum* spp.; Plasp: *Planktothrix* sp.

the fully mixed water column. **M** representatives are able to regulate their buoyancy and are accompanied by floating colonies of *Botryococcus* or large colonies of *Pseudosphaerocystis lacustris*. *Ceratium furcoides* is considered also as a $\mathbf{L}_M$ representative, despite its smaller dimensions (46 000 μm^3) compared to *C. hirundinella* and its different shape. Nevertheless, it develops under z_{mix}/z_{eu} conditions comparable to those occurring when *C. hirundinella* dominated except for higher temperature values. Its sudden and abundant appearing is due to the fact that it came from Lake Garcia, where it was blooming. Contemporary, organisms belonging to **S1** association (undetected filamentous cyanoprokaryota and *Planktothrix agardhii* (Gomont) Anagnostidis & Komárek) occurred, even though with rather low relative biomass values. Members of the **P** association dominated the assemblage during winter period.

Figure 2 summarises the C-, S-, and R-tendencies of a selection of planktic algae recorded in Lake

Arancio. Most of them show an R- and S-tendency. S-species are defined by Reynolds (1997) as *acquisitive strategists*; they are generally large unicells or large colonies which, through compounding motility or the ability to control buoyancy, may self-regulate their position in the water column. These organisms are typically K-selected strategists. R-species are defined *acclimating strategists*, tolerating or depending upon entrainment within deeply circulating water layers. They include both r-selected and K-selected organisms and are pre-adapted to maximise suspension opportunities as well as low irradiance values.

It has to be pointed out that despite the fact that clear responses in structure were observed and, arguably, were triggered by changes in insolation wrought by depth-transparency-stability changes, however, such clear drivers were producing S or R strategists but not as theory predicts. Actually, according to the classical work by Margalef (1978), where the linkages between environment, morphology and life-cycle strategies were for the first time clearly underlined, K-species should be more strongly favoured when weaker mixing and a diminished resource base prevails. Accordingly, acquisitive strategists are favoured in a low resources/high energy environmental pattern whereas acclimating strategists under high resources and low energy conditions (see Reynolds, 1997).

In Lake Arancio, in the studied period, a general scenario occurred in which high nutrient concentrations led to high biomass values and to a consequent reduction of light availability (low energy) as described by the trend of z_{mix}/z_{eu} ratio. Thus, a much more pronounced R-tendency would have been apparently more suitable in this environment. Nevertheless, *Ceratium* spp. showed high biomasses in early spring and autumn when the water column was fully mixed (lower light availability) and nutrient values were higher. An explanation to this event could be linked to the drought that Sicily experienced in the last two years, which lead to several empty reservoirs.

The total biomass trend in the studied period showed a continuous decrease throughout spring 2001 due to the negative growth rates of *Ceratium hirundinella*. Unfortunately, data are lacking in the period immediately preceding the beginning of the investigation and we do not have an exact idea of the previous structure of the assemblage. Moreover, we have to consider that *Microcystis*, having a more pronounced S-tendency than *Ceratium*, reached rather high biomass values under a condition of stable stratification and when epilimnetic nutrient values tended to be exhausted. From the beginning of March to half-June RP passed from 50 to 16 μg l^{-1} and N-NO_3 from 2 to 0.4 mg l^{-1} (Naselli-Flores, 2003). Even though these values are far to be considered limiting in 'normal' conditions they can not sustain the early-spring biomass values.

Accordingly, a progressive decrease in total phytoplankton biomass was recorded in this period due to the shift from the large *Ceratium* to the small *Microcystis*. Moreover, once that *Microcystis* well established its dominance, a more or less subtle change in the environmental conditions may make another alga grow faster than *Microcystis*, but it will not cause a change in the dominance if the environment is already knee-deep in *Microcystis*. The change effectively occurred rather suddenly in mid-October when the addition of a relatively large amount of waters, transporting a *Ceratium furcoides* bloom, 'diluted' *Microcystis* in Lake Arancio. This event was able to overcome the 'inoculum effect' quite rapidly. But also in this case the high biomass-forming *C. furcoides*, inoculated in Lake Arancio, showed negative growth rates (likely due to the decrease in temperature), even though this organism showed a higher R-tendency than *C. hirundinella*. As a consequence, it was rather rapidly outcome by winter **P** association members. Thus, in early spring 2002 we observed the re-starting of *Ceratium hirundinella* growth in a more 'normal' condition of deep mixing and lower nutrient concentration, due to the consumption operated by winter assemblage not balanced by an adequate income from the catchment due to a dry winter. Since also winter in the precedent year was rather dry we can hypothesise that the explanation of the anomaly recorded in the development of R- and S- species may depend on the drought that the island was experiencing since 1999.

Among S-species, two morphotypes of *Microcystis* were common in Lake Arancio in the studied period: *M.* cf. *panniformis* Komárek et al., and *M. aeruginosa* (Kützing) Kützing. As Reynolds et al. (1981) pointed out, *Microcystis* is a genus showing a big variability in structure, nomenclature and some plasticity among forms.

Both the morphotypes started growing when the lake was still homothermal at the beginning of March and phytoplankton assemblage was dominated in terms of biomass by *Ceratium hirundinella*. This period, despite the high *Ceratium* biomass, was characterised by relatively high transparency with euphotic depth ranging between 4 and 5.5 m (Fig. 1B). The deep penetration of light, especially along the shores, allowed the recruitment of *Microcystis* colonies from the sediment. Reynolds & Bellinger (1992) showed that *Microcystis* can form vast deposits on the sediments where they may remain viable for years. Their successful recruitment depends upon a preceding phase of water-clarity. When the colonies lying on the bottom receive sufficient light, it activates anoxygenic photosynthesis and the cells start producing gas-vesicles that allow them to become buoyant. These traits of *Microcystis* life-history are supported by the microscopic observations carried out on Lake Arancio phytoplankton. Actually, several colonies, at the beginning of their growth period, were observed to show a nucleus of green cells, lacking gas vesicles, surrounded by a multi-layer crown of black cells, filled with gas vesicles.

Microcystis cf. *panniformis* formed large (3 – 4 mm), spherical colonies with cells, 3 μm in diameter, tightly arranged. These colonies tended to preserve their arrangement after Lugol's solution addition. On the contrary, *M. aeruginosa* was characterised by smaller spherical colonies (0.25 – 0.30 mm in diameter) with larger cells (6 μm), which easily tended to disaggregate when fixed in Lugol's solution. A third morphotype, *Microcystis protocystis* Crow, forming flat colonies with cells, about 6 μm in diameter, loosely arranged in a diffuse, irregular mucilage, was recorded sporadically with a few colonies except for a pulse in October when it reached a biomass of 80 mg l^{-1}.

During spring and till mid-July, the increasing biomass of these canopy species led to a progressive increase of z_{mix}/z_{eu} ratio, despite the onset of a stable thermal stratification and the reduction of mixing depth. *M.* cf. *panniformis* colonies showed a decrease in their dimensions (Fig. 3A) accompanied by a progressive increase of biomass. Also *M. aeruginosa* showed a progressive increase in biomass, even more conspicuous than that of *M.* cf. *panniformis* due to the greater cell dimensions, but the diameter of the colonies also increased (Fig. 3B). July was characterised by high values of z_{mix}/z_{eu} due to the strong contraction of the euphotic depth and *M.* cf. *panniformis* colonies reached their smaller size. After the re-establishing of lower values of z_{mix}/z_{eu}, *M.* cf. *panniformis* colonies tended again to become larger and those of *M. aeruginosa* smaller. From mid-July to the beginning of October, *M. aeruginosa* was the dominant species in the lake accompanied by *Pseudosphaerocystis lacustris*, and a few colonies of *Botryococcus braunii* and *B. terribilis*. At the end of this period, the water column

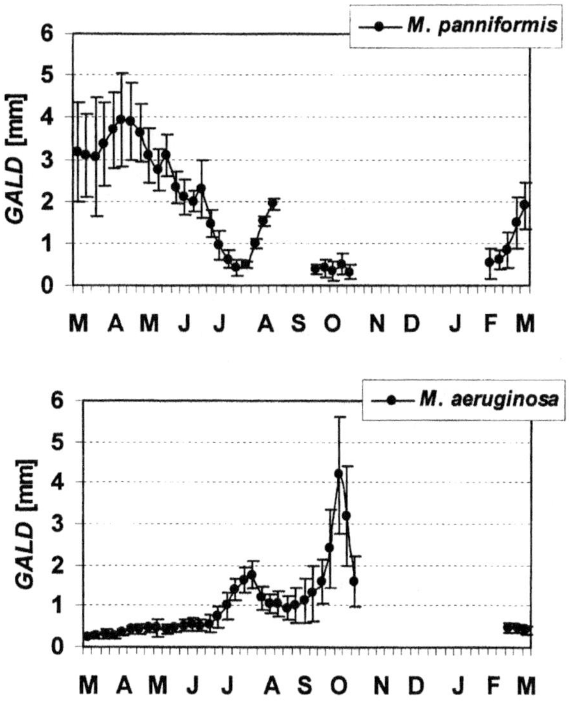

Figure 3. Variation of maximal linear dimension (GALD) in *Microcystis* cf. *panniformis* (**A**) and *M. aeruginosa* (**B**) in Lake Arancio in the studied period. Vertical bars indicate standard deviation.

was characterised by low values of z_{mix}/z_{eu} and the colonies of *Microcystis* reached the largest dimensions recorded in the study period.

Not only the morphology but also the physiological processes seems to follow different pathways in these two morphotypes. Actually, *M.* cf. *panniformis* showed an average growth rate of 0.18 day^{-1}, with a maximum of 0.73 and a minimum of 0.06, whereas *M. aeruginosa* showed lower values ranging between 0.04 and 0.45 day^{-1} with an average value of 0.14 day^{-1}.

Buoyancy regulation most likely occurs through two mechanisms in natural populations of *Microcystis*: regulation of gas vesicle synthesis and cell ballast (Wallace et al., 2000). Regulation of cell ballast promotes the frequently observed pattern of cyanobacteria accumulation during the night and early morning and their sinking away from the surface during late morning and afternoon (Van Rijn & Shilo, 1985). Moreover, the position of colonies in the water column depends also on their shape and dimensions and on the nature and the extent of water movements according to wind intensity and thermal stability of the water body. In addition, Wallace & Hamilton (1999) recorded that in some *Microcystis* populations there is an initial non-linear response period in the rate of carbohydrate accumulation, responsible for ballast, after an increase in light. For the organisms able to form large colonies, a longer response time in a dark environment may ensure the possibility to sink and float at greater speeds and so quickly return to the surface after a mixing event. A short response time might represent an adaptation of a species that is frequently exposed to a fluctuating light climate experienced when entrained by turbulent motions. Such a species might forms small colonies and its response time should be in accordance with the time scale of the light fluctuation.

The lower growth rates showed by *Microcystis aeruginosa* suggest a longer response time compared to that of *M.* cf. *panniformis*. As observed by Wallace et al. (2000), the advantages offered to *M. aeruginosa* by an increased carbohydrate dynamics to create an efficient buoyancy regulation and the control over vertical position in the water column are counterbalanced by lower growth rate.

As a result, *M.* cf. *panniformis*, with a higher growth rate, seems to be advantaged when the water column is fully and deeply mixed. A shorter response time also permits to the large colonies of this morphotype a rapid sinking away from the surface and an easier entrainment in spring when the lake is deeply mixed because of the wind intensity, generally higher than in summer, and a not severe light limitation. On the contrary, as the environment becomes darker and the mixing zone shallower, a reduction in colony dimensions may represent an attempt of *M.* cf. *panniformis* to adapt to the frequent fluctuations of the light climate.

Nevertheless, as the environment becomes darker, *Microcystis aeruginosa* colonies increase their dimensions and a longer response time ensures them greater speeds in sinking and floating, which makes them advantaged over *M.* cf. *panniformis*. The largest colonies were found at the end of September – beginning of October when the lake, at its minimum holding, tended to form diurnal thermoclines located in the first 50 cm but the euphotic depth was reduced to a few centimetres. These environmental conditions, characterised by low wind intensities, may enhance the formation of large over-buoyant colonies, possibly because in this way they are subjected to fewer light fluctuations than small colonies (Wallace & Hamilton, 2000). In addition, under calm conditions, small colonies may clump to form fewer large ones, and as a result light penetrates deeper. Strong winds disperse the large clumps into smaller colonies. This mechanism, according to Robarts & Zohary (1984) gives *Microcystis aeruginosa* an advantage.

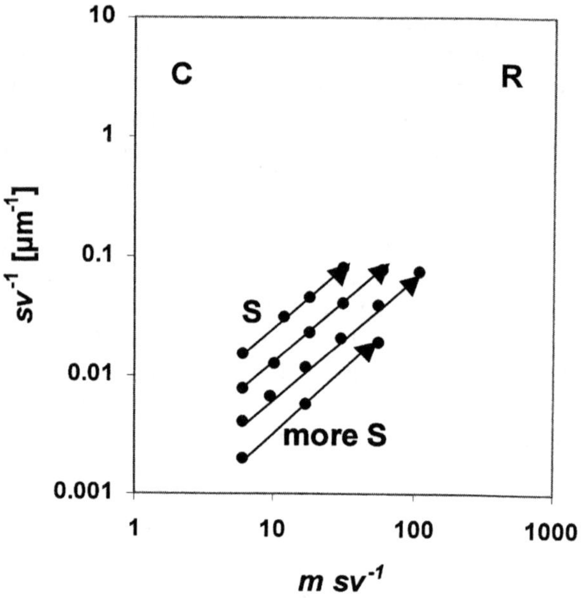

Figure 4. Morphological ordination of some *Microcystis* colonies, plotted against maximal linear dimension (*m*), surface area (*s*) and volume (*v*) of colonies, and the ecological C-, S-, and R-tendencies. The arrows indicate the "direction" that a colony follows assuming a more elongated elliptic shape but maintaining invariate its maximal linear dimension.

It has also to be considered that *M. aeruginosa* is able to modify not only the diameter of its colonies but also their shape assuming complicated clathrate forms,

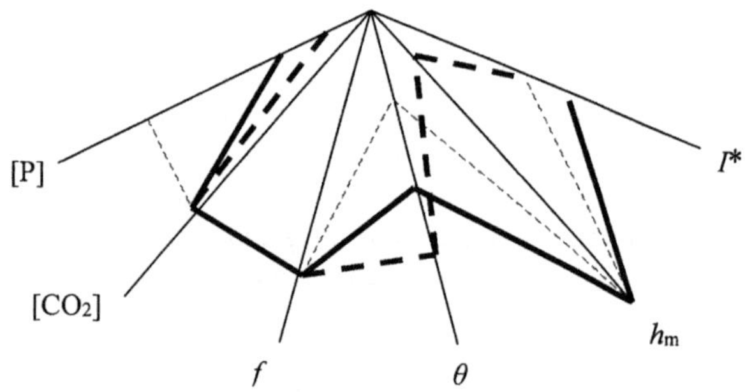

Figure 5. Hexacle illustrating some environmental features recorded when *Microcystis* cf. *panniformis* (thick solid line) and *M. aeruginosa* (thick dashed line) reached their maximum growth rate in Lake Arancio in the studied period. The lines are compared with that (thin line) reported by Reynolds (1997) for *Microcystis*. I*: integrated light [E m^{-2} d^{-1}]; h_m: mixed depth [m]; θ; temperature of water [°C]; *f*: volume of water processed by filter-feeding zooplankton [ml $l^{-1}d^{-1}$], [CO_2]: availability of carbon [mmol C l^{-1}]; [P]: availability of phosphorus [μmol P l^{-1}]. For scales see Reynolds (1997).

which likely represent a response to the changing turbulent and light regime of the water column. Figure 4 summarises the different strategies of the *Microcystis* colonies as they modify their dimensions and pass from a spherical to a more or less elongated elliptic form. According to Reynolds (1997), as the diameter of large and spherical colonies decreases, they move from a 'more S' to an 'S' strategy. As they tend to become narrower and the proportion between their maximal linear dimension and their width increases, they also exhibit a shift toward R-tendency. Actually, colony thinning increases the surface to volume ratio and thus the light harvesting performance, which may result more useful in a dark or in a strongly fluctuating light environment.

The morphological plasticity showed by *Microcystis* does not represent an exception. Different strategies are known for several phytoplankton from within single taxa *Planktothrix* is outstanding. *Dictyosphaerium* and *Scenedesmus* are sufficiently plastic to alter from generation to generation. The marine *Phaeocystis* does it routinely. Naselli-Flores & Barone (2000) showed that environments differing in their hydrology and trophic state may share the same phytoplankton taxa but they generally show different morphologies.

Although the morphological properties are merely surrogates of a more detailed template, their fidelity is remarkable. Nevertheless, an attempt to draw the *Microcystis* hexacles is made in Figure 5. The values on the template are those reached by the two *Microcystis* morphotypes when the higher growth rates were recorded. Compared to the hexacle reported in Reynolds (1997) for *Microcystis* the most evident differences deal with concentration of nutrient, temperature, and to mixed depth for *M. aeruginosa*. Anyway, mixed depth *per se* is not so important if it is uncoupled from light conditions. A shallow mixed depth under poor light conditions can be compared to a deeper mixing zone in a more lightened environment. Nutrient concentrations might be meaningless in this context since they are well above limiting thresholds. Also volumes of water processed by filter-feeding zooplankton were rather low. Actually, very few large filter feeders such as *Daphnia* were recorded in Lake Arancio in the studied period compared to previous investigations. Zooplankton assemblage was mostly formed by rotifer, cyclopoid copepods, and calanoids (Naselli-Flores, unpublished data). This last group of organisms was never recorded previously in lake Arancio and underlines the scarcity of small edible phytoplankton.

If we start with a null hypothesis that everything can live (almost) anywhere and that the template is not prescriptive, since it describes a pattern through probability, we can consider the template itself as a filter. To one side the mesh is coarse and corresponds to 'benign environments' (Reynolds, pers. com.). Most species pass the filter so any could dominate. The only qualification is that the species is there, probably in substantial numbers (the 'memory' *sensu* Padisák, 1992) but it is also r-selected. Templates can not predict what species will be found at given co-ordinates. What it will do is to suggest the functional adaptations of the species which will perform best (or is locally the fittest) at that locus. Carving the biota up into functional associations provides a way of giving an identity to those adaptations. But it will never predict what species will be found there. Nevertheless, away form the benign corner, the selective filter tightens, demanding more specialist survival traits. With fewer species able to survive, prediction gets easier as life becomes more difficult. At the extreme (R- and S-tendencies) we may predict quite well the kind of 'monoculture' that we experience. Of course there are more dimensions than those depicted and the template may have depth as an intaglio, but at the present level of knowledge the view of Grime (see Grime et al., 1996) on stress and disturbance can represent a good compromise.

Conclusions

Lake Arancio may be considered an 'extreme' environment with regard to its underwater light availability and hydraulic regime. Nevertheless, previous investigations on its phytoplankton dynamics and structure (Barone & Naselli-Flores, 1994; Naselli-Flores & Barone, 1998) highlighted rich assemblages characterised by period of high diversity mainly corresponding to the emptying phases of the reservoir. Nutrient concentrations, as well as TP and TN, showed in 2001–2002 values two times higher than those recorded in previous investigations. In addition, a stronger nutrient depletion due to phytoplankton uptake would have been expected in epilimnetic layers during summer periods. By contrast, this did not take place, the reason being likely in the depth decrease and in the increased internal loading (Naselli-Flores, 2003).

According to traditional ecological theories, a 'closed' ecosystem with a regular energy input tend to assume a steady state of maximum attainable biomass. Energy transfer between species is energetically expensive. It follows that the greater the diversity, the greater the number of energy transfers and the greater the energetic costs in maintaining the system. Thus in systems of low diversity, biomass accumulates despite low productivity and in those of high diversity biomass is relatively low compared to energy input.

If we consider Lake Arancio a closed ecosystem (and it is quite realistic during summer when there is not any water income from the catchment), it may be argued that dewatering, and its impacts on nutrient dynamics allowing a continuous nutrient supply, maintains a regular energy input to this ecosystem. This led to a shift toward a higher trophic state and to a steady state of maximum attainable biomass. The most evident consequence was the virtual extinction of several phytoplankton taxa, and ultimately the reduction of species richness (especially of r-selected taxa). The species recorded in 2001–2002, according to C-S-R tendencies (Reynolds, 1997), belong to S- or R-strategic groups and exhibit K-selected features such as motility, large dimensions, low light adaptation, buoyancy regulation and long time span of occurrence. The adding of one more characteristic could make the list of K-selected features more complete: the ability to modify the morphology (dimensions and shape) within a relatively short interval of time to best fit and counteract environmental variability. Nevertheless, species populations are dissipative units *per se* and diversity may be considered at different levels, from genes to communities. The morphological variability likely includes a genetic diversity among the members of a population which allows the survival, and even the growth, of that population under a certain degree of sub-optimal conditions. Ecologists often refer to this characteristic as 'tolerance'. This word anyway includes a certain passive meaning whereas some species demonstrate to react actively to environmental variability. They accumulated information in their evolutive process and stored it in their genetic equipment. As much information they accumulated as much diversity they can express within a population, until the formation of new 'species'. The implausibility of steady state (and even more of equilibrium) according to Margalef (2000) has been often discussed by philosophers. It would be interesting to start discussing it among ecologists.

Acknowledgements

L.N.-F. is deeply indebted to Colin Reynolds for his comments on a previous version of the manuscript and his willingness and patience in discussing several aspects dealing with C-S-R tendencies, templates, hexacles, their 'predictive' value and applicability. Most of his e-suggestions are reported in this paper. We wish also to thank Tamar Zohary for her helpful suggestions on a previous draft of the manuscript.

References

Allen, T. F. H. & T. B. Starr, 1982. Hierarchy: perspectives for ecological complexity. University of Chicago Press, Chicago.

Barone, R. & L. Naselli-Flores, 1994. Phytoplankton dynamics in a shallow, hypertrophic reservoir (Lake Arancio, Sicily). Hydrobiologia 289: 199–214.

Grime, J. P., J. G. Hodgson & R. Hunt, 1996. Comparative plant ecology. A functional approach to common British species. Chapman & Hall, New York

Harris, G. P., 1986. Phytoplankton Ecology. Strucure, function and fluctuation. Chapman and Hall, London.

Hillebrand, H., C.-D. Dûrselen, D. Kirschtel, U. Pollingher & T. Zohary, 1999. Biovolume calculation for pelagic and benthic microalgae. J. Phycol. 35: 403–424.

Kennedy, R. H., T. Cole, W. Boyd & K. Barko, 2002. Operational influences on the limnological characteristics of reservoirs: a model study. Extended Abstracts of the 4th International Conference on Reservoir Limnology and Water Quality, Èeské Budìjovice, Czech Republic, August 12–16, 2002. pp. 168–170.

Komárek, J. & P. Marvan, 1992. Morphological differences in natural populations of the genus Botryococcus (Chlorophyceae). Archiv für Protistenkunde 141: 65–100.

Komárek, J., J. Komárkova-Legnerová, C. Sant'Anna, M.T. de Paiva Azevedo & P.A.C. Senna, 2002. Two common Microcystis species from tropical America. Cryptogamie/Algologie 23: 159–177.

Margalef, R., 1978. Life forms of phytoplankton as survival alternatives in an unstable environment. Oceanologica Acta 1: 493–509.

Margalef, R., 2000. Exosomatic structures and captive energies relevant in succession and evolution. In Jørgensen S. E. (ed.), Thermodynamics and ecological modelling. Lewis Publishers, Boca Raton: 3–15.

Naselli-Flores, L. 2003. Man-made lakes in Mediterranean semiarid climate: The strange case of Dr Deep Lake and Mr Shallow Lake. Hydrobiologia (in press).

Naselli-Flores, L. & R. Barone, 1998. Phytoplankton dynamics in two reservoirs with different trophic state (Lake Rosamarina and Lake Arancio, Sicily, Italy). Hydrobiologia 369/370: 163–178.

Naselli-Flores, L. & R. Barone, 2000. Phytoplankton dynamics and structure: a comparative analysis in natural and man-made water bodies of different trophic state. Hydrobiologia 438: 65–74.

Naselli-Flores, L., J. Padisák, M. T. Dokulil & I. Chorus, 2003. Equilibrium/steady-state concept in phytoplankton ecology. Hydrobiologia 502 (Dev. Hydrobiol. 172): 395–403.

Naselli-Flores, L., R. Barone & R. Mosello, 2003b. Eutrophication control by lime addition: a preliminary approach in Sicilian reservoirs. Hydrobiologia (in press).

Padisák, J., 1992. Seasonal succession of phytoplankton in a large, shallow lake (Balaton, Hungary) – a dynamic approach to ecological memory, its possible role and mechanisms. J. Ecol. 80: 217–230.

Padisák, J., C. S. Reynolds & U. Sommer (eds), 1993. Intermediate Disturbance Hypothesis in Phytoplankton Ecology. Developments in Hydrobiology 81. Kluwer Academic Publishers, Dordrecht, The Netherlands.

Padisák, J., F. Barbosa, R. Koschel & L. Krienitz, 2003. Deep layer cyanoprokaryota maxima in temperate and tropical lakes. In Koschel R. & D. Adams (eds), Lake Stechlin. An Approach to Understanding an Oligotrophic Lowland Lake. Archiv für Hydrobiologie/Advances in Limnology (in press).

Reynolds, C. S., 1997. Vegetation processes in the pelagic: A model for ecosystem theory, Ecology Institute, D-21385 Oldendorf, Germany.

Reynolds, C. S. & E. G. Bellinger, 1992. Patterns of abundance and dominance of the phytoplankton of Rostherne Mere, England: evidence from an 18-year data set. Aquat. Sci. 54: 10–36.

Reynolds, C. S., G. H. M. Jaworski, H. A. Cmiech & G. F. Leedale, 1981. On the annual cycle of the blue-green alga, Microcystis aeruginosa Kütz. emend. Elenkin. Phil. Trans. r. Soc. Lond. Ser B 293: 419–477.

Reynolds, C. S., V. L. Huszar, C. Kruk, L. Naselli-Flores & S. Melo, 2002. Towards a functional classification of the freshwater phytoplankton. J. Plankton Res. 24: 417–428.

Robarts, R. D. & T. Zohary, 1984. *Microcystis aeruginosa* and underwater light attenuation in a hypertrophic lake (Hartbeespoort Dam, South Africa). J. Ecol. 72: 1001–1017.

Scheffer, M., S. Reinaldi, J. Huisman & F. J. Weissing, 2003. Why plankton communities have no equilibrium: solutions to the paradox. Hydrobiologia 491: 9–18.

Senese, F. A., 2003. http://antoine.frostburg.edu/chem/senese/javascript/water-density.html.

Sommer, U., J. Padisák, C. S. Reynolds & P. Juhász-Nagy, 1993. Hutchinson's heritage: the diversity-disturbance relationship in phytoplankton. Hydrobiologia 249: 1–7.

Straškraba, M., I. Dostálková, J. Hejzlar & V. Vyhnálek, 1995. The effect of reservoirs on phosphorus concentration. Int. Rev. ges. Hydrobiol. 80: 403–413.

Van Rijn, J. & M. Shilo, 1985. Carbohydrate fluctuations, gas vacuolation, and vertical migration of scum-forming cyanobacteria in fishponds. Limnol. Oceanogr. 30: 1219–1228.

Wallace, B. B. & D. P. Hamilton, 1999. The effect of variations in irradiance on buoyancy regulation in Microcystis aeruginosa. Limnol. Oceanogr. 44: 1127–1138.

Wallace, B. B. & D. P. Hamilton, 2000. Simulation of water-bloom formation in the cyanobacterium Microcystis aeruginosa. J. Plankton Res. 22: 1127–1138.

Wallace, B. B., M. C. Bailey & D. P. Hamilton, 2000. Simulation of vertical position of buoyancy regulating Microcystis aeruginosa in a shallow eutrophic lake. Aquat. Sci. 62: 320–333.

Whittaker, R. H., 1975. Communities and Ecosystems. Macmillan, New York.

Whittaker, R. H. & S. A. Levin, 1977. The role of mosaic phenomena in natural communities. Theor.Pop. Biol. 12: 117–139.

Hydrobiologia **502:** 145–155, 2003.
L. Naselli-Flores, J. Padisák, M. T. Dokulil (eds),
Phytoplankton and Equilibrium Concept: The Ecology of Steady-State Assemblages.

Algal assemblage types of bog-lakes in Hungary and their relation to water chemistry, hydrological conditions and habitat diversity

Gábor Borics[1], Béla Tóthmérész[2], István Grigorszky[3], Judit Padisák[4], Gábor Várbíró[1] & Sándor Szabó[5]
[1]*Environmental Protection Inspectorate for Trans-Tiszanian Region, Debrecen Piac u. 9/b H-4025 Hungary*
E-mail: boricsg@freemail.hu
[2]*Department of Ecology and Hydrobiology, University of Debrecen, Debrecen, Egyetem-tér 1, H-4010 Hungary*
[3]*Department of Botany, University of Debrecen, Debrecen, Egyetem-tér 1, H-4010 Hungary*
[4]*Department of Limnology, University of Veszprém, Veszprém, Egyetem u. 10, H-8200. Hungary*
[5]*Department of Botany, Nyíregyháza College, H-4401 Nyíregyháza, P.O. Box 166, Hungary*

Key words: bog lakes, habitat heterogeneity, algal flora, pseudovicarism, association types

Abstract

Algal flora of 12 bog lakes was investigated during the period of March 1995 to August 1999 in Hungary. Of the 129 samples, 624 taxa of algae were identified. Species richness of individual samples ranged between 8 and 107 except the extraordinarily species rich Baláta-tó where 533 algal taxa were observed. Ordination of the samples resulted in five groups: (1) assemblages dominated by chlorococcalean algae and planktonic Cyanoprokaryota; (2) assemblages dominated by flagellates (Dinophyta, Cryptophyta, Euglenophyta, Chrysophyceae, Raphidophyceae); (3) chlorococcalean algae and cyanoprokaryotic assemblages with desmids, cryptophytes, dinoflagellates or euglenophytes as subdominants; (4) assemblages dominated by diatoms and (5) a group of samples where other taxa belonging to Xanthophyta and filamentous green algae dominated. The bogs were typically rich in inorganic N and P, moreover, their water chemical characteristics (including pH and conductivity) were rather uniform. Therefore, other factors than chemical properties were responsible for different flora. The above groups were characteristic to certain types of habitats. In the first group, plankton samples from relatively large pools with considerable open water can be found. The second group included samples taken from small bog pools. The third group contained the periphyton samples from macrophytes, living in bogs with constantly reliable water supply. Samples of group four and five comprised small bogs that occasionally dry up. Periphyton of lakes with *Sphagnum* belonged exclusively to the fifth group. This study has shown that small bog-pools are often inhabited by different species of flagellates and desmids are not as important as it has been widely believed. Hydrological properties and habitat diversity are the major factors influencing species richness of Hungarian bog-lakes.

Introduction

Peat bogs represent one of the major vegetation type at higher latitudes of the Northern hemisphere and they are, even though not that characteristic, abundant in the cooler and wet temperate regions. The climate of Hungary is neither cold nor humid and therefore peat bogs (with at least some *Sphagnum*) occur sporadically and due to edaphic reasons. However, as occurring at the margins of geographic distribution of this type of vegetation they deserve protection and distinguished scientific attention.

The beauty and large diversity of the algae living in the peat bogs has always attracted the attention of the algologists. The first reports of peat bog algae in Hungary were published in the late 19th century (Hazslinszky, 1868; Schaarschmidt, 1882; Csató, 1885) and early data were followed by dozens of other works in the 20th century (for detailed literature review see Borics, 2001). In these works, it was recognized that they are similar to other bog systems in the Alps, in the Carpathians and in the northern part of Europe. Uherkovich et al. (1994) investigated the algal flora of several peat bogs and considered first environmental factors like water balance, water chemistry, geomorphology, etc. that can be important for understanding assembly of algal flora in these ecosystems. These early publications, despite very low international publicity, are of great importance as being the only sources of information about the Hungarian peat bogs, and therefore they provide a solid basis for comparative studies (Stollmayer-Boncz, 1992).

The early investigators of peat bogs focused on detailed description of the desmid flora, while other taxa were largely neglected even though these groups include many interesting and unique species. More recent investigations have reported that many species of silica-scaled Chrysophyceae, (Kristiansen, 1986; Péterfi & Momeu, 1996; Kristiansen & Stoyneva, 1998; Péterfi et al., 1998; Padisák et al., 2000), Volvocales, Chlorococcales (Cambra & Hindák, 1998; Borics et al., 1998a, b) and dinoflagellates (Borics et al., 2000) can be useful in estimation of ecological status of the peat bogs. Other investigations have demonstrated that the rich desmid flora is not exclusive for acidic environments, rather, it can be remarkably rich in alkaline conditions. For example, Kiss (1977) found 40 species of desmids in the reed-belt of a salt marsh (dominant ions: HCO_3^- and Na^+, pH between 8.5 and 10) in southern Hungary. Fehér (2000) described 43 taxa of desmids in one sample originating from a side-branch of Danube (pH: 8.2–8.4) some of which were considered previously as typical acidophylous organisms. The checklist of the Kiskunság National Park and the surrounding areas (with no acidic localities) contains 169 species of Desmidiales (Padisák, 1998). After investigating many types of water bodies, Coesel (1984) revealed that the alkaline seepage-lakes have the richest desmid flora in the Netherlands.

Quite often, algal flora of such small lakes has been more rich than that in deep lakes and they have continued to attract attention of florists but not of ecologists. They are difficult objects for ecologists because, due to their variable water balance, small water volume and spatial heterogeneity most are rather changeable with no clear patterns to estimate major processes. As a consequence, ecological knowledge of small lakes, representing the bulk of the World's biodiversity, is incomparably poorer than that of large ones.

Between 1995 and 1999 we performed a comprehensive research on 12 bogs in Hungary with the following main aims:

(i) What kind of algal assemblages characterize the small habitats that are typical in the bog systems?

(ii) Which taxa can be considered typical bog-dwellers?

(iii) What is the main group of environmental factors (hydrology, water chemistry, habitat diversity) that influences the species richness of the bogs?

(iv) Are there any differences in the composition of the algal flora in comparison to results of early investigators?

(v) What considerations are necessary when studying algal assemblages of such lakes?

Materials and methods

Study area

Plankton, metaphyton and periphyton samples were collected from the largest Hungarian bog-lake, Baláta-tó, between 1995 and 1999. In 1999, we extended our investigations to 11 other bogs of different type located in the western part of Hungary (Fig. 1).

1. Fekete-tó (300 m a.s.l.) is a *Sphagnum*-bog, which is situated in a depression of a hill plateau. The surface area is about 0.06 ha, without any open water area. The whole basin is filled with *Sphagnum*. Typical macrophyte-association is the *Carici echinatae-Sphagnetum*. The bog is fed by precipitation.

2. Huszászi-patak lápja. This bog is situated in a depression along the Huszászi-streamlet (210 m a.s.l.). It was a small pond 30 years ago with considerable open water area, but as a result of a rapid natural succession the whole basin was invaded by macrophytes, mainly by *Carex elata*. Among the tussocks of this species several small pools are hidden. The stocks of the tussocks are covered by *Sphagnum*. The most of the small pools are densely covered by *Utricularia neglecta*. Surface area is 0.5 ha, mean depth is about 0.8 m. Water supply: precipitation, floods of the streamlet and groundwater seepage.

3. Grajka-patak lápja is a swampy area, which is situated in the westernmost part of Hungary. Thirty years ago the bog of Grajka-streamlet (about 200 m a.s.l.) was a typical mesotrophic spring-bog. After giving up the cutting of this area characteristic swamp vegetation (dominated by *Carex* species) developed during the last decades and parallel the *Sphagnum* decreased. Water is on groundwater level. Water is supplied by precipitation and groundwater seepage.

4. Ördög-tó (approx. 200 m a.s.l.) is a small, eutrophic pool of unknown origin in the westernmost

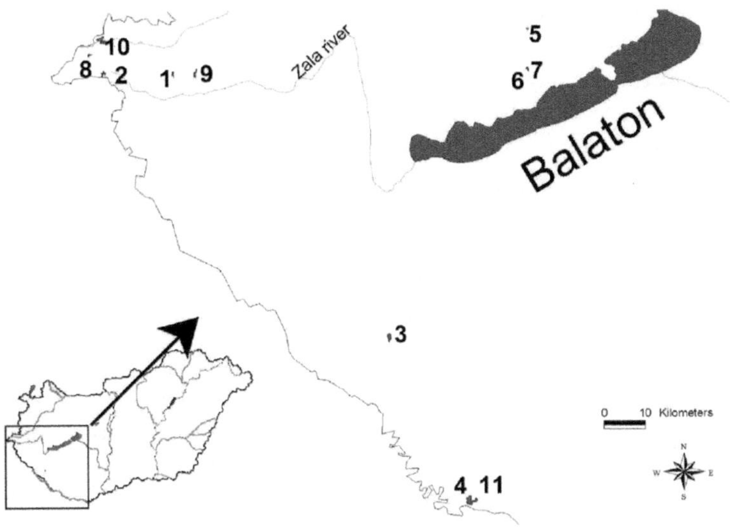

Figure 1. Map of the investigated localities 1. Vadkacsás-tó 2. Huszászi-patak lápja, 3. Baláta-tó, 4. Szûrûhely-folyás, 5. Öcsi Nagy-tó, 6. Barkás-tó, 7. Monostori-tó, 8. Grajka-patak lápja, 9. Szõcei láp, 10. Ördög-tó, 11. Tiva-tó; 12 Fekete-tó.

part of Hungary. Surface area is 0.04 ha, mean depth is 0.4 m. *Sphagnum*-stands are present in the southern shore. The basin is divided by *Carex* tussocks. The water supply is exclusively by precipitation.

5. Vadkacsás-tó (about 180 m a.s.l.) is a transition bog, which is in an advanced successional phase. Almost the whole area of the bog (0.5 ha) is covered by *Salix cinerea*. Among the trees and bushes small pools (0.1–0.2 m depth) are found, with *Sphagnum* on the margin. Submerged aquatic macrophytes were not found. The bog is supplied by water from an artificial ditch that collects the runoff from the surrounding area.

6. Szõcei láp. In the vicinity of Szõce village (approx. 140 m a.s.l.) a small meadow is lying in the valley of a streamlet. Surface area is approx. 1 ha. Patches of *Sphagnum* species can be found in the small depressions of the area. *Drosera rotundifolia* is also present. Although open water-surfaces appear only temporarily, the meadow is always wet because of the relatively high precipitation and some leaking sources around.

Bogs of Fekete hegy. Several basalt hills can be found in the Balaton Upland. Some of these hills have a large plateau of irregular shape. In the depressions small swamps and bogs developed. Fekete-hegy (350–371 m a.s.l.) is a hill, which has five pools, of which Monostori-tó and Barkás-tó are the most interesting.

7. Monstori-tó is a relatively large (8–10 ha) bog-lake. Max depth is 2 m. In the central area, a typical floating *Sphagnum*-island is developed which is quite unusual in Hungary. The pond is supplied exclusively by the precipitation. At the time of the sampling, the whole surface was covered by *Typha angustifolia*, hence we were able to take samples from the small pools, at the margin.

8. Barkás-tó is a small (0.5 ha), slightly acid, ephemeral pond, that is situated at the edge of the plateau. Mean depth is 0.4 m. Patches of *Schoenoplectus lacustris* and *Sparganium erectum* divide the basin. At the margin *Ceratophyllum submersum* and *Utricularia* sp. are dominating. The pond is fed by the precipitation.

9. Öcsi Nagy-tó (330 m a.s.l.) is a bog-lake that is situated in a depression of the southern slope of the Kab-hegy, the highest peak of the Bakony-mountains. The pond is divided into two different regions. The upper part is a typical transition bog with birch and *Sphagnum* in the centre and *Salix cinerea* in the edge. The lower part is an artificially dammed pond with considerable open water area, which collects the leaking water of the upper bog-area. Besides the precipitation, the pond is supplied by a source.

10. Lake Baláta (160 m a.s.l.) is the largest acid bog-lake in Hungary having a surface area of 174 ha. Various vegetation types cover approximately 80% of the surface. The lake has a mean depth of 1 m, maximum depth is 3–4 m with bottom deposits of 1–2 m thickness. The lake is fed primarily by precipitation.

11. Tiva-tó (120 m a.s.l.) is a small, artificial pond in the area of Barcs Nature Reserve, and during the

last decades it became similar to the natural transition bogs. Surface area is 1.5 ha, mean depth is 1 m. At the time of the sampling, macrovegetation was restricted to the shoreline of the basin. The pond is fed by precipitation and a small canal which originates from another marshland.

12. Szûrûhely-folyás (120 m a.s.l.) is a small, ephemeral forest-pool in the area of Barcs Nature Reserve. Surface area is about 0.14 ha, mean depth is 0.4 m. Earlier the basin was edged by *Sphagnum* (Uherkovich, 1981), but at the time of sampling *Sphagnum* was not observable. *Riccia fluitans* and *Lemna minor* covered the whole water surface. The pool is fed by precipitation.

Sampling and sample processing

Water samples were taken by filling bottles approximately 40 cm below the surface for phytoplankton and water-chemical analyses. For investigating the periphyton, parts of macrophytes, peat and other organic remnants were collected in 100 ml plastic bottles. In the laboratory 100–250 ml distilled water was added to them. After shaking they were allowed to settle in 500 ml cylinders, and were decanted some days later. The remnants were investigated with LEICA DMRB microscope equipped by brightfield, phase-contrast and Nomarski-contrast techniques. Both types of samples were fixed with Lugol's solution with acetic acid.

Water chemical analyses were performed according to the technical guidelines of the Hungarian Environmental Protection Inspectorates. These guidelines apply the internationally accepted analytical methods.

Data-analysis

For comparison of the microflora of the bog lakes hierarchical cluster analysis (Euclidean distance, Nearest neighbour methods; Florek et al., 1951) was applied. Non-metric multidimensional scaling was used to display the similarity of the algal flora of the bog-lakes (Legendre & Legendre, 1998). The similarity was measured by the Matusita similarity, which is defined in the following way:

$$s(\mathrm{p},\mathrm{q}) = \sum_{i=1}^{S} \sqrt{p_i \cdot q_i},$$

where p_i is the relative frequency of the species i in the first compared bog-lake (p), and q_i is the relative frequency of the species i in the second compared bog-lake (q), and S is the total number of species. We preferred to use this similarity, because it is based on the relative frequency of the species. Calculations were done by the NuCoSA program package (Tóthmérész, 1993).

Results

Water chemistry data (Table 1) allow describing these water-bodies as slightly acidic ones with very low conductivity and ionic concentrations. The lowest pH and conductivity values were measured in the samples taken from *Sphagnum* sites. In the majority of localities the PO_4^{3-}-P concentrations were fairly high (95–1240 $\mu g\ l^{-1}$) with exception of a small pool among *Carex* tussocks (Huszászi-patak lápja), and twice among *Sphagnum* hummocks (Fekete-tó, Szõcei láp) where concentrations $<10\ \mu g\ l^{-1}$ were found. The inorganic-nitrogen ($N_{inorg} = NH_4^+$-N + NO_3^--N + NO_2^--N) concentration changed between 850 and 7540 $\mu g\ l^{-1}$.

A total of 624 taxa were identified during the study: Cyanobacteria – 50, Bacillariophyceae – 80, Chrysophyceae – 44, Xanthophyceae – 14, Chlorococcales – 99, Volvocales – 34, Desmidiales – 115 , other Chlorophyta – 26, Euglenophyta – 126, Dinophyta – 23, Cryptophyta – 10, Raphidophyceae – 2 and Haptophyta – 1. Floristic composition both at level of main algal groups and species showed considerable differences among the localities (Table 2). Four groups of bog systems could be distinguished based on hierarchical cluster of the floristic compositions (Fig. 2):

I. Desmids, diatoms, and euglenophytes are dominant. Chlorococcalean algae also appear in these localities in relatively high number. The bogs that belong to the first group possess a large diversity of habitats (open-water areas, small pools, rich aquatic macroflora) and have relatively good water supply.

II. 'Typical' bog flora, with overwhelming dominance of desmids and diatoms, as well as epiphytic Cyanobacteria. These can also be characterised by sufficient and permanent water supply and rich macroflora but they do not have open surface areas.

III. The contribution of flagellated algae, mainly the euglenophytes, is high. This group comprises two bog-lakes. Both are quite nutrient rich and have relatively large open water areas and good water supply. Macrovegetation is of minor importance.

Table 1. Some physico-chemical parameters of the investigated localities. 1. Vadkacsás-tó 2. Huszászi-patak lápja, 3. Baláta-tó, 4. Szûrûhely-folyás, 5. Öcsi Nagy-tó, 6. Barkás-tó, 7. Monostori-tó, 8. Grajka-patak lápja, 9. Szõcei láp, 10. Ördög-tó, 11. Tiva-tó; 12 Fekete-tó

	1	2	3	4	5	6	7	8	9	10	11	12
NH_4^+ mg l^{-1}	1.69	0.56	0.66	0.84	0.56	0.52	0.63	0.71	0.56	5.56	0.9	0.49
NO_2^- mg l^{-1}	0.06	0.02	0.03	0.07	0.04	0.03	0.05	ø0.04	0.02	0.08	0.04	0.02
NO_3^- mg l^{-1}	1.9	0.9	1	1.1	3	0.3	1.5	1.1	0.9	1.9	2.6	1.1
PO_4^{3-} μg l^{-1}	1042	<10	95	220	252	427	552	337	<10	1240	172	<10
pH	5.99	6.28	6.35	5.61	5.75	6.77	6.3	6.34	6.03	6.12	6.11	4.75
Conductivity. μS cm^{-1}	70	92	92	43	89	142	75	60	52	102	73	25
Ca^{2+} mg l^{-1}	11.43	15.61	9.55	10.40	11.43	21.44	5.72	9.21	9.21	10.81	11.40	10.62
Mg^{2+} mg l^{-1}	5.2	3.04	7.53	3.65	3.47	6.07	3.47	3.29	3.81	4.74	3.04	0.97
Na^+ mg l^{-1}	1.9	3.2	2.8	2.90	4.7	5.0	4.0	3.45	6.44	5.75	4.61	1.30
K^+ mg l^{-1}	<1.0	<1.0	3.5	0.6	2.0	2.0	14.3	6.01	5.47	1.19	6.4	1.96
Cl^- mg l^{-1}	7.09	7.9	14.18	7.10	10.64	14.18	7.09	12.05	21.62	12.42	7.13	5.31
SO_4^{2-} mg l^{-1}	33.6	16.81	28.8	24.01	24.0	23.89	24.0	12.00	9.59	19.25	19.25	24.02
HCO_3^- mg l^{-1}	18.31	44.18	22.56	24.42	12.20	60.98	30.88	30.53	30.46	30.46	42.68	7.32

IV. Extraordinarily poor flora, desmids did not occur. This group comprises only one locality: the *Sphagnum*-bog of Fekete-tó. (This bog-system cannot be considered typical: thirty years ago it was a transitory *Sphagnum*-bog, but as a result of the arid decades it desiccated which led to the impoverishment of the microflora.)

Despite the fact that the investigated localities differ from each other in their flora and habitat characteristics, quite a large number of common species were found. Species as *Eunota bilunaris, Gomphonema gracile, G. parvulum, Nitzschia palea, Closterium dianae, C. venus, Cosmarium regnellii, C. polymorphum* and *Euglena oxyuris* were present in almost each sample. These are species not exclusively associated with bogs and acidic environments. Well-known bog-dwellers like *Actinotaenium turgidum, A. cucurbita, Netrium digitus, Micrasterias truncata, Tetmemorus granulatus, T. laevis*, etc. were observable only in one, or two sites. Nevertheless we have found three taxa which appeared in the majority of the localities, but not being widespread. *Heimiochrysis actinotrychia* (Chrysophyceae), *Rhipidodendron huxlei* (Haptophyta), and *Peridinium bipes* tab. *γ travectum* fa. *tabulatum* (Dinophyta) are acidophylous organisms and proved to be constant dwellers of the investigated localities.

Physiognomies of certain types of habitats were very similar in the different bogs. In order to investigate whether this resemblance appears at microscopic level, similarity of the flora of each sample has been analysed quantitatively. The basis for the comparison was the relative abundance of the specimens that belong to the following categories:

1. Diatoms that frequently occur in acidic environments (mainly *Eunotia* and *Pinnularia* species)
2. Diatoms not so specific, with wide ecological range
3. Chrysophyceae spp.
4. Xanthophyceae spp.
5. Unidentified cyst-like Xanthophyceae species
6. Chlorococcales and planktonic *Ulotrichales* species (*Koliella* sp.)
7. Volvocales spp.
8. Tetrasporales spp.
9. Filamentous green algae and *Vaucheria* sp.
10. Desmids
11. Euglenophytes
12. Dinophytes
13. Cryptophytes and Raphidophyceae spp.
14. Epiphytic Cyanobacteria
15. Planktonic Cyanobacteria

The ordination of the semi-quantitative data based on Matusita-simililarities (Fig. 3) allowed distinguishing of five groups:

1. Group-C included the samples in which the chlorococcalean algae (*Radiococcus nimbatus, Scenedesmus ovalternus, Scenedesmus* spp., *Dictyosphaerium pulchellum, Ankistrodesmus bibraianus*) and planktonic Cyanobacteria (*Aphanocapsa* sp., *Snowella lacustris*) were the dominants.

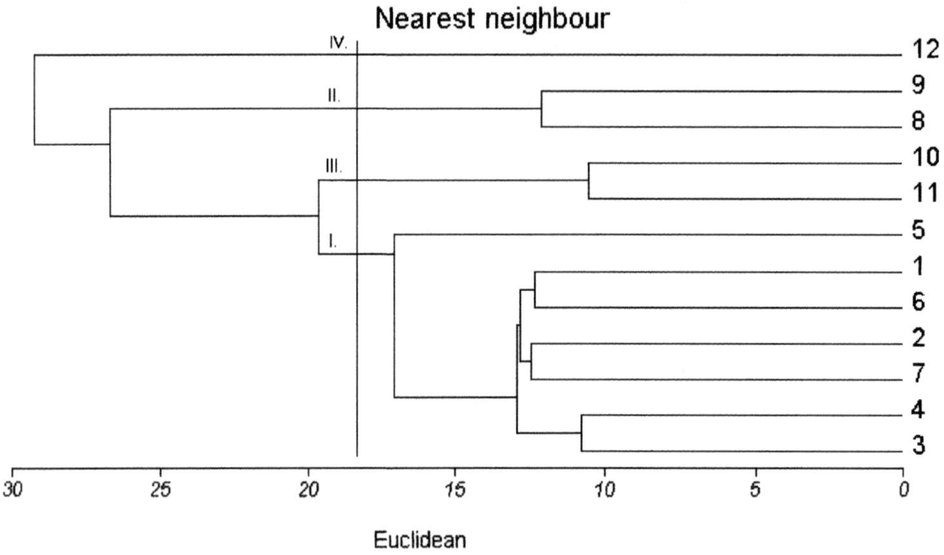

Figure 2. Dendrogram of the investigated localities on the basis of the floristic composition. 1. Vadkacsás-tó 2. Huszászi-patak lápja, 3. Baláta-tó, 4. Szûrûhely-folyás, 5. Öcsi Nagy-tó, 6. Barkás-tó, 7. Monostori-tó, 8. Grajka-patak lápja, 9. Szõcei láp, 10. Ördög-tó, 11. Tiva-tó; 12 Fekete-tó.

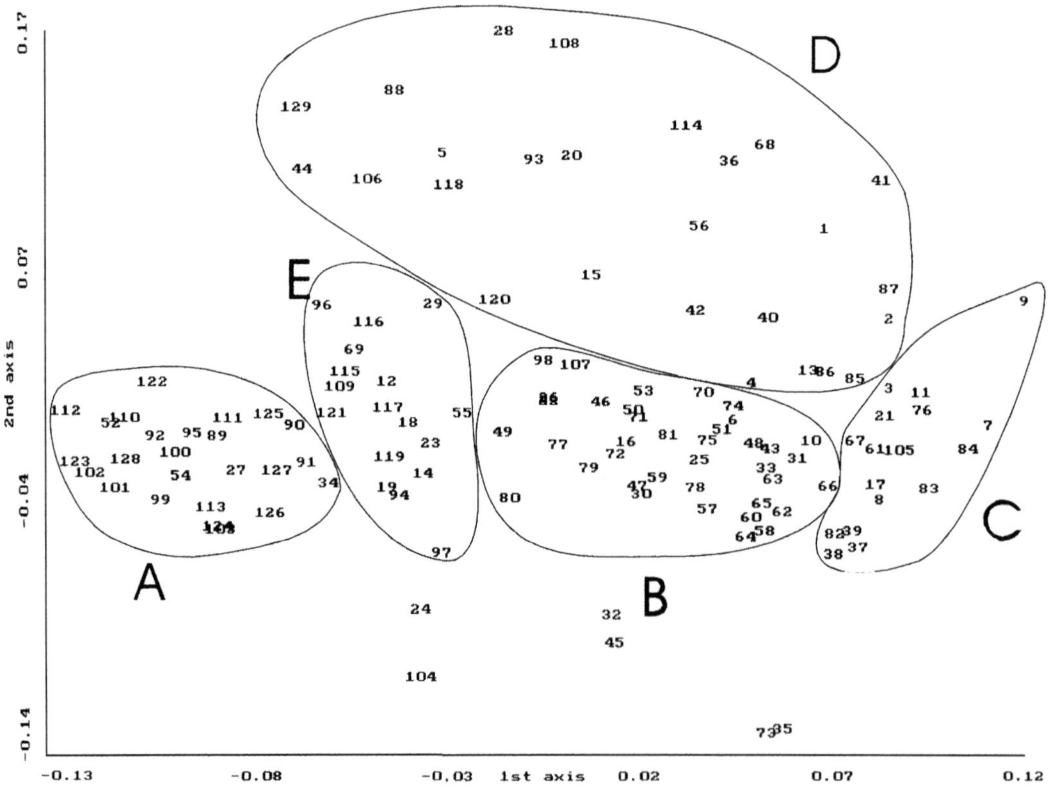

Figure 3. Ordination of the samples on the basis of relative abundance of the specimens that belong to the ecological categories mentioned above.

Table 2. Floristic composition of the localities arranging in accordance with the basic types of microvegetation. 1. Vadkacsás-tó 2. Huszászi-patak lápja, 3. Baláta-tó, 4. Szûrûhely-folyás, 5. Öcsi Nagy-tó, 6. Barkás-tó, 7. Monostori-tó, 8. Grajka-patak lápja, 9. Szõcei láp, 10. Ördög-tó, 11. Tiva-tó; 12 Fekete-tó

Taxa	Group I							Group II		Group III		Group IV
	1	2	3	4	5	6	7	8	9	10	11	12
Cyanobacteria	1	6	47	2	12	6	7	6	4	2	4	1
Bacillariophyceae	8	13	54	6	29	23	13	38	19	7	10	2
Chrysophyceae	1	2	44	3	1	3	2	1	0	3	3	0
Xanthophyceae	0	2	12	2	0	2	3	0	0	1	3	0
Raphidophyceae	1	1	1	0	0	1	0	0	0	0	0	0
Chlorococcales	1	5	85	13	2	11	9	0	0	5	17	2
Volvocales	0	3	28	2	4	3	7	2	1	6	8	1
Desmidiales	4	18	99	12	13	12	18	20	7	4	6	0
Other Chlorophyta	1	1	23	3	1	2	2	0	1	0	4	1
Euglenophyta	9	14	109	17	17	25	27	2	1	22	44	1
Dinophyta	1	6	22	3	5	6	6	0	1	2	7	0
Cryptophyta	2	2	8	3	4	6	6	0	1	1	4	0
Haptophyta	1	1	1	1	1	1	1	0	0	1	1	0
Total number of taxa	30	74	533	67	89	101	101	69	35	54	111	8

2. Samples in Group-D were dominated by large sized flagellates (Dinophyta: *Peridinium umbonatum, P. bipes*; Cryptophyta: *Cryptomonas ovata*; Euglenophyta: *Euglena sanguinea, Lepocinclis* spp., *Phacus* spp.; Chrysophyceae: *Mallomonas* spp. and Raphidophyceae: *Gonyostomum semen, G. intermedium*).

3. In Group-B the above mentioned chlorococcalean algae and Cyanobacteria were dominants and desmids (*Cosmarium subprotumidum, Closterium gracile, C. dianae, Staurastrum tetracerum, Staurodesmus dejectus, S. indentatus*), cryptophytes, dinoflagellates or euglenophytes appeared as subdominants.

4. Diatom dominated samples can be found in Group-A. (Characteristic species: *Eunotia bilunaris, Pinnularia* spp., *Gomphonema gracile, Nitzschia* cf. *fonticola*).

5. The fifth group (Group-E) contains all the samples in which other taxa, such as *Vaucheria dichotoma*, filamentous green algae: *Oedogonium* spp., *Cladophora* spp., *Microspora* spp. and *Microthamnion kuetzingianum*. prevailed.

Numbers, situated outside the encircled areas of Figure 2 indicate samples in which only one species (*Cladophora* sp. or filamentous blue green algae: *Anabaena solitaria, Nostoc* spp., *Microchaete tenera*), was found.

The groups mentioned above are closely linked with certain types of habitats.

In Group-A one can find samples dominated by planktonic species and they were collected from large pools with considerable open water area. Group-B contains the samples derived from small bog pools. Group-C includes sampled dominated by periphytic samples taken from bogs of good water supply. In Group-D also periphytic species dominate but these samples were collected from bogs that occasionally dry up. Exclusively samples collected in *Sphagnum* stands belong to Group-E.

Some of these bogs had been investigated previously. Therefore we tried to establish the basic trends that characterise the changes of the algal flora. In general (Table 3), there is a significant decrease of the total number of taxa that was observed at almost each sample-site. This decreasing tendency is attributable to the impoverishment of species rich algal groups such as diatoms, desmids and Cyanobacteria. The richer these taxa were in the past, the bigger is the drop in species number. Inverse trend has been found in case of Ördög-tó, that is an eutrophicated bog-pool. In this case, the larger species number is due to the increase of the species number of *Euglenophyta*.

Table 3. Floristic results (species numbers) from the previous and the present investigations. Abbreviations: 4: Szûrûhely-folyás; 5: Öcsi Nagy-tó; 7: Monostori-tó; 8: Grajka patak lápja; 9: Szõcei-láp; 10: Ördög-tó; 11: Tiva-tó; 12: Fekete-tó; a: Kol (1967); b: Vízkelety (1987); c: Uherkovich (1981); d: Uherkovich (1979); e: Uherkovich (1978); f: Kol (1970); g: Uherkovich et al. (1994); h: Uherkovich (1982); p: present study

locality	4	4	5	5	7	7	8	8	9	9	10	10	11	11	12	12	12
reference	c	p	d	p	h	p	f	p	g	p	b	p	e	p	a	b	p
Cyanobacteria	12	2	25	12	21	7	15	6	3	4	3	8	5	4	15	2	1
Bacillariophyceae	67	6	100	29	41	13	0	38	54	19	10	7	23	10	4	20	2
Chrysophyceae	9	3	7	1	6	3	0	1	0	0	0	9	15	3	0	1	0
Xanthophyceae	5	2	6	0	6	3	1	0	0	0	3	1	3	3	0	2	0
Raphidophyceae	0	0	0	0	0	0	0	0	0	0	0	0	0	0	0	0	0
Chlorococcales	19	13	41	2	13	9	2	0	6	0	1	5	15	17	10	0	2
Volvocales	4	2	8	4	6	7	0	2	0	1	2	6	6	8	3	2	1
Desmidiales	23	12	27	13	74	18	78	20	14	7	6	4	37	6	64	0	0
other Chlorophyta	11	3	5	1	9	2	14	0	3	1	2	0	13	4	8	2	1
Euglophyta	47	17	61	17	32	27	2	2	8	1	8	22	30	44	1	1	1
Dinophyta	3	3	4	5	6	6	0	0	0	1	1	2	8	7	0	0	0
Cryptophyta	7	3	6	4	1	6	0	0	0	1	0	1	2	4	0	0	0
Haptophyta	0	1	0	1	0	1	0	0	0	0	0	1	0	1	0	0	0
Number of taxa	**207**	**67**	**290**	**89**	**212**	**102**	**112**	**69**	**88**	**35**	**36**	**66**	**157**	**111**	**105**	**30**	**8**

Discussion

The expression 'bog systems' includes several types of wetlands, which differ from each other in many respects. Despite remarkable differences in origin, morphology, physiognomy and macroflora, the water-chemical properties of the localities proved to be rather uniform especially in context of nutrients. As threshold values for P_{inorg} and N_{inorg} were set as 10 μg l^{-1} and 100 $\mu g\, l^{-1}$, respectively (Sas, 1989), no nitrogen limitation can arise in these bogs and P_{inorg} might be limiting in only three bogs. Nevertheless, the low phosphorus concentrations did not coincide necessarily with poor algal flora. For example, the species richness and the species composition of Huszászi-patak lápja was very similar to that of Szûrûhely-folyás, although the phosphorus content in the latter locality was much higher. These records allow the conclusion that water chemistry, and especially nutrient supply, cannot be considered as the major factor for the observed differences. It is worth to note that in large lakes with dominant phytoplankton water chemistry and its changes along different scales can be the major selecting force for species and a chief driving force for annual and interannual changes (Padisák & Dokulil, 1994; Sommer et al., 1996; Reynolds, 1997).

The total number of identified taxa per localities ranged from 8 to 107. The occurrence of algae was quite variable within and among localities. These differences are due to several factors, especially water supply and habitat diversity. The richest flora was found in those localities where the water supply was stable and the encroachment of macrophytes resulted in various microhabitats.

Flora of Baláta-tó was neglected in the previous comparison because this transitional bog was investigated more in detail and for a longer period. Despite the Baláta-tó is a relatively small water body its algal flora is exceptionally rich (533 taxa). Such species richness is usually typical only for large lakes or rivers. As comparison, Kiss & Schmidt (1998) recorded 594 taxa from the Hungarian section of the Danube river during the last twenty years. The experienced large number of taxa is inconsistent with the generally accepted view, that the algal flora of dystrophic lakes is rather poor (Maitland & Morgan, 1997).

It has been widely accepted that the acidic environment is one of the most important factors that can determine the species pool of a bog. In this study, similarity of the water chemical parameters (including pH) of the investigated localities and the large differences of their algal flora suggest that other elements of the environment are also important. It seems that the character of these elements is rather physical than chemical. Because of their hypothesized terrestrial origin (Stebbins & Hill, 1980) most of the desmids are benthic or metaphytic forms. Proportion of desmids, indeed, was relatively high in those localities where the lake basin was dominated by macrophytes. According to observations by Fehér (in press) desmids seem to be associated with certain species of macrophytes like *Utricularia vulgaris* and *Miryophyllum spicatum*. Taxonomic composition of those localities where the open surface areas are also present is characterised by the high ratio of flagellates. Dimensions of these open areas seem to be the most important factor. Because of the sheltering of the terrestrial macrovegetation, wind cannot induce significant water turbulences. In these largely undisturbed, lentic environments organisms that do not have capability of effective locomotion will sink to the sediments. It justifies that possession of flagella is the most advantageous strategy in this kind of environment. Additional factors might be that acidic waters are usually richer in dissolved organic material which can fulfil requirements of potentially mixotrophic species. The benthic or metaphytic life of chlorococcalean algae that are believed to be principally planktonic is another strategy of survival. They might be also favoured by the relatively high amount of available inorganic nutrients. The relevant physico-chemical parameters of the small pools of different localities proved to be very similar, but only in a few of them was observable the dominance of the same assemblage of flagellated species. In the majority of cases different species of Raphidophyceae, Cryptophyta or Euglenophyta groups prevailed in the small pools, which is a good example of the pseudovicarism of algae. It is important to note here that the large number of species in several groups (like Chlorococcales) is probably due to evolution in such environments rather than in large lakes where they are usually the most numerous in terms of species number but mostly insignificant in terms of biomass (Padisák & G.-Tóth, 1991; Dokulil, 1991; Padisák & Dokulil, 1994; Padisák et al., 1998).

In the large bog-pools the wind induced water turbulences enable permanent presence of chlorococcalean algae and Cyanoprokaryota species in the plankton. This type of microflora only appeared in the largest Hungarian bog lake, Baláta-tó.

Comparing our results with that of the previous investigations a drop in algal species number has been observed. Although it cannot be excluded that the differences in sampling techniques or the different number of the samplings can influence the results, impoverishment of the flora can be most likely attributed to other factors. On the basis of an algal floristic investigation on a peat bog, which had lost considerable part of its catchment as a consequence of basalt-mining area in Balaton Upland region, Uherkovich (1984) concluded that the desmids are more sensitive to drying out than diatoms, consequently the impoverishment of the flora starts with the disappearance of desmids. The most dramatic changes have been shown in case of the *Sphagnum*-bog of Fekete-tó. This locality was described earlier as a very desmid rich habitat in Hungary (Kol, 1970). Impoverishment of the desmid flora was observed in the 1980s (Vízkelety, 1987). Very likely, the decreasing species number of the investigated bog systems is a reflection of the long lasting dry period that started in the early eighties, and lasted for one and half decade in Hungary. These observations indicate that such algal assemblages are sensitive indicators of both various human impacts at local scale or climate changes at global scales, moreover, that indication at species level is sometimes possible (*Heimiochrysis actinotrychia* [Chrysophyceae] *Rhipidodendron huxlei* [Haptophyta], *Peridinium bipes* tab. *γ travectum* f. *tabulatum* [Dinophyta]), but careful establishment of a functional classification system with different association types (Reynolds et al., 2002) has been probably more promising as evidenced by Padisák et al. (2003), Stoyneva (2003) and O'Farrell et al. (2003).

In accordance with results of this floristic studies and their ordination profile we may outline five basic types of the microvegetation that appear in the different habitats of a hypothetical bog system (Fig. 4).

(1) Littoral algal vegetation in forested shores (trees are many times rooting in water) is rich in cryptomonads most probably because of its ability for chromatic adaptation (phycobiliproteins), good photosynthetic parameters in low-light environments and its good spatial orientation in small scale environments (utilization of varying sunshine-spots penetrating through the canopy of the trees). *Mallomonas* spp. might also be characteristic.

(2) In dense, emerged macrophyte stands periphytic diatoms are the most characteristic type of association sometimes with periphytic filamentous algae (*Chlorophyta*).

(3) In habitats with rich submerged higher plant vegetation, however with good light climate, species of *Euglenophyta* appear in large numbers. They are probably favoured because they are good swimmers and their need for organic substances are usually sufficed in such environments. In this as-

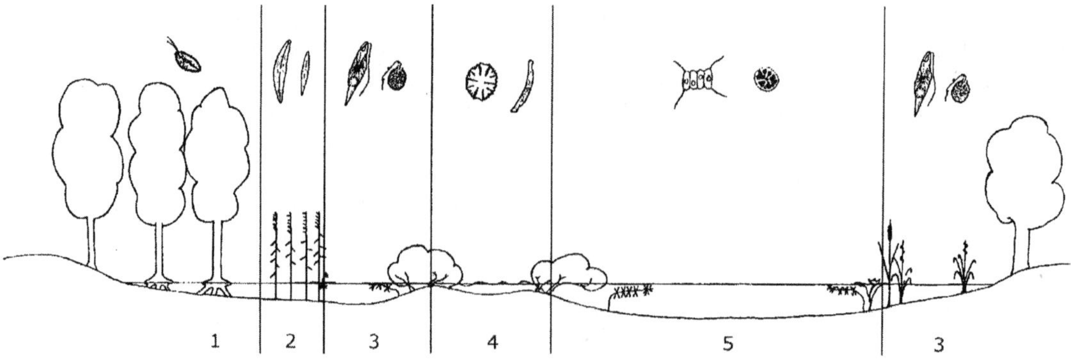

Figure 4. Vegetation profile of a hypothesised bog-lake with the characteristic types of habitats and the relevant members of the microflora. 1. Shallow, shaded and lentic area without macrophytes; 2. Emerged macrophytes; 3. Small, lentic bog pools with good light conditions and submerged macrophyte; 4. *Sphagnum* tussocks; 5. Large open water areas.

sociation type many of well-known planktonic organisms, such as *Scenedesmus, Dictyosphaerium, Eutetramorus*, species or a heterocytic cyanoprokaryotes like *Anabaena solitaria*, instead of being in the plankton, can sufficiently establish populations in the metaphyton where settling on leaves of macrophytes prevent them of sinking to aphotic layers.

(4) *Sphagnum*-patches are inhabited by typical bog dwellers (desmids, diatoms like *Eunotia*).

(5) Large open surfaces allow for typical planktonic associations of chiefly chlorococcalean, sometimes cyanobacterial types. In such habitats wind-driven mixing can keep non-motile planktonic green algae in suspension and their high demand for inorganic nutrients are mostly sufficed.

In more alkaline waters with similar physiognomy filamentuos species of type two can be more characteristically periphytic *Oscillatoriales*. In such lakes a 6th association type may be present: dense benthic algal mats of oscillatorealean filaments with large benthic diatoms.

When dealing with such small-scale canopy systems one should consider spatial heterogeneity. With 'rough' sampling methods we may easily mix different association types which makes impossible to recognize them even though they are equilibrated (*sensu* Sommer et al., 1993) one by one.

Acknowledgements

This work was founded by various sources, including Environmental Protection Inspectorate for Trans-Tiszanian Region, Hungarian National Science Foundation (OTKA T 029636, F 31802), Hungarian Higher Education Found (0195).

References

Borics, G., 2001. A Baláta-tó és néhány más hazai lápvíz algaflórájának főbb jellemvonásai. [Characteristic features of algal flora of Lake Baláta and some other Hungarian bog lakes] PhD Thesis, University of Debrecen [in Hungarian with English summary].

Borics G., J. Padisák, I. Grigorszky, I. Oldal, L. I. Péterfi &L. Momeu, 1998a. Green algal flora of the acidic bog-lake, Baláta-tó SW Hungary. Biol. Bratislava 53: 457–465.

Borics G., I. Oldal, I. Grigorszky, J. Padisák, L. I. Péterfi & L. Momeu, 1998b. Adatok a Baláta-tó algaflórájához [Data to the algal flora of Lake Baláta]. Hidrológiai Közlöny 78 5–6: 276–278 [in Hungarian with English summary].

Borics G., I. Grigorszky, J. Padisák & S. Szabó, 2000. A Baláta-tó Dinoflagellata flórája és annak sajátságai [Characteristic features of the dinoflagellate flora of Lake Baláta]. Hidrológiai Közlöny 80: 282–284 [in Hungarian with English summary].

Cambra, J. & F. Hindák, 1998. Green algae from mountain peat-bogs in the Eastern Pyrenees (Catalonia, Spain). Biol. Bratislava 53: 467–480.

Coesel, P. F. M., 1984. The significance of desmids as indicators of the trofic status of freshwaters. Swiss J. Hydrol. 45: 388–393.

Csató, J., 1885. A Mluha nevû tó (Teu Mluhi) és viránya [Algae of Lake Mluha]. Magy. Növénytani Lapok 9: 1–8 [in Hungarian].

Dokulil, M., 1991. Contribution of green algae to the phytoplankton assemblage in a mesotrophic lake, Mondsee, Austria. Archiv für Protistenkunde 139: 212–223.

Fehér, G., 2000. Desmidiales elõfordulások a szeremlei Duna-ágban [Occurrence of desmid taxa in the Szeremlei Duna (Gemenc Area, Duna-Dráva National Park, South of Hungary]. Acta Agronomica Óváriensis. 41: 197–210 [in Hungarian with English summary].

Fehér, G., in press. Szikes területek vizeinek Desmidiales flórája [Desmidiales flora of alcaline salt marshes]. Természetvédelmi Közlemények [in Hungarian].

Florek, K., J., Lukaszewicz, J., Perkal, H., Steinhaus & S. Zubrzycki, 1957. Sur la liaison et la division des points d'un ensemble fini. Colloquium Mathematicum 2: 282–285.

Hazslinszky, F., 1868. Magyarhon s társországai moszatviránya [Algae of Hungary and the Carpatian Basin]. Mathematikai és Természettudományi Közlemények 5: 163–181 [in Hungarian].

Kiss, I., 1977. A Kiskunsági Nemzeti Park III-as számú területén lévõ szikes tavak algológiai vizsgálata [Algological investigation of salt ponds in Kiskunság Nature Reserve]. Unpublished manuscript [in Hungarian].

Kiss, K. T. & A. Schmidt, 1998. Changes of the Chlorophyta species in the Hungarian section of the Danube river during the decades (1961–1997). Biol. Bratislava 53, 509–518.

Kol, E., 1967. Algologische und hydrobiologische Untersuchungen im *Sphagnum*-moor 'Fekete-tó' bei Farkasfa. Acta Bot. Acad. Sci. Hungariae 13: 113–131.

Kol, E., 1970. Algológiai és hydrobiológiai vizsgálatok a Grajkapatak forrásláp jain Vas megyében [Algological and hydrobiological investigations of the bog of Grajka-streamlet, Vas-county]. Savaria (Szombathely) 4: 9–29 [in Hungarian with English summary].

Kristiansen, J., 1986. Silica-scale bearing Chrysophytes as Environmental Indicators. British Phycol. J. 21: 425–436.

Kristiansen, J. & M. Stoyneva, 1998. Silica-scaled Chrysophytes in Bulgaria. Cryptogamie, Algologie 19: 19–28.

Legendre, P. & L. Legendre, 1998. Numerical Ecology. 2nd ed. Elsevier 272 p.

Maitland, P. S. & N. C. Morgan, 1997. Conservation Management of Freshwater Habitats. Chapman & Hall. 130 p.

O'Farrell, I., R. Sinistro, I. Izaguirre & F. Unrein, 2003. Do steady state assemblages occur in shallow lentic environments from wetlands? Hydrobiologia 502 (Dev. Hydrobiol. 172): 197–209.

Padisák, J., 1999. Checklist of aquatic algae found in the Kiskunság National Park and in the Danube-Tisza Interfluve. In Lõkös, L. & M. Rajczy (eds), The Flora of the Kiskunság National Park II. Cryptogams. Natural History of the National parks of Hungary. Magyar Természettudományi Múzeum, Budapest. 9: 9–146.

Padisák, J., G. Borics, G. Fehér, I. Grigorszky, I. Oldal, A. Schmidt & Z. Zámbóné-Doma, 2003. Dominant species, functional assemblages and frequency of equilibrium phases in late summer phytoplankton assemblages in Hungarian small shallow lakes. Hydrobiologia 502 (Dev. Hydrobiol. 172): 157–168.

Padisák, J. & M. Dokulil, 1994. Contribution of green algae to the phytoplankton assemblage in a large, turbid, shallow lake (Neusiedlersee, Austria/Hungary). Biol. Bratislava 49: 571–579.

Padisák, J. & L. G.-Tóth, 1991. Some aspects of the ecology of the subdominant green algae in a large nutrient limited shallow lake (Balaton, Hungary). Archiv für Protistenkunde 139: 225–242.

Padisák, J., L. Krienitz, W. Scheffler, R. Koschel, J. Kristiansen & I. Grigorszky, 1998. Phytoplankton succession in the oligotrophic Lake Stechlin (Germany) in 1994 and 1995. Hydrobiologia 369/370: 179–197.

Padisák, J., S. L. Péterfi & L. Momeu, 2000. Silica scaled chrysophytes from Hungary. Verh. int. Ver. Limnol. 27: 131–134.

Péterfi, L. S. & L. Momeu, 1996. Three Rumanian *Mallomonas* species (Synurophyceae), with special distribution patterns. Hydrobiologia 336: 143–150.

Péterfi, L. S., J. Padisák, L. Momeu & G. Borics, 1998. Silica-scaled chrysophytes from the bog-lake Baláta-tó, Sw Hungary. Nordic J. Bot. 18: 727–733.

Reynolds, C. S., 1997. Vegetation processes in the pelagic: a model for ecosystem theory. Ecology Institute, Oldendorf/Luhe.

Reynolds, C. S., V. Huszar, C. Kruk, L. Naselli-Flores & S. Melo (2002): Towards a functional classification of the freshwater phytoplankton. Journal of Plankton Research 24: 417–428.

Sas, H., 1989. Lake Restoration by Reduction of Nutrient Loading: Expectations, Experiences, Extrapolations. Academia Verlag Richarz, St. Augustin. 212 p.

Schaarschmidt, Gy., 1882. Tanulmányok a magyarhoni Desmidiaceákról [Researh of the Hugarian desmid flora]. Mathemaikai és Természettudományi Közlemények 18: 257–280 [in Hungarian].

Sommer, U., J. Padisák, C. S. Reynolds & P. Juhász-Nagy, 1993. Hutchinsons heritage: the diversity-disturbance relationship in phytoplankton. Hydrobiologia 249: 1–8.

Sommer U, M. Gliwicz, W. Lampert & A. Duncan, 1986. The PEG model of seasonal succession of planktonic events in fresh waters. Archiv für Hydrobiol. 106: 433–471.

Stebbins, G. L. & G. J. C. Hill, 1980. Did multicellular plants invade the land? Am. Nat. 115: 342–353.

Stollmayer-Boncz, E., 1992. The alga species of the Csömör pool II. Studia Bot. Hungarica, 23: 17–48.

Stoyneva, M. P., 2003. Steady-state phytoplankton assemblages in shallow Bulgarian wetlands. Hydrobiologia 502 (Dev. Hydrobiol. 172): 169–176.

Tóthmérész, B., 1993. NuCoSA 1.0: Number cruncher for Community Studies and other ecological applications. Abstracta Bot. 7: 283–287.

Uherkovich, G., 1978. A Tíva tó és a Nagyberek (Barcsi õsborókás) algáiról [Algae from Tiva-tó and Nagyberek in Barcs Nature Revserve]. Dunántúli Dolgozatok Természettudományi Sorozat 1: 9–35 [in Hungarian with English summary].

Uherkovich, G., 1979. Az öcsi Nagy-tó limnológiája [Die Limnologie der Grossen See von Öcs]. A Veszprém Megyei Múzeumok Közleményei 14: 25–53 [in Hungarian with German summary].

Uherkovich, G., 1981. A Szürühely-folyás (Barcsi Borókás) tõzegmohás tavacskájának algái [Algae of the pond "Szürühely-folyás" in the Juniper Woodland of Barcs, Hungary]. Dunántúli Dolgozatok Természettudományi Sorozat 2: 5–23 [in Hungarian with English summary].

Uherkovich, G., 1982. A Fekete-hegy (Balaton-felvidék) Kerek tava algavegetációja [Die Algenvegetation des Kerek (Rund)-Sees vom Fekete (Schwarz)-Berg (Balaton-Oberland)]. Folia Musei Historico-naturalis Bakonyiensis 1: 81–110 [in Hungarian with English summary].

Uherkovich, G., 1984. A Vad-tó (Kovácsi hegy, Zala megye) algavegetácójáról [Microvegetation of the bog-lake 'Vad-tó']. Folia Musei Historico-naturalis Bakonyiensis 3: 43–56 [in Hungarian with English summary].

Uherkovich, G., L. Szilágyi & É. Vízkelety, 1994. Die Algenvegetation von zwei kleinen Stillgewassern Westungarns: Vadása See und Moor von Szõce. Janus Pannonus Múzeum Évkönyve 38: 5–17.

Vízkelety, É., 1987. Az õrségi Fekete-tó és Ördög-tó algológiai vizsgálata. [Algological research of the Sphagnum bog "Fekete-tó" and the bogpool "Ördög-tó"]. Praenorica 2: 59–62 [in Hungarian with English summary].

Hydrobiologia **502:** 157–168, 2003.
L. Naselli-Flores, J. Padisák, M. T. Dokulil (eds),
Phytoplankton and Equilibrium Concept: The Ecology of Steady-State Assemblages.

Dominant species, functional assemblages and frequency of equilibrium phases in late summer phytoplankton assemblages in Hungarian small shallow lakes

Judit Padisák[1], Gábor Borics[2], Gizella Fehér[3], István Grigorszky[4], Imre Oldal[5], Antal Schmidt[3] & Zsuzsa Zámbóné-Doma[1,6]
[1]*University of Veszprém, Department of Limnology, H-8200 Veszprém, P. O. Box 158, Hungary*
E-mail: padisak@tres.blki.hu
[2]*Trans-Tiszanian Environmental Protectorate, H-4025 Debrecen, Piac u. 9/b, Hungary*
[3]*South Danubian Environmental Protectorate, University of Debrecen, H-6501 Baja, P.O. Box 113, Hungary*
[4]*Botanical Department, H-4010 Debrecen Egyetem tér 1, Hungary*
[5]*South Transdanubian Environmental Protectorate, H-7615 Pécs, Cserkút, Csárdakörnyék 6, Hungary*
[6]*Balaton Limnological Institute of the Hungarian Academy of Sciences, H-3237 Tihany, Hungary*

Key words: phytoplankton composition, equilibrium phase, stress, trophic states, species number, phytoplankton functional groups

Abstract

Late summer phytoplankton associations were studied qualitatively and quantitatively in 80 Hungarian lakes altogether (mostly shallow salt lakes, reservoirs, oxbows, gravel pit lakes). Equilibrium phases *sensu* Sommer et al. (1993) were found only in 17 lakes. Most of them were under some kind (high salt content or very low level of nutrients) of stress factor. It is concluded that environmental stress forces phytoplankton communities towards equilibrium. No relationship between occurrence of equilibria and trophic state was found. Species number of non-equilibrated lakes was almost three times as high as those in equilibrium. Of the 31 recently described (Reynolds et al., 2002) phytoplankton assemblages most of those were recognized that are likely to occur in shallow lakes. Separation of a functional group $\mathbf{W_S}$ from **W2** for *Synura* dominated lakes is suggested. It seemed also necessary to raise a group ($\mathbf{Y_{Ph}}$) for lakes dominated by *Phacotus*. Sorting of Dinophyta species into different already described functional groups is desirable.

Introduction

Phytoplankton research on small lakes has been controversial. While they have been preferred targets for taxonomic and floristic work, the bulk of our knowledge on ecology of phytoplankton originates from middle-sized or large lakes.

The number of small lakes is much higher than that of large ones and, at a global scale, the volume of freshwater that is stored in small lakes is comparable to that in large lakes. They are used as drinking water supply in many regions of the world and their importance is increasing as recreational areas (water sports, swimming, angling, eco-tourism etc.). The socio-economic importance of small lakes in itself calls for more detailed scientific knowledge about their limnology and phytoplankton ecology. Moreover, many of them exhibit a large habitat diversity, therefore, they are very important in conservation biology. As a consequence of their small water volume and often unstable hydrological balance small lakes react quickly to human impacts like increased N and P loadings on the watershed, acidification or climatic changes even at small scales. Their common feature is that historical data are largely absent. Interest towards understanding driving forces that govern their spatial and temporal phytoplankton patterns has just started to increase (O'Farrell et al., 2003; Ortega-Mayagoitia et al., 2003; Stoyneva, 2003).

During its annual development, the phytoplankton

passes several quite distinct successional stages during which equilibrial compositions develop and pertain for shorter or longer periods. One of the few attempts to quantify criteria for equilibrium phase of phytoplankton communities was provided by Sommer et al. (1983) as: *"In natural phytoplankton communities, it is often difficult to determine whether a given 'phase' in a seasonal sequence can be considered to be in an equilibrium state or not, due either to a lack of chemical data, or to insufficient sampling frequency, or to any other cause. For practical purposes, provided that (i) 1, 2 or 3 species of algae contribute more than 80% of total biomass, (ii) their existence or coexistence persists for long enough (more than 1–2 weeks) and (iii) during that period the total biomass does not increase significantly, then that phase can be considered to be at equilibrium."*

Restriction of species number in this definition was based on Hardin's Competitive Exclusion Theory (1960) (only as many species can exist as there are limiting factors) which was supported for phytoplankton by modelling and in chemostat experiments (Tilman, 1982). Since it is not likely that more than 3 basic resources (N, P, light silica for diatoms; micronutrients are very rarely limiting) are in limiting amounts at a given space and time, species number for persistent coexistence was maximized at 3. The 80% additive dominance was based merely on empirical field observations. Persistence of such a structure was set at 1-2 weeks because disturbance frequency in natural lakes operates at such scale. Constancy of biomass was needed as measure of functional persistence of the ecosystem (top-down and bottom-up controls are probably equilibrated or bottom-up dominates). Equilibria as defined above may occur during the spring phytoplankton peak (usually dominated by diatoms), in clear-water phases (usually dominated by one or several species of chrysophytes and cryptophytes), during winter stagnation, when species are largely selected according to their tolerance of stress conditions (cold temperatures, mixing, low light) and most prominently during summers when the stagnation period allows the development of competitively stabilized associations.

In this paper, we present data from a survey of small lakes in Hungary. We investigate how often equilibria occur during summer stagnation (July–September). Since this study is based on one sampling per lake, duration of equilibria cannot be considered. We discuss how phytoplankton composition of mostly shallow oxbows, gravel pit lakes, salty lakes and small reservoirs fit to functional classification of phytoplankton assemblages recently described by Reynolds et al. (2002).

Study sites, sampling and methods

The studied lakes are located in four different regions of Hungary (Fig. 1). I. One oxbow and 5 small reservoirs were sampled in Eastern Hungary. Their surface areas range from 20 to 91 ha and their depths are in the range of 1.5–5.5 m. II. There are a number of continental salt lakes (conductivity range: 1600–32 000 μS cm^{-1}) in the Kiskunság National Park located between the rivers Duna and Tisza. Samples were taken from 7 lakes having surface areas from 35 to 456 ha and their depth is 0.5–1 m. III. In the Danube region of South Hungary 15 lakes, oxbows and deadarms were sampled. Surface range from 4 ha to 74 ha and their depth was 1.5–6 m. IV. On the territority of the Duna-Dráva National Park 39 oxbows (surface area: 1–100 ha; depth: 0.5–3 m) and 13 gravel pit lakes (surface area: 4–400 ha; depth: 3–12 m) were sampled.

Large volume surface samples were taken preferably in the open water (in areas as free as possible of macrophytes) of the lakes in July–middle October of 1999–2001. They were preserved in Lugol's solution and kept in the dark until analyses. Inverted microscopes were used for establishing cell numbers and biomass was obtained from geometric approximation. In each case except the southern 39 oxbows and gravel 13 pit lakes (category IV) a minimum of 1000 individuals were counted in each sample providing a counting error of <5%. In the southern 39 oxbows and 13 gravel pit lakes phytoplankton species were counted in one transect independently from density. Counting accuracy in these cases was between 10 and 20% but species occurring outside the transect were also recorded and included in the number of species. Thus, species numbers provided by the two different counting methods were roughly comparable.

According to chemical analyses, soluble reactive P concentrations were $>10\ \mu$g l^{-1} and soluble inorganic N was $>100\mu$g l^{-1} except the gravel pit lakes that were apparently seriously nutrient deficient. Conductivity ranged between 200 and 1500 μS cm^{-1}for most lakes except the salt lakes where values between 3000 and 25 000 μS cm^{-1} were registered.

Table 1. Dominant species, phytoplankton biomass and species number in 80 Hungarian shallow lakes. Abbreviations: S – salt lake; RE – reservoir; GP – gravel pit lake; OX – oxbow; SA – side arm (at one edge has connection to the mainflow even though it is rarely, if at all, washed through); NL – natural lake of other origin (mostly wind erosion)

Nr.	Name of the locality	Type	1st dominant species	%	2nd dominant species	%	3rd dominant species	%	Biomass ($\mu g\,l^{-1}$)	Species number
1	Böddi-tó, NY3	S	*Euglena ehrenbergii*	95	*Stenopterobia pelagica*	5			19	2
2	Látókepi tározó	RE	*Cryptomonas* sp.	93					4067	20
3	Fancsika I.	RE	*Tetraedron trigonum*	69	*Lepocinclis* sp.	21			7028	24
4	Fehér szék	S	unidentified green	50	*Phacus acuminatus*	40			4141	21
5	Fancsika II.	RE	*Aphanocapsa* sp.	82	*Treurbaria triappendiculata*	6			4745	28
6	Kavicsbánya-tó III, Gyékényes	GP	*Cyclotella comta*	54	*Rhodomonas pusilla*	19	*Cryptomonas marssonii*	15	2066	16
7	Kavicsbánya-tó I, Gyékényes	GP	*Zygnema* sp.	82	*Cryptomonas marssonii*	5			21574	24
8	Kishobogy, Dráva holtág	OX	*Synura uvella/pettersenii*	74	*Cryptomonas rostrata*	8	*Melosira varians*	5	13247	27
9	Mosztonga	OX	*Cryptomonas* sp.	86					3756	22
10	Sugovica	SA	*Synura uvella/pettersenii*	73	*Pseudokephyrion* sp.	10			6277	22
11	Kavicsbánya-tó Ib, Gyékényes	GP	*Cryptomonas erosa*	50	*Phacotus lenticularis*	23	*Rhodomonas nannoplanctica*	10	115	9
12	Szeremlei Duna	SA	*Trachleomonas volvocina*	51	*Anabaena* sp.	20	*Centrales* sp.	12	12559	44
13	Szilháti Dráva-holtág, Cun	OX	*Pandorina morum*	48	*Synura uvella*	35			14093	31
14	Böddi-tó, NY1	S	*Navicula* sp.	82					513	17
15	Roza-tó, Dráva holtág, Cun	OX	*Synura uvella*	66	*Stephanodiscus* sp.	8	*Cryptomonas obovata*	6	7478	32
16	Kavicsbánya-tó Ic, Gyékényes	GR	*Cyclotella comta*	64	*Rhodomonas pusilla*	11	*Cryptomonas marssonii*	5	3140	23
17	Kavicsbána-tó I., Berzence	GR	*Phacotus lenticularis*	47	*Dinobryon divergens*	26	*Fragilaria ulna*	7	705	13
18	Sziki-tó	NL	*Rhodomonas* sp.	39	*Chrysochromulina parva*	25	*Other Chrysophyceae* sp.	15	273	14
19	Riha-tó (kelet)	OX	*Aphanizomenon ovalisporum*	63	*Pseudanabaena limnetica*	10	*Other Oscillatoriales* sp.	5	217053	41
20	Holt-Tisza, Tiszadob	OX	*Cylindrospermopsis raciborskii*	54	*Aphanizomenon aphanizomenoides*	17	*Aphanizomenon issatschenkoi*	7	10476	32
21	Böddi-tó K2	S	*Navicula* sp.	48	*Botryococcus braunii*	20	*Cryptomonas* sp.	8	513	17
22	Támosós Dráva-holtág, Cun	OX	*Pinnularia maior*	47	*Cymatopleura solea*	17	Nitzschia sigmoidea	12	3753	20
23	Dráva holtág, Zákány I.	OX	*Cryptomonas rostrata*	33	*Salpingoeca ringens*	28	*Cryptomonas ovata*	14	22714	19
24	Priszpa	OX	unidentified green	56	*Botryococcus braunii*	10	*Euglena* sp.	8	784	27
25	Babócsa II. holtág	OX	*Ceratium hirundinella*	73					8265	29
26	Mérnökházi Dráva-holtág	OX	*Didymosphaenia geminata*	36	*Cymatopleura solea*	26	*Coelastrum astroideum*	11	5839	30
27	Dráva-holtág, Recske	OX	*Pandorina morum*	43	*Synura uvella*	18	*Dictyosphaerium pulchellum*	11	26548	45
28	T ozeges-tó, Bolhó	NL	*Pediastrum simplex*	51	*Ceratium hirundinella*	11	*Pinnularia maior*	9	34289	45
29	Kelemen-szék	S	*Navicula* sp.	27	*Nitzschia* sp.	27	unidentified green	15	278	15
30	Babócsa I. holtág	OX	*Ceratium hirundinella*	68					1670	25
31	Dráva-holtág, Zákány II.	OX	*Peridinium cinctum*	33	*Cryptomonas rostrata*	19	*Cryptomonas ovata*	16	2453	27
32	Boros-Dráva, Dráva-holtág	OX	*Stephanodiscus hantzschii*	48	*Coelastrum astroideum*	10	*Pandorina morum*	9	26379	43
33	Kavicsbánya-tó I., Bélavár	GR	*Cryptomonas ovata*	35	*Cryptomonas marssonii*	16	*Cryptomonas rostrata*	16	4825	30
34	Dráva-holtág, Révfalu	OX	*Scenedesmus obliguus*	52	*Cryptomonas ovata*	7	*Cyclotella comta*	7	163782	47
35	Fekete-tó	NL	*Euglena ehrenbergii*	38	*Centrales* sp.	14	*Phacus* sp.	14	3019	23
36	Vájás Dráva-holtág, Drávasztára	OX	*Botryococcus braunii*	33	*Ceratium hirundinella*	26	*Melosira varians*	7	30672	47
37	Dráva-holtág, Bélavár	OX	*Ceratium hirundinella*	31	*Melosira varians*	26	*Dibobryon sociale*	7	7402	39
38	Kisinci Dráva-holtág, Cun	OX	*Ceratium hirundinella*	48	*Peridinium aciculiferum*	8	*Nitzschia vermicularis*	7	5118	37
39	Szelidi-tó	S	*Goniochloris mutica*	41	*Cylindrospermopsis raciborskii*	13	*Aphanizomenon ovalisporum*	9	7151	31
40	Mézeshegyi-tó	RE	*Tetraedron trigonum*	25	*Snowella lacustris*	23	*Coelastrum pseudomicroporum*	13	1685	44
41	Vadkerti-tó	NL	*Cryptomonas* sp.	33	*Tetraedron minimum*	20	*Lobocystis planktonica*	6	759	27
42	Kati-tó	RE	*Aphanizomenon flos-aquae*	32	*Navicula* sp.	16	*Nitzschia* sp.	11	13966	41
43	Böddi-tó NY1	S	*Cryptomonas* sp.	33	*Navicula* sp.	16	*Nitzschia* sp.	9	947	31
44	Kavicsbányató III. Bélavár	GR	*Phacotus lenticularis*	26	*Peridinium aciculiferum*	20	*Cryptomonas marssonii*	12	974	20
45	Szilháti Dráva-holtág, Cun	OX	*Euglena proxima*	31	*Synura uvella*	16	*Trachelomonas planktonica*	10	922	19
46	Kishobogy, Dráva-holtág, Cun	OX	*Pinnularia viridis*	34	*Cryptomonas marssonii*	15	*Amphora ovalis*	7	13247	27
47	Böddi-tó NY2	S	*Botryococcus braunii*	27	*Chrysochromulina parva*	14	*Cryptomonas* sp.	13	121	17
48	Klágya-Duna	OX	*Cylindrospermopsis raciborskii*	28	*Oscillatoriales* sp.	25			9210	50

Table 1. contd.

Nr.	Name of the locality	Type	1st dominant species	%	2nd dominant species	%	3rd dominant species	%	Biomass ($\mu g\,l^{-1}$)	Species number
49	Kavicsbányató, Örtilos	GR	*Peridinium inconspicuum*	33	*Rhodomonas minuta* var. *nannoplanktica*	10	*Dictyosphaerium pulchellum*	9	410	24
50	Dráva-holtág, Vízvár	OX	*Synura uvella*	29	*Peridinium cinctum*	13	*Peridinium aciculiferum*	10	3984	25
51	Sóséri csatorna	RE	*Volvox* sp.	28	*Rhodomonas* sp.	12	*Cryptomonas* sp.	11	516	25
52	Kondor-tó	S	*Scenedesmus quadricauda*	27	*Mougeotia* sp.	13	*Unidentified green*	11	373	25
53	Kavicsbányató I., Berzence	GR	*Cryptomonas ovata*	22	*Dinobryon divergens*	19	*Aphanizomenon issatschenkoi*	10	1156	18
54	Földvári-tó	OX	*Crucigenia tetrapedia*	33	*Oocystis marina*	9	*Scenedesmus quadricauda*	7	60802	53
55	Kavicsbányató, Somogyudvarhely	GR	*Peridinium aciculiferum*	27	*Crucigenia tetrapedia*	12	*Cyclotella comta*	10	2070	27
56	Dázsonyi-tó	NL	*Pinnularia viridis*	38	*Cryptomonas rostrata*	10			1504	22
57	Kavicsbányató II., Gyékényes	GR	*Rhodomonas pusilla*	22	*Cryptomonas obovata*	14	*Pseudanabaena catenata*	12	771	22
58	Riha-tó (nyugat)	OX	*Cryptomonas marssonii*	22	*Cylindrospermopsis raciborskii*	15	*Aphanizomenon ovalisporum*	11	2011	31
59	Ménökházi Dráva-holtág	OX	*Melosira varians*	23	*Trachleomonas oblonga*	15	*Navicula cryptocephala*	9	2515	35
60	Kisinci Dráva-holtág, Cun	OX	*Pandorina morum*	16	*Pinnularia viridis*	16	*Staurastrum manfeldtii*	15	14446	37
61	Hótedra, Dráva-holtág, Gordása	OX	*Cryptomonas marssonii*	32	*Coelastrum astroideum*	8	*Stephanodiscus hantzschii*	7	15020	59
62	Mrtvica, Dráva-holtág	OX	*Synura uvalla*	28	*Ceratium hirundinella*	11	*Rhopalodia gibba*	7	34991	79
63	Boros-Dráva, Dráva holtág	OX	*Ceratium hirundinella*	21	*Aphanizomenon issatschenkoi*	13	*Peridinium aciculiferum*	12	9461	41
64	Hótedra, Dráva-holtág, Gordása	OX	*Rhopalodia gibba*	24	*Peridinium cinctum*	11	*Closterium aciculare*	7	4969	51
65	Bresztik, Dráva-holtág, Drávasztára	OX	*Melosira varians*	24	*Phacus maximus*	11	*Pseudanabaena limnetica*	7	23217	70
66	Támosós Dráva-holtág, Cun	OX	*Cryptomonas rostrata*	22	*Phacus pleuronectes*	12	*Trachelomonas oblonga*	8	2373	20
67	Aldrivica, Dráva-holtág, Zaláta	OX	*Cymatopleura solea*	21	*Cyclotella comta*	13	*Euglena acus*	8	21833	52
68	Aldrivica, Dráva-holtág, Zaláta	OX	*Anabaenopsis arnoldii*	16	*Aphanizomenon issatschenkoi*	13	*Euglena proxima*	11	18034	44
69	Roza-tó, Dráva-holtág, Cun	OX	*Peridinium cinctum*	14	*Euglena texta*	13	*Cryptomonas rostrata*	12	3691	41
70	Sós-tó	S	*Oscillatoriales* sp.	15	*Chrysophyceae* sp.	12	unidentified green	10	224	24
71	Kis-Bok, Dráva-holtág, Barcs	OX	*Microcystis* sp.	21	*Peridinium aciculiferum*	9	*Euglena proxima*	6	4445	48
72	Recske, Dráva-holtág	OX	*Tetrastrum glabrum*	22	*Cryptomonas rostrata*	6	*Microcystis* sp.	7	4344	52
73	Vájás, Dráva-holtág, Drávasztára	OX	*Botryococcus braunii*	13	*Cryptomonas rostrata*	11	*Coelastrum pseudomicroporum*	11	10259	66
74	Kavicsbányató III., Berzence	GR	*Rhodomonas pusilla*	14	*Peridinium cinctum*	12	*Cryptomonas marssonii*	9	1774	31
75	Kavicsbányató II., Berzence	GR	*Oscillatoria granulata*	14	*Cryptomonas rostrata*	12	*Dinobryon sociale*	8	9140	32
76	Ó-Dráva holtág, Barcs	OX	*Trachelomonas oblonga*	13	*Cryptomonas erosa*	11	*Peridinium aciculiferum*	9	4531	44
77	Bresztik, Dráva-holtág, Drávasztára	OX	*Staurastrum paradoxum*	11	*Mougeotia* sp.	11	*Anabaenopsis arnoldii*	10	12266	65
78	Kavicsbányató, Õrtilos	GR	*Euglena texta*	14	*Euglena acus*	5	*Anabaena solitaria*	5	3437	56
79	Mrtvica, Dráva-holtág	OX	*Crucigenia tetrapedia*	10	*Stephanodiscus hantzschii*	7	*Cyclotella comta*	7	5719	44
80	Dráva-holtág, Majláthpuszta	OX	*Planktolyngbya limnetica*	6	*Anabaenopsis arnoldii*	6	*Coelastrum astroideum*	6	72571	76

Results

The investigated 80 lakes covered all trophic states and the distribution of the lakes was rather equal among the different trophic categories. Dominance was estimated from percentage contribution of individual species to total biomass.

The first three dominant species in each of the 80 lakes are listed in Table 1. Only those species which had a minimum of 5% contribution to total biomass were considered as dominants. For this reason, for Lake nr 2 only *Cryptomonas* is mentioned as dominant although there were 19 other species present. Biomass contribution of none of these 19 reached the 5% limit and therefore for this lake only *Cyryptomonas* was listed. Numbers in square brackets in the following sections of this paper refer to serial number of the given lakes as listed in Table 1.

As demonstrated on Figure 2, equilibrium phase (additive dominance of the three most abundant species reached 80%) was detected in only 17 of the 80 lakes. Among the lakes within or close to equilibrium stressed (salt lakes and nutrient poor gravel pit lakes) lakes were represented in a larger amount than their share among the lakes. In 26 of the 80 lakes additive dominance did not exceed 50% which indicates that

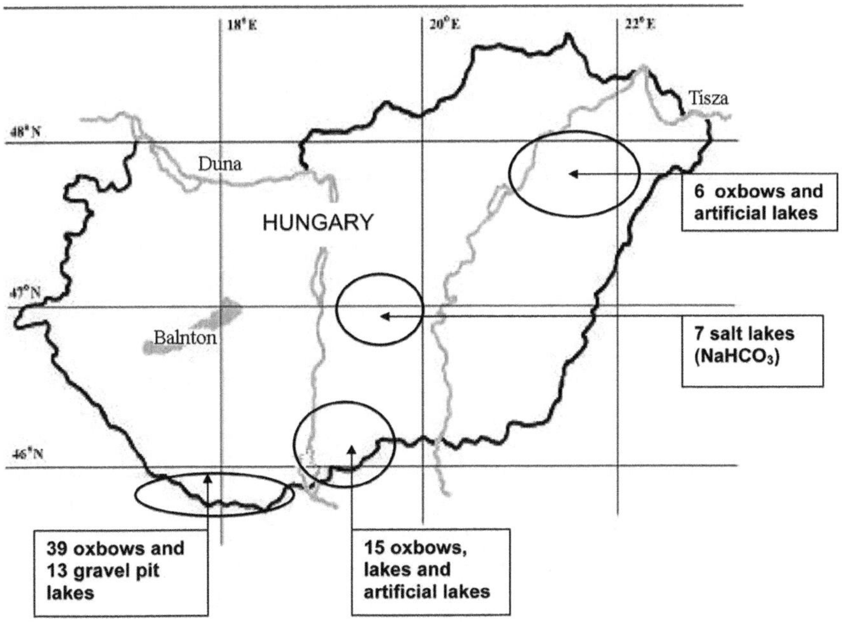

Figure 1. Location of the small lakes within Hungary. I: Eastern Hungarian sampling site; II. Salt lakes in the territory of the Kiskunság National Park; III: South-Danubian sampling district; IV: oxbows and gravel pit lakes on the territory of the Duna-Dráva National Park.

phytoplankton of many systems are far from being competitively equilibrated.

In 6 cases, one single species alone made up more than 80% in terms of biomass [1, 2, 5, 7, 9, 14]. In 5 lakes two species together reached the 80% dominance [3, 4, 8, 10, 13]. This means that equilibrium states, as defined in the introduction, frequently represent monodominance or codominance of only 2 species.

Green algae were relatively frequent among dominant species. These groups provided the first dominant species in 21 of the 80 cases, the second dominant in 12 of the total of 75 cases and the third dominant in 14 of the altogether 66 cases (Table 2). The involved species among the 3 most dominant were *Pandorina morum* (5x – numbers after species names here show how many times the given taxon was either as first, second or third dominant), *Botryococcus braunii* (5x), *Coelastrum astroideum* (5x), *Phacotus lenticularis* (2x), *Coelastrum pseudomicroporum* (2x), *Crucigenia tetrapedia* (2x), *Dictyosphaerium pulchellum* (2x), *Volvox* sp., *Coelastrum microporum, Crucigenia quadrata, Lobocystis planktonica, Oocystis marina, Pediastrum duplex, P. simplex, Scenedesmus obliquus, S. quadricauda, Tetraedron mimimum, T. trigonum, Treurbaria triappendiculata, Tetrastrum glabrum, Mougeotia* sp., *Staurastrum manfeldtii, Zygnema* sp., and species that were registered as 'unidentified green'. Most of the above species are typically planktonic, however in some cases species of metaphytic (*Mougeotia, Zygnema, Staurastrum*) origin were abundant.

Cyanoprokaryote (Table 1) dominants included all the three main groups of the division. Of Chroococcales, *Microcystis* spp. (2x), *Snowella lacustris* and '*Aphanocapsa* sp.' were found among dominants. The latter includes colonial forms with cells in picoalgal range ($<2\ \mu m^3$) that may dominate in salt lakes. Most frequent planktonic species of Oscillatoriales were *Pseudanabaena limnetica* and *Planktolyngbya limnetica*, the others (unidentified Oscillatoriales sp. and *Pseudanabaena catenata*) can be considered as periphytic or benthonic. Nostocales were represented by species *Cylindrospermopsis raciborskii, Aphanizomenon flos-aquae, A. issatschenkoi, A. ovalisporum, A. aphanizomenoides, Anabaenopsis arnoldi* and an unidentified *Anabaena* sp.

Euglenophyta rarely reach high biomass in large lakes, however, in the small lakes they often occurred in large proportion (Table 1). The most frequent taxa were: *Euglena proxima* (3x), *Trachelomonas oblonga* (3x), *E. ehrenbergii* (2x), *E. texta* (2x), *E. acus, Euglena* sp., *Lepocinclis* sp., *Phacus acuminatus, P. maximus, P. pleuronectes, Phacus* sp., *Trachelomonas planktonica* and *T. volvocina.*

Table 2. Number of cases when taxa belonging to the indicated higher taxonomic groups were the first, second or third dominants in 80 Hungarian lakes in late summers

	1st dominant	2nd dominant	3rd dominant	Total
Chlorophyta	21	12	14	47
Cryptophyta	14	15	13	42
Bacillariophyceae	14	12	16	42
Cyanoprokaryota	10	11	10	31
Dinophyta	9	9	3	21
Euglenophyta	6	6	7	19
Chrysophyceae	5	8	3	16
Haptophyta	0	2	0	2
Xanthophyceae	1	0	0	1
total	80	75	66	

Diatoms are normally not characteristic members of phytoplankton associations in late summers. In this survey, diatoms contributed significantly to total biomass more frequently than was expected (Table 1), but benthic or periphytic taxa were more abundant than planktonic ones. Of planktonic species, *Cyclotella comta* (5x), *Stephanodiscus hantzschii* (3x), Centrales spp. (2x), *Cyclotella* sp., and *Stephanodiscus* sp. were prominent (more precise identification was impossible in lack of additional TEM/SEM studies). Among benthic, periphytic or meroplantktonic forms (c.f. Padisák & Dokulil, 1994) the following species were most characteristic: *Melosira varians* (5x), *Navicula* sp. (5x), *Cymatopleura solea* (3x), *Nitzschia* sp. (3x), *Pinnularia viridis* (3x), *P. major* (2x), *Rhopalodia gibba* (2x), *Amphora ovalis, Didymosphaenia geminata, Navicula cryptocephala, Nitzschia sigmoidea, N. vermicularis* and *Stenopterobia pelagica.*

Cryptophyta were found many times (Table 1): *Cryptomonas marssonii* (9x), *C. rostrata* (9x), *Cryptomonas* sp. (7x), *Rhodomonas pusilla* (4x), *Cryptomonas ovata* (4x), *C. erosa* (2x), *C. obovata* (2x), *Rhodomonas minuta* var. *nannoplanktica* (4x) and *Rhodomonas* sp. (2x).

Of Chrysophyceae (Table 1), *Synura* spp. (7x; most probably *S. uvella* and *S. petersenii*), *Dinobryon divergens* (2x), *D. sociale* (2x) *Pseudokephyrion* sp., *Salpingoeca* sp. were the most important and twice species that were registered as 'other Chrysophyceae'. Four species of Dinophyta (Table 1) contributed significantly to late summer phytoplankton: *Ceratium hirundinella* (8x), *Peridinium aciculiferum* (6x), *P. cinctum* (5x) and *P. inconspicuum.* Only one species of Xanthophyceae (*Goniochlorys mutica*) was represented among dominants and of Haptophyta *Chrysochromulina parva* was found among the 3 most important species twice.

Discussion

One of the basic conclusions of this survey is that equilibrium phases in phytoplankton assemblages are relatively rare. Of the 80 assemblages investigated in this study only in 21% of the cases did we find equilibrium composition. This number may be even smaller if we consider persistence of dominance as a serious factor of qualifications (and it, in fact, is a serious factor). Another important observation is that if equilibrium occurred at all it frequently manifested as monodominance or codominance of only two species. This observation supports the preconception of Sommer et al. (1993) that equilibrium phases of phytoplankton assemblages are not likely to be dominated by more than a few species. In other words, phytoplankton assemblages with a particular, overdriving force (it can be limitation by a particular nutrient or light) are pressed towards the selection of the most adapted population, i.e. the one that has the lowest threshold to the given driving force.

Water chemistry data (not shown in this study) of the 80 studied lakes were in the 'average range' in most cases. Extremes were the high conductivity of salt lakes. It was also apparent that most (not all) gravel pit lakes were very poor in nutrients. Phytoplankton assemblages of these lakes were more often found in equilibrium than those of other lakes in the 'average range'. It tells us that stressed ecosystems are quite likely to support equilibrated phytoplankton communities. Causal mechanisms behind are relatively easy to trace: in such lakes evolutionary adaptation for stress conditions is the primary selection factor and therefore their species pool is limited to those that can survive under the given circumstances. Competitive interactions are only of secondary importance.

Hypertrophic conditions often lead to water blooms comprising only one species; a striking example is described by Borics et al. (2000) where *Cylindrospermopsis raciborskii* made up 100% of the assemblage thus setting the diversity to zero. In such cases, the equilibrium is strongly apparent. On this basis, it is often supposed that eu- or hypertrophic phytoplankton are more likely to reach an equilibrium state than others. This study did not support this preconception: lakes with high, low or medium biomasses

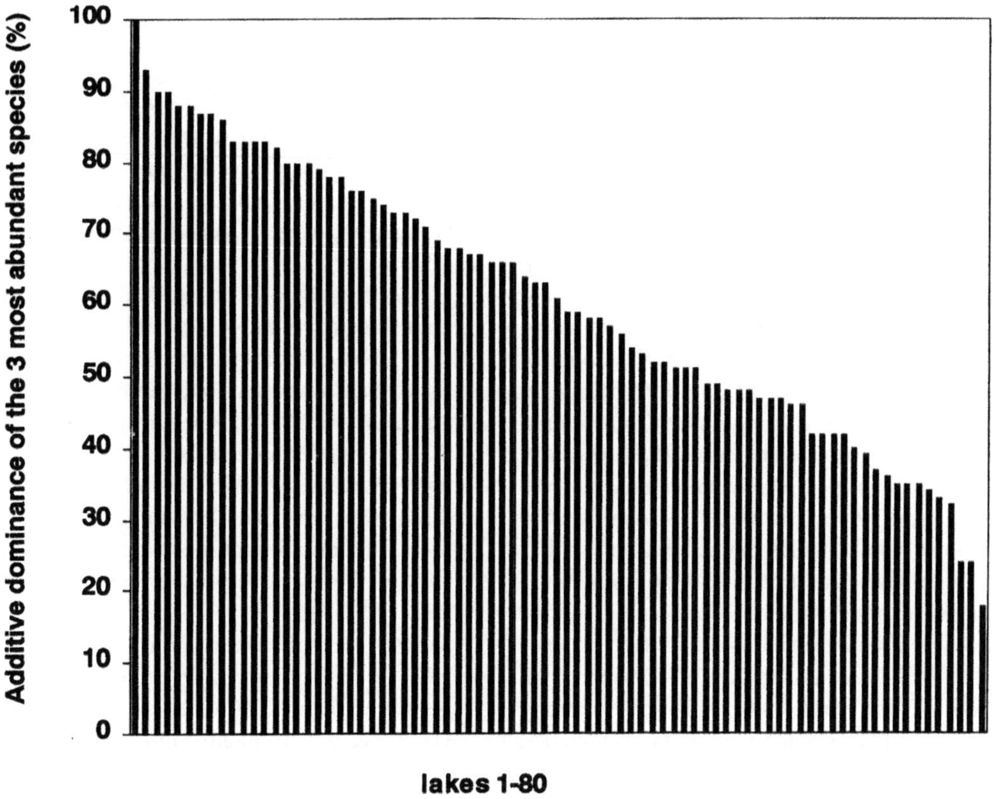

Figure 2. Additive dominance of the 3 most dominant (dominance eas based on percentage contribution of individual species to total biomass) species in 80 Hungarian small lakes late summers. Sequence of lakes correspond to that in Table 1.

are rather randomly distributed in the equilibrium spectrum sequences in Table 1 and this conclusion remains unchanged if stressed lakes (salty ones and gravel pits) are excluded.

Species number, however, exhibits a clear increase with decreasing equilibrium: species number of lakes far from equilibrium was considerably higher as averages by 10-lakes increments show (Fig. 3). Whether we regard or disregard stressed lakes (salty or gravel pit) this relationship remains apparent. This result does not mean that lakes with phytoplankton in equilibrium are less species rich. However, it does mean that in lakes far from the equilibrium many more species have a sufficient relative frequency to become dominants than in others (c.f. Padisák, 1992, 1993a).

As mentioned in the introduction, lakes with small water volumes are increasingly subjected to changes originating from climactic fluctuations or from human activities. Moreover, they are frequently overgrown by macrophytes. The latter originate for at least three reasons: (1) they are shallow (2) their small surface area is less exposed to wind action (Padisák & Reynolds, in press) and (3) they are often surrounded by dense forest further decreasing wind effects (Padisák,

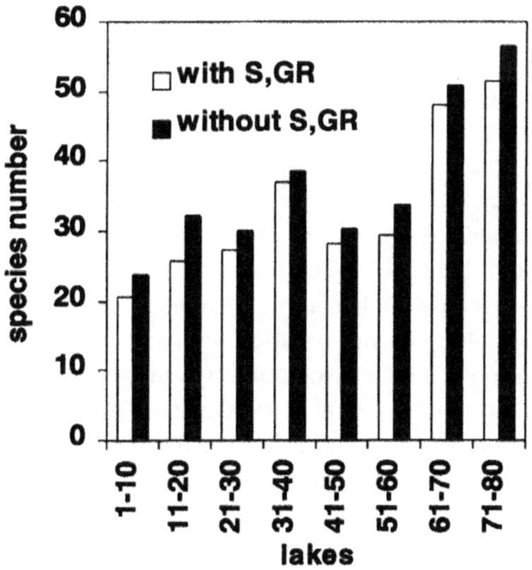

Figure 3. Species number in lakes averaged in 10 lakes increments. Sequence of lakes correspond to that in Table 1.

in press). A recent study (Borics et al., 2003) established different kinds of algal assemblages in a hypothesized bog-lake and phytoplankton is only one of them. The influence of metaphytic, periphytic or benthic assemblages on composition of phytoplankton samples can be supposed in many cases: *Zygnema* sp. [7], *Melosira varins* [8, 36, 37, 59, 65], *Navicula* sp. [14, 21, 29], *Fragilaria (Synedra) ulna* [17], *Pinnularia maior* [22, 28], *Cymatopleura solea* [22, 26], *Nitzschia sigmoidea* [22], *Didymosphaenia geminata* [26], *Nitzschia* sp. [29], *Nitzschia vermicularis* [38], *Pinnularia viridis* [46, 56, 60], *Amphora ovalis* [46], *Mougeotia* sp. [52, 77], unidentified green algal sp. [52], *Pseudanabaena catenata* [57], *Navicula cryptocephala* [59], *Staurastrum manfeldtii* [60], *Rhopalodia gibba* [62, 64], *Closterium aciculare* [64], *Oscillatoriales* sp. [70], *Staurastrum paradoxum* [77]. Since the above listed species occurred among the 3 dominants it can be concluded, that phytoplankton samples taken in these 80 lakes were heavily influenced by other associations than plankton at least in 23 cases (29%). As concluded by Borics et al. (2003) when dealing with such small-scale canopy systems one should consider spatial heterogeneity. With 'rough' sampling methods we may easily mix different association types which makes it impossible to recognize if they were equilibrated (*sensu* Sommer et al., 1993) or not.

It is a challenging exercise to compare algal associations to those that were recently reviewed by Reynolds et al. (2002). Those associations (**A**, **B**, **C** – see abbreviations for associations in the afore cited paper) which are characteristic for spring diatom development or typically characterize epilimnetic (**N**, **P**, **T**, **H2**, **U**, $\mathbf{L_o}$) or metalimnetic (**Z**, **R**, **V**) layers of stratified lakes are not represented at all in Table 1. Assemblage **X1** might have occurred but did not probably because such shallow systems that were included in this study do not require tolerance to onset of stratification. The **X3** association was also not represented in this sample set and tolerance of low base status has not been a basic selective force in the lakes investigated. The **S2** type was not found although might have occurred; at least with *Rhaphidiopsis* dominace. This species (unlike planktonic *Spirulina* and *Arthrospira*) has been recorded in the Hungarian flora (Tamás, 1974). The small shallow lakes in Carpathian Basin in the end of the summer are typically warm (up to 30 °C), they are alkaline, moreover, they often represent the northernmost distribution edge of many tropical species (Péterfi & Momeu, 1996). The **S2** association is probably severely limited to low latitudes. Many of the 80 lakes, especially the oxbows, are humic lakes therefore the **Q** assemblage characterized by *Gonyostomum* might have occurred. The sensitivity of this species to preservation materials is well known and it is quite probable that it was present in some lakes but disintegrated. Presence of *G. latum* in south Hungarian lakes is well documented (Schmidt et al., 1990).

Each of the other described phytoplankton associations occurred, however, with different frequency.

The **S1** assemblage was rather poorly represented. Its species were subdominants in a natural lake [19] and in an oxbow [48] where otherwise species of **H1** or $\mathbf{S_N}$ were more characteristic. The S_N functional group with *Cylindrospermopis raciborskii* was significant in three lakes [20, 39, 48] and each of these lakes are used for recreation purposes including angling. Members of the H1 assemblage were significant in many lakes [6, 11, 16, 18, 47, 49, 51, 57], however, they contributed to total biomass significantly in only one case [18]; in the others their percentage dominance was between 10 and 22%. Species belonging to groups **S1**, **H1** or $\mathbf{S_N}$ are frequently in competition and seasonal sequences often appear as $\mathbf{H1} \rightarrow \mathbf{S_N} \rightarrow \mathbf{S1}$. Whether this sequence can complete or not it largely depends on abundance of inorganic N, disturbance pattern or sudden weather changes (cooling). Summer succession in Lake Balaton provides a clear example (Padisák & Reynolds, 1998). The **H1**, $\mathbf{S_N}$ and **S1** species often co-dominated [19, 20, 39, 48, 58, 80] and among them, the **H1** species exhibited the highest affinity to occur as subdominants other associations (like **D** in [12, 42], **W1** in [68, 78], **J** in [39, 80], **Y** [53, 58], **E** [53] or even when otherwise metaphytic algae were the most abundant [77]). The $\mathbf{S_N}$ assemblage showed no tendency to co-occur with any other assemblage but **H1** and **S1**. It is also worth noting that in lakes where species representing as **H1**, $\mathbf{S_N}$ and **S1** alone or in any combination, biomass was considerably higher (34 mg l^{-1}) than the average of the 80 lakes (13 mg l^{-1}). Exceptions were lakes [53] and [58] where **H1** members contributed 10 and 11% to the overall low biomass. *Microcystis* spp. representing the **M** association were only subdominant is lakes [71, 72] with moderate biomass (4.3 and 4.4 mg l^{-1}, respectively).

Assemblages of **D**, **J** and **K** are expected in shallow, nutrient rich, turbid lakes. The assemblage **K** can be recognized in composition of one lake [5, reservoir with fish] where it was represented by *Aphanocapsa* sp. with 82% contribution to total biomass and the second dominant was a **J** species, *Treubaria triappendiculata*. Probably the most evident representatives of **K** assemblages are described in Vörös et al. (1991) when reporting about hypertrophic conditions in a reservoir (with considerable fish-stock) dominated by algae of picoalgal cell sizes. The **D** assemblage typically occurs in spring, however it may be recurrent in late summers in shallow lakes (and also rivers). Such assemblages dominated by *Cyclotella comta* were found in two gravel pit lakes [6, 16] with low biomass (2–3 mg l^{-1}) and in an oxbow [32] where *Stephanodiscus hantzschii* contributed 48% to the rather high biomass (26 mg l^{-1}). In all other lakes where species of group **D** occurred among the 3 dominants they made a small fraction (7–14%) of the total biomass. It is a widely accepted fact that coccal green algae occur in large species numbers, however, usually with small (<20%) contribution to total biomass (Happey-Wood, 1988; Dokulil, 1991; Padisák & G.-Tóth, 1991; Padisák & Dokulil, 1994) in most lakes. Examples in this study are lakes [55, 61, 70, 73, 79 and 80]. However, in some cases coccal green algae may be the characteristic group as exemplified here by *Tetraedron trigonum* (69%) in lake [3]; *Scenedesmus obliquus* (52%) in lake [34]; *Crucigenia tetrapedia* (33%) + *Oocystis marina* (9%) + *Scenedesmus quadricauda* (7%) in lake [54]; the xanthohycean *Goniochlorys mutica* (41%) in lake [39]; *Tetraedron trigonum* (25%) + *Coelastrum pseudomicroporum* (13%) in lake [40]; *Tetraedron minimum* (20%) and *Lobocystis planktonica*, a very rare species among Chlorococcales (6%) in lake [41] and *Tetrastrum glabrum* (22%) in lake [72]. Species of assemblage **J** were found in association with representatives of almost any other assemblage, however, co-occurrence of other 'shallow, turbid, nutrient rich' groups was frequent (examples are lakes [5, 32, 33, 39, 55,61, 79, 80].

The most characteristic forms of phytoplankton in the studied 80 lakes were flagellates belonging to several functional groups. Of these assemblages the **X2** with *Plagioselmis* (*Rhodomonas*) and/or *Chrysochromulina* is the easiest to detect. This assemblage typically occurred either as dominant [18] or subdominant [6, 11, 16, 47, 49, 51, 57] in lakes with very low (range: 121 – 3140 $\mu g\ l^{-1}$) phytoplankton biomass. The assemblage **F** occurred in 3 cases, two of them were 'low biomass' lakes [24, 47] and one [36] was a 'high biomass' lake with no clear pattern of recognition.

Assemblage **E** is described (Reynolds et al., 2002) to be characteristic for small, oligotrophic base-poor lakes. Its representatives (*Dinobryon divergens, D. sociale, Salpingoeca* sp.) occurred with 8–28% biomass contribution in lakes [17, 23, 53, 75]. In two cases [17, 53; both with *Dinobryon divergens*] biomass was indeed low (705 and 1156 $\mu g\ l^{-1}$, respectively) but in the two other cases [22714 and 9140 $\mu g\ l^{-1}$, respectively) it was rather high. None of the four lakes was base-poor. For example, in lake [53] *D. divergens* coexisted with *Aphanizomenon issatschenkoi*, a species which is not at all likely to occur in base-poor lakes. The case of lake [17] is even more striking: *Dinobryon divergens* was subdominant to *Phacotus lenticularis*, a species, that needs HCO_3^- carbon source dominance to build up its thick $CaCO_3$ loricae. It is argued in Reynolds et al. (2002) that members of assemblage **E** need the presence of free CO_2 since they cannot use HCO_3^- as carbon source. An alternative nutritional option to suffice their carbon demand might be phagocytosis of bacteria (Bird & Kalff, 1986, 1987). However, an aberrant bloom in Lake Balaton in 1993 have shown that *Dinobryon sociale* could reach extremely high densities (33 mg l^{-1} biomass) while apparently neither free CO_2 nor high amount of grazeable bacteria were present (Reynolds et al., 1993).

The most abundantly represented assemblage in the 80 lakes was the **W1** characteristic for small organic ponds with euglenoids, *Synura* and *Gonium* as main representatives. *Gonium* did not occur in these lakes but euglenoids and *Synura* did. It seems pertinent to distinguish between *Synura* and euglenoid dominance, especially, because coexistence (as dominants) of these two forms was observed once only [45]. *Synura* occurred in 8 cases of which it was the first dominant in 5 lakes [8, 10, 15, 50, 62], the second in 4 lakes [13, 27, 45] and in no case occurred as 3rd dominant. Of the 8 cases when *Synura* was significant 3 were equilibrium states *sensu* Sommer et al. (1993). Average biomass with *Synura* (disregarding [45] when it codominated with euglenoids) was 15.2 mg l^{-1}. Euglenoids (except *Trachelomonas* that belongs to association **W2**) were found among the 3 dominants in 12 cases of which they were the first dominants in 4 cases, 2nd in 6 cases and 3rd in 5 cases. The only example when an euglenoid was found in equilibrium case was a salty lake [1] that is a hardly representative as an example. Average biomass with euglenoids (also disregarding [45] when they codominated with *Synura*) was 7.7 mg l^{-1}. Euglenoids, at least in the Carpathian basin, are regularly found in organically rich, however typically small volume basins like temporary ponds, oxidation ponds of sewage treatment plants, etc. where they might co-occur with *Gonium* or *Eudornia* (Padisák, 1993a). In lakes with larger volumes (like those in this study), they are usually subdominants. Eugleonoids can be dominant in small lakes influenced abundant waterfowl (Padisák, 1993b). *Synura* occurs typically in permanent lakes, very often in oxbows, well sheltered by forests from direct sunshine. The water is often colored by humic material deriving from decomposition of leaves from the surrounding forests. As the above comparison shows, *Synura* is apparently a good competitor: if once established it drives the assemblage to equlibrium. Almost monodominant *Synura* blooms in oxbows are decribed (Kiss & Kristiansen, 1994) from the region. Euglenoids in lakes of similar characteristics may be subdominants. For these reasons hereby we suggest to establish a new functional group, $\mathbf{W_S}$, for associations dominated by *Synura.* This functional group is sensitive to high irradiances, tolerates coloration of the water due to humic substances and natural organic richness of the water, but it is not likely to establish if organic 'richness' results primarily from excretion products of waterfowl or sewage processing. These criteria do not hold for **W1** with euglenoids and colonial Volvocales.

Trachleomonas, as representative of **W2**, occurred among the 3 dominants in 5 lakes, each oxbows [12, 45, 59, 66, 76]. In one of them [59], it occurred with *Melosira varians* that often grows on the bottom of small lakes thus supporting the supposed 'bottom dweller' feature of *Trachelomonas*. The relationship of this functional group to others or to environmental patterns that enhance its occurrence have remained rather unclear.

Cryptomonad flagellates as representatives of functional group **Y** were found in many cases. As first dominants they occurred in 11 cases [2, 9, 11, 23, 33, 41, 43, 53, 58, 61, 66], it was the second dominant in 9 lakes [31, 33, 56, 57, 72, 73, 75, 76] and the third in 11 cases [2, 16, 21, 23, 31, 33, 44, 46, 51, 69, 74]. These lakes in most cases supported low phytoplankton biomass: the average was 4.4 mg l^{-1} (average of the 80 lakes was 13 mg l^{-1}). Members of the **Y** group can co-exist with almost any other assemblage. The lakes where *Cryptomonas* spp. were present among the 3 dominants were usually small oxbows or gravel pit lakes with no fish. This observation allows to suppose that functional group **Y**, and also group **X2**, can become dominant if grazing pressure of zooplankton is quite high since fish, as main predators of zooplankton are largely absent. Under such circumstances only fast growing species, like *Cryptomonas* and *Rhodomonas*, can survive and maintain a rather low standing crop. This supposition is supported by the observation that in the clear water phase of phytoplankton succession they are often the most numerous kind of phytoplankton (Sommer et al., 1986). Moreover, after fish kills or in biomanipulation experiments when fish are removed (McQueen et al., 1984; Korponai et al., 1997; Borics et al., 2000) mostly sudden biomass decrease is observed with dominance of *Cryptomonas* and *Rhodomonas*. Consequently, these groups become dominant under high level of top-down control either if it arises as temporary event in the seasonal successional of plankton (clear-water phase) or as a consequence of removal of predators of zooplankton. This would also explain why group **Y** can be find associated with almost any kind of assemblages as can be evidenced by data provided in Table 1.

Species of group **G** occurred in 4 lakes and was represented by *Pandorina morum* [13, 27, 32] and *Volvox* sp. [51]. Based on data from the *Pandorina* lakes it seems to be justified that nutrient rich environments are necessary for this group to establish (average biomass of these lakes were 22 mg l^{-1}). In the sample set used in this study, functional groups **G** and the recently proposed $\mathbf{W_S}$ seem to be closely related.

The association $\mathbf{L_M}$ was traced in 7 sites [25, 30, 36, 37, 38, 62, 63], each were oxbows. This group was represented by *Ceratium* in the above lakes and *Microcystis* in no case co-dominated. Data indicate that *Ceratium* can become the only quantitatively important species [see lakes 25 and 30] probably due to its excellent migration abilities in such shallow but stratified water columns in which it can even form 'deep layer' maxima (Grigorszky et al., 2003a).

Two groups of species need separate discussion: *Peridinium* spp. and *Phacotus lenticularis.*

Phacotus with its thick $CaCO_3$ loricae has been quite a phytoplankton enigma. However, it is widespread and can be found in lakes from large to small, from stratified to very shallow, from oligotrophic to hypertrophic, from organically rich to poor. In any case, typical form of *Phacotus* can occur only in lakes where HCO_3^- is the dominant carbon source, moreover the lake is calcareous. High density populations are typically found in smaller, clear water, meso-eutrophic, calcareous ponds (Schlegel et al., 1998, 2000). In this study, *Phacotus* contributed to equilibrium phases in lakes [11, 17] and was dominant in lake [44]. In each of the three cases it was associated with other flagellates being members of groups **Y**, **X2** or **E**. All the three localities were nutrient poor gravel pit lakes with very low phytoplankton biomass (115 - 974 μg l^{-1}). Since description of none of the functional groups insists on abundance of HCO_3^- and Ca^{2+}, it seems to be necessary to raise a separate group, $\mathbf{Y_{Ph}}$, for *Phacotus*. Necessity of this $\mathbf{Y_{Ph}}$ functional group can be justified by the rather high frequency of *Phacotus* in small lakes and by its enigmatic features and requirements.

In the list by Reynolds et al. (2002) *Peridinium* is mentioned, together with *Woronichinia*, as a typical representative of group $\mathbf{L_o}$, characteristic for summer epilimnia and in **W1** where peridinoids co-occur with euglenoids, *Synura* and *Gonium*. Since the bulk of the 80 lakes in this study was not stratified, the term 'epilimnion' is difficult to define and *Peridinium* occurrences cannot be considered as $\mathbf{L_o}$. Moreover, publications on winter blooms of Dinophyta are available (Grigorszky & Padisák, 1998). Therefore neither 'epilimnion', nor 'summer' are appropriate terms to describe their dominance. *Peridinium* contributed to dominants in this study quite often: *Peridinium cinctum* in [31, 50, 64, 69, 74], *P. aciculiferum* in [38, 44, 50, 55, 63, 71], and *P. inconspicuum* in [49]. *P. cinctum* was associated with members of **Y** [31, 69, 74], W1 [50, 69], X2 [74] and non-planktonic algae [64]; except the latter each were flagellate group. *P. inconspicuum* occurred with **X2** and **F** species [49]. *P. aciculiferum* co-occurred with very various groups as $\mathbf{L_H}$ [38], $\mathbf{X_{Ph}}$ [44], **Y** [44, 74], **W1** [44], **J** [55], **D** [55], $\mathbf{L_M}$ [63], **H1** [63], **M** [74] and non-plaktonic algae [38, 64]. This very limited example is sufficient to demonstrate that high species-specific differences may occur among species of Dinophyta regarding their environmental requirements that largely determines with which groups will they co-exist and with which ones not. A recent study (Grigorszky et al., 2003b) explored quite distinct differences between environmental preferences of different species of Dinophyta especially regarding their relation to ambient temperatures and organic matter content of the locality. For the above reasons, species of Dinophyta (*Peridinium, Peridiniopsis, Woloszynskia*, etc.), similarly to species of Nostocales, have to be sorted to various phytoplankton associations and cannot be handled as a guild, like most chlorococcalean algae as mentioned also by Reynolds et al. (2002).

Acknowledgements

This project was supported by the Hungarian National Science Foundation (OTKA T029636, F31802), the Higher Education Found (FKFP 0195) and the Békessy Science Foundation.

References

Bird, D. F. & J. Kalff, 1986. Bacterial grazing by planktonic algae. Science 231: 493–495.

Bird D. F. & J. Kalff, 1987. Algal phagotrophy: regulating factors and importance relative to photosynthesis in *Dinobryon* (Chrysophyceae). Limnol. Oceanogr. 32: 277–284.

Borics, G., I. Grigorszky, S. Szabó & J. Padisák, 2000. Phytoplankton associations under changing pattern of bottom-up vs. top-down control in a small hypertrophic fishpond in East Hungary. Hydrobiologia 424: 79–90.

Borics, G., B. Tóthmérész, I. Grigorszky, J. Padisák, G. Várbíró & S. Szabó, 2003. Algal assemblage types of bog-lakes in Hungary and their relation to water chemistry, hydrological conditions and habitat diversity. Hydrobiologia 502 (Dev. Hydrobiol. 172): 145–155.

Dokulil, M., 1991. Contribution of green algae to the phytoplankton assemblage in a mesotrophic lake, Mondsee, Austria. Archiv für Protistenkunde 139: 212–223.

Grigorszky I. & J. Padisák, 1998. Jég alatti *Peridinium aciculiferum* Lemmermann (Dinophyta) populáció a Balatonban [*Peridinium aciculiferum* Lemmermann (Dinophyta) under-ice population from the shallow lake Balaton, Hungary]. Hidrológiai Közlöny 78: 282–284. [in Hungarian with English summary].

Grigorszky, I., J. Padisák, G. Borics, C. Schnitchen & G. Borbély, 2003a. Description of a deep chlorophyll maximum due to *Ceratium hirundinella* (O. F. Müller) Bergh. Hydrobiologia 506–509 (in press).

Grigorszky I., G. Borics, J. Padisák, B. Tótmérész, G. Vasas, S. Nagy & G. Borbély, 2003b. Factors controlling occurrence of Dinophyta species in Hungary. Hydrobiologia 506–509 (in press).

Happey-Wood, C. M., 1988. Ecology of freshwater planktonic algae. In Sandgren, C. D. (ed.), Growth and Reproductive Strategies of Freshwater Phytoplankton. Cambridge: 175–226.

Hardin, G., 1960. The competitive exclusion theory. Science 131: 1292–1297.

Kiss, K. T. & J. Kristiansen, 1994. Silica-scaled chrysophytes (Synurophyceae) from some rivers and shallow lakes. Hydrobiologia 289: 157–162.

Korponai, J., K. Mátyás, G. Paulovits, I. Tátrai & N. Kovács, 1997. The effect of different fish communities on the cladoceran plankton assemblages of the Kis-Balaton Reservoir, Hungary. Hydrobiologia 360: 211–221.

McQueen, D. J., J. R. Post & E. L. Mills, 1986. Trophic relationship in freshwater pelagic ecosystems. Can. J. Fish. aquat. Sci. 43: 1571–1578.

O'Farrell, I., R. Sinistro, I. Izaguirre & F. Unrein, 2003. Do steady state assemblages occur in shallow lentic environments from wetlands? Hydrobiologia 502 (Dev. Hydrobiol. 172): 197–209.

Ortega-Mayagoitia, E., C. Rojo & M. A. Rodrigo, 2003. Controlling factors of phytoplankton assemblages in wetlands: an experimental approach. Hydrobiologia 502 (Dev. Hydrobiol. 172): 177–186.

Padisák, J., 1992. Seasonal succession of phytoplankton in a large shallow lake (Balaton, Hungary) – a dynamic approach to ecological memory, its possible role and mechanisms. J. Ecol. 80: 217–230.

Padisák, J., 1993a. The influence of different timescale disturbances on the species richness, diversity and equitablity of phytoplankton in shallow lakes. Hydrobiologia 249: 135–156.

Padisák, J., 1993b. Dynamics of phytoplankton in brown-water lakes enclosed with reed-belt (Fertö/Neusiedlersee; Hungary/Austria). Verh. int. Ver. Limnol. 25: 675–679.

Padisák, J., in press: Chapter 10. The Phytoplankton. In O'Sullivan, P. & C. S. Reynolds (eds), The Lakes Handbook. Blackwell Science Ltd. Oxford.

Padisák, J. & M. Dokulil, 1994. Contribution of green algae to the phytoplankton assemblage in a large, turbid, shallow lake (Neusiedlersee, Austria/Hungary). Biológia Bratislava 49: 571–579.

Padisák, J. & C. S. Reynolds, 1998. Selection of phytoplankton associations in Lake Balaton, Hungary, in response to eutrophication and restoration measures, with special reference to cyanoprokaryotes. Hydrobiologia 384: 41–53.

Padisák, J. & C. S. Reynolds, in press. Shallow lakes: the absolute, the relative the functional and the pragmatic. Hydrobiologia.

Padisák, J. & L. G.-Tóth, 1991. Some aspects of the ecology of the subdominant green algae in a large nutrient limited shallow lake (Balaton, Hungary). Archiv für Protistenkunde 139: 225–242.

Péterfi, L. S. & L. Momeu, 1996. Three Roumanian *Mallomonas* species (Synurophyceae), with special distribution patterns. Hydrobiologa 336: 143–150.

Reynolds, C. S., V. Huszar, C. Kruk, L. Naselli-Flores & S. Melo, 2002. Towards a functional classification of the freshwater phytoplankton. J. Plankton Res. 24: 417–428.

Reynolds, C. S., J. Padisák & I. Kóbor, 1993. A localized bloom of *Dinobryon sociale* in Lake Balaton: Some implications for the perception of patchiness and the maintenance of species richness. Abstracta Bot. 17: 251–260.

Schlegel, I., R. Koschel & L. Krienitz, 1998. On the occurrence of *Phacotus lenticularis* (Chlorophyta) in lakes of different trophic state. Hydrobiologia 369/370: 353/361.

Schlegel, I., L. Krienitz & D. Hepperle, 2000. Variability of calcification of *Phacotus lenticularis* (Chlorophyta, Chlamydomonadales) in nature and culture. Phycologia 39: 318–322.

Schmidt, A., F. Vasas & L. Dobler, 1990. Adatok a *Gonyostomum latum* Ivanov magyarországi elôfordulásáról. (Data on the presence of *Gonyostomum latum* Ivanov in Hungary.) - Botanikai Közlemények 77/1–2: 39–46. [in Hungarian with English summary]

Sommer, U., M. Gliwicz, W. Lampert & A Duncan, A., 1986. The PEG model of seasonal succession of planktonic events in fresh waters. Archiv Hydrobiol. 106: 433–471.

Sommer, U., J. Padisák, C. S. Reynolds & P. Juhász-Nagy, 1993. Hutchinson's heritage: the diversity-disturbance relationship in phytoplankton. Hydrobiologia 249: 1–8.

Stoyneva, M. P., 2003. Steady-state phytoplankton assemblages in shallow Bulgarian wetlands. Hydrobiologia 502 (Dev. Hydrobiol. 172): 169–176.

Tamás, G., 1974. The occurrence of *Rhaphidiopsis mediterranea* Skuja in the plankton of Lake Balaton. Annales Instituti Biologici Academia Scientarium Hungariae, Tihany 41: 317–321.

Tilman, D., 1982. Resource competition and community structure. Princeton Univ. Press.

Vörös, L., P. Gulyás & F. Németh, F., 1991. Occurrence, dynamics and production of picoplankton in Hungarian shallow lakes. Int. Rev. ges. Hydrobiol. 76: 617–629.

Hydrobiologia **502:** 169–176, 2003.
L. Naselli-Flores, J. Padisák, M. T. Dokulil (eds),
Phytoplankton and Equilibrium Concept: The Ecology of Steady-State Assemblages.

Steady-state phytoplankton assemblages in shallow Bulgarian wetlands

Maya P. Stoyneva
Department of Botany, Faculty of Biology, Sofia University "St Kliment Ohridski". 8 Dr. Zankov, 1164 – Sofia, Bulgaria
E-mail: mstoynev@techno-link.com

Key words: phytoplankton, steady-state assemblages, shallow lakes, trophic state, cyanoprokaryote blooms

Abstract

The present paper reports the results on the occurrence and composition of steady state phytoplankton assemblages in the shallow water bodies of 18 Bulgarian natural wetlands. They all differ in their morphometry, hydrology and trophic state. Steady states were identified according to Sommer & Padisák (1993) when (i) 1, 2 or 3 species of algae contribute more than 80% of total biomass, (ii) their existence or coexistence persists for long enough (more than 1–2 weeks) and (iii) during that period the total biomass does not increase significantly. These situations were rare and were dominated by 7 stress-tolerant, colony-forming and ruderal disturbance-tolerant cyanoprokaryotes (*Microcystis aeruginosa, Microcystis wesenbergii, Anabaena spiroides, Anabaenopsis arnoldii, Aphanizomenon flos-aquae, Aphanizomenon gracile* and *Cylindrospermopsis raciborskii*). The equilibrium states had been registered only during hypertrophic conditions in three stressed and frequently disturbed shallow water bodies and were related to the summer – late summer season.

Introduction

It has been well recognized that natural lakes can be very rich in phytoplankton species and the most widely quoted statement on the apparently anomalous diversity of algal communities is that by Hutchinson (1961) in his paper on the "paradox of the plankton". Over the last decades, different equilibrium and non-equilibrium paradigms tried to explain the maintenance of this diversity. Among them, Connells' (1978) Intermediate Disturbance Hypothesis (IDH) represents a reasoned expression of the interactions among the externally exposed stochasticity of environmental variability and internally driven progress towards community organization and energetic equilibrium (Reynolds et al., 1993). Disturbance is recognized as an important factor that increases heterogeneity at all scales (Naselli-Flores et al., 2003). And still "the magnitude of temporal variability evidenced by lake communities continues to surprise limnologists and defies prediction" (Carpenter & Kitchell, 1988). The temporal dynamics of phytoplankton species is, in many respects, analogous to the succession sequences of other plant species in other habitats (Margalef, 1961; Harris, 1983; Reynolds et al., 1993; Sommer et al., 1993). Since that recognition, the existence of steady state assemblages in the field has been discussed. Most opinions were that these states are an exception in nature and that commonly there is a continuous shift in their structure, which is continuously readjusted to a changing environment (Harris, 1986; Alimov, 1990). Considering the microscopic sizes of planktic algae and their brief generation time, it is well recognized that both stress and disturbances represent very important selective forces to which phytoplankton respond under natural conditions (Reynolds, 1988; Olrik, 1994). After the identification of equilibrium phases for practical purposes in relation with IDH was proposed (Sommer & Padisák, 1993) the question as to whether we really find steady-state phytoplankton assemblages in nature strongly rose again and has been given different and sometimes even contradictory answers. This problem is of particular interest if concerning the unpredictable, fragile and vulnerable ecosystems of shallow water bodies (Moss, 1980) with fluctuating conditions.

Our knowledge of phytoplankton ecology in the Bulgarian wetlands is far from complete: although algology in Bulgaria has been developing since the end of the 19th century, information on phytoplankton is quite scanty and is generally based on spot qualitative observations. There are few studies on the year-round variations of phytoplankton composition and abundance and these are quite recent (detailed review has been already provided by Stoyneva, 2000a, b).

This paper reports the results on the occurrence and composition of steady state phytoplankton assemblages in the shallow water bodies of 18 Bulgarian natural wetlands. They all differ in their morphometry, hydrology and trophic state.

Description of sites studied

Detailed descriptions of the wetlands studied and disturbances and stress they are subject to have been provided in Stoyneva & Valchanova (1997), Stoyneva & Michev (1998), Stoyneva (1998b, 2000a, b) and Stoyneva et al. (2001). The sites differ in their area (ranging from 2 to 1930 ha), depth (from 0.5 to 9.5 m), altitude (from −0.5 m b.s.l. to 860 m a.s.l.), hydrological regime (from rain and snow fed or karst spring inflow to seawaters input), salinity (from <0.5 to 250‰), nutrient loading and trophic state (from meso- to hypertrophic). The main features of the water bodies are summarized in Table 1. In spite of their peculiarities, all of the sites could be generalized as shallow ones, which are subject to a disturbed hydrological regime and are undergoing accelerated succession, as described in the publications mentioned above.

Materials and methods

The present study is based mainly on quantitative samples taken weekly and bi-weekly from the surface water layer (0–0.5 m) and at 2 m intervals depth, when depth exceeds 2 m. Phytoplankton sampling was conducted almost simultaneously in the coastal wetlands during the last 8 years and in the Danubian wetlands during the last 20 years. The small mountain water bodies Choklyovo Blato and Boyansko Blato were sampled during 1995–1996 and 1997–1999, respectively. All the samples were processed according to the sedimentation method after fixation in 2–4% formalin. In some cases, for more correct identification, simultaneous living samples were collected and some extra samples were fixed with Lugols' solution.

All determinations and measurements were done on an Amplival microscope with a magnification up to 1200 ×. Detailed taxonomic background followed Stoyneva (1998a), Hindák (2000), Hegewald et al. (2001) and John et al. (2002). Counting was done across 64 small squares of the Thoma blood-counting chamber in eight reiterations. Cells were the main counting unit. Cell numbers were converted to volumes by reference to the closest volumetric shape with the help of BIODATA computer program (Stoynev, pers.). Species were considered as dominant if they contributed more than 25% to the total biomass. Steady states were identified according to Sommer and Padisák (1993) when (i) 1, 2 or 3 species of algae contribute more than 80% of total biomass, (ii) their existence or coexistence persists for long enough (more than 1–2 weeks) and (iii) during that period the total biomass does not increase significantly.

The functional phytoplankton groups were classified according to: (1) the ecological concept of adaptive strategists (Grime, 1979; Reynolds, 1988, Olrik, 1994); (2) the behavioral grouping by Mur et al. (1993) and (3) the trait-separated functional groups (Reynolds et al., 2002). The limiting nutrient was estimated according to Overbeck (1988). Wetlands are defined according to Ramsar definition (Davis, 1994). Their trophic state was evaluated through the average annual biomass according to the open-boundary OECD system (Vollenweider & Kerekes, 1982; Vollenweider, 1990).

Results

In total, more than 1300 algal taxa were found and more than 75% of them were cyanoprokaryotes and chlorophytes, which contributed notably to the phytoplankton composition and structure of all studied wetlands (Fig. 1). The general strong increase in total biomass accompanied by severe cyanoprokaryote blooms has already been reported for most of them (Stoyneva & Valchanova, 1997; Stoyneva, 1998b, 2000a, b; Stoyneva et al., 2001) and was recently detected for Vaya Lake also. Cyanoprokaryotes comprised up to 99% of the phytoplankton biomass during the summer periods and reached up to 88% of average biomass values in water bodies, which have reached a decidedly critical trophic level (Table 1, Fig. 1).

The phytoplankton of the all water bodies under investigation was dominated by more than 200 species altogether. A period of dominance for a given species longer than one week occurred only in a few occasions (e.g. 1% in Boyansko Blato and in Dourankoulak; 3% in Srebarna and 5% in Vaya). Situations in which one species keeps a dominant position for one week,

Table 1. Main characteristics of the investigated Bulgarian wetlands. Abbreviations: Sr – Srebarna, DO – Orlovo Blato, D – Dourankoulak, E – Ezerets, Sh – Shabla, ST – Shablenska Touzla, Af – freshwater part of Atanassovsko Lake, As – salinas of Atanassovslo Lake, Ac – canal of Atanassovsko Lake, VL – Vaya Lake, Al – Alepou, Ar – Arkoutino, VV – Velyov Vir, St – Stamopolou, R – Ropotamo River, BB – Boyansko Blato and CB – Choklyovo Blato

Parameter Sites	SR	DO	D	E	Sh	ST	Af	As	Ac	VL	Al	Ar	VV	St	R	BB	CB
Altitude (m a.s.l.)	0.51	0.5	0.5	0.8	0.8	0.7	−0.5–1	−1	−1–1	−0.5–1.5	0.4	0.2	1	0.4	0–1	580	860
Total area (ha)	902	20	350	72	79	19	8	1930	12	2462.7	167	36	13.6	7	1001	24	180
Open water surface (ha)	150	20	250	70	72	19	2	1930	12	2000	164	36	13.6	7	1001	3	150
Maximum depth (m)	3	1.5	4	8.5	9.5	0.8	1.3	0.8	1.5	1.5	1	0.5	1.2	0.5	3	1.3	2.5
Water temperature (°C)**	28	28	26–27	25–26	25–26	30	26	28	26	27	28	26	27	28	29	28	28
Salinity range (%‰)	<0.05	0.6–1.3	0.3–0.4	0.4–0.6	0.5	4–5.3	0.12–0.18	18–280	0.12–169	0.25–18	>0.5	<0.5	>0.5	<0.5	4.1–18.2	<0.05	<0.05
pH	8.6–9	8.6	8.8	8.6	8.6	8	7.5–8.8	8.4–8.8	7.4–8.3	7.4–9.1	6.5–7.6	6.8	7.3	7.9–9.2	8.1–8.5	7.5–8.2	7.5–8
NH_4^+ (mg l^{-1})*	4	2.8	2.3	1.5	1.3	3.7	0.39	0.24	0.13	1.3	0.5	0.5	0.3	0.2	0.4	0.5	1.2
NO_3^- (mg l^{-1})*	1	0.2	0.1	0.05	0.04	0.5	0.42	0.26	0.12	0.04	0.3	0.4	0.17	0.3	4.2	0.4	0.12
NO_2^- (mg l^{-1})*	1.5	0.004	0.004	1.5	0.64	0.09	0.05	0.01	0.01	0.64	0.05	0.02	0.01	0.02	0.003	0.01	0.04
HPO_4^{3-} (mg l^{-1})*	0.6	0.12	0.18	0.37	0.52	0.3	0.43	0.15	0.19	0.52	0.1	0.012	0.05	0.9	0.15	0.06	0.2
Total phytoplankton biomass (mg l^{-1})	63.7	14.7	8.4	13.7	8.2	6.8	30.53	27.39	10.79	74.5	80.78	21.16	6.0	32.82	17.58	33.2	12.81

* – average annual values, ** – average summer temperature.

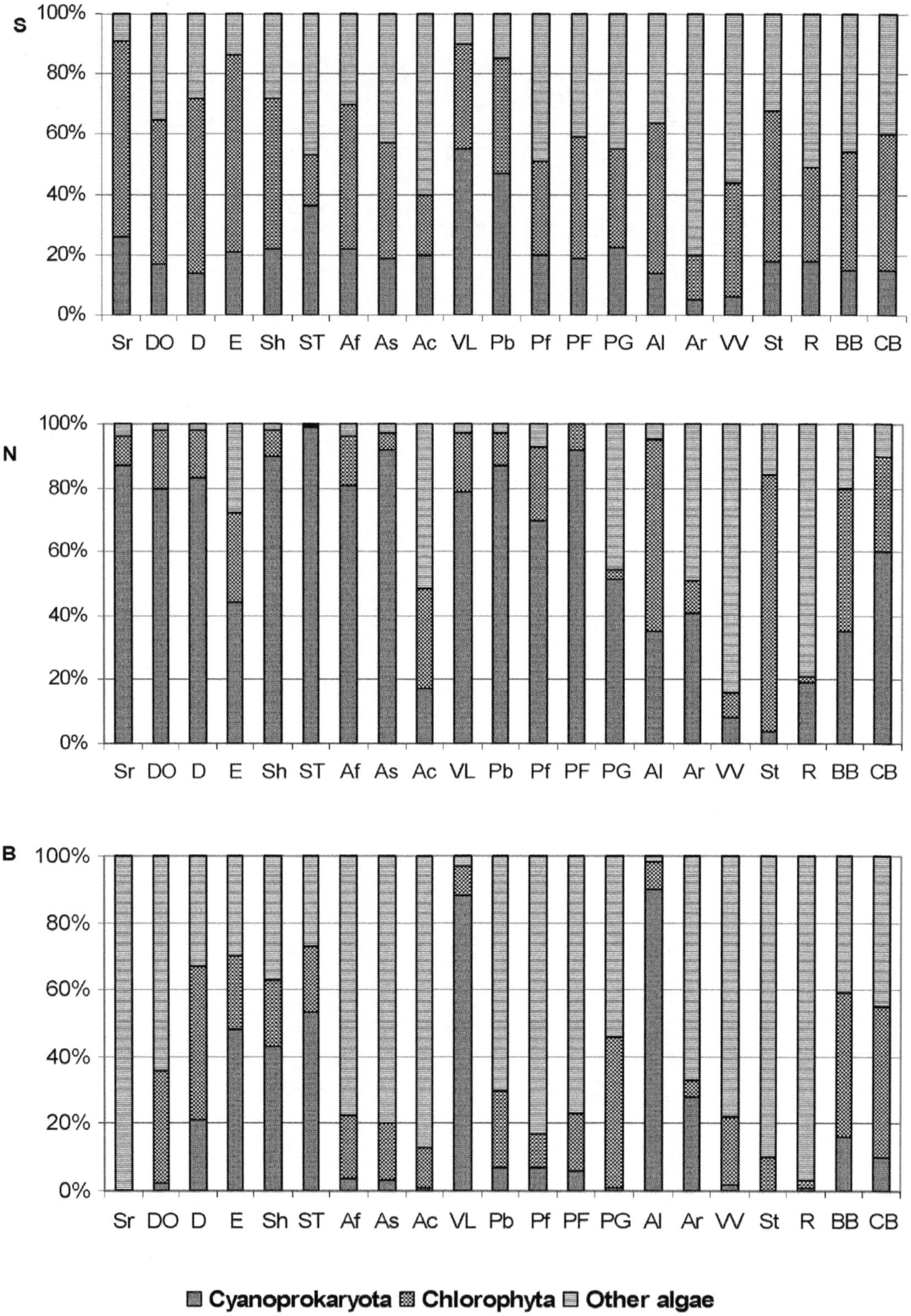

Figure 1. Average contribution of Cyanoprokaryota, Chlorophyta and other algae to the total species composition (S), to the total phytoplankton numbers (N) and to the total phytoplankton biomass (B) in Bulgarian wetlands. Abbreviations: Sr – Srebarna, DO – Orlovo Blato, D – Dourankoulak, E – Ezerets, Sh – Shabla, ST – Shablenska Touzla, Af – freshwater part of Atanassovsko Lake, As – salinas of Atanassovsko Lake, Ac – canal of Atanassovsko Lake, VL – Vaya Lake, Pb – brackish small water bodies of Poda wetland, Pf – freshwater small water bodies of Poda wetland, PF – Foros Bay, PG – Ouzoungeren, Al – Alepou, Ar – Arkoutino, VV – Velyov Vir, St – Stamopolou, R – Ropotamo River, BB – Boyansko Blato and CB – Choklyovo Blato.

but after some short periods appears as dominant again were also detected (*Phacus orbicularis* Hübn. in the summer period of 1997 in Boyansko Blato; *Microcystis aeruginosa* Kütz. emend. Elenk. in Dourankoulak, Shabla and Ezerets in summers of 1997 and 1998). When higher taxonomical levels are taken into account, representatives of the same taxonomical group could dominate for a longer period, but with an alteration at the species level (e.g. 7 weeks dominance of Euglenophyta during September–October 1999 in Boyansko Blato with alteration of *Strombomonas acuminata* (Schm.) Defl., *Trachelomonas hispida* (Perty) Stein emend. Defl. and *Euglena acus* Ehrenb. and with small variations in total biomass).

Periods of dominance of 1–3 species lasting longer than one week in most occasions were accompanied by pronounced changes of biomass. An example could be given by the dominance of *Aphanizomenon flos-aquae* (L.) Ralfs in June–July 1996 in Shabla. In the beginning of its dominance (4.06.1996) this species comprised 87% of the total biomass. During the following 7 weeks, both its own biomass and total phytoplankton biomass gradually decreased from 9.6 to 0. 3 mg l^{-1} (or 21% of total biomass) and from 11.7 to 1.7 mg l^{-1}, respectively. This event was combined with the introduction of *Microcystis aeruginosa* in the dominant complex, which reached 70% of the biomass on 31.07.1996.

A 2 week period of dominance without biomass change (both total biomass and relative representation of species) was detected occasionally (e.g. 2 weeks dominance of *Microcystis wesenbergii* (Kom.) Kom. in Kondr., *Aphanizomenon gracile* Lemm. and *Anabaena spiroides* Klebahn in August–September 2001 in Vaya lake; 2 weeks dominance of *Cylindrospermopsis raciborskii* (Wolosz.) Seenaya et Subba Raju and *Aphanizomenon gracile* in September 1999 in Srebarna; 2 weeks dominance of *Phacus orbicularis* in July 1998 and *Trachelomonas hispida* in July–August 1998 in Boyansko Blato).

The relative participation of dominants in biomass also changed rapidly in most cases (e.g. in Ezerets, 88% of biomass by *Anabaenopsis elenkinii* Müll. and 11% by *Microcystis wesenbergii* for 2 weeks (24.07–7.08.1997), but was followed by 30% *A. elenkinii* and 56% *Microcystis wesenbergii* in the next week). An often-detected event was that dominants generally form between 25 and 65% of the total biomass. For instance, *Peridinium* cf.*cinctum* (O. F. Müll.) Ehrenb. dominated the phytoplankton of Boyansko Blato between 5 August and 2 September 1998 but formed only 27% of its biomass; *Microcystis aeruginosa* and *Coelosphaerium* sp. dominated during 7–26 July 1995 in Choklyovo Blato, but formed only 32% of the total biomass without changing their proportional representation. The dominant contribution about 80% and more to the biomass was detected only in summer periods in hypertrophic conditions (*Cylindrospermopsis raciborskii* formed 87% of the biomass in August 1992 in Srebarna; *Aphanizomenon flos-aquae* in June–July 1996 in Shabla, etc.).

Steady-state situations (when all the three criteria for a theoretical equilibrium state were fulfilled) were detected only 7 times: 3 weeks dominance (97%) of *Microcystis aeruginosa* in July–August 1990 in Srebarna (Fig. 2), 3 weeks dominance (98%) of *Microcystis wesenbergii* in July 1991 in Srebarna (Fig. 2), 3 weeks dominance of *M. wesenbergii* and *Anabaenopsis arnoldii* (87%) in August–September 1995 in Shabla; 3 weeks dominance (97%) of *Cylindrospermopsis raciborskii* in August–September 1999 in Srebarna (Fig. 2); 3 weeks dominance (89%) of *M. wesenbergii, Aphanizomenon flos-aquae* and *Anabaena spiroides* in August–September 2001 in Vaya; 4 weeks dominance (92%) of *C. raciborskii* and *Aphanizomenon gracile* in July–August 1998 in Srebarna (Fig. 2); 4 weeks dominance (98%) of *M. wesenbergii* and *A. gracile* in Vaya Lake in August–September 2002. A relatively long (4 weeks) dominance of *Peridinium* cf. *umbonatum* Stein (99% of the total biomass without changes) was detected in September 1993 in Srebarna Lake, concomitant with the lowest water level (down to 0.2–0.5 m) in the whole history of this wetland (Stoyneva & Michev, 1998) and it resembled much more a temporary pool than a real lake.

Discussion

This study showed that steady states, as Sommer & Padisák (1993) have defined them, occur quite rarely in Bulgarian shallow water bodies. This result is absolutely predictable considering the quite fluctuating conditions in this type of water bodies (see also Borics et al., 2003; Ortega-Mayagoitia et al., 2003; Padisák et al., 2003; O'Farrell et al., 2003). At the same time, it seems that the concept that environmental and internal disturbances can prevent the establishment of equilibrium conditions (Sommer et al., 1993) has to be applied carefully to shallow systems. It could also be proposed tentatively that sometimes the disturbance(s)

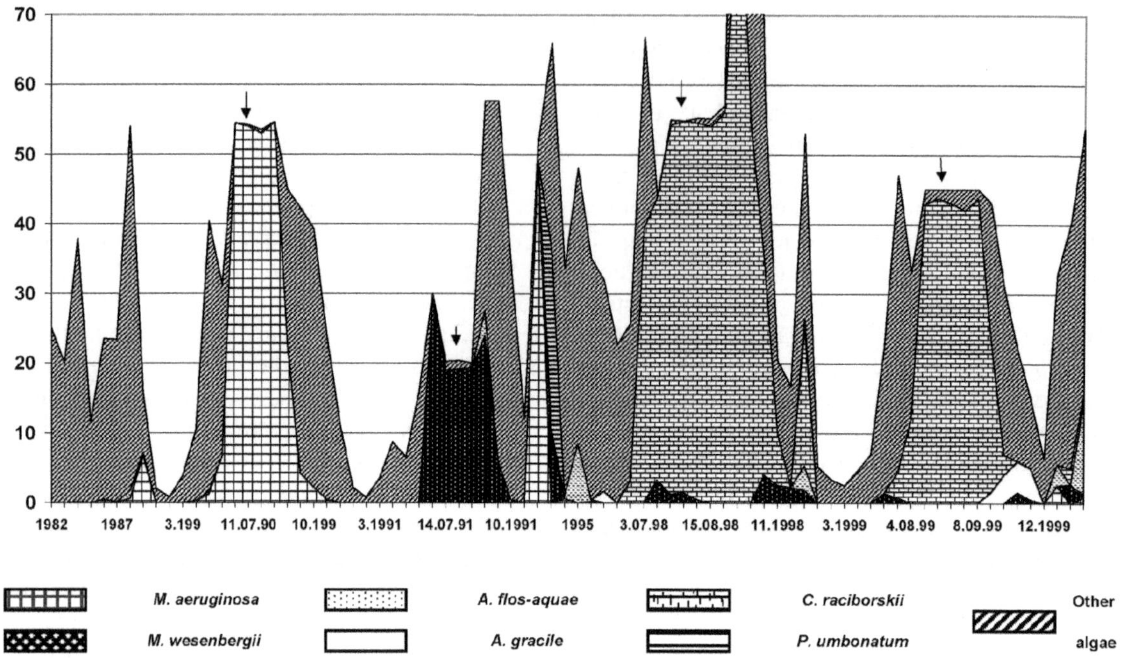

Figure 2. Phytoplankton biomass dynamics (mg l^{-1}) in Srebarna during 1982–2002 with dynamics of dominants in steady-state assemblages. [Data for 1982–1989, 1992–1997 and 1999–2002 are represented as annual averages; data for years 1990, 1991, 1998 and 1999 are represented as monttly averages combined with weekly data on steady-state situations; arrows indicate the steady-state conditions].

and stress factors themselves could lead to establishment of steady states. All cases of steady states occurrence in Srebarna and in Vaya in 2002 were detected after drastic changes in hydrological regime (significant and rapid decrease in water level in Srebarna (1990–1993, 1998) and seawater input through a canal in Vaya) combined with extremely high temperatures (cases 1990 and 1998 in Srebarna and 2002 in Vaya). However, since these cases are quite infrequent and they occurred on a background of accelerated eutrophication (Stoyneva, 1998b; Stoyneva & Michev, 1998) and since there is no clear explanation for the appearance of stable assemblages in Shabla and in Vaya (2001), it is still impossible to prove this proposal.

The detected steady state assemblages in Bulgarian shallow water bodies were dominated by 7 cyanoprokaryotes, which belonged to two morphological types – coccal and filamentous. They could be referred generally to two of the four ecotypes proposed by Mur et al. (1993) according to their behaviour in the water column: to group 1 of nitrogen-fixing species (*Aphanizomenon gracile, Anabaena spiroides, Anabaenopsis arnoldii, Cylindrospermopsis raciborskii*) and to group 4 of colony or aggregate forming species (*Microcystis aeruginosa, Microcystis wesenbergii*). The bundle-forming *Aphanizomenon flos-aquae* with its ability to fix nitrogen could be placed in both groups. According to the concept of adaptive strategists (Grime, 1979; Reynolds, 1988; Olrik, 1994) dominating cyanoprokaryotes belong to two groups – of stress–tolerant, colony-forming specialists (S) and ruderal, disturbance-tolerant filamentous (R) strategists (e.g. *Microcystis* and *Aphanizomenon*, respectively). In the classification scheme of trait-separated phytoplankton groups (Reynolds et al., 2002) the species found are placed separately in groups $\mathbf{S_N}$, $\mathbf{H_1}$ and $\mathbf{L_M}$. The combinations of dominants found in the studied sites were not represented in this scheme. *Microcystis* and *Cylindrospermopsis* dominated stable phytoplankton phases both separately and in combination with *Aphanizomenon* and *Anabaena*, while the last genera were not detected as single dominants during equilibrium phases. All these taxa have been widely recognized as typical and abundant members of the hypertrophic phytoplankton (Álvarez-Cobelas & Jacobsen, 1992). Published ecological requirements, physiological abilities and behaviour of species mentioned above are not only different but also could be regarded from different angles (Padisák, 1997; Bulgakov & Levich, 1999; Reynolds, 1999; Smith & Bennett, 1999; Sommer, 1999; Teubner & Dokulil, 2002). Many lakes show a similar pattern of phytoplankton response to enhanced nutrient loading which was supposed to be implicated directly in the selection of dominant species. However, still the problem of why certain species should dominate mixed assemblages under certain circumstances and which are the driving forces for this remains (Reynolds, 1999). After the underlying mechanisms of cyanoprokaryote dominance (Dokulil & Teubner, 2000; Komárková & Tavera, 2003; Mischke, & Nixdorf, 2003; O'Farrell et al., 2003; Padisák et al., 2003) were analyzed and discussed using data from various shallow mixed and deep stratifying lakes, it was stated that this dominance can not be viewed independently from all other members of phytoplankton and is strongly affected by complex interactions between lake morphometry, water temperature, underwater light availability, nutrient supply and total food-web structure (Harris, 1986; Reynolds, 1987; Dokulil & Teubner, 2000). These results have been obtained in almost all studies on phytoplankton dynamics and structure and are relevant to most algal species (e.g. Reynolds & Reynolds, 1985; Rhew et al., 1999; Naselli-Flores & Barone, 2000 among the many others). The interplay of factors was outlined for hypertrophic environments particularly (Álvarez-Cobelas & Jacobsen, 1992). The results from a recent study did not allow us to determine all the ecological conditions which lead to achieving of steady states. However, it has to be underlined that equilibrium states dominated by cyanoprokaryotes were seen only during hypertrophic conditions in stressed and frequently disturbed shallow water bodies and were related with the summer – late summer periods.

Acknowledgements

The author wishes to thank Mrs V. Chainadjieva, Mr M. Dimitrov, Mr I. Botev, Mrs L. Tsvetkova and Mr Chr. Naydenov for obtaining and providing the chemical characteristics of the studied lakes. Special thanks are due to Mr T. Michev, Mr N. Mikhov, Mrs V. Chajnadjieva, Mrs M. Valchanova, Ms M. Nedelkova, Dr S. Andreev, Dr D. Kozhuharov and to the late Dr S. Kovachev for their inestimable help in collecting the samples. The biggest part of the study was done within the framework of BSBCP, scientific program of Le Balkan Foundation and monitoring program of Bulgarian Ministry of Environment and Waters. Special thanks are due to Dr L. Naselli-Flores for the invitation to present this work at the 13th IAP-meeting in Castelbuono (Italy), to Prof. J.-P. Descy and to both unknown referees for their helpful comments on the manuscript.

References

Alimov, A. F., 1990. The main principles of the theory of functioning of aquatic systems. Gidrobiologicheskiy Zhournal 26: 3–12 (in Russian, English summ).

Álvarez-Cobelas, M. & B. A. Jacobsen, 1992. Hypertrophic phytoplankton: An overview. Freshwat. Forum 2: 184–199.

Borics, G., B. Tóthmérész, I. Grigorszky, J. Padisák, G. Várbíró & S. Szabó, 2003. Algal assemblage types of bog-lakes in Hungary and their relation to water chemistry, hydrological conditions and habitat diversity. Hydrobiologia 502 (Dev. Hydrobiol. 172): 145–155.

Bulgakov, N. G. & A. P. Levich, 1999. The nitrogen: phosphorus ratio as a factor regulating phytoplankton community structure. Archiv für Hydrobiol. 146: 3–22.

Carpenter, S. R. & J. F. Kitchell, 1988. Introduction. In Carpenter, S. R. (ed.), Complex Interactions in Lake Communities. New York, Springer Verlag: 1–7.

Connell, J., 1978. Diversity in tropical rain forests and coral reefs. Science 199: 1304–1310.

Davis, T. D. (ed.), 1994. The Ramsar Convention Manual: A Guide to the Convention on Wetlands of International Importance especially as Waterfowl Habitat. Ramsar Convention Bureau, Gland, Switzerland. 207 pp.

Dokulil, M. T. & K. Teubner, 2000. Cyanobacterial dominance in lakes. In: Hamilton, P. B., H. Kling & M. T. Dokulil (eds), Cyanoprokaryotes and Chlorophytes across Lake Trophic States. Hydrobiologia 438: 1–12.

Grime, J. P., 1979. Plant Strategists and Vegetation Processes. New York. Willey.

Harris, G. P., 1983. Mixed layer physics and phytoplankton populations; studies in equilibrium and non-equilibrium ecology. In Round, F. E. & D. Chapman (eds), Progress in Phycological Research 2. Elsevier, Amsterdam: 1–52.

Harris, G. P., 1986. Phytoplankton Ecology. Structure, Function and Fluctuation. Chapmann and Hall, London, New York. 384 pp.

Hegewald, E., D. Hepperle, M. Wolf & L. Krienitz, 2001. Phylogenetic placement of *Chlorotetraedron incus*, *C. polymorphum* and *Polyedriopsis spinulosa* (Neochloridaceae, Chlorophyta). Phycologia 40: 399–402.

Hindák, F., 2000. Morphological variation of four planktic nostocalean cyanophytes – members of the genus *Aphanizomenon* or *Anabaena*? In Hamilton, P. B., H. Kling & M. T. Dokulil (eds), Cyanoprokaryotes and Chlorophytes across Lake Trophic States. Hydrobiologia 438: 107–116.

Hutchinson, G. E., 1961. The paradox of the plankton. Am. Nat. 95: 137–145.

John, D. M., B. A. Whitton & A. J. Brook, 2002. The Freshwater Algal Flora of the British Isles. An Identification Guide to Freshwater and Terrestrial Algae. Cambridge University Press. 702 pp.

Komárková, J. & R. Tavera, 2003. Steady state of phytoplankton assemblage in the tropical Lake Catemaco (Mexico). Hydrobiologia 502 (Dev. Hydrobiol. 172): 187–196.

Margalef, R., 1961. Communication of structure in planktonic populations. Limnol. Oceanogr. 6: 124–128.

Mischke, U. & B. Nixdorf, 2003. Equilibrium phase conditions in shallow German lakes: How Cyanoprokaryota species establish a steady state phase in late summer. Hydrobiologia 502 (Dev. Hydrobiol. 172): 123–132.

Moss, B., 1980. Ecology of fresh waters. Blackwell ScientificPublications, Oxford. 332 pp.

Mur, L. R., H. Shreurs & P. Viser, 1993. How to control undesirable cyanobacterial dominance. In Giussani, G. & C. Callieri (eds), Strategies for Lake Ecosystems Beyond 2000. Proceedings of the 5th International Conference on Conservation and Management of Lakes, Stresa, Italy: 565–569.

Naselli-Flores, L. & R. Barone, 2000. Phytoplankton dynamics and structure: a comparative analysis in natural and man-made water bodies of different trophic state. In Hamilton, P. B., H. Kling & M. T. Dokulil (eds), Cyanoprokaryotes and Chlorophytes across Lake Trophic States. Hydrobiologia 438: 65–74.

Naselli-Flores, L., J. Padisák, M. Dokulil & I. Chorus, 2003. Equilibrium/steady-state concept in phytoplankton ecology. Hydrobiologia 502 (Dev. Hydrobiol. 172): 395–403.

O'Farrell, I., R. Sinistro, I. Izaguirre & F. Unrein, 2003. Do steady state assemblages occur in shallow lentic environments from wetlands? Hydrobiologia 502 (Dev. Hydrobiol. 172): 197–209.

Olrik, K., 1994. Phytoplankton – Ecology. Determining Factors for the Distribution of Phytoplankton in Freshwater and the Sea. Ministry of the Environment, Denmark, Danish Environmental Protection Agency, Nofo Print, Elsinore. 183 pp.

Ortega-Mayagoitia, E., C. Rojo & M. A. Rodrigo, 2003. Controlling factors of phytoplankton assemblages in wetlands: an experimental approach. Hydrobiologia 502 (Dev. Hydrobiol. 172): 177–186.

Overbeck, J., 1988. Ecosystem concepts. In: Jorgensen, S. E. & R. A. Vollenweider (eds), Guidelines of Lake Management. Volume 1. Principles of Lake Management, ILEC: 19–34.

Padisák, J., 1997. *Cylindrospermopsis raciborskii* (Woloszynska) Seenayya et Subba Raju, an expanding, highly adaptive cyanobacterium: worldwide distribution and review ot its ecology. Archiv für Hydrobiol./Supplement 107 (Monographic Studies), 4: 563–593.

Padisák, J., G. Borics, G. Fehér, I. Grigorszky, I. Oldal, A. Schmidt & Z. Zámbóné-Doma, 2003. Dominant species, functional assemblages and frequency of equilibrium phases in late summer phytoplankton assemblages in Hungarian small shallow lakes. Hydrobiologia 502 (Dev. Hydrobiol. 172): 157–168.

Rhew, K., R. M. Baca, C. A. Ochs & S. T. Threlkeld, 1999. Interaction effects of fish, nutrients, mixing and sediments on autotrophic picoplankton and algal composition. Freshwat. Biol. 42: 99–109.

Reynolds, C. S., 1987. Cyanobacterial water blooms. Adv. Bot. Res. 13: 67–143.

Reynolds, C. S. 1988. Functional morphology and the adaptive strategies of freshwater phytoplankton. In Sandgren, C. D. (ed.), Growth and Reproductive Strategies of Freshwater Phytoplankton. Cambridge University Press: 388–433.

Reynolds, C. S., 1999. Non-determinism to Probability, or N:P in the community ecology of phytoplankton. Archiv für Hydrobiol. 146: 23–35.

Reynolds, C. S. & J. B. Reynolds, 1985. The atypical seasonality of phytoplankton in Crose Mere 1972: an independent test of the hypothesis that variability in the physical environment regulates community dynamics and structure. British Phycol. J. 20: 227–242.

Reynolds, C. S., J. Padisák & U. Sommer, 1993. Intermediate disturbance in the ecology of phytoplankton and the maintenance of species diversity: a synthesis. In Padisák, J., C. S. Reynolds & U. Sommer (eds), Intermediate Disturbance Hypothesis in Phytoplankton Ecology. Hydrobiologia 249: 183–188.

Reynolds, C. S., V. Huszar, C. Kruk, L. Naselli-Flores & S. Melo, 2002. Towards a functional classification of the freshwater phytoplankton. J. Plankton Res. 24: 417–428.

Rhew, K., R. M. Baca, C. A. Ochs & S. T. Threlkeld, 1999. Interaction effects of fish, nutrients, mixing and sediments on autotrophic picoplankton and algal composition. Freshwat. Biol. 42: 99–109.

Smith, V. H. & S. J. Bennet, 1999. Nitrogen:phosphorus supply ratios and phytoplankton community structure in lakes. Archiv für Hydrobiol. 146: 37–53.

Sommer, U., 1999. A comment on the proper use of nutrient ratios in microalgal ecology. Archiv für Hydrobiol. 146: 55–64.

Sommer, U. & J. Padisák, 1993. Appendix 1. Identification of equilibrium. In Sommer, U., J. Padisák, C. S. Reynolds & P. Juhász-Nagy, 1993. Hutchinson's heritage: the diversity-disturbance relationship in phytoplankton. Hydrobiol. 249: 5–6.

Sommer, U., J. Padisák, C. S. Reynolds & P. Juhász-Nagy, 1993. Hutchinson's heritage: the diversity-disturbance relationship in phytoplankton. In Padisák, J., C. S. Reynolds & U. Sommer (eds), Intermediate Disturbance Hypothesis in Phytoplankton Ecology. Hydrobiologia 249: 1–7.

Stoyneva, M., 1998a. Algae. In Michev, T. M., B. B. Georgiev, A. V. Petrova & M. P. Stoyneva (eds), 1998. Biodiversity of the Srebarna Biosphere Reserve. Checklist and Bibliography. Sofia, Co-publ. Context & Pensoft: 10–37.

Stoyneva, M. P., 1998b. Development of the phytoplankton of the shallow Srebarna Lake (North-Eastern Bulgaria) across the trophic gradient. In Álvarez-Cobelas, M., C. S. Reynolds, P. Sanchez-Castillo & J. Kristiansen (eds), Phytoplankton and Trophic Gradients. Hydrobiologia 369/370: 259–367.

Stoyneva, M. P., 2000a. Algological studies of Bulgarian coastal wetlands. I. Species composition of the phytoplankton of Dourankoulak and Shabla-Ezeretz lakes. – Ann. Univ. Sof., 91, Book 2 – Botany: 27–48.

Stoyneva, M. P., 2000b. Planktic green algae of Bulgarian coastal wetlands. In Hamilton, P. B., H. Kling & M. T. Dokulil (eds), Cyanoprokaryotes and Chlorophytes across Lake Trophic States. Hydrobiologia 438: 25–41.

Stoyneva, M. P. & T. M. Michev, 1998. Srebarna: general characteristics and brief history. In Michev, T. M., B. B. Georgiev, A. V. Petrova & M. P. Stoyneva (eds), Biodiversity of the Srebarna Biosphere Reserve. Checklist and bBbliography. Sofia, Co-publ. Context & Pensoft: 1–8.

Stoyneva, M. & M. Valchanova, 1997. Pilot studies on annual alteration of various dominant life strategists in the phytoplankton of the peat-bog Tschokljovo (South-western Bulgaria). Annual of Sofia University "St Kliment Ohridski", Faculty of Biology, Book 2 – Botany 89: 23–33.

Stoyneva, M. P., D. N. Temniskova-Topalova, K. G. Uzunova & M. G. Nedelkova, 2001. Long-term compositional response of the algal flora of the Boyana swamp (western Bulgaria) to the environmental changes. Annual of Sofia University "St Kliment Ohridski", Faculty of Biology, Book 2 – Botany 93: 43–72.

Teubner, K. & M. T. Dokulil, 2002. Ecological stoichiometry of TN:TP:SRSI in freshwaters: nutrient ratios and seasonal shifts in phytoplankton assemblages. Archiv für Hydrobiol. 154: 625–646.

Vollenweider, R. A., 1990. Eutrophication: conventional and non-conventional considerations and comments on selected topics. In de Bernardi, G. Giussani & L. Barbanti (eds), Scientific Perspectives in Theoretical and Applied Limnology. Mem. Ist. ital. Idrobiol. 47: 77–137.

Vollenweider, R. & J. Kerekes, 1982. Eutrophication of water: monitoring, assessment and control. OECD, Paris: 154 pp.

Hydrobiologia **502:** 177–186, 2003.
L. Naselli-Flores, J. Padisák, M. T. Dokulil (eds),
Phytoplankton and Equilibrium Concept: The Ecology of Steady-State Assemblages.

Controlling factors of phytoplankton assemblages in wetlands: an experimental approach

Elizabeth Ortega-Mayagoitia[1], Carmen Rojo* & María A. Rodrigo
[1]*Laboratory of Limnology, CyMA, UIICSE, FES-Iztacala Universidad Nacional Autónoma de México. Av. de los Barrios s/n, Los Reyes Iztacala, 54090 Tlalnepantla, Estado de México, México*
Laboratory of Limnology, Institute "Cavanilles" of Biodiversity and Evolutionary Biology, University of Valencia, Apdo. Oficial 22085, E-46071 Valencia, Spain
E-mail: Carmen.Rojo@uv.es
*(*Author for correspondence)*

Key words: steady state, sediment, Tablas de Daimiel National Park, perturbations, nutrient pulses, food web

Abstract

The aim of this work is to answer some questions like: what factors control the phytoplankton assemblage? What factor or factors are perturbing the assemblage? What factors are driving or maintaining the stability? Are the different responses to the control factors dependent on the hierarchy level? For that, we tested experimentally the influence of herbivory, planktivory, nutrients and sediment on phytoplankton assemblages and its stability from a hypertrophic wetland (*Las Tablas de Daimiel* National Park, Spain) in three microcosm experiments. The study of the steady state phytoplankton assemblages in this perturbed system could point out some underlying processes instead of competition. The presence of planktivorous fish and the different composition of zooplankton have minor importance in phytoplankton composition. Conversely, sediment is of paramount importance, promoting a more diverse phytoplankton assemblage. When sediment or nutrient pulses are absent, phytoplankton become dominated by slow-growing algae, present but not dominant in the studied wetland community. We suggest that alternate states of phytoplankton assemblages in a eutrophic wetland occur as a sequence of substitutions persistently altered by perturbations, thanks to the close coupling with sediment, and that possibly trophic relationships are irrelevant.

Introduction

Changes in species composition and dominance of phytoplankton can be mediated by a variety of mechanisms, including ambient temperature, light income, nutrient supply, as well as through settling out, removal by zooplankton, etc. (Reynolds, 1993). A great part of this research has focused on lakes, where the alternation between mixing of the water column (abiotic factors) and stability periods (autogenic mechanisms) promotes changes in the composition and structure of lake phytoplankton (Reynolds, 1993).

Some general patterns on phytoplankton assembly have been widely recognised and frequently corroborated. Examples are the seasonal PEG-model (Sommer et al., 1986), the algal associations corresponding each environment (Reynolds, 1988; Olrik, 1994) and a reasonably clear sequence of states when autogenic succession occurs (Reynolds, 1997). Autogenic succession needs an unperturbed environment to occur, in such conditions it is probably that a competition by resources results in a steady state assemblage of few species, these best competitors will be dominants during some weeks and their biomass quite constant. This process is similar to an experience about competition in microcosms in a laboratory without grazing or other stressors (Lampert & Sommer, 1997, Cottingham, 1999).

This whole landscape about phytoplankton structural change may apply less well to wetlands (Padisák et al., 2003), where a different and weak pattern has been established (Chow-Fraser, 1998; Rojo et al., 2000a). In this way, for example, only when we give up species (as variable) and use the main taxonomic groups some vague seasonal and inter-annual sequences are observed (Rojo et al., 2000b) and periods with autogenic trajectories are minimum along the year (Rojo et al., 2000a). The study of the steady state phytoplankton assemblages in these perturbed systems could point out some underlying processes instead of competition. Phytoplankton assemblages (one, two, three, or more species) with a constant composition, not only the species list rather the biomass proportion, during some weeks can be due to predation, different niches, initial pool of species, etc (Rojo & Álvarez-Cobelas, 2003). In shallow environments – like wetlands are –, the relative importance of abiotic and biotic factors change, increasing the influence of both the sediment layer and the commonly abundant planktivorous fish (Jeppesen, 1998), among others. The importance of these factors should be stressed when we try to understand the sequence of substitutions of phytoplankton structure in wetlands. Unfortunately, the background about these possible relationships is scarce. Thus, despite the increasing interest on the benthic-pelagic coupling in all kind of aquatic ecosystems in recent years (e.g. Hansson, 1996; Marcus & Boero, 1998; Boero et al., 2000; Borics et al., 2003), the relationship between complex sediment system and phytoplankton composition remains poorly known. In addition, the effects of planktivory and grazing on phytoplankton are usually focused on the size structure of predators and prey, but less information exists on the effects of trophic interactions on the species composition of phytoplankton (Komárková & Tavera, 2003). This gap can be important when it is known that differences in resource use by zooplankton are dependent as much on food quality, or on taxonomic differences in the food, as on size (reviewed in Rothhaupt, 1990). Differences in resource use can depend on food particle sizes (Neill, 1973; Irvine, 1986), on food quality (Hoenicke & Goldman, 1987), or on taxonomic differences in the food (DeMott & Kerfoot, 1982).

Finally, our work is being developed in a hypertrophic system with high phosphorus load, weak water inputs and high temperature, however, its algal composition never reached their expected potential steady state (Reynolds, 1984) and for example, cyanophycean algae were not dominant (Rojo et al., 2000 a).

Consequently, the influence of other environmental factors changing the outcome autogenic succession in phytoplankton assemblage is presumed. Therefore, our main goal is to test the influence of sediment, nutrient load and trophic relationships in phytoplankton taxonomical structure in wetlands, seeking, in particular, answers to the following questions:

What factors control the phytoplankton assemblage? What factor or factors are perturbing the assemblage? What factors are driving or maintaining the stability? Are the different responses to the control factors dependent on the hierarchy level (populations or main taxonomic groups)?

With these objectives in mind, we performed microcosm experiments with plankton from a wetland deeply studied by us (Rojo et al., 2000a; Ortega-Mayagoitia et al., 2000; Sánchez-Carrillo & Álvarez-Cobelas, 2001).

Methods

Plankton community

The plankton community studied is from a floodplain wetland (Las Tablas de Daimiel National Park, central Spain, TDNP hereafter) that is undergoing a severe eutrophication process and continuous water shortages. When the experiment was performed TDNP flooding area was approximately 14 km^2. Mean depth of the wetland is 0.5 m. The light climate when water was sampled is characterised by high incident PAR irradiance 1209–1645 μE m^{-2} s^{-1}. The uppermost water column strata (0.3–0.4 m) received 200–450 μE m^{-2} s^{-1}. During 1996–1998, total phosphorus ranged from 0.01 to 2.3 mg l^{-1}, and total nitrogen ranged from 1 to 10 mg l^{-1} (Sánchez-Carrillo & Álvarez-Cobelas, 2001). Phytoplankton biomass is highly variable both temporally and spatially in the wetland, averaging annually 4–14 mg l^{-1} (fresh weight) (Rojo et al., 2000a). Cryptophytes and diatoms are the dominant groups most of times, followed by chlorophytes and cyanophytes as dominant or co-dominant taxa. Species composition of phytoplankton has been described in detail in Rojo et al. (1999) and in Ortega-Mayagoitia & Rojo (2000a, b). Zooplankton biomass mostly consists of ciliates; the largest zooplankton components during most part of the year are cyclopoid copepods except in spring, when cladocerans dominate; rotifer biomass is scarce (Ortega-Mayagoitia et al., 2000). Introduced mosquitofish (*Gambusia holbrooki* Girard), carp (*Cyprinus carpius* Linneo) and sunfish pumpkinseed (*Lepomis gibbosus* Linneo) are the dominant fish species. The main area of the wetland is covered by the emergent macrophytes saw grass (*Cladium mariscus*) and reed (*Phragmites australis*). For further information on TDNP, the reader is referred to Álvarez-Cobelas & Cirujano (1996).

Experimental design

Three experiments were performed in 40-l microcosms placed in an isolated room-chamber at constant temperature (20 ± 1 °C), with a 12 × 12 h light-dark cycle. The design of Experiments 1 and 2, a 2 × 2 factorial design (two factors × two treatment levels each) resulted in four treatment combinations, whereas Experiment 0 consisted in only two treatments. Each treatment combination had three replicates, and were randomly assigned to microcosms. Experiments were performed as follows.

Experiment 0 (E0)

Designed to test if planktivory can be a controlling factor of phytoplankton composition by changing zooplankton abundance and composition. Treatments: (1) Fish present (F), (2) Fishless (nF). Twenty L of water from TDNP with a spring plankton community was placed on each microcosm and three mosquitofish (2.2–2.5 cm standard length) were added as appropriate. The experiment ran for 4 weeks.

Experiment 1 (E1)

We examined the single and combined effects of planktivory and sediment layer on phytoplankton structure. Sediment was expected to influence phytoplankton structure by changing the phosphorus concentration in water. Treatment combinations: (1) no sediment and fishless (nS/nF), (2) no sediment and fish present (nS/F), (3) sediment present and fishless (S/nF), and (4) both sediment and fish present (S/F). Water and sediment were collected in autumn, in a shallow area of the wetland. Sediment was distributed on the bottom of the microcosms, 20 l of water from TDNP was poured, and after a while, four mosquitofish (1.3–2 cm standard length) were added as appropriate. The experiment ran for 3 weeks.

Experiment 2 (E2)

Designed to assess the effects of zooplankton composition and nutrient level on phytoplankton, and whether the effects of both factors are co-dependent. As the TDNP wetland is eu-hypertrophic, instead of increasing nutrient concentrations as a treatment, we diluted the natural concentration of wetland water. Treatment combinations: (1) diluted nutrients and reduced zooplankton (dN/rZ), (2) diluted nutrients and normal zooplankton (dN/Z), (3) normal nutrients and reduced zooplankton (N/rZ), and (4) normal nutrients and normal zooplankton (N/Z). Water was collected in the TDNP in autumn. To test for the effects of reduced nutrient concentration, while keeping a similar abundance of plankton among combinations, we mixed the wetland water with de-chlorinated tap water in all microcosms. Then, a solution of 12 mg $NaNO_3$/l and 1 mg $NaH_2PO_42H_2O$/l was added in the 'normal' nutrient treatment to give an initial nutrient concentration of 12.5 ± 0.9 mg TN/L and 0.82 ± 0.02 mg TP/L. These concentrations were significantly ($P = 0.002$ and $P < 0.001$, respectively) higher than the 7.9 ± 0.6 mg TN/L and 0.24 ± 0.01 mg TP/L measured in the diluted treatment. The zooplankton structure was manipulated by filtering the wetland water through a 50-μm mesh to remove most meta-zooplankton. This operation eliminated copepods very efficiently and reduced rotifer and protozoan abundances remarkably. Aquaria were filled with 14 l of the filtered or unfiltered wetland water, mixed with 8 l of tap water, and then the nutrient solution was added. This experiment ran for 6 weeks.

A more detailed explanation of the methods can be found in Ortega-Mayagoitia et al. (2002).

Sampling and analyses

Water samples for nutrient concentrations were only taken the first and last days of the experiments. We measured total phosphorus (TP) in E1, and TP and total nitrogen (TN) in E2. TP was analysed following Standard Mehtods (APHA, 1985). TN was measured according to the method of Bachman & Canfield (1996).

To monitor plankton response to manipulations, we sampled microcosms every 6 day in E0, and every 7 day in E1 and E2. As plankton is very abundant in the TDNP wetland, and to avoid an excesive reduction of water within experimental units, only 50 ml of water were withdrawn for phytoplankton samples. Algal populations were estimated according to the Utermöhl method, counting at least 400 individuals of the most abundant species from each sample with a 10% error (Lund et al., 1958). Phytoplankton was divided in higher taxonomic categories according to Reynolds (1984). Ciliate and heterotrophic flagellate abundances were estimated simultaneously to phytoplankton, counting at least 100 individuals of each group. Algal and protozoan biovolumes were calculated by associating cellular shapes to geometrical bodies (Rott, 1981), using measurements of at least 20 individuals of each population.

For metazooplankton, 500 ml of water were filtered with a 50 μm Nytal net, returning the filtered water carefully to the corresponding microcosm. Rotifers and microcrustaceans were counted and measured with an inverted microscope, observing the entire sample or a subsample if organisms were too abundant. Microcrustacean biomass was estimated by length-weight regressions, and rotifer biomass was estimated by biovolume (McCauley, 1984). Body density was assumed at 1 g/cm^3. All biomass data are expressed as fresh weight.

Statistical analyses

Initial and final nutrient concentration data were analysed by means of one-way ANOVA in E1 (factor: sediment), and by two-way ANOVA in E2 (factors: zooplankton treatment, nutrient level). To determine the effects of the main factors and their interactions on plankton over the course of each experiment, two-way repeated measures analyses of variance (ANOVAR) were performed. The relative contribution (X) of each grouping to total phytoplankton biomass was transformed (arcsin $\sqrt{X}$) when necessary to reduce variance heterogeneity among groups (Underwood, 1997). All except the initial sampling data were used in the analysis. ANOVAR was performed mainly adhering to von Ende (1993). We evaluated the effects of each one of the inter-subject factors, its interaction, the effect of time and the time $\times$ factors interaction. Effects were considered significant at $P < 0.05$. All statistical analyses were run using SPSS release 9.0.1 General Linear Model (GLM) routine. A more detailed explanation can be found in Ortega-Mayagoitia et al. (2002).

Results

Experiment 0

Total zooplankton biomass at the beginning of the experiment was 13.8 $\pm$ 3.7 mg l^{-1}, with the copepod *Acanthocyclops robustus* as the main contributor (83 $\pm$ 4%), followed by protozoans (heterotrophic flagellates and ciliates, 12 $\pm$ 2%) and nauplii (4 $\pm$ 1%). Significant effects of mosquito fish presence on zooplankton were a noticeable copepod and nauplii decrease, and an increase in rotifers abundance. Mean total phosphorus was 0.23 $\pm$ 0.01 mg l^{-1}, similar among combinations.

In the beginning of this experiment, mean phytoplankton biomass was 6.4 $\pm$ 0.7 mg l^{-1} and consisted of a diverse array of species (47 taxa), among which an unidentified oscillatorial cyanophyte, a group of small flagellates and *Monoraphidium komarkovae* Nygaard were the main contributors (1.7, 0.8 and 0.4 mg l^{-1}, respectively). The only significant effects of the fish presence during the course of the experiment were a slight enhancement of the proportion of cryptophytes and euglenophytes (groups of minor relevance), and a change on the trajectory of the small flagellates group (Table 1, Fig. 1A).

Moreover, we observed several changes in phytoplankton composition which occurred in both treatments, so they were not related to fish presence: a phytoplankton richness decrease, whereas some species substitutions like *Cryptomonas erosa* Ehrenberg by *Cryptomonas marsonii* Skuja and *Chroomonas* sp., and *Katodinium fungiforme* (Anisimova) Loeblich III and *Peridinium umbonatum* Stein by *Gymnodinium* cf. *wawrikae* Schiller occurred. Only *M. komarkovae* persisted as an important species along the experiment.

Experiment 1

At the begining of the experiment, TP in water was 0.56 $\pm$ 0.03 mg l^{-1}, similar among combinations ($P = 0.470$). By the end of the experiment, TP was not affected by treatments, averaging 0.46 $\pm$ 0.7 mg l^{-1}.

Initially, total zooplankton biomass was 4.25 $\pm$ 0.40 mg l^{-1}. *A. robustus* contributed 60 $\pm$ 5% of zooplankton biomass, followed by protozoans (27 $\pm$ 3%) and rotifers (8 $\pm$ 2%). Most important significant effects of fish presence were the decrease of copepods abundance and increase of rotifers and heterotrophic flagellates. Sediment presence decreased significantly protozoan and rotifer abundance but favoured higher numbers of copepod populations (Ortega-Mayagoitia et al., 2002).

At the beginning of the experiment, mean phytoplankton biomass was 8.0 $\pm$ 0.8 mg l^{-1} and was mainly composed by the cryptophytes *C. erosa* (1.8 mg l^{-1}) and *Chroomonas* spp. (1.0 mg l^{-1}), a group of small flagellates (1.1 mg l^{-1}), the diatoms *Cyclotella atomus* Hustedt (1.0 mg l^{-1}) and *C. meneghiniana* Kützing (0.5 mg l^{-1}), and the oscillatorial

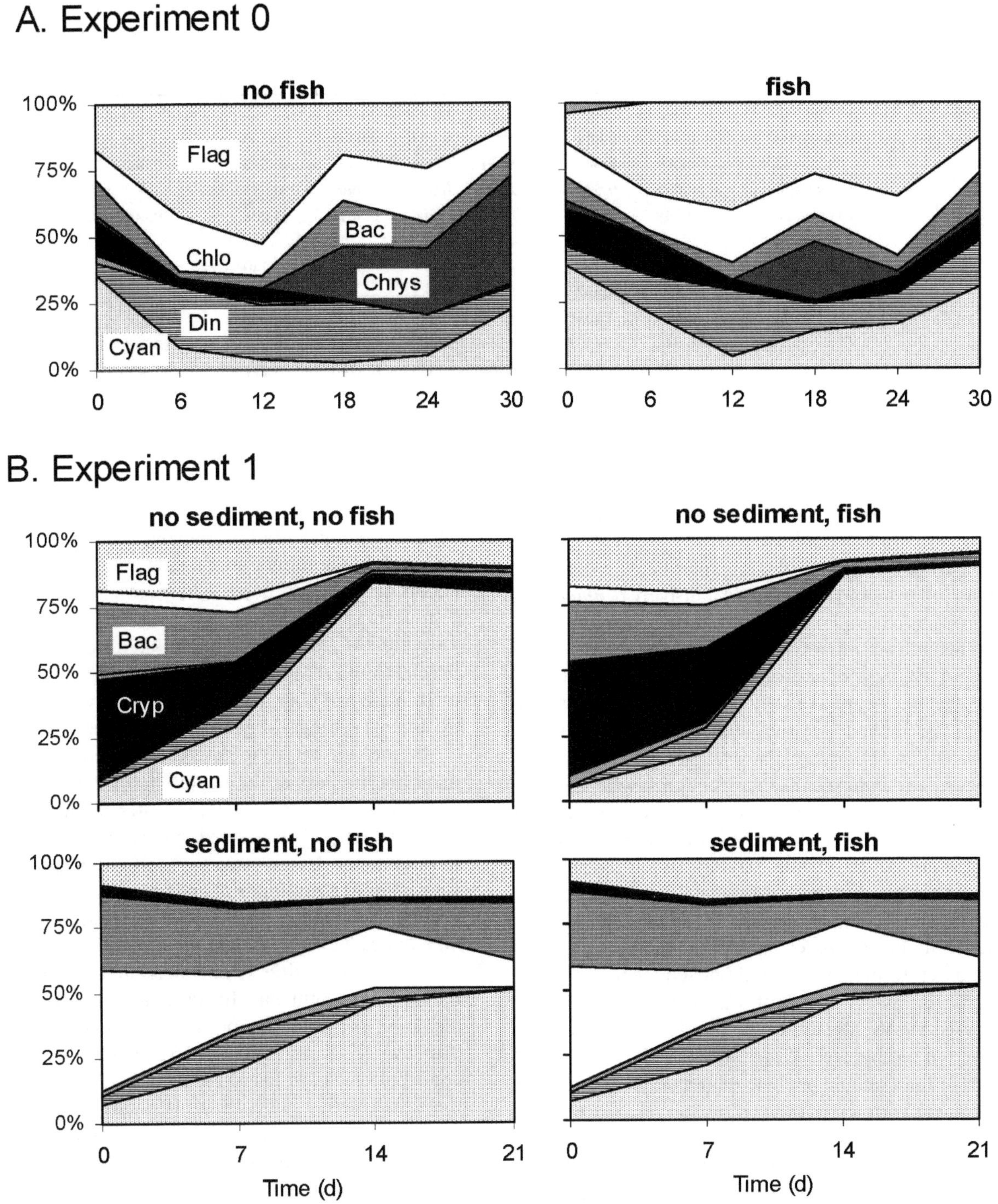

Figure 1. Examples of the time course of the relative biomass of phytoplankton taxonomic groups. Highest mean between-replicate difference was ± 10% (Chrysophyta in the fishless treatment). Cyan: Cyanophyta; Din: Dinophyceae; Cryp: Cryptophyceae; Chrys: Chrysophyceae; Bac: Bacillariophyceae; Chlo: Chlorophyta; Flag: small phytoflagellates.

Table 1. Results of repeated measures analysis of variance (ANOVAR) on Experiments 0, 1 and 2. Only the probabilities (*P*) of the main effects and their interactions are shown. (^{T}) indicates if the interaction with Time was significant. Effects were considered significant at $P < 0.05$

	Experiments						
	E0		E1		E2		
	Fish	Sediment	Fish	S × F	Nutrients	Zooplankton	N × Z
Cyanophyta	0.445	**<0.001**^{T}	0.932	0.451	0.053^T	0.446	0.779
Dinophyceae	0.609	**0.005**^{T}	**0.031**	**0.020**	0.244	0.598	0.580
Euglenophyta	**0.039**	**0.020**	0.507	0.091	**0.000**^{T}	0.075^T	
Cryptophyta	**0.040**^{T}	**<0.001**^{T}	**0.005**	**0.004**	**0.001**^{T}	**0.008**^{T}	**0.009**
Chrysophyceae	0.250	–	–	–	–	–	–
Haptophyta	–	–	–	–	**<0.001**^{T}	0.740^T	0.634
Xanthophyceae	–	**0.032**^{T}	0.091	0.681	–	–	–
Bacillariophyceae	0.640	**<0.001**^{T}	0.09	0.389	0.238^T	0.023	0.706
Chlorophyta	0.644	0.656^T	0.719	0.457	**<0.001**^{T}	**0.001**	0.340
Small flagellates	0.232^T	**0.043**	**0.009**	0.095	**0.020**	0.871	0.153

cyanophyte *Pseudanabaena limnetica* (Lemmermann) Komárek (0.5 mg l^{-1}). Other 43 algal taxa were present in minor quantities.

During the course of the experiment, a clear increase of cyanophytes, almost exclusively *P. limnetica*, was observed in all microcosms, most remarkably and significantly in microcosms without sediment, where the contribution of the other taxa, except small flagellates that were constant, decreased significantly (Fig. 1B, Table 1). In the presence of sediment, the contribution of dinoflagellates, euglenophytes, cryptophytes, xanthophyceans, and diatoms was significantly higher. Thus, in the microcosms with sediment, there were no dominant species or taxonomical group, so a more heterogeneous phytoplankton assemblage was observed over the experimental period (Fig. 1B). At the end of the experiment, species richness was higher in the microcosms with sediment (37 spp.) compared to the aquaria without sediment (28 spp.). Moreover, when sediment was not present, *G.* cf. *wawrikae* was substituted by *K. fungiforme* and *C. marsonii* disappeared. Other species substitutions occurred, for example pennate diatoms (*Nitzschia* spp.) alternated with the initial central species, but without a clear effect of any of the manipulated factors.

The effects of fish were less appreciable, but their presence increased the contribution of dinoflagellates and cryptophytes, and reduced the percentage of the small flagellates. The Sediment × Fish interaction had significant effects only in the dinoflagellate and the cryptophyte contribution.

Experiment 2

TP and TN concentrations remained higher in the normal than in the diluted treatments until the end of the experiment. In the normal nutrients treatment, concentrations were 0.23 ± 0.02 mg P/L and 10.12 ± 1.61 mg N/L, whereas in the diluted treatments 0.07 ± 0.001 mg P/L and 4.51 ± 0.91 mg N/L were measured. Neither TN pr TP were affected by zooplankton treatments (Ortega-Mayagoitia et al., 2002).

In the normal zooplankton treatment, total zooplankton biomass at the beginning of the experiment was 4.7 ± 0.3 mg l^{-1}, significantly higher than the 2.2 ± 0.09 mg l^{-1} estimated in the reduced zooplankton treatment. In the same way, all zooplankton taxa were significantly more abundant in the Z treatment (Ortega-Mayagoitia et al., 2002). These differences maintained along the experiment except in the case of ciliate and rotifer abundance, that became similar among treatments. In the non-manipulated zooplankton, the main component of biomass from Day 0 to Day 42 were copepods (51 ± 5%), with lower contributions of ciliates, nauplii and heterotrophic flagellates (16 ± 4%, 15 ± 4% and 15 ± 3%, respectively), and rotifers as the minor contributor (3 ± 1). In contrast, in the reduced zooplankton treatment copepods were observed in only one microcosm (6 ± 3%), so the contribution of heterotrophic flagellates, ciliates and rotifers was higher (40 ± 5%, 33 ± 4%, and 20 ± 4%, respectively).

The initial mean phytoplankton biomass was 7.6 ± 0.4 mg l^{-1}. The inoculum mostly consisted of the central diatoms *C. meneghiniana* (4.9 mg l^{-1}) and the brackish-water indicator *Chaetoceros muelleri* Lemmermann (1.2 mg l^{-1}). A minor quantity of the chlorophytes *Monoraphidium* cf. *minutum* (Nägeli) Komárková-Legnerová (0.4 mg l^{-1}), *Tetraselmis cordiformis* Carter (0.3 mg l^{-1}) and *M. komarkovae* (0.2 mg l^{-1}) was observed.

In the diluted nutrient treatment, the *Prymnesium* sp. (Haptophyta) contribution, which developed in the second part of the experiment, was higher and dominated the phytoplankton biomass at the end of the experiment (Fig. 2, Table 1). In contrast, in the normal nutrient treatment, chlorophytes maintained a higher contribution, and a development of cyanophytes (*P. limnetica*) and cryptophytes (*Chroomonas* sp.) was observed at the middle of the experiment.

In microcosms with large zooplankton (copepods), the average contribution of diatoms and cryptophytes was slightly higher, and the chlorophyte percentage slightly lower, than in the modified zooplankton treatment (Fig. 2).

Some species replacements were observed in all treatments: chlorophytes exhibited a change from the initial structure (*M.* cf. *minutum, M. komarkovae* and *T. cordiformis*) to a final state where *Stichococcus* sp. was dominant.

Discussion

Sediment and nutrient effects

These results clearly indicate that the biotic factors or trophic relationships cause weak effects, because the presence of planktivorous fish and the different composition of zooplankton seem to have minor importance in this phytoplankton assemblage and its dynamics. Moreover, in our results, only the treatments with sediment or non-diluted nutrients showed a phytoplankton assemblage trajectory very similar to that which was observed in the field during the same period (see Rojo et al., 2000a).

In the situation most similar to natural conditions, the initial composition was maintained, however, when sediment was removed an important loss of richness occurred. Direct effects of sediment on phytoplankton dynamics are difficult to evaluate properly (Beklioglu & Moss, 1996; Ortega-Mayagoitia et al., 2002) and have scarcely been studied. The sediment layer may be a reservoir of resistance stages of planktonic organisms (which can act like founder controller, Keddy, 2000) and, among others, an actively nutrient exchanger. Its reservoir role can be discarded in this case because species with resistance stages were not relevant in TDNP, or in our experiments. With respect to the nutrient exchange capacity of sediment, Jensen et al. (1994) suggest, for example, that small, fast growing species as chlorococcalean algae, can take advantage and outcompete species with lower growth rates, such as filamentous cyanophytes, in shallow lakes where sediments could be producing nutrient pulses. In this case, the controlling factor is not the nutrient concentration rather its frequency of variation, so variability of resource is like a new resource in the competitive system (Levins, 1979). Therefore, the complex relationship between sediment and plankton community allows the stable state of a multi-specie assemblage.

In our case, *P. limnetica* became the dominant species when sediment was not present. It is a filamentous cyanophyte common but not dominant compared to other filamentous cyanobacteria in eutrophic systems (Berger & Bij de Vaate, 1983), such as those occuring, for example, in the TDNP (Ortega-Mayagoitia & Rojo, 2000a) where *Planktothrix agardhii* (Gomont) Komárek & Anagnostidis was more important. One possible explanation for its explosive development could be that, in the absence of sediment, the growth of phytoplankton reduces available nutrients, giving a competitive advantage to the initially small *P. limnetica* population. This would be in accordance to the *P. limnetica* peaks occurring in the field when the other filamentous cyanobacteria disappear, coincidental to overall nutrient depletion in TDNP and elsewhere (Rojo & Álvarez-Cobelas, 1993; Ortega-Mayagoitia & Rojo, 2000a). Concluding, in this case two factors can be observed: the importance of sediment and the inocula or founder controlled assemblage.

When the nutrients were lower (nutrients were not added), species richness decreased and the flagellated *Prymnesium* sp. became dominant. *Prymnesium* is a widely distributed marine/brackish genus of potentially toxic species (Guo et al., 1996). Species like *Prymnesium parvum* Carter often occur in eutrophic, warm, standing environments like aquaculture ponds (Collins, 1978), though its ecophysiology is still poorly known (Johansson & Granéli, 1999). *P. parvum* is difficult to control once it becomes the dominant species, but this situation may be prevented by adding nutrients, as this species is a relatively slow-growing algae and does not compete well with other phytoplankton if nutrient pulses are produced (Guo et al.,

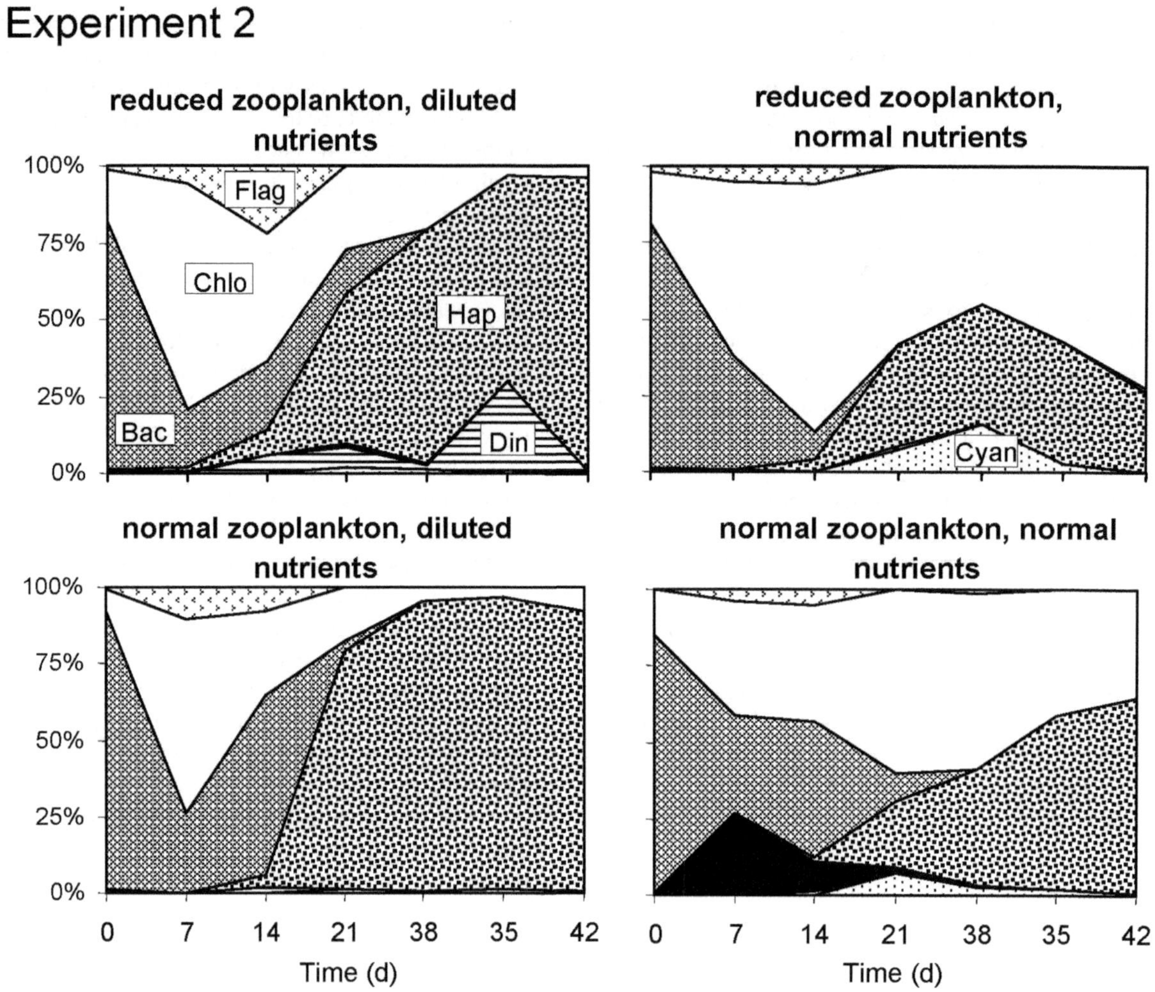

Figure 2. Examples of the time course of the relative biomass of phytoplankton taxonomic groups. Hap: Haptophyta; rest of legend as in Figure 1.

1996) which, in contrast, favour coccal chlorophytes (Sommer, 1985) as observed in our results. In this case, the variability of nutrient concentration (either pulsed or not) is more important than their concentration, in accordance to the Levins suggestion (1979). Consequently, a stable warm period in a eutrophic brackish area as TDNP, would favour this toxic species, becoming dominant and dangerous to fish. As an example: in a plankton culture from the wetland TDNP in spring, *Prymnesium* sp. became dominant in 2 weeks at 20 °C provoking a fish-kill episode (unpublished data). Fortunately, in the natural conditions of the wetland, nutrient pulses could be frequent.

Planktivory and grazing effects

Despite the strong changes in zooplankton structure and biomass attributable to *Gambusia,* eliminating cyclopoid copepods and increasing the abundance of small zooplankton as expected in hypertrophic ecosystems (reviewed by Jeppesen, 1998), the cascading effect on phytoplankton biomass was weak (Ortega-Mayagoitia et al., 2002). In contrast to the classic trophic cascade experiences, where a large daphnid was present (Carpenter et al., 1985; Komárková & Tavera, 2003). This may be a result of the copepod's omnivorous role compared to the highly efficient grazing exerted by *Daphnia* species. However, an interesting reflection on the trophic cascade effects on phytoplankton taxonomic structure can be made: the same taxonomic group, for example cryptophytes, can be affected positively (e.g. Beklioglu & Moss, 1996; Stephen et al., 1998) or negatively (Kurmayer & Wanzenböck, 1996) independently of whether the larger zooplankters were *Daphnia* or cyclopoid copepods (as in this case). The reason could be that zooplankton of very different body size can feed on algae from the same taxonomical group (Klaveness, 1988), which is also constituted by a wide range of algae size. In this way, changes in the relative abundance of populations (even extinction) are not noticeable in taxonomical groups. This is a possible explanation for the responses of populations, main taxonomical groups and biomass of phytoplankton to environmental changes at different time scales (Rojo & Álvarez-Cobelas, 2001). Then a sort of steady state assemblage can be observed in the taxonomical groups or functional groups at the same time that continuous substitutions of population are occurring.

In general, we conclude that influence of sediment on phytoplankton composition was of paramount importance. The intermediate perturbations due to the complex features of sediment hinder long stable periods. Consequently, the sequence of phytoplankton composition expected as a result of competence, when grazing factor is over, is not observed. Then, we suggest that alternate states of phytoplankton taxonomic structure in a eutrophic wetland occur as a sequence of substitutions, persistently altered by perturbations thanks to the close relationship with sediment; the autogenic trajectory is quite impossible, and the trophic relationships are possibly irrelevant. So, the degree of stochasticity of phytoplankton composition in wetlands is very high as a result of sediment interactions and the steady state episodes are related to it.

Acknowledgements

The authors would like to express their gratitude to M. Álvarez-Cobelas for his valuable comments on the manuscript, to P. Riolobos who kindly performed the nutrient analyses, to P. Sánchez-Castillo for his help with some species identification, to the *Tablas de Daimiel* National Park staff, and to the Spanish Ministry of Science and Technology to support this work with a grant (BOS2002-02003). D. Clifton-Sewell reviewed the English language of the manuscript.

References

Álvarez-Cobelas, M. & S. Cirujano (eds), 1996. Las Tablas de Daimiel, Ecología Acuática y Sociedad. Ministerio Español de Medio Ambiente, Madrid. 368 pp.

APHA, AWWA & WPCF, 1985. Standard Methods for the Examination of Water and Wastewater (16th ed.). American Public Health Association, Washington. 1268 pp.

Bachmann, R. W. & D. E. Canfield, 1996. Use of an alternative method for monitoring total nitrogen concentrations in Florida lakes. Hydrobiologia 323: 1–8.

Beklioglu, M. & B. Moss, 1996. Mesocosms experiments on the interaction of sediment influence, fish predation and aquatic plants with the structure of phytoplankton and zooplankton communities. Freshwat. Biol. 36: 315–325.

Berger, C. & A. Bij de Vaate, 1983. Limnological studies on the eutrophication of Lake Wolderwijd. A shallow hypertrophic *Oscillatoria* dominated lake in the Netherlands. Schweizerische Zeitschrift für Hydrol. 45: 458–479.

Boero, F., G. Belmonte, G. Fanelli, S. Piraino & F. Rubino, 2000. The continuity of living matter and the discontinuities of its constituents: do plankton and benthos really exist? Trends Ecol. Evol. 11: 117–180.

Borics, G., B. Tóthmérész, I. Grigorszky, J. Padisák, G. Várbíró & S. Szabó, 2003. Algal assemblage types of bog-lakes in Hungary and their relation to water chemistry, hydrological conditions and habitat diversity. Hydrobiologia 502 (Dev. Hydrobiol. 172): 145–155.

Carpenter, S., J. F. Kitchell & J. R. Hodgson, 1985. Cascading trophic interactions and lake productivity. BioScience 35: 634–639.

Chow-Fraser, P., 1998. A conceptual ecological model to aid restoration of Cootes Paradise Marsh, a degraded coastal wetland of Lake Ontario, Canada. Wetlands Ecol. Manage. 6: 43–57.

Collins, M., 1978. Algal toxins. Microbiol. Rev. 42: 725–746.

Cottingham, K. L. Nutrients and zooplankton as multiple stressors of phytoplankton communities: evidence from size structure. Limnol. Oceanogr. 44: 810–827.

Guo, M., P. J. Harrison & F. J. Taylor, 1996. Fish kills related to *Prymnesium parvum* Carter (Haptophyta) in the People's Republic of China. J. Appl. Phycol. 8: 111–117.

Jensen, J. P., E. Jeppesen, K. Olrik & P. Kristensen, 1994. Impact of nutrients and physical factors on the shift from cyanobacterial to chlorophyte dominance in shallow Danish lakes. Can. J. Fish. aquat. Sci. 51: 1692–1699.

Jeppesen, E., 1998. The ecology of shallow lakes. Trophic interactions in the pelagial. Doctor's dissertation (DSc). NERI Technical report No. 247. National Environmental Research Institute, Silkeborg. 358 pp.

Johansson, N. & E. Granéli, 1999. Influence of different nutrient conditions on cell density, chemical composition and toxicity of *Prymnesium parvum* (Haptophyta) in semi-continuous cultures. J. exp. mar. Biol. Ecol. 239: 243–258.

Klaveness, D., 1988. Ecology of the Cryptomonadina: a first review. In Sandgren C. D. (ed). Growth and Reproductive Strategies of Freshwater Phytoplankton. Cambridge University Press, Cambridge: 105–133.

Komárková, J. & R. Tavera, 2003. Steady state of phytoplankton assemblage in the tropical Lake Catemaco (Mexico). Hydrobiologia 502 (Dev. Hydrobiol. 172): 187–196.

Kurmayer, R. & J. Wanzenböck, 1996. Top-down effects of underyearling fish on a phytoplankton community. Freshwat. Biol. 36: 599–609.

Hansson, L.-R. , 1996. Algal recruitment from lake sediments in relation to grazing, sinking, and dominance patterns in the phytoplankton community. Limnol. Oceanogr. 41: 1312–1323.

Lampert W. & U. Sommer, 1997. Limnoecology: the ecology of lakes and streams. Oxford University Press, New York.

Levins, R., 1979. Coexistence in a variable environment. Am. Nat. 114: 765–783.

Lund, J. W. G., C. Kipling & E. D. Le Cren, 1958. The inverted microscope method of estimating algal numbers and the statistical basis of estimations by counting. Hydrobiologia 11: 143–170.

McCauley, E., 1984. The estimation of the abundance and biomass of zooplankton in samples. In Downing, J. A. & F. H. Rigler (eds), A Manual on Methods for the Assessment of Secondary Productivity in Fresh Waters. Blackwell Scientific Publications, Oxford: 228–265.

Marcus, N. H. & F. Boero, 1998. Minireview: the importance of benthic-pelagic coupling and the forgotten role of life cycles in coastal aquatic systems. Limnol. Oceanogr. 43: 763–768.

Olrik, K., 1994. Phytoplankton ecology. Miljøministeriet, København. 183 pp.

Ortega-Mayagoitia, E. & C. Rojo, 2000. Fitoplancton del Parque Nacional Las Tablas de Daimiel. II. Las cianofíceas, los dinoflagelados, las criptofíceas, las crisofíceas y las xantofíceas. Anales del Jardín Botánico de Madrid 57: 251–266.

Ortega-Mayagoitia, E. & C. Rojo, 2000. Fitoplancton del Parque Nacional de Las Tablas de Daimiel. III. Las diatomeas y las clorofíceas. Anales del Jardín Botánico de Madrid 58: 17–37.

Ortega-Mayagoitia, E., X. Armengol & C. Rojo, 2000. Zooplankton structure and dynamics in a semiarid wetland, the National Park "Las Tablas de Daimiel" (Spain). Wetlands 24: 629–638.

Ortega-Mayagoitia, E., M. A. Rodrigo & C. Rojo, 2002. Factors masking the trophic cascade in shallow eutrophic wetlands: evidence from a microcosm study. Archiv für Hydrobiol. 155: 43–63.

Padisák, J., G. Borics, G. Fehér, I. Grigorszky, I. Oldal, A. Schmidt & Z. Zámbóné-Doma, 2003. Dominant species, functional assemblages and frequency of equilibrium phases in late summer phytoplankton assemblages in Hungarian small shallow lakes. Hydrobiologia 502 (Dev. Hydrobiol. 172): 157–168.

Reynolds, C. S., 1984. The ecology of freshwater phytoplankton. Cambridge University Press, Cambridge. 384 pp.

Reynolds, C. S., 1988. Functional morphology and the adaptive strategies of freshwater phytoplankton. In Sandgren C. D. (ed.), Growth and Reproductive Strategies of Freshwater Phytoplankton. Cambridge University Press, Cambridge: 388–433.

Reynolds, C. S., 1993. Scales of disturbance and their role in plankton ecology. Hydrobiologia 249: 157–171.

Reynolds, C. S., 1997. Vegetation processes in the pelagic: a model for ecosystem theory. Ecology Institute, Oldendorf-Luhe. 371 pp.

Reynolds, C. S., J. Padisák & U. Sommer, 1993. Intermediate disturbance in the ecology of phytoplankton and the maintenance of species diversity: a synthesis. Hydrobiologia 249: 183–188.

Rojo, C. & M. Álvarez-Cobelas, 1993. Population dynamics of *Limnothrix redekei, Oscillatoria lanceaeformis, Planktothrix agardhii* and *Pseudanabaena limnetica* (Cyanobacteria) in a shallow hypertrophic lake (Spain). Hydrobiologia 275/276: 165–171.

Rojo, C. & M. Álvarez-Cobelas, 2001. Phytoplankton structure and dynamics at a daily temporal scale: response to the thermal overturn. Archiv für Hydrobiol. 151: 549–569.

Rojo, C., E. Ortega-Mayagoitia & V. Conforti, 1999. Fitoplancton del Parque Nacional de las Tablas de Daimiel. I. Las Euglenofitas. Anales del Jardín Botánico de Madrid 57: 15–23.

Rojo, C., E. Ortega-Mayagoitia, M. Álvarez-Cobelas & M. A. Rodrigo, 2000a. Phytoplankton structure and dynamics in a semiarid wetland, the National Park "Las Tablas de Daimiel" (Spain). Archiv für Hydrobiol. 148: 397–419.

Rojo, C., E. Ortega-Mayagoitia & M. Álvarez-Cobelas, 2000b. Lack of pattern among phytoplankton assemblages. Or, what does the exception to the rule mean? Hydrobiologia 424: 133–139.

Rojo, C. & M. Álvarez-Cobelas, 2003. Are there steady-state phytoplankton assemblages in the field? Hydrobiologia 502 (Dev. Hydrobiol. 172)): 3–12.

Rothhaupt, K. O., 1990. Resource competition of herbivorous zooplankton: a review of approaches and perspectives. Archiv für Hydrobiol. 118: 1–29.

Rott, E., 1981. Some results from phytoplankton counting intercalibrations. Schweizerische Zeitschrift für Hydrol. 43: 34–62.

Sánchez-Carrillo, S. & M. Álvarez-Cobelas, 2001. Nutrient dynamics and eutrophication patterns in a semi-arid wetland: the effects of fluctuating hydrology. Wat. Air Soil Pollut. 127: 12–27.

Stephen, D., B. Moss & G. Phillips, 1998. The relative importance of top-down and bottom-up control of phytoplankton in a shallow macrophyte-dominated lake. Freshwat. Biol. 39: 699–713.

Sommer, U., 1985. Comparison between steady state and non-steady state competition: experiments with natural phytoplankton. Limnol. Oceanogr. 30: 335–346.

Sommer, U., M. Z. Gliwicz, W. Lampert & A. Duncan, 1986. The PEG-model of seasonal succession of planktonic events in fresh waters. Archiv für Hydrobiol. 106: 433–471.

Underwood, A. J., 1997. Experiments in ecology: their logical design and interpretation using analysis of variance. Cambridge University Press, Cambridge. 504 pp.

von Ende, C. N., 1993. Repeated measures analysis: growth and other time-dependent measures. In Scheider, S. M. & J. Gurevitch (eds), Design and Analysis of Ecological Experiments. Chapman and Hall, New York: 113–137.

Hydrobiologia **502:** 187–196, 2003.
L. Naselli-Flores, J. Padisák, M. T. Dokulil (eds),
Phytoplankton and Equilibrium Concept: The Ecology of Steady-State Assemblages.

Steady state of phytoplankton assemblage in the tropical Lake Catemaco (Mexico)

Jaroslava Komárková[1] & Rosaluz Tavera[2]
[1]*Hydrobiological Institute AS CR and University of South Bohemia, Na sádkách 7, CZ- 379 01, České Budějovice E-mail: jarkakom@hbu.cas.cz*
[2]*Ecology of Algae, Ecology and Natural Resources Department, School of Sciences, UNAM, Apdo. Postal 70-620, C.U.Coyoacán 04510, México City, México*

Key words: tropical lake, shallow lake, phytoplankton, steady state, herbivorous fish, competition, *Cylindrospermopsis*

Abstract

Phytoplankton of the tropical lake Catemaco (Veracruz, Mexico) showed similar species composition during samplings from 1993 to 1995. There were two small dominant cyanobacterial species *Cylindrospermopsis catemaco* Kom.-Legn. et Tavera and *Cylindrospermopsis philippinensis* (Taylor) Kom., and a group of larger algae and cyanobacteria that were always present, however in smaller numbers: *Aulacoseira granulata* (Ehr.) Simons morphotype *curvata*, *A.* cf. *italica* (Ehr.) Simons mf. *curvata*, *Fragilaria construens* (Her.) Grun., *Achnanthes minutissima* Kütz., *Planktolyngbya circumcreta* (G.S.West) Anagn. et Kom., *Chroococcus microscopicus* Kom.-Legn. et Cronberg. Moreover we found several other scarcely present species. The percentage of total biomass of the two dominant species of *Cylindrospermopsis* varied between 34 and 81%, but they accounted for 80 to 95% of abundance. Apart from geomorphological features and climate conditions, biological variables played an important role. Fish-stock was formed by filter-feeding native herbivorous species of fish *Dorosoma petenense* (Günther), *Bramocharax caballeroi* (Contreras et Rivera), *Astyanax mexicanus* (Filippi), and an introduced, also herbivore *Oreochromis niloticus* (L.). Feeding activity of fish removed large species of algae and cyanobacteria as well as detrital remnants and zooplankton from the water. Smaller, inedible cyanobacteria remained in the water and formed the stable portion of the phytoplankton, dominant both in biomass and abundance. CANOCO analysis of samples and species variability demonstrated results of competition between two species of *Cylindrospermopsis*: steady state during the dominance of *C. catemaco* lasting probably for the whole year 1993 (one dry and one wet season) and steady state during the dominance of *C. philippinensis* in 1994 and 1995. According to the functional classification of phytoplankton suggested by Reynolds et al. (2002), Catemaco dominant assemblage would belong to the functional group $\mathbf{S_N}$.

Introduction

According to Reynolds et al. (2002) and Rojo & Álvarez-Cobelas (2003), "a steady-state assemblage means the invariance of its species composition over some time period". Thus, the dominant species should represent at least 50–80% of total abundance and the steady state period should last at least 2 weeks.

The definition of a steady state must also account for the stability of biomass concentration of the assemblage. Rojo & Álvarez-Cobelas (2003) suggested not insisting on its invariability in case the percent share of main dominant species remains the same. However a recommendation was accepted by attendants of the 13th IAP symposium (Naselli-Flores et al., 2003) to look for steady state situations maintained by competition between the potential dominants and not by biomanipulation or random events. The steady state is understood as the result of a dynamic equilibrium.

A phytoplankton steady state in temperate regions under completely natural conditions, i.e. without biomanipulation, is not easy to find. Disturbance of the system due to seasonal changes and weather variability is too great and variable. Shallow, polymictic lakes or ponds in summer are more promising biotopes (Nixdorf et al., 2003, Mischke & Nixdorf, 2003). Among deep water-bodies situated in temperate zones, lakes with a long retention time have a greater chance to have a steady state in warm, quiet summers (Salmaso, 2003), especially when the conditions are favourable for filamentous cyanobacteria (Reynolds, 1994). A permanent limitation by some of the important nutrients and a stable temperature course can reduce the assemblage to one or two competing dominants. Not only physical factors but also biotic events (spring outburst of calanoids, increasing abundance of filtering herbivorous *Daphnia* species etc.) influence exchange of the phytoplankton species and interrupt a potential steady state soon after its occurrence. Temperate zones are moderately stable in terms of physical and biological conditions in comparison with other geographical zones. Steady state conditions would be more probable and longer lasting in tropical (Ganf, 1974; Rott, 2002) and polar regions (Allende & Izaguirre, 2003) because of the smaller ranges in physical changes.

The geomorphology of water-bodies and their water discharge are other important arguments for the appearance of the steady state of phytoplankton. Shallow polymictic lakes are promising localities for the establishment of a phytoplankton steady state. Their everyday overturn (atelomixis, Barbosa & Padisák, 2002) provides a 'dynamic equilibrium' state for the phytoplankton by its regularity. Based on the preconditions for a steady state, shallow tropical lakes with long retention times and regular mixing due to wind appear to be the most convenient biotope for the emergence and susceptibility of the steady state (Ganf, 1974; Reynolds, 1989; Rott, 2002).

Considering the complex of biota that influence the phytoplankton of any of water bodies, top down control plays a more and more decisive role with increasing eutrophy. However, highly eutrophic and hypertrophic water bodies are exceptions to this point of view. The high bulk of inedible algal biomass in the water results in this potential pool of organic matter and nutrients starts to have a driving role for the water body (Reynolds, 1984). Especially when composed of cyanobacteria, such assemblages can remain for a long time in the lake even if they may not be photosynthetically active. In temperate zones such populations may disappear only with autumnal cold weather and mixing of the water column (Reynolds et al., 1991; Komárková & Hejzlar, 1998). Maintenance of such a steady state situation can be predicted from the previous behavior of the epilimnion, and should be carefully followed especially by water work authorities in reservoirs used for drinking water production. Production of toxic substances by the dominant cyanobacterial assemblage is very probable (Chorus & Bartram, 1999). In our opinion, a situation where the remnants of summer populations survive during the autumnal months cannot be considered as a steady state.

The aim of this paper is to describe the steady state conditions in a shallow tropical lake with small physical and chemical disturbances during the dry and rainy seasons and with a long retention time. This steady state is the result of a cycling food web and the existence of a special composition of phytoplankton due to competition between the two dominant cyanobacteria.

Lake Catemaco

The lake is situated between 18° 27′–18° 21′ N and 95° 01′–95° 07′ W in Los Tuxtlas region, State Veracruz, in an extensive volcanic area, at 332 m above sea level (Fig. 1). Climate is humid and hot (air temperatures between 12 and 38 °C), with a long lasting rainy season in the summer. The region is regarded as one of the most humid in Mexico (Fig. 2).

The primary vegetation is evergreen tropical forest (Miranda & Hernandez, 1963). However, a great portion of the land around the lake has been deforested recently and used for agriculture and livestock purposes. The small town of Catemaco (about 60 000 inhabitants, Tavera & Castillo, 2000) is situated on the west shore and more than about 10 000 people, mainly fishermen with families, live around the lake. The town releases waste water into the lake.

The lake is a rich source of fish-stock, composed of native Poecilidae and Characidae (*Dorosoma petenense, Bramocharax caballeroi, Astyanax mexicanus*), and an introduced cichlid *Oreochromis niloticus* (tilapia). An adequate portion of fishstock is netted every day by fishermen living in the surroundings. Catemaco lake is catalogued as one of the most productive in Mexico.

Water in lake has a permanently slightly blue-green color and transparency is low. The species composition of phytoplankton and appearance of specific spe-

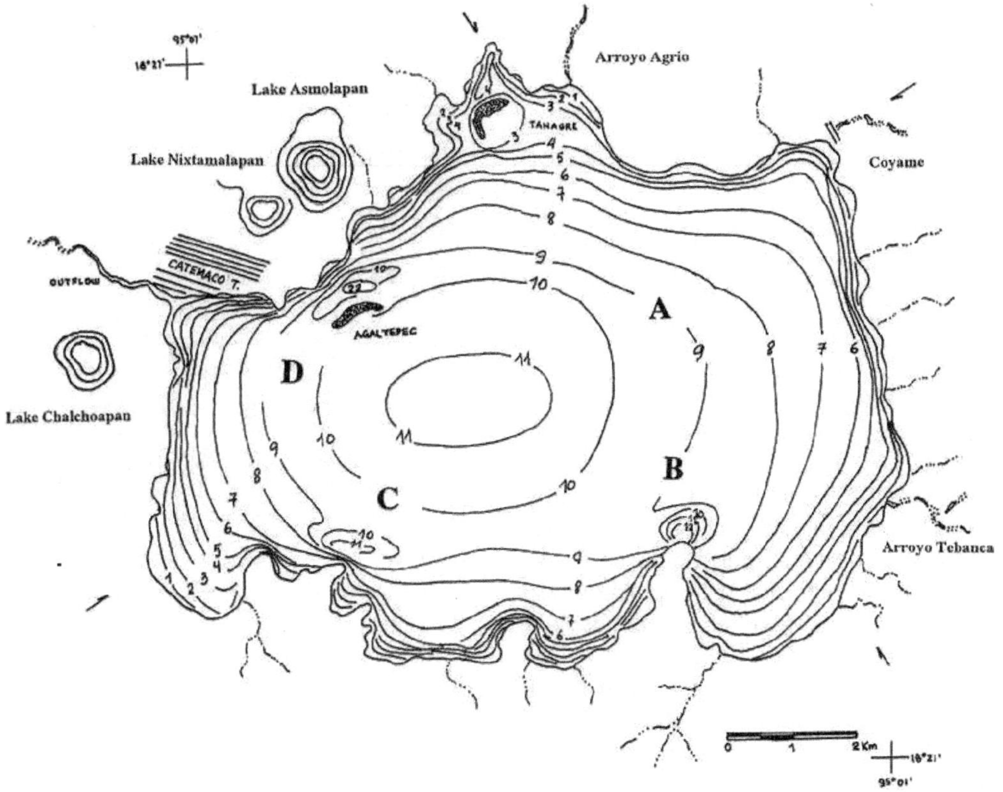

Figure 1. Lake Catemaco, Veracruz State, Mexico. Morphometry and bathymetry from Pérez-Rojas & Torres-Orozco (1992). Sampling sites are marked with capitals.

cies of Cyanobacteria were studied in detail in 1993 (Komárková & Tavera, 1996).

Morphometrically the bed is a rounded shallow bowl without deep bays (Fig. 1). The surface of the lake is 72.5 km^2 and the area of the watershed is 244 km^2. Maximum depth is 11 m and mean depth 7.6 m. Relative depth (z_R, Hutchinson, 1957) is 0.229%, so that there exist conditions for a permanent mixture of the water column. Average retention time is 0.9 year. The bottom is formed of clays and silt clays and covered with voluminous layer of organic matter. There is a transition to coarse sand with pebbles and gravel toward the littoral.

The region is rather windy, with a mean wind speed of 3–4.2 $m.s^{-1}$ (Atlas Nacional de México 1990). Maximal wind speed is reached either during the winter winds ('nortes') from November to February, or during cyclones in the Gulf of Mexico which can occur from July to September.

Methods

Sampling of phytoplankton, zooplankton and measuring of the main chemical and physical variables (pH, alkalinity, oxygen concentration, conductivity, temperature, transparency) were performed at four pelagic sites (A–D, see Fig. 1). We chose six sampling periods from the total sampling set (Table 1), representing dry and rainy periods in the three consecutive years: May and August 1993 (DRYMay3, RAIAug3), May and September 1994 (DRYMay4, RAISep4), and May and August 1995 (DRYMay5, RAIAug5). Each of them is a mean of samples taken at four sites in the lake (A–D, Fig. 1).

Samples for more detailed chemistry were taken only in May and August in 1993 and were analyzed in the Chemical Laboratory of the IMTA, Juitepec, Morelos. As the other analyses from 1995 were fewer and were processed in another laboratory, we decided to use only the data from 1993 to determine the ranges of nutrient concentrations in the lake.

Table 1. Frequency of phytoplankton sampling during three successional years. Rain season is marked in black

	J	F	M	A	M	J	J	A	S	O	N	D
1993		x	x	x	x	x		x				
1994	x				x				x	x	x	
1995	x	x	x	x	x	x		x				

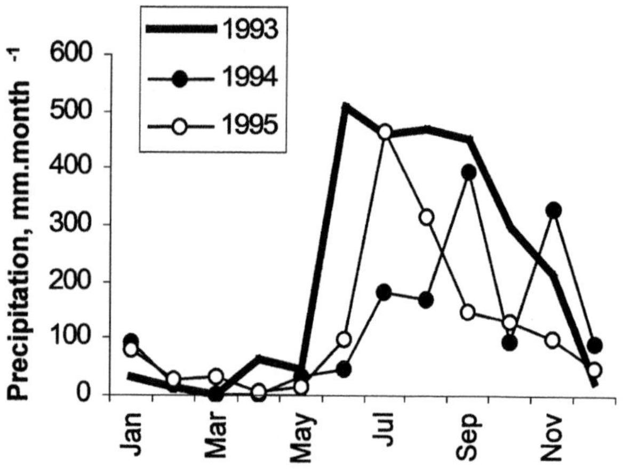

Figure 2. Yearly courses of precipitation in the years studied (Observatorio Nacional de México, Tavera & Castillo, 2000). Extremely heavy rains were in June and July 1993.

Samples from all four sites were taken at the surface, 3m, and the horizon above the bottom (mostly 7 or 9 m) using a Van Dorn sampler. Qualitative samples for zooplankton were taken by hauling an Epstein plankton net (70 and 200 μm mesh) from the bottom to the surface at each site. A chlorophyll concentration was estimated according to Lorenzen (1967) and corrected for pheopigments. Phytoplankton was estimated by counting the cells after sedimentation in Utermöhl chambers (Lund, 1951; Sournia, 1978). The abundance of each species is presented as the number of organisms (or units) per milliliter; filamentous and coenobial algae were considered as a unit. Cell volumes were calculated from measures of their shapes and applied to the relevant geometrical solids. Biovolume was converted to biomass and expressed as mg l^{-1} fresh mass (FM). The samples were preserved in Lugol solution. Other biological samples were preserved in 1.5–3% formalin. Analyses were conducted on the biomass data (fresh mass, – FM), as they are better comparable than the numbers of cells and coenobia or colonies (filament biovolumes of dominant *C. catemaco* and *C. philippinensis* differed 7 times, coenobia of *Pediastrum* can be several thousand times more voluminous than a filament of *Cylindrospermopsis*).

Phytoplankton composition was evaluated from two points of view using detrended correspondence analysis (DCA, Ter Braak & Šmilauer, 2000). Each sample and species biomass were taken separately and submitted to analysis for similarity between the assemblages.

Zooplankton samples taken using the 200 μm mesh Epstein plankton net were determined only informally to species.

Samples of adult fish specimens were obtained in August 1993 and October 1994 from fishermen from their catch in the morning at the shore of the lake. In October 1994 the fish were immediately killed and kept on ice before the examination, in August 1993 they were preserved by formalin. Their guts were examined for their contents.

Results

Concentrations of the main nutrients and elements in samples from the dry and wet periods in 1993 are given in Table 2. Concentrations of the forms of nitrogen, sulfates and silica were higher in the rainy season. Even if the water column during the dry season should have been more stable, the daily stratification was better developed in August. The discrepancy is a consequence of only one sampling in the period. However, we checked all of the available data in 1993 and found that the character of the chemical environment did not change substantially over the periods. Temperature and chlorophyll *a* levels were similar between the wet and dry seasons (Table 2).

Yearly courses of precipitation are shown in Figure 2. Rather short dry periods from January to May were followed by longer lasting warm rains that culminated in September. In 1993, high precipitation started already in June and continued until November. That year, the June to November precipitation was almost twice as high as the long-term average. Therefore, the chemical data in Table 2 represent a maximal concentrations of nutrients, as the nutrient load was much higher than the average values (Tavera & Castillo, 2000).

Species composition in units of FM is given in Figure 3a, b. Diatoms, Chlorophyceae, and the group of 'other cyanobacteria' were not abundant. In the

Table 2. Average data on the concentration of nutrients and other chemical and physical variables in May and August 1993. Each number is a mean for 4 sampling sites (A–D). Differences between dry and rainy seasons are marked in black. No difference was found in temperature

		May (dry season)			August (rainy season)		
		surface	3 m	9 m	surface	3 m	9 m
Alkalinity	meq l^{-1}	0.97	1.00	1.00	0.85	0.80	0.85
$P\text{-}PO_4$	$\mu g\, l^{-1}$	10	10	10	10	10	10
N_{org}	$\mu g\, l^{-1}$	77	93	93	140	290	380
$N\text{-}NO_3$	$\mu g\, l^{-1}$	35	30	31	50	45	52
$N\text{-}NH_4$	$\mu g\, l^{-1}$	154	154	158	140	142	140
$N_{tot.}$	$\mu g\, l^{-1}$	268	280	291	330	430	580
SO_4^{2-}	mg l^{-1}	2.6	2.3	2.2	4.3	4.1	5.6
COD	mg l^{-1}	20.4	51.2	30.7	45.0	40.5	63.0
Mg^{2+}	mg l^{-1}	6.9	6.9	6.9	7.2	6.7	7.4
K^{+}	mg l^{-1}	2.3	2.3	2.4	2.6	2.9	2.9
Na^{+}	mg l^{-1}	15	14	18	20	21	20
Si^{3+}	mg l^{-1}	1.0	1.1	1.1	1.3	1.2	1.7
Ca^{2+}	mg l^{-1}	6.2	6.4	6.4	6.6	6.3	6.9
pH		8.9	8.6	7.8	8.6	8.9	8.8
Temperat.	°C	28.9	27.3	27.0	28.0	28.0	28.0
Transp.	m	0.6			0.55		
Chlor. *a*	$\mu g\, l^{-1}$	46.7	48.3	43.0	69.0	51.2	29.7

first sample, *Cylindrospermopsis catemaco* was dominant. The same composition was found roughly in other samples from this period (Tavera, unpublished results). The second sample from the rainy season of 1993 still maintained this pattern, even if the numbers of *C. philippinensis* were increasing. The dominance of *C. philippinensis* increased beginning with the dry period of 1994, and remained so up to the last sampling in summer 1995. The same pattern is seen when given as percentage of total fresh mass (Fig. 3b). Tavera (1996) studied the number of heterocytes in the filaments of both *Cylindrospermopsis* species. She found much more heterocytes in the population of *C. philippinensis* than in *C. catemaco*. *C. philippinensis* responded successfully to low nitrogen concentration, which decreased greatly due to low precipitation levels in the following quiet and drier years of 1994 and 1995 (NO_3-N dropped from 30–52 to 5.4 $\mu g\ l^{-1}$, NH_4-N from 140–158 to 14 $\mu g\ l^{-1}$, $P\text{-}PO_4$ remained almost unchanged, Tavera, 1996). This pattern of phytoplankton composition lasted for the following two years, with small variations in the occurrence of additional species.

Species biomass data were also analyzed by DCA (Fig. 4). Together with analysis of the species, individual samples were plotted. Even if the number of the samples is very low for this type of analysis, the distribution corroborated our results. The samples were distributed into two groups, situated in the same place as the dominant and attendant species. While *C. catemaco* was related to the 1993 samples, the other four samples were joined to *C. philippinensis* (see white arrows and the indication).

It is interesting that there was no difference between the samples from the rainy and dry periods. The main reason for the shift of dominance and the appearance of new species in the lake was evidently the change in the summer precipitation levels between 1993 and 1994, as mentioned above. Heavy flooding rains in 1993 enabled *C. catemaco* to dominate (vegetation cells have no aerotopes) along with the appearance of heavy planktonic diatoms (*Aulacoseira granulata, A. italica)* and coenobia of *Pediastrum boryanum* (Turp.) Menegh., *P. simplex* Meyen, *P. duplex* Meyen, and *P. simplex* var. *biwaense* Fukush. The rains helped in mixing the whole water column and probably lifted up a lot of sedimented nutrients. Runoff from the surrounding agricultural lands also introduced high concentrations of nutrients to the lake. In the next two years, precipitation returned to mean

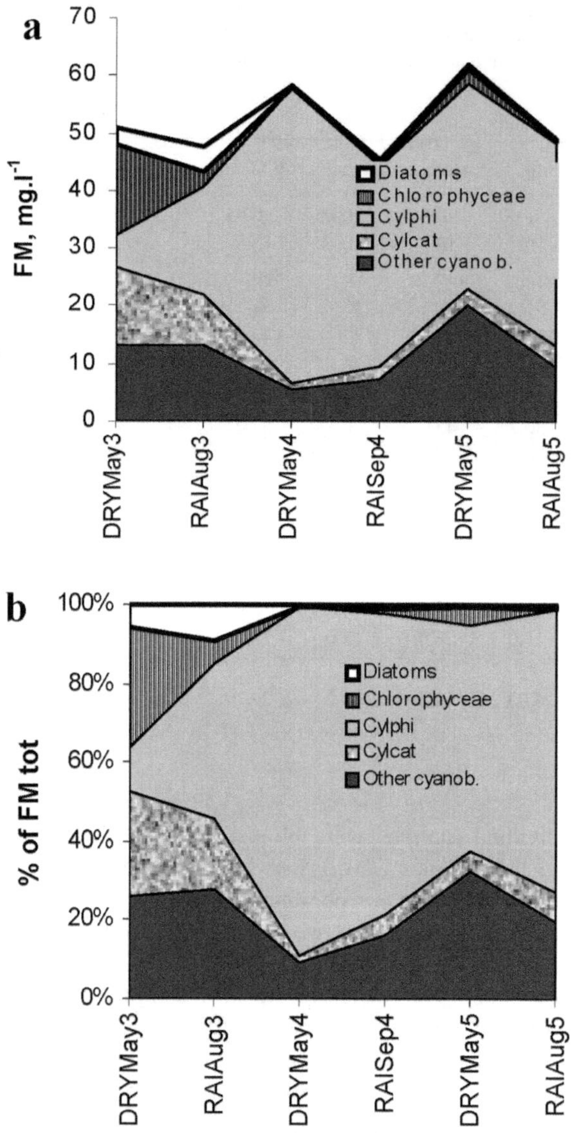

Figure 3. Fresh mass (FM, biovolume, 3a) and share of individual dominants on the total FM (3b) during six samplings in dry and rainy seasons in three subsequent years.

values and the phytoplankton used up most of the nutrient load in primary production.

Stomach and gut contents of different herbivorous fish were studied in summer 1993 and 1994 from the fish yield of the Catemaco fishermen. A high percentage of the contents consisted of the rests of large diatoms and green algae, while small sized *Cylindrospermopsis* filaments were relatively few. There was a large difference between the composition of the phytoplankton assemblage and the gut content of the native and dominant clupeid *Dorosoma petenense* (Fig. 5a,

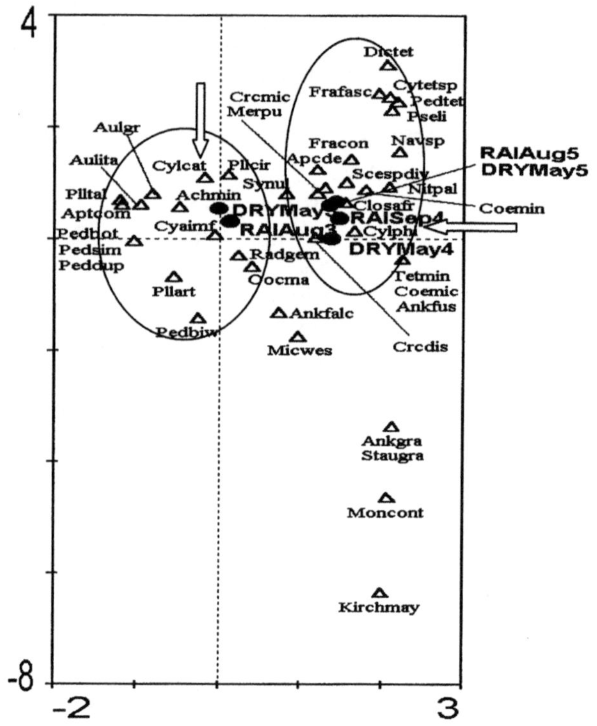

Figure 4. Detrended correspondence analysis (DCA) of FM of individual species and their distribution over first and second DCA axes. Comparison is done for whole samples taken in 6 seasons in three subsequent years. Abbreviations: Achmin – *Achnanthes minutissima* Kütz.; Ankfalc – *Ankistrodesmus falcatus* (Corda) Ralfs; Ankfus – *A.fusiformis* Corda; Ankgra – *A. gracilis* (Reinsch) Korš; Apcde – *Aphanocapsa delicatissima* W. et G.S. West; Aptcom – *Aphanothece comasii* Kom.-Legn. et Tavera; Aulgr – *Aulacoseira granulata* (Ehr.) Simonsen mf. *curvata*; Aulita – *Aulacoseira italica* (Ehr.) Simonsen; Closafr – *Closteriopsis acicularis* (G.M. Smith) Belcher et Swale var. *africanum*; Coemic – *Coelastrum microporum* Näg. in A.Br.; Coemin – *Coelomoron minimum* (Bernard) Kom.-Legn. et Tavera; Crcdis – *Chroococcus distans* (G.M. Smith) Kom.-Legn. et Cronb.; Crcmic – *Ch. microscopicus* Kom.-Legn. et Cronb.; Cyaimf – *Cyanodictyon imperfectum* Cronb. et Weibull; Cylcat – *Cylindrospermopsis catemaco* Kom.-Legn. et Tavera; Cylphi – *C. philippinensis* (Taylor) Kom.; Cytetsp – *Cyanotetras* sp.; Dictet – *Dictyosphaerium tetrachotomum* Printz; Fracon – *Fragilaria construens* var. *construens* (Ehr.) Grunow; Frafasc – *F. fasciculata* (Agardh) Lange-Berth.; Kirchmay – *Kirchneriella mayori* (G.S.West) Kom.-Legn.; Merpu – *Merismopedia punctata* Meyen; Micwes – *Microcystis wesenbergii* (Kom.) Kom. in Kondr.; Moncont – *Monoraphidium contortum* (Thur.) Kom.-Legn.; Navsp – *Navicula* sp.; Nitpal – *Nitzschia palea* (Kütz.) W.Smith; Oocma – *Oocystis marssonii* Lemm.; Pedbiw – *Pediastrum simplex* var. *biwaense* Fukush.; Pedbor – *P. boryanum* (Turp.) Menegh.; Peddup – *P. duplex* Meyen; Pedsim – *P. simplex* Meyen; Pedtet – *P. tetras* (Ehrenb.) Ralfs; Pllart – *Planktolyngbya arthrospiroides* Kom.-Legn. et Tavera; Pllcir – *P. circumcreta* (G.S.West) Anagn.et Kom.; Plltal – *P. tallingii* Kom. et Kling; Pseli – *Pseudanabaena limnetica* (Lemm.) Kom.; Radgem –*Radiocystis geminata* Skuja; Scespdiv – *Scenedesmus* sp.div.; Staugra – *Staurastrum gracile* Ralfs; Synul – *Synedra ulna* (Nitzsch) Ehr.; Tetmin – *Tetraedron minimum* (A.Br.) Hansg.

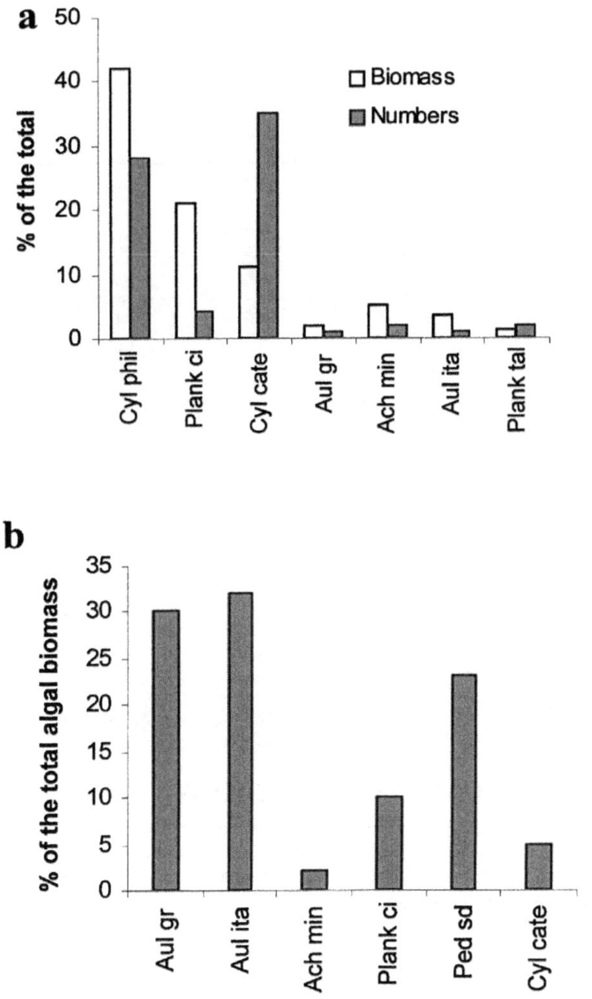

Figure 5. Percentual composition of the planktonic algae and cyanobacteria in the Catemaco lake in August 1993 (5a) and corresponding shares in the guts of 12 cm long specimens of *Dorosoma petenense* (5b).

b). While the gut content consisted mostly of debris from algae larger than 40–50 μm, *Cylindrospermopsis* occupied the largest percentage in the phytoplankton.

Discussion

A simplified food web model of Lake Catemaco is presented in Figure 6. The main predators (white shad, *Bramocharax caballeroi* and tilapia, *Oreochromis niloticus*) of the phytoplankton were mostly dependent on the large fraction of the phytoplankton (>20 μm). Large zooplankton such as *Daphnia* sp., *Macrocyclops albidus* (Jurine), *Mastigodiaptomus albuquerquensis* (Herrick), and *Diaphanosoma* sp. were not taken into account, as they have much lower concentrations compared with the fishstock (Tavera & Castillo, 2000). However, in case large daphnias are present, they filter off their portion of the phytoplankton.

Zooplankton and all larger particles of organic detritus and plant debris were greatly consumed by herbivorous fish (Drenner et al., 1982). Drenner et al. (1982) studied prey escape concerning the gizzard shad in experiments. They found that large daphnias had a small chance to sustain their populations under high predation pressure of fish, while *Diaphanosoma* and *Diaptomus* were better able to avoid swallowing. Larvae and juvenile fish were probably dependent on zooplankton very much, as was found for closely related gizzard shad from North America (*Dorosoma cepedianum*, Pierce et al., 1981; Dettmers & Stein, 1992). Small zooplankton as *Bosmina longirostris* (O.F.Müller), *Brachionus havanaensis* (Rousselet), and *Keratella americana* (Carlin) were more frequent. These can consume small particles and bacteria and can remove partly also *Cylindrospermopsis* and small algae. Except for *Diaphanosoma*, all of the species were found in Lake Catemaco already in 1985 by Suárez et al. (1986). They also found a 90% dominance of cyanobacteria.

Drenner et al. (1984) measured the distances between the rake gills of the related species *Dorosoma cepedianum*. They found distances between 1 and 85 μm in five studied specimens of the same size as was a majority of *Dorosoma petenense* specimens from Lake Catemaco (body length 136–163 mm). Also other species of fish present in the lake were consumers of phytoplankton: *Bramocharax caballeroi*, *Astyanax mexicanus* and the introduced species *Oreochromis niloticus* (tilapia). The steady state based on competition between the dominant species (see Fig. 6 right bottom) concerns here the assemblage already modified by long lasting and permanent predation pressure of the fish, or fish and zooplankton. Even if many other algae of such small size existed in the lake, none of them was able to grow under such poor nitrogen concentration and low light conditions. In comparison with cyanobacteria, green algae were well digested.

Bodola (1966) observed that *Dorosoma cepedianum* consumed large amounts of *Microcystis aeruginosa* colonies and filaments of *Anabaena spiroides*; however, the cells remained viable after passing through the guts. The vitality of blue-greens and algae isolated from the fish guts can be tested also by

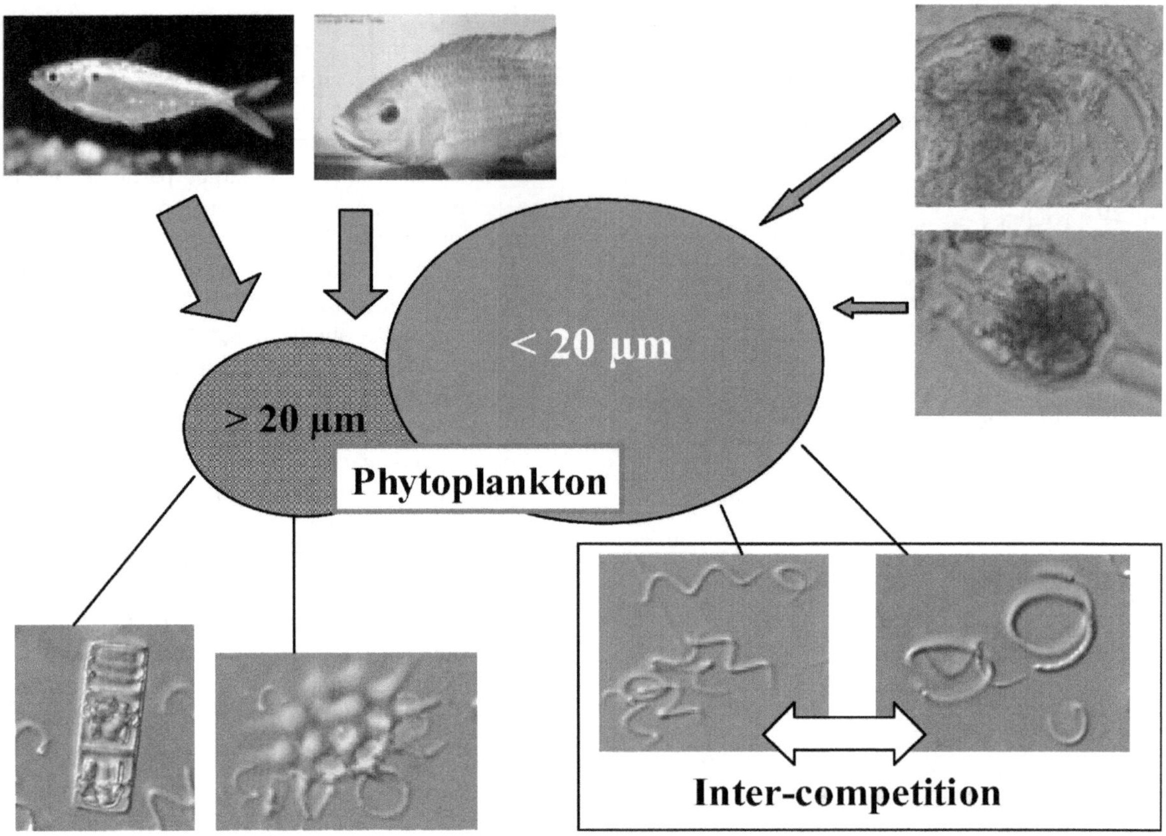

Figure 6. Simplified food net in the lake Catemaco. Dark arrows: predation on the phytoplankton. Large (>20 μm) and small (<20 μm) phytoplankton: while small phytoplankton is supported by predation activity of herbivorous fish, large particles are removed. Predation pressure on small phytoplankton is negligible. Inter-competition can exist between the inedible species, *Cylindrospermopsis catemaco* and *C. philippinensis* (see white double-arrow to the right bottom).

cultivation (Dokulil, 1983, on *Sarotherodon mosambicus*). However, adult specimens of tilapia have a pH of about 1.4 in their digestive organs (Moriarty, 1973; Payne, 1978; Northcott et al., 1991; Tavera, 1996), so cyanobacteria could be digested by tilapia.

The lake probably has enough nutrients for growth of diatoms only in rainy and windy summers, when the concentration of nutrients increases and heavy filaments are able to counter sedimentation (Tavera & Castillo, 2000). Everyday fishing of herbivorous fish on the lake provides a regulation of the fish-stock which keeps the food web functioning. Small sized cyanobacteria were probably dominant in the lake already for many years (analysis of the sample collected by E. Rott in 1970 detected dominance of *C. catemaco* and *C. philippinensis*). Such a long lasting steady state of the phytoplankton composition is mostly conditioned by a sort of biomanipulation, as other potential competitors (large buoyant colonies, flagellates, filamentous buoyant species) are removed mechanically from the environment (equivalent of pasture). Drenner et al. (1984) found a similar modification of the phytoplankton assemblage due to the presence of herbivorous fish as we did. Larger species (there *Ceratium*) were suppressed while the smallest algae and bacteria were enhanced. A convenient niche for the development of cyanobacteria was supported by high and stable temperature and low transparency, as the cyanobacteria irradiance demand is low (Oliver & Ganf, 2000). Also the Redfield ratio (Reynolds, 1984) was appropriate for nostocalean cyanobacteria which can supply the missing concentration of N by nitrogen fixation activity of heterocytes.

Tavera & Castillo (2000) explained the further eutrophication of the lake by an increase in the percentage of total cyanobacteria forming the phytoplankton (from 64–83% in 1993 to 94–99% in 1994 and 1995). However, phytoplankton biomass did not increase substantially. The important change was diminution of the portion of large phytoplankton, which was probably a consequence of increasing predation pressure by herbivorous fish and increased sedimentation processes during the quiet 1994 and 1995 years.

Stable physical conditions and certain morphometrical features of the lake are necessary characteristics of the lakes occupied by a phytoplankton assemblage enrolled to the S_N group. Such lakes should be warm (tropical) and shallow enough as to be polymictic, mostly mixed every day. Tavera (1996) calculated the relative depth (Z_R) that indicated the instability of the water column (Hutchinson, 1957). The value for Catemaco is very low (0.228), which indicated a high instability and daily mixing. Diurnal measurement in the lake during one day proved also the establishment of short lasting stratifications of oxygen concentration and temperature; however, pH values remained unchanged. Climatic conditions in the region ensured small seasonal variations in water temperature (22–26 °C). The retention time of the lake was computed to be about one year (except for the extremely rainy year 1993), which is also one of the preconditions for development of a stable equilibrium system.

Maximum biomass of phytoplankton in a shallow polymictic eutrophic lake (4–5 m deep) with available nutrients was assumed to be about 200 μg l^{-1} chlorophyll *a* (Reynolds, 1984). Tavera (1996) recorded concentrations between 17 and 60 μg l^{-1} of chlorophyll *a*, which also proved that the phytoplankton grew under permanent limitation of nutrients. The main factors for maintenance of a stable steady state, including permanent limitation by physical factors and nutrients (Reynolds, 1997), were therefore present in Catemaco Lake during all of the study.

Conclusions

Phytoplankton biomass in Lake Catemaco consisted from 36 to 81% of two very small cyanobacterial species, *Cylindrospermopsis catemaco* and *C. philippinensis* during all three years studied.

Algae larger than 20 μm (in GALD) and zooplankton were scarce but always present. They were mainly filtered by phytophagous species of fish: *Dorosoma petenense, Bramocharax caballeroi, Astyanax mexicanus* and *Oreochromis niloticus.*

Everyday fishing forced the fish-stock to rejuvenate, even if the food supply for fish (large particles) was kept low.

Strong predation pressure by filter-feeding herbivorous fish, permanent mixing and low transparency and limitation by lack of nitrogen manifested in small differences in biomass of total phytoplankton during the yearly cycles and in permanent dominance of two species of the genus *Cylindrospermopsis.* Exchange of the two dominant species, *Cylindrospermopsis catemaco* Kom.-Legn. et Tavera and *C. philippinensis* (Taylor) Kom., took part after extremely high precipitation in the rainy period of 1993. Before and after the change, a probably long lasting steady state of the phytoplankton assemblage existed in the lake as a result of inter-species competition.

The eutrophic character of the lake, its shallowness and permanent mixing of the water column well fit the characteristic dominance of *Cylindrospermopsis* belonging to the functional group $\mathbf{S_N}$ (Reynolds et al., 2002).

Acknowledgements

We are grateful to Facultad de Ciencias and DGIA, Universidad Nacional Autónoma de México, for possibility to realize the investigation in the Catemaco Lake in collaboration of Czech and Mexican scientists. We thank Prof. J. Komárek and all Mexican colleagues for help during the field sampling. For identification of zooplankton species we are indebted to Dr Seda and Mgr. Devetter from Hydrobiological Institute AS CR. Elaboration of this paper was supported by the project GA CR No. K6005114 (Biodiversity and functioning of ecological systems) and project MSMT No. 123100004 (Structure and functioning of the system water-soil-plant).

References

Allende, G. & I. Izaguirre, 2003. The role of physical stability on the selection of steady state assemblages in two maritime Antarctic lakes. Hydrobiologia 502 (Dev. Hydrobiol. 172): 211–224.

Atlas Nacional de México, 1990. Instituto de Geografía, UNAM. Vol. II. Naturaleza.

Barbosa, A. R. & J. Padisák, 2002. The forgotten lake stratification pattern: atelomixis, and its ecological importance. Int. Ver. theor. angew. Limnol. Verh. 28: 1385–1395.

Bodola, A., 1966. Life history of gizzard shad, *Dorosoma cepedianum* (Le Sueur), in Western Lake Erie. Fish. Bull. (US), 65: 391–425.

Chorus, I. & J. Bartram (eds), 1999. Toxic Cyanobacteria in Water. A guide to their Public Health Consequences, Monitoring and Management. E. & F. N. Spon, London & N. York. 416 pp.

Dettmers J. M. & R. A. Stein, 1992. Food consumption by larval gizzard shad: zooplankton effects and implications for reservoir communities. Trans. am. Fish. Soc. 121: 494–507.

Dokulil, M., 1983. Aspects of gut passage of algal cells in *Sarotherodon mosambicus* (Peters), (Pisces, Cichlidae). In Schiemer, M. (ed), Limnology of Parakrama Saamudra: 165–170.

Drenner, R.W., F. de Noyelles & D. Kettle, 1982. Selective impact of filter-feeding gizzard shad on zooplankton community structure. Limnol. Oceanogr. 27: 965–968.

Drenner, R. W., J. R. Mummert, F. de Noyelles Jr. & D. Kettle, 1984. Selective particle digestion by a filter-feeding fish and its impact on phytoplankton community structure. Limnol. Oceanogr. 29: 941–948.

Ganf, G. G., 1974. Diurnal mixing and the vertical distribution of phytoplankton in a shallow equatorial lake (Lake George, Uganda). J. Ecol. 62: 611–629.

Hutchinson, G. E., 1957, A treatise on Limnology I. Geography, Physics and Chemistry. John Willey & sons, N.Y. 1015 pp.

Komárková, J., 1998. Fish stock as a variable modifying trophic pattern of phytoplankton. Hydrobiologia 367/370: 139–152.

Komárková, J. & J. Hejzlar, 1996. Summer maxima of phytoplankton in the Římov reservoir in relations to hydrologic parameters and phosphorus loading. Archiv für Hydrobiol. 136: 217–236.

Komárková, J. & R. Tavera, 1996. Cyanoprokaryota (Cyanobacteria) in the phytoplankton of the lake Catemaco (Veracruz, Mexico). Archiv für Hydrobiol./Algological Studies 83: 403–422.

Lorenzen, C. J., 1967. Determination of chlorophyll *a* and pheopigments: spectrofotometric equations. Limnol. Oceanogr. 12: 343–346.

Lund, J. W. G., 1951. A sedimentation technique for counting algae and other organisms. Hydrobiologia 3: 390–394.

Miranda, F. & E. Hernandez, 1963. Los tipos de vegetación de México y su classificación. Boletín de la Sociedad Botanica de México 28: 29–179.

Mischke, U. & B. Nixdorf, 2003. Equilibrium phase conditions in shallow German Lakes: How Cyanoprokaryota species establish a steady state phase in late summer. Hydrobiologia 502 (Dev. Hydrobiol. 172): 123–132.

Moriarty, D. J., 1973. The physiology of digestion of blue-green algae in the cichlid fish, *Tilapia nilotica*. J. Zool. 171: 25–39.

Naselli-Flores, L., J. Padisák, M. T. Dokulil & I. Chorus, 2003. Equilibrium/steady-state concept in phytoplankton ecology. Hydrobiologia 502 (Dev. Hydrobiol. 172): 395–403.

Nixdorf, B., U. Mischke & J. Rücker, 2003. Phytoplankton assemblages and steady state in deep and shallow eutrophic lakes – an approach to differentiate the habitat properties of Oscillatoriales. Hydrobiologia 502 (Dev. Hydrobiol. 172): 111–121.

Northcott, M. A., M. C. Beveridge & L. G. Ross, 1991. A laboratory investigation of the filtration and ingestion rates of the tilapia, *Oreochromis niloticus*. Environ. Biol. Fish. 31: 75–85.

Oliver, R. L. & G. G. Ganf, 2000. Freshwater bloom. In Whitton B. A. & M. Potts (eds), The Ecology of Cyanobacteria. Their Diversity in Time and Space. Kluwer Academic Publishers, Dodrecht, London, Boston: 149–194.

Payne, A. I., 1978. Gut pH and digestive strategies in estuarine pray mullet (Mugilidae) and tilapia (Cichlidae). J. Fish Biol. 13: 627–629.

Pérez-Rojas, A., R. Torres-Orozco & A. Z. Márquez-García, 1993. Los sedimentos recientes del lago di Catemaco, Veracruz, México. Anales del Instituto de Ciencias del Mar y Limnologia, UNAM, México 20: 67–76.

Pierce, R. J., T. E. Wissing & B. A. Megrey, 1981. Aspects of the feeding ecology of gizzard shad in Acton lake, Ohio. Trans. am. Fish. Soc. 110: 391–395.

Reynolds, C. S., 1984, The ecology of freshwater phytoplankton. Cambridge University Press, Cambridge, London, New York, New Rochelle, Melbourne, Sydney. 384 pp.

Reynolds, C. S., 1989. Physical determinants of phytoplankton succession. In Sommer U. (ed.), Plankton Ecology. Springer Verlag, N.Y.: 9–56.

Reynolds, C. S., 1994. The long, the short and the stalled. On the attributes of phytoplankton selected by physical mixing in lakes and rivers. In Descy, J.-P., C. S. Reynolds & J. Padisák (eds), Phytoplankton in Turbid Environments: Rivers and Shallow Lakes. Hydrobiologia 289: 9–21.

Reynolds, C. S., 1997. Vegetation processes in the Pelagic: a Model for Ecosystem Theories. – Ecology Institute, Oldenburg/Luhe. 371 pp.

Reynolds, C. S., G. H. Jaworski, H. A. Cmiech & F. G. Leedale, 1991. On the annual cycle of the blue-green alga *Microcystis aeruginosa* Kütz. emend. Elenkin. Phil. Trans. r. Soc. Lond. B 293: 419–477.

Reynolds, C. S., V. L. Huszar, C. Kruk, L. Naselli-Flores & S. Melo, 2002. Towards a functional classification of the freshwater phytoplankton. J. Plankton Res. 24: 417–428.

Rojo, C. & M. Álvarez-Cobelas, 2003. Are there steady-state phytoplankton assemblages in the field? Hydrobiologia 502 (Dev. Hydrobiol. 172): 3–12.

Rott, E., 2002. Is the persistence of the *Cylindrospermopsis raciborskii* and *Fragilaria acus* dominated phytoplankton in a shallow tropical reservoir indicating equilibrium conditions? In Naselli-Flores, L. (ed.), Abstracts of the 13th Workshop IAP, Castelbuono, Italy, 31 August–8 September, 2002. p. 49.

Salmaso, N., 2003. Interactions of physical, chemical and biotic factors affecting the turnover rate of phytoplankton assemblages in large and deep lakes. Hydrobiologia 502 (Dev. Hydrobiol. 172): 13–36.

Sournia, A. (ed.), 1978, Phytoplankton Manual. UNESCO, Paris. 337 pp.

Suárez, E., L. Segura & M. A. Fernandez, 1986. Diversidad y abundancia del Plankton en la laguna de Catemaco, Veracruz, durante el ciclo annual. Anales del Instituto de Ciencias del Mar y Limnologia, UNAM, México 13: 313–316.

Tavera, R. L., 1996. Phytoplankton of the tropical Lake Catemaco. Ph.D. Thesis, University of South Bohemia, Faculty of Biological Sciences, České Budějovice, Czech Republic. 62 pp.

Tavera, R. & S. Castillo, 2000. An eutrophication-induced shift in the composition, frequency and abundance of the phytoplankton in Lake Catemaco, Veracruz, Mexico. In Munawar, M., S. G. Lawrence, I. F. Munawar & D. F. Malley, (eds), Aquatic Ecosystems in Mexico: Status and Scope, Ecovision World Monograph Series. Backhuys Publ., Leiden, The Netherlands: 103–117.

Ter Braak, C. J. F. & P. Šmilauer, 2000. CANOCO Reference Manual and User's Guide to Canoco for Windows, Software for Canonical Community Ordination (version 4). Centre for Biometry, Wageningen, Microcomputer Power (Ithaca, NY, USA). 352 pp.

Hydrobiologia **502:** 197–209, 2003.
L. Naselli-Flores, J. Padisák, M. T. Dokulil (eds),
Phytoplankton and Equilibrium Concept: The Ecology of Steady-State Assemblages.

Do steady state assemblages occur in shallow lentic environments from wetlands?

Inés O'Farrell, Rodrigo Sinistro, Irina Izaguirre & Fernando Unrein
Laboratorio de Limnología, Departamento de Ecología, Genética y Evolución, Facultad de Ciencias Exactas y Naturales, Universidad de Buenos Aires, C1428EHA, Buenos Aires, Argentina, Conicet
E-mail: ines@bg.fcen.uba.ar

Key words: wetland, algal associations, steady state, disturbance, macrophyte, South America

Abstract

This study was conducted in a wetland on the Lower Paraná River (Argentina). Algal assemblages from a permanent shallow lake, three relict oxbow lakes (ROL) and an isolated small pond were sampled fortnightly during the spring-summer time of 1998–99. The algal assemblages strongly differed among the permanently vegetated ROLs and the shallow lake and small pond, both temporarily covered by floating macrophytes. Anoxic conditions and poor light penetration have been pointed out in previous studies to be the leading factors in the selection of algal species in this wetland. Two or three Cyanobacteria species and some large pennate diatoms dominated the algal biomass in the ROLs, whereas only the former taxa contributed significantly to density. Assemblages were characterised by their low diversity and species richness; the highest SIMI values between consecutive samplings were registered in the ROLs. Even though some characteristics of the **Z** and $\mathbf{S_2}$ Reynolds associations describe the ROLs assemblages, they do not strictly match with them. Changes in algal density and biomass were more pronounced in the shallow lake. An alternative dominance of *Cryptomonas* spp., *Peridinium* spp., *Euglena* spp., *Monoraphidium* spp., Oscillatoriales and Chroococcales occurred, with a consequent replacement of the **Y**, $\mathbf{L_o}$, $\mathbf{W_1}$ $\mathbf{X_1}$,$\mathbf{S_2}$, **K**, and **F** Reynolds associations. Variation in the macrophytic cover of the small pond accounted for the fluctuations in its algal composition. When the cover was complete, the assemblages resembled those of the ROLs, whereas when its surface was free of plants, they were more similar to those of the shallow lake. Steady state assemblages related to the persistence of a dense macrophyte cover may be found in wetlands in environments with constant and extreme ecological conditions such as those recorded in the ROLs. In contrast, 'unpredictable' assemblages can be expected in shallow lakes with fluctuating conditions.

Introduction

Holling (1973) characterised stability as persistence of a system near or close to an equilibrium state. According to equilibrium theory, the observed diversity of living organisms has arisen as a result of both niche specialization and diversification resulting from current or past competition, respectively. An alternative explanation to the observed diversity is that there never has been any competition at all and that the organisms merely evolved separate strategies by utilizing different resources in a multivariate environment. On the other hand, non-equilibrium theory establishes that competitive interactions are minimal and community structure is dominated by the vagaries of the environment. A non-equilibrium model requires that environmental disturbances occur with sufficient frequency that the course of competitive exclusion is disrupted (Harris, 1986).

Resilience was introduced in contrast to stability to indicate the behaviour of dynamic systems far from equilibrium. The concept of ecological resilience presumes the existence of multiple stability domains and a tolerance of the system to perturbations that facilitate transition among stable states (Gunderson, 2000). Multiple stable states and transition among them were described by Scheffer et al. (2001) for the alternative clear and turbid states observed in shallow lakes.

The principle of community assembly assumes that most species have a cosmopolitan or even ubiquitous distribution, but their relative sensitivity to certain properties and environmental circumstances weaken their performance in comparison with those that are less sensitive or positively tolerant to such conditions (Reynolds et al., 2002). These authors promote a functional classification of the freshwater phytoplankton based upon functional associations of species that share common adaptive features. Phytoplankton information given in terms of functional associations is scarce, even in floodplain systems where the pulsing of river discharge induces morphological, anatomical and physiological adaptations (Junk et al., 1989). Two studies have focussed on tropical or subtropical shallow systems (Huszar et al., 2000; Unrein, 2001), and in particular, the latter described the temporal variability of the associations of a floodplain shallow lake on the Lower Paraná River. Simultaneously, Izaguirre et al. (unpublished) compared the algal assemblages of several shallow systems across this wetland including the previously mentioned lake. These authors found a spatial segregation of species regulated by the macrophyte cover that determined the optical properties of the water, oxygen and nutrient concentrations and sediment load along with other parameters. The aim of this paper is to define the occurrence, nature and properties of 'equilibrial' species, which dominate the assemblages under equilibrium or stable conditions as well as of those taxa characteristic of non-equilibrium conditions.

Study area

Most of the area of the Otamendi Natural Reserve is a wetland, which is delimited by the Paraná de las Palmas and Luján Rivers, Argentina (34° 10′–34° 17′ S; 58° 48′–58° 53′ W) (Fig. 1). It comprises two shallow lakes and several semi-permanent water bodies, some of which are relict oxbow lakes. All of them are surrounded by marshy vegetation and, especially the latter, are permanently covered by floating macrophytes. The 'laguna Grande', the largest water body (about 28 ha) is colonized by rooted and floating aquatic plants. This region has a temperate sub-humid climate, with mean summer and winter temperatures of 22 °C and 9.5 °C, respectively and a mean annual precipitation of 950 mm. The area has been described in detail in Izaguirre et al. (unpublished).

Materials and methods

Sampling points were chosen following a transect from three relict oxbow lakes (ROLs hereinafter) of the Paraná Basin to the largest shallow lake, where two sampling sites were selected, littoral (LSL) and pelagial (PSL) zones (Fig. 1). Along this transect, an additional small pond (SP) was also sampled; the macrophyte cover of this pond exhibited a wide range of variation throughout the study period, which allowed us to assess the influence of the floating vegetation on the algal assemblages. Samples were collected fortnightly from October 1998 to January 1999. Temperature, pH and conductivity were measured *in situ* using Hanna HI 8314 and HI 8033 portable electronic meters. Dissolved oxygen was determined by the Winkler method.

Samples for nutrient analyses were filtered through Whatman GF/F. The concentration of the main inorganic compounds (phosphates, nitrates, sulfates, chlorides, potassium, sodium, calcium and magnesium) were determined by ionic chromatography using DIONEX DX-100 instrument with a conductivity detector and DIONEX AS-4 or CS-10 cromatographic columns. Ammonia was analyzed following the indophenol method (APHA, 1975). The humic substances were estimated by spectrophotometry, reading the absorbance at 250 nm (Kronberg, 1999). Total phosphorus and total nitrogen were determined by ionic chromatography after alkaline digestion according to Koroleff (1983). Suspended solids were evaluated drying the non-filterable residue at 103–105 °C until constant weight (APHA, 1975). Incident and underwater irradiance were measured with a Li-Cor PAR sphaerical quantum sensor (Li-250).

Qualitative algal samples were obtained with a 15 μm pore net and fixed in 4% formalin. Water samples for quantitative analyses of the algal communities were preserved in PVC flasks with 1% Lugol's iodine solution. Counts were performed according to Utermöhl (1958). Replicate chambers of 5 or 10 ml each were left to sediment for at least 24 h. Counting errors were estimated according to Venrick (1978), accepting a maximum of 20% for the most frequent species.

The diversity of the phytoplankton community was estimated according to the Shannon-Weaver index. Phytoplankton biomass was estimated calculating species biovolumes following the criteria and formulae of Hillebrand et al. (1999). A minimum of 10 individuals were measured for each taxon; for less abundant

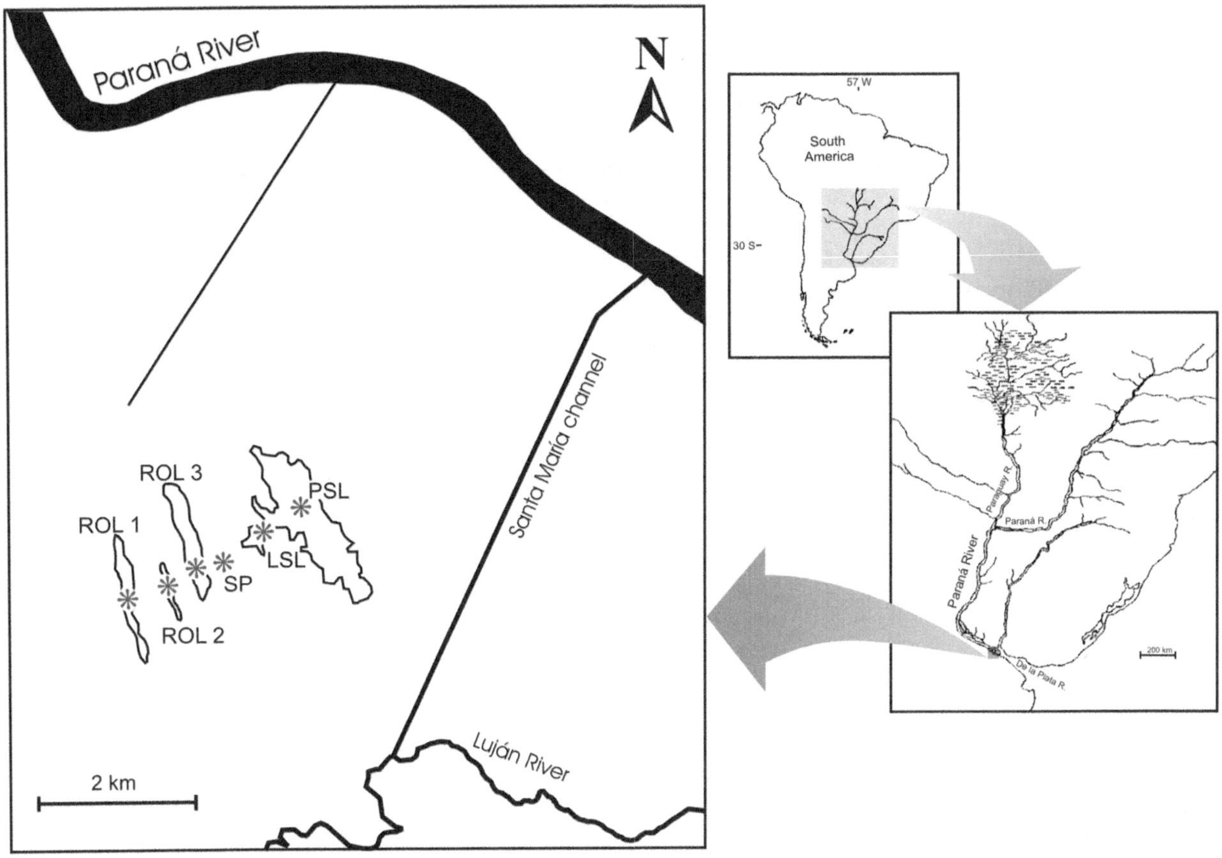

Figure 1. Location of the Otamendi Reserve in the Lower Paraná floodplain, showing the oxbow lakes (ROLs), the small pond (SP) and the littoral (LSL) and pelagic (PSL) areas of the shallow lake.

species, we also considered measures of specimens found in the Lower Paraná during previous studies. The taxa listed were classified in functional groups according Reynolds et al. (2002). The SIMI index between samples was calculated according to Elber & Shantz (1988).

Results

Table 1 summarizes the statistics of the main hydrological, physical and chemical variables corresponding to the studied systems. Figure 2a shows the water level fluctuations of the Paraná River in a site adjacent to the reserve Otamendi and the rainfall levels for the study period. Figure 2b illustrates the depth variation at each sampling site. The greatest depth among sampling sites was recorded in the open water area of the shallow-lake while the small pond was not only the shallowest site but it also exhibited the largest water level variations as seen by the variation coefficient (VC).

Water temperature in the shallow lake was approximately 10 °C higher than in the oxbow lakes. A high thermal gradient characterized both ROL2 and the small pond; the latter showed values between the lake and the oxbow-lakes. pH exhibited a temporal and a spatial pattern; lowest values, corresponding to acid and neutral waters were observed from October to December and increased steadily to the end of January. Even though lowest values were recorded from the small pond, mean pH in the oxbow lakes never exceeded 6. pH in the pelagic area was higher than in the littoral zone. Once again, the highest pH gradient occurred in the small pond. pH was inversely correlated with the presence of humic substances measured as absorbance at 250 nm ($r = -0.67$; $p < 0.05$). A clear spatial pattern was observed in the dissolved oxygen concentrations; this was illustrated by either anoxia or extremely low oxygen values in the oxbow

Table 1. Limnological variates at the six sampling sites. Mean, minimum and maximum values are given

	ROL1	ROL2	ROL3	SP	LSL	PSL
Water level	48.2	32.6	50.7	35.7	52.2	77.2
(cm)	36.5–63	21–50	35–82	14–60	29–77	65–94
Temperature	15.7	16.8	16.9	19.5	24.7	25.9
(°C)	9.1–20.1	9.8–23.5	9.9–21.7	12.1–25.6	20.1 –30.7	21.4–30.3
pH	6	5.8	5.9	5.7	6.3	7.7
	5.0–7.8	4.4–7.7	4.9–7.7	3.8–7.2	5.2–8.1	6.3–9.1
Dis. oxygen	ND	0.05	ND	1.4	4.77	8.74
(mg l^{-1})		ND–0.39		ND–3.59	2.88–7.2	1.95–4.05
Conductivity	2375	2806	2169	3758	2180	1451
(μS cm^{-1})	1310–3400	1231–3980	1440–2790	1300–7100	1230–3150	902–2170
Susp. solids	34	175	113	767	33	12
(mg l^{-1})	10–94	114–227	30–203	34–3493	8–53	2–26
Humic acids	3.21	2.23	2.04	1.94	1.35	0.87
(Abs. 250 nm)	2.42–3.87	1.77–3.25	2.58–3.49	1.11–2.38	0.88–1.80	0.7–1.27
T. phosphorus	831	756	844	1504	621	447
(μg l^{-1})	490–1187	219–1095	387–1239	315–5010	12–1263	138–824
P–PO_4	601	579	532	748	545	268
(μg l^{-1})	385–958	160–894	203–832	169–1880	172–1705	24–493
T. nitrogen	1391	1425	1847	2341	1583	1739
(μg l^{-1})	118–3256	45–3597	10–3586	83–8857	234–2824	54–4482
N–NO_3	143	272	560	155	66	42
(μg l^{-1})	10–304	23–1424	10–3260	20–390	10–170	10–93
N–NH_4	148	204	170	2166	156	38
(μg l^{-1})	10–387	24–709	26–555	23–7290	10–864	15–100

lakes to increasing concentrations in the limnetic area of the shallow-lake (Table 1).

The highest conductivities were recorded at sites exhibiting high water depth variation; thus, maximum conductivity values were found in the small pond. Lowest conductivities were registered in the limnetic area of the shallow lake where this variable showed fairly constant values. Conductivity was inversely correlated to water depth and pH ($r = -0.71, r = -0.67$; $p < 0.05$) and directly correlated to ammonia and phosphate concentrations ($r = 0.59$, $r = 0.52$; $p <$ 0.05). Suspended solid concentrations were always lower and more constant in the shallow lake than at the other sites. Maximum suspended solids values were recorded from the small pond, which almost dried up towards the end of the study period. This variable was inversely correlated with water level ($r = -0.64$; $p < 0.05$). A clear spatial pattern of absorbance at 250 nm was observed across this system. Estimated humic acid content was lowest in the limnetic area of the shallow lake and increased towards the oxbow lakes. Absorbance was inversely correlated to temperature and total nitrogen ($r = -0.7$, $r = -0.6$; $p < 0.05$). While only 4% of the incident radiation reached at 5 cm depth in the ROLs, nearly 50% of the radiation attained this depth in the lake.

Both total phosphorus and phosphate concentrations were more stable in the oxbow lakes while in the small pond there was a high degree of variation. The phosphorus content of this system was quite high exceeding 20 μg l^{-1}. Total nitrogen concentrations were quite similar in all sites with the exception of the values found in the small pond. Following rain events the total nitrogen concentrations in the small pond were found to increase two or threefold above the values recorded in the oxbow lakes and in the shallow-lake (3 November 1998 and 22 January 1999). Dissolved fractions were usually lower in the shallow lake, especially in the limnetic zone.

The analysis of the algal flora revealed 238 species and varieties. Most taxa belonged to the Bacillariophyceae, Cyanobacteria and Chlorophyceae while a small proportion belonged to the classes Zygo-

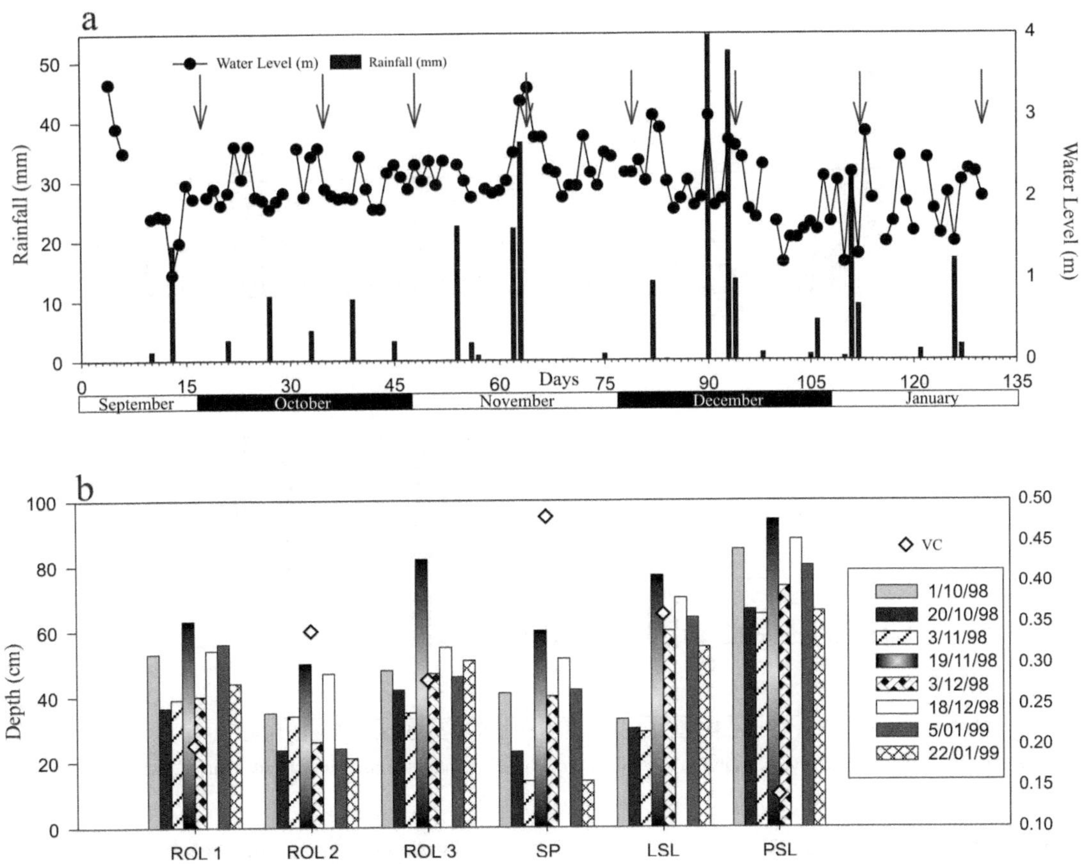

Figure 2. (a) Fluctuations of the water level of the Paraná River and its relation to rainfall during the study period. (b) Depth variations at the sampling sites.

phyceae, Chrysophyceae, Cryptophyceae, Euglenophyceae and Dinophyceae.

The abundance of Cyanobacteria was directly correlated with humic acids ($r = 0.6$, $p < 0.05$). This group dominated the algal communities of ROL1 with a high contribution of *Synechococcus nidulans*, an un-identified oscillatorian species (illustrated and described by Izaguirre et al. (2001b)) and *Synechocystis* sp.. Total algal density ranged between 797 and 45233 ind ml^{-1} with a peak at the end of October (Fig. 3a). Densities at ROL2 and ROL3 were lower than at ROL1 varying between 326 and 8762 ind ml^{-1}. The Cyanobacteria species mentioned above also occurred at these sites, but here there was a higher contribution of Bacillariophyceae represented by *Achnanthes hungarica, Cocconeis placentula, Fragilaria construens* and *Neidium iridis* (Fig. 4a).

Algal biomass in the ROLs varied from 1.3 to 18.5 $mm^3\ l^{-1}$ (Fig. 3b). ROL3 exhibited the highest values at the beginning of December due to the contribution of some large benthic diatoms. In ROL1 algal biomass was alternatively dominated by the un-identified oscillatorian, *Synechococcus nidulans, Woronichinia elorantae* and *Neidium iridis*. In ROLs 2 and 3 the non-identified oscillatorian was also important in terms of algal biomass, but in this case heavy diatoms such as *Pinnularia maior, Neidium iridis* and *Amphiprora alata* contributed more significantly to the bulk of the biomass (Fig. 4b). In the ROLs the variation coefficient of algal biomass ranged from 0.66 to 0.77. The SIMI index calculated using species density values showed a strong similarity (>0.8) between consecutive samplings at ROL1, whereas when calculated using biovolumes, figures were slightly lower.

None of the phytoplankton associations defined by Reynolds et al. (2002) strictly matched the dominant taxa recorded from the ROLs. Nevertheless, **Z** (picophytoplankton) and **S_1/S_2** (solitary filamentous cyanoprokaryotes) associations seem the most appropriate classification for the functional groups encountered.

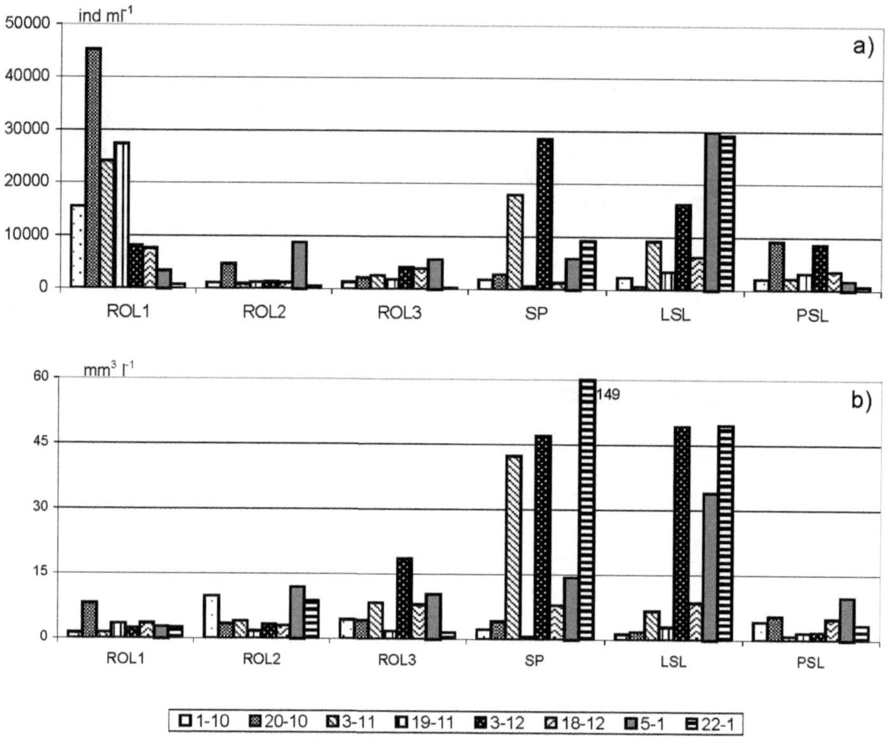

Figure 3. Total algal density variation (a) and algal biovolume (b) fuctuations at the six sites during the study period.

The **Z** association (prokaryote picoplankton) dominated in terms of density at ROL1, $\mathbf{S_1/S_2}$ (unidentified oscillatorian) prevailed at ROL3. An intermediate situation was observed at ROL2. *Woronichinia elorantae,* classified as $\mathbf{L_0}$, appeared in the three oxbow-lakes (Fig. 5).

The algal community of the small pond presented a slightly different compositional pattern. During the first months it resembled that of the ROLs, with blue-green dominance as well as an important diatom biomass component (Fig. 4). A gradual succesion occurred as species belonging to the Chlorophyceae, Cryptophyceae, Euglenophyceae and Chrysophyceae increased in abundance. At the end of the study, diatoms dominated the small pond. This site represented a high degree of variability, not only in species composition, but also in total algal density that ranged from 657 to 53 334 ind ml^{-1} (Fig. 3a). Algal biomass in the small pond ranged between 0.59 and 149 mm^3 l^{-1} (Fig. 3b). Smaller peaks observed were mainly a result of increases in the occurrence of *Euglena variabilis* and *Amphiprora alata* (Fig. 4b). The latter large diatom was responsible for the highest peak registered in this pond at the end of the study period. These large biomass fluctuations were responsible for the highest variation coefficient among all sites (1.50). The SIMI index fluctuated between very high and low values throughout the period studied. During the first period of this study, the algal associations resembled that of the ROLs (mainly $\mathbf{S_1/S_2}$ and $\mathbf{L_0}$, which is typically represented by *Peridinium, Woronichinia, Merismopedia*). Nevertheless, other algal associations such as $\mathbf{W_1}$ (typically represented by euglenoids, *Synura* and *Gonium*), $\mathbf{X_1}$ (defined as an eutrophic-hypertrophic association represented by small Chlorococcales) and **Y** (represented by Cryptomonad species) became important during the study. Towards the end of spring and beginning of summer, $\mathbf{S_1/S_2}$ once again became the dominant functional group, but at this stage it was represented by a greater species diversity (*Leptolyngbya fragilis, Oscillatoria sancta, Planktolyngbya* sp.) (Fig. 5).

The littoral area of the shallow lake was characterized by the co-occurrence of species from Chlorophyceae, Cyanobacteria, Cryptophyceae, Euglenophyceae and Dinophyceae. Total density ranged from 673 to 29 869 ind ml^{-1}. Green algae dominated the phytoplankton at this site, reaching numbers up to 23 200 ind ml^{-1} (Fig. 3a). Algal biomass ranged between 1.31 and 49.54 mm^3 l^{-1} (Fig. 3b).

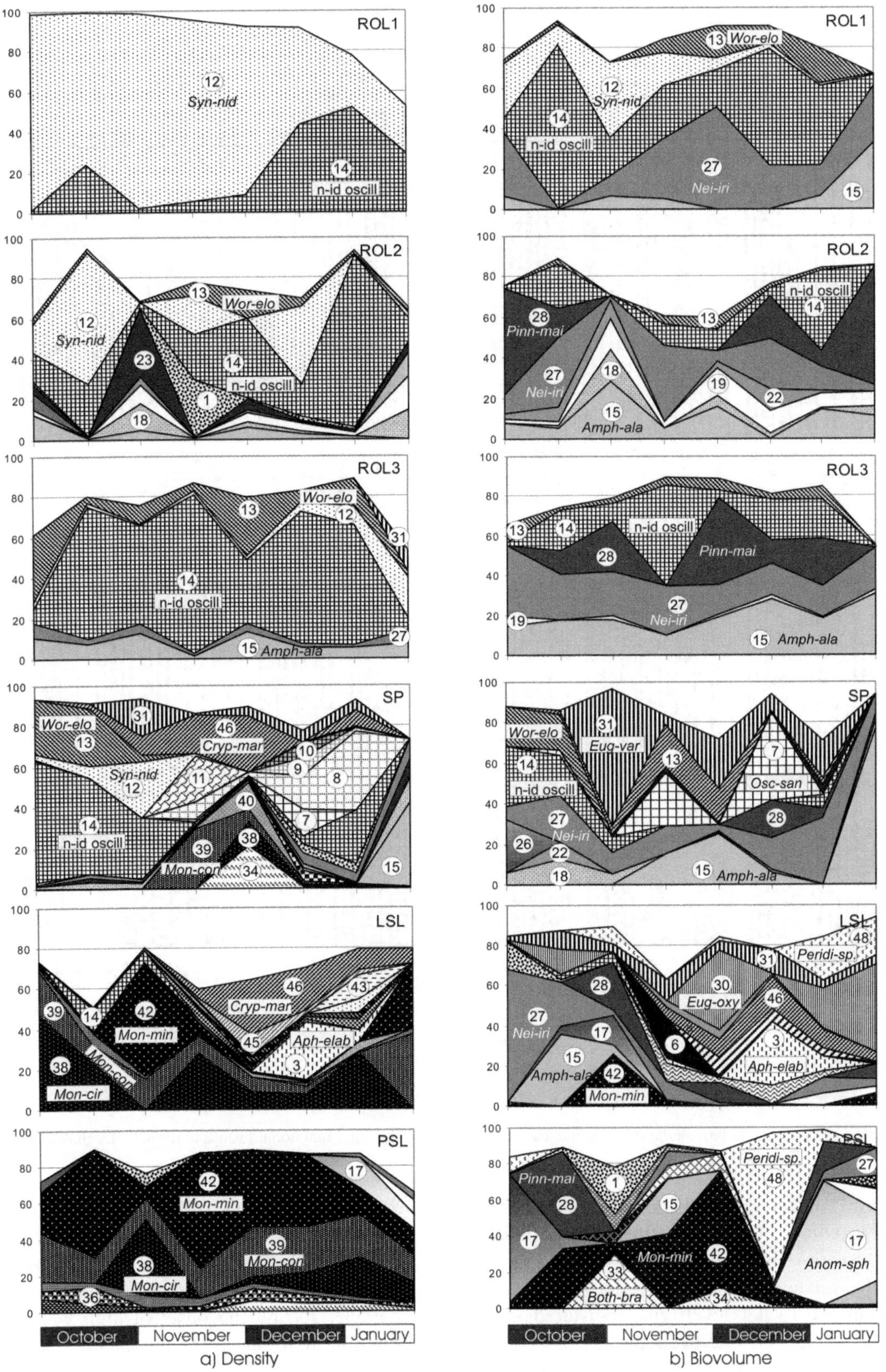

Figure 4. Species total density (a) and biovolume (b) percentages at the six sites during the study period.

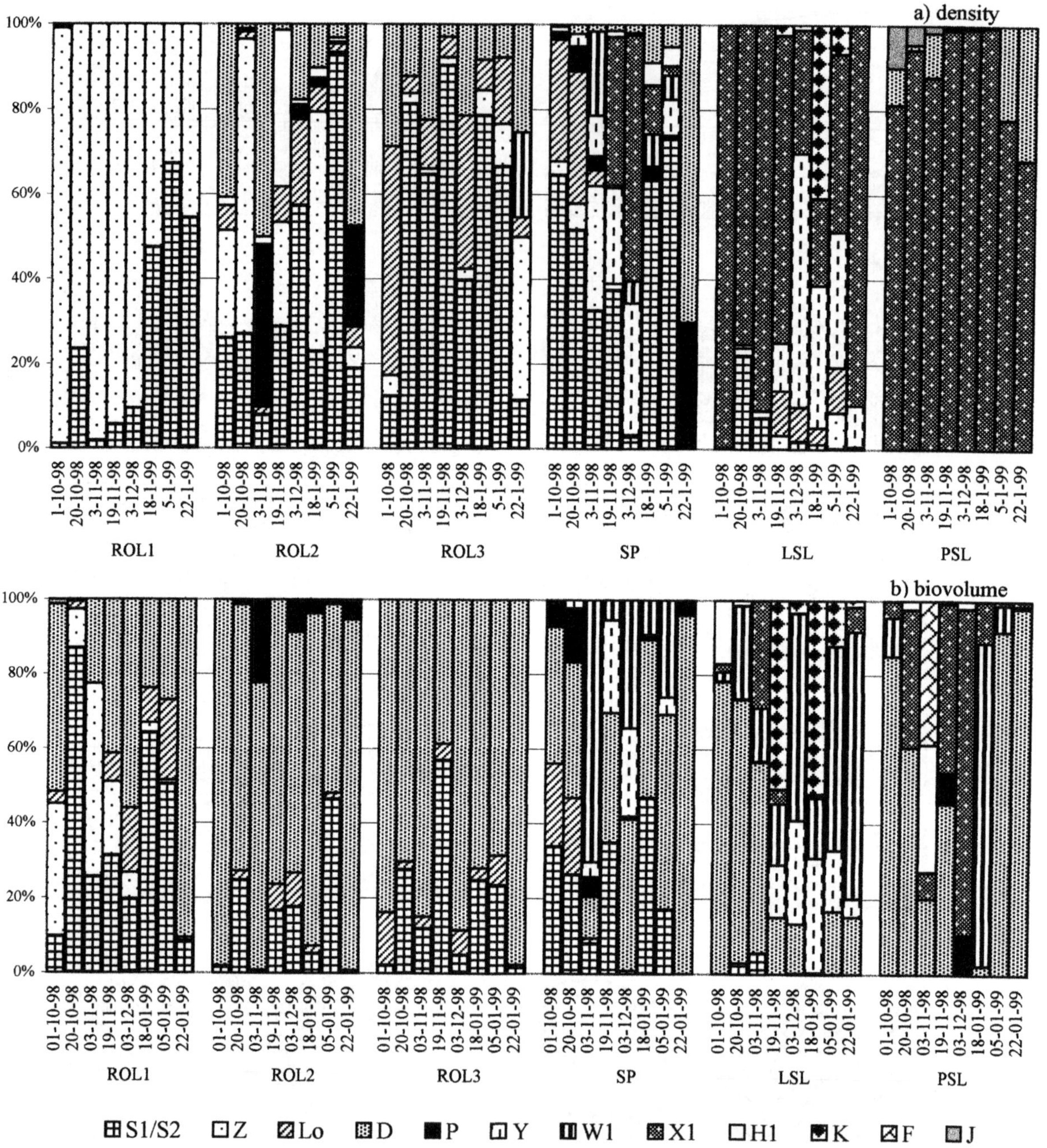

Figure 5. Total density (a) and biovolume (b) percentages corresponding to the algal functional groups at the six sites during the study period.

The SIMI index revealed brief periods of relatively stable composition alternating with periods of radical changes. The latter dominated the biomass and density fluctuations following the most important hydrometric (water level) rise registered in the shallow lake (Fig. 3b) in November. $\mathbf{X_1}$ association, represented by several *Monoraphidium* species, dominated the initial 'stable' stage in terms of density (Fig. 4a), whereas large non-planktonic diatoms also contributed to the bulk of biomass (Fig. 4b). A second stage of the littoral area of the lake may have been characterized by the onset of high waters (Fig. 2b), which brought on the development of several species of *Cryptomonas* and *Euglena/Peridinium*, as well as some colonial Chroococcales, belonging to the **Y**, $\mathbf{W_1}$ and **K** associations respectively. The latter has been defined by Reynolds et al. (2002) as an association of small-celled colonial Cyanoprokaryotes. Due to their large size, euglenoids and dinoflagellates were relevant only in terms of biomass. Towards the end of the study period, and coinciding with a slight decrease in the water level, *Monoraphidium* populations increased again and the $\mathbf{X_1}$ association acquired new significance (Fig. 5).

The pelagic area showed lower and less variable algal densities ranging from 741 to 9184 ind ml^{-1} (VC = 0.81) (Fig. 3a), as well as a more constant phytoplankton composition. A net dominance of Chlorophyceae species was observed throughout the study period. The most representative species were *Crucigenia quadrata*, *Didimocystis bicellularis*, *Monoraphidium circinale*, *M. contortum* and *M. minutum* ($\mathbf{X_1}$ association). Algal biomass values were also comparatively lower than in the littoral zone and ranged between 0.93 and 9.79 $mm^3\ l^{-1}$ and represented low temporal variations (VC = 0.70) (Fig. 3b). Large benthic diatoms were responsible for the major contribution to the standing crop during several periods. These alternated with associations dominated either by *Monoraphidium minutum* ($\mathbf{X_1}$) and *Peridinium* sp. ($\mathbf{W_1}$) or by *Anabaena sphaerica* ($\mathbf{H_1}$) and *Botryococcus braunii* (**F**) (Figs 4 and 5).

Species diversity calculated using density values varied between 0.16 in ROL1 and 3.21 in the littoral area of the shallow lake. In general terms, the littoral area showed the highest values throughout the study period. The small pond exhibited intermediate values.

Discussion

In general terms, the aquatic systems of this wetland may be differentiated into two groups: the profusely vegetated ROLs and the shallow lake with large areas free of floating macrophytes. The ROLs are habitats with an unvarying limiting condition, such as the low light penetration, which seems to have selected for continued dominance of the most appropriate species. Contrarily, the shallow lake represents a more variable environment, where the limiting constraint is frequently varied and the identity of the best adapted species changes. In this system, diversity is kept high as different species respond to alternations of relative difficulty and relative benefit, so no particular features are consistently selected.

The species conforming to the most frequent associations in the ROLs displayed phylogenetic affinities, mainly Cyanobacteria, and similar adaptive strategies. Cyanobacteria show several physiological adaptations to the extreme conditions that prevailed in the ROLs. Photosynthesis in Cyanobacteria, especially in solitary filamentous forms, saturates at low light intensity, and thus displays a high affinity for light and their maintenance requirements are extremely low (van Liere & Mur, 1979). These properties allow photosynthesis even under unfavourable light conditions. Moreover, in addition to normal aerobic photosynthesis, several species are capable of sulfide-dependent anoxygenic photosynthesis (Garlick et al., 1977; Sorokin, 1999). Cyanobacteria are well adapted for life under anoxic conditions for, in addition to the normal aerobic dark respiration, virtually all species constituting microbial mats are capable of fermentation (Stal & Moezelaar, 1997).

Reynolds et al. (2002) grouped several solitary filamentous Cyanoprokaryotes into association **S** with subdivisions $\mathbf{S_1}$ and $\mathbf{S_2}$, which are composed of the genera, *Planktothrix*, *Limnothrix*, *Pseudoanabaena*, *Spirulina*, *Arthrospira* and *Raphidiopsis.* These mainly inhabit warm and shallow systems. The unidentified oscillatorian species recorded in the ROLs is surely related to these genera, and the environmental conditions match. Even though there is a clear similarity, the anoxic conditions recorded in these environments seem to be significant enough to necessitate, either the creation of a new subdivision for association S, or enlargement of the habitat features of this group.

The other dominant association in the ROLs is referred to as **Z**. We think this is a more adequate designation for the high abundance of *Synechococcus nidulans* and *Synechocystis* spp. observed and it is typically represented by very small cyanobacteria. Nevertheless, there are some features, such as the sensitivity to light deficiency and low nutrient tolerance, which do not adequately describe the situation encountered. There are several explanations for this association success in colored waters under the profuse macrophyte cover. Picocyanoplankton may be classified into two cell types according to their major light harvesting pigments (Stockner et al., 2000): PE cells with yellow autofluorescing phycoerythrin and PC cells with red autofluorescing phycocyanin, respectively. Vörös et al. (1998) related the dominance of phycocyanin-rich APP in eutrophic waters to the qualitative changes in underwater photosynthetically active radiation. In hypertrophic lakes, wavelengths in the blue-green region are rapidly attenuated by chlorophyll pigments, detritus and dissolved yellow substances, while those in the green and red regions penetrate deeper. Callieri et al. (1996) demonstrated that phycocyanin-rich APP exploit red light better and have higher growth rates than phycoerythrin-rich APP. Moreover, Waterbury et al. (1986) showed that the optimum growth rate of *Synechococcus* occurs under low light conditions. On the other hand, Huszar et al. (2000) suggested assigning the synechococcoid phytoplankton to a more eutrophic association ($\mathbf{X_1}$) after a high biomass was observed in some coastal lagoons in Brazil. Nevertheless, we think it might be better to preserve the **Z** association to denote prokaryote plankton, as other authors also found very high concentrations of picocyanobacteria in very enriched lakes (Carrick & Schelske, 1997; Vörös et al., 1998; Izaguirre et al., 2001a), demonstrating that there is not always an inverse correlation between picoplankton abundance and nutrient concentrations.

The food web in the ROLs could be characterized as typically heterotrophic. The absence of grazing by cladocerans, copepods and rotifers referred to as one of the factors controlling phytoplankton in shallow lakes, is much less important in these environments. In anoxic waters of eutrophic and distrophic lakes, protozooplankton usually prevails. Metazooplankton, and protozoans become dominant (Straskrabová & Šimek, 1993; Callieri & Stockner, 2002). It has been repeatedly stated, under these conditions nano-sized ciliates and heterotrophic flagellates are the primary picoplankters grazers (Azam et al., 1983). We hypothesize that the grazing pressure in the ROLs would be relatively constant in relation to the low variation in algal biomass throughout the study period. Likewise, Ortega-Mayagoitia et al. (2003) suggested that in eutrophic wetlands trophic relationships are possibly irrelevant.

Contrary to the phylogenetic and physiological affinities between the dominant algal associations encountered in the ROLs, the shallow lake displays a more diverse taxonomic composition of organisms with multiple adaptive strategies. Typically phototrophic small chlorococcalean taxa alternate with mixotrophic flagellates (Cryptophyceae, Dinophyceae and Euglenophyceae) or phototrophic colonial cyanobacteria. Large diatoms appear as accompanying taxa, primarily derived from benthic and periphytic communities.

Obligate photosynthesis was the dominant algal strategy in the shallow lake. This was seen in the high densities of the eutrophic nanoplankton represented by non-motile chlorophytes (Chlorococcales) throughout the study period, especially in the pelagial area of this lake. The prevailing conditions consisting of high light and nutrient availability, enabled a successful development of the $\mathbf{X_1}$ association. The increased rainfall during the month of November produced significant environmental changes in the littoral region: the water column level increased by twice its depth, and there was an initial decrease followed by a sustained increase in conductivity and humic acids concentration. These fluctuations brought about a change not only in the algal composition, but also in the prevailing nutrition strategies. After this rain event, nearly 40% of the plankton density was represented by two species of *Cryptomonas*. The opportunistic behaviour of Crytophyceae has been described by Tolotti et al. (2003). These species from association **Y** use a mixotrophic strategy to survive (Urabe et al., 2000). Conditions of unfavorable light and relatively high dissolved organic carbon (DOC) should present a disadvantage to obligate phototrophs, because of light limitation and increased competition with bacteria for inorganic nutrients. On the other hand, mixotrophy provides nutrient supplementation through the ingestion of bacteria or the uptake of DOC (Jones, 2000). The *Cryptomonas* density maximum, coincides with a biomass pulse of *Euglena* spp. (association $\mathbf{W_1}$) which accounts for approximately 45% of the phytoplankton in the littoral area. Both functional groups $\mathbf{W_1}$ and **Y** share this mixotrophic strategy and the flagellar mobility not withstanding their other taxonomical similarities. Unrein (2001) pointed out that the abundance of mixotrophs, during periods when nitrogen was not limiting, suggests that these organisms have an alternative adaptive strategy. It may be that the possession of flagella allows organisms to keep themselves in suspension in non-turbulent areas such as the rich macrophytes zones. Moreover, **K** association occurred simultaneously with the mobile **Y** and $\mathbf{W_1}$ groups. The non-motile small colonial cyanobacteria in this association also display characteristics that favor buoyancy and are maintained in the water column around macrophyte environments such as the littoral area. As Reynolds (1997) points out, the average algal density can be reduced by the presence of peripheral mucilage, provided that the increase in particle size does not enhance the sinking velocity as postulated by the Stokes equation. However, small cyanobacteria are also known for their capacity to absorb DOC (Sorokin, 1999) and a *Synechococcus* pulse was recorded at the beginning of January simultaneously with the occurrence of the **K** species.

Pithart (1999), Zalocar de Domitrovic (2000) and Unrein (2001) have already shown that flagellates such as euglenoids and *Cryptomonas* spp. are more prevalent in shallow lakes with a high dominance of aquatic vegetation compared to a predominance of the $\mathbf{X_1}$ association in less vegetated systems. Stickney et al. (2000) stated that if light conditions and nutrient concentrations are optimal, autotrophs outcompete mixotrophs. Mitsch & Gosselink (2000) argued that even though nitrate and ammonia are the main nitrogen sources for phytoplankton, other fractions such as dissolved and particulate organic nitrogen are not taken into account and they can be potentially very important in wetlands. The abundance of mixotrophs suggests that this alternative N source may be a significant source of N. The replacement of mixotrophic species by small Chlorococcales in the littoral area by the end of the study period (22 January 1999) could be cited as evidence of the relative importance of the various alternative nitrogen fractions. The low levels of dissolved nitrogen encountered at this site recover and, the $\mathbf{X_1}$ association once again constitutes about 70% of the phytoplankton density. Although we did not perform experiments to determine nutrient limitation during the study period, Unrein (2001) carried out bioassays that identify N as the limiting factor in the littoral area of the lake. He postulated that concentrations $>100\ \mu g\ l^{-1}$ would be sufficient to satisfy phytoplankton requirements, whereas concentration $<60\ \mu g\ l^{-1}$ would indicate a potential limitation.

Most of the species encountered were satisfactorily assigned to the phytoplankton associations defined in Reynolds et al. (2002). Nevertheless, we think that considering nutrition strategies enriches the characterization of these associations as functional groups. These are closely related to physiological adaptations and as Reynolds et al. (2002) pointed out are not constrained by phylogenetic affinities. In particular, Nixdorf et al. (2001) stated that in wetlands with a heavy macrophyte development it is very important to consider the roles of algal cell size, shape, motility, changes in metabolism and the ability to shift between heterotrophy and mixotrophy. Previous results obtained by studying the spatial gradient in this wetland using canonical correspondence analysis, support the relationship between the environmental parameters and the species assemblages encountered at each sampling site (Izaguirre et al., unpublished).

Having defined the characteristics of the assemblages present at each site, it is pertinent to analyse whether they correspond either to a steady state or, as expected from shallow lakes, to non-equilibrium conditions. In temperate lakes equilibrium is currently described by one to three species contributing with more than 80% of the standing crop, with a coexistence persisting for more than 2 weeks and with no significant changes in total biomass during this period (Sommer et al., 1993). In relation to this, we suggest that additional species should be considered for floodplain shallow systems. A description of the acquisition of a steady state in floodplain shallow systems should take into account the high biodiversity of these systems. The shallow depth of these environments implies that there will be a sporadic contribution of benthic organisms with no ecological significance. This is especially evident in the analysis of biomass data where large heavy diatoms from the bottom contribute a significant proportion to the biomass but cannot yet be classified into any of the functional groups.

We hypothesise that macrophyte cover and consequently water column stability are the main ecological conditions leading to the steady state assemblage in this wetland. The existence of such 'equilibrium or steady state' in ROL1 and ROL3, and partially in ROL2, was associated with a stable water column with persistent anoxia and poor light conditions, as consequence of profuse and dense floating vegetation. On one hand, we could assume that the constancy of extreme ecological conditions provided a chance for exclusion mechanisms to take place or, in other words, enough time has passed for competition to exclude relevant species. On the other hand, based on the equilibrium theory, we seriously consider and support the possibility that competition may not have been a factor and that the existing species evolved separate strategies using the resources encountered in the ROLs. The extreme conditions of these systems permits few physiologically adapted species to flourish. Stoyneva (2003) and Padisák et al. (2003) also found equilibrium states in stressed ecosystems, which were dominated by cyanoprokaryotes just as in the ROLs. As a consequence, we hypothesise that the ecological resilience (*sensu* Gunderson, 2000) of the ROLs will be very low, and disturbance will change the self-organized processes and structures. In this sense, the different situations observed in the small pond suggest the existence of alternative scenarios. This change corresponds to the loss of the macrophyte cover as a consequence of the heavy rains. Here we conclude that the potential of a steady state or equilibrium situation is frequently disrupted as a consequence of major hydrologic events such as flushing.

The species replacement recorded in the shallow lake and the small pond revealed recurrent oscillations of primarily the same taxa. These assemblages were characterised by the coexistence of many species with no single dominant taxa. According to Sommer et al. (1993) the phytoplankton in these systems do not represent the characteristics that can give rise to an equilibrium phase. The variability in individual species was high, and only in the pelagial area was there near stability in total community biomass. In the latter situation, the pattern of oscillation resembled the planktonic biodiversity model generated by Huissman & Weissing (1999). These authors postulate that such a scenario supports the hypothesis that competition in high diversity ecosystems increases variability at the species level while at the same time stabilizing the global ecosystem properties such as total community biomass. Nevertheless, the diversity and periodicity of dominant species observed in the shallow lake and the small pond may probably be explained by frequent allogenic disturbances such as postulated in the Intermediate Disturbance Hypothesis (Sommer et al., 1993). As Reynolds (1997) pointed out, the IDH assists the understanding of the reactions of the biotic structures but does not provide a record of the resilience of a system. There are several facts suggesting that although the phytoplankton of these systems is not close to equilibrium, it has the capacity to withstand abrupt external disturbances and at the same time retain a pattern of its basic composition and organisation. In this sense, the diversity of nutritional strategies allows the enhancement of potential energy sources and/or a consistency of the energy flux. In spite of the fluctuations in the dominant species, the dominating functional groups displayed a strong persistence. Unrein (2001) describes a comparable situation in a nearby swamp, where he found strong differences at the species level (up to 80% of similarity using association density values) between consecutive samplings. Thus, in both floodplain systems, the dominant species changed but the prevailing associations were always the same (**Y**, **W** and $\mathbf{X_1}$), suggesting that succession in these functional groups might account for a type of stability in the phytoplankton community. In this sense, Stewart & Wetzel (1986) pointed out that the intermittent and rapid increases of cryptophytes act as an important community-stabilizing feature that can have profound effects upon pelagic metabolic processes.

Finally, we conclude that the extreme ecological conditions leading to the steady state assemblages encountered in the ROLs requires physiological adaptations to low light penetration and nocturnal anoxia. These were characteristic of certain Cyanobacteria, whereas the changing conditions of the shallow lake are related to the high degree of fluctuation in several algal functional groups representing diverse nutrition strategies with no phylogenetic affinities. Nevertheless, the former assemblages appear to be less resilient than the latter because of the effect of the extreme conditions resulting from the dense macrophyte cover.

Acknowledgements

We thank the staff of the Otamendi Reserve (Parques Nacionales) for their assistance in the field. We are grateful to CONICET, UBA and ANCYPT for their financial support.

References

APHA, 1975. Standard methods for the examination of water and wastewaters. American Public Health Association, Washington. 1193 pp.

Azam, F., T. Fenchel, J. G. Field, J. S. Gray, L. A. Meyer-Reil & F. Thingstad, 1983. The ecological role of water-column microbes in the sea. Mar. Ecol. Prog. Ser. 10: 257–263.

Callieri, C., E. Amicucci, R. Bertoni & L. Vörös, 1996. Fluorimetric characterization of two picocyanobacteria strains from lakes of different underwater light quality. Int. Rev. ges. Hydrobiol. 81: 1–11.

Callieri, C. & J. G. Stockner, 2002. Freshwater picoplancton: a review. J. Limnol. 61: 1–14.

Carrick H. J. & C. L. Schelske, 1997. Have we overlooked the importance of small phytoplankton in productive waters? Limnol. Oceanog. 42: 1613–1621.

Elber, F. & F. Shanz, 1988. The causes of change in diversity and stability of phytoplankton community in small lakes. Freshwat. Biol. 21: 237–251.

Garlick S., A. Oren & E. Padan, 1977. Occurrence of facultative anoxygenic photosynthesis among filamentous and unicellular cyanobacteria. J. Bacteriol. 129: 623–629.

Gunderson, L. H., 2000. Ecological Resilience – In Theory and Application. Ann. Rev. Ecol. Syst. 31: 425–439.

Harris, G. P., 1986. Phytoplankton ecology. Structure, function and fluctuation. Chapman & Hall, London. 384 pp.

Hillebrand H., C-D. Durselen, D. Kirshtel, U. Pollingher & T. Zohary, 1999. Biovolume calculation for pelagic and benthic microalgae. J. Phycol. 35: 403–424.

Holling, C. S., 1973. Resilience and stability of ecological systems. Ann. Rev. Ecol. Syst. 4: 1–23.

Huisman, J. & F. J. Weissing, 1999. Biodiversity of plankton by species oscillations and chaos. Nature 402: 407–411.

Huszar, V. L. M., L. H. S. Silva, M. Marinho, P. Domingos & C. L. Sant'Anna, 2000. Cyanoprokaryote assemblages in eight productive tropical Brazilian waters. Hydrobiologia 424: 67–77.

Izaguirre, I., G. Mataloni, L. Allende & A. Vinocur, 2001a. Summer fluctuations of microbial planktonic communities in a eutrophic lake Cierva Point, Antarctica. J. Plankton Res. 23: 1095–1109.

Izaguirre, I. R. Sinistro, I. O'Farrell, F. Unrein & G. Tell, 2001b. Algal assemblage in anoxic relictual oxbow lakes from the Lower Paraná floodplain (Argentina). Nova Hedwigia 123: 95–106.

Izaguirre, I., I. O'Farrell, F. Unrein, R. Sinistro, M. dos Santos Afonso & G. Tell. Algal assemblages and limnological features across a wetland, from a shallow lake to the relict oxbow lakes (Lower Paraná River, South America), ms.

Jones R. I., 2000. Mixotrophy in planktonic protists: an overview. Freshwat. Biol. 45: 219–226.

Junk, W., P. B. Bayley & R. E. Sparks, 1989. The flood pulse concept in river floodplain system. Spec. Publ. Can. J. Fish. aquat. Sci. 106: 110–127.

Koroleff, F., 1983. Simultaneous oxidation of nitrogen and phosphorus compounds by persulfate. In Grosshoff, K., M. Eberhodt & K. Kremling (eds), Methods of Seawater Analysis. Verlag Chemie, Wainheimer: 168–169.

Kronberg, L., 1999. Content of Humic Substances in Freshwater. In Keskitalo, J. & P. Eloranta (eds), Limnology of Humic Waters. P. Backhuys Publishers, Leiden 2: 9–10.

Mitsch W. J. & J. G. Gosselink, 2000. Wetlands. J. Wiley & Sons, N. Y. 610 pp.

Ortega-Mayagoitia, E., C. Rojo & M. A. Rodrigo, 2003. Controlling factors of phytoplankton assemblages in wetlands: an experimental approach. Hydrobiologia 502 (Dev. Hydrobiol. 172): 177–186.

Nixdorf, B., A. Fyson & H. Krumbeck, 2001. Review: plant life in extremely acidic waters. Environ. Exp. Bot. 46: 203–211.

Padisák, J., G. Borics, G. Fehér, I. Grigorszky, I. Oldal, A. Schmidt & Z. Zámbóné-Doma, 2003. Dominant species, functional assemblages and frequency of equilibrium phases in late summer phytoplankton assemblages in Hungarian small shallow lakes. Hydrobiologia 502 (Dev. Hydrobiol. 172): 157–168.

Pithart, D., 1999. Phytoplankton and water chemistry of several alluvial pools and oxbows after the flood event – a process of diversification. Algol. Stud. 95: 93–113.

Reynolds, C. S., 1997. Vegetation processes in the Pelagic: a model for ecosystem theory. Excellence in Ecology, Ecology Institute, Oldendorf. 371 pp.

Reynolds, C. S., V. Huszar, C. Kruk, L. Naselli Flores & S. Melo, 2002. Towards a functional classification of the freshwater phytoplankton. J. Plankton Res. 24: 417–428.

Scheffer, M., S. Carpenter, J. A. Foley, C. Folkes & B. Walker, 2001. Catastrophic shifts in ecosystems. Nature 413: 591–596.

Sommer, U., J. Pádisak, C. S. Reynolds & P. Juhász-Nagy, 1993. Hutchinson's heritage: the diversity–disturbance relationship in phytoplankton. Hydrobiologia 249: 1–7.

Sorokin, Y. I., 1999. Aquatic Microbial Ecology. Backhuys Publ., Leiden. 247 pp.

Stal L. J. & R. Moezelaar, 1997. Fermentation in cyanobacteria. FEMS Microbiol. Rev. 21: 179–211.

Stewart, A. J. & R. G. Wetzel, 1986. Cryptophytes and other microflagellates as couplers in plankton community dynamics. Archiv für Hydrobiol. 106: 1–19.

Stickney, H. L., R. R. Hood & D. K. Stoecker, 2000. The impact of mixotrophy on planktonic marine ecosystems. Ecol. Model. 125: 203–230.

Stockner J., C. Callieri & G. Cronberg, 2000. Picoplankton and other non-bloom formers. In Whitton, B. A. & M. Potts (eds), The Ecology of Cyanobacteria. Their Diversity in Time and Space. Kluwer Academic Publishers, Dordrecht: 195–231.

Stoyneva, M. P., 2003. Steady-state phytoplankton assemblages in shallow Bulgarian wetlands. Hydrobiologia 502 (Dev. Hydrobiol. 172): 169–176.

Straskrabová V. & K. Šimek, 1993. Microbial loop in lakes and reservoirs related to trophy and mezooplankton development. Int. Ver, theor. angew. Limnol: Verh. 25: 1183–1186.

Tolotti, M., H. Thies, M. Cantonati, C. M. E. Hansen & B. Thaler, 2003. Flagellate algae (Chrysophyceae, Dinophyceae, Cryptophyceae) in 48 high mountain lakes of the Northern and Southern slope of the Eastern Alps: biodiversity, taxa distribution and their driving variables. Hydrobiologia 502 (Dev. Hydrobiol. 172): 331–348.

Unrein, F., 2001. Efecto de los nutrients y el pH sobre el crecimiento y la estructura del fitoplancton en ambientes de la llanura alluvial del Paraná Inferior. Tesis Doctoral. Universidad de Buenos Aires. 162 pp.

Urabe, J., T. B. Gurung, T. Yoshida, T. Sekino, M. Nakanishi, M. Mauro & E. Nakayama, 2000. Diel changes in phagotrophy by *Cryptomonas* in Lake Biwa. Limnol. Oceanog. 45: 1558–1563.

Utermöhl, M., 1958. Zur Vervollkommung der quantitativen Phytoplankton Methodik. Mitt. int. Ver. theor. angew. Limnol. 9: 1–38.

Ven Liere, L.& L. R. Mur, 1979. Growth kinetics of *Oscillatoria agardhii* Gomont in continuous culture, limited in its growth by the light energy supply. J. Gen. Microbiol. 115: 153–160.

Venrick, E. L., 1978. How many cells to count?. In: Ed. Sournia A., Phytoplankton Manual. UNESCO, Paris: 167–180.

Vörös L, C. Callieri, K. V. Ballogh & R. Bertoni, 1998. Freshwater picocyanobacteria along trophic gradient and light quality range. Hydrobiologia 369/370: 117–125.

Waterbury, J. B., S. W. Watson, F. W. Valois & D. G. Franks, 1986. Biological and ecological characterisation of the marine unicellular cyanobacterium *Synechococcus*. Can. J. Fish. aquat. Sci. 214: 17–120.

Zalocar de Domitrovic Y., 2000. Estructura y dinámica del fitoplancton en la cuenca del eje potámico Paraguay-Paraná (Argentina). Tesis Doctoral. Universidad Nacional de Córdoba, Córdoba. 371 pp.

Hydrobiologia **502:** 211–224, 2003.
L. Naselli-Flores, J. Padisák, M. T. Dokulil (eds),
Phytoplankton and Equilibrium Concept: The Ecology of Steady-State Assemblages.

The role of physical stability on the establishment of steady states in the phytoplankton community of two Maritime Antarctic lakes

Luz Allende & Irina Izaguirre
Facultad de Ciencias Exactas y Naturales, Departamento de Ecología, Genética y Evolución, Universidad de Buenos Aires, (C1428 EHA), Buenos Aires, Argentina. E-mail: iri@bg.fcen.uba.ar

Key words: Antarctic lakes, functional groups, picocyanobacteria, thermal patterns

Abstract

Two Antarctic lakes near Hope Bay were studied during summers 1998 and 1999. One of the lakes (Boeckella) is located near Esperanza Station and exhibits a meso-eutrophic condition due to the input of nutrients of a nearby penguin rookery. Its surface generally remains ice-free during the Antarctic summer (December–March). The other lake (Chico) is situated on the Mount Flora shelf, is typically oligotrophic and its surface is ice-free only during brief periods in the summer season. The difference in the duration of the ice-cover insures that the wind mixes the former lake continuously throughout the summer, while the latter remains almost always stratified. $\mathbf{X_2}$, $\mathbf{X_3}$ and **Z** functional groups defined by Reynolds dominated phytoplankton in both lakes. In Lake Boeckella, *Chlamydomonas* spp. followed by *Ochromonas* sp. were the most frequently encountered taxa in the nano-phytoplankton fraction, however the latter species was dominant when the lake froze. In Lake Chico, the major contribution to this fraction was due to different genera of flagellated Chrysophyceae (*Ochromonas* sp., *Chromulina* spp., cf. *Chrysidalis*). In terms of density and biomass in both lakes picocyanobacteria represented a large proportion of the phytoplankton. Probably due to the typically low algal biodiversity of Antarctic lakes, both water bodies showed periods of more than 2 weeks when a maximum of only three species comprised more than 80% of the standing crop. In spite of this, Chico Lake was the only one in which no significant change was recorded in total biomass. Thus, we were able to identify equilibrium phases in the latter lake, which were confirmed by a low coefficient of variation. The presence of an almost permanent ice cover in Chico Lake generated more stable ecological conditions, allowing the development of steady state assemblages. On the contrary, the wind influence in the shallow Antarctic ice free lake (Lake Boeckella) provided continuous mixing events, disrupting the possibility of establishing a steady state.

Introduction

Two main regions are usually recognized in Antarctica: the Maritime and the Continental. Both regions are rich in different types of water bodies, including deep and shallow permanent lakes, and a great variety of temporary ponds and pools. The lakes described in this paper are located in the Maritime region, which extends from the South Sandwich Island and Bouvetøya, through the South Orkney and South Shetland Islands and down the western side of the Antarctic Peninsula to approximately 70° S (Ellis-Evans, 1996).

The lakes involved in this study are located at Hope Bay (Antarctic Peninsula), where several investigations were carried out since 1991. Previous research has shown that most of the water bodies in this region are relatively shallow. As in other regions of the Maritime Antarctica, the coastal lakes in the Hope Bay area are exposed to natural eutrophication, since birds and seals often frequent them. This is the most important factor determining their total phytoplankton biomass. The most enriched lakes clearly show an increased development of phytoplankton, usually dominated by volvocalean taxa, whereas the oligotrophic water bodies are very clear and contain dense benthic algal mats (Izaguirre et al., 1993, 1998). Similar conditions were described for other places in Maritime Antarctica (Hawes, 1983; Hawes, 1990; Butler, 1999; Mataloni et al., 2000, Vinocur & Pizarro, 2000; Vinocur & Unrein, 2000).

Contrary to lakes in Continental Antarctica, where ice-free periods are limited and lakes generally develop a thick ice-cover (Heywood, 1977; Ellis-Evans, 1996; Zipan & Deprez, 2000), lakes from the Antarctic Peninsula usually are free of ice during the summer period, except for those located at relatively higher altitudes.

Antarctic lakes are subject to extreme conditions typical of polar environments: strong variations in irradiance, seasonal changes in day length, and very low temperatures. Thus, freshwater algae must be adapted to these extreme conditions. For example, polar cyanobacteria have developed adaptive strategies such as, adjusting their photosynthetic apparatus by decreasing their photosynthetic capacity at low temperatures (Tang & Vincent, 1999). Another important adaptation of some Antarctic planktonic algae is the capacity to survive for long periods at low light intensities, or even during the winter complete darkness (Priddle et al., 1986). Various authors have mentioned the success of certain algal groups such as Chrysophyceae as being amongst the best adapted to living at low light irradiances due to their mixotrophic behaviour (Sandgren, 1988). In particular, *Ochromonas* sp., one of the dominant taxa in the lakes of Hope Bay, was included by Jones (2000) in the group of the protists whose primary mode of nutrition is heterotrophy and which uses phototrophy only when prey concentrations limit their heterotrophic growth.

Despite the common extreme conditions, Antarctic lakes can exhibit important physical and chemical differences. On one hand, the degree of enrichment determines their phytoplankton abundance and composition. On the other hand, the water stability and the hydrologic changes strongly affect the phytoplankton dynamics. Pose and Izaguirre (1997) described the drastic decrease in phytoplankton abundance and the changes of its composition during the melting and concomitant high water period in Boeckella Lake.

The aim of this paper is to analyse the influence of the physical stability on the phytoplankton assemblages of two lakes at Hope Bay. The difference in the duration of ice-cover between the lakes allows for continuously mixing throughout the summer, while the other one remains almost always inversely stratified.

According to Sommer et al. (1993), an equilibrium state in the phytoplankton community is a period when one, two or a maximum of three species coexist, together contributing more than 80% of the standing biomass, and coexisting for more than two weeks with no significant changes in total biomass. Thus, an equilibrium state is approached as competitive exclusion operates in favour of the dominant species. Contrary to this, disturbances are considered to be the counterparts of an equilibrium state and usually prevent competitive exclusion (Sommer, 1989).

We hypothesise that the presence of a semi-permanent ice cover will generate more stable ecological conditions in a lake, thus allowing steady state assemblages to develop. On the other hand, disturbance generated by wind, promoting frequent mixing events, will discourage the possibility of the establishing of a steady state.

In this paper, we compare the phytoplankton structure and dynamics of two Antarctic lakes throughout two summer seasons. We discuss the algal composition in terms of density and biomass for each lake in relation to the phytoplankton associations proposed by Reynolds et al. (2002).

The lakes

Two lakes of Hope Bay (Antarctic Peninsula), which have been studied since 1991, were selected for this study: Boeckella and Chico. Both lakes have a glacial origin and are situated near the Argentinean Esperanza Station, but differ in their size (surface), altitude, distance to the sea, permanence of their ice-cover and trophic status. Their morphometric features have been previously reported in Izaguirre et al. (1998), while the main differences between them are summarized in Table 1.

Methods

The lakes were sampled weekly on the same day, during the Antarctic summers of 1998 and 1999. Various physical and chemical variables were measured *in situ*: temperature, pH and conductivity with Hanna CT-1 portable meters. Dissolved oxygen was estimated by the Winkler method. Samples for chlorophyll *a* and nutrient analyses were taken in acid-washed PVC flasks and immediately filtered through Whatman GF/F.

Filters for chlorophyll *a* analyses were preserved frozen at −20 °C. Concentrations of the pigment corrected for phaeopigments were determined by spec-

Table 1. Ranges of the main limnological features registered in Lake Chico and Lake Boeckella, during summers 1998 (a) and 1999 (b)

	Chico	Boeckella
(a)		
Temperature (°C)	0–0.3 (**0.04**)	0–4.1 (**2.14**)
pH	7–7.7 (**7.19**)	5.7–7.34 (**6.94**)
Conductivity (μS cm^{-1})	19–62 (**40.78**)	15–49 (**26.56**)
Dissolved oxygen (mg l^{-1})	12.51–15.01 (**13.50**)	11.96–13.31 (**12.57**)
Secchi depth (m)	Bottom	0.59–1.2 (**0.76**)
DRP (mg l^{-1})	0.01–0.17 (**0.06**)	0.1–0.54 (**0.24**)
Nitrites + Nitrates (mg l^{-1})	0.04–1.44 (**0.42**)	0.05–4.85 (**1.25**)
Ammonia (mg l^{-1})	ND–0.12 (**0.06**)	0.13–1.3 (**0.56**)
Chlorophyll *a* (μg l^{-1})	0–2.4 (**0.9**)	1.25–9.44 (**4.86**)
(b)		
Temperature (°C)	0–1.5 (**0.42**)	0–4 (**2.36**)
pH	6.62–7.26 (**6.97**)	6.5–6.79 (**6.68**)
Conductivity (μS cm^{-1})	34–62 (**47**)	25–67 (**48**)
Dissolved oxygen (mg l^{-1})	10.78–13.81 (**12.83**)	11.95–13.9 (**12.8**)
Secchi depth (m)	Bottom (>1)	0.43–0.6 (**0.46**)
DRP (mg l^{-1})	0.06–0.465 (**0.198**)	0.237–1.053 (**0.657**)
Nitrites + Nitrites (mg l^{-1})	0.183–0.835 (**0.382**)	0.36-3.11 (**1.256**)
Ammonia (mg l^{-1})	ND–0.34 (**0.059**)	0.36–2.3 (**1.299**)
$Si0_2$ (mg l^{-1})	0.28–1.29 (**0.73**)	0.58–1.08 (**0.79**)
Hardness (Mg) (mg l^{-1})	2.38–4.4 (**3.34**)	2.27–4.07 (**3.17**)
Hardness (Ca) (mg l^{-1})	0–5.36 (**1.51**)	0–1.71 (**0.59**)
OCD (mg l^{-1})	0–6 (**2.62**)	0–10 (**6.12**)
Chlorophyll *a* (μg l^{-1})	ND–2.45 (**1.71**)	ND–8.71 (**2.66**)

Mean values are indicated between brackets.

trophotometry before and after acidification (HCl 0.1 N), using hot ethanol (60–70 °C) as solvent. Equations published by Marker et al. (1980) were used for the calculations.

Soluble Reactive Phosphorus (SRP), nitrates, ammonia, silicates, oxygen chemical demand (OCD), and hardness were determined at the research station laboratory, using a Hach DR/890 colorimeter and their corresponding kits of reagents.

Taxonomic analyses of the phytoplankton communities were based on microscopic observation *in situ* of living specimens. Previous floristic reports published for these lakes (Izaguirre et al., 1993; Vinocur & Izaguirre, 1994; Tell et al., 1995; Izaguirre et al., 1998) were used for the identifications of the taxa present.

Samples for quantitative analyses of phytoplankton were separated into two fractions. One of them, for net and nano-phytoplankton enumeration, was preserved with acidified 1% Lugol's iodine solution. The one for picoplankton counts was fixed with 2% ice-cold glutaraldehyde.

Counts of micro and nanoplanktonic fractions were performed using an inverted microscope according to Utermöhl (1958). Replicate chambers were left to sediment for at least 24 h, and counting errors were estimating following recommendations given by Venrick (1978). Using epifluorescence microscopy, autotrophic picoplankton (APP) cells were observed and counted from the fluorescence given off by photosynthetic pigments (Callieri & Pinolini, 1995). Samples were filtered on 0.2 μm black polycarbonate filters Isopore GTPB 02500, which were then mounted on a microscope slide with a drop of immersion oil for fluorescence (Immersol 518 F). Each filter was examined for pigment autofluorescence with a Zeiss Axioplan Microscope equipped with an HBO 50W lamp, a plan-Apochromat 100X objective and a filter set for blue light excitation (BP 450–490 nm, FT 510 nm, LP 520 nm) and green light excitation (BP 546 nm, FT 580 nm, LP 590 nm) (Wynn-Williams, 1992; Rankin et al., 1997). Cell volume was calculated for each species on the basis of the average cell dimensions using appropriate volume formulae (Hillebrand et al., 1999). In all cases, individuals were counted and for colonial or filamentous algae the size and/or number of cells corresponding to a standard individual were established.

Phytoplankton diversity was calculated using the Shannon-Wiener index (Magurran, 1988) using the biovolume data.

Steady states were studied for each lake during summer 1999, analysing the fluctuations of specific and total biomass as well as calculating the variation coefficient.

Results

The range and mean values of limnological features for the water bodies surveyed are summarized in table 1. In both cases waters were well oxygenated ($>$12.57 mg O_2 ml^{-1}), showing conditions of oversaturation throughout the two summer seasons. Mean temperatures were similar during the two studied periods, with Lake Chico values near 0 °C exhibiting only brief ice-free periods. Lake Boeckella remained unfrozen throughout most of the survey, but exhibited an important ice-cover event at the beginning of summer 1998 and towards the end of summer 1999. Variations in conductivity values throughout the summer were associated with the hydrological events such as melting and freezing periods.

Differences in trophic states between these lakes were evident from the differences in nutrients (SRP, nitrates and ammonia) and chlorophyll *a* concentrations. Values of all nutrients were much lower in Lake Chico, and although some figures exceeded the upper limit of the oligotrophic range according OECD criteria, the historical records and the chlorophyll *a* concentrations allow us to classify this water body as oligo-mesotrophic with a recent trend to natural eutrophication (Izaguirre et al., 1998; Izaguirre et al., in press). Lake Boeckella in turns, can be classified as meso-eutrophic according to the chlorophyll *a* data, but clearly eutrophic regarding its nutrient load.

A total of 48 taxa were identified from Lake Chico and 40 from Lake Boeckella (Table 2). In both lakes apparently different species of the genera *Ochromonas* and *Chlamydomonas* mainly represented the nanophytoplankton. Tychoplanktonic species were very important in both water bodies and were mainly represented by epilithic Bacillariophyceae and filamentous Cyanobacteria. Cryobiontic Chlorophyceae were only important during brief periods associated with the melt events.

Micro and nano-phytoplankton abundance and biovolume exhibited differences between the lakes surveyed. In Lake Chico, abundances varied between 51–214 ind. ml^{-1} during 1998 and 118–682 ind. ml^{-1} during 1999, and remained rather constant throughout the two study periods. Changes in biovolumes ranged between 1890–60045 μm^3 ml^{-1} in 1998, and 15811–99299 μm^3 ml^{-1} in 1999 (Fig. 1a, b).

During the first study period, micro and nanophytoplankton density in Lake Boeckella was similar to that recorded from Lake Chico, ranging from 87–187 ind. ml^{-1}. Biovolumes, were also similar varying from 3583–50198 μm^3 ml^{-1}. On the contrary, during summer 1999 the differences between the lakes were more pronounced. The density in Lake Boeckella ranged from 44–3690 ind. ml^{-1}, and the biovolumes between 3524–553881 μm^3 ml^{-1}. Both parameters continually increases towards the end of the summer (Fig. 2a, b).

Pico-phytoplankton was also more abundant in Lake Boeckella than in Lake Chico, ranging from 5.04 $\times$ 10^4 to 3.64 $\times$ 10^5 ind. ml^{-1} in the former and from 3.66 $\times$ 10^4 to 8.79 $\times$ 10^4 ind. ml^{-1} in the latter.

Since the patterns of density and biovolume of pico-phytoplankton fluctuations were almost identical, we illustrated only the latter (Fig. 3). Clear differences were observed between the lakes. The peak in picoplankton abundances in Lake Boeckella appeared from 11/2/99 to 25/3/99. In Lake Chico, more constant values were seen and no peaks were recorded. This picoplankton fraction was mostly dominated by picocyanobacteria, which according to Callieri & Stockner (2002) probably belong to the genus *Cyanobium* sp. and *Synechococcus* sp.

Figure 4, shows the difference in contribution of each size fraction to the total phytoplankton biovolume between the water bodies. In Lake Boeckella the pico-phytoplankton contribution showed important changes along summer 1999, constituting more than 80% during the first five sampling dates, and diminishing in importance towards the end of the summer as the lake became frozen. However, in Lake Chico the contribution of the pico-phytoplankton was rather constant and its biovolume never exceeded 70%.

The phytoplankton biodiversity in both lakes showed the typical low values for high latitude water systems (Fig. 5).

Taking into account the associations proposed by Reynolds et al. (2002), it is evident that both lakes

Table 2. Floristic composition of the autotrophic plankton at each study site during the surveyed summers

	Summer 1998		Summer 1999	
	Chico	Boeckella	Chico	Boeckella
Heterokontophyta				
Bacillariophyceae				
Achnanthes laevis var. *ninckei* (Guerm.et Mang.) L.-Bert.	x			x
Achnanthes subatomoides (Hust.) L.-Bert. et Archibald	x	x		x
Achnanthes sp. 1	x			
Achnanthes spp.	x			x
Gomphonema sp.				x
Luticola muticopsis (V. Heurck) D. G. Mann		x		x
Navicula laticeps Husdedt	x			
Navicula soehrensis var. *hassiaca* (Krasske) L.- Bertalot	x			
Navicula tabellariaeformis Krasske	x			
cf. *Navicula contenta* Grun.	x			
Navicula spp.	x			
Nitzschia gracilis Hantz.			x	x
Nitzschia homburgiensis L.-Bert.	x	x	x	x
Nitzschia paleacea Grun.			x	x
cf. *Nitzchia acicularis* (Kützing) W. Smith	x			
Nitzschia sp. 1	x		x	x
Nitzschia sp. 2			x	x
Pinnularia borealis (Ehr.)		x		
Pinnularia crucicola Freng.	x			
Pinnularia microstauron (Ehr.) Cl.	x	x	x	x
Chrysophyceae				
Chromulina sp. 1			x	x
Chromulina sp. 2				x
cf. *Chrysidalis*			x	x
Ochromonas sp.	x			x
Chrysophyceae (n.i)	x	x		x
Chrysophyceae cysts	x	x	x	x
Tribophyceae				
Tribonema australis Vinocur *et* Pizarro				x
Tribonema monochloron Pascher et Geitler	x	x	x	x
Tribonema utriculosum (Kütz.) Hazen	x	x		x
Chlorophyta				
Carteria sp.			x	
Chlamydomonas sp. 1	x	x		x
Chlamydomonas sp. 2	x		x	
Chlamydomonas sp. 3				x
Chlamydomonas aff. *braunii* Goroschankin		x	x	x
Chlorogonium sp.				x
Cylincrocystis brebisonii (Menegh. ex Ralfs) De Bary	x			
Cylindrocystis crassa De Bary	x		x	x
cf. *Hyalogonium* sp.			x	
Trochiscia sp.			x	x
Monoraphidium contortum (Thur.) Kom. Legm	x			
Palmella miniata var. *aequalis* Näg sensu G.M. Smith	x			
Prasiococcus calcarius (Boye Petersen) Vischer	x			
Prasiola callophylla (Carm.) Menegh.	x		x	x
Raphidonema nivale Lagerh.	x		x	x
Ulothrix sp.			x	x

Table 2. Continued

Cyanobacteria				
Aphanocapsa elachista W. et G. S. West				x
Chroococcus minutus (Kütz.) Näg.	x		x	x
Eucapsis minuta Fritsch.				x
Gloeocapsa ralfsiana (Harv.) Kütz.	x		x	x
Isocystis pallida Woronichin			x	x
Leptolyngbya antarctica (W. et. G. S. West) Anag. et Kom.	x	x		x
Leptolyngbya fragilis (Gomont) Anag. et Kom.	x	x	x	x
Leptolyngbya frigida (Fritsch) Anag. et Kom.	x	x	x	x
Leptolyngbya tenuis (Gom.) Anag. et Kom.				x
Myxosarcina concinna Printz				x
Phormidium autumnale (Ag.) Gomont	x		x	x
Phormidium spp.	x	x		
Pseudoanabena catenata Lautern.				x

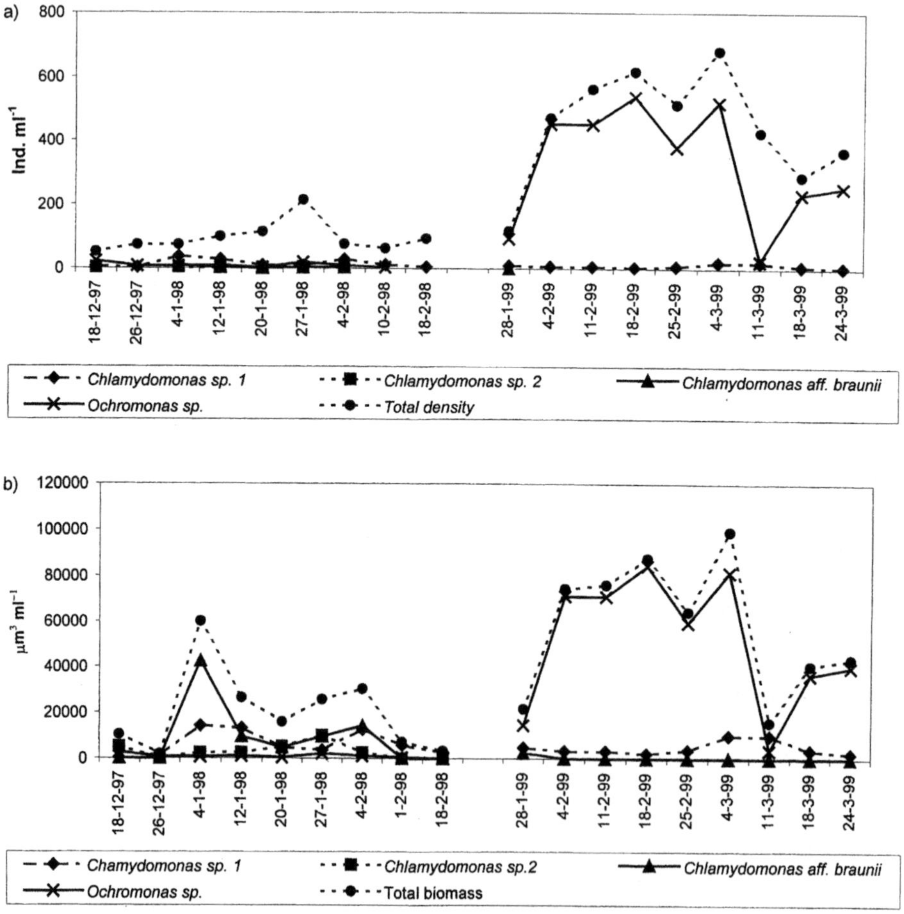

Figure 1. Summer fluctuations in phytoplankton abundances (a) and phytoplankton biovolumes (b) in Lake Chico, during summers 1998 and 1999.

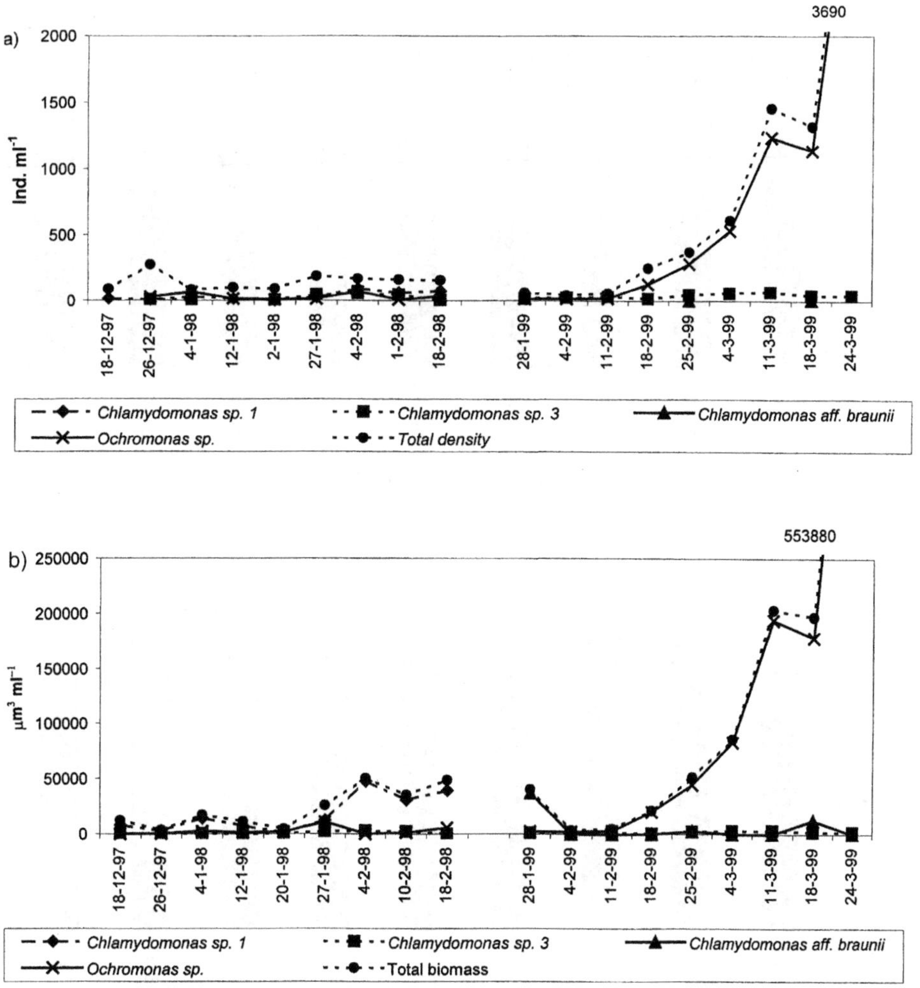

Figure 2. Summer fluctuations in phytoplankton abundances (a) and phytoplankton biovolumes (b) in Lake Boeckella, during summers 1998 and 1999.

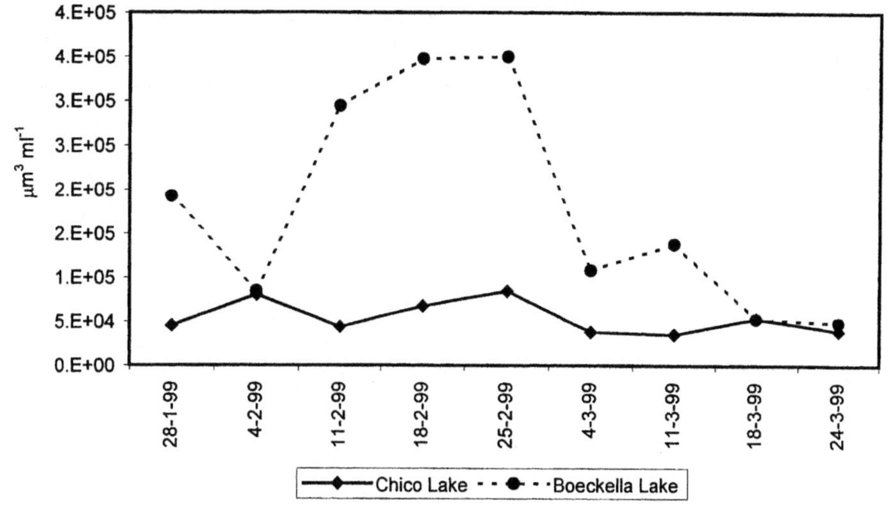

Figure 3. Autotrophic picoplankton biovolume in Lake Chico and Lake Boeckella during summer 1999.

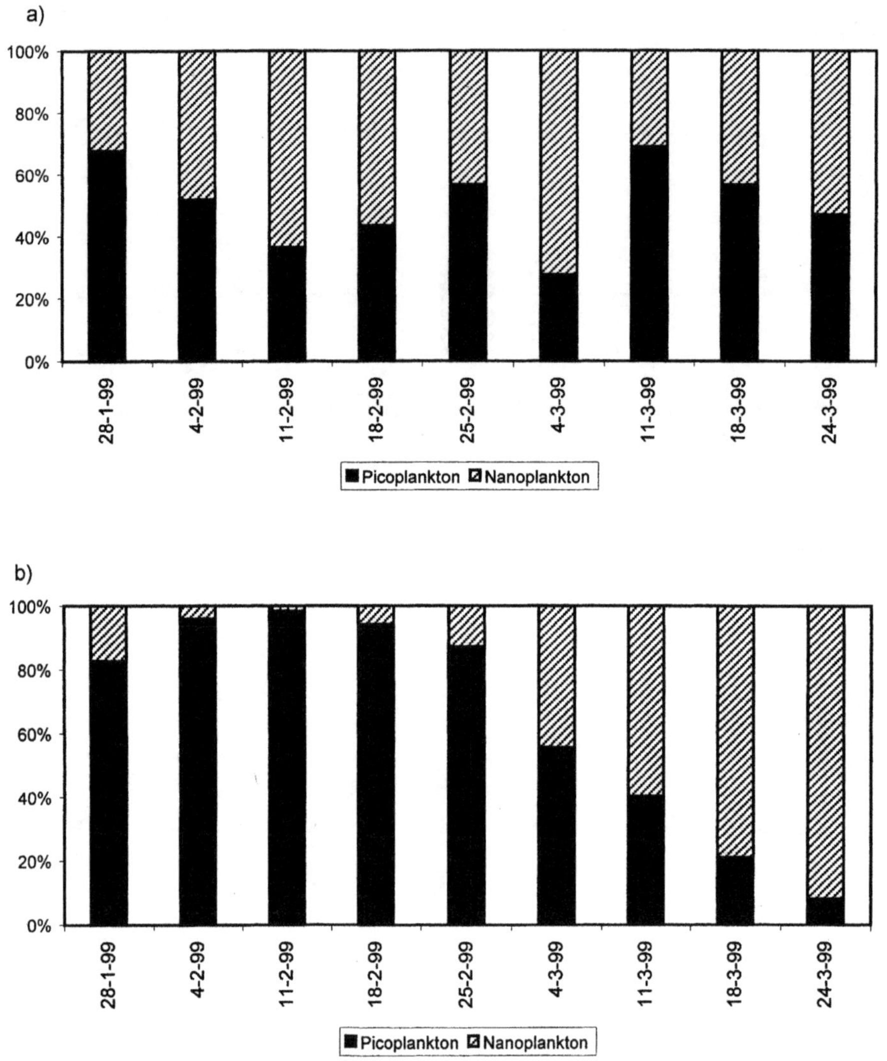

Figure 4. Proportion of autotrophic nanoplankton and picoplankton size fractions in Lake Chico (a) and Lake Boeckella (b), during summer 1999.

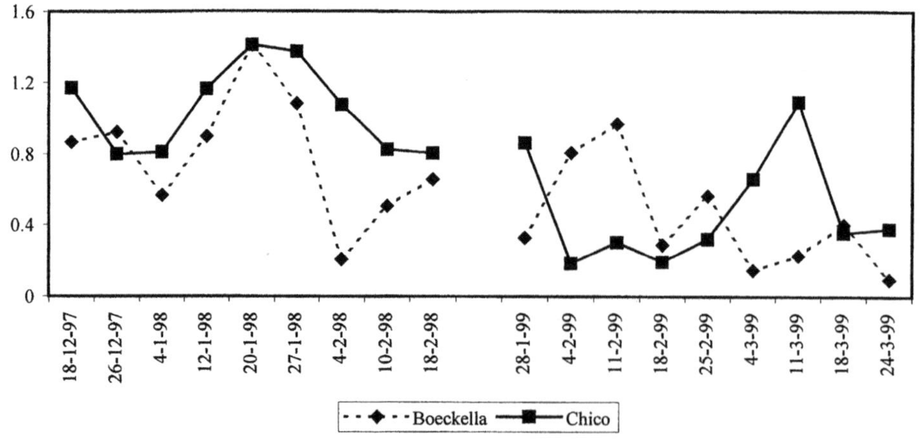

Figure 5. Shannon-Wiener index for phytoplankton biodiversity in Lake Chico and Lake Boeckella, during summers 1998 and 1999.

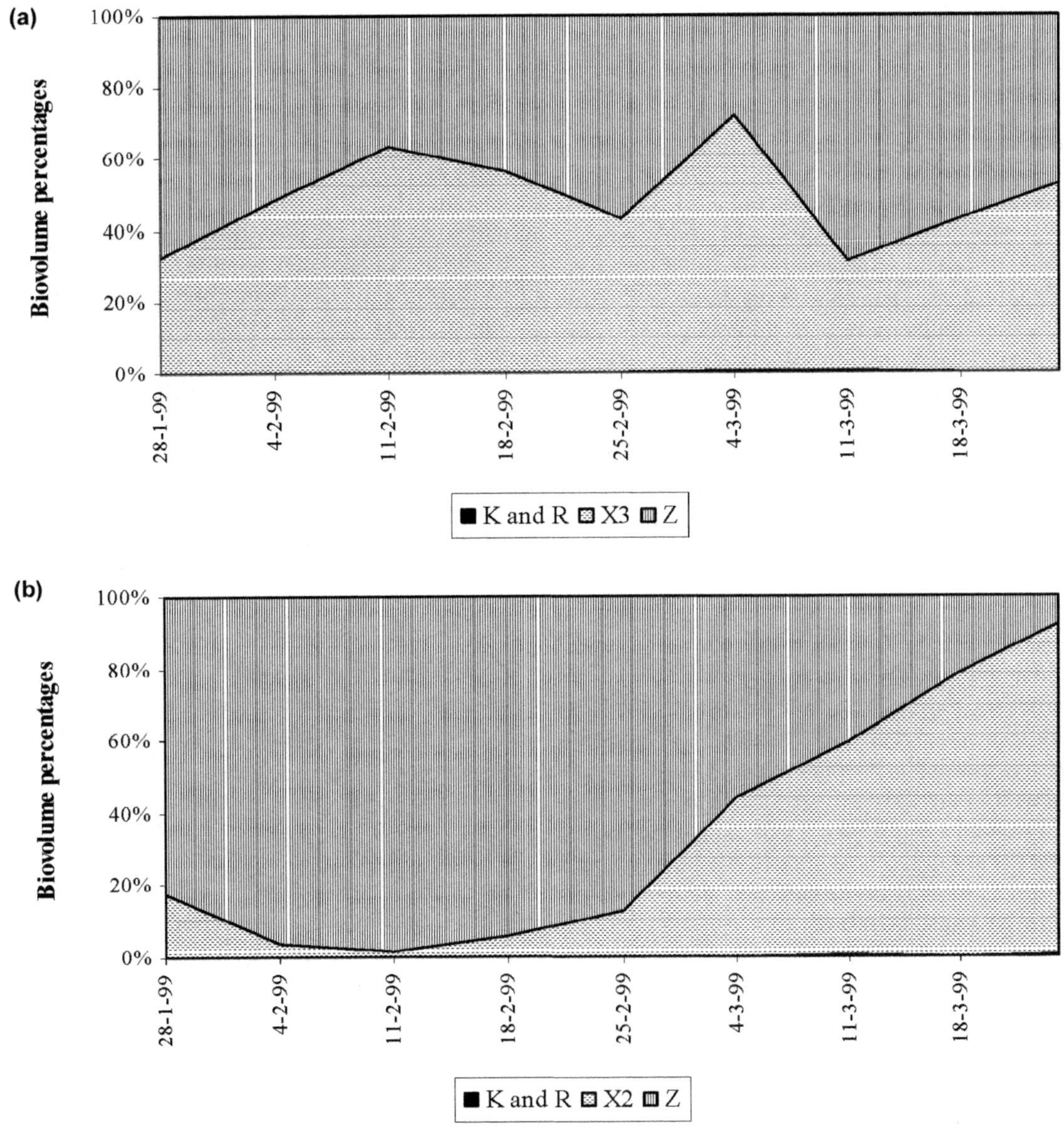

Figure 6. Reynold's functional groups of phytoplankton in Lake Chico (a) and Lake Boeckella (b).

shared some of them, although noticeable differences could be noted in their duration (Fig. 6a, b). The oligotrophic lake is represented by the associations **K**, **R**, $\mathbf{X_3}$ and **Z**, while in Lake Boeckella the functional groups were **K**, **R**, $\mathbf{X_2}$ and **Z**. The **K** association was mainly represented by *Eucapsis minuta*, *Choococcus minutus* and *Aphanocapsa elachista*. On the other hand, the **R** association was composed of *Isocystis pallida*, *Leptolyngbya antarctica*, *Leptolyngbya tenue* and *Pseudoanabaena catenata*. In both cases, the most relevant associations were **X** and **Z**. The former is composed of small nanoplanktonic species mostly unicellular (*Chlamydomonas* spp., *Chlorogonium* sp., *Raphidonema nivale, Carteria* sp., *Ochromonas* sp., *Chromulina* spp. and cf. *Chrysidalis*) that can live along the trophic spectrum. Reynolds et al. (2002) proposed three associations ($\mathbf{X_1}$, $\mathbf{X_2}$ and $\mathbf{X_3}$) on the basis of a decreasing nutrient gradient. On the other hand, the **Z** association includes the prokaryotic picophytoplankton.

In Lake Chico, the proportion of the two main algal functional groups remained rather constant, whereas in Lake Boeckella a clear shift between them was observed. Thus, from the end of January until late

February 1999 **Z** association dominated, followed by the X_2 association which replaced it.

Discussion

Differences in the trophic status between both lakes were reported in previous published and unpublished work from this region (Izaguirre et al., 1998; Izaguirre et al., unpublished). Data concerning the concentration of nutrients, conductivity and chlorophyll *a* clearly provide evidence of the oligo-mesotrophic condition of Lake Chico with a recent trend to natural eutrophication. This fact is probably related to the relatively higher temperatures recorded during the last years in some regions of Antarctica, and particularly in Hope Bay, which provokes the thawing of neighboring areas to the lakes. As this region is subjected to the influence of sea animals in summer time, the runoff increases the inputs of nutrients to the water bodies. On the other hand, Lake Boeckella exhibits an advanced eutrophication level, which results from the continuous supply of nutrients derived from a large nearby penguin rookery every summer. This lake showed a progressive nutrient enrichment, which became more evident as its water level diminishes. This was particularly noticeable in this lake when a dramatic natural decrease occurred during summer 2001. A clear peak was observed in most of the eutrophication indicators (Izaguirre & Almada, 2001).

Despite the differences in trophic status between the lakes above described, we stress in this paper the role of the persistent ice-cover as a physical characteristic affecting the phytoplankton structure and dynamics. The differences in the ice cover between the high altitude Lake Chico with only short ice free periods and Lake Boeckella with a long ice free season, resulted in contrasting behavior related to mixing events. Lake Boeckella shows a typical cold polymictic regime, whereas Lake Chico can be described as an oligomictic water body. As pointed out by Andreoli et al. (1996), life in Antarctic lakes is influenced by the presence of an ice cover. Ice cover isolates the water body from the external environment, by protecting the water profile from the wind action, as well as reducing light penetration.

Regarding their algal communities, these lakes exhibit important differences. A sparse phytoplankton and well developed perennial phytobenthos characterize oligotrophic Lake Chico while Lake Boeckella shows a more abundant phytoplankton community as a result of its higher trophic state. As has been reported by other authors (Priddle, 1980; Hawes, 1990; Izaguirre et al., 1998) when the Antarctic lakes are subjected to increasing nutrient inputs, the dominant communities shift from phytobenthos to phytoplankton.

In relation to the nano-phytoplankton composition, the dominant taxa recorded during this study coincide with those reported previously for these lakes (Izaguirre et al., 1993; Pose and Izaguirre, 1997; Izaguirre et al., 1998; Izaguirre et al., unpublished). Lake Chico is continually dominated by small flagellated chrysophytes whereas Lake Boeckella shows a succession from volvocaleans taxa during the open water season to chrysophytes when the lake freezes. This under-ice prevalence of small flagellates has been reported in the literature (Priddle et al., 1986; Burch, 1988, Sandgren, 1988, Izaguirre et al., 1998), and their success is attributed to several causes. For example, the small flagellated chrysophytes, which frequent in this type of environments, are well adapted to deal with low light intensities, and have been reported as nutritionally opportunistic (Sandgren, 1988). Priddle et al. (1986) pointed out that the presence of flagella in algae from Antarctic or Arctic freshwater environments gives them the capability for migration and allow them maintain their position in a water column. This is an advantage in stratified lakes.

Regarding the algal biovolume values, the interannual variations observed in both lakes can be attributed to the sampling period, which was different in 1998 and 1999. The second year samplings started at the end of January, when temperatures were higher, probably favoring a greater algal growth. In particular, in a previous study (Izaguirre et al., 1998), the peak of total phytoplankton biovolume was also reported for a similar period of the summer season.

The mode of nutrition of the dominant species is also a key factor during some periods of the year, and probably results in their success in frozen Antarctic lakes. The phytoplankton community cannot be viewed strictly as autotrophic as many authors (Bird & Kalff, 1986; Olrick & Nauwerck, 1993; Jones, 1994; Laybourn-Parry et al., 2000) provide evidence to the contrary. Apparently some natural phytoplankton are also phagotrophic and are capable of supplying their energy requirements by ingesting picoplankton. It is important to point out that previous studies revealed a high abundance of both autotrophic and heterotrophic picoplankton in these lakes (Izaguirre et al., unpublished). In particular, bacterioplankton is supported by sufficient levels of DOC, which were measured by Rocco et al. (2002).

One typical mixotrophic taxon in these lakes is *Ochromonas* sp., which was described by Jones (2000) as a protist whose primary mode of nutrition is heterotrophy and which employs phototrophy only when prey concentration limit the heterotrophic growth. Thus, mixotrophy can be an adaptive advantage under two different situations. On one hand, under conditions of limiting nutrients such as those found in Lake Chico, mixotrophs survive by ingesting particles or uptaking dissolved organic carbon. On the other hand, in low light environments, or even in darkness, these organisms may switch from autotrophic to heterotrophic metabolism. The latter situation occurs in both systems when they freeze. Tolotti et al. (2003) discussed the ability of Chrysophyceae to live in enviroments with extreme temperatures and light availability in detail. Species of *Chlamydomonas* constitute another group of phytoplankton well-adapted to these environments. This genus was reported several times from maritime Antarctic lakes (Mataloni et al., 1998, 2000; Izaguirre et al., 2001), and its increase in abundance has been correlated with enrichment of the water bodies (Hawes, 1990; Ellis-Evans, 1991). In agreement with this statement, *Chlamydomonas* spp. was found to be more relevant in the meso-eutrophic lake.

Low phytoplankton diversity values might be related to the extreme environmental characteristics of Antarctic lakes. The analysis of the Reynolds' associations revealed that both lakes share almost the same functional groups. It is important to point out that, due to the complexity of dominant algae and difficulty in species identification, it could be useful that in the future taxonomic studies involve new molecular techniques. In Lake Chico many different species of Chrysophyceae were observed. These species are nearly impossible to identify using the traditional microscopic techniques. In spite of this, the ecological adaptations and requirements of the dominant taxa allow us to include them in the categories proposed by Reynolds et al. (2002). To explain the similarities in the functional groups recorded in each of the two lakes of contrasting trophic status and physical stability, we hypothesise that it is the very extreme conditions of Antarctica that force the community to select for only those taxa that can develop in this extreme environment. The effect of stress factors, such as harsh climate, accounts for the dominance of only a few species in lakes (Naselli-Flores et al., 2003; Padisák et al., 2003).

Tychoplanktonic species found in these lakes do not belong to any of the functional groups proposed by Reynolds and were not taken into account in this study. We consider that further research should include this fraction as it could be an important component of the plankton community and might influence the structure of the phytoplankton assemblages.

In terms of the temporal phytoplankton succession, in lakes of Maritime Antarctica the most important changes occurred during the warmer period (spring–summer). In winter, many species form vegetative resting stages that may persist in the water column or germinate from the sediments in spring, or they use an over wintering strategy such as the capacity of a heterotrophic mode of nutrition (Mc Knight et al., 2000). In spring, as light and temperature constraints relax, photosynthesis begins (Priddle et al., 1986), and algal succession takes place with all the important population changes occurring over a period of only 3–5 months. During the summer period, the melting, and mixing events, produce drastic physical changes that influence phytoplankton. In the case of Antarctic lakes, the changes described for autogenic succession in the PEG MODEL (Sommer et al., 1986) can occur in a relatively shorter time compared to other latitudes. Nevertheless, these shallow lakes are subjected to frequent mixing episodes and in these cases the succession may be continuously disrupted, as stated in the IDH (Intermediate Disturbance Hypothesis) applied to phytoplankton ecology (Reynolds et al., 1993). Due to its shallowness, the absence of an ice-cover and the strong winds of Hope Bay, this situation is found in Lake Boeckella during the summer period, while the longer periods of physical stability are associated with the persistent ice-cover on Lake Chico, during the same period. In spite of this, it must be taken into account that even under the presence of an ice-cover, convectional movements can occur in the water profile due to warming-cooling cycles which generate density changes.

If we consider the definition of 'equilibrium' given by Sommer et al. (1993), we can recognize periods of at least 2 weeks, with a dominance of one, two or a maximum of three algal species which together make up more than 80% of the total standing biomass composition in both lakes. Lake Chico showed periods of more than two weeks without significant changes in phytoplankton biomass, supported by low variation coefficient values. Of the two lakes, it appears that steady state only occurred in this water body. Dokulil & Teubner (2003) examined the existence of steady-state conditions during thermal stratification in alpine lakes, Although these authors found that always more than three species dominated the phytoplankton, the persistence of the dominant species for more than two weeks suggests that equilibrium could occur under stratification conditions.

Considering the ratio of nano-phytoplankton to pico-phytoplankton, a greater constancy was observed in Lake Chico. However, in Lake Boeckella there was an interesting succession from autotrophic picoplankton to flagellated nanoplankton (mixotrophs) that occurred towards the end of the summer, as the lake froze. Although future studies on the nutritional mode of the species in these lakes would be advisable, there is a clear indication that the reduction in light favored the growth of mixotrophic species. McKnight et al. (2000), study of phytoplankton dynamics in Lake Fryxell (Victoria Land - Antarctica) during winter darkness, supports the increased heterotrophic growth in winter and the dominance of a mixotrophic behavior throughout the year.

Autotrophic picoplankton, were observed to be very important in the both studied lakes, however the abundance differed between them. Lake Chico had lower numbers that remained rather constant over the summer while Lake Boeckella showed variation exceeding one order of magnitude the values reported for the oligotrophic lake. This finding coincides with previous reports for Antarctic lakes. These organisms have a high nutrient-scavenging ability and can be an important part of the pelagic community in terms of biomass, particularly in more oligotrophic conditions (Stockner, 1988; Vincent, 2000; Callieri & Stockner, 2000; Stockner et al., 2000). According to different authors, in temperate lakes the pico-phytoplankton importance in terms of biomass diminishes as the trophic level increases (Sommaruga & Robarts, 1997; Vörös et al., 1998; Agawin et al., 2000). On the other hand, studies carried out in lakes at higher latitudes, revealed that even in eutrophic conditions the pico-phytoplankton can be an important fraction of the phytoplankton community (Ellis-Evans, 1991; Andreoli et al., 1996; Izaguirre et al., 2001). Thus, trophic level alone cannot account for the importance of this fraction in a lake (Callieri & Stockner, 1997). Other variables such as light availability and grazing have to be considered as factors affecting the pico-phytoplankton fraction (Stockner et al., 2000; Reynolds et al., 2002).

Acknowledgements

This project was supported by grants of the "Agencia de Promociones Científicas y Técnicas" (FONCYT), and CONICET (Argentina). The Antarctic expeditions were supported by a cooperative project between Argentinean Antarctic Institute and the University of Buenos Aires. We also wish to thank the members of the Argentinean Esperanza Station for on site logistic support.

References

Agawin, N. S. R., C. M. Duarte & S. Agusti, 2000. Nutrient and temperature control of the contribution of picoplankton to phytoplankton biomass and production. Limnol. Oceanogr. 45: 591–600.

Andreoli, C., N. Rascio, I. Moro, L. Scarabel, C. Tolomio, F. Dalla Vecchia & L. Maseiro, 1996. Ecological, ultrastructural and physiological features of Antarctic microalgae. In Prisco G., S. Focasdi & P. Liporini (eds.), Proceedings of the Third Meeting on Antarctic Biology: 275–285.

Bird, D. F. & J. Kalff, 1986. Bacterial grazing by planktonic lake algae. Science 231: 493–494.

Burch, M. D., 1988. Annual cycle of phytoplankton in Ace Lake, an ice covered, saline meromictic lake. Hydrobiologia 165: 59–75.

Butler, H. G., 1999. Seasonal dynamics of the planktonic microbial community in a maritime Antarctic lake undergoing eutrophication. J. Plankton Res. 21: 2393–2419.

Callieri, C. & J. Stockner, 1997. Pico-blues success across freshwater trophic gradient. In Proceedings of 7th International Conference on Lakes Conservation and management. S. M. de los Andes.

Callieri, C. & M. L. Pinolini, 1995. Picoplankton in Lake Maggiore, Italy. Int. Rev. ges. Hydrobiol. 80: 491–501.

Callieri, C. & J. Stockner, 2000. Picocyanobacteria success in oligotrophic lakes: fact or fiction? J. Limnol. 59: 72–76.

Callieri, C. & J. Stockner, 2002. Freshwater autotrophic picoplankton: a review. J. Limnol. 61: 1–14.

Dokulil, M. T. & K. Teubner, 2003. Steady state phytoplankton assemblages during thermal stratification in deep alpine lakes. Do they occur? Hydrobiologia 502 (Dev. Hydrobiol. 172): 65–72.

Ellis-Evans, J. C., 1991. Numbers and activity of bacterio- and phytoplankton in contrasting Maritime Antarctic lakes. Int. Ver. für theor. angew. Limnol.: Verh. 24: 1149–1154.

Ellis-Evans, J. C., 1996. Microbial diversity and function in Antarctic freshwater ecosystems. Biodiversity and Conservation 5: 1395–1431.

Hawes, I., 1983. Nutrients and their effects on phytoplankton populations in lakes on Signy Island, Antarctica. Polar Biol. 2: 115–126.

Hawes, I., 1990, Eutrophication and vegetation development in Maritime Antarctic Lakes. In Kerry, K. R. & G. Hempel (eds), Antarctic Ecosystems. Ecological Change and Conservation. Springer-Verlag, Berlin, Heidelberg, New York: 83–90.

Heywood, R. B., 1977. Antarctic freshwater ecosystems – a review and synthesis: In Llano G. A. (ed.), Adaptations within Antarctic Ecosystems. Gulf Publishing, Houston: 801–828.

Hillebrand H., C-D. Durselen, D. Kirshtel, U. Pollingher & T. Zohary, 1999. Biovolume calculation for pelagic and benthic microalgae. J. Phycol. 35: 403–424.

Izaguirre, I., G. Mataloni, A. Vinocur & G. Tell, 1993. Temporal and spatial variations of phytoplankton from Boeckella Lake (Hope Bay, Antarctic Peninsula). Antarctic Sci. 5: 137–141.

Izaguirre, I., A. Vinocur, G. Mataloni & M. Pose, 1998. Comparison of phytoplankton communities in relation to trophic status in lakes from Hope Bay (Antarctic Peninsula). Hydrobiologia 369/370: 73–87.

Izaguirre, I., G. Mataloni, L. Allende & A. Vinocur, 2001. Summer fluctuations of microbial planktonic communities in a eutrophic lake – Cierva Point, Antarctica. J. Plankton Res. 23: 1095–1109.

Izaguirre, I., L. Allende & C. Marinone. Comparative study of the planktonic communities from lakes of contrasting trophic status at Hope Bay (Antarctic Peninsula). J. Plankton Res., 25 (9).

Izaguirre, I. & P. Almada, 2001. Cambios en las características limnológicas y en biomasa fitoplanctónica del lago Boeckella (Bahía Esperanza) asociados al brusco descenso en su nivel hidrométrico. Contribución Instituto Antártico Argentino 553: 1–6.

Jones, R. I., 1994. Mixotrophy in planktonic protists as a spectrum of nutritional strategies. Mar. Microbial Food Webs 8: 87–96.

Jones, R. I., 2000. Mixotrophy in planktonic protists: an overview. Freshwat. Biol. 45: 219–226.

Laybourn-Parry, J., E. C. Roberts & E. M. Bell, 2000. Mixotrophy as a survival strategy among planktonic protozoa in Antarctic lakes. In Davison, W. C. Howard-Williams & P. Broady (eds), Antarctic Ecosystems: Model for a Wider Ecological Understanding. The Caxton Press, Christchurch: 33–40.

Magurran, A. E., 1988. Ecological Diversity and Its Measurements. Chapman and Hall, London. 179 pp.

Marker, A. F. H., A. Nusch, H. Rai & B. Riemann, 1980. The measurement of photosynthetic pigments in freshwater and standardization of methods: conclusions and recommendations. Archiv für Hydrobiol./Ergeb. Limnol. 14: 91–106.

Mataloni, G., G. Tesolín & G. Tell, 1998. Characterization of a small eutrophic Antarctic lake (Otero Lake, Cierva Point) on the basis of algal assemblages and water chemistry. Polar Biol. 19: 107–114.

Mataloni, G., G. Tesolin, F. Sacullo & G. Tell, 2000. Factors regulating summer phytoplankton in a highly eutrophic Antarctic lake. Hydrobiologia 432: 65–72.

Mc Knight, D. M., B. L. Howes, C. D. Taylor & D. D. Goehringer, 2000. Phytoplankton dynamics in a stably stratified Antarctic lake during winter darkness. J. Phycol. 36: 852–861.

Naselli-Flores, L., J. Padisák, M. T. Dokulil & I. Chorus, 2003. Equilibrium/steady-state concept in phytoplankton ecology. Hydrobiologia 502 (Dev. Hydrobiol. 172): 395–403.

Olrik, K. & A. Nauwerck, 1993. Stress and disturbance in phytoplankton community of a shallow, hypertrohic lake. Hydrobiologia 249: 15–24.

Padisák, J., G. Borics, G. Fehér, I. Grigorszky, I. Oldal, A. Schmidt & Z. Zámbóné-Doma, 2003. Dominant species, functional assemblages and frequency of equilibrium phases in late summer phytoplankton assemblages in Hungarian small shallow lakes. Hydrobiologia 502 (Dev. Hydrobiol. 172): 157–168.

Pose, M. & I. Izaguirre, 1997. Sucesión estival de las poblaciones fitoplanctónicas en el lago Boeckella (Bahía Esperanza, Antártida). Contribución Instituto Antártico Argentino 449: 1–27.

Priddle, J., 1980. The production ecology of benthic plants in some Antarctic lakes. 1. *In situ* production studies. J. Ecol. 68: 141–153.

Priddle, J., I. Hawes, J. C. Ellis-Evans & T. J. Smith, 1986. Antarctic aquatic ecosystems as habitats for phytoplankton. Biol. Rev. 61: 199–238.

Rankin, L. M., P. D. Franzmann, T. A. Michelin & H. R. Burton, 1997. Seasonal distribution of picoCyanobacteria in Ace Lake, a marine- derived Antarctic lake. In Battaglia, B., J. Valencia & D. W. H. Walton (eds), Antarctic Communities. Species, Structure and Survival. Cambridge University Press, Cambridge: 178–184.

Reynolds, C. S., J. Padisák & U. Sommer, 1993. Intermediate disturbance in the ecology of phytoplankton and the maintenance of species diversity: a synthesis. In Padisák, J., C. S. Reynolds & U. Sommer (eds), Intermediate Disturbance Hypothesis in Phytoplankton Ecology. Developments in Hydrobiology 81, Kluwer Academic Publishers, Dordrecht: 157–171.

Reynolds, C. S., V. Huszar, C. Kruk, L. Naselli-Flores & S. Melo, 2002. Towards a functional classification of the freshwater phytoplankton. J. Plankton Res. 24: 417–428.

Rocco, V. E., O. Oppezzo, R. Pizarro, R. Sommaruga, M. Ferraro & H. E. Zagarese, 2002. Ultraviolet damage and counteracting mechanisms in the copepod *Boeckella poppei* from Antarctic freshwater lakes. Limnol. Oceanogr. 47: 837–847.

Sandgren, C. D., 1988. Growth and reproductive strategies of freshwater phytoplankton. Cambridge University Press, Cambridge: 442 pp.

Sommaruga, R. & R. D. Robarts, 1997. The significance of autotrophic and heterotrophic picoplankton in hypereutrophic ecosystems. FEMS Microbiol. Ecol. 24: 187–200.

Sommer, U., 1989. Maximal growth rates of Antarctic phytoplankton: only weak dependence on cell size. Limnol. Oceanogr. 34: 1109–1112.

Sommer, U., Z. M. Gliwicz, W. Lampert & A. Duncan, 1986. The PEG Model of seasonal succession of planktonic events in freshwaters. Archiv für Hydrobiol. 106: 433–471.

Sommer, U., J. Padisák, C. S. Reynolds & P. Juhász-Nagy, 1993. Hutchinson's heritage: the diversity disturbance relationship in phytoplankton. Hydrobiologia 249: 1–7.

Stockner, J. G., 1988. Phototrophic picoplankton: an overview from marine and freshwater ecosystems. Limnol. Oceanogr. 33: 765–775.

Stockner, J. G, C. Callieri & G. Cronberg, 2000. Picoplankton and other non-bloom-forming Cyanobacteria in lakes. In Whitton, B.A. & M. Potts (eds), The Ecology of Cyanobacteria. Kluwer Academic Publishers, Dordrecht: 195–231.

Tang, E. P. & W. F. Vincent, 1999. Strategies of thermal adaptation by high-latitude Cyanobacteria. The New Phycol. 142: 315–323.

Tell, G., A. Vinocur & I. Izaguirre, 1995. Cyanophyta of lakes and ponds of Hope Bay, Antarctic Peninsula. Polar Biol. 15: 503–509.

Tolotti, M., H. Thies, M. Cantonati, C. M. E. Hansen & B. Thaler, 2003. Flagellate algae (Chrysophyceae, Dinophyceae, Cryptophyceae) in 48 high mountain lakes of the Northern and Southern slope of the Eastern Alps: biodiversity, taxa distribution and their driving variables. Hydrobiologia 502 (Dev. Hydrobiol. 172): 331–348.

Utermöhl, H., 1958. Zur vervollkommnung der quantitativen Phytopankton Methodik. Mitt. int. Ver. theor. angew. Limnol. 9: 1–38.

Venrick, E. L., 1978. How many cells to count? In von Sournia, A. (ed.), Phytoplankton Manual. Unesco, Paris: 167–180.

Vincent, W. F., 2000. Cyanobacterial dominance in polar regions. In Whitton, B. A. & M. Pots (eds), The Ecology of Cyanobacteria. Kluwer Academic Publishers, Dordrecht: 321–340.

Vinocur, A. & H. Pizarro, 2000. Microbial mats of twenty-six lakes from Potter Peninsula, King George Island, Antarctica. Hydrobiologia 37: 71–185.

Vinocur, A. & F. Unrein, 2000. Typology of lentic water bodies of Potter Peninsula (King George Island, Antarctica) based on physical-chemical characteristics and phytoplankton communities. Polar Biol. 23: 858–870.

Vinocur, A. & I. Izaguirre, 1994. Freshwater algae from nine lakes and pools of Hope Bay, Antarctic Peninsula. Antarctic Sci. 6: 483–489.

Vörös, L., C. Callieri, K V-Balogh & R. Bertoni, 1998. Freshwater picoCyanobacteria along a trophic gradient and light quality range. Hydrobiologia 369/370: 117–125.

Wynn-Williams, D. D., 1992. Direct quantification of microbial propagules and spores. In Wynn-Williams, D. D. (ed.), Biotas Manual of Methods. Scientific Committee on Antarctic Research, Cambridge: 34 pp.

Zipan, W. & P. P. Deprez, 2000. Study on ecological structures of coastal lakes in Antarctic continent. Chinese J. Polar Sci. 11: 147–160.

Hydrobiologia **502:** 225–270, 2003.
L. Naselli-Flores, J. Padisák, M. T. Dokulil (eds),
Phytoplankton and Equilibrium Concept: The Ecology of Steady-State Assemblages.

A companion to the identification of cryptomonad flagellates (Cryptophyceae = Cryptomonadea) *

Gianfranco Novarino
Department of Zoology, The Natural History Museum, Cromwell Road, London SW7 5 BD, UK
E-mail: gn@nhm.ac.uk

Key words: cryptomonads, Cryptomonadea, cryptophytes, Cryptophyta, Cryptophyceae, identification, morphology, *Pyrenomonas duplex* comb. nov., taxonomy, ultrastructure

Abstract

This paper provides an annotated synopsis of cryptomonad morphology, ultrastructure, taxonomy and nomenclature. Its aim is to aid cryptomonad identification down to whatever taxonomic level is realistically possible, based on currently used taxonomic criteria. It is intended as a 'companion', i.e. a tool to be used in conjunction with the existing literature, rather that a self-contained guide making it possible to identify all known cryptomonads. A nominalistic philosophy is adopted as far as possible, whereby taxa are considered to be acceptable as long as their names have been published in accordance with the nomenclatural rules and prevailing taxonomic practice. A new combination is proposed, *Pyrenomonas duplex* comb. nov. (basyonym: *Rhodomonas duplex* Hill & Wetherbee). Although identification keys are provided whenever possible, an empirical, multilateral approach is followed with a view to facilitating cryptomonad identification by non-specialists using only what equipment is available to them at any one time. This approach is based on the fact that different degrees of taxonomic resolution can be achieved by using different combinations of observation techniques. Identification rationales are based on the following techniques (in ascending order of complexity, time-consumption, and financial cost): light microscopy, scanning electron microscopy, and transmission electron microscopy, supplemented by freeze-fracture investigations and spectrophotometry. Information given under any one section is cross-referenced to other relevant sections by means of italicized links. It is envisaged that this paper will constitute the core of an interactive identification tool to be published electronically on the Internet in due course. Amongst the factors contributing to the difficulties of cryptomonad identification, the most important appear to be the following: the state of flux of cryptomonad classification systems; typification difficulties, especially the historical disregard for the type method and the incorrect belief that preserved type materials of cryptomonads do not exist; difficulties in relating 'classical' taxa and newer ultrastructural descriptions; the description of monospecific genera; inadequate methodology; and the long-standing gap between ecological studies on the one hand and cryptomonad taxonomy and systematics on the other. A wider use of scanning electron microscopy is advocated wherever possible because it appears to be a highly informative and cost-effective tool for discovering, describing, and mapping cryptomonad biodiversity. It is also to be hoped that collaborative projects between ecologists and taxonomic specialists will start to fluorish soon.

Introduction

Cryptomonad flagellates are of great evolutionary importance owing especially to the presence of a nucleomorph, a subcellular structure which has attracted much interest ever since it was first detected in ultrathin sections (Lucas, 1970a) and later recognized as a distinct cell organelle resembling an eukaryotic nucleus (Greenwoood, 1974; Greenwood et al., 1977). Studies using cytochemical (Hansmann et al., 1985; Ludwig & Gibbs, 1985; Hansmann et al., 1986) and molecular methods (Douglas et al., 1991; Maier, 1992; Sitte & Maier, 1992; Maier & Sitte, 1994) have strongly supported the hypothesis of Gillot & Gibbs

(1980) that the nucleomorph is the vestigial nucleus of a red algal-like photosynthetic endosymbiont from which the present-day cryptomonad chloroplast has evolved. Molecular methods have also made it possible to investigate the phylogenetic relationships of cryptomonads as a group (Maerz et al., 1992; Cavalier-Smith et al., 1994; McFadden et al., 1994; Cavalier-Smith et al., 1996; Liaud et al., 1997; Van Der Auwera et al., 1998; Ben Ali et al., 2001), and the phylogeny of recognised cryptomonad genera and suprageneric taxa (Cavalier-Smith et al., 1996; Marin et al., 1998; Hoef-Emden et al., 2002). Thanks to the cryptomonads it is now accepted that eukaryotic chloroplast-containing protists have evolved not only through the ingestion and retention of a photosynthetic prokaryote endosymbiont, but also through the establishment of a secondary endosymbiotic association with a photosynthetic eukaryote (e.g. Douglas, 1992; Maier & Sitte, 1994; Cavalier-Smith et al., 1996; Van Der Auwera et al., 1998).

In addition to being of much evolutionary interest cryptomonads are also of great ecological significance. It has long been known that they are widely distributed in marine and freshwater ecosystems (Wailes, 1939; Pratt, 1959; Ruttner, 1959; Lackey & Lackey, 1963; Findenegg, 1971; Lack, 1971; Ilmavirta, 1975; Eriksson et al., 1977; Reynolds, 1978; Seaburg et al., 1979; Beers et al., 1980; Willén et al., 1980; Hecky & Kling, 1981; Booth et al., 1982; Cronberg, 1982; Chang, 1983; Schanz, 1985; Novarino et al., 1997; Roberts et al., 2000; Barone & Naselli-Flores, 2003; Borics et al., 2003; Menezes & Novarino, 2003; Lepistö & Holopainen, 2003; Padisák et al., 2003; Tolotti et al., 2003; Willén 2003). They have also been found in terrestrial subsurface or subaerial habitats such as soil (Sandon, 1927; Morrall, 1980; Paulsen et al., 1992), groundwater (Novarino et al., 1994) and snow (Javornický & Hindák, 1970). In the marine plankton, cryptomonads may often reach very high population densities (e.g. Haigh et al., 1992), and it is especially in this environment that photosynthetic forms may contribute significantly to primary production (e.g. Robinson et al., 1999). In freshwater environments cryptomonads are often present in constant but relatively low numbers throughout the year, although sudden population increases are also common. The vast literature on cryptomonad ecology is scattered in numerous journals, and the reader is referred to two comprehensive reviews for further information (Klaveness, 1988, 1988/1989).

Most cryptomonads can be detected easily in unfixed environmental samples thanks to their characteristic swimming behaviour. This has been described as 'recoiling' (Christensen, 1980), a term which has also provided the group with its vernacular name in the Danish language ('rekylalger': e.g. Thomsen, 1992). However, identification at or below the generic level is notoriously complex. An up-to-date monographic revision of the group is not available and therefore it is necessary to have access to an abundant taxonomic literature, including many of the older, hard to find publications. A range of observation methods may be indispensable but not all may be available to the investigator. The quality of some of the early taxonomic descriptions, which were based necessarily on the small number of characters visible in living cells using light microscopy, is not always adequate and therefore it is often difficult to use those descriptions in a modern taxonomic context. Furthermore, cryptomonad systematics is in a state of flux, not least because molecular sequencing studies are providing a wealth of new phylogenetic information. In some cases the degree of congruence between molecular phylogenies and classifications based on ultrastructural characters such as the nucleomorph position (Novarino & Lucas, 1993b, 1995; Clay et al., 1999) is high (Cavalier-Smith et al., 1996; Deane et al., 2002; Hoef-Emden et al., 2002), while in others the situation is still unclear and requires further investigation (Hoef-Emden et al., 2002).

The difficulties outlined above are such that they may discourage non-specialist investigators from attempting to identify cryptomonads altogether. The situation is made worse by the existence of an authoritative viewpoint whose influence still seems to persist nowadays: namely, that cryptomonads are simply impossible to classify (Pringsheim, 1968). As a result only a few named taxa are mentioned frequently in the non-taxonomic literature, often with little information on the actual specimens observed. However, there are several reasons why it is still important to identify unknown field specimens as far as is reasonably practicable:

1. Cryptomonad diversity may be far from having been described in full. Estimates of the total number of exisiting species range from 300 (Novarino, unpublished data) to 1200 (Andersen, 1992), i.e. one and a half to six times the number of currently known species. Therefore it is always possible that environmental samples will contain previously undescribed taxa, perhaps with novel morphological, ecological,

physiological or other features of interest within the context of cryptomonad systematics and phylogeny.
2. Our knowledge of how diversity varies in relation to geographical and environmental variability is still very scant, and the very possibility of addressing these questions depends entirely on the ability to recognize species in the first place.
3. Cryptomonads may form nuisance blooms both in marine (Andreoli et al., 1986; Dame et al., 2000) and freshwater ecosystems (Nishijima, 1990). Sound taxonomic identifications are a necessary preamble to any modelling studies of such blooms if the models are to be sufficiently predictive.
4. There is evidence that freshwater cryptomonad species may differ considerably from one another in their physiological and ecological characteristics, and therefore they may occupy distinct, well-defined ecological niches (Willén et al., 1980; Sommer, 1992; Gervais, 1997). Although there is scant information on the subject in marine environments, *a priori* it cannot be excluded that the situation is analogous to that in freshwaters. Here again, the ability to address this sort of question will depend on the availability of reliable taxonomic identifications.

Work carried out in in the last three or four decades has shown that progress can be made towards cryptomonad identification if the taxonomic value of all observable characters is evaluated closely, particularly in the case of ultrastructural features (Hill & Wetherbee, 1986, 1988, 1989, 1990; Hill, 1991a, b; Novarino, 1991a, b; Novarino & Lucas, 1993a; Novarino et al., 1994; Kugrens et al., 1999). Recently, several ultrastructural characters have been incorporated in two formal classification systems of the group (Novarino & Lucas, 1993b; Clay et al., 1999).

The aim of this paper is to aid identification down to whatever taxonomic level is realistically possible, based on morphological criteria adopted in more recent years by several independent workers. It is intended as a supplement to the existing taxonomic literature rather than a self-contained identification guide. It is by no means exhaustive and it does not make it possible to identify all described cryptomonad species. Although simple identification keys are provided where possible, an empirical, multilateral approach is followed in order to facilitate identification by non-specialists using only what equipment is available to them at any one time. This approach is based on the fact that different degrees of taxonomic resolution can be achieved by using different combinations of observation techniques: light microscopy, scanning electron microscopy, transmission electron microscopy, freeze-fracture investigations, and spectrophotometry. The most recently described cryptomonad genus (*Hanusia* Deane et al., 1998) has been separated from morphologically similar genera based on 18S rRNA sequences. In future it is likely that more and more molecular sequence data will be incorporated into taxonomic protologues. However, in the current practice molecular sequence data do not play a major role in cryptomonad identification.

From a taxonomic viewpoint, this paper adopts a nominalistic philosophy, whereby taxa are considered to be acceptable as long as their names have been published in accordance with the nomenclatural rules and prevailing taxonomic practice. A number of poorly known genera whose affinities with the cryptomonads are uncertain and genera which are taxonomically synonymous with others are also included for information purposes. Other uncertain genera are listed elsewhere (Novarino & Lucas, 1993b). Owing to the fact that cryptomonad systematics is outside the scope of this guide, no particular classification system is followed. Cryptomonads are ambiregnal protists (Patterson, 1986), and therefore they are considered here to be under the dual jurisdiction of the botanical (ICBN) and the zoological (ICZN) Codes of Nomenclature (Novarino & Lucas, 1993b, 1995).

It is hoped that using this companion will be as flexible a process as possible. Readers may choose to go through all the paragraphs sequentially, or else they may prefer to go directly to the parts most pertinent to the specimens at hand and the observation techniques available. For this purpose, links to related text in this paper are given wherever possible. All links are italicized and they indicate the related paragraph(s), e.g. *2.2.2.3*. Abbreviations used in the text are as follows: EPC – external periplast component; IPC – internal periplast component; LM – light microscopy; PC – phycocyanin; PE – phycoerythrin; SEM – scanning electron microscopy; TEM – transmission electron microscopy.

1. Cryptomonads and light microscopy (Figs 1–11)

Light microscopical observations are best carried out on live, unfixed cells, and the morphological details visible to the observer will depend largely on the kind of equipment and illumination techniques available. The quality of the observations will depend also on the personal skills of the observer – there is some evidence to suggest that a few of the early light microscopists had observed some ultrastructural features of cryptomonads many years before their official discovery (Novarino, 1993). However, this is probably an unlikely result for the average worker, and the most realistic achievements appear to be the following:

- to observe swimming behaviour;
- to observe details of cell shape and measure cell size parameters;
- to observe the large ejectosomes;
- to tell apart heterotrophic (colourless) from photosynthetic cryptomonads or those with photosynthetic endosymbionts (coloured);
- to identify all known heterotrophs to species or genus level;
- to tell apart phycoerythrin - from phycocyanin-containing cryptomonads;
- to count the number of chloroplasts and pyrenoids in photosynthetic cryptomonads;
- to identify *Hemiselmis* at the generic and subgeneric level; and
- to observe certain other cellular features of possible taxonomic value.

Light microscopical observations on fixed cells (especially in phase-contrast or differential intereference contrast illumination) may yield additional information on flagellar length and ratio. This may be important for species identification in some genera (*1.2*). Examination of fixed samples may also be necessary to measure cells accurately, since cell size may be taxonomically significant at or below the species level (*1.3*). If fixation is suspected to alter cell size or shape significantly, living cells can be measured on photomicrographs obtained using a microflash apparatus or digital images extracted from digital film clips (Novarino & Roberts, 1994). Some morphological details may be easier to see in inverted (negative) photographic prints rather than positive ones (Figs 6 and 8).

1.1. Swimming behaviour (Figs 1, 3 and 4)

With the exception of the genus *Goniomonas* (see *Goniomonas* under *1.6*), the vast majority of cryptomonads swim according to a characteristic pattern, described by Christensen (1980) as 'recoiling'. The cell moves along a spiral with the flagella directed forwards, whilst also revolving about its longitudinal axis (Figs 1, 3 and 4). Changes of direction are frequent, and analogous to many other flagellates it is also common for the cell to come periodically to a halt and revolve on itself, with no translational movement.

1.2. Flagella (Figs 2, 5, 6 and 7–9a)

The motile unicell (Figs 7–9a) is the most common life-form in the cryptomonads, but non-motile cells such as palmelloid cells (single, paired or forming irregular colonies) and resting stages (cysts) may also occur (Fig. 9b). Pascher (1913) believed that the ability to form these stages was systematically significant, and therefore he divided the cryptomonads into two major groups based upon the relative importance of the motile and non-motile stages throughout the life-cycle. However, this view is untenable because in culture many cryptomonads tend to assume a non-motile palmelloid state (Pringsheim, 1944).

Motile cells have two flagella, which are readily visible with the LM (Figs 7–9a). Lugol-fixed material is preferable, and contrast illumination (phase-contrast, differential interference contrast) may be an advantage but in general it is not absolutely indispensable. Flagella may be equal, unequal or subequal in length and this may be useful for identification purposes at the specific level. The position of the flagella may be significant at the generic level (laterally attached flagella: go to *1.3.1*).

1.3. Cell size and shape

Cell size parameters are the cell length, width, and thickness. These terms, as applied to cryptomonad cells, refer to the length of the longitudinal, perlateral, and dorso-ventral axes, respectively (see Klaveness, 1985). The ventral face is defined as the face from which the flagella appear to emerge. When a cryptomonad cell is seen from the ventral face, the right and left sides of the cell are to the left- and right-hand sides of the observer, respectively. Differences in cell size have been used extensively in traditional taxonomic descriptions and therefore they may be useful for identification purposes at the specific level.

Morphological parameters such as the shape of the cell apex and antapex, dorso-ventral or lateral compression, and torsion or curvature of the cell along its longitudinal axis may be taxonomically significant at the specific level (Hill & Wetherbee, 1986, 1988, 1989; Hill, 1991a, b; Novarino, 1991a, b; Gervais, 1997; Kugrens et al., 1999; Javornický, 2003). In addition, all members of the heterotrophic genus *Goniomonas* are characteristically flattened along the

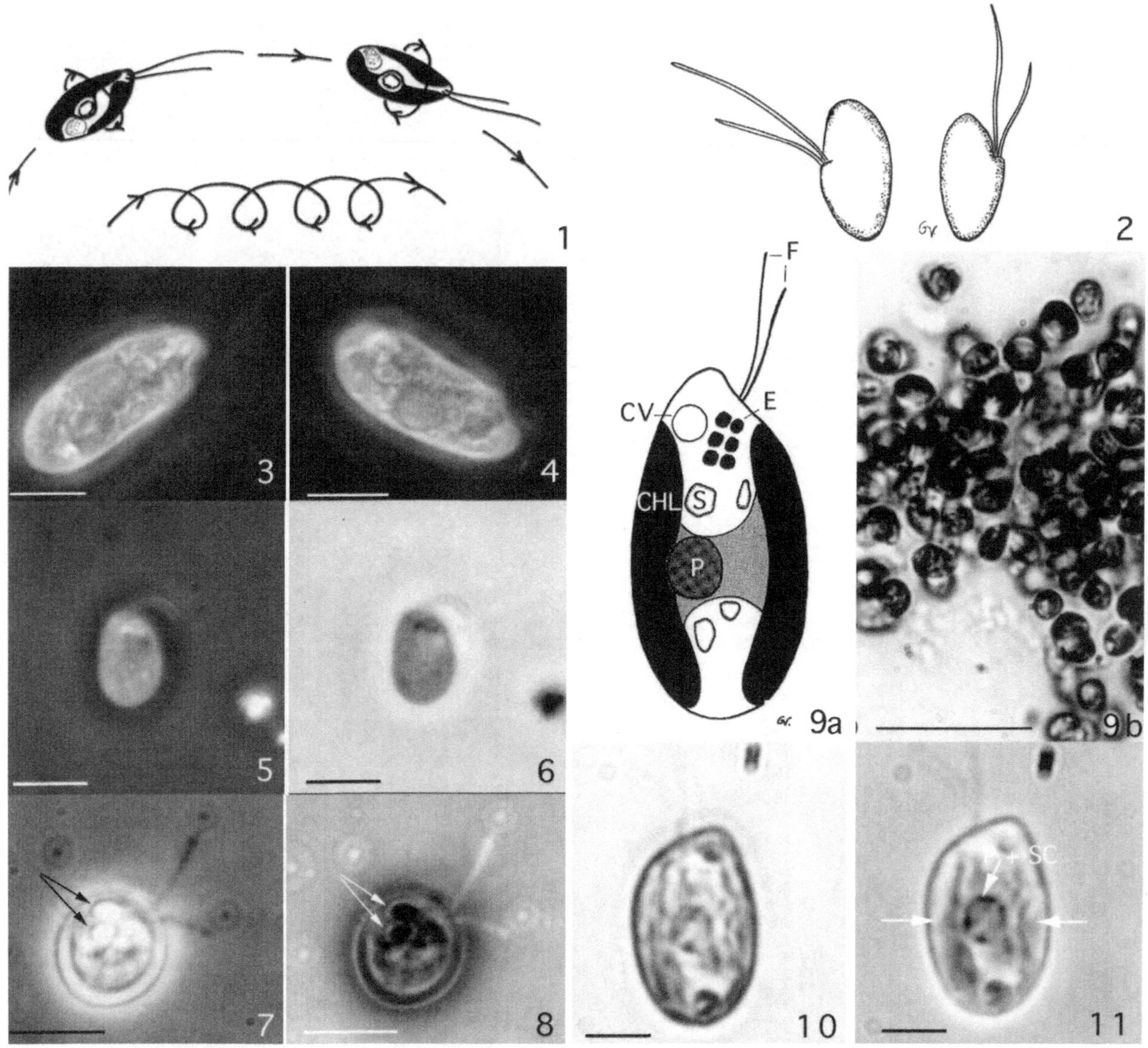

Figure 1–11. Some features of cryptomonads visible under the light microscope. *Figure 1.* Diagrams of swimming behaviour. *Figure 2.* Cell shape and point of flagellar insertion in *Hemiselmis*. *Figures 3 and 4.* Sequential flash micrographs of an actively swimming cell of *Chilomonas paramecium*. Scale bars = 5 μm. *Figures 5 and 6.* A living, unstained cell of *Goniomonas truncata* photographed with a microflash, in positive and reverse (negative) print, showing the characteristic flagella and the anterior row of ejectosomes. Scale bars = 5 μm. *Figures 7 and 8.* A living, unstained cell of *Chroomonas placoidea*, phase-contrast illumination, in positive and reverse (negative) print, showing the paired refringent bodies (arrows). Scale bars = 5 μm. *Figures 9a and 9b.* Basic cell types of cryptomonads. *Figure 9a.* Flagellate cell; F = flagella, CV = contractile vacuole, E = large ejectosomes, S = starch granules, Chl = chloroplast, P = pyrenoid. *Figure 9b.* Group of non-motile palmelloid cells of *Chroomonas ligulata*. Scale bar = 25 μm. *Figures 10 and 11.* An osmium-fixed cell of *Protemonas pseudobaltica*, before and after treatment with dilute Lugol's solution, which stains the starch cap (SC) around the pyrenoid (Figure 11). The two lobes of the chloroplast are also visible (arrows). Scale bars = 5 μm.

perlateral axis (see *Goniomonas* under *1.6*). Cell morphology can be observed with LM but SEM (go to *2*) is preferable owing to the fact that it provides additional three-dimensional information. Descriptions of cell morphology in published taxonomic protologues can be interpreted with the aid of a standard terminology (Systematics Association Committee for Descriptive Biological Terminology, 1962; Klaveness, 1985).

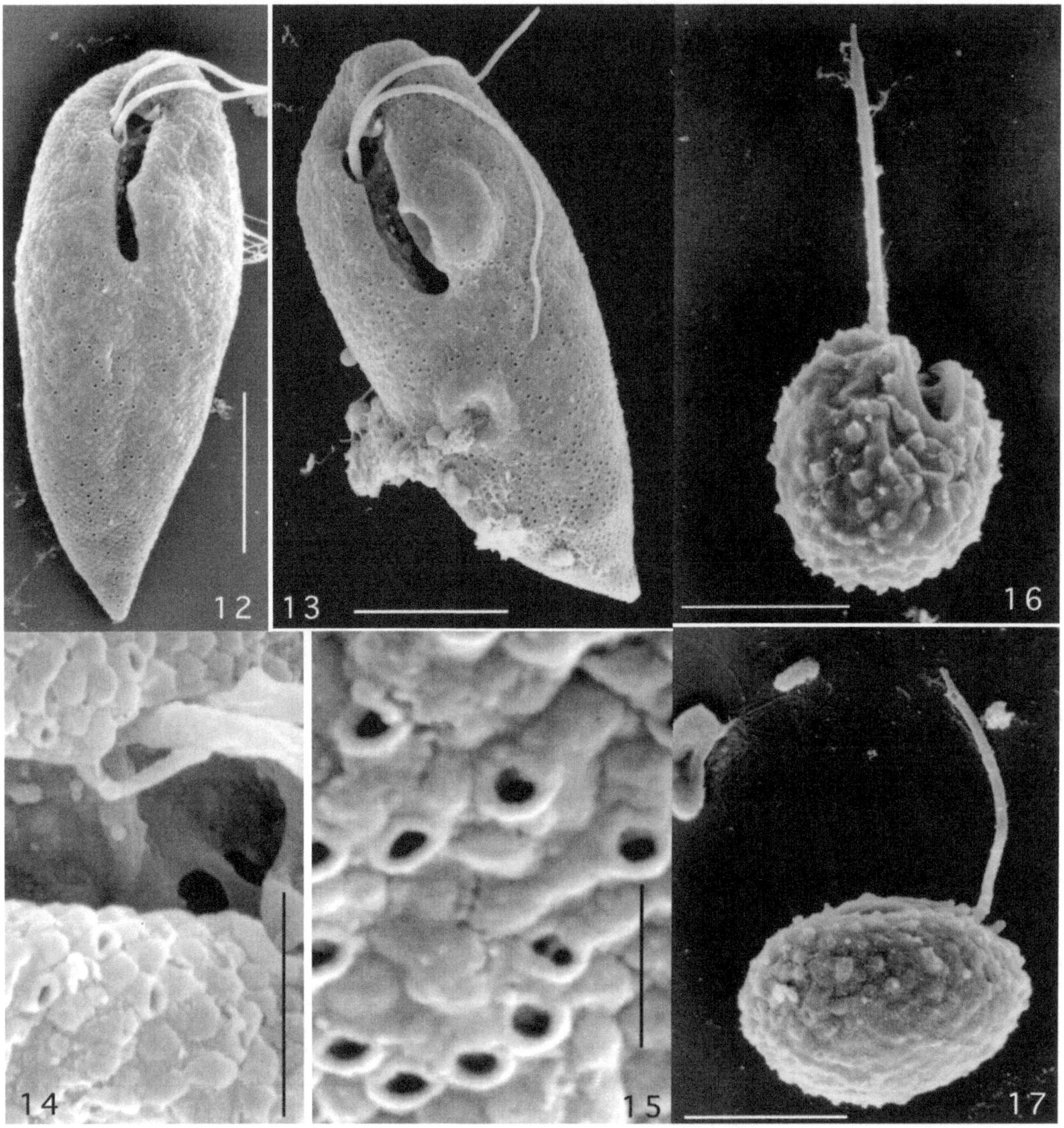

Figure 12–17. SEM micrographs of cryptomonads. *Figures 12–15. Campylomonas* sp. Figures 12 and 13 show the ventral cell face with the furrow, whose left-hand margin carries the discharge site of the contractile vacuole, and the flagella inserted at its top right-hand side. Figure 14 shows a detail of the furrow region, and Figure 15 shows the periplast with ejectosome pores. Scale bars = 1 μm (Figs 12 and 13) or 500 nm (Figs 14 and 15). *Figures 16 and 17. Cryptochloris* sp., showing the ventral face with furrow and flagella (Fig. 16), and the dorsal face (Fig. 17). Scale bars = 2.5 μm.

The reniform cell shape, together with the position of the flagella, is taxonomically significant at the generic level (*1.3.1*).

1.3.1. The reniform cryptomonads: **Hemiselmis** *and its allies (***Planonephros** *and* **Protochrysis***) (Fig. 2)*

Genus *Hemiselmis* Parke (ICBN, ICZN)
M. Parke 1949. *Journal of the Marine Biological As-*

sociation of the United Kingdom vol. 28, p. 279, pl. II figures 11–16, text-figures 69–73.
Type species: *Hemiselmis rufescens* Parke 1949, p. 279, pl. II figures 11–16, text-figures 69–73.
Subgenus *Hemiselmis* Parke, 1949 p. 279 (ICZN), equivalent to *Hemiselmis* Butcher, 1967 p. 13 (ICBN).
Subgenus *Plagiomonas* Butcher, 1967 p. 16 (ICZN, ICBN).

Genus *Planonephros* Christensen (ICBN, ICZN)
T. Christensen 1978. *Botanisk Tidsskrift* vol. 73, p. 67. See also Christensen 1967, p. 92.
Synonyms: *Nephroselmis* sensu Pascher (1913, p. 111), *Sennia* sensu Skuja (1948, p. 366).
Type species: *Planonephros dispar* Christensen 1978, p. 67; synonyms: *Nephroselmis olivacea* sensu Pascher (1913, p. 111), *Sennia commutata* sensu Skuja (1948, p. 366).

Genus *Protochrysis* Pascher (ICBN, ICZN)
A. Pascher 1911. *Ber. deut. Bot. Ges.* vol. 29, p. 191, figures II 1, 2.
Type species: *Protochrysis phaeophycearum* Pascher, 1911, p. 191, figures II 1, 2.

It is possible to identify members of the genus *Hemiselmis* to the generic level by using LM alone, thanks to the characteristic reniform (kidney-shaped) cells and the laterally attached flagella (Fig. 2). Species of *Hemiselmis* belong to one of two subgenera (*Hemiselmis* and *Plagiomonas*) based on the phycobiliprotein pigment (phycoerythrin – PE – in subgen. *Hemiselmis*, versus phycocyanin – PC – in subgen. *Plagiomonas*: see also *1.5*, *6*). Therefore they can be readily identified at the subgeneric level based on cell colour (blue-green in subgen. *Plagiomonas* versus 'red' in subgen. *Hemiselmis*, i.e. shades of red, brown, red-brown, yellowish-brown or yellowish-green, but never blue-green). There are numerous described species, many of which are due to Butcher (1967). Traditional species distinctions have relied largely on cell shape and size, but electron microscopy has provided additional taxonomic information (see *Hemiselmis* under *2.2.2.3*).

Until recently, this genus was known only from marine habitats but there is a recent report of its presence in a oligohaline environment (Domingos & Menezes, 1998) and a new freshwater species has also been described (Clay & Kugrens, 1999a). Genera allied to *Hemiselmis* are *Planonephros*, which is considered synonymous with *Hemiselmis* (see Novarino & Lucas, 1995) and only contains the type species; and *Protochrysis*, which shows a strong overall resemblance to *Hemiselmis*. Both these genera have hardly been reported after the original descriptions and no ultrastructural studies are available.

Selected species:

Hemiselmis amylosa Clay & Kugrens
Clay & Kugrens, 1999a, p. 298, figures 1–4 (holotype), 2–12.
Note: This is a PC-containing freshwater species. The description is based largely on ultrastructural features (see under *2.2.2.3*). See also *Planonephros dispar* below.

Hemiselmis brunnescens Butcher
Butcher, 1967, p. 14, pl. I figure 4, pl. XIII figure 2.
Note: This is a PE-containing marine species. Butcher omitted to designate a type, which was required at his time of writing, and therefore a lectotype is necessary in order to fulfill the nomenclatural rules. Some ultrastructural information is available (see under *2.2.2.3*).

Hemiselmis rufescens Parke
Parke, 1949, p. 279, pl. II figures 11–16, text-figures 69–73.
Note: The type species contains PE and is marine. Some ultrastructural information is available (see under *2.2.2.3*).

Hemiselmis virescens Droop
Droop, 1955, p. 238, figures 7–11.
Note: This is a PC-containing marine species. Some ultrastructural information is available (see under *2.2.2.3*).

Planonephros dispar Christensen
Christensen, 1978, p. 67.
Note: This is a freshwater form presumably containing PC. There is a complex nomenclatural and taxonomic question – see Christensen (1967, 1978). If *Planonephros* is effectively a synonym of *Hemiselmis*, then *Planonephros dispar* is allied to *Hemiselmis amylosa* (see above), the only *Hemiselmis* species known from freshwater environments.

Protochrysis phaeophycearum Pascher
Pascher, 1911, p. 191, figures II 1, 2.
Note: This is a freshwater form. As suggested by the specific binomen, the cell colour is mostly yellow-brown and therefore *P. phaeophycearum* may contain PE. This is also supported by the observation that the

colour can also be red (Pascher, loc. cit.), but it is in contrast with the fact that Pascher also found blue-greenish cells (Pascher, loc. cit.). It is possible that Pascher was unknowingly observing two species at the same time, one containing PE and the other containing PC. If this inference is correct then *Protochrysis* would be very much like *Hemiselmis*, which also has PE- and PC-containing species. *Protochrysis phaeophycearum* was originally described has having two chloroplasts (compare *Protochrysis vinosa* and see *1.7.2*).

Protochrysis vinosa Conrad

Conrad, 1939, p. 5, figures 10, 11.
Note: This brackish-water form is red in colour, and therefore it may contain PE. It was described has having a single chloroplast (*1.7.2*).

1.4. Ejectosomes (Figs 5, 6 and 9a)

Cryptomonads typically possess peculiar ejectile organelles termed ejectosomes, ejectisomes or trichocysts (Dragesco, 1951; Morrall, 1980; Grim & Staehelin, 1982, 1984). Smaller ejectosomes may be present at the cell periphery. The larger ejectosomes, which are usually relatively easy to observe using LM, are found in the vestibular region of the cell from which the flagella emerge (Fig. 9a). The number and arrangement of the large ejectosomes has been used extensively for systematic purposes (Butcher, 1967), but this view has largely fallen in disuse. However, thanks to the large ejectosomes it may be possible to detect the position and extent of the furrow/gullet system, which is taxonomically important (see under SEM section, *2.3*, and TEM section, *4.3*).

1.5. Cell colour as a rough indicator of phycobilin pigment identity

Photosynthetic cryptomonads contain phycobilins as the main accessory pigments. These water-soluble compounds exist in two main forms, phycoerythrin (red, PE) and phycocyanin (blue-green, PC). Within each category there are numerous sub-types, distinguished on the basis of their absorption spectra (Hill & Rowan, 1989) (go to *6*). In the cryptomonads, the identity of the particular phycobilin spectral type present is now widely accepted as being taxonomically significant at the generic level. The genus *Hemiselmis* has both PE- and PC-containing species, but this is now thought to be the result of an evolutionary character reversal producing a phycoerythrin from a phycocyanin thanks to a relatively simple biochemical transformation (Marin et al., 1998).

The accurate identification of phycobilin spectral types requires the pigment to be extracted and analysed using spectrophotometry (Hill & Rowan, 1989) (go to *6*). However, simple visual observation of live cells using LM is usually sufficient to establish whether a PC or a PE is present. PC-containing forms are blue-green in colour, while PE-containing ones can be red, brown, red-brown, yellowish-brown or yellowish-green, but never blue-green. Subtle variations in cell colour have been used amply in taxonomic accounts of cryptomonads from British coastal waters (Butcher, 1967), but such variations are now known to reflect varying concentrations of phycobilin pigment relative to chlorophylls. If a cell appears grass-green in colour it may have accidentally lost the phycobilin pigment completely.

- Cells are blue-green in colour, i.e. they contain PC .... *Chroomonas, Cryptochloris, Falcomonas, Hemiselmis* subgen. *Plagiomonas, Hillea* sensu stricto, *Komma, ?Protochrysis* (partim) (go to *1.7*)

- Cells are red, brown, red-brown, yellowish-brown or yellowish-green, but never blue-green in colour, i.e. they contain PE. *Campylomonas, Cryptomonas, Geminigera, Guillardia, Hanusia, Hemiselmis* subgen. *Hemiselmis, Hillea* sensu Butcher, *Plagioselmis* and *Isoselmis, Proteomonas, Protochrysis* (partim), *Pyrenomonas/Rhodomonas* emend., *Rhinomonas, Storeatula, Teleaulax* (go to *1.7*)

- Cells lack colour altogether. . . Heterotrophs (go to *1.6*)

Note: There is another photosynthetic genus (*Cryptochrysis* Pascher 1911, p. 190, figures I 1–4), but the cell colour was not mentioned in the original description. It has not been reported frequently and very few species have been described (Butcher, 1952; Norris, 1964). No ultrastructural information is available. The genus was reduced to a subgenus of *Chroomonas* by Butcher (1967, p. 29), based on cell colour (not blue-green in *Cryptochrysis* versus blue-green in *Chroomonas*). This is no longer accepted and *Chroomonas* only includes blue-green forms. Santore (1984) based his genus *Pyrenomonas* on a species of *Chroomonas*, subgenus *Cryptochrysis: C. salina* (Wislouch) Butcher (go to *1.7*).

1.6. Heterotrophic cryptomonads: Chilomonas, Goniomonas, Platychilomonas, Protocryptomonas *(Figs 3–6)*

If a live cell lacks colour in bright-field light microscopy, chloroplasts are absent and therefore the cell is heterotrophic. There are two well-established heterotrophic genera, *Goniomonas* von Stein and *Chilomonas* Ehrenberg. The affinities of *Platychilomonas* Larsen & Patterson, a third heterotrophic genus which is not widely reported, are unclear because no ultrastructural studies are available. It is also possible that *Platychilomonas* is not a cryptomonad at all. The heterotrophic genera *Katablepharis/Kathablepharis* Skuja (1939), *Leucocryptos* Butcher (1967) and *Cryptaulax* Skuja (1948) (also known as *Cryptaulaxoides* Novarino, 1996 under the ICZN) have been assigned in the past to the cryptomonads but ultrastructural studies have shown that they are not cryptomonads and their systematic position is uncertain (Vørs, 1992; Clay & Kugrens, 1999b). A further heterotrophic genus has been described which has been considered as a cryptomonad (*Protocryptomonas* Skvortzov, 1960), but the poor original descriptions make it very difficult to establish its true affinities (see below).

- Small, bacterivorous, laterally flattened cells, lacking starch granules; swimming does not occur in the usual spiral pattern; ejectosomes (*1.4*) arranged transversely in the anterior region of the cell .... *Goniomonas*

- Larger, osmotrophic cells with starch granules; ejectosomes (*1.4*) arranged vertically in the anterior – median part of the cell ...... *Chilomonas*

- Cells appear similar to *Chilomonas* but are laterally compressed; in cells at rest the ventral flagellum (*1.2*) is held in a curl .... *Platychilomonas*

- Cells appear similar to *Chilomonas* but lack a furrow/gullet (see also SEM section *2.3*, TEM section *4.3*) and ejectosomes (*1.4*) .... *Protocryptomonas*

Genus *Goniomonas* von Stein (ICBN, ICZN) (Figs 5 and 6)

F. R. von Stein, 1878. *Der Organismus der Infusionsthiere. III Abteilung. Die Naturgeschichte der Flagellaten oder Geisselinfusoria. I Halfte:* pl. II, figure II, 1–6.
Synonym: *Cyathomonas* sensu Bütschli 1884 (p. 844, pl. 45, figures 8a, b); sensu Fisch 1885 (p. 74, pl. I, figures 25–38); sensu Ùlehla 1911 (p. 284, figures lc,d); sensu auctorum; non *Cyathomonas* Fromentel 1874 (206, pl. 27, figures 27, 28, 30, 31).
Type species: *Goniomonas truncata* (Fresenius) von Stein 1878 (pl. II, figure II, 1-6); basionym: *Monas truncata* Fresenius 1858 (p. 227, pl. X figure 42).

This genus is often referred to by the name of *Cyathomonas* Fromentel, but this is incorrect (see Larsen & Patterson, 1990). Members of *Goniomonas* are readily recognizable in the light microscope thanks to the laterally flattened cell shape, the large ejectosomes (*1.4*) arranged in a horizontal row in the anterior region of the cell (Figs 5 and 6), and their unusual swimming behaviour. Unlike all other cryptomonads, cells of *Goniomonas* move along irregular tracks close to the substrate, which they face with one of the lateral sides (Figs 5 and 6). This genus, which is known from both marine and freshwater environments, contains a further two species in addition to the type: *G. amphinema* Larsen & Patterson and *G. pacifica* Larsen & Patterson (Larsen & Patterson, 1990). These species, which have been described based on LM only, are distinguished from one another based mostly on subtle size differences, the behaviour of the flagella, and the presence or absence of longitudinal striations, but the species boundaries are not entirely clear-cut and require further investigation. *Goniomonas truncata*, which has been found both in freshwater and marine habitats, is by far the most widely reported species, although most of the early reports refer to it incorrectly as "*Cyathomonas truncata* Fromentel". Bacterivory, which has been described based on TEM observations, seems to be present throughout the genus.

Goniomonas truncata (Fresenius) von Stein
von Stein, 1878, pl. II, figure II, 1–6; basionym: *Monas truncata* (Fresenius, 1858), p. 227, pl. X figure 42.
Note: This is the most widely reported species. It has been found both in freshwater and marine habitats.

Goniomonas amphinema Larsen & Patterson
Larsen & Patterson, 1990, p. 809, figures 1f, 2d (type figure), 2e.
Note: This marine species is characterized by its swimming behaviour, which takes place with the shorter flagellum directed forwards and the other one backwards.

Goniomonas pacifica Larsen & Patterson
Larsen & Patterson, 1990, p. 810, figures 1g, 2a, 2b (type figure), 2c.
Note: This marine species is characterized by its swimming behaviour, which takes place with the flagellum situated closest to the dorsal face pointing forwards, and the other flagellum usually bending towards the ventral face. Another characteristic feature is given by the presence of longitudinal striations. These were not commented on further in the description although it is likely that they correspond to the periplast ridges or grooves visible using electron microscopy (see *Goniomonas* in SEM section *2.2.2.2*, and Figs 25 and 26).

Genus *Chilomonas* Ehrenberg (ICBN, ICZN) (Figs 3 and 4)

Vide C. G. Ehrenberg, 1838. *Die Infusionsthierschen als volkommene Organismen. Ein Blick in das tiefere organische Leben der Natur*: p. 30, pl. II figures V–VII.
Type species: *Chilomonas paramaecium* Ehrenberg. This was not designated by Ehrenberg himself – see Senn (1900, p. 168). Ehrenberg also used the spelling *Chilomonas paramecium* (Ehrenberg 1838, p. 300). The original illustrations relating to this species are to be found in Ehrenberg (1838, pl. II, figure VI).

This genus was introduced by Ehrenberg on several occasions. The earliest appearances of the generic name in the literature (Ehrenberg, 1831a, b) are sanctioned by the on-line version of the Index Nominum Genericorum Plantarum (http://rathbun.si.edu/botany/ing/) as the original publications for nomenclatural purposes. In Ehrenberg (1831a) there are only succinct synoptic tables without descriptions or illustrations. More detailed information is available in Ehrenberg (1831b) but without illustrations. These were provided in a later work (Ehrenberg, 1838) together with a bilingual (Latin and French) formal diagnosis and several illustrations. For this reason the publication place and date adopted here is that of Ehrenberg (1838) rather than the others.

Members of the genus *Chilomonas* (Figs 3 and 4) are readily recognizable by LM based on the lack of colour and the presence of starch granules. Nutrition appears to be entirely osmotrophic. There are several described species but the genus is known mostly on the basis of the widely reported *Chilomonas paramaecium/paramecium*, which is also the only species examined using electron microscopy (see *Chilomonas* under *2.2.1*). So far *Chilomonas paramaecium/paramecium* is known mainly from freshwater environments, although there are a few reports from estuarine and marine habitats (Kent, 1880–82; Pratt, 1959; Hada, 1974; Jones, 1974). The second most reported species is *Chilomonas oblonga* Pascher. Taxonomic descriptions have relied mostly on such characters as cell shape and size but in most cases the species boundaries are unclear, making species-level identification uncertain.

The following is an annotated *bona fide* list of published species names in addition to *Chilomonas paramaecium/paramecium*. Many of these have hardly been used again after their first publication.

Chilomonas abrupta Skvortzow
Skvortzow, 1957, p. 184, pl. 6 figure 52.
Note: This was described as being similar to *Chilomonas oblonga* Pascher.

Chilomonas acuta Schiller
Schiller, 1929, p. 456, text-figure 22 a–e.
Note: This was later recombined as *Cyanomonas acuta* Schiller (1957) based on the presence of four or more grey-blueish plastids (see *Cyanomonas* under *1.7.1*).

Chilomonas acuta Schiller var. *insignis* Skuja
Skuja, 1956, p. 349, pl. LX figures 41–45.
Note: This has been raised to the level of a species by Javornický (1967) – see *Chilomonas insignis* below.

Chilomonas amygdalum Kent
Kent, 1880–82, p. 426, pl. XXIV, figure 49.
Note: based on the original description there is a strong resemblance with *Platychilomonas psammobia* Larsen & Patterson (see below).

Chilomonas bacillaris Javornický
Javornický, 1967, p. 47, pl. 3 figures 4–6.

Chilomonas coniformis Ondratschek
Ondratschek, 1941, p. 286, text-figure 3.

Chilomonas cryptomonadoides Skuja
Skuja , 1956, p. 348, pl. LX figures 25–36.

Chilomonas curta Pringsheim *"nom. provis."*
Pringsheim, 1963, text-figures 30.2, 30.3.

Note: This name is nomenclaturally illegitimate since no formal taxonomic protologue was provided.

Chilomonas curvata Pringhseim *nomen nudum*
Pringsheim, 1963, text-figure 28.1.
Note: This name is nomenclaturally illegitimate since no formal taxonomic protologue was provided.

Chilomonas cylindrica (Ehrenberg) Kent
Kent, 1880–82, p. 425, pl. XXIV, figure 50.
Basionym: *Cryptomonas cylindrica* Ehrenberg. *Vide* Ehrenberg 1838 (p. 42, pl. II figure XIX).
Note: *Cryptomonas cylindrica* was originally described by Ehrenberg (1838) as being green in colour.

Chilomonas destruens Ehrenberg
Ehrenberg, 1838, p. 31, pl. II figure VII.

Chilomonas globosa Ondratschek
Ondratschek, 1941, p. 288, text-figure 5.

Chilomonas granulosa Dujardin
Dujardin, 1841, p. 295, pl. III, figure 15.

Chilomonas ?inflata Fromentel
Fromentel, 1874, p. 332, pl. XXIII figure 54.

Chilomonas insignis (Skuja) Javornický
Javornický, 1967, p. 48, pl. 3 figures 7, 8
Basyonym: *Chilomonas acuta* Schiller var. *insignis* Skuja 1956, p. 349, pl. LX figures 41–45.

Chilomonas longata Ondratschek
Ondratschek, 1941, p. 287, text-figure 4.

Chilomonas marina (Braarud) Halldal
Halldal, 1953, p. 58, text-figure 21.
Basionym: *Bodo marina* Braarud, 1935, p. 115, text-figure 36.
Note: This species has been transferred to the genus *Leucocryptos* Butcher (1967). Ultrastructural investigations have shown that this genus does not to belong to the cryptomonads (Vørs, 1992).

Chilomonas minor Czosnowski
Czosnowsky, 1948, p. 13, pl. V figures 12, 13.

Chilomonas naviculaeformis Schiller
Schiller, 1957, p. 36, pl. XII figure 56a–e.

Chilomonas obliqua Dujardin
Dujardin, 1841, p. 295, no illustrations.

Chilomonas oblonga Pascher
Pascher, 1913, p. 109, figure 153 (?).
Note: This is the second most reported species and the only one recognized by Pascher (1913) in addition to *Chilomonas paramaecium/paramecium.* There are some difficulties concerning its typification because Pascher's Figure 153, indicated in the text (p. 109) as being relative to *Chilomonas oblonga*, is said to show *Chilomonas ovata* (no authority given) in the figure legends (p. 97). However, this is clearly not the *C. ovata* of Stokes (see below), which Pascher himself (1913, pp. 97, 109) justly removed from the cryptomonads. Furthermore, Pascher's Figure 153 is not an original one since it reproduces Figure 1A of Ùlehla (1911), which Ùlehla referred to as "*Chilomonas paramaecium, forma obtusa*".

Chilomonas oblonga Pascher forma *minor* (Czosnowsky) Javornický
Javornický, 1967, p. 46, pl. 3 figures 1–3.
Basyonym: *Chilomonas minor* Czosnowsky 1948, p. 13, pl. V figures 12, 13.

Chilomonas ovata Stokes
Stokes, 1888, p. 123, pl. II figures 27–29.
Note: This is very unlikely to be a cryptomonad – see notes for *Chilomonas oblonga* above.

Chilomonas sp. (?) Conn
Conn, 1905, p. 35, figure 95.

Chilomonas volvox Ehrenberg
Vide Ehrenberg, 1838, p. 30, pl. II figure V.

Genus *Platychilomonas* Larsen & Patterson (ICBN, ICZN)
J. Larsen & D. J. Patterson 1990, *Journal of Natural History* vol. 24, p. 811, figures 1e, 2f–g.
Type species: *Platychilomonas psammobia* Larsen & Patterson, 1990, p. 811, figures 1e, 2f–g.

The genus *Platychilomonas* Larsen & Patterson is known from its authors' light microscopical observations on *P. psammobia* Larsen & Patterson, a flagellate described from tropical marine sediments. As its name implies there is a broad resemblance to *Chilomonas*. Cells are laterally compressed and when they are at rest the flagella are held in a characteristic position, the ventral one forming a curl. This is strongly reminiscent of some observations by Kent (1880–82) on the 'lower' flagellum of *Chilomonas paramaecium*, which "was thrown into a loose spiral coil" (Kent, 1880–82, p. 425), and also some observations on *Chilomonas amygdalum* which "anchored itself by a spiral coil of the posterior of the two flagella" when at rest (Kent, 1880–82, p. 426). Identical observations on *Chilomonas paramaecium* were made by Edmonson, (1906, p. 50), who saw that the 'lower' flagellum was "often thrown into a coil and serves as an anchor to temporarily attach the organism to some support". Cells of *Chilomonas amygdalum* are also laterally flattened ('almond-shaped' according to Kent), as is *Platychilomonas,* whereas Edmonson's cells of *Chilomonas paramecium* have the uncompressed shape generally believed to be typical of *Chilomonas*. However, in the original description (Ehrenberg, 1838), *Chilomonas paramaecium/paramecium* was described as having keeled cells ("corps cariné longitudinalement", p. 30), suggesting a laterally compressed shape, as observed by later investigators (Conn, 1905).

Platychilomonas has never been examined using electron microscopy and therefore it is still unknown whether it is a true cryptomonad or whether its phylogenetic affinities lie elsewhere. If *Platychilomonas* is effectively a cryptomonad the generic boundaries between it and *Chilomonas* will require further investigation. In the meantime, for identification purposes it is advisable to refer laterally flattened cells with a coiling ventral flagellum to *Platychilomonas*, bearing in mind however that *Platychilomonas psammobia, Chilomonas amygdalum* and *Chilomonas paramecium* sensu Kent and sensu Edmonson could effectively be congeneric.

Genus *Protocryptomonas* Skvortzov ex Bicudo *sensu* Castro, Bicudo & Bicudo (ICBN, ICZN)
Vide Skvortzov (Skvortzow), 1960, p. 5; Skvortzov (Skvortzow), 1968, p. 12. *nom. non rite publicatum* (*vide* Bicudo, 1989, p. 82; Castro et al., 1991, pp. 93–96, figures 78–88).
Lectotype species: *Protocryptomonas mukdenensis* Skvortzov 1960, p. 5, figures 14–18 (*vide* Bicudo, 1989, p. 82).

This rare freshwater genus is known from the original descriptions and those of Castro et al. (1991). There are some complexities regarding its typification (Bicudo, 1989) and it is unclear whether the selected lectotype species is the most representative one of its genus. The original descriptions are very poor and they contain little to suggest that this genus is a true member of the cryptomonads. The concept of *Protocryptomonas* adopted by later workers (Castro et al., 1991) is better defined and more circumscribed, However, apart from the (putative) presence of starch as a reserve product there is still little to confirm that *Protocryptomonas* is effectively a cryptomonad. Cells lack ejectosomes (*1.4*) and a furrow/gullet (see under SEM section *2.3*), and species descriptions are based largely on differences in the cell shape and size. The nutrition mode (osmotrophy versus phagotrophy) is unknown. No ultrastructural information is available and this is urgently required in order to clarify the systematic position of this genus.

Protocryptomonas acuta Castro, Bicudo & Bicudo
Castro et al., 1991, p. 94, figures 86–88.

Protocryptomonas chilomonoides Skvortzov ex Castro, Bicudo & Bicudo
Castro et al., 1991, p. 94, figures 78–80.

Protocryptomonas ellipsoidea Skvortzov ex Castro, Bicudo & Bicudo
Castro et al., 1991, p. 95, figure 81.

Protocryptomonas sygmoidea Castro, Bicudo & Bicudo
Castro et al., 1991, p. 95, figures 82–85.

1.7. Photosynthetic cryptomonads (Figs 7–11)

Photosynthetic cryptomonads have one or two chloroplasts, which are variable in shape but they tend to be elongated with one or two conspicuous parietal portions occupying a large part of the cell (Fig. 11). A few genera (known mostly from the original descriptions) possess multiple chloroplasts, which have never been documented using electron microscopy. It is possible that these organelles are effectively photosynthetic endosymbionts or sequestered chloroplasts of ingested photosynthetic prey, harboured by phagotrophic cryptomonad cells. In this respect it is interesting to note that a few genera have effectively been described which possess blue-green endosymbionts (cyanelles), but here again there is very little information in addition to the original descriptions. It is possible that some of these genera may not belong with the cryptomonads at all. For instance, *Strobilomonas* Schiller

(1954) is reminiscent of the heterotrophic genus *Katablepharis/Kathablepharis* Skuja (1939), which does not belong with the cryptomonads (see under *1.6*).

The number of chloroplasts has been used as a taxonomic character in specific and generic descriptions. Although the chloroplasts can be seen easily using LM, it is not always possible to count them unequivocally using ordinary illumination. There is a well-documented LM study in which cryptomonad chloroplasts have been observed in isolation from the containing cells (Ettl, 1980). This showed that a single bilobed chloroplast was in effect present rather than two independent ones. If fluorescence microscopy is available (either by transmitted light or epifluorescence) the natural red autofluorescence of chlorophyll can be used to better visualize the chloroplasts, especially in unfixed, actively swimming cells. However, electron microscopy may be still be indispensable for counting the chloroplasts unambiguously. In this respect, discrepancies have occurred between the original taxomic descriptions based on LM and ultrastructural re-examinations of certain taxa (*1.7.2*). Therefore some caution is necessary when counting the chloroplasts using only one method (LM or TEM) at any one time.

Photosynthetic cryptomonads may also have one, two, four or more pyrenoids within the plastidial complex (see Klaveness, 1985). Their number and position may be taxonomically significant at the specific or even the generic level. For instance, in the genus *Pyrenomonas* (= *Rhodomonas* emend.) the single pyrenoid occurs in a characteristic position within the bridge connecting the two lobes of the single chloroplast. Pyrenoids are often surrounded by a starch cap, which stains with Lugol's solution thereby making the number and position of the pyrenoid(s) clearly visible (Figs 10 and 11).

1.7.1. Cryptomonads with multiple chloroplasts or with photosynthetic endosymbionts: Cryptella, Cryptochrysis polychrysis, Cyanomonas, Peliaina, Pseudocryptomonas, Strobilomonas

All but one of the genera with multiple chloroplasts/photosynthetic endosymbionts are blue-green in colour. The exception is *Pseudocryptomonas* (see below). In addition to the cryptomonads mentioned in the key below, another form with small, multiple chloroplasts/photosynthetic endosymbionts has been described under the name of *Cryptochrysis polychrysis* Pascher (1913, p. 102, figure 155). This has never been reported again after the original description and the colour of the chloroplasts/photosynthetic endosymbionts is unknown since it was not mentioned in the diagnosis. The type species of *Cryptochrysis* has two chloroplasts, but once again their colour was not mentioned in the diagnosis (*1.7.2*).

1. Chloroplasts/photosynthetic endosymbionts small and elliptical or polygonal in shape 2
   Photosynthetic endosymbionts (cyanelles) large, discoid 4
2. *Cryptomonas*-like cells about 16–60 μm long with 'chestnut' coloured chloroplasts . . *Pseudocryptomonas*
   Cells with blue-green chloroplasts/photosynthetic endosymbionts 3
3. *Chroomonas*-like cells with 6–10 discoid chloroplasts/photosynthetic endosymbionts; flagella not notably longer than the cells . . . . *Cyanomonas*
   Katablepharis-like cells with numerous small elliptical chloroplasts/photosynthetic endosymbionts, flagella twice as long as the cells . . . . *Strobilomonas*
4. Cyanelle single, ejectosomes present . . . *Cryptella*
   3–5 cyanelles, ejectosomes absent . . . . . . *Peliaina*

Genus *Cyanomonas* Oltmanns (ICBN, ICZN)

F. Oltmanns 1904, *Morphologie und Biologie der Algen*, p. 30, figures 16.1, 16.2 (reproduced from Davis, 1894, pl. XI, figures 1, 2). Non *Cyanomonas* Büttner 1911, p. 129, figure 9, *nom. illegit.*

Type species: *Cyanomonas americana* (Davis) Oltmanns, 1904, p. 30, figures 16.1, 16.2; basionym: *Cryptoglena americana* Davis, 1894, p. 101, pl. XI figures 1–5.

The distinctive feature of *Cyanomonas* is given by the presence of multiple (4–10) discoid blue-green chloroplasts in cells which are otherwise similar to *Chroomonas*. No ultrastructural observations are available and therefore it is unknown whether they are effectively chloroplasts rather than photosynthetic endosymbionts.

This is a very rare genus, known mostly from observations on the type species (Davis, 1894, as *Cryptoglena americana*; Skuja, 1948), a flagellate which has been found in brackish and freshwater habitats. There is one additional species, *Cyanomonas acuta* (Schiller) Schiller, introduced by Schiller (1957, p. 17) based on *Chilomonas acuta* which he had described on a previous occasion (Schiller, 1929). Patterson & Hedley (1992, figure 124) also provided an original line-drawing of a cell of *Cyanomonas* but they referred this genus to *Chroomonas* Hansgirg with no further explanation. Perhaps this was due to the fact that Hill (199lb, p. 134) had recombined *Cryptoglena americana* under *Chroomonas americana* (Davis) Hill, with *Cyanomonas americana* (Davis) Oltmanns as a synonym, thus making the entire genus *Cyanomonas* a taxonomic synonym of *Chroomonas*. However, *Chroomonas* was described by Hansgirg (1885) as having a single parietal chloroplast of a different shape: "chromatophoris (cyanophoris) simplicibus, laminaeformis, parietalibus aerugineo-coerulis" (Hansigirg, 1885, p. 230). This agrees with the currently accepted concept of *Chroomonas* based on electron microscopy (see *Chroomonas* under *2.2.2.2*, *2.2.2.3*). Furthermore, no re-examination of *Cyanomonas americana* was provided when it was recombined (Hill, 1991b).

Genus *Cryptella* Pascher (ICBN, ICZN)
A. Pascher 1929, *Jahrbücher für wissenschaftliche Botanik* vol. 71, p. 459, figures 19A, 27a, b.
Type species: *Cryptella cyanophora* Pascher, 1929, p. 459, figures 19A (holotype), 27a.

Cryptella is a rare monospecific genus which is known exclusively from Pascher's original observations on *Cryptella cyanophora*. This freshwater cryptomonad is characterized by the presence of a single discoid blue-green endosymbiont placed centrally in an otherwise colourless cell possessing starch granules and two longitudinal rows of ejectosomes (*1.4*) in the ventral region. No ultrastructural information is available.

Genus *Peliaina* Pascher (ICBN, ICZN)
A. Pascher 1929, *Jahrbücher für wissenschaftliche Botanik* vol. 71, p. 458–459, figures 8–17, 27b.
Type species: *Peliaina cyanea* Pascher, 1929, p. 459, figures 8-17, 27 b.

This very rare freshwater flagellate was described on the same occasion as *Cryptella*, from which it differs in the presence of several (up to about six) cyanelles and the absence of ejectosomes. Pascher (loc. cit.) also observed cells without cyanelles, which were capable of dividing outside the host cell. No other species are known in addition to the type. According to Pascher (loc. cit.) there are no phylogenetic relationships between *Cryptella* and the cryptomonads, and the morphological similarities are the result of evolutionary convergence. No ultrastructural information is available.

Genus *Pseudocryptomonas* Bicudo & Tell (ICBN, ICZN)
C. E. de M. Bicudo & G. Tell 1988, *Nova Hedwigia* vol. 46, p. 408, figures 1–11.
Type species: Pseudocryptomonas subcylindrica Bicudo & Tell 1988, p. 409, figure 1.

As its name implies, *Pseudocryptomonas* shows a broad resemblance to *Cryptomonas* on the basis of light-microscopical features. However, there are numerous disc-shaped chloroplasts as opposed to one or two in *Cryptomonas*. The colour of the chloroplasts was not mentioned in the original description but a later paper reports that it is 'variable' and various 'tones of chestnut' (Castro et al., 1991, p. 96, in Portuguese). Should this effectively reflect the presence of phycoerythrin, then *Pseudocryptomonas* would appear to be the phycoerythrin-containing counterpart of *Cyanomonas* (see above), particularly *Pseudocryptomonas parrae* which shows a general resemblace to *Cyanomonas americana*. No ultrastructural information is available.

This is a very rare genus which is known mostly from the original description. It has only been reported from two freshwater habitats in South America (Brazil and Chile: Bicudo & Tell, 1988; Castro et al., 1991). It contains three species, all of which were originally described by the authors of the genus themselves.

Pseudocryptomonas americana Bicudo & Tell
Bicudo & Tell, 1988, p. 408, figures 7–11.
Note: This has been distinguished from the other two species based on the presence of 'naviculoid' ejectosomes (*1.4*) which are "distributed in a single row lining the gullet". The two other species have 'granular' ejectosomes.

Pseudocryptomonas parrae Bicudo & Tell
Bicudo & Tell, 1988, p. 409, figures 2–6.
Note: This is the smallest species. It is reminiscent of *Cyanonomonas americana* (see above) but it has 'chestnut' rather than blue-green chloroplasts/photosynthetic endosymbionts.

Pseudocryptomonas subcylindrica Bicudo & Tell
Bicudo & Tell, 1988, p. 409, figure 1.
Note: this is the largest species (ca. 63 × 23 μm).

Genus *Strobilomonas* Schiller (ICBN, ICZN)
J. Schiller 1954, *Archiv für Protistenkunde* vol. 100, p. 119, figures 2a–e.
Type species: *Strobilomonas cyanea/cyaneus* Schiller, 1954, p. 119, figures 2a–e.

This is a very rare monospecific genus from freshwater habitats. Its distinctive feature is given by the presence of numerous small, elliptical, blue-green chloroplasts/photosynthetic endosymbionts in cells which are otherwise similar to the heterotrophic genus *Katablepharis/Kathablepharis* Skuja (1939), which is not a cryptomonad (Vørs, 1992). No ultrastructural information is available.

1.7.2. Genera with one or two chloroplasts: Campylomonas, Chroomonas, Cryptochloris, Cryptochrysis, Cryptomonas, Falcomonas, Geminigera, Guillardia, Hanusia, Hemiselmis, Hillea, Komma, Plagioselmis *and* Isoselmis, Proteomonas, Pyrenomonas/Rhodomonas *emend.,* Rhinomonas, Storeatula, Teleaulax

With the exception of those mentioned in *1.7.1*, all photosynthetic cryptomonads have one or two chloroplasts of the type mentioned in *1.7.* Their number appears to be constant within each genus but there are some important exceptions. The original diagnoses of the genera *Hillea* and *Protochrysis* both mention that two chloroplasts are present, but later species descriptions have mentioned a single chloroplast (*Protochrysis: 1.3.1*; *Hillea:* see under *2.2.1*). In other cases, the number of chloroplasts mentioned in the original taxonomic descriptions based on LM has not been confirmed by later ultrastructural studies. For instance, the original description of *Plagioselmis prolonga* mentioned two chloroplasts (Butcher, 1967, p. 18), although an ultrastructural re-examination has shown that the chloroplast is single (Novarino et al., 1994) (see *Plagioselmis* under *2.2.2.3*). In the genus *Pyrenomonas* (= *Rhodomonas* emend.) one species has two chloroplasts while all the others have a single chloroplast (see *Pyrenomonas* under *2.2.2.4*). Finally, the case of the genus *Cryptomonas* is still complex because generic descriptions based on electron microscopy mention two chloroplasts (Santore, 1984; Hill, 1991a), but cells with a single bilobed chloroplast are also known from the LM observations of Ettl (1980). As far as is possible to infer, the cells observed by Ettl (loc. cit.) have a periplast composed of a single sheet (*2.2.1*), and therefore they closely resemble *Campylomonas*, although this was originally described as having two chloroplasts rather than one only (Hill, 1991a) (*2.2.1*).

Perhaps some of the discrepancies about the number of chloroplasts within a particular species or genus could be partly explained by the fact that different investigators may have observed cells during different stages of the life-cycle. This could account for the presence of cells with two chloroplasts (i.e., cells about to divide) in taxa described as having a single chloroplast.

- One chloroplast .... *Chroomonas, Cryptomonas* (partim), *Falcomonas, Geminigera, Guillardia, Hanusia, Hemiselmis, Hillea* (partim), *Komma, Plagioselmis* and *Isoselmis, Proteomonas, Pyrenomonas/Rhodomonas* emend. (partim), *Protochrysis* (partim), *Rhinomonas, Storeatula, Teleaulax*

- Two chloroplasts .... *Campylomonas, Cryptochloris, Cryptochrysis, Cryptomonas* (partim), *Hillea* (partim), *Protochrysis* (partim), *Pyrenomonas/Rhodomonas* emend. (partim)

1.8. Other LM characters of possible taxonomic value: Corps de Maupas and other refringent inclusions, eyespots (Figs 7 and 8)

Several cryptomonad species (especially within the genera *Cryptomonas* and *Chilomonas*) have been described which possess paired structures called 'Corps de Maupas' (CdM), a term used in the past for paired refringent bodies seen also in the ciliates (Deflandre, 1938). The CdM can be detected thanks to their location (usually in the anterior part of the cell) and their characteristic birefringence when viewed using polarized light microscopy. There is very little information on the nature of the CdM. Santore (1985) quoted information derived from Pringsheim, stating that in cultured cryptomonad cells the CdM disappeared after subculturing. This suggested that the CdM were neither specialized inclusions or organelles, and therefore the term 'Corps de Maupas' should have been abandoned (Santore, 1985). By contrast, other ultrastructural observations suggest that they are reserve materials (Joyon, 1963) or have a lysosomal function (Lucas, 1970a, b). Structures termed 'Corps de Maupas' were studied ultrastructurally also by Taylor & Lee (1971) and Kugrens & Lee (1990), who saw that they contained degenerating ejectosomes. Overall the status of the CdM as a reliable taxonomic character is still very unclear and requires further investigation.

Cryptomonads may have other refringent inclusions, for instance the 'refractive bodies' (RBs) seen by Butcher (1967), especially in species of *Chroomonas* (Figs 7 and 8). Butcher (loc. cit.) made no suggestions as to their nature and function although he did use the number of RBs as a specific character. Although there are several discrepancies in Butcher's descriptions of the RBs (see Novarino & Lucas, 1993a), accurate RB counts have been used as a diacritical character between two species of *Chroomonas* sensu auctorum (Novarino & Lucas, 1993a). Ultrastructural observations also suggest that the RBs may correspond to the 'Corps de Maupas', but further studies are necessary to confirm their identity and function (Novarino & Lucas, 1993a).

Several light-microscopical descriptions of cryptomonads mention the presence of putative eyespots, especially in blue-green forms. Examples can be found in Butcher (1967), who noticed that some species of *Chroomonas* possessed a 'stigma' and/or 'carotin grains', which may or may not represent true eyespots of the kind documented later using electron microscopy (Lucas, 1982). True eyespots have been found also in other genera, e.g. *Hemiselmis* (Clay & Kugrens, 1999a), and therefore their taxonomic value is at the specific level. TEM is necessary in order to confirm the identity of putative eyespots seen using LM (*4.5*).

2. Cryptomonads and scanning electron microscopy (Figs 12–37 and 39–52)

Except for the characters considered above, the author believes that LM does not reveal further characters which can be observed easily and used unequivocally in a taxonomic context. Electron microscopy is now necessary to further identify cryptomonads at and below the generic level.

Beginning in the mid-1960s electron microscopy has revealed a multitude of previously unknown cellular features, many of which have been included in a comprehensive review of taxonomically valuable characters within the cryptomonads (Klaveness, 1985). Since then, several ultrastructural characters have been used for describing new genera and species, and amending existing generic diagnoses (Hill & Wetherbee, 1986, 1988, 1989, 1990; Hill, 1991a, b; Novarino, 1991a, b; Novarino & Lucas, 1993a; Novarino et al., 1994; Kugrens et al., 1999). Subsequently, two attempts have been made to erect formal classification systems of the cryptomonads based on ultrastructural characters (Novarino & Lucas, 1993a, 1995; Clay et al., 1999).

There is still some discussion about which ultrastructural characters are taxonomically significant, and at what level. The characters listed below are visible using SEM and are accorded a taxonomic value at the generic and/or specific levels by various workers. However, in spite of the great taxonomic usefulness of SEM not all cryptomonads can be identified conclusively using this observation method alone because further information might be necessary on the identity of the accessory photosynthetic phycobilin pigment (go to *6*), the detailed architecture of the periplast (as revealed by freeze-fracture/etch methods; go to *5*) and the internal cell ultrastructure, especially the position of the nucleomorph (as revealed by transmission electron microscopy; go to *4.1*). The possible occurrence of preparation artefacts should also be borne in mind (*2.9*), but certain artefacts are not necessarily undesirable at all times since they may still provide some information of taxonomic interest (*2.9.1*, *2.9.2*).

2.1. Cell size and shape

Strictly speaking these are not 'ultrastructural' characters, but SEM makes it possible to observe the cell shape much more accurately than is possible using LM alone (*1.3*). The SEM also makes it possible to measure specimens very accurately. Shrinkage (up to about 25% linear) may occur during specimen preparation for SEM using critical point-drying (Novarino, 1991b; Novarino & Lucas, 1993a), and this should be borne in mind when comparing LM- and SEM-based measurements.

2.2. Kind of periplast present (Figs 12, 13, 15–23, 25, 26, 28–32, 34–37, 39–48)

Cryptomonads possess a typical cell covering termed the 'periplast'. There is evidence to suggest that the periplast was detected during two early investigations which were based on light microscopy only (Novarino, 1993). However, such cases are exceptional and only electron microscopy has made it possible to elucidate the periplast structure in detail. Since the first electron microscopical observations (Lucas, 1970a; Gantt, 1971; Hibberd et al., 1971; Faust, 1974), our knowledge of periplast architecture has progressed consid-

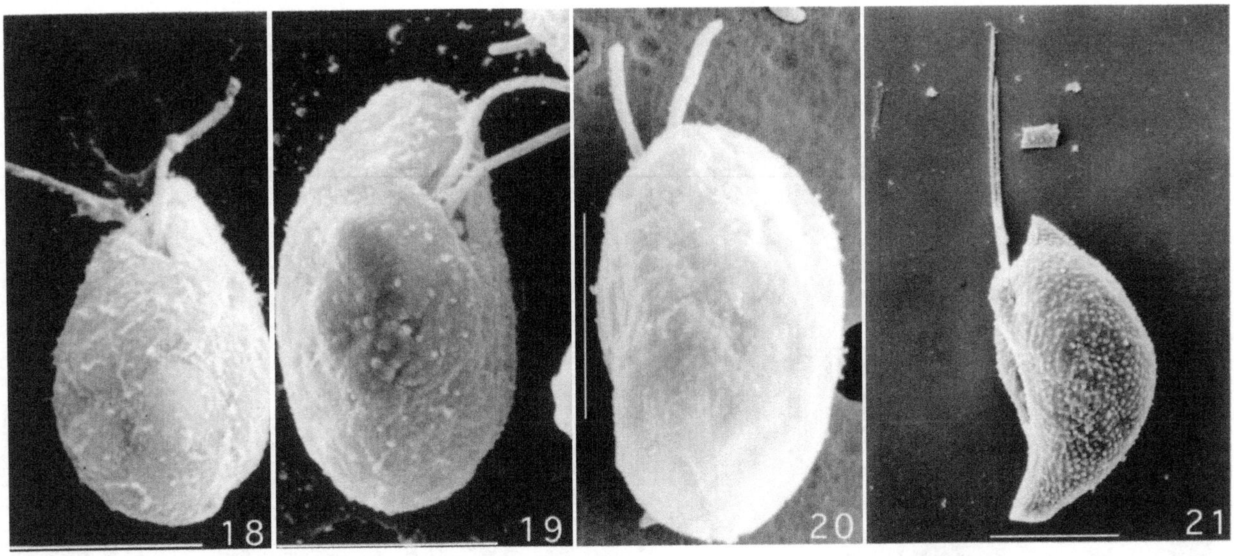

Figure 18–21. SEM micrographs of cryptomonads. *Figures 18 and 19.* Two cells of *Proteomonas pseudobaltica* seen from the ventral face. Scale bars = 5 μm. *Figure 20.* A cell of *Storeatula* sp. Seen from th dorsal face. Scale bar = 5 μm. *Figure 21. Teleaulax amphioxeia* seen in latero-ventral view, showing the long furrow and flagella. Scale bar = 5 μm.

erably thanks especially to freeze-fracture studies (*5*). Periplast construction is very diverse (*5*), but there is always a protein layer underneath the cell membrane (internal periplast component, IPC). The IPC can be made up either of a single element (i.e. it is a continuous sheet), or a series of discrete plates. There is also an external periplast component (EPC), which is very variable in structure. Some cryptomonads with a plated IPC have an EPC composed of plates similar to the IPC. One exception is *Guillardia*, which has a sheet-like IPC and an EPC composed of large plates (see *Guillardia* under *2.2.2.5*).

Plated versus non-plated periplasts are accepted as a taxonomically significant character at the generic level (Santore, 1984; Kugrens & Lee, 1987; Hill & Wetherbee, 1988, 1989, 1990; Hill, 1991a, b; Novarino, 1991a, b; Novarino & Lucas, 1993a, b; Novarino et al., 1994; Novarino & Lucas, 1995; Clay et al., 1999; Kugrens et al., 1999). The exception is *Proteomonas* (see below), whose type species has two life-stages, one with a non-plated periplast (see *2.2.1*) and the other one with a plated periplast (see *2.2.2*). The SEM is usually capable of revealing the sheet-like appearance of the IPC in taxa with a non-plated IPC, and the plated appearance of the IPC or the EPC in those taxa with a plated IPC and/or EPC.

Among the possible preparation artefacts affecting the appearance of the periplast in the SEM, cell shrinkage is the one which is most likely to occur (*2.9.2*). Another possible artefact is given by the discharge of the ejectosomes present at the cell periphery (*2.9.1*).

2.2.1. Genera whose IPC is non-plated and whose periplast appears sheet-like under the SEM: Chilomonas, Campylomonas, Cryptochloris, Cryptomonas *(partim)*, Geminigera, Hanusia, Hillea *(?)*, Proteomonas *(diplomorph stage)*, Storeatula, Teleaulax

- Heterotrophs .... *Chilomonas*
- PC-containing genera *Cryptochloris, Hillea* sensu stricto (?)
- PE-containing genera ..... *Campylomonas, Cryptomonas* (partim), *Geminigera, Hanusia, Hillea* sensu lato (?), *Proteomonas* (diplomorph stage), *Storeatula, Teleaulax*

Apart from *Chilomonas*, all known cryptomonads with a non-plated (sheet-like) periplast are photosynthetic and most contain PE. Morphologically they form a closely knit cluster and further information on the position of the nucleomorph (go to *4.1*) and the type of PE (*6*) may be necessary for a conclusive identification at the generic level.

Genus *Chilomonas* Ehrenberg
(For publication details see *Chilomonas* under *1.6*).

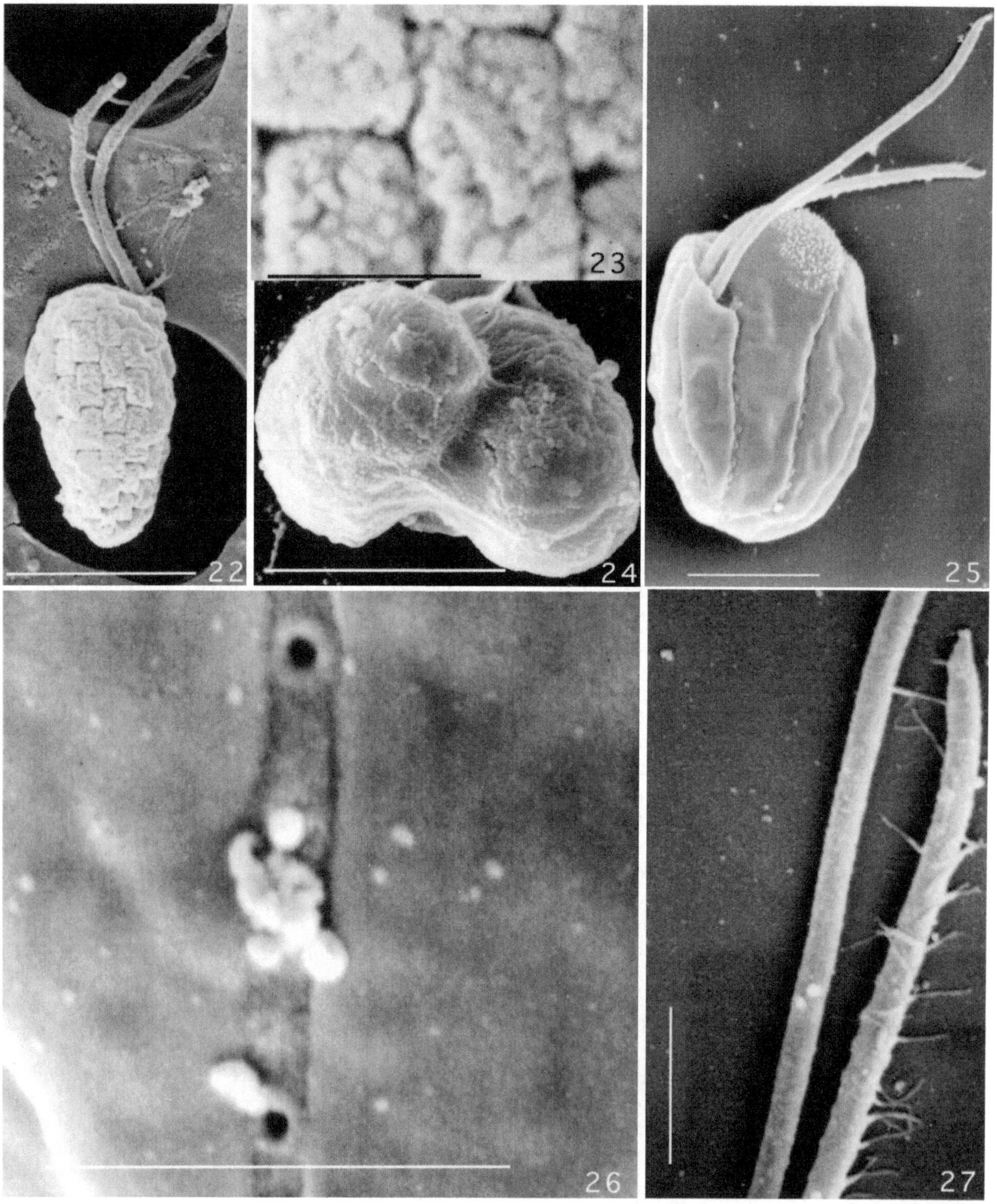

Figure 22–27. SEM micrographs of cryptomonads. *Figures 22–24.* A species of *Chroomonas* sensu auctorum. Figure 22 shows a flagellate cell, Figure 23 the periplast plates, and Figure 24 a pair of palmelloided cells covered with a layer of mucilage, which obscures the periplast. Scale bars = 5 μm (Figs 22 and 24) or 1 μm (Fig. 23). *Figures 25–27. Goniomonas* cf. *amphinema*. Figure 25 shows a whole cell, Figure 26 the periplast and Figure 27 the flagellar appendages. Scale bars = 2 μm (Fig. 25) or 1 μm (Figs 26 and 27).

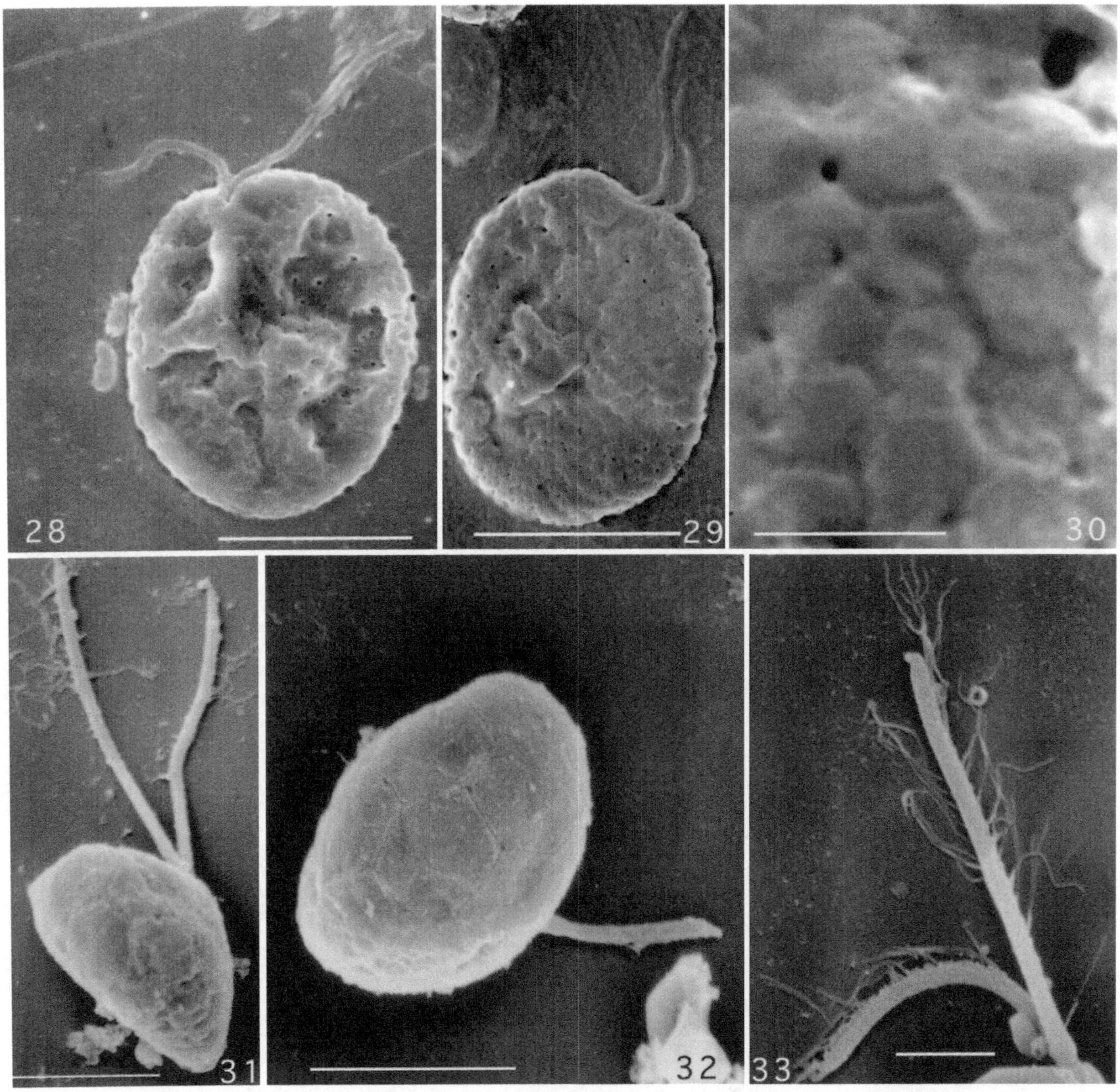

Figure 28–33. SEM micrographs of cryptomonads. *Figures 28–30.* Hansgirg's original type specimens of *Chroomonas nordstedtii*, the type species of its genus. Figures 28 and 29 show cells without and with discharged ejectosomes visible in the background. Figure 30 shows a detail of the periplast; hexagonal plates are clearly visible. Scale bars = 5 μm (Figs 28 and 29) or 1 μm (Fig. 30). *Figures 31–33. Hemiselmis* sp. Figures 31 and 32 show different cells and in Figure 32 the periplast plates are clearly visible. *Figure 33.* shows the flagellar appendages arranged in the 'normal' pattern. Scale bars = 2 μm (Figs 31 and 32) or 1 μm (Fig. 33).

Chilomonas has a non-plated periplast which is very similar to *Campylomonas* (see below) (Roberts et al., 1981; Grim & Staehelin, 1982, 1984; Heywood, 1988; Kugrens & Lee, 1991). *Chilomonas* and *Campylomonas* also share an unusual arrangement of flagellar appendages (see *Campylomonas* below and *2.8*, *3.1.1*).

Genus *Campylomonas* Hill (ICBN, ICZN) (Figs 12–15)

D. R. A. Hill 1991, *Phycologia* vol. 30, p. 173, figures 3, 7–15.

Type species: *Campylomonas reflexa* (Skuja) Hill, 1991a, p. 173, figures 3, 7–15.

Basionym: *Cryptomonas reflexa* Skuja, 1939, p. 93-

Figure 34–38. SEM and TEM micrographs of *Plagioselmis prolonga*. Figures 34 and 35 show different cells in lateral (Fig. 34) and dorso-lateral (Fig. 35) views. The flagella with typically arranged appendages are clearly visible, as are the hexagonal periplast plates on the main portion of the cell body, and the sheet-like (non-plated) periplast on the posterior tail. Scale bars = μm. *Figures 36 and 37.* Details of the posterior tail. Figure 36 shows the mid-ventral band (MVB) extending from the ventral furrow to the tail. Figure 37 shows another tail in dorsal view. Scale bars = 2.5 μm (Fig. 36) or 1 μm (Fig. 37). *Figure 38.* Transverse section showing the ventral furrow (F) lined with ejectosomes, and the nucleomorph (arrow) in the periplastidial compartment between the chloroplast (Chlp) and the nucleus (N). Scale bar = 1 μm.

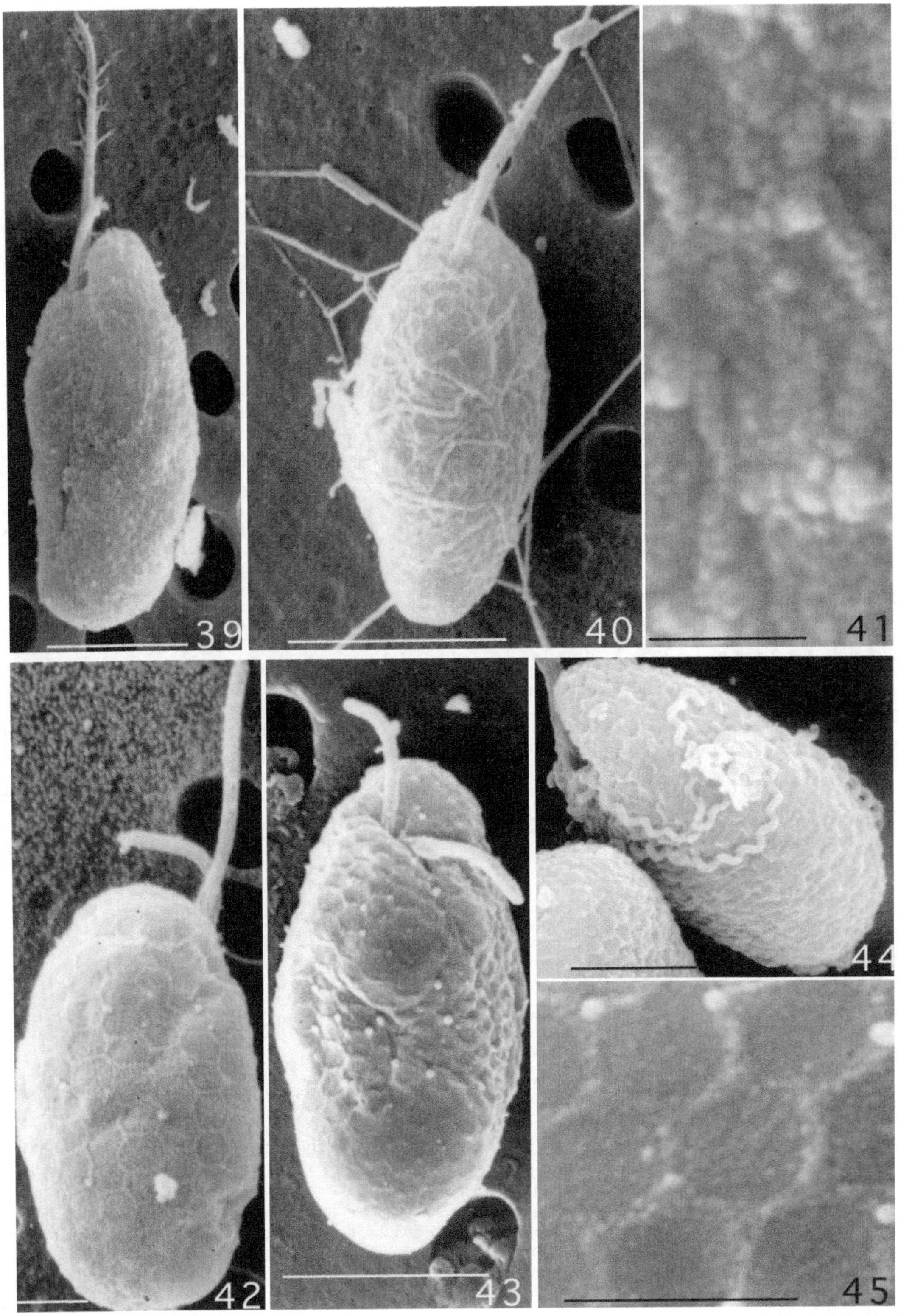

Figure 39–45. SEM micrographs of cryptomonads. *Figures 39–41. Pyrenomonas* sp. Figure 39 is a cell in lateral view and Figure 40 a cell in ventral view with several discharged ejectosomes. Figure 41 shows the periplast plates. Scale bars = 5 μm (Figs 39 and 40) or 1 μm (Fig. 41). *Figures 42–45. Rhinomonas reticulata.* Figures 42 and 43 show two cells in dorso-lateral and ventral view, respectively. Figure 44 shows another cell with spirochaete bacteria in the gullet. Figure 45 shows the periplast. Scale bars = 5 μm (Fig. 43), 2.5 μm (Fig. 44), or 1 μm (Fig. 45).

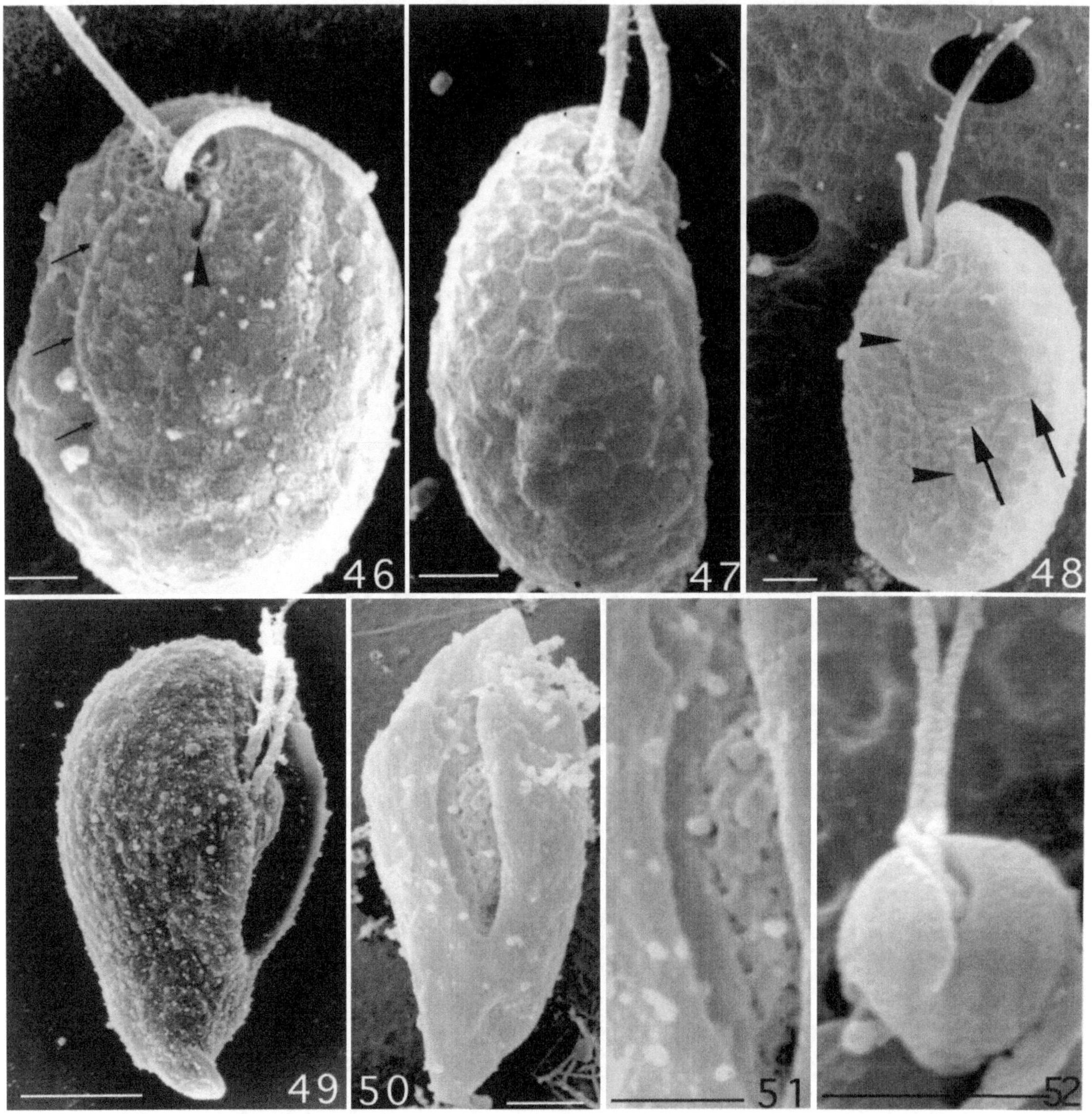

Figure 46–52. SEM micrographs of artefactual and non-artefactual ventral furrows in cryptomonads. *Figure 46.* A cell of *Rhinomonas reticulata* showing a short, notch–like artefactual furrow produced by cell shrinkage, whose effects are visible also elsewhere in the cell (arrows, showing a lateral ridge-like fold). Figure 47 is another cell of the same strain from the same SEM preparation, showing no signs of shrinkage and lacking a ventral notch-like furrow. Figure 48 is another cell of the same strain from the same SEM preparation, showing a narrow furrow-like ventral fold which also branches towards the left-hand ventro-lateral region of the cell (arrows). Scale bars = 1 μm. *Figures 49–52.* True, non-artefactual furrows. Figure 49. *Plagioselmis* sp. inedit., showing a wide furrow with flagella inserted at the top end and slightly displaced towards the right-hand side. Scale bar = 2 μm. Figures 50 and 51. The furrow of *Teleaulax acuta* at different magnifications. Note the ejectosome pores on the internal surface (Fig. 51). Scale bars = 2 μm. *Figure 52. Hillea* (?) sp. with a short, curved furrow from which the flagella emerge. Scale bar = 2 μm.

94, pl. 5 figure 9; synonym: *Cryptomonas erosa* Ehrenberg var. *reflexa* Marsson (non vidi).

Campylomonas has been separated from *Cryptomonas* based on the presence of a sheet-like periplast (Figs 12, 13 and 15) and a long keeled rhizostyle (see Hill, 1991a). It was described based on the type species only, a new combination for a *Cryptomonas* species which has been reported moderately frequently from freshwater environments. It is a PE-containing cryptomonad with two chloroplasts and two pyrenoids, a pointed posterior end which is dorsally curved, and a short ventral furrow, which is also found in other species of the genus (Fig. 14). Further information on the flagella is available from the study of Gromov et al. (1998), where a newly isolated strain of *Campylomonas reflexa* was seen to possess a single row of hairs on each flagellum (see *2.8*, *3.1.1*). This contrasts with *Cryptomonas*, where one flagellum has two rows of hairs and the other one a single row (Hill, 1991a) (see *3.1*, *3.1.1*). So far *Campylomonas* only includes the type species, but is is possible that several species of *Cryptomonas* may have to be transferred here (see *Cryptomonas* below).

Genus *Cryptochloris* Schiller (ICBN, ICZN) non Bentham (ICBN) non Shortridge & Carter (ICZN) (Figs 16 and 17)

J. Schiller 1925, *Archiv für Protistenkunde* vol. 53, p. 88, pl. 3 figure 10.
Type species: *Cryptochloris vittata* Schiller, 1925, p. 88, pl. 3 figure 10.

This is a poorly known marine genus which has been reported very rarely. To date it only includes the type species. An unidentified species was observed with the SEM (Novarino et al., 1997), which showed a non-plated periplast with a papillate appearance caused by the underlying ejectosomes bulging towards the cell surface (Figs 16 and 17). Under the ICBN, the generic name is pre-occupied by *Cryptochloris* Bentham, a name of a genus of grasses, but a proposal has been made to reject Bentham's name in the interest of ambiregnal nomenclature (Novarino & Gilbert, 2002).

Genus *Cryptomonas* Ehrenberg (partim) (ICBN, ICZN)

Vide C. G. Ehrenberg, 1838. *Die Infusionsthierschen als volkommene Organismen. Ein Blick in das tiefere organische Leben der Natur*: p. 40, pl. II figures XVI–XXII.
Type species: not designated by Ehrenberg (1831a, b, 1838).

This genus was introduced by Ehrenberg on several occasions. The earliest appearances of the generic name in the literature (Ehrenberg, 1831a, b) are sanctioned by the on-line version of the Index Nominum Genericorum Plantarum (http://rathbun.si.edu/botany/ing/) as the original publications for nomenclatural purposes. In Ehrenberg (1831a) there are only succinct synoptic tables without descriptions or illustrations. More detailed information is available elsewhere (Ehrenberg, 1831b) but without illustrations. These were provided in a later work (Ehrenberg, 1838) together with a bilingual (Latin and French) formal diagnosis and several illustrations. For this reason the publication place and date adopted here is that of Ehrenberg (1838) rather than the others.

Cryptomonas and the other Ehrenbergian genus (*Chilomonas*) are the oldest described cryptomonad genera. Since Ehrenberg's early work, which allowed for the existence of some seven species, a great many species have been described and *Cryptomonas* is the largest genus in terms of published species names. *Cryptomonas* is also the most difficult genus from a taxonomic point of view. Ehrenberg never designated a type species, and common usage has established a consensus that the 'typical' Ehrenbergian species are *C. erosa* and *C. ovata*. This was sanctioned in particular by the designation of *C. ovata* as the lectotype species by Butcher (1967), although this has not been incorporated in the on-line version of the Index Nominum Genericorum Plantarum (http://rathbun.si.edu/botany/ing/).

Cryptomonas erosa and *C. ovata* are the most frequently reported species, not only within this genus but also within the cryptomonads as a whole. Electron-microscopical and biochemical studies (Santore, 1984, 1985; Hill & Rowan, 1989; Hill, 1991a) have established a concept of *C. ovata* based on the number of chloroplasts, the identity of the PE pigment, the architecture of the periplast, and the number and position of the nucleomorphs (*4.1*). Within this concept, *C. ovata* has two chloroplasts (each with its own associated nucleomorph placed outside the pyrenoid, *4.1.3*), PE III (go to *6*), and a periplast with an IPC composed of discrete hexagonal plates (see Hill, 1991a, *2.2.2.3*). The presence of a ventral furrow (*2.3*) in *C. ovata* is supported by SEM micrographs (see Hill, 1991a and references therein) although other workers have rejected it (Santore, 1985).

Available SEM micrographs of some species (e.g. *Cryptomonas rostratiformis* Skuja: Klaveness, 1985, pl II figure E; *Cryptomonas undulata* Gervais: Gervais, 1997, figures 1–5), show a sheet-like periplast. Overall those species are strongly reminiscent of *Campylomonas* (see above) and therefore they may have to be recombined under that genus. However, there are discrepancies regarding the number of chloroplasts because *Cryptomonas rostratiformis* and *Cryptomonas undulata* both have a single chloroplast rather than two (Ettl, 1980; Gervais, 1997). This casts further uncertainties on the generic value of the number of chloroplasts (see *1.7.2*). A concept of *Cryptomonas* has been proposed which includes forms with both types of periplast (Novarino & Lucas, 1993b), and it has been suggested that the differently constructed periplasts might be useful for defining subgenera (Novarino, 1991b). An additional diacritical character between *Campylomonas* and *Cryptomonas* is given by the arrangement of the flagellar hairs (go to *2.8*, *3.1*, *3.1.1*).

Genus *Geminigera* Hill (ICBN, ICZN)

D. R. A. Hill 1991, *Phycologia* vol. 30, p. 181, figure 2.

Type species: *Geminigera cryophila* (Taylor & Lee) Hill, 1991a, p. 181, figure 2.

Basionym: *Cryptomonas cryophila* Taylor & Lee, 1971, p. 270, figure 1.

Geminigera was also separated from *Cryptomonas* based on the presence of a sheet-like periplast and a long keeled rhizostyle, as in *Campylomonas* (see Hill, 1991a), but it has a single chloroplast rather than two, with two pyrenoids. *Geminigera cryophila*, which is the only described species, is based on *Cryptomonas cryophila*, a marine cryptomonad from Antarctica possessing a short furrow and peculiar electron-dense bodies underneath the cell surface which give the cell a 'warty' appearance. Similar bodies have been seen also in a species of *Chroomonas* sensu auctorum (*2.2.2.2*) cultured at a high irradiance (Novarino & Lucas, 1993a), whereas cells cultured at a low irradiance lacked these bodies altogether. It was hypothesized that their function is to shield the chloroplast from irradiance levels which are too high for optimal growth (Novarino & Lucas, 1993a).

Genus *Hanusia* Deane, Hill, Brett & McFadden (ICBN, ICZN)

J. A. Deane, D. R. A. Hill, S. J. Brett & G. I. McFadden 1998, *European Journal of Phycology* vol. 33, p. 153, figures 1, 4–10, 13.

Type species: *Hanusia phi* Deane, Hill, Brett & McFadden, 1998, p. 153, figures 1, 4–10, 13 (holotype).

Hanusia is a monospecific genus based on a PE-containing marine cryptomonad with a single chloroplast and a sheet-like periplast. The diagnosis states that *Hanusia* is "similar in morphology to *Teleaulax* Hill and the *Proteomonas* Hill & Wetherbee diplomorph but differing in 18S ribosomal RNA gene sequence". (See *Proteomonas* and *Teleaulax* below). *Hanusia* shares molecular and karyotypic features with *Guillardia* (see *Guillardia* under *2.2.2.5*), but this has a different periplast. Therefore molecular and genetic techniques are necessary in order to identify correctly any newly observed specimens as *Hanusia phi*.

Genus *Hillea* Schiller (ICBN, ICZN) (Fig. 52)

J. Schiller 1925, *Archiv für Protistenkunde* vol. 53, p. 87, pl. 3 figures 7–9.

Type species: *Hillea fusiformis* (Schiller) Schiller 1925, p. 87, pl. 3 figures 7–9.

Basionym: *Chlamydomonas fusiformis* Schiller, 1913, p. 625, figure 7.

This is a poorly known marine genus which has been reported very rarely. The type species was described as being blue-green in colour, probably reflecting the presence of PC. However, Butcher (1952) described a species with a 'dull yellow' chloroplast, *Hillea marina* (Butcher, 1952, p. 184, pl. 2 figures 39–41). Information on the periplast is available for *Hillea marina* (Chang, 1983, figure 7B) and an unidentified species (Novarino et al., 1997 , figure 11; this paper, Fig. 52). In the SEM both those species show a non-plated periplast (Fig. 52) but further details are necessary, including additional taxonomic information on the illustrated specimens because the systematic affinities of *Hillea* are still very unclear.

Genus *Proteomonas* Hill & Wetherbee (ICBN, ICZN) non Podlipaev, Frolov & Kolesnikov (ICBN, ICZN) (Figs 18 and 19)

D. R. A. Hill & R. Wetherbee 1986, *Phycologia* vol. 25, p. 522, figures 1–64. Type species: *Proteomonas sulcata* Hill & Wetherbee, 1986, p. 181, figures 1 (holotype), 2–64.

The type species of *Proteomonas* is a marine cryptomonad with a single chloroplast containing PE. Its life-cycle includes two morphologicallly distinct stages: a larger 'diplomorph' stage with a sheet-like periplast (Figs 18 and 19), and a smaller 'haplomorph' stage with a plated periplast (see *Proteomonas* under *2.2.2.3*).

To date, only two taxa have been described in addition to the type species:

Proteomonas pseudobaltica (Butcher) Novarino (Figs 18 and 19)
Novarino, 1991b, p. 604, figures 5–7, 10, 11, 13–15.
Basionym: *Cryptomonas pseudobaltica* Butcher, 1967, p. 44, pl. VI fig 2, pl. XIX fig 5, typ. non desig. Lectotype (designated by Novarino, 1991b): Butcher, 1967, pl. VI figure 2.
Note: In this species the 'haplomorph' stage has not been observed, causing Deane et al. (2002) to cast a doubt on its assignment to *Proteomonas*. If *Proteomonas* is effectively an unsuitable generic vehicle, then this cryptomonad could represent an hitherto undescribed genus.

Proteomonas pseudobaltica (Butcher) Novarino var. *leonardiana* Novarino
Novarino, 1991b, p. 606, figures 17 (holotype), 18, 19.
Note: This differs from the type variety in the lateral compression of the cells.

Genus *Storeatula* Hill (ICBN, ICZN) (Fig. 20)
D. R. A. Hill 1991, *Phycologia* vol. 30, p. 181, figures 5, 34–40.
Type species: *Storeatula major* (Butcher) Hill 1991a, p. 181, figures 5, 34–40.
Basionym: *Cryptomonas major* Butcher, 1967, p. 39, typ. non. desig. Lectotype (designated by Hill, 1991a): pl. V. figure 3.

Storeatula was described based on a marine cryptomonad with a single PE-containing chloroplast and a sheet-like periplast (Fig. 20). For a conclusive identification, information is necessary on the position of the nucleomorph (*4.1*, *4.1.2*).

Only one species has been described in addition to the type species:

Storeatula rhinosa Kugrens, Clay & Lee
Kugrens et al., 1999, p. 1080, figures 1, 2, 3 (holotype), 4–12.
Note: This is a freshwater species.

Genus *Teleaulax* Hill (ICBN, ICZN) (Figs 21, 50 and 51)
D. R. A. Hill 1991, *Phycologia* vol. 30, p. 177, figures 6, 16–24, 25, 26.
Type species: *Teleaulax acuta* (Butcher) Hill, 1991a, p. 177, figures 6, 16–24.
Basionym: *Cryptomonas acuta* Butcher, 1952, p. 188, pl. II, figures 51–53.

Teleaulax was first described based on *T. acuta*, a new combination introduced for a frequently reported marine cryptomonad with a single chloroplast containing PE, an acute posterior end and a wide furrow on the ventral face reaching well into the median region of the cell. So far this genus is known exclusively from marine habitats.

Two species have been described in addition to the type species:

Teleaulax amphioxeia (Conrad) Hill (Fig. 21)
Hill, 1992b, pp. 175–176, figures 1A–N.
Basionym: *Rhodomonas amphioxeia* Conrad, 1939, p. 4, figures 3–6.
Note: This is a new combination based on *Rhodomonas amphioxeia* Conrad (1939). Hill (1992b) examined specimens from the Baltic Sea using light and electron microscopy, which showed that the main distinctive feature of this species is given by the presence of unequal flagella, of which the longer one is the ventral flagellum (Hill, 1992b, p. 176). However, Conrad's original description of *Rhodomonas amphioxeia* makes no mention of this feature and specifies instead that the flagella are equal ("les fouets sont égaux", Conrad, 1939, p. 4), as shown also by the illustrations (Conrad, 1939, figures 3–6). Therefore, it is very unlikely that *Teleaulax amphioxeia* effectively represents the cryptomonad studied by Conrad under the name of *Rhodomonas amphioxeia*. For this reason, the cryptomonad studied by Hill (1992b) might be better considered as belonging to an entirely new species, but no formal taxonomic protologue is available.

Teleaulax merimbula Hill
Hill, 1991a, p. 177, figures 25, 26 (holotype).
Note: This species is distinguished from *Teleaulax acuta* by its larger size, the more rostrate anterior end and the chloroplast, which extends to the posterior of the cell.

***2.2.2. Genera whose IPC and/or EPC is plated and therefore the periplast appears plated when viewed using SEM:* Chroomonas, Cryptomonas *(partim)*, Falcomonas, Goniomonas, Guillardia, Hemiselmis, Komma, Plagioselmis, Proteomonas *(haplomorph stage)*, Pyrenomonas/Rhodomonas *emend.*, Rhinomonas**

- Heterotrophs .... *Goniomonas*
- PC-containing genera . *Chroomonas, Falcomonas, Hemiselmis* subgen. *Plagiomonas*, *Komma*
- PE-containing genera .... *Cryptomonas* (partim), *Guillardia, Hemiselmis* subgen. *Hemiselmis, Plagioselmis, Proteomonas* (haplomorph stage), *Pyrenomonas/Rhodomonas* emend., *Rhinomonas*

2.2.2.1. Shape and size of the periplast plates when plates are present

The shape of the periplast plates is a well-established taxonomic character at the generic level (Santore, 1984; Kugrens & Lee, 1987; Hill & Wetherbee, 1988, 1989, 1990; Hill, 1991a, b; Novarino, 1991a, b; Novarino & Lucas, 1993a, b; Novarino et al., 1994; Novarino & Lucas, 1995; Clay et al., 1999; Kugrens et al., 1999), while the size of the plates has been used to delimit species within particular genera (Meyer, 1984; Novarino, 1991a, b; Novarino & Lucas, 1993a; Novarino et al., 1994). When measuring the periplast plates, the possible occurrence of shrinkage artefacts (*2.9.2*) should be borne in mind and may require some interpretation on the part of the observer.

2.2.2.2. Periplast plates rectangular:* Chroomonas *sensu auctorum*, Goniomonas *(Figs 22, 23, 25 and 26)

There are only two genera with rectangular plates, although in the SEM the periplast of *Guillardia* appears as a series of longitudinally oriented plate-like strips (see *Guillardia* under *2.2.2.5*). These may somewhat recall the truly rectangular plates of *Goniomonas* but the resemblance is only superficial.

Genus *Chroomonas* sensu auctorum (Figs 22–24)
Vide Hill, 1991b, pp. 135–137.

The original description of the genus *Chroomonas* Hansgirg, the first blue-green (PC-containing) genus of cryptomonads ever to be published, was based necessarily on light microscopy only (Hansgirg, 1885) (see *Chroomonas* under *2.2.2.3*). In more recent years, many studies have provided this genus with a well-defined ultrastructural identity (for a summary see Hill, 1991b). In the SEM the most prominent feature is the presence of a periplast with relatively large, rectangular plates, arranged in offset longitudinal rows which give the appearance of a brick wall. However, an SEM examination of Hansgirg's original type material of *Chroomonas nordstedtii*, the type species, has revealed the presence of a periplast with hexagonal plates (Novarino & Oliva, 1998; *2.2.2.3*; Figs 28–30). Therefore, based on the taxonomic significance which is currently accorded to the periplast, *Chroomonas* is a heterogeneous assemblage. Species with rectangular periplast plates are not *Chroomonas* in the original sense of Hansgirg and will be referred to here as "*Chroomonas* sensu auctorum". Species with hexagonal plates will be referred to as "*Chroomonas* sensu stricto" (go to *2.2.2.3*). The fact that the type material of *Chroomonas nordstedtii* has hexagonal plates also raises some difficulties in relation to the existence of recently described genera also containing PC and possessing a periplast with hexagonal plates (*Falcomonas, Komma:* go to *2.2.2.3*).

The species listed below have been examined with the SEM and fit the concept of *Chroomonas* sensu auctorum because they have rectangular periplast plates. There are interspecific differences in the size of the plates. In palmelloid cells (see under *1.2*) the periplast may be difficult or impossible to visualize in the SEM owing to the presence of an external mucilage layer (Fig. 24). Additional interspecific differences visible with the SEM include the periplast raphe (see *2.7*).

Chroomonas africana Meyer & Pienaar
Meyer & Pienaar, 1984, p. 307, fis 2a–c, 3a, b, 3c (holotype), 4–32.
Note: The periplast plates range from ca. 0.2×0.15 to 0.6×0.5 μm in size.

Chroomonas coerulea (Geitler) Skuja
Skuja, 1948, p. 350, pl. XXXVII, figures 24, 25.
Basionym: *Cryptomonas coerulea* Geitler, 1922, p. 684, figure a, b.
Note: For ultrastructural information, see Hill (1991b), p. 135 et seq. The periplast plates are ca. 0.4×0.3 μm in size and cells have a periplast raphe (*2.7*).

Chroomonas collegionis Butcher ex Hill
Hill, 1991b, p. 134.

Synonym: *Chroomonas collegionis* Butcher, 1967, p. 28, typ. non desig. Lectotype (designated by Hill, 1991b): Butcher, 1967, pl. II, figure 2.
Note: The size of the periplast plates is not known.

Chroomonas ikaitensis Kristiansen & Kristiansen
Kristiansen & Kristiansen, 1999, p. 755, figures 1–6, 7 (holotype), 8–21.
Note: The size of the periplast plates is not known.

Chroomonas ligulata Novarino & Lucas
Novarino & Lucas, 1993a, p. 584, figures 1 (holotype), 2, 5–8.
Note: The size of the periplast plates is 0.89±0.070 × 0.62±0.054 μm.

Chroomonas placoidea Butcher ex Novarino & Lucas
Novarino & Lucas, 1993a, p. 585, figures 3, 4, 9–14.
Synonym: *Chroomonas placoidea* Butcher, 1967, p. 29, typ. non desig. Lectotype (designated by Novarino & Lucas, 1993a): Butcher, 1967, pl. XVI, figure 3r, first cell from bottom.
Note: The size of the periplast plates is 0.69±0.101 × 0.43±0.043 μm.

Chroomonas sp. Meyer & Pienaar
Meyer & Pienaar, 1987, p. 129 et seq., figures 1–30.
Note: The periplast plates range from 0.6 × 0.4 to 2 × 1 μm in size.

Genus *Goniomonas* von Stein (Figs 25–27)
(For publication details see *Goniomonas* under *1.6*).

Goniomonas (Figs 25–27) has a characteristic periplast composed of rather large rectangular plates forming few longitudinal, non-offset rows (Kugrens & Lee, 1991). Longitudinal ridges or grooves may be apparent between the plate rows (Fig. 26). For a species list see *Goniomonas* in the light-microscopical section under *1.6*.

***2.2.2.3. Periplast plates distinctly hexagonal:* Chroomonas *(sensu stricto)*, Cryptomonas *(partim)*, Falcomonas, Hemiselmis, Komma, Plagioselmis, Proteomonas *(haplomorph stage)*, Rhinomonas**

Genus *Chroomonas* Hansgirg sensu stricto (ICBN, ICZN) (Figs 28–30)

A. Hansgirg 1885, *Botanisches Centralblatt* vol. 23, p. 230, no figures.
Type species: *Chroomonas nordstedtii* Hansgirg, 1885, p. 230.

Genus *Falcomonas* Hill (ICBN, ICZN)
D. R. A. Hill 1991, *Journal of Phycology* vol. 27, p. 141, figures 3, 32–44.
Type species: *Falcomonas daucoides* (Conrad & Kufferath) Hill, 1991b, p. 141, figures 3, 32–44.
Basionym: *Chroomonas daucoides* Conrad & Kufferath, 1954, p. 136, pl. X figure 5.
Synonym: *Hillea daucoides* (Conrad & Kufferath) Butcher, 1967, p. 10.

Genus *Komma* Hill (ICBN, ICZN)
D. R. A. Hill 1991, *Journal of Phycology* vol. 27, p. 137, figures 2, 18–31.
Type species: *Komma caudata* (Geitler) Hill, 1991b, p. 137, figures 2, 18–31.
Basionym: *Chroomonas caudata* Geitler, 1924, p. 246, figures a–h.

Chroomonas Hansgirg sensu stricto (Figs 28–30), *Falcomonas* and *Komma* all contain PC and form a closely knit morphological cluster because they all have a periplast with hexagonal plates. The main difference between these genera and *Chroomonas* sensu auctorum is that the latter has rectangular periplast plates (see *2.2.2.2*). *Falcomonas* can be identified unequivocally thanks to its unique phycocyanin (*6*). *Komma* is very similar to *Chroomonas* sensu stricto. Both *Falcomonas* and *Komma* only include the type species. *Chroomonas* has numerous described species, but most of those examined using electron microscopy possess a periplast with rectangular plates and therefore they do not fit the concept of *Chroomonas* Hansgirg sensu stricto (*2.2.2.2*).

The following blue-green cryptomonads all show hexagonal periplast plates when examined using SEM:

Chroomonas nordstedtii Hansgirg
Hansgirg, 1885, p. 230, no figures.
Note: The hexagonal periplast plates of Hansgirg's original type material are illustrated here (Figs 28–30).

Chroomonas pochmanni Huber-Pestalozzi
Huber-Pestalozzi, 1950, p. 31, figures 16a–f.
Note: An ultrastructural examination of a cryptomonad assigned to this species with some hesitation is available in the literature (Kugrens & Lee, 1990).

Chroomonas sp. inedit. Novarino
Novarino, 2002, p. 181, figure 44 I.

Chroomonas vectensis Carter
Carter, 1937, p. 56, pl. VIII figures 1–3.
Note: An SEM micrograph of a cryptomonad assigned to this species was provided by Hill (1992a).

Falcomonas daucoides (Conrad & Kufferath) Hill
Hill, 1991b, p. 141, figures 3, 32–44.
Basionym: *Chroomonas daucoides* Conrad & Kufferath, 1954, p. 136, pl. X figure 5.
Synonym: *Hillea daucoides* (Conrad & Kufferath) Butcher, 1967, p. 10.

Komma caudata (Geitler) Hill
Hill, 1991b, p. 137, figures 2, 18–31.
Basionym: *Chroomonas caudata* Geitler, 1924, p. 246, figures a–h.

Genus *Cryptomonas* Ehrenberg (partim)
(For publication details see *Cryptomonas* under *2.2.1*).

As mentioned above (see *2.2.1*), numerous species of *Cryptomonas* fit the generic concept defined by several biochemical and ultrastructural studies (Santore, 1984, 1985; Hill & Rowan, 1989; Hill, 1991a). Within this concept the periplast is composed of hexagonal plates. Other species have a sheet-like periplast, as described for the genus *Campylomonas* (see under *2.2.1*).

Genus *Hemiselmis* Parke (Figs 31–33)
(For publication details see *Hemiselmis* under *1.3.1*).

Hemiselmis has a characteristic periplast composed of large hexagonal plates forming few longitudinal rows (Wetherbee et al., 1986). The size of the plates (Figs 31 and 32) varies between species and therefore it may be significant for identification purposes.

- Length of the sides of the hexagon ca. 1–1.25 μm *Hemiselmis amylosa* Clay & Kugrens (*1.3.1*)
- Length of the sides of the hexagon ca. 0.6 μm *Hemiselmis brunnescens* Butcher (*1.3.1*)
- Length of the sides of the hexagon ca. 0.4–0.65 μm, tail-like posterior end (compare *2.5*) *Hemiselmis* sp. inedit. Novarino

Genus *Plagioselmis* Butcher emend. Novarino Lucas & Morrall (ICBN) = *Plagioselmis* Butcher (ICZN) (Figs 34–38 and 49)
R. W. Butcher 1967, *Fishery Investigations, London, Series IV*, p. 18, pl. I figures 9, 10, pl. XIV figures 2, 3. *Emend.* G. Novarino, I.A.N. Lucas & S. Morrall 1994, *Cryptogamie-Algologie* vol. 15, p. 90, figures 1–26.
Lectotype species: *Plagioselmis prolonga* Butcher ex Novarino Lucas & Morrall 1994, p. 90, figures 1–18. (ICBN); nomenclatural equivalent: *Plagioselmis prolonga* Butcher 1967, p. 18, pl. I figure 9, pl. XIV figure 2 (ICZN).
Synonym (ICBN): *Plagioselmis prolonga* Butcher 1967, p. 18, pl. I figure 9, pl. XIV figure 2, typ. non desig. Lectotype (designated by Novarino et al., 1994): Butcher 1967, pl. I figure 9.

Plagioselmis includes both marine and freshwater cryptomonads with a characteristically acute cell posterior (tail, *2.5*, Figures 36 and 37), a periplast composed of hexagonal plates on the main portion of the cell body (Figs 34 and 35), and a non-plated periplast (*2.2.1*) on the tail (*2.5*, Figures 36 and 37). All confirmed members of this genus contain PE but it was inferred that at least one species could contain PC (Novarino et al., 1994). SEM characters which make it possible to tell apart species of *Plagioselmis* include the size of the periplast plates, the length of the flagella, the presence or absence of a true, non-artefactual ventral furrow (*2.3*, *4.3*), the length of the tail (*2.5*), and the presence or absence of a mid-ventral band (*2.6*, Fig. 36).

Plagioselmis nannoplanctica (Skuja) Novarino, Lucas et Morrall
Novarino et al. 1994, p. 97, figures 21–26
Basionym: *Rhodomonas minuta* var. *nannoplanctica* Skuja 1948, p. 347, pl. XXXVII, figures 11–15 (excl. figure 16):
Lectotype (designated by Novarino et al., 1994): Skuja 1948, pl. XXXVII, figure 13.
Note: This is a freshwater species with unequal flagella 3/4 and 1/2 the cell length, periplast plates on average 0.46$\pm$0.053 μm in size, a tail 1/5 to 1/4 the cell length, no ventral furrow and no mid-ventral band.

Plagioselmis prolonga Butcher ex Novarino Lucas & Morrall (ICBN) = *Plagioselmis prolonga* Butcher (ICZN) (Figs 34–38)
Novarino et al., 1994, p. 90, figures 1–18; Butcher

1967, p. 18, pl. I figure 9, pl. XIV figure 2, typ. non desig. (ICBN). Lectotype (designated by Novarino et al., 1994): Butcher 1967, pl. I figure 9.
Note: The lectotype species (Figs 34–38) is marine and has subequal flagella with the longer (dorsal) one about as long as the cell, periplast plates on average 0.39±0.087 μm in size, a tail (*2.5*, Figs 36 and 37) 1/7 to 1/3 the cell length, a furrow (*2.3*, *4.3*, Fig. 38) extending to about 1/2 the cell length, and a mid-ventral band (*2.6*, Fig. 36) extending from the tip of the tail to the base of the furrow.

Plagioselmis prolonga var. *nordica* Novarino Lucas & Morrall (ICBN)
Novarino et al., 1994, p. 92, figures 19 (holotype), 20.
Note: Differences with the type variety include the smaller size, the smaller periplast plates (0.31±0.041 μm), the absence of a furrow (*2.3*) and the absence of a mid-ventral band (*2.6*). This taxon is only recognized by the ICBN because taxa at the variety level have no status under the ICZN.

Plagioselmis sp. inedit. Novarino (Fig. 49)
Note: This is a marine species with equal–subequal flagella about 1/3 to 3/3 the cell length, very delicate periplast plates 0.15–0.25 μm in size, a tail (*2.5*, Fig. 49) 1/10 the cell length, a ventral furrow (*2.3*, *4.3*, Fig. 49) extending to about 2/3 the cell length, and a very short mid-ventral band (*2.6*, Fig. 49) which does not extend from the tail towards the furrow.

Genus *Proteomonas* Hill & Wetherbee non Podlipaev, Frolov & Kolesnikov
(For publication details see under *2.2.1*).

The 'haplomorph' stage of *Proteomonas sulcata* has a periplast composed of hexagonal plates, rather than the non-plated periplast found in the 'diplomorph' stage (see under *2.2.1*). Together with the ventral furrow and the mid-ventral band found on the slightly pointed posterior end, this may be a cause for mistaking the *Proteomonas sulcata* haplomorph for a *Plagioselmis*. However, in *Plagioselmis* the periplast on the tail is non-plated rather than plated (see *Plagioselmis* under *2.2.2.3*).

Genus *Rhinomonas* Hill & Wetherbee (ICBN, ICZN) (Figs 42–48)
D. R. A. Hill & R. Wetherbee 1988, *Phycologia* vol 27, p.356, figures 1–18.
Type species: *Rhinomonas pauca* Hill & Wetherbee, 1988, p. 359, figures 1 (holotype), 2–18.

Rhinomonas (Figs 42–48) was described based on a single marine species containing PE and possessing a periplast with hexagonal plates (Fig. 45). There is an overall resemblance to *Pyrenomonas/Rhodomonas* emend. (see under *2.2.2.4*) but the shape of the periplast plates is different. Furthermore, many (but not all) species of *Pyrenomonas/Rhodomonas* emend. possess a ventral furrow (*2.3*), but this is always absent in *Rhinomonas* (Fig. 47). From a taxonomic point of view there is some confusion, because several 'classic' cryptomonad species which were originally described using light microscopy only and which were subsequently recombined under *Rhinomonas* have also been included elsewhere, e.g. in *Pyrenomonas/Rhodomonas* emend. (see under *2.2.2.4*) by other authors. The main feature in common between *Rhinomonas* and *Pyrenomonas/Rhodomonas* emend is the position of the nucleomorph, which is most easily observed using trasmission electron microscopy (*4.1*, *4.1.2*). In all the other genera with a plated periplast the nucleomorph is in a different position (*4.1*, *4.1.3*). The nucleomorph is found in the same position also in *Storeatula*, which has a non-plated periplast (see under *2.2.1*, *4.1*, *4.1.2*).

Rhinomonas includes a small number of species in addition to the type and all known species are marine. Three additional species were introduced by the authors of the genus (Hill & Wetherbee, 1988) based exclusively on pre-existing light-microscopical observations (Butcher, 1952, 1967). All of the other members of the genus have been described based on electron microscopy and are as follows:

Rhinomonas reticulata (Lucas) Novarino (Figs 42–48)
Novarino, 1991a, p. 244, figures 1–4, 6–11, 14–17, 19, 20.
Basionym: *Cryptomonas reticulata* Lucas, 1968, p. 535, figures 1b (holotype), 6, 7 (paratypes).
Note: Prior to its recombination under *Rhinomonas*, this cryptomonad had also been considered as a member of *Pyrenomonas/Rhodomonas* emend., i.e. *Pyrenomonas reticulata* (Lucas) Santore (1986), *Rhodomonas reticulata* (Lucas) Hill & Wetherbee (1989), but no ultrastructural re-examination was provided on either one of those occasions.

Rhinomonas reticulata (Lucas) Novarino var. *atrorosea* (Butcher ex Hill & Wetherbee) Novarino

Novarino, 1991b, p. 606, figures 24, 25.
Basionym: *Rhodomonas atrorosea* Butcher ex Hill & Wetherbee (1989), p. 156.
Synonym: *Chroomonas atrorosea* Butcher, 1967, p. 33, pl. II figure 11, pl. XVII figure 1, typ. non desig.
Lectotype (designated by Hill & Wetherbee, 1989): Butcher 1967, pl. II figure 11.
Note: This differs from the type variety in the slight degree of lateral compression of the cells.

Rhinomonas reticulata (Lucas) Novarino var. *compressa* Novarino
Novarino, 1991b, p. 607, figures 26, 27 (holotype).
Note: This variety resembles var. *atrorosea* but the cells are strongly compressed laterally.

Rhinomonas reticulata (Lucas) Novarino var. *eleniana* Novarino
Novarino, 1991a, p. 244, figures 5 (holotype), 18.
Note: This differs from the type variety in the smaller size of the periplast plates.

2.2.2.4. Periplast plates rectangular-hexagonal (square with bevelled edges): **Pyrenomonas/Rhodomonas *emend*** .

Genus *Pyrenomonas* Santore (ICBN, ICZN) (Figs 39–41)
U. J. Santore 1984, *The New Phytologist* vol. 98, p. 637, figure 3.
Type species: *Pyrenomonas salina* (Wislouch) Santore, 1984, p. 637.
Basionym: *Cryptomonas salina* Wislouch, 1924, p. 115, pl. III, figures 5a–d; synonym: *Chroomonas salina* (Wislouch) Butcher, 1967, p. 33, pl. III figures 1–9, pl. IV figures 1–4, pl. X figures 1, 7, pl. XII figure 4, pl. XVII figures 2–6, pl. XVIII figures 1–3.

Pyrenomonas (Figs 39–41) is a PE-containing genus with a characteristic plated periplast. The shape of the plates (Fig. 41), which has been variously described as approximately rectangular, rectangular-hexagonal, elongated-hexagonal, or square with bevelled edges, is clearly different from the regularly rectangular shape found in *Chroomonas* sensu auctorum (see under *2.2.2.2*) and the distinctly hexagonal shape found in numerous other genera (*2.2.2.3*). The genus resembles most closely *Rhinomonas* but some of its species have ventral furrows, which have never been documented in *Rhinomonas*. Most species are marine but a freshwater one has been described recently.

The nomenclature and taxonomy of this genus are complex and have attracted some controversy. A few years after its original description, which was based on electron microscopy, an emended description of the genus *Rhodomonas* Karsten (1898) was proposed based mainly on electron microscopy (Hill & Wetherbee, 1989). Because the examined strains showed the characteristic ultrastructural features of *Pyrenomonas*, it was argued (Hill & Wetherbee, 1989) that *Pyrenomonas* was a later (junior) synonym of *Rhodomonas* and therefore it was to be abandoned. Elsewhere it was strongly contented that the name *Rhodomonas* emend. Hill & Wetherbee (1989) was the later synonym and therefore it was to be abandoned in favour of *Pyrenomonas*, which name ought to have been adopted by Hill & Wetherbee (1989) rather than *Rhodomonas* (Novarino, 1991a; Novarino & Lucas, 1993b). In essence, the name *Rhodomonas* emend. Hill & Wetherbee (1989) is a later unnecessary name for *Pyrenomonas* Santore (1984), i.e. a *nomen superfluum* (Novarino & Lucas, 1993b). There is no evidence whatsoever that the genus *Rhodomonas* in the original sense of Karsten (1898) has the ultrastructural identity of the cryptomonads examined by Hill & Wetherbee (1989), or of any other cryptomonads which have been examined ultrastructurally, simply because the ultrastructure of the specimens examined by Karsten (1898) is totally unknown. Because there appears to be no original type material (i.e. that examined by Karsten himself) for *Rhodomonas*, nobody will ever know exactly which cryptomonad Karsten was looking at when he described the genus, i.e. what was its ultrastructural identity. Virtually *any* 'red' cryptomonad fits Karsten's original diagnosis of the genus to a greater or lesser degree (Novarino, 1991a; Novarino & Lucas, 1993b), and therefore the use of the name *Rhodomonas* in the original sense of Karsten or any other sense is to be strongly discouraged. This view has also been adopted more recently by other workers (Kugrens et al., 1999).

In *Pyrenomonas* the nomenclatural difficulties are made worse by the existence of another emended description of *Rhodomonas* (Erata & Chihara, 1989) in which all of the strains examined using electron microscopy showed several features characteristic of *Pyrenomonas*. It appears that the authors of each one of the emended descriptions of *Rhodomonas* were unaware of the existence of the other and therefore they introduced several new combinations with homonymous names (Erata & Chihara, 1989; Hill & Wetherbee, 1989). However, no SEM information was provided on the strains examined by Erata & Chihara (1989), and therefore it is impossible to establish with certainty whether those strains effectively possessed the kind of periplast which is typical of *Pyrenomonas/Rhodomonas* emend. Hill & Wetherbee.

Numerous cryptomonads have been assigned to this genus in addition to the type species, including several 'classic' cryptomonad species which have been recombined here at a later stage. However, few of these have been examined ultrastructurally and therefore it is unclear whether or not all of them effectively belong here. All of the taxa in the list which follows have been examined using electron microscopy and are unequivocal members of this genus.

Pyrenomonas duplex (Hill & Wetherbee) comb. nov. (ICBN, ICZN)
Basionym: *Rhodomonas duplex* Hill & Wetherbee 1989, p. 151, figures 32–34, 35 (holotype), 36, 37.
Note: This is an unusual species with two chloroplasts and two pyrenoids. There is also a ventral furrow (*2.3*, *4.3*) and a mid-ventral band (*2.6*). See also *Pyrenomonas maculata* below.

Pyrenomonas heteromorpha (Butcher) Santore
Santore, 1986, p. 75, figures 1–7.
Basionym: *Chroomonas heteromorpha* Butcher 1967, p. 32, pl. V figure 1, pl. XVI figure 5, typ. non desig..
Lectotype (designated by Hill & Wetherbee, 1989): Butcher 1967, pl. V figure 1.
Synonym: *Rhodomonas heteromorpha* Butcher ex Hill & Wetherbee, 1989, p. 157.
Note: This has no furrow (*2.3*) and no mid-ventral band (*2.6*).

Pyrenomonas maculata (Butcher) Santore
Santore, 1986, p. 80.
Basionym: *Cryptomonas maculata* Butcher 1967, p. 45, pl. V figure 6, pl. XIX figure 6, typ non desig.
Lectotype (designated by Hill & Wetherbee, 1989): Butcher 1967, pl. V figure 6.
Synonym: *Rhodomonas maculata* Butcher ex Hill & Wetherbee, 1989, p. 149, figures 22–31.
Note: This is similar to *Pyrenomonas duplex* (see above) but it has a single chloroplast with one pyrenoid.

Pyrenomonas ovalis Kugrens, Clay & Lee
Kugrens et al., 1999, p. 1081, figures 13–22.
Note: This is the only freshwater species described to date.

Pyrenomonas salina var. *curvata* Novarino
Novarino, 1991b, p. 607, figures 26, 27 (holotype).
Note: This species lacks a furrow and a mid-ventral band. It is easily recognized because the cells are curved along the longitudinal axis.

***2.2.2.5. Long periplast "plates" oriented longitudinally:* Guillardia**

Genus *Guillardia* Hill & Wetherbee (ICBN, ICZN)
D. R. A. Hill & R. Wetherbee 1990, *Can. J. Bot.* vol. 68, p. 1873, figures 1–7.
Type species: *Guillardia theta* Hill & Wetherbee, 1990, p. 1873, figure 1 (holotype), 2–7.

Guillardia is a marine PE-containing genus which only includes the type species. The IPC is sheet-like while the EPC is composed of "... irregularly shaped crystalline plates" (Hill & Wetherbee, 1990, diagnosis on p. 1873). In the SEM the EPC appears as a series of "longitudinally oriented plate-like strips" (Hill & Wetherbee, 1990), although a later re-examination (Deane et al., 1998, p. 152, figures 2, 3) mentioned longitudinally oriented 'plates'. This kind of periplast architecture is very different from that of the other genera with a sheet-like IPC. Freeze-fracture investigations (*5*) may be necessary in order to conclusively characterize the periplast of putative *Guillardia* specimens.

***2.2.2.6. Genera whose periplast structure is totally unknown:* Cryptella, Cryptochrysis, Cyanomonas, Isoselmis, Peliaina, Planonephros, Platychilomonas, Protochrysis, Protocryptomonas, Pseudocryptomonas, Strobilomonas**
Cryptella (see under *1.7.1*)
Cryptochrysis (see under *1.5*)
Cyanomonas (see under *1.7.1*)

Genus *Isoselmis* Butcher (ICBN, ICZN)
R. W. Butcher 1967, *Fishery Investigations, London, Series IV*, p. 19, pl. I figure 11, pl. XII figure 1, pl. XIV figure 4.
Type species: *Isoselmis obconica* Butcher, 1967, p. 12, pl. I figure 11, pl. XII figure 1, pl. XIV figure 4, typ. non desig.

The description of *Isoselmis* was based on a single red marine species, which Butcher (1967) distinguished from *Plagioselmis* "with some hesitation" based on the behaviour of the flagella and the num-

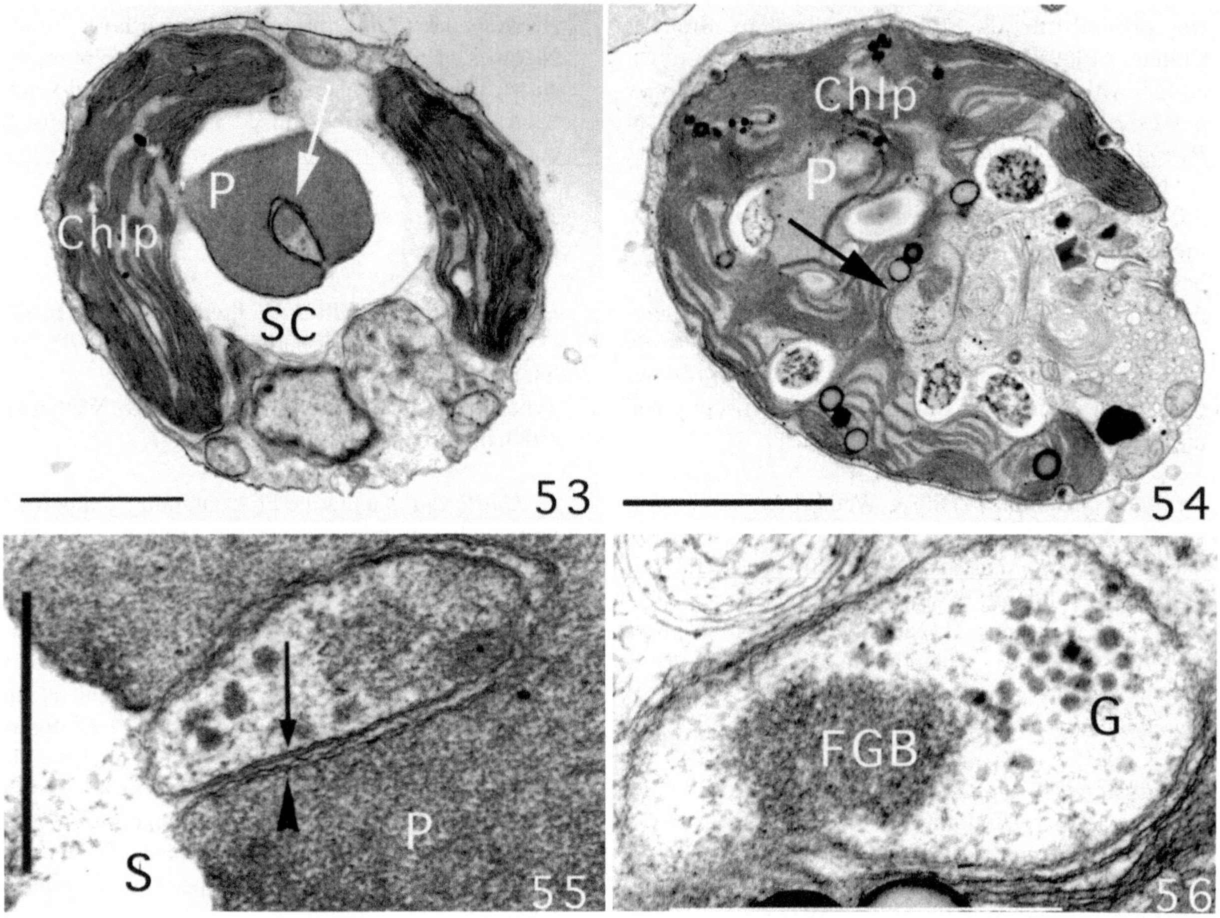

Figure 53–56. TEM micrographs of ultrathin sections of cryptomonads, showing some features of the nucleomorph. *Figure 53.* Transverse section of *Rhinomonas reticulata* with its nucleomorph (arrow) characteristically placed in a groove within the pyrenoid (P). Chlp = chloroplast, SC = starch cap. Scale bar = 2 μm. *Figure 54.* Oblique longitudinal section of *Chroomonas ligulata* with its nucleomorph (arrow) characteristically placed outside the pyrenoid (P). Scale bar = 2.5 μm. *Figure 55.* Detail of the intrapyrenoidal nucleomorph of *Rhinomonas reticulata*, showing the double membrane of the nucleomorph (arrow) in relation to the double membrane of the chloroplast envelope (arrowhead) in the pyrenoid (P) region of the chloroplast; S = starch. Scale bar = 0.5 μm. *Figure 56.* Detail of the extrapyrenoidal nucleomorph of *Chroomonas ligulata*, with its characteristic granular inclusions (G) and fibrillo-granular body (FGB). Scale bar = 1 μm.

ber of large ejectosomes. These are both difficult to observe with certainty, particularly the behaviour of the flagella owing to the cells' rapid swimming. The status of this genus is very unclear since no ultrastructural information is available. Specimens which could be tentatively identified as possible members of the genus *Isoselmis* under the light microscope were briefly examined elsewhere (Novarino et al., 1994). In the SEM the periplast appeared sheet-like, similar for instance to the *Proteomonas* 'diplomorph', *Storeatula* and *Teleaulax* (*2.2.1*). However, it is impossible to know with certainty if those specimens effectively belong to *Isoselmis* because the original culture of *I. obconica* is no longer available, and no other strains bearing the generic name *Isoselmis* are available for comparative purposes. The status of *nomen dubium* has been suggested for *Isoselmis* (Novarino et al., 1994).

Peliaina (see under *1.7.1*)
Planonephros (see under *1.3.1*)
Platychilomonas (see under *1.6*)
Protochrysis (see under *1.3.1*)
Protocryptomonas (see under *1.6*)
Pseudocryptomonas (see under *1.7.1*)
Strobilomonas (see under *1.7.1*)

2.3. Morphology of the vestibular region of the cell from which the flagella arise, especially the presence or absence of a true, non artefactual ventral furrow

This structure has been the object of much controversy ever since the early light-microscopical descriptions. The term 'furrow' refers to a shallow groove on the ventral face of cell, of variable length and width (Figs 49–52). Some of the early light microscopists described cryptomonads with open furrows, but the exact architecture of these structures (especially their spatial relationships with other vestibular structures such as the closed, tubular gullet) was virtually impossible to establish with certainty using light microscopy only. Since the advent of electron microscopy, scanning electron microscopy has been the tool of choice for studying these structures. It has been hypothesized (Santore, 1984) that furrows are always preparation artefacts (*2.9*), and therefore they have no taxonomic value at all, or that they are never artefactual (Kugrens et al., 1986; Hill & Wetherbee, 1989) and are taxonomically significant at the generic level (Clay et al., 1999).

A more complex view is that the term 'furrow' may encompass artefactual and non-artefactual structures alike (Novarino, 1991b), and it is possible to judge whether or not an observed furrow is an artefact based on a few simple considerations:

- presence or absence of obvious signs of shrinkage or collapse in other regions of the cell;
- frequency of the occurrence of observed furrows in a sample of cells (say 30–40) belonging to the same species;
- and, in cryptomonads with a plated periplast, presence or absence of periplast plates on the internal face of the furrow, because true furrows are never lined internally with discrete periplast plates of the same kind as those found on the rest of the cell.

It has been argued that true furrows are taxonomically significant at the specific rather than generic level because they are not necessarily present in all of the species of a particular genus (Novarino, 1991b; Novarino et al., 1994). Further comments on artefacts associated with specimen preparation for SEM are given below (*2.9*). Furrows can also be observed on sectioned material for TEM (*4.3*). The function of the furrow/gullet system of cryptomonads is still unknown. Cryptomonads with a closed tubular gullet may occasionally harbour bacteria in the gullet itself, e.g. spirochaete bacteria in *Rhinomonas* (Novarino, unpublished observations; and this paper, Fig. 44).

2.3.1. Genera where non-artefactual furrows have been observed using SEM: **Campylomonas, Chilomonas, Cryptomonas, Falcomonas, Hillea, Plagioselmis, Proteomonas, Pyrenomonas/Rhodomonas** ***emend.,*** **Teleaulax**

Campylomonas (see under *2.2.1*)
Chilomonas (see under *2.2.1*)
Cryptomonas (see under *2.2.1*, *2.2.2.3*)
Falcomonas (see under *2.2.2.3*)
Hillea (see under *2.2.1*)
Plagioselmis (see under *2.2.2.3*)
Proteomonas (see under *2.2.1*, *2.2.2.3*)
Pyrenomonas/Rhodomonas emend. (see under *2.2.2.4*)
Teleaulax (see under *2.2.1*)

2.4. Arrangement, absolute and relative length of the flagella

Flagella can be observed also with the LM (see *1.2*), but their length and point of insertion can be detemined much more accurately with the SEM. Features such as the median/subapical versus apical insertion point of the flagella and their absolute and relative length have been considered to be taxonomically significant since the times of the early light-microscopical descriptions (at the genus and species level, respectively). This view has been upheld also in more recent ultrastructural and taxonomic investigations ((Hill & Wetherbee, 1988; Novarino & Lucas, 1993b; Clay & Kugrens, 1999a).

2.5. Presence or absence of a posterior tail

'Tails' – i.e. acute posterior ends which are often curved ventrally or dorsally – can be observed also with the light microscope, but the SEM can show whether or not the periplast is of the same kind present on the rest of the cell surface (Figs 36 and 37). This feature is taxonomically significant at the generic level because it is diagnostic of *Plagioselmis* (see under *2.2.2.3*).

2.6. Presence or absence of a mid-ventral band in the region between the posterior end of the furrow and the posterior end of the cell

Mid-ventral bands (Fig. 36) are present in a few genera. They are easily observed with the SEM, where they appear as a cord-like structure on the ventral cell surface in the posterior region of the cell. Within individual genera they do not appear to be present in all

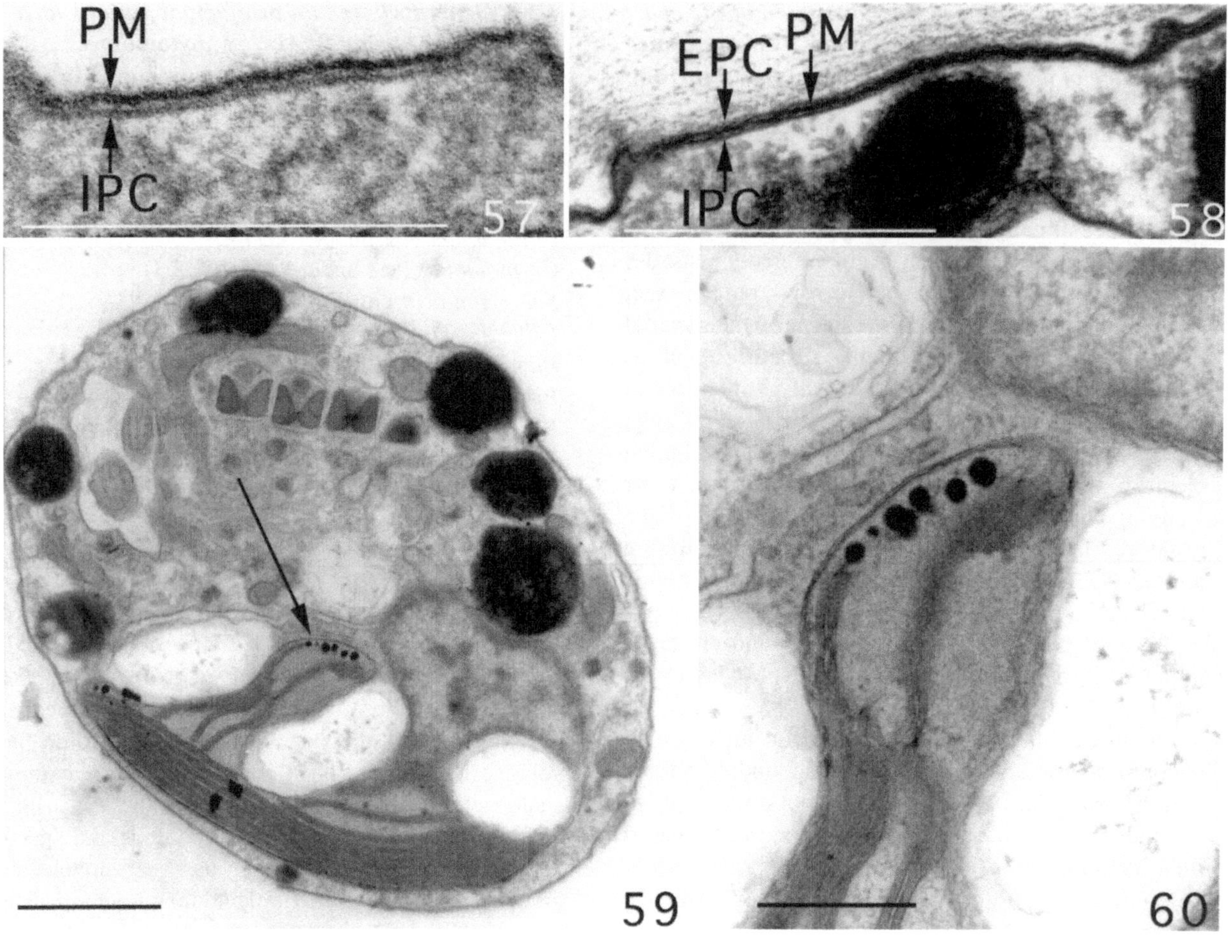

Figure 57–60. TEM micrographs of ultrathin sections of cryptomonads, showing some features of the periplast and eyespot. *Figure 57.* Periplast of *Rhinomonas reticulata*, showing the plated internal periplast component (IPC) in relation to the plasma membrane (PM); note also the fibrillar component on the external face of the PM. Scale bar = 500 nm. *Figure 58.* Plated periplast of *Chroomonas ligulata*, showing the characteristic 'sandwich' composed of an external (EPC) and internal (IPC) layer with the plasma membrane (PM) in between; note also the thick mucilage layer external to the EPC. Scale bar = 500 nm. Figures 59 and 60. The eyespot (arrow) of *Chroomonas ligulata*. Figure 60 shows the globular inclusions in the stalked extension of the pyrenoid making up the eyespot. Scale bars = 1 μm (Fig. 59) or 350 nm (Fig. 60).

of the species (Hill & Wetherbee, 1989; Hill, 1991b; Novarino et al., 1994), suggesting that their taxonomic value is at the specific level.

2.6.1. Genera where mid-ventral bands have been observed using SEM: Falcomonas, Plagioselmis, Pyrenomonas/Rhodomonas *emend*

Falcomonas (see under *2.2.2.3*)
Plagioselmis (see under *2.2.2.3*)
Pyrenomonas/Rhodomonas emend. (see under *2.2.2.4*)

2.7. Presence or absence of a periplast raphe

The 'raphe' – a line at the posterior end of the cell where the periplast plates seem to converge – is a unique feature of some species of the genus *Chroomonas* sensu auctorum (see under *2.2.2.2*). It may be visible using SEM although freeze-fracture methods for TEM are preferable (*5*).

2.8. Arrangement of flagellar appendages

The nature and arrangement of flagellar appendages in cryptomonads are quite diverse (Morrall, 1980; Kugrens et al., 1987). The taxonomic significance of these appendages is not entirely clear but some genera always show a characteristic pattern. Whole mounts for transmission electron microscopy are the choice observation method (see *3.1*) but the SEM may still show the appendages in adequately fixed cells (Figs

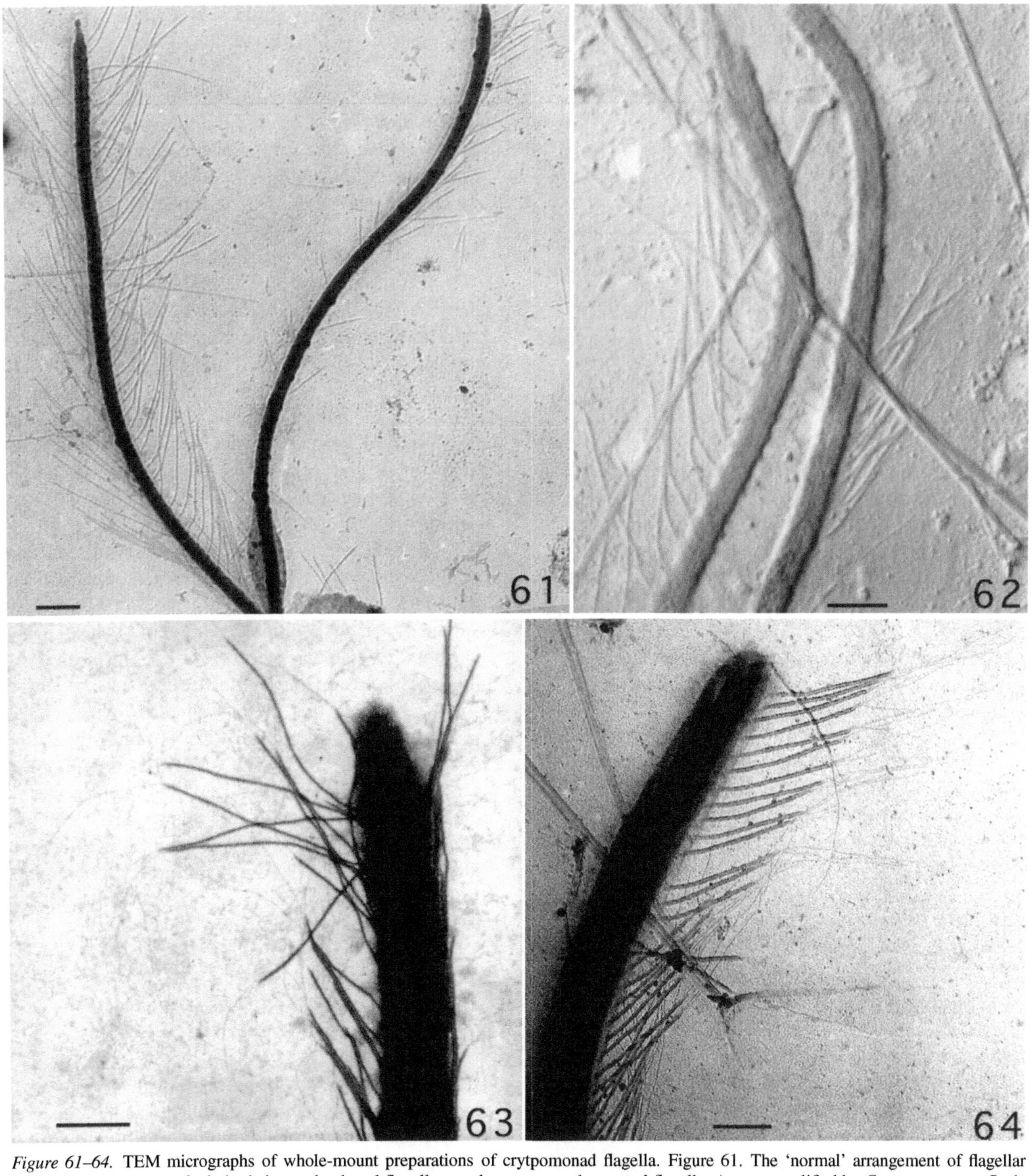

Figure 61–64. TEM micrographs of whole-mount preparations of crytpomonad flagella. Figure 61. The 'normal' arrangement of flagellar appendages (two rows of tubular hairs on the dorsal flagellum and one row on the ventral flagellum), as exemplified by *Cryptomonas* sp. Scale bar = 600 nm. *Figure 62. Chilomonas paramecium*, with a single row of tubular hairs on each flagellum. This arrangement is found also in *Campylomonas*. Scale bar = 400 nm. *Figures 63 and 64.* Details of the dorsal (Fig. 63) and ventral (Fig. 64) flagellum of *Cryptomonas* sp., showing the tubular hairs with terminal filaments. Scale bars = 200 nm.

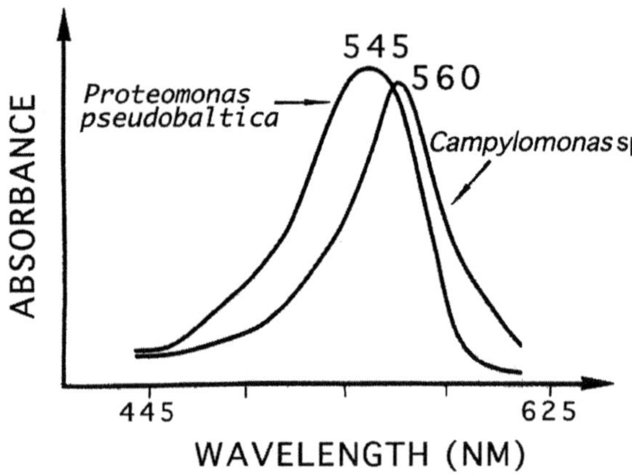

Figure 65. Absorbance spectra of phycoerythrin extracts from *Proteomonas pseudobaltica* and *Campylomonas* sp. The absorbance peaks (at 545 and 560 nm, respectively), identify the *P. pseudobaltica* pigment as Cr-PE I and the *Campylomonas* pigment as Cr-PE III.

27, 33 and 34) revealing whether or not they are arranged according to the 'normal' pattern (see Morrall, 1980; Kugrens et al., 1987). The 'normal' pattern is two rows of tubular hairs on the dorsal (usually longer) flagellum, and one row on the ventral (usually shorter) flagellum (*3.1*).

2.9. Some common artefacts associated with the preparation of cryptomonad cells for scanning electron microscopy

The possible occurrence of preparation artefacts should always be borne in mind when observing cryptomonad cells with the SEM. In addition to cell shrinkage, which may produce artefactual furrows (*2.3*, *4.3*), some common artefacts are the following:

2.9.1. Artefacts arising from peripheral ejectosome discharge or bulging

The discharge of ejectosome vesicles present underneath the cell membrane produces distinctly visible pores. As surprising as this may seem, this artefact may turn out to be a useful one because it provides information on periplast morphology when the normal appearance is not seen. In cryptomonads with a plated periplast the peripheral ejectosomes are placed at the corners of the plates, so their discharge immediately reveals the plate shape. For instance, ejectosome pores arranged in an hexagonal pattern indicate the presence of a periplast composed of hexagonal plates. When the periplast is sheet-like (= non-plated) the peripheral ejectosomes are frequently arranged in a spiral pattern. Bulging of the ejectosomes towards the external surface immediately reveals this arrangement, showing in turn that the periplast is not plated.

2.9.2. Artefacts arising from inadequate fixation of the cell membrane

These artefacts are very variable and it is difficult to give an unequivocal interpretation of how they affect the appearance of the periplast because cryptomonad cells may have periplast components on one, the other or both sides of the cell membrane (Figs 58 and 59, *4.2*). In the case of periplast plates present only on the inside of the membrane, the complete loss of the membrane may give rise to plates with a smooth, more angular or deformed appearance. At times the plates may be bordered by thickened ridges but adequately fixed cells usually lack these ridges and have thin grooves instead, so the ridges could be a result of inadequate fixation of the membrane over the intra-membrane particles present between adjoining periplast plates (see Kugrens & Lee, 1987). Shrinkage of the membrane over the underlying (internal) periplast plates may cause either a more distinct or indistinct appearance of the plates, depending on the degree of shrinkage itself. An amount of shrinkage up to about 25% linear is to be considered normal during specimen preparation for SEM (Novarino, 1991b), so it is possible that membrane shrinkage itself is a main factor contributing to the 'normal' appearance of the periplast in SEM. In turn, if the membrane of a particular cell has not shrunk sufficiently over the underlying IPC, a plated IPC may appear to be non-plated. Such cases may cause some difficulties with identification, so it is always preferable to examine a sufficient number of cells (e.g. 30) in order to judge on the plated or non-plated nature of the periplast.

3. Cryptomonads and transmission electron microscopy whole-mounts (Figs 61–64)

Whole-mount preparations for transmission electron microscopy are a long-established and relatively simple technique for studying the external micromorphology of flagellates. They are useful for studying the nature and arrangement of flagellar appendages and scales.

3.1. Flagellar appendages

The typical flagellar appendage of cryptomonads is a tubular hair (mastigoneme) with terminal filament(s) (Figs 63 and 64). The typical arrangement, which may also be visible using SEM (*2.8*, Figures 27 and 34), is two rows on the dorsal (usually longer) flagellum, and one row on the ventral (usually shorter) flagellum (Figs 61, 63, 64); this is referred to here as the '2+1' pattern. However, there is considerable variation on this basic pattern (Morrall, 1980; Kugrens et al., 1987). Three 'atypical' genera are listed below for illustration purposes.

3.1.1. Genera with flagellar appendages not arranged in a typical '2+1' pattern: **Campylomonas, Chilomonas, Goniomonas**

Campylomonas and *Chilomonas* (Fig. 62) both have a single row of mastigonemes on each flagellum (Kugrens et al., 1987; Gromov et al., 1998). *Goniomonas* (Fig. 27, SEM) has been reported as having a similar arrangement of mastigonemes (Morrall, 1980), or peculiar, non-tubular 'spike' appendages on one flagellum only (Kugrens et al., 1987).

3.2. Scales

Cryptomonads may possess characteristic scales over the cell body and/or the flagella. (e..g. Santore, 1983; Lee & Kugrens, 1986). The typical size of the body scales is ca. 150–200 nm, although scales about 80 nm in size have been found in sectioned material of *Plagioselmis* (Novarino et al., 1994). Freeze-fracture investigations (*5*) often reveal more detail than whole-mounts. The taxonomic distribution and significance of cryptomonad scales is not entirely clear at present and therefore it requires further investigation.

4. Cryptomonads and ultrathin sections for transmission electron microscopy (Figs 53–60)

Studies using ultrathin sections for TEM have provided much ultrastructural information. Some ultrastructural features of taxonomic usefulness are listed here.

4.1. Occurrence and position of the nucleomorph

The cryptomonad nucleomorph (Figs 55 and 56) is a distinct cell organelle resembling an eukaryotic nucleus (Greenwoood, 1974; Greenwood et al., 1977) and representing the vestigial nucleus of a red algal-like photosynthetic endosymbiont from which the present-day cryptomonad chloroplast has evolved (Hansmann et al., 1985; Ludwig & Gibbs, 1985; Hansmann et al., 1986). It is located in the periplastidial compartment and within a particular genus it occurs in one of two basic positions, i.e. embedded in a groove within the pyrenoid ('intrapyrenoidal nucleomorph', Figs 53 and 55) or outside the pyrenoid ('extrapyrenoidal', Figures 54 and 56). The intra- versus extrapyrenoidal position was first considered to be systematically significant at the order level elsewhere (Novarino & Lucas, 1993b) and later expanded upon and incorporated into a newer classification system (Clay et al., 1999). It is sometimes possible to locate the nucleomorph by light (fluorescence) microscopy if the cells have been stained with the DNA-specific fluorochrome DAPI (Ludwig & Gibbs, 1985). There is also some evidence suggesting that in the 1950s the nucleomorph may have been observed unknowingly using conventional histochemical staining (Novarino, 1993). However, TEM is the choice method.

4.1.1. Genera without a nucleomorph: **Goniomonas**

Goniomonas is the only known cryptomonad genus where a nucleomorph is lacking (Morrall, 1980; Kugrens & Lee, 1991). As such it is believed to be closely related to the ancestral cryptomonad cell which acquired a photosynthetic endosymbiont and led to the evolution of the group. This view is supported by molecular evidence (McFadden et al., 1994).

4.1.2. Genera with an intrapyrenoidal nucleomorph: **Pyrenomonas/Rhodomonas *emend.*, Rhinomonas, Storeatula**

The three known genera with an intrapyrenoidal nucleomorph (e.g. Fig. 53, *Rhinomonas*) are all grouped in a single order (Novarino & Lucas, 1993b; Clay et al., 1999), whose separation is supported by molecular studies (Cavalier-Smith et al., 1996; Hoef-Emden et al., 2002; Deane et al., 2002). From the point of view of differential taxonomic identification, information on periplast ultrastructure is necessary (*2.2.1*, *2.2.2.3*, *2.2.2.4*):

- Periplast plates hexagonal .... *Rhinomonas*
- Periplast plates elongated-hexagonal (square with bevelled edges) ...... *Pyrenomonas/Rhodomonas* emend.

- Periplast non-plated (sheet-like) ...... *Storeatula*

4.1.3. Genera with an extrapyrenoidal nucleomorph: Campylomonas, Chilomonas, Chroomonas *sensu auctorum,* Chroomonas *sensu stricto,* Cryptomonas *sensu lato,* Cryptomonas *sensu stricto,* Falcomonas, Geminigera, Guillardia, Hanusia, Hemiselmis, Komma, Plagioselmis, Proteomonas, Teleaulax

All the other nucleomorph-bearing genera have the nucleomorph in an extrapyrenoidal position (e.g. Fig. 54, *Chroomonas* sensu auctorum). With the exception of *Chilomonas* they are all photosynthetic. From the point of view of taxonomic identification they can be told apart based on a combination of pigment (*1.5*, *6*) and periplast (*2.2.1*, *2.2.2*) features.

4.1.4. Genera in which the presence or absence of a nucleomorph is not known: Cryptella, Cryptochloris, Cyanomonas, Hillea, Isoselmis, Peliaina, Platychilomonas, Protochrysis, Protocryptomonas, Pseudocryptomonas, Strobilomonas

4.2. Number and general kind of periplast layers

TEM can confim or refute SEM observations on plated versus non-plated periplasts (*2.2*, *2.2.1*, *2.2.2*). In cross-section the latter appear as a discontinuous line whereas the former appear a continuous line. It might also be possible to count the number of periplast layers; this could be important for identification at the specific level because there are reports of cryptomonad species missing one of the plate-layers expected for their particular genus (see Novarino, 1991a; Novarino & Lucas, 1993a) (Figs 57 and 58).

4.3. Furrow

TEM can be used as a complementary technique to SEM for observing ventral furrows (*2.3*). In cross-section non-artefactual furrows are usually wide and distinctly visible, and can be seen to be lined internally with large ejectosomes (Fig. 38).

4.4. Number of chloroplasts

Uncertainties about the number of chloroplasts (as revealed by LM: *1.7.2*) can benefit from the observation of TEM sections (Figs 38, 53 and 54). For instance, the original description of *Plagioselmis prolonga* mentioned two chloroplasts (Butcher, 1967, p. 18), but an ultrastructural re-examination has shown that the chloroplast is single (Novarino et al., 1994) (Fig. 38). See also *Plagioselmis* under *2.2.2.3*.

4.5. Occurrence of a true eyespot

Several species of *Chroomonas* (*2.2.2.2*) may be seen to possess putative eyespots in the LM (*1.8*). TEM can confirm or refute whether these are true eyespots rather than other structures. True eyespots (Figs 59 and 60) appear as a stalked extension of the pyrenoid containing several electron dense globules at the periphery, i.e. in the direction facing the cell surface (Lucas, 1982).

5. Cryptomonads and freeze-fracture investigations

Numerous freeze-fracture studies have been carried out on the cryptomonads (Brett & Wetherbee, 1986; Wetherbee et al., 1986; Kugrens & Lee, 1987; Wetherbee et al., 1987). These investigations have provided more detailed insight into the intimate architecture of the cell surface, showing that cryptomonad periplasts (*2.2*, *4.2*) are highly diversified and at the same time providing more general information on the eukaryotic cell surface (Brett et al., 1994). As a consequence there has been an increasing use of freeze-fracture investigations in purely taxonomic studies (e.g. Hill & Wetherbee, 1986, 1988, 1989, 1990; Hill, 1991a, b; Clay & Kugrens, 1999a), although these observations are rather difficult to carry out during routine taxonomic surveys. One of the most interesting cases studied using freeze-fracture is that of the periplast of *Guillardia* (*2.2.2.5*), which has been shown to be composed of a sheet-like internal component and a plated external component (Hill & Wetherbee, 1990).

Freeze-fracture investigations also provide better information on cryptomonad scales but their taxonomic distribution is not entirely clear at present. (*3.2*). Finally, the periplast raphe (*2.7*) found in some species of *Chroomonas* sensu auctorum is better revealed by freeze-fracture studies (Hill, 1991b) rather than SEM.

6. Cryptomonads and spectrophotometry (Fig. 65)

Hill & Rowan (1989) reviewed the spectral characteristics and taxonomic distribution of the different spectral types of phycoerythrin and phycocyanin pigments found in photosynthetic cryptomonads (see also *1.5*). For the purposes of taxonomic identification it may be sufficient to establish whether a phycocyanin or a phycoerythrin is present, and this may be done approximately by looking at the colour of living cells under the light microscope (*1.5*). However, more detailed information on the pigment's spectral characteristics may be necessary because a few cryptomonads have unique phycobilins (see below). The technical protocol involved is well-established (Hill & Rowan, 1989) and not excessively demanding from a technical point of view since the main instrument required is a conventional laboratory spetrophotometer. However, cultured material is necessary to provide a sufficient quantity of extracted pigment, although Klaveness (1985) successfully obtained spectral scans from whole cells.

Cryptomonad phycobilins are designated by the prefix 'Cr' followed by the type of phycoerythrin (PE) or phycocyanin (PC) in Roman numerals. Each type has a characteristic absorbance curve, identified by one or two absorbance peaks at given wavelengths. For instance, 'Cr-PE I (545)' refers to cryptomonad type I phycoerythrin with a single absorbance peak located at a wavelength of 545 nm. For illustration purposes, Figure 65 shows the absorbance curves of Cr-PE I and Cr-PE III from the author's isolates of *Proteomonas pseudobaltica* and *Campylomonas* sp., respectively.

The following is a key to Cr-phycobilin spectral types, based on the characteristics of the absorbance curves, together with their taxonomic distribution (based mainly on data in Hill & Rowan, 1989).

- Peak in the region 540–550 nm .... Cr-PE I
  This is the most widespread type in red marine cryptomonads. Genera *Geminigera, Plagioselmis, Proteomonas, Pyrenomonas/Rhodomonas* emend., *Rhinomonas, Storeatula, Teleaulax.*

- Peak in the region 555 nm .... Cr-PE II
  This is a relatively rare type, found mainly in *Hemiselmis* subgen. *Hemiselmis.*

- Peak in the region 560–566 nm .... Cr-PE III
  This is the most widespread type in red freshwater cryptomonads. Genera *Campylomonas, Cryptomonas* sensu lato, *Cryptomonas* sensu stricto.

- Peaks at 569 (major) and 630 nm ..... Cr-PC 569
  This pigment is diagnostic of the genus *Falcomonas.*

- Major peak in the region of 615 nm and minor peak in the region 577–585 nm .... Cr-PC 615
  This is found in *Hemiselmis* subgen. *Plagiomonas.*

- Major peak in the region of 625–630 nm and minor peak in the region of 580–584 nm .... Cr-PC 630
  This is a relatively rare pigment found in a few species of *Chroomonas.*

- Major peak in the region of 640-650 nm and minor peak in the region of 580–585 nm .... Cr-PC 630
  This is found in the majority of species of *Chroomonas* and *Komma caudata.*

Concluding remarks

Non-specialists are often overawed by the complexities of cryptomonad identification. Some of the difficulties may be real – i.e., they may be due to the high degree of diversity and phenotypic variability present within the cryptomonads – but others may depend purely on the historical and cultural development of our knowledge of the group.

State of flux of cryptomonad classification systems

This is an obvious obstacle to identification but it is probably inevitable, at least until the full extent of cryptomonad diversity will have been described. In future, classification systems may have to be modified even further in order to accommodate more recent molecular information. However, this should not act as a deterrent to identifying cryptomonads in as much detail as possible using what equipment is available at any one time. Biological classification relies entirely on the 'descriptive basement level' of biology, which is 'the foundation on which all else is built' (Mayr, 1974). Inexorably, classification systems will have to change until the foundation will have been described entirely: in the botanical words of Arnold (1948), "No system can be accepted as final so long as a single fact concerning any kind of plant remains unknown".

Typification

The type method is the fundamental principle of the Codes of Nomenclature which apply to the cryptomonads. It also suggests the ideal identification practice because comparisons with nomenclatural types offer a theoretically unequivocal method for deciding which name should be applied to a particular specimen. However, within the cryptomonads the type method has been largely disregarded. This is shown by the lack of the required designation of holotypes in taxonomic protologues, particularly by Butcher (1967); the poor quality of many of the early taxonomic descriptions and illustrations; and the incorrect assumption that preserved, original type materials of cryptomonads do not exist and therefore it is acceptable to amend generic descriptions in the absence of any information on the type – the case of *Rhodomonas* emend. is striking (see under *2.2.2.4*). On the contrary, original type materials may well exist and it may be possible to re-examine them ultrastructurally. This is shown by the case of the preserved type specimens observed by Hansgirg (1885) during the description of the genus *Chroomonas*, which have been examined by the author using SEM (Novarino & Oliva, 1998; this paper, Figs 28–30). Such a case may have important consequences for establishing the true ultrastructural identity of the genus, since the currently accepted concept of *Chroomonas* does not fit with Hansgirg's type specimens (*2.2.2.2*, *2.2.2.3*).

Relationships between 'classical' taxa and newer ultrastructural descriptions

Several 'classical' taxa have been re-described using electron microscopy but the relationship between the light- and electron-microscopical taxon concepts is not always clear. The type method offers the only reliable rationale with which such cases can be addressed. The availability of ultrastructural re-descriptions of certain taxa might also be a source of confusion regarding later reports of the taxa in question. If a given species has been re-described or recombined in a different genus based on electron microscopy, in later reports it is undesirable to report that species under the emended or recombined name unless electron microscopy has been used for identifying the specimens. For instance, *Chroomonas caudata* was first described using light microscopy, and later transferred to the new genus *Komma* based on electron microscopy (see under *2.2.2.3*). In order to be fully substantiated, all later reports of *Komma caudata* should be based on electron microscopy because the distinctive features of the genus *Komma* are not visible under the light microscope. In other words, it is important for all later reports of cryptomonad species to be based on the same observation methods used to describe the species in the original taxonomic protologue. This will become more and more imperative as molecular sequence data are incorporated in taxonomic protologues, as in the case of *Hanusia* (see under *2.2.1*).

Monospecific genera

Several genera have been described based on the type species only. This might lead to an incorrect appreciation of the variability of characters included in the generic diagnosis. In turn, this might make it difficult to compare newly found specimens with the diagnosis of a given monospecific genus if the specimens deviate to some extent from the diagnosis itself. Another possible consequence is that the diagnosis of a particular monospecific genus might have to be emended at a later stage if new, 'deviant' species of that genus are eventually discovered.

Methodology

As more and more ultrastructural information is incorporated into cryptomonad taxonomy and systematics, some of the identification difficulties could be reduced by adopting electron microscopy on a more routine basis during ecological surveys. The most practicable method appears to be SEM, which is relatively cost-effective, poses relatively few technical difficulties, and reveals much information of taxonomic usefulness. SEM has already been used successfully during several nanoplankton surveys (e.g. Booth et al., 1982; Kuylenstierna & Karlson, 1994; Novarino et al., 1997; Bérard-Therriault et al., 1999). It is the belief of the author that, as technology progresses and becomes more affordable, SEM investigations will become part of the routine toolkit of the practising phytoplankton and nanoplankton worker. From the point of view of the dissemination of taxonomic information, electronic publication offers an inexpensive, convenient and easily accessible solution. It is envisaged that the present paper will constitute the core of an interactive identification tool to be published electronically on the Internet in due course. Electronic publication will also ensure that the information can be modified and updated as required and with minimal disruption.

The electronic version will also provide a number of detailed cards for the most common or otherwise interesting cryptomonads.

Relationships between ecological studies and cryptomonad taxonomy and systematics

There is a wide, long-standing gap between the study of cryptomonad ecology on the one hand and taxonomy and systematics on the other. This is unfortunate because these disciplines could benefit greatly from one another, particularly in the study of diversity in relation to geographical and environmental variability, bloom ecology and modelling, and niche partitioning. Another topic which might benefit from a greater input from ecology is that of the cryptomonad species concept, which is largely understood in a morphological-typological sense rather than a biological one (Novarino & Lucas, 1993b). In the case of an essentially asexually-reproducing group such as the cryptomonads, it could be hypothesized that local ecological and/or geographical factors could confer recognizable morphological discontinuities to a pre-existing genetic continuum – such is the essence of the ecospecies concept as applied to the cryptomonads (Klaveness, 1985). In order to establish whether or not the ecospecies concept is applicable to this group, it becomes of paramount importance to relate ecological data to taxonomic information. It is probably unrealistic for ecologists to carry out detailed taxonomic investigations at the same time, and vice-versa, but ongoing collaborative projects involving both ecologists and taxonomist specialists could represent a possible solution. Such collaborations are hitherto rare, but in the author's view they are to be advocated on a much more frequent basis than normally takes place today.

Postscript

While publication in electronic form is in the making the author would be very grateful to receive criticisms, comments and suggestions. This companion is aimed at the end-users, and therefore their constructive feedback would be most welcome.

Acknowledgements

I thank my Natural History Museum colleagues Merche Martin-Cereceda for supplying Figures 25–27, Susan Morrall for supplying Figures 61, 63 and 64, and Emilia Oliva for taking several SEM photos. Part of the observations leading to the present work were carried out at the School of Ocean Sciences, University of Wales Bangor, Menai Bridge, U.K., thanks mainly to a special contribution from *Consiglio Nazionale delle Ricerche*, Italy (1987–88), and later an Award and a Fellowship from the Royal Society, London, within the European Science Exchange Programme jointly with *Accademia Nazionale dei Lincei*, Rome (1990 and 1991, respectively). I am grateful to the anonymous referees for their comments. I also thank Paul Kugrens for an afternoon conversation on cryptomonads high up in the Colorado mountains in August 1999.

References

Arnold, C. A., 1948. Classification of the gymnosperms from the viewpoint of palaeobotany. Bot. Gaz. 110: 2–12.

Barone, R. & L. Naselli-Flores, 2003. Distribution and seasonal dynamics of cryptomonads in Sicilian water bodies. Hydrobiologia 502 (Dev. Hydrobiol. 172): 325–329.

Beers, J. R., F. M. H. Reid & G. L. Stewart, 1980. Microplankton population structure in Southern California nearshore waters in late spring. Mar. Biol. 60: 209–226.

Ben Ali, A., R. De Baere, G. Van der Auwera, R. De Wachter & Y. Van de Peer, 2001. Phylogenetic relationships among algae based on complete large- subunit rRNA sequences. Int. J. Syst. Evol. Microbiol. 51: 737–749.

Bérard-Therriault, L., M. Poulin & L. Bossé, 1999. Guide d'identification du phytoplancton marin de l'estuaire et du golfe du Saint-Laurent incluant également certains protozoairs. Publication spéciale canadienne des sciences halieutiques et aquatiques 128: 387 pp.

Bicudo, C. E. d. M., 1989. Validation of the generic name *Protocryptomonas* (Cryptophyceae: Cryptomonadaceae). Taxon 38: 82–83.

Bicudo, C. E. d. M. & G. Tell, 1988. *Pseudocryptomonas*, a new genus of Cryptophyceae from southern Brazil. Nova Hedwigia 46: 407–411.

Booth, B. C., J. Lewin & R. E. Norris, 1982. Nanoplankton species predominant in the subarctic Pacific in May and June 1978. Deep-Sea Res. 29: 185–200.

Borics, G., B. Tóthmérész, I. Grigorszky, J. Padisák, G. Várbíró & S. Szabó, 2003. Algal assemblage types of bog-lakes in Hungary and their relation to water chemistry, hydrological conditions and habitat diversity. Hydrobiologia 502 (Dev. Hydrobiol. 172): 145–155.

Braarud, T., 1935. The 'Øst' expedition to the Denmark Strait 1929. II. The phytoplankton and its conditions of growth (including some qualitative data from the Arctic in 1930). Hvalrådets Skrifter 10: 1–173.

Brett, S. J., L. Perasso & R. Wetherbee, 1994. Structure and development of the cryptomonad periplast – a review. Protoplasma 181: 106–122.

Brett, S. J. & R. Wetherbee, 1986. A comparative study of periplast structure in *Cryptomonas cryophila* and *C. ovata* (Cryptophyceae). Protoplasma 131: 23–31.

Butcher, R. W., 1952. Contributions to our knowledge of the smaller marine algae. J. mar. biol. Ass. U.K. 52: 175–191.

Butcher, R. W., 1967. An introductory account of the smaller algae of British coastal waters. Part IV: Cryptophyceae. Fishery Investigations London Ser. 4: 1–54.

Bütschli, O., 1884. Erster band. Protozoa. II Abteilung: Mastigophora. In Bronn H. G. (ed.), Klassen und Ordnungen des Thier-Reichs 1883–1887. Leipzig und Heidelberg: 617–1097.

Büttner, J., 1911. Die farbigen Flagellaten des Kieler Hafens. Wissenschaftliche Meeresuntersuchungen, Abteilung Kiel, N. F. 12: 121–132.

Carter, N., 1937. New or interesting algae from brackish water. Archiv für Protistenkunde 90: 1–68.

Castro, A. A. J. d., C. E. d. M. Bicudo & D. d. C. Bicudo, 1991. Criptógamos do Parque Estadual das Fontes do Ipiranga, Sao Paulo, SP. Algas, 2: Cryptophyceae. Hoehnea 18: 87–106.

Cavalier-Smith, T., M. Allsopp & E. E. Chao, 1994. Chimeric conundra – Are nucleomorphs and chromists monophyletic or polyphyletic. Proc. Natl Acad. Sci. U.S.A. 91: 11368–11372.

Cavalier-Smith, T., J. A. Couch, K. E. Thorsteinsen, P. Gilson, J. A. Deane, D. R. A. Hill & G. I. McFadden, 1996. Cryptomonad nuclear and nucleomorph 18S rRNA phylogeny. Eur. J. Phycol. 31: 315–328.

Chang, F. H., 1983. Winter phytoplankton and microzooplankton populations off the coast of Westland, New Zealand, 1979. New Zealand J. mar. Freshwat. Res. 17: 279–304.

Christensen, T., 1967. Two new families and some new names and combinations in the algae. Blumea 15: 91–91.

Christensen, T., 1978. Annotations to a textbook of phycology. Botanisk Tidsskrift 73: 65–70.

Christensen, T., 1980. Algae. A taxonomic survey. Vol I. Aio Tryk as, Odense.

Clay, B. L. & P. Kugrens, 1999a. Characterization of *Hemiselmis amylosa* sp. nov. and phylogenetic placement of the blue-green cryptomonads *H. amylosa* and *Falcomonas daucoides*. Protist 150: 297–310.

Clay, B. & P. Kugrens, 1999b. Systematics of the enigmatic kathablepharids, including EM characterization of the type species, *Kathablepharis phoenikoston*, and new observations on *K. remigera* comb. nov. Protist 150: 43–59.

Clay, B. L., P. Kugrens & R. E. Lee, 1999. A revised classification of Cryptophyta. Bot. J. linn. Soc. 131: 131–151.

Conn, H. W., 1905. A preliminary report on the Protozoa of the fresh waters of Connecticut. State Geological and Natural History survey of Connecticut 1: 1–69.

Conrad, W., 1939. Notes protistologiques. XII.- Sur quatre Cryptomonadines rouges. Bulletin du Musée royal d'Histoire naturelle de Belgique 15: 1–5.

Conrad, W. & H. Kufferath, 1954. Recherches sur les eaux saumâtres des environs de Lilloo. II. Partie descriptive. Algues et protistes. Considérations écologiques. Mémoires de l'Institute royal des Sciences Naturelle de Belgique 127: 1–346.

Cronberg, G., 1982. Phytoplankton changes in Lake Trummen induced by restoration. Long-term whole-lake studies and food-web experiments. Folia Limnologica Scandinavica 18: 1–119.

Czosnowsky, J., 1948. Materialy do flory wiciowcow polski. Poznanskie Towarzystwo Przyjaciol Nauk Wydzial matematyczno-przyyrodniczy, Prace Komisji Biologicznej 11: 1–40.

Davis, B. M., 1894. Notes on the life-history of a blue-green motile cell. Bot. Gazette 19: 96–102.

Deane, J. A., D. R. A. Hill, S. J. Brett & G. I. McFadden, 1998. *Hanusia phi* gen. et sp. nov. (Cryptophyceae): characterization of 'Cryptomonas sp. Phi'. Eur. J. Phycol. 33: 149–154.

Deane, J. A., I. M. Strachan, G. W. Saunders, D. R. A. Hill & G. I. McFadden, 2002. Cryptomonad evolution: Nuclear 18S rRNA phylogeny versus cell morphology and pigmentation. J. Phycol. 38: 1236–1244.

Deflandre, G., 1938. Les corpuscules biréfringents des Ciliés et des cryptomonadines. Bullettin de la Société Française de Microscopie 7: 110–129.

Domingos, P. & M. Menezes, 1998. Taxonomic remarks on planktonic phytoflagellates in a hypertrophic tropical lagoon (Brazil). Hydrobiologia 370: 297–313.

Douglas, S. E., 1992. Eukaryote-Eukaryote Endosymbioses – Insights from studies of a cryptomonad alga. Biosystems 28: 57–68.

Douglas, S. E., C. A. Murphy, D. F. Spencer & M. W. Gray, 1991. Cryptomonad algae are evolutionary chimeras of 2 phylogenetically distinct unicellular eukaryotes. Nature 350: 148–151.

Dragesco, J., 1951. Sur la structure des trichocystes du flagellé cryptomonadine *Chilomonas paramecium*. Bulletin de Microscopie Appliquée 2e série, 1: 172–175.

Dujardin, F., 1841. Histoire naturelle des Zoophytes. Infusoires, comprenant la physiologie et la classification de ces Animaux, et la maniere de les etudier a l'aide du microscope. Librairie Encyclopedique de Roret, Paris. 295 pp.

Edmonson, C. H., 1906. Protozoa of Iowa. Proc. Davenport Acad. Sci. 11: 1–124.

Ehrenberg, C. G., 1831a. Animalia Evertebrata exclusis Insectis. Series Prima cum Tabularum Decade Prima. In Hemprich, F. & C. G. Ehrenberg (eds), Symbolae Physicae seu Icones et Descriptiones Animalium Evertebratorum sepositis Insectis. Berolini ex Officina Academica, Berlin: 1–126.

Ehrenberg, C. G., 1831b. Über die Entwicklung und Lebensdauer der Infusionsthiere: nebst ferneren Beiträgen zu einer Vergleichung ihrer organischen Systeme. Abhandl. Akad. Wissens. Berlin 1832: 1–154.

Ehrenberg, C. G., 1838. Die Infusionsthierschen als volkommene Organismen. Ein Blick in das tiefere organische Leben der Natur. Leopold Voss, Leipzig. 548 pp. + atlas

Erata, M. & M. Chihara, 1989. Re-Examination of *Pyrenomonas* and *Rhodomonas* (Class Cryptophyceae) through ultrastructural survey of red pigmented cryptomonads. Bot. Magazine-Tokyo 102: 429–443.

Eriksson, S., C. Sellei & K. Wallström, 1977. The structure of the plankton community of the Öregrundsgrepen (Souhwest Bothnian Sea). Helgolander Wiss. Meeresunters. 30: 582–597.

Ettl, H., 1980. Die Zweiteiligkeit der Chromatophoren bei Cryptomonaden. Plant Syst. Evol. 135: 227–234.

Faust, M. A., 1974. Structure of the periplast of *Cryptomonas ovata* var. *palustris*. J. Phycol. 9: 489–495.

Findenegg, I., 1971. Die Produkmonsleistungen einiger planktischer Algenarten in ihren naturlichen Milieu. Archiv für Hydrobiol. 69: 272–293.

Fisch, C., 1885. Untersuchungen über einige Flagellaten und verwandte Organismen. Zeitschrift fur Wissenschaftliche Zool. 42: 47–125.

Fresenius, G., 1858. Beiträge zur Kenntniss Mikroskopicher Organismen. Abhandlungen herausgegeben von der Senckerbergischen Naturforschenden Gesellschaft 2: 211–242.

Fromentel, E. D., 1874. Etudes sur les microzoaires ou Infusoires proprement dits. Masson, Paris. 364 pp.

Gantt, E., 1971. Micromorphology of the periplast of *Chroomonas* sp. J. Phycol. 7: 177–184.

Geitler, L., 1922. Die Mikrophyten-Biocoenose der Fontinalis-Bestände des Lunzer Untersees und ihre Abhängigkeit vom Lichts. Int. Rev. Hydrobiol. 10: 683–691.

Geitler, L., 1924. *Chroomonas caudata*, nov. spec. Österreichische Botanische Zeitschrift 73: 246–247.

Gervais, F., 1997. *Cryptomonas undulata* spec. nov., a new freshwater cryptophyte living near the chemocline. Nova Hedwigia 65: 353–364.

Gillot, M. A. & S. P. Gibbs, 1980. The cryptomonad nucleomorph: Its ultrastructure and evolutionary significance. J. Phycol. 16: 558–568.

Grim, J. N. & L. A. Staehelin, 1982. Ejectisomes of *Chilomonas paramecium* – a freeze-fracture study of their IMPs and docking sites. J. Protozool. 29: 474–474.

Grim, J. N. & L. A. Staehelin, 1984. The ejectisomes of the flagellate *Chilomonas paramecium* – visualization by freeze-fracture and isolation techniques. J. Protozool. 31: 259–267.

Gromov, B. V., K. A. Mamkaeva & O. V. Gavrilova, 1998. Ultrastructure of the flagella of *Campylomonas reflexa* (Cryptophyceae equals Cryptomonadea). Nova Hedwigia 66: 197–204.

Hada, Y., 1974. The Flagellata examined from polluted water of the Inland Sea, Setonaikai. Bull. Plankton Soc. Japan 20: 112–125.

Haigh, R., F. J. R. Taylor & T. F. Sutherland, 1992. Phytoplankton ecology of Sechelt Inlet, a fjord system of the British Columbia coast. 1. General features of the nanoplankton and microplankton. Mar. Ecol. Prog. Ser. 89: 117–134.

Halldal, P., 1953. Phytoplankton investigations from Weather Ship M in the Norwegian Sea, 1948–49 (including observations during the 'Armauer Hansen' cruise, July 1949. Hvalrådets Skrifter 38: 1–91.

Hansgirg, A., 1885. Anhang zu meiner Abhandlung – Ueber den Polymorphismus der Algen. Botanisches Centralblatt 23: 229–233.

Hansmann, P., H. Falk, U. Scheer & P. Sitte, 1986. Ultrastructural localization of DNA in two *Cryptomonas* species by use of a monoclonal DNA antibody. Eur. J. Cell Biol. 42: 152–160.

Hansmann, P., H. Falk & P. Sitte, 1985. DNA in the nucleomorph of *Cryptomonas* demonstrated by DAPI fluorescence. Zeitschrift für Naturforschung 40c: 933–955.

Hecky, R. E. & H. J. Kling, 1981. The phytoplankton and protozooplankton of the euphotic zone of Lake Tanganyika: Species composition, biomass, chlorophyll content, and spatio-temporal distribution. Limnol. Oceanogr. 26: 564–564.

Heywood, P., 1988. Ultrastructure of *Chilomonas paramecium* and the phylogeny of the Cryptoprotists. BioSystems 21: 293–298.

Hibberd, D. J., A. D. Greenwood & H. B. Griffiths, 1971. Observations on the ultrastructure of the flagella and periplast in the Cryptophyceae. British Phycol. J. 6: 61–72.

Hill, D. R. A., 1991a. A revised circumscription of *Cryptomonas* (Cryptophyceae) based on examination of Australian strains. Phycologia 30: 170–180.

Hill, D. R. A., 1991b. *Chroomonas* and other blue-green cryptomonads. J. Phycol. 27: 133–145.

Hill, D. R. A., 1992a. *Chroomonas vectensis* Carter (Cryptophyceae). Baltic Sea identification Sheet no. 6. Ann. Bot. Fenn. 29: 161–162.

Hill, D. R. A., 1992b. *Teleaulax amphioxeia* (Conrad) Hill, comb. nov. (Cryptophyceae). Baltic Sea Identification Sheet no. 13. Ann. Bot. Fenn. 29: 175–176.

Hill, D. R. A. & K. S. Rowan, 1989. The biliproteins of the Cryptophyceae. Phycologia 28: 455–463.

Hill, D. R. A. & R. Wetherbee, 1986. *Proteomonas sulcata* gen. et sp. nov. (Cryptophyceae), a cryptomonad with two morphologically distinct and alternating forms. Phycologia 25: 521–543.

Hill, D. R. A. & R. Wetherbee, 1988. The structure and taxonomy of *Rhinomonas pauca* gen. et sp. nov. (Cryptophyceae). Phycologia 27: 355–365.

Hill, D. R. A. & R. Wetherbee, 1989. A reappraisal of the genus *Rhodomonas* (Cryptophyceae). Phycologia 28: 143–158.

Hill, D. R. A. & R. Wetherbee, 1990. *Guillardia theta* gen. et sp. nov. (Cryptophyceae). Can. J. Bot. 68: 1873–1876.

Hoef-Emden, K., B. Marin & M. Melkonian, 2002. Nuclear and nucleomorph SSU rDNA phylogeny in the cryptophyta and the evolution of cryptophyte diversity. J. Mol. Evol. 55: 161–179.

Huber-Pestalozzi, G., 1950. Das Phytoplankton des Süßwassers. 3. Teil. Cryptophyceen, Chloromonadinen, Peridineen. E. Schweizerbart'sche Verlagsbuchhandlung (Erwin Nägele), Stuttgart. 310 pp.

Ilmavirta, V., 1975. Diel periodicity in the phytoplankton of the oligotrophic Lake Pääjärvi, Southern Finland. II. Late summer phytoplankton biomass. Ann. Bot. Fenn. 12: 37–44.

Javornický, P., 1967. Some interesting algal flagellates. Folia Geobotanica et Phytotaxonomica 2: 43–67.

Javornický, P., 2003. Taxonomic notes on some freshwater planktonic Cryptophyceae based on light microscopy. Hydrobiologia 502 (Dev. Hydrobiol. 172): 271–283.

Javornický, P. & F. Hindák, 1970. *Cryptomonas frigoris* spec. nova (Cryptophyceae), the new cyst-forming flagellate from the snow of the High Tatras. Biológia (Bratislava) 25: 241–250.

Jones, E. E., 1974. The Protozoa of Mobile Bay, Alabama. University of South Alabama Monographs 1: 1–113.

Joyon, L., 1963. Contribution à l'étude cytologique de quelques Protozoaires flagellés. Annales de la Faculté de Sciences de l'Université de Clermont 22 (Biologie Animale 1er Fasc.): 1–96.

Karsten, G., 1898. *Rhodomonas balica* N. g. et sp. Wiss. Meeresunters. Abteilung Kiel, N. F. 3: 15–17.

Kent, W. S., 1880–82. A Manual of the Infusoria. Vols I–III. David Bogue, London.

Klaveness, D., 1985. Classical and modern criteria for determining species of Cryptophyceae. Bull. Plankton Soc. Japan 32: 111–123.

Klaveness, D., 1988. Ecology of the Cryptomonadida: a first review. In Sandgren, C. D. (ed.), Growth and Reproductive Strategies of Freshwater Phytoplankton. Cambriddge University Press, Cambridge: 105–133.

Klaveness, D., 1988/1989. Biology and ecology of the Cryptophyceae: status and challenges. Biol. Oceanogr. 6: 257–270.

Kristiansen, J. & A. Kristiansen, 1999. A new species of *Chroomonas* (Cryptophyceae) living inside the submarine ikaite columns in the Ikkafjord, Southwest Greenland, with remarks on its ultrastructure and ecology. Nordic J. Bot. 19: 747–758.

Kugrens, P., B. L. Clay & R. E. Lee, 1999. Ultrastructure and systematics of two new freshwater red cryptomonads, *Storeatula rhinosa*, sp nov and *Pyrenomonas ovalis*, sp nov. J. Phycol. 35: 1079–1089.

Kugrens, P. & R. E. Lee, 1987. An ultrastructural survey of cryptomonad periplasts using quick-freezing freeze-fracture techniques. J. Phycol. 23: 365–376.

Kugrens, P. & R. E. Lee, 1990. Ultrastructural evidence for bacterial incorporation and mixotrophy in the photosynthetic cryptomonad *Chroomonas pochmanni* (Cryptomonadida). J. Protozool. 37: 263–267.

Kugrens, P. & R. E. Lee, 1991. Organization of cryptomonads. In Patterson D. J. & J. Larsen (eds), The biology of free-living heterotrophic flagellates. Clarendon Press, Oxford: 219–233.

Kugrens, P., R. E. Lee & R. A. Andersen, 1986. Cell form and surface patterns in *Chroomonas* and *Cryptomonas* cells (Cryptophyta) as revealed by scanning electron microscopy. J. Phycol. 22: 512–525.

Kugrens, P., R. E. Lee & R. A. Andersen, 1987. Ultrastructural variations in cryptomonad flagella. J. Phycol. 23: 511–518.

Kuylenstierna, M. & B. Karlson, 1994. Seasonality and composition of pico- and nanoplanktonic cyanobacteria and protists in the Skagerrak. Bot. Mar. 37: 17–33.

Lack, T. J., 1971. Quantitative studies on the phytoplankton of the rivers Thames and Kennet at Reading. Freshwat. Biol. 1: 213–224.

Lackey, J. B. & E. W. Lackey, 1963. Microscopic algae and protozoa in the waters near Plymouth in August 1962. J. mar. biol. Ass. U.K. 43: 797–805.

Lee, R. E. & P. Kugrens, 1986. The occurrence and structure of flagellar scales in some fresh-water Cryptophytes. J. Phycol. 22: 549–552.

Lepistö, L. & A.-L. Holopainen, 2003. Occurrence of Cryptophyceae and katablepharids in boreal lakes. Hydrobiologia 502 (Dev. Hydrobiol. 172): 307–314.

Liaud, M. F., U. Brandt, M. Scherzinger & R. Cerff, 1997. Evolutionary origin of cryptomonad microalgae: two novel chloroplast/cytosol-specific GAPDH genes as potential markers of ancestral endosymbiont and host cell components. J. Mol. Evol. 44: S28–S37.

Lucas, I. A. N., 1968. Three new species of the genus *Cryptomonas*. British Phycol. Bull. 3: 535–541.

Lucas, I. A. N., 1970a. Observations on the ultrastructure of the Cryptophyceae. I. The genus *Cryptomonas*. J. Phycol. 6: 30–38.

Lucas, I. A. N., 1970b. Observations on the ultrastructure of representatives of the genera *Hemiselmis* and *Chroomonas*. British Phycol. J. 5: 29–37.

Lucas, I. A. N., 1982. Observations on the fine structure of the Cryptophyceae. II. The eyespot. British Phycol. J. 17: 13–19.

Ludwig, M. & S. P. Gibbs, 1985. DNA is present in the nucleomorph of cryptomonads: Further evidence that the chloroplast evolved from a eukaryotic endosymbiont. Protoplasma 127: 9–20.

Maerz, M., S. Rensing, G. L. Igloi & U. G. Maier, 1992. Evolutionary analysis of the plastid-encoded gene for the alpha subunit of the DNA-dependent RNA-polymerase of *Pyrenomonas salina* (Cryptophyceae). Current Gen. 22: 479–482.

Maier, U. G., 1992. The 4 genomes of the alga *Pyrenomonas salina* (Cryptophyta). BioSystems 28: 69–73.

Maier, U. G. & P. Sitte, 1994. Cryptomonad evolution – insights into a eucyte within a eucyte. Endocytobiosis and Cell Res. 10: 129–135.

Marin, B., M. Klingberg & M. Melkonian, 1998. Phylogenetic relationships among the Cryptophyta: analyses of nuclear-encoded SSU rRNA sequences support the monophyly of extant plastid-containing lineages. Protist 149: 265–276.

McFadden, G. I., P. R. Gilson & D. R. A. Hill, 1994. *Goniomonas* – Ribosomal-RNA sequences indicate that this phagotrophic flagellate is a close relative of the host component of cryptomonads. Eur. J. Phycol. 29: 29–32.

Mayr, E., 1974. The challenge of diversity. Taxon 23: 3–9.

Menezes, M. & G. Novarino, 2003. How diverse are planktonic cryptomonads in Brazil? Advantages and difficulties of a taxonomic-biogeographical approach. Hydrobiologia 502 (Dev. Hydrobiol. 172): 297–306.

Meyer, S. R. & R. N. Pienaar, 1984. The microanatomy of *Chroomonas africana* n. sp. (Cryptophyceae). South African J. Bot. 3: 306–319.

Meyer, S. R. & R. N. Pienaar, 1987. The taxonomic implications of the ultrastructure and cell division of a stigma-containing *Chroomonas* sp. (Cryptophyceae) from Rocky Bay, Natal, South Africa. South African J. Bot. 53: 129–139.

Morrall, S., 1980. A comparison of the periodic substructure of the trichocysts of the Cryptophyceae and Prasinophyceae. BioSystems 12: 71–83.

Morrall, S., 1980. Ultrastructure of the Cryptophyceae, including cyst formation and trichocyst substructure. PhD Thesis, Imperial College, University of London.

Nishijima, T., 1990. Growth characteristics of *Plagioselmis* sp. (strain 87) causing freshwater red tide in the lower part of the Nakasuji River, Japan. Bull. Jap. Soc. Sci. Fish. 56: 353–359.

Norris, R. E., 1964. Studies on phytoplankton in Wellington Harbour. New Zealand J. Bot. 2: 258–278.

Novarino, G., 1991a. Observations on *Rhinomonas reticulata* comb. nov. and *Rhinomonas reticulata* var. *eleniana* var. nov. (Cryptophyceae), with comments on the genera *Pyrenomonas* and *Rhodomonas*. Nordic J. Bot. 11: 243–252.

Novarino, G., 1991b. Observations on some new and interesting Cryptophyceae. Nordic J. Bot. 11: 599–611.

Novarino, G., 1993. Possible detection of the periplast areas and the nucleomorph of cryptomonads by light microscopy: some early observations by Künstler, Skuja and Hollande. Quekett J. Microsc. 37: 45–51.

Novarino, G., 1996. Notes on flagellate nomenclature. I. *Cryptaulaxoides* nom. nov., a zoological substitute for *Cryptaulax* Skuja, 1948 (Protista *incertae sedis*) non *Cryptaulax* Tate, 1869 (Mollusca, Gastropoda) non *Cryptaulax* Cameron, 1906 (Insecta, Hymenoptera), with remarks on botanical nomenclature. Acta Protozool. 35: 235–238.

Novarino, G., 2002. Phylum Cryptophyta (Cryptomonads). In John, D. M., B. A. Whitton & A. J. Brook (eds), The Freshwater Algal Flora of the British Isles. An Identification Guide to Freshwater and Terrestrial Algae. Cambridge University Press, Cambridge: 180–185.

Novarino, G. & M. Gilbert, 2002. Proposal to conserve the name *Cryptochloris* J. Schiller (Cryptophyceae) against *Cryptochloris* Bentham (Poaceae). Taxon 51: 181–182.

Novarino, G. & I. A. N. Lucas, 1993a. A comparison of some morphological characters in *Chroomonas ligulata* sp. nov. and *C. placoidea* sp. nov. (Cryptophyceae). Nordic J. Bot. 13: 583–591.

Novarino, G. & I. A. N. Lucas, 1993b. Some proposals for a new classification system of the Cryptophyceae. Bot. J. linn. Soc. 111: 3–21.

Novarino, G. & I. A. N. Lucas, 1995. A zoological classification system of cryptomonads. Acta Protozool. 34: 173–180.

Novarino G. & E. Oliva, 1998. Typification and ultrastructural characterization of long-described flagellate taxa based on original type material. In The Flagellates Symposium, University of Birmingham, 7–11 September 1998 (abstract).

Novarino, G., I. A. N. Lucas & S. Morrall, 1994. Observations on the genus *Plagioselmis* (Cryptophyceae). Cryptogamie, Algologie 15: 87–107.

Novarino, G., D. K. Mills & F. Hannah, 1997. Pelagic flagellate populations in the southern North Sea, 1988–89.1. Qualitative observations. J. Plankton Res. 19: 1081–1109.

Novarino, G. & D. M. Roberts, 1994. Digital film as an aid for studying live protists. Binary – Computing in Microbiology 6: 197–203.

Novarino, G., A. Warren, N. E. Kinner & R. W. Harvey, 1994. Protists from a Sewage-Contaminated Aquifer on Cape-Cod, Massachusetts. Geomicrobiol. J. 12: 23–36.

Ondratschek, K., 1941. Uber die 'Variabilität' farbloser Flagellaten. Beihefte Botanisches Centralblatt 61 (Abteil. A): 277–309.

Padisák, J., G. Borics, G. Fehér, I. Grigorszky, I. Oldal, A. Schmidt & Z. Zámbóné-Doma, 2003. Dominant species, functional assemblages and frequency of equilibrium phases in late summer phytoplankton assemblages in Hungarian small shallow lakes. Hydrobiologia 502 (Dev. Hydrobiol. 172): 157–168.

Pascher, A., 1911. Zwei braune Flagellaten. Berichte der deutschen Botanischen Gesellschaft 29: 190–192.

Pascher, A., 1913. Cryptomonadinae. In Pascher, A. (ed.), Die Süsswasser-Flora Deutschlands, Österreichs und der Schweiz. Verlag von Gustav Fischer, Jena: 96–114.

Patterson, D. J., 1986. Some problems of ambiregnal taxonomy and a possible solution. Symp. Biol. Hungarica 33: 87–93.

Patterson, D. J. & S. Hedley, 1992. Free-living freshwater Protozoa – a colour guide. Wolfe, London. 223 pp.

Paulsen, B. S., A. A. H. Vieira & D. Klaveness, 1992. Structure of extracellular polysaccharides produced by a soil *Cryptomonas* sp. (Cryptophyceae). J. Phycol. 28: 61–63.

Pratt, D. M., 1959. The phytoplankton of Narragansett Bay. Limnol. Oceanogr. 4: 425–440.

Pringsheim, E. G., 1944. Some aspects of taxonomy in the Cryptophyceae. New Phytol. 43: 143–150.

Pringsheim, E. G., 1963. Farblose Algen. Ein Beitrag zur Evolutionsforchung. Gustav Fischer Verlag, Stuttgart.

Pringsheim, E. G., 1968. Zur Kenntnis der Cryptomonaden des Susswassers. Nova Hedwigia 16: 367–401.

Reynolds, C. S., 1978. Notes on the phytoplankton periodicity of Rostherne Mere, Cheshire, 1967–1977. British Phycol. J. 13: 329–335.

Roberts, E. C., J. Laybourn-Parry, D. M. McKnight & G. Novarino, 2000. Stratification and dynamics of microbial loop communities in Lake Fryxell, Antarctica. Freshwat. Biol. 44: 649–661.

Roberts, K. R., K. D. Stewart & K. R. Mattox, 1981. The flagellar apparatus of *Chilomonas paramecium* (Cryptophyceae) and its comparison with certain zooflagellates. J. Phycol. 17: 159–167.

Robinson, C., S. D. Archer & P. J. L. Williams, 1999. Microbial dynamics in coastal waters of East Antarctica: plankton production and respiration. Mar. Ecol. Prog. Ser. 180: 23–26.

Ruttner, F., 1959. Einige Beobachtungen ueber das Phytoplankton norditalienischer Seen. Mem. Ist. ital. Idrobiol. Dott. Marco de Marchi 11: 73–111.

Sandon, H., 1927. The composition and distribution of the protozoan fauna of the soil. Oliver & Boyd, Edinburgh & London.

Santore, U. J., 1983. Flagellar and body scales in the Cryptophyceae. British Phycol. J. 18: 239–248.

Santore, U. J., 1984. Some aspects of the taxonomy in the Cryptophyceae. New Phytol. 98: 627–646.

Santore, U. J., 1985. A cytological survey of the genus *Cryptomonas* (Cryptophyceae) with comments on its taxonomy. Archiv für Protistenkunde 130: 1–52.

Santore, U. J., 1986. The ultrastructure of *Pyrenomonas heteromorpha* comb. nov. (Cryptophyceae). Bot. Mar. 29: 75–82.

Schanz, F., 1985. Vertical light attenuation and phytoplankton development in Lake Zurich. Limnol. Oceanogr. 30: 299–310.

Schiller, J., 1913. Vorläufige Ergebnisse der Phytoplanktonuntersuchungen auf den Fahrten S.M.S. 'Najade' in der Adria. Sitzungsberichte Kais Akademie der Wissenschaften, Mathematisch-Naturwissenschaften Klasse, Abt. I 122: 621–630.

Schiller, J., 1925. Die planktontischen Vegetationen des adriatischen Meeres. B. Chrysomonadina, Heterokontae, Cryptomonadina, Eugleninae, Volvocales. I. Systematischer Teil. Archiv für Protistenkunde 53: 59–123.

Schiller, J., 1929. Neue Chryso- und Cryptomonaden aus Altwässern der Donau bei Wien. Archiv für Protistenkunde 66: 436–458.

Schiller, J., 1954. Neue Mikrophyten aus künstlichen betonierten Wasserbehältern. Archiv für Protistenkunde 100: 116–126.

Schiller, J., 1957. Untersuchungen an den planktischen Protophyten des Neusiedlersees 1950–1954, II. Teil. Wiss. Arbeit. Burgenland 18: 1–44.

Seaburg, K. G., B. C. Parker, G. W. Prescott & L. A. Whitford, 1979. The algae of Southern Victorialand, Antarctica. Bibliotheca Phycologica 46, J. Cramer, Vaduz. 169 pp.

Senn, G., 1900. Flagellaten. In Engler A. & K. Prantl (eds), Die naturliche Pflanzenfamilien. Verlag von Wilhelm Engelmann, Leipzig: 93–192.

Sitte, P. & U. G. Maier, 1992. Evolution of complex plastids from eukaryotic endosymbionts. Endocytobiosis and Cell Res. 8: 223–225.

Skuja, H., 1939. Beitrag zur Algenflora Lettlands II. Acta. Horti Botanici Universitatis Latviensis 11/12: 41–169.

Skuja, H., 1948. Taxonomie des Phytoplanktons einiger Seen in Uppland, Schweden. Symbolae Bot. Ups. 9: 1–399.

Skuja, H., 1956. Taxonomische und biologische Studien über das Phytoplankton Schwedischer Binnengewasser. N. Acta Regiae Societatis Scientiarum Upsaliense, Ser. IV 16: 1–404.

Skvortzov (Skvortzow), B. V. (B. W), 1960. Algae novae et minus cognitae Chinae boreali-orientalis. 2. Flagellatae prope oppidum mukden in anno 1957 collectarum. Bull. Herb. North-east. Forest. Acad. Harbin 2: 1–8.

Skvortzov (Skvortzow), B. V. (B. W), 1968. On new genera of Cryptomonadinae recorded from North-Eastern China. J. Jap. Bot. 43: 8–16.

Skvortzov (Skvortzow), B. V. (B. W), 1957. New and rare Flagellatae from Manchuria, Eastern Asia. Philippine J. Sci. 86: 139–202.

Sommer, U., 1982. Vertical niche separation between two closely related planktonic flagellate species (*Rhodomonas lens* and *Rhodomonas minuta* v. *nannoplanctica*). J. Plankton Res. 4: 137–142.

Stokes, A. C., 1888. A preliminary contribution toward a history of the fresh-water Infusoria of the United States. J. Trenton Nat. Hist. Soc. 1: 71–344.

Systematics Association Committee for Descriptive Biological Terminology, 1962. II. Terminology of simple symmetrical plane shapes (chart 1). Taxon 11: 145–156.

Taylor, D. L. & C. C. Lee, 1971. A new cryptomonad from Antarctica: *Cryptomonas cryophila* sp. nov. Archiv für Mikrobiologie 75: 269–280.

Thomsen, H. A. (ed.), 1992. Plankton i de Indre Danske Farvande. Havforsnkning fra Miljøstyrelsen, Miljøministeriet Miljøstyrelsen, Copenhagen. 331 pp.

Ùlehla, V., 1911. Die Stellung der Gattung *Cyathomonas* From. im System der Flagellaten. Berichte der Deutschen Botanischen Gesellschaft 29: 284–292.

Tolotti, M., H. Thies, M. Cantonati, C. M. E. Hansen & B. Thaler, 2003. Flagellate algae (Chrysophyceae, Dinophyceae, Cryptophyceae) in 48 high mountain lakes of the Northern and Southern slope of the Eastern Alps: biodiversity, taxa distribution and their driving variables. Hydrobiologia 502 (Dev. Hydrobiol. 172): 331–348.

Van der Auwera, G., C. J. B. Hofmann, P. De Rijk & R. De Wachter, 1998. The origin of red algae and cryptomonad nucleomorphs: A comparative phylogeny based on small and large subunit rRNA sequences of *Palmaria palmata, Gracilaria verrucosa*, and the *Guillardia theta* nucleomorph. Mol. Phylogen. Evol. 10: 333–342.

Vørs, N., 1992. Ultrastructure and autecology of the marine, heterotrophic flagellate *Leucocryptos marina* (Braarud) Butcher 1967 (Katablepharidaceae/Kathablepharidae), with a discussion of the genera *Leucocryptos* and *Katablepharis/Kathablepharis*. Eur. J. Protistol. 28: 369–389.

Wailes, G. H., 1939. The plankton of Lake Windermere, England. Annals and Magazine of Natural History, Series 11, 3: 401–404.

Wetherbee, R., D. R. A. Hill & S. J. Brett, 1987. The structure of the periplast components and their association with the plasma membrane in a cryptomonad flagellate. Can. J. Bot. 65: 1019–1026.

Wetherbee, R., D. R. A. Hill & G. I. McFadden, 1986. Periplast structure of the cryptomonad flagellate *Hemiselmis brunnescens*. Protoplasma 131: 11–22.

Willén, E., 2003. Dominance patterns of planktonic algae in Swedish forest lakes. Hydrobiologia 502 (Dev. Hydrobiol. 172): 315–324.

Willén, E., M. Oké & F. Gonzàlez, 1980. *Rhodomonas minuta* and *Rhodomonas lens* (Cryptophyceae) – Aspects of form-variation and ecology in lakes Mälaren and Vättern, Central Sweden. Acta Phytogeographica Suecica 68: 163–172.

Wislouch, S., 1924. Przyczynek do biologji solnisk i genezy szlamów leczniczych na Krymie (Beiträge zur Biologie und Entstehung von Heilschlamm der Salinen der Krim) [in Polish and German]. Acta Societatis Botanicorum Poloniae 2: 99–129.

Hydrobiologia **502:** 271–283, 2003.
L. Naselli-Flores, J. Padisák, M. T. Dokulil (eds),
Phytoplankton and Equilibrium Concept: The Ecology of Steady-State Assemblages.

Taxonomic notes on some freshwater planktonic Cryptophyceae based on light microscopy

Pavel Javornický
Běloveská 1674, CZ-54701 Náchod, Czech Republic
E-mail: pjavornicky@quick.cz

Key words: freshwater phytoplankton, *Cryptomonas*, *Rhodomonas*, *Plagioselmis*, light microscopy

Abstract

The cryptomonads sampled frequently from pelagial of large freshwater bodies, lakes and ponds, but also found in littoral regions and in small water bodies covered with vegetation, are often characterized by the sigmoid (S-shaped) form of their cells. According to the quoted findings of electron microscopy it seems that these species should be incorporated into the new genus *Campylomonas* Hill. Because the EM characteristics have not been proved so far for all species, and the following nomenclatural combinations accomplished, the traditional classification into the genus *Cryptomonas* is kept here. The correct name for the largest of sigmoid cryptomonads is *Cryptomonas curvata* Ehr. em. Penard. Contrarily, *C. rostrata* Troitz. em. Kisel. is to be held for later synonyms. The existence of the species C. *rostratiformis* Skuja remains uncertain. The smaller species of sigmoid shape, i.e. *C. reflexa* (Marss.) Skuja and *C. marssonii* Skuja, may also easily be discerned under the light microscope. From the small puddles with H_2S in water, shortened forms of both these species are documented, one of them under the published name *C. anas* Javorn. The pelagic assemblage of the above sigmoid cryptomonads frequently is completed by *Plagioselmis nannoplanctica* (Skuja) Novar., Lucas et Morr. and by *P. lacustris* (Pasch. et Ruttn.) Javorn. These flagellates so far are currently determined as the species of the genus *Rhodomonas* Karsten. In addition to the EM characteristic, they differ from *Rhodomonas* by the absence of a true gullet (pseudopharynx) having only the ventral furrow with rows of superficial ejectosomes (similar to the genus *Cryptochrysis* Pascher). The ellipsoidal or ovoid cryptomonads are sampled more frequently from peat pools and small water bodies covered with vegetation than from open pelagial of lakes. An erroneous determination of them causes confusion. Because some strains are wrongly labeled, electron-microscopic characteristics are vaguely determined species. For example, some EM features of *Cryptomonas ovata* in fact belong to *C. pyrenoidifera* Geitl. or to *C. phaseolus* Skuja. *Cryptomonas ovata* Ehr. em. Stein is not a collective species with a wide dimensional range of ellipsoid cells. It is a large species the typical morphology which is described here in detail. *C. splendida* Czosn. differs from *C. ovata* only by the transversal orientation of the cell. Similar to *C. ovata* but smaller species is *C. tatrica* Czosn. These species are further compared with the well-defined species *C. obovata* Skuja and the particularly small *C. phaseolus* Skuja.

Abbreviations: bg – basal refractive grain; cv – contractile vacuole; f+t – furrow with trichocysts; g+t – gullet with trichocysts; Mc – Maupas corpuscles; n+n – nucleus with nucleolus; p – pyrenoid; rm – reserve material; sp – striated periplast

Introduction

In opinion of Klaveness (1989: 259), much more taxonomic work based upon the full range of available characters is necessary before new, mainly electron microscopic, characters are ascribed to the taxons. In the present paper, the long term light microscopic observations of the common freshwater *Cryptophyceae* are made available.

According to the observations of electron microscopists, the genus *Cryptomonas* Ehr. is not homogenous but breaks up into several groups, possibly genera. Two of these groups are frequent in fresh water bodies. After some hesitation if the tubular gullet (pseudopharynx) is present in these flagellates at all it was confirmed that the species of the genus *Cryptomonas* possess both furrow and gullet (Brett et al., 1994). There is a difference between the two types of furrow (Kugrens et al., 1986). Members of both groups may or may not possess pyrenoids; if present, they are not traversed by thylakoids. Another difference appears in the flagellar apparatus. The conspicuously keeled rhizostyle of some species is believed to cause the sigmoid shape of the cells (Hill, 1991). Moreover the periplast structure of both groups is different.

On the basis of the above stated differences, Hill (1991) established the new genus *Campylomonas*. The type species is *Campylomonas reflexa* (Marsson) Hill. Some more species belong to this group according to Kugrens et al. (1986): *Cryptomonas curvata, C. rostratiformis, C. marssonii, C. platyuris, C. caudata, C. ozolinii.* Hill (1991), however, wrote that the variability in size and shape of the cells in his culture was wide and his organism could be identified with *C. reflexa, C. marssonii*, and *C. gracilis.*

The cells of most of these species are sigmoid, S-shaped, with the apex curved ventrally and the antapex (posterior end) curved dorsally. If such a simple visible characteristric were joined with the ultrastructures of the cell, the LM determination of *Campylomonas* would be easy. This would be a very useful solution of the problem but more evidence is necessary. Before more cultures and natural materials are studied in detail by EM, the old wide concept of the genus *Cryptomonas* is used in the present contribution.

Results

All figures have been drawn by the autor according to his own light-microscopical observations. Figures 1–8 and 12–17 are published here for the first time. The other figures are reprinted from the following publications: 9–11 (Javornický, 1957); 18–27 (Javornický, 1976); 28–42 (Javornický, in print).

Sigmoid species of the genus Cryptomonas *Ehr.*

Several sigmoid species of *Cryptophyceae* are frequently found in plankton of freshwater lakes, dam reservoirs and fishponds in temperate zones. They may be sampled also among littoral vegetation, as well as in small water bodies overgrown with vegetation.

The most apparent is a robust species which was described by Ehrenberg (1838) as *Cryptomonas curvata.* His pictures are very schematic but the total habitus of the cells as well as their dimensions (39–44 μm) are clear enough. The species was emended first by Stein (1878) and later by Penard (1922) both of whom provided good pictures of it. Nevertheless, there still exists nomenclatoric confusion about this species. For example Skuja (1948) writes that it is not clear which *"reflected Cryptomonas"* is to be understood under *C. curvata* Ehr. Evidently the same species was schematically pictured and described by Troitzkaja (1922) as *C. rostrata* (emended by Kiselev, 1931, and distinctly drawn by Czosnowski, 1948). Very similar species was described as *C. rostrata* by Skuja (1948), later renamed as *C. rostratiformis* (Skuja) Skuja in Huber-Pestalozzi (1950).

All the above named cryptomonads are rather large; some authors give their maximum length at 60–65 μm. The cell length varies between 42–52 μm in my materials (Figs 1–5). The elongated cells have the conspicuous apical finger-shaped rostrum, sometimes with a cut off peak (Fig. 1a), and funnel-shaped mouth of the pseudopharynx. The wide, laterally flattened pseudopharynx is inlaid with numerous ejectisomes (trichocysts). The antapex is bluntly pointed, curved to the cell dorsum, and sometimes keel-shaped. The cells are frequently deformed and have a screw-like distortion. Therefore, the cell orientation is hard to define precisely. Most cells are somewhat laterally flattened and contain two lateral chloroplasts, sometimes joined with narrow dorsal isthmus. Maupas corpuscles are not present. No pyrenoids are visible under a light microscope.

In spite of the above given nomenclatorical confusion, the species *Cryptomonas curvata* Ehr. emended by Penard is well defined and may be easily identified. Ettl (1980) makes the difference between *C. curvata* and *C. rostratiformis.*

C. curvata is frequently accompanied in phytoplankton by *Cryptomonas reflexa* (Marsson) Skuja (1939).

C. reflexa is smaller than *C. curvata.* Its cells were 22–30 μm long (average 26, 9 μm) in my materials (Figs 6–8). They were of a regular sigmoid (S) shape with the blunt apical rostrum curved ventrally and the rounded or bluntly pointed antapex curved dorsally. The cells were slightly laterally flattened with two lateral brown chloroplasts sometimes joined with dorsal isthmus. Two Maupas corpuscles were most often present but in some cells they were absent. This species may be easily differentiated from *C. curvata.*

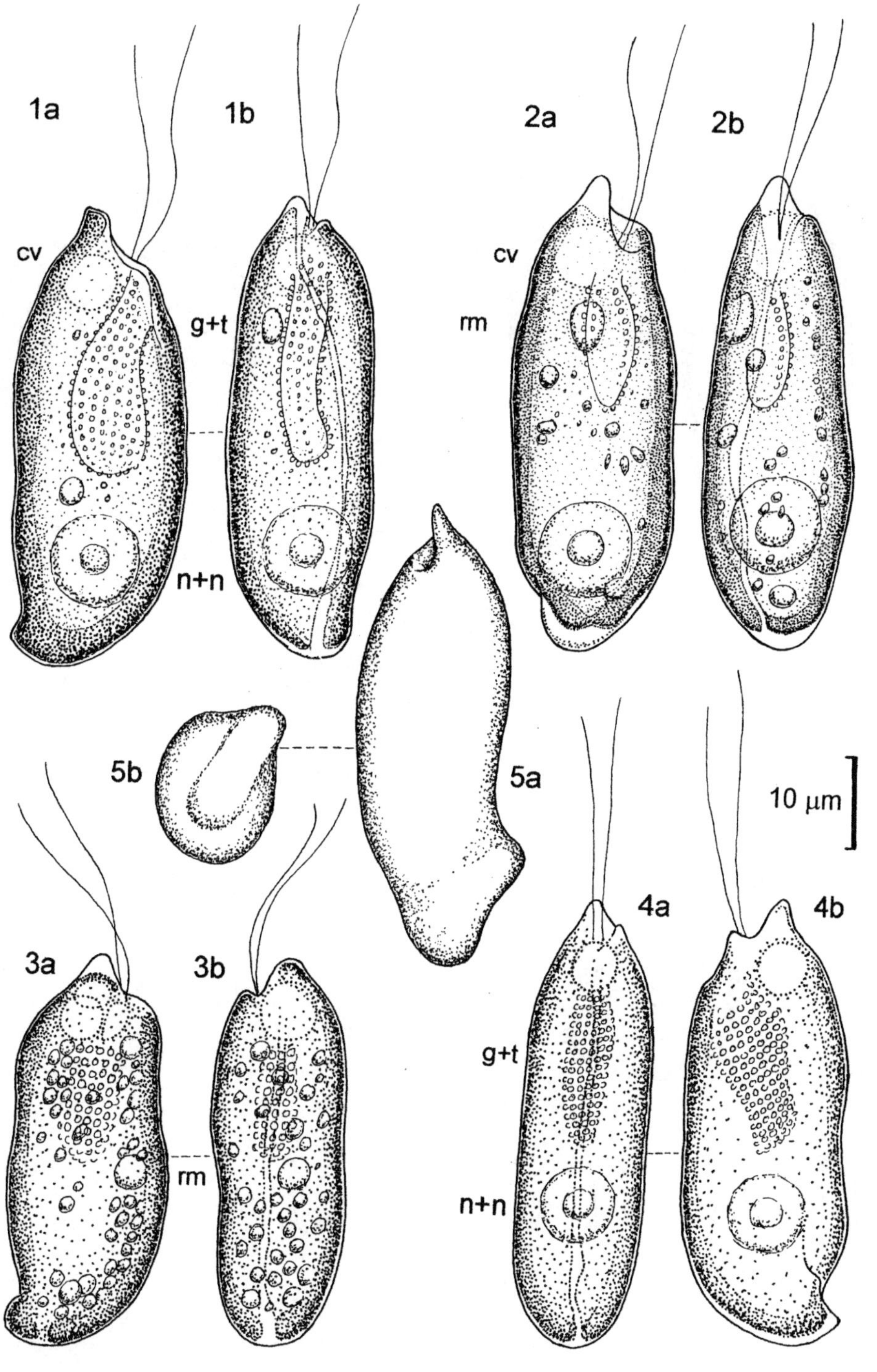

Figure 1–5. Cryptomonas curvata Ehr. 1a, 2a, 3a, 4b – lateral views; 1b, 2b, 4a – ventral views; 3b – dorsal view; 5 – contour of a keel-shaped cell, 5a – lateral view, 5b – antapical view.

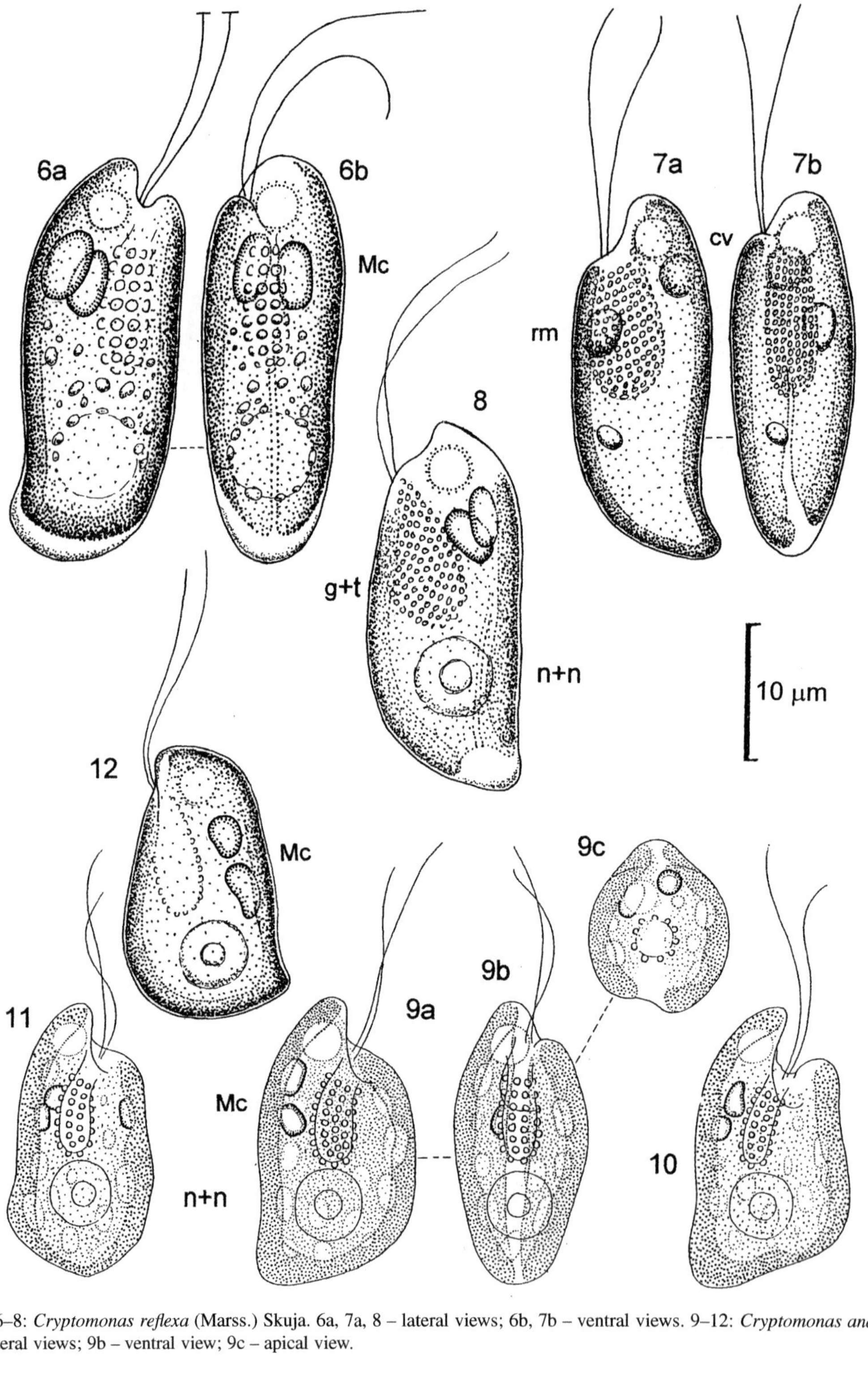

Figure 6–12. 6–8: *Cryptomonas reflexa* (Marss.) Skuja. 6a, 7a, 8 – lateral views; 6b, 7b – ventral views. 9–12: *Cryptomonas anas* Javorn. 9a, 10, 11, 12 – lateral views; 9b – ventral view; 9c – apical view.

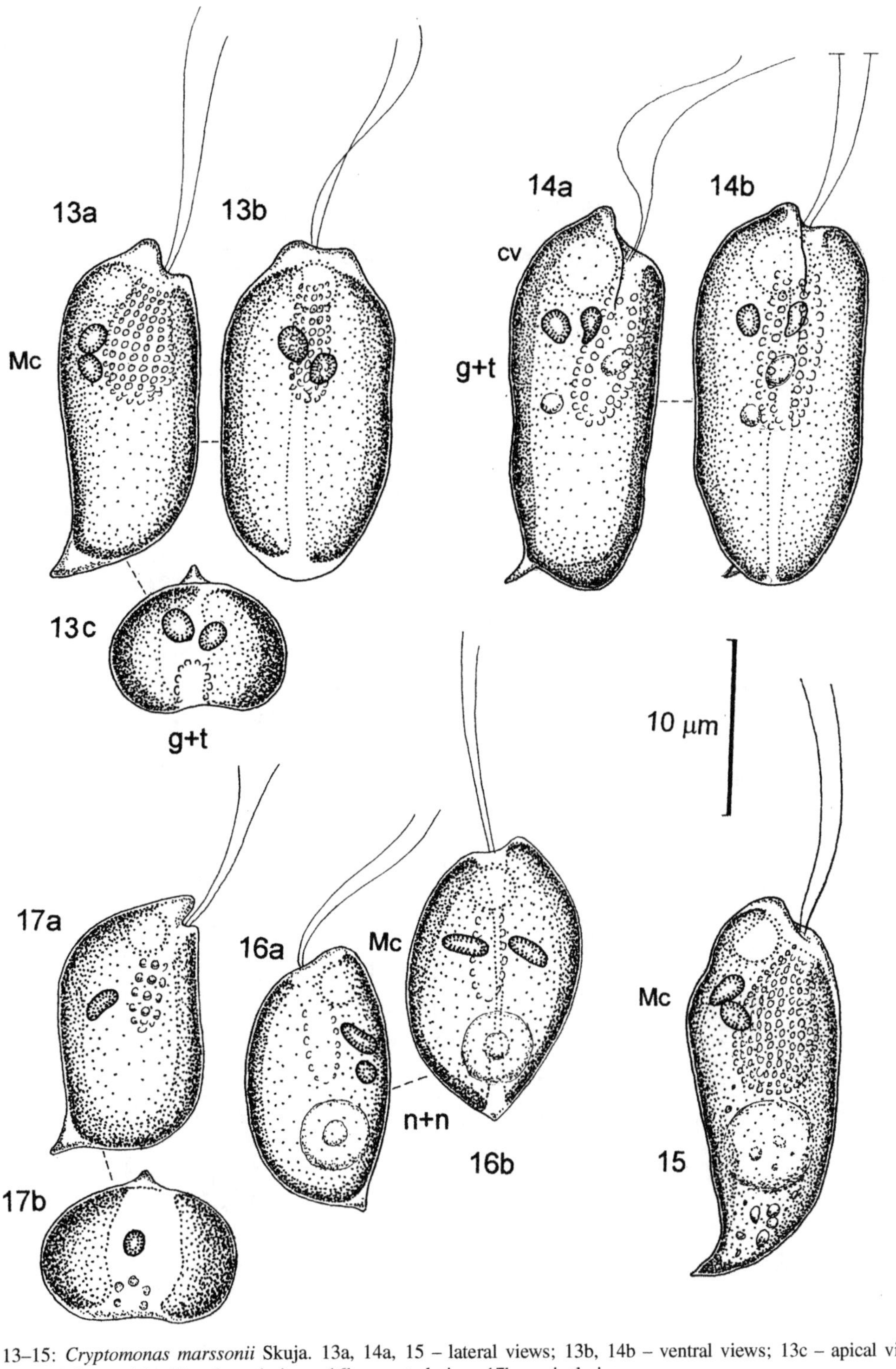

Figure 13–17. 13–15: *Cryptomonas marssonii* Skuja. 13a, 14a, 15 – lateral views; 13b, 14b – ventral views; 13c – apical view. 16 – 17: *Cryptomonas marssonii* forma. 16a, 17a – lateral views; 16b – ventral view; 17b – apical view.

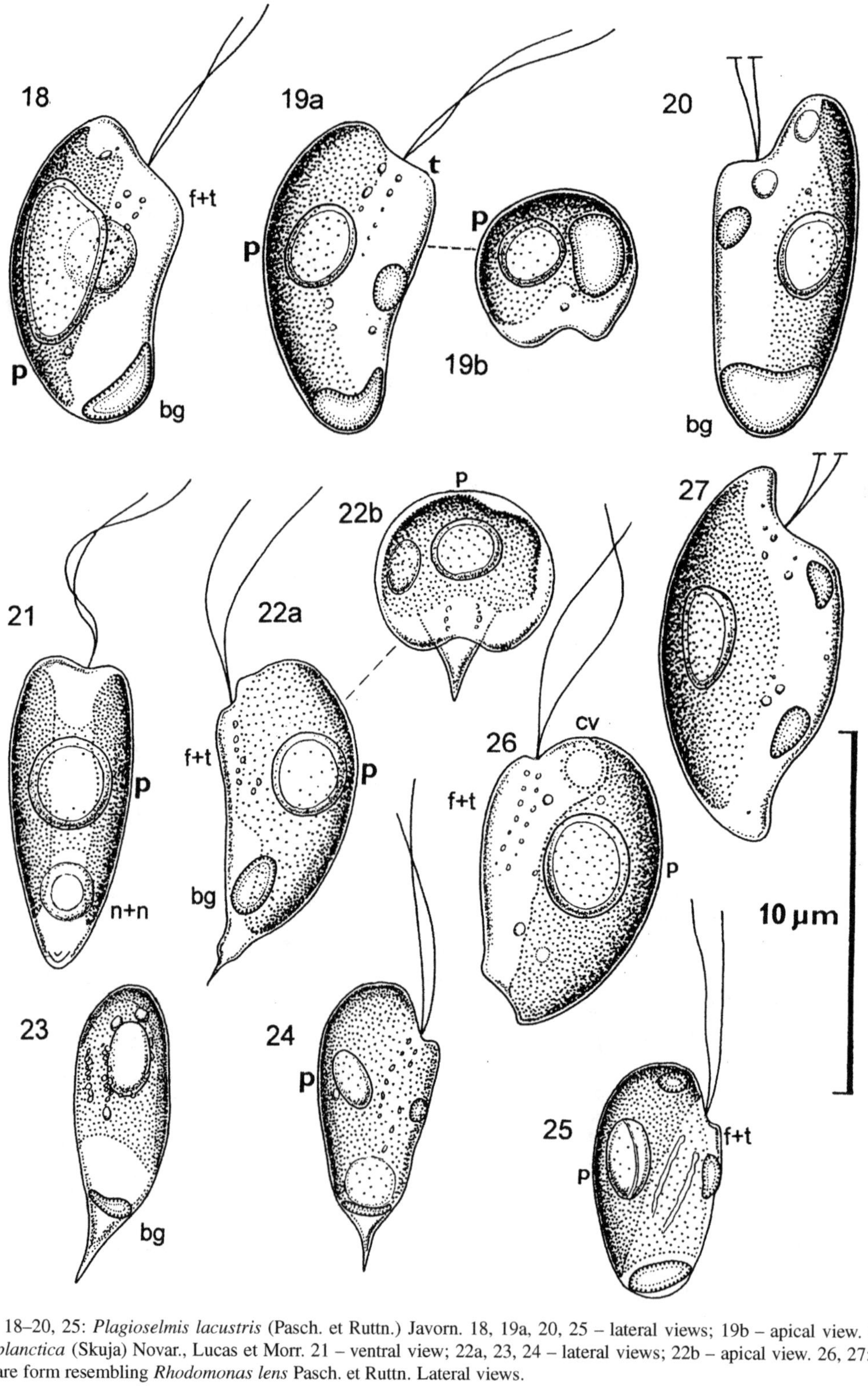

Figure 18–27. 18–20, 25: *Plagioselmis lacustris* (Pasch. et Ruttn.) Javorn. 18, 19a, 20, 25 – lateral views; 19b – apical view. 21–24: *Plagioselmis nannoplanctica* (Skuja) Novar., Lucas et Morr. 21 – ventral view; 22a, 23, 24 – lateral views; 22b – apical view. 26, 27: *Plagioselmis lacustris*, the rare form resembling *Rhodomonas lens* Pasch. et Ruttn. Lateral views.

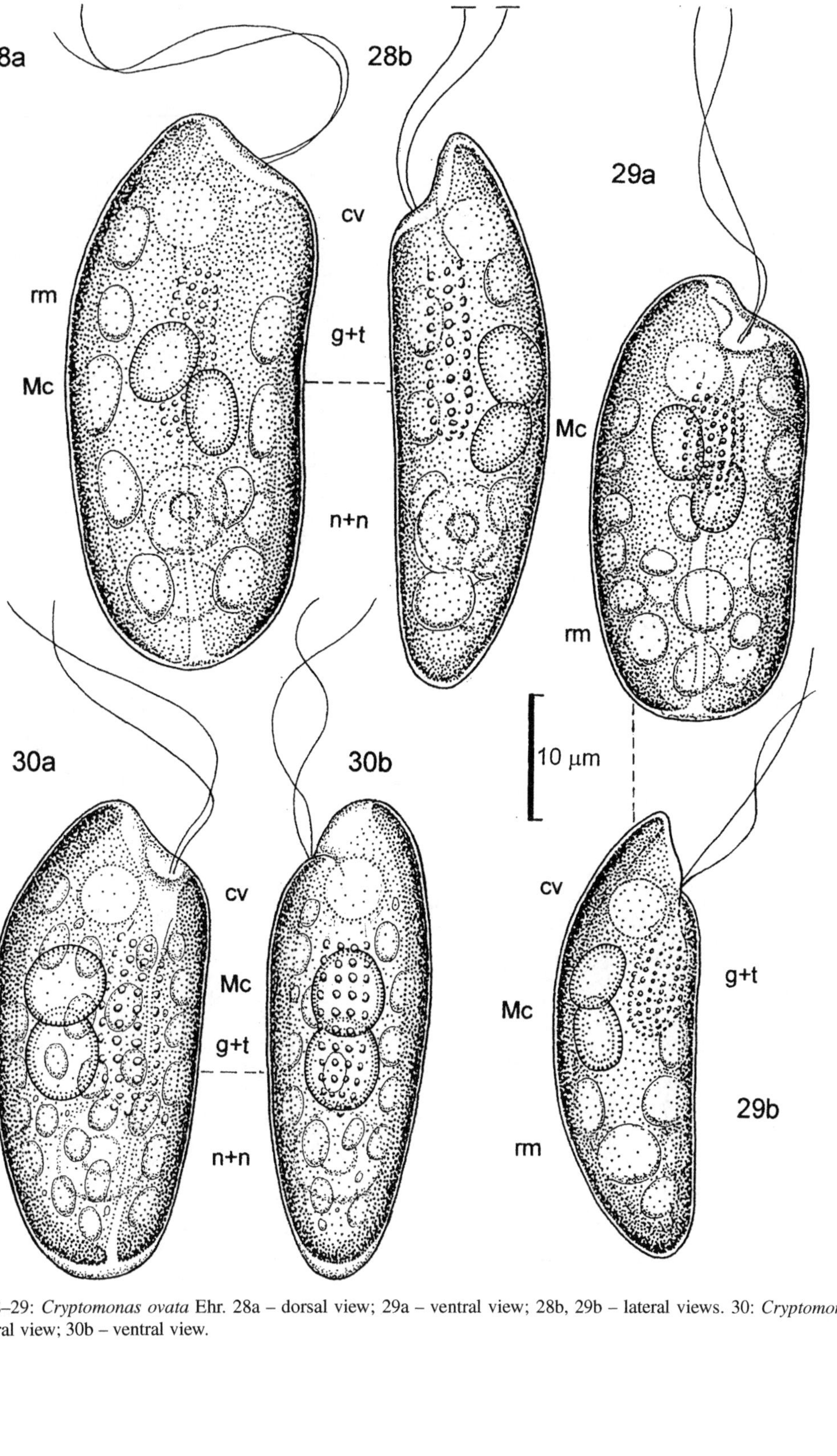

Figure 28–30. 28–29: *Cryptomonas ovata* Ehr. 28a – dorsal view; 29a – ventral view; 28b, 29b – lateral views. 30: *Cryptomonas splendida* Czosn. 30a – lateral view; 30b – ventral view.

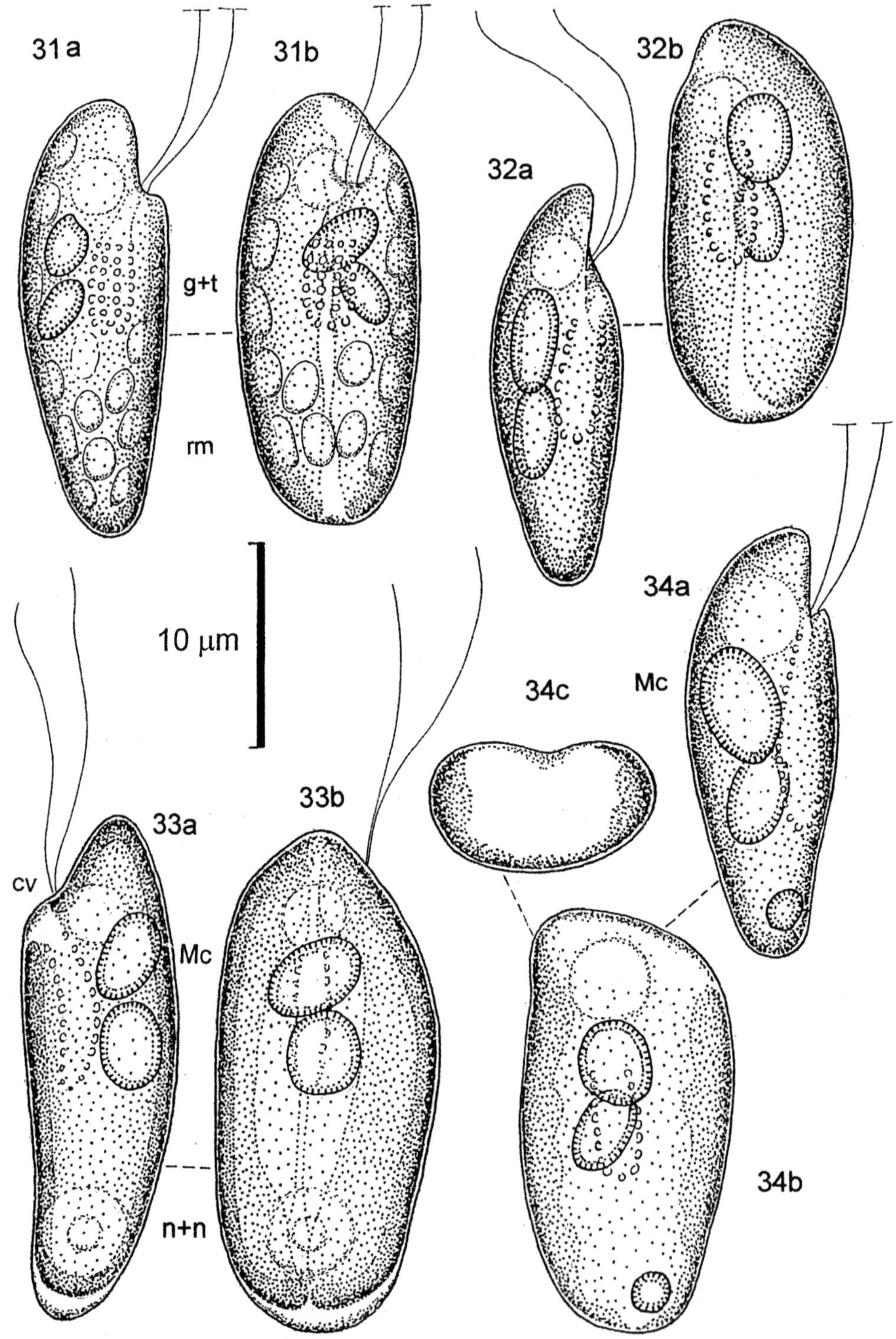

Figure 31–34. Cryptomonas tatrica Czosn. 31a, 32a, 33a, 34a – lateral views; 31b, 32b, 33b – ventral views; 34b – dorsal view; 34c – equatorial cross section.

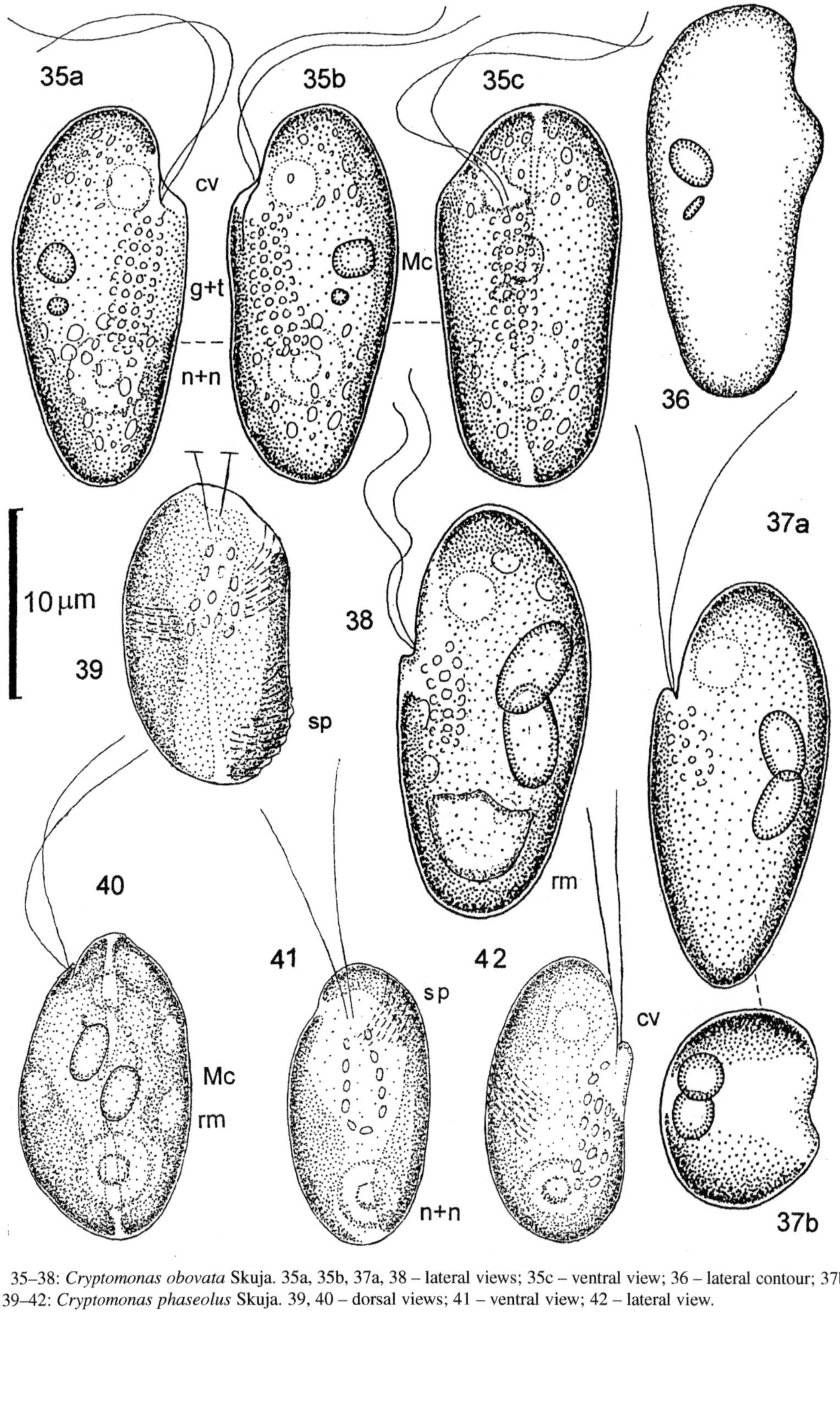

Figure 35–42. 35–38: *Cryptomonas obovata* Skuja. 35a, 35b, 37a, 38 – lateral views; 35c – ventral view; 36 – lateral contour; 37b – equatorial cross section. 39–42: *Cryptomonas phaseolus* Skuja. 39, 40 – dorsal views; 41 – ventral view; 42 – lateral view.

Except for the size, the cells of *C. reflexa* are not deformed or spirally distorted.

Two pyrenoids have been described for the type species *Campylomonas reflexa* (Marsson) Hill. Hill (1991) commented that they are inconspicuous under a light microscope. Limnologists who frequently observe phytoplankton preserved with iodine (KJ + J solution) may sometimes find the sigmoid cryptomonad with pyrenoids along with the vast majority of the specimens without any visible pyrenoids, as I did. This may suggest the existence of two taxons of *Cryptomonas (Campylomonas) reflexa*.

Another member of the frequent plankton assemblage is *Cryptomonas marssonii* Skuja. Some authors doubt this species is justifiable but one can easily recognize it. It is smaller, even though the dimensions of *C. reflexa* and *C. marssonii* overlap: the cells of *C. marssonii* in my materials were 13.5–24 μm (average 19 μm) long (Figs 13–15). Above all, however, the antapices of the cells are different: that of *C. marssonii* is again curved dorsally but sharply pointed particularly in the side projection. Maupas corpuscles are almost always present; they are relatively small, sometimes of irregular shape. Pyrenoids were not observed.

There is one more diacritical characteristic: while the cells of *C. reflexa* are slightly laterally flattened, or are not transversally flattened at all, those of *C. marssonii* are flattened dorsoventrally. This can be easily determined because of the dorsal curvature of the posterior tail.

C. reflexa and *C. marssonii* may be found in open pelagial, in littoral growth or in shallow, overgrown pools. Beside the typical specimens, however, there are special shortened forms of both these species sampled from shallow pools with decaying leaves and water containing humic acids and sulfan (H_2S). The form related to *C. reflexa* was described as *Cryptomonas anas* Javornický (1957) (Figs 9–12 in the present publication). The short and relatively wide cells of this species are similar to those of *C. reflexa*. They are sigmoid, but at the same time bag-shaped, resembling the body of a duck (*Anas* in Latin). They were 12.5–21 μm (average 18.2 μm) long; for *C. reflexa* was the average length 26.9 μm. The width (dorsoventral) was practically the same for both species, so that the ratio of length: width was 2.52 for *C. reflexa* and 1.93 for *C. anas*. The cells of both species were laterally slightly flattened or they were not flattened at all.

Similar 'small-water pendant' was found also for *C. marssonii* (Figs 16–17). Again the cells of this form are relatively wider. They are sigmoid and possess more or less sharp, dorsally curved antapices. One to two 2 Maupas corpuscles may be disposed also at the ventral inner wall of the cell. The cells were on average 16.6 μm long, dorsoventrally flattened, as in typical form. The ratio of length: width was on average 2.42 for *C. marssonii* and 1.95 for the small water form.

Freshwater species of the genus Plagioselmis (Rhodomonas *p.p.)*

The above named cryptomonads are in the pelagic waters of large reservoirs regularly accompanied by the minute autotrophic flagellates belonging to the genus *Plagioselmis* Butcher (1967). These organisms were initially described by Pascher and Ruttner in Pascher (1913) as *Rhodomonas lacustris* and *R. lens*. Later the similar species obtained different names. The most accepted were the excellent drawings by Skuja (1948) and consequently the names *R. minuta* Skuja and *R. minuta* var. *nannoplanctica* Skuja prevailed. Since the original description in 1913, the presence of the ventral furrow with superficial rows of trichocysts (ejectisomes) and the absence of any gullet were clear. Nevertheless, all authors (including Javornický, 1967, 1976) kept the wrong original classification of this organism as *Rhodomonas*.

Klaveness (1981) published a thorough electron-microscopic study on *Rhodomonas lacustris*. His specimens lacked a gullet but had a deep ventral furrow in which more than 20 relatively large ejectisomes were located. Novarino et al. (1994) studied and compared the ultrastructures of three strains and accepted for the genus the name *Plagioselmis* Butcher emended Novarino, Lucas et Morrall. The quoted authors used for the freshwater species the name of Skuja's (1948) variety and combined it as *Plagioselmis nannoplanctica* (Skuja) Novar., Lucas et Morr. comb. et stat. nov.

Butcher's (1967) separating the rhodomonads with superficial rows of trichocysts but without a gullet from the genus *Rhodomonas* (according to him from *Cryptomonas*) into an independent genus *Plagioselmis* is certainly correct, even though he erroneously stated the absence of any sulcus for the other generic feature.

On the other hand, the narrowed concept of the genus emended by Novarino et al. "with a posterior, often ventrally bent tail" and the "discrete periplast areas absent in the region of the tail" excludes the specimens without any distinct tail. Such species were described as *Rhodomonas lacustris* Pasch. et Ruttn., *R. minuta* Skuja, and *Plagioselmis punctata* Butch.

I expressed the opinion that the existence of an additional genus for specimens without any tail would be superfluous (Javornický, 2001). Therefore, I proposed, (1) to keep the wider definition of the genus *Plagioselmis* as given by Butcher except his "sine sulco" (without furrow); and (2) to use the presence or absence of the sharp hyaline tail for discerning the freshwater species of *Plagioselmis*.

Consequently, the common minute freshwater rhodomonads belong to the species *Plagioselmis nannoplanctica* (Skuja) Novar., Lucas et Morr. and *Plagioselmis lacustris* (Pasch. et Ruttn.) Javorn. (Javornický, 2001). These species are not discernible by their dimensions varying within the range 5.0–14.0 μm of length. The difference between them, visible by means of a light microscope, is in their cell shape: cells of *P. nannoplanctica* possess the sharp hyaline tail (Figs 21–24), those of *P. lacustris* have a narrowed but rounded antapex without any tail (Figs 18–20 and 25). The well visible single shelled pyrenoid is present in all cells.

Some phycologists make a difference between *Rhodomonas lacustris* and *R. lens* Pasch. et Ruttn. (Willén et al., 1980). The latter organism was so rare in my materials that it has not been possible to decide if the species *R. lens* really exists or if it belongs to the range of variability of *P. lacustris* (Figs 26 and 27).

The chloroplasts of the freshwater species of *Plagioselmis* are yellow, brownish or amber; I never found them wine red as it is typical for *Rhodomonas* Karsten.

Concerning ecology, both *P. nannoplanctica* and *P. lacustris* are very common in plankton of lakes and dam reservoirs in Europe and North America (for localities and authors see e.g. Lund, 1962; Javornický, 1976). In my experience, the two species in question occur frequently in the same phytoplankton assemblage with the prevalence of one of them, mostly of *P. nannoplanctica*.

Cryptomonas ovata *Ehr. and some affiliated species*

Cryptomonas ovata Ehrenberg (1838) and related species belong very probably to "the true" genus *Cryptomonas* concerning their ultrastructures. *C. ovata*, type of the genus *Cryptomonas* Ehr., is not a collective species, as is supposed by different authors.

C. ovata is not so frequent in nature as is indicated in the literature and one can hardly find this species in pelagial phytoplankton. It lives in peat bogs, overgrown pools and little ponds. It is apparent by the large cells which have a slightly peculiar shape. It is a rather sizable species with conspicuously crosswise flattened cells.

I was finding the large cryptomonads (31–45 μm long) corresponding to the concept of *Cryptomonas ovata* Ehr. as pictured by Skuja (1939) in the above mentioned biotopes (Javornický, in print). The chloroplast covered nearly the whole inner wall of the periplast, being divided only with narrow slots. Its colouring varied from yellow-brown to olive green. Two conspicuous relatively large Maupas corpuscles are present in every cell. No shelled pyrenoid was observed.

The cells were conspicuously dorsoventrally flattened, in dorsal or ventral views broadly oval (Figs 28a and 29a), their long axis not being straight, but slightly bowed to the right or left side. The cell apex was slightly notched, the antapex widely rounded. The mouth of the vestibule may be shifted to the right or to the left side of the cell. In the lateral view (Figs 28b and 29b) the cells were not only more slender, but they also carried the pointed apical rostrum slightly curved over the ventral vestibule. Consequently, they resemled the slender half-moon and the shape of the whole cell may be compared to the content of a full spoon.

Almost identical species but flattened laterally (Fig. 30) was described by Czosnowski (1948) as *Cryptomonas splendida*. This species very probably belongs to the variability of *C. ovata*.

Some authors give the range of the dimensions of *C. ovata* much wider than is given above. I have never found in my materials specimens 50 μm long or longer. On the other hand, I frequently found cryptomonads of similar morphology being only 20–24.5 μm long (Figs 31–34). The cells always had two conspicuos Maupas corpuscles but no pyrenoid. Because the cell lengths of the two groups in my materials did not overlap, and taking some details of their habitus into account, it is proposed to consider them for two different species (Javornický, in print). The flattened cryptomonads, ellipsoidal in dorsoventral views and slender in the lateral views, particularly smaller than *C. ovata* (or *C. splendida)* were described by Czosnowski (1948) under the name *C. tatrica*. It is proposed to use the name *C. tatrica* Czosn. for "the small *C. ovata*". The species was named according to the finding place, i.e. swamps in the Polish Tatra Mountains. I sampled this cryptomonad from peat or forest pools and watered ditches.

Cryptomonas obovata Skuja is another cryptomonad with the habitus similar to *C. ovata*. I found this species repeatedly in pools, swamps and small fishponds (Figs 35–38). It is a middle-sized cryptomonad, not flattened or only slightly flattened dorsoventrally. Its size variability was narrow in my materials: 19–25.6 μm long, 9–13 μm wide and thick. The cell form is obovoid from the side projection, widest under the front apex, forming sometimes a 'belly' on the ventral contour (Fig. 36). The apex is rounded, sometimes in the form of a helmet. The mouth of the gullet is conspicuously subapical. The Maupas corpuscles are of different size but always present. No shelled pyrenoid was observed.

The spherical palmelloid stages of *C. obovata* were found to make part of the neustonic membrane on the surface of a fishpond.

Skuja (1948) described another species of an ellipsoidal cryptomonad. For *Cryptomonas phaseolus* Skuja he gave small dimensions (12–15 μm long, 7–8 μm wide), two lateral chloroplasts, no true pyrenoid and only sometimes 2–3 Maupas corpuscles. Skuja wrote: "*C. phaseolus* gehört zu der Kolektivspezies C. erosa... auf der Grenze zwischen *C. erosa* und *C. ovata*..."

I studied a very similar cryptomonad isolated by F. Hindák from the soil in the inundation area of the Dyje (Thaya) River close to the town of Lednice, Moravia, Czech Republic (Javornický, in print). The ellipsoidal cells were not flattened (Figs 39–42). The pseudopharynx (gullet) was rather deep, inlaid with relatively large ejectisomes (trichocysts). Olive-green up to dark brown chloroplast was a single one in each cell, being split on the dorsal part almost in two, but always joined by a narrow isthmus. No visible pyrenoid was present and also Maupas corpuscles could be seen only on exception (Fig. 40). When treated with iodine, the surface striate structure was visible in some parts of cells. It could cause the slightly undulated contour of the cell (Figs 39–42). The cells of this soil cryptomonad were 13.5–16.5 (average 15.0) μm long, 7–10 (average 8.9) μm wide and thick.

The minute species oval *C. phaseolus* may hardly be confused with *C. ovata*, but it is sometimes the case. Therefore, it is presented here for comparison.

Discussion

The main problem of the comparison of cryptophycean taxa based upon light microscopy with those derived from observations by electron microscopy may be the uncertain sources of name labels of culture strains. EM studies sometimes attribute their exact findings to vaguely identified strains. For example, some authors worked with the so-called *Cryptomonas ovata* var. *palustris* Pringsheim kept in "Sammlung von Algenkulturen in Göttingen" as the strain B 979-3 (Schlösser, 1982: 195). The strains used e.g. by Faust (1974), Roberts (1984), Santore (1978, 1984, 1985), Kugrens et al. (1986), Kugrens & Lee (1987) originated probably from this old culture of Pringsheim (1968: 388).

The cells of Pringsheim's strain are middle sized (18–24 μm long) "mit breit-elliptischem Querschnitt", bearing two conspicuous shelled pyrenoids in each cell. The short inspection of Pringsheim's drawing (1968: Tab. 148,1) reveals that it shows *Cryptomonas pyrenoidifera* Geitler (1922), excellently re-pictured by Skuja (1948).

Hill (1991) compiled the electron-microscopic data on the so-called *Cryptomonas ovata*. He takes *C. ovata* for lectotype of the genus and the characteristics, he found, should determine the genus from the viewpoint of ultrastructures. They are: The inner component of the periplast "consists of discrete plates that are roughly oval in shape" – "the superficial component consists of a thin layer of very fine fibrils" – "the flagellar apparatus of *C. ovata* possesses a rhizostyle with a peculiar structure" (p.173).

All these features are discernible only under an electron microscope. Hill gave also two rather large shelled pyrenoids attached to the lateral chloroplasts. This phenomenon could hardly have been overlooked by so many light-microscopists when in some other species (*C. pyrenoidifera* Geitl., *C. parapyrenoidifera* Skuja, *C. tetrapyrenoidosa* Skuja, let alone other genera like *Rhodomonas* or *Chroomonas*), the shelled pyrenoids are very clearly visible.

Hill did not mention the source of his schematic summarizing drawing of *C. ovata* (p. 172, Fig. 1). He did not even give its dimensions. The 5 μm-scale in his table indicates cell length at about 23 μm, which again, together with two shelled pyrenoids, indicates *Cryptomonas pyrenoidifera* Geitl., i.e. the isolate by Pringsheim (*C. pyrenoidifera*, erroneously ascribed to Skuja, is listed by Hill in "allied species").

The flagellates examined by some authors are too small to represent *C. ovata* at all (Brett & Wetherbee, 1986: 11 μm long, 9 μm wide). Their organism was very probably related to *Cryptomonas phaseolus* Skuja.

Many other details found by electron microscopists for the so-called *C. ovata* Ehr. may be quoted.

Of course, the value of these data for characterizing this species is problematic so far since the authors worked with vaguely identified strains. Their findings should be attributed mostly to *Cryptomonas pyrenoidifera* Geitl. or to *C. phaseolus* Skuja. The type species of the genus, *Cryptomonas ovata* Ehr. first needs to be defined unambiguously with the basic characters based upon light microscopy and then studied in detail. Conceivably some ultrastructures found for the above mentioned strains of the so-called "*C. ovata*" could appear to be valid also for the correctly interpreted species *C. ovata* Ehr. em. Stein and for its affiliated species, but so far we cannot be sure of it.

References

Brett, S. J., L. Perasso & R. Wetherbee, 1994. Structure and development of the cryptomonad periplast: a review. Protoplasma 181: 106–122.

Brett, S. J. & R. Wetherbee, 1986. A comparative study of periplast structure in *Cryptomonas cryophila* and *C. ovata* (Cryptophyceae). Protoplasma 131: 23–31.

Butcher, R. W., 1967. An introductory account of the smaller algae of British coastal waters. 4. Cryptophyceae. Fishery Investigations, Ser. 4: London. 54 pp.

Czosnowski, J., 1948. Materialy do flory wiciowców Polski (Materials for the flora of flagellates of Poland). Poznañskie Towarzystwo Przyjaciol Nauk 11. 57 pp.

Ehrenberg, Ch. G., 1838. Die Infusionsthiere als vollkommene Organismen, Berlin und Leipzig. 547 pp.

Ettl, H., 1965. Untersuchungen an Flagellaten. Österreichische Botanische Zeitschrift 112: 701–745.

Ettl, H., 1980. Beitrag zur Kenntnis der Süsswasseralgen Dänemarks. Botaniska Tidsskrift 74: 179–223.

Faust, M. A., 1974. Structure of the periplast of *Cryptomonas ovata* var. *palustris*. J. Phycol. 10: 121–124.

Geitler, L., 1922. Die Microphytenbiocoenose der Fontinalis-Bestände des Lunzer Untersees und ihre Abhängigkeit vom Licht. Int. Rev. ges. Hydrobiol. 10: 683–691.

Hill, D. R. A., 1991. A revised circumscription of *Cryptomonas* (Cryptophyceae) based on an examination of Australian strains. Phycologia 30: 170–188.

Huber-Pestalozzi, G., 1950. Das Phytoplankton des Süsswassers. Systematik und Biologie 3. Cryptophyceen, Chloromonadinen, Peridineen. In Thiemannn, A. (ed.), Die Binnengewässer 16, 3. Schweizerbart'sche Verlagsbuchhandlung, Stuttgart. 310 pp.

Javornický, P., 1957. Několik nových a malo známých bičíkovců z kmene Pyrrhophyta (Some new and scarcely known flagellata from the phyllum Pyrrhophyta). Universitas Carolina, Biologica 3: 251–268.

Javornický, P., 1967. Some interesting algal flagellates. Folia geobotanica et phytotaxonomica 2: 43–67.

Javornický, P., 1976. Minute species of the genus *Rhodomonas* Karsten (Cryptophyceae). Archiv für Protistenkunde 118: 98–106.

Javornický, P., 2001. Freshwater Rhodomonads (Cryptophyceae). Algol. Stud. 102: 93–116.

Javornický, P., 2003. *Cryptomonas ovata* Ehrenberg (Cryptophyceae) and some related species. Algological Studies (in print).

Kiselev, I. A., 1931. Zur Morphologie einiger neuer und seltener Vertreter des pflanzlichen Mikroplanktons. Archiv für Protistenkunde 73: 246–260.

Klaveness, D., 1981. *Rhodomonas lacustris* (Pascher and Ruttner) Javornický (Cryptomonadida): ultrastructure of the vegetative cell. J. Protozool. 28: 83–90.

Klaveness, D., 1989. Biology and ecology of the Cryptophyceae: Status and challenges. Biol. Oceanogr. 6: 257–270.

Kugrens, P. & R. E. Lee, 1987. An ultrastructural survey of cryptomonad periplasts using quick-freezing freeze fracture techniques. J. Phycol. 23: 365–376.

Kugrens, P., R. E. Lee & R. A. Andersen, 1986. Cell form and surface patterns in *Chroomonas* and *Cryptomonas* cells (Cryptophyta) as revealed by scanning electron microscopy. J. Phycol. 22: 512–522.

Lund, J. W. G., 1962. A rarely recorded but very common British alga, *Rhodomonas minuta* Skuja. British Phycol. Bull. 2: 133–139.

Marsson, M., 1904. Die Abwasser-Flora und -Fauna einiger Kläranlagen bei Berlin. Mitteilungen an der Königlichen Prüfungsanstalt für Wasserversorgung und Abwässerbeseitigung zu Berlin, 4.

Novarino, G., I. A. N. Lucas & S. Morrall, 1994. Observations on the genus *Plagioselmis* (Cryptophyceae). Cryptogamie, Algologie 15: 87–107.

Pascher, A., 1913. Cryptomonadineae. In Pascher, A. & E. Lemmermann (eds), Süsswasserflora Deutschlands. Österreichs und der Schweiz 2, Flagellatae 2: Jena. 96–114.

Penard, E., 1922. Studies on some Flagellata. Proc. Acad. Nat. Sci. Phil. 73: 105–168.

Pringsheim, E. G., 1968. Zur Kenntnis der Cryptomonaden des Süsswassers. Nova Hedwigia 16: 367–401.

Roberts, K. R., 1984. Structure and significance of the cryptomonad flagellar apparatus. I. *Cryptomonas ovata* (Cryptophyta). J. Phycol. 20: 590–599.

Santore, U. J., 1978. Light- and electron-microscopic observations of the palmelloid phase in members of the genus *Cryptomonas* (Cryptophyceae). Archiv für Protistenkunde 120: 420–435.

Santore, U. J., 1984. Some aspects of taxonomy in the Cryptophyceae. New Phytol. 98: 627–646.

Santore, U. J., 1985. A cytological survey of the genus *Cryptomonas* (Cryptophyceae) with comments on its taxonomy. Archiv für Protistenkunde 130: 1–52.

Schlösser, U. G., 1982. Sammlung von Algenkulturen, Pflanzenphysiologisches Institut der Universität Göttingen. Berichte der Deutschen Botanischen Gesellschaft 95: 181–276.

Skuja, H., 1939. Beitrag zur Algenflora Lettlands 2. Acta Horti botanici Universitatis Latviensis 11/12: 41–169.

Skuja, H., 1948. Taxonomie des Phytoplanktons einiger Seen in Uppland, Schweden. Symb. Bot. Ups. 9: 1–399.

Stein, F., 1878. Der Organismus der Infusionsthiere 3, 1. Leipzig. 154 pp.

Troitzkaja, O. V., 1922. O novoj kryptomonade (De nova Cryptomonadinearum species). Botanièeskie materialy Instituta sporovych rastenij 1, 8.

Willén, E., M. Oké & F. Gonzáles, 1980. *Rhodomonas minuta* and *Rhodomonas lens* (Cryptophyceae) – aspects on form-variation and ecology in Lakes Mälaren and Vättern, central Sweden. Acta phytogeographica Suecica 68: 163–172.

Hydrobiologia **502:** 285–296, 2003.
L. Naselli-Flores, J. Padisák, M. T. Dokulil (eds),
Phytoplankton and Equilibrium Concept: The Ecology of Steady-State Assemblages.

Ecological and taxonomic observations on the flagellate algae characterising four years of enclosure experiments in Lake Tovel (Southern Alps)

Massimiliano Tardio[1], Monica Tolotti[1,2], Gianfranco Novarino[3] & Marco Cantonati[1]
[1]*Museo Tridentino di Scienze Naturali, Limnology and Phycology Section, Via Calepina 14, I-38100 Trento, Italy*
E-mail: tardio@mtsn.tn.it
[2]*Institute of Zoology & Limnology, University of Innsbruck, Technikerstrasse 25, A-6020 Innsbruck, Austria*
[3]*Department of Zoology, The Natural History Museum, Cromwell Road, London SW7 5BD, U.K.*

Key words: Lake Tovel, flagellate algae, enclosures, ecology, taxonomy

Abstract

Lake Tovel (Italian Alps, Brenta Dolomites, 1178 m a.s.l.), with a dimictic mixolimnion, is a meromictic and oligotrophic mountain lake, once notorious for the summer reddening of its waters due to dinoflagellate blooms that ceased suddenly in 1964. Since 1997 the Museo Tridentino di Scienze Naturali has been carrying out *in situ* experiments in order to understand the factors regulating the blooms. Experiments were carried out in different types of enclosures to test the reaction of phytoplankton to phosphorus enrichment and light reduction, with special reference to *Glenodinium sanguineum* Marchesoni, the dinoflagellate responsible for the reddening. The aim of the present contribution is to study the ecology of other flagellate algae typical of Lake Tovel, by analysing the data from the enclosure experiments. Particular attention is given to competition for resources, the comparison with *G. sanguineum* and some interesting taxonomic observations. The four flagellate taxa selected for this investigation were *Gymnodinium uberrimum* (Allman) Kofoid & Swezy, *Dinobryon* cf. *sociale* var. *americana* (Brunnthaler) Bachmann, *Campylomonas* sp. and *Tetraselmis* sp. Number of individuals and biovolume calculated for each taxon during the experiments were statistically related to NO_3–N and TP concentrations, water temperature, pH and oxygen saturation. A prompt and positive response to P-enrichment was found in all experiments for both *G. sanguineum* and *G. uberrimum*. However, the first taxon showed a pronounced development only when nutrients were coupled with high light intensities, while the latter appeared to be a better competitor in low light intensities and low water temperature. The behaviour of *Tetraselmis* sp. was similar to that of *G. sanguineum*, while the development pattern of *Campylomonas* sp. was similar to that of *G. uberrimum*. *D. sociale* did not show a clear relationship either to nutrient enrichment or weather conditions.

Introduction

Lake Tovel (Italian Alps) is notorious because of the summer reddening of its waters – due to blooms of the dinoflagellate *Glenodinium sanguineum* Marchesoni – that ceased suddenly in 1964 (Baldi, 1941; Tomasi, 1989; Paganelli, 1992). Since 1997, the Museo Tridentino di Scienze Naturali has been carrying out *in situ* experiments in order to understand the factors regulating the blooms (Tolotti et al., 1999; Tardio et al., 2001; Cantonati et al., 2003). In addition to field observations and laboratory experiments based on algal cultures, *in situ* experiments may be a useful method for better understanding several aspects of phytoplankton ecology. These experiments are usually carried out in isolated portions of water called enclosures. The use of enclosures has become widespread in several countries during the last few decades (e.g. Riemann & Sondergaard, 1986; Riemann et al., 1988; Chorst & Riemann, 1994; Halac et al., 1997; Vinebrooke & Leavitt, 1999). However, the use of this technique is still limited in Italy (e.g. Ravera, 1989, 1990; Premazzi et al., 1999), particularly in mountain lakes (e.g. Paris et al., 1995). In these lakes small flagellates usually dominate the phytoplankton because of their shorter replication time (Tilzer, 1972, 1973) and their ability to migrate in the water column between the euphotic zone during the day and the nutrient richer hypolimnion during the night (Salonen & Rosenberg, 2000). Several species of Chrysophyceae, Dinophyceae, and probably Cryptophyceae are able to improve their nutrient uptake by mixotrophy (Porter, 1988; Tranvik et al., 1989), which allows them to develop in lakes limited by low nutrient concentrations and low light intensity (Jansson, et al. 1996). Owing to these physiological characteristics, flagellate algae are particularly well adapted to the extreme environmental conditions typical of mountain lakes (e.g. thermal condition, light intensity and nutrient availability, cf. Pechlaner, 1971), as shown by the fact that they are usually the most important algal groups in mountain lakes from a quantitative point of view (Rott, 1988). Therefore, flagellate algae must be considered closely in any study aimed at understanding ecological processes in mountain lentic environments.

The aims of the present contribution are: (i) to analyse the ecological requirements of a selection of flagellate phytoplankton algae of Lake Tovel, with particular attention to inter-specific competition for resources and a comparison with *G. sanguineum;* (ii) to provide some taxonomic observations on the Cryptophyceae, whose taxonomy is undergoing constant revision (Hill, 1991; Novarino et al., 1994; Novarino, 2003; Javornický, 2001; Javornický, 2003; Novarino, 2003; Javornický, 2003).

Description of sites studied

Lake Tovel is located at 1178 m a.s.l. in the northern part of the Brenta Dolomites (also known as Western Dolomites) in the Tovel Valley, one of the most renowned sites of the Adamello-Brenta Natural Park. The lake surface is ca. 0.39 km^2, while the drainage basin is slightly larger than 39 km^2 (Baldi, 1941). The lake consists of one main deep NE basin (max. depth = 38.5 m) and a shallow SW basin also called 'Red Bay' (max. depth = 5 m). The water renewal time is variable and short (from 18.5 to 100 days) and the lake is characterised by marked seasonal water level fluctuations (Paganelli, 1992). Lake Tovel is a meromictic (monimolimnion at about 30 m with dimictic mixolimnion) and oligotrophic lake: TP = 3–6 $\mu g\ l^{-1}$, NO_3-N = 280–420 $\mu g\ l^{-1}$, mean Secchi disk = about 10 m (Paganelli, 1992; I.A.S.M.A., 1998; Corradini et al., 2000).

Materials and methods

The enclosure experiments considered in this contribution were carried out in the 'Red Bay' from 1998 to 2001. Enclosures of two different shapes were used (Cantonati et al., 2002), both made of flexible and transparent polyethylene (0.5 mm thickness): 'cylinders' (1998, 2000 and 2001) open at the top and bottom, with a diameter of 3 m and a length of 4 m, and a capacity of about 25 m^3; and 'pools' (1999) shaped as an irregular parallelepiped (3 × 1 m), with a closed bottom which follows the inclination of the shore bottom and with only the upper part opening towards the atmosphere. The capacity is ca. 3 m^3. The experiment features are summarised in Table 1. Temperature and solar radiation values were recorded by the meteorological station run by the Agrarian Institute of S. Michele all'Adige and available on the Internet site http://relay.ismaa.it:12164/meteo/dati.html. Physical and chemical parameters (water temperature, oxygen saturation, pH, conductivity and redox potential) were measured with a Hydrolab H20 probe. Phosphorus enrichments were obtained by adding a concentrated Na_2HPO_4 solution 35÷40 mM, while nitrogen additions were obtained by means of a $mg(NO_3)_2{*}6H_2O$ solution 0.36 M. In 2000, enclosure C was covered with a darkening sheet in order to produce a reduction of light intensity of about 60%. Conversely, in 2001 an attempt to increase incident light was carried out using 16 neon lamps (OSRAM L 18W/77 FLUORA). Light extinction in the enclosures was measured with a LI-COR (LI-1400) quantum sensor. Water samples for chemical and biological analyses were taken with a 5 l Patalas-Schiendler sampler at 0.5 m and 2.5 m in all enclosures. Laboratory chemical analyses were carried out following I.R.S.A. – C.N.R. (1994). Water samples for chlorophyll *a* analyses were filtered in the field within a few hours on Whatman GF/C 0.45 μm filters, which were then stored at –20 °C until extraction in 90% acetone and spectrophotometric analyses (Lorenzen, 1967). Phytoplankton samples were preserved with Lugol's solution acidified with 10% of CH_3COOH. Counts were carried out using an OLYMPUS IX70 inverted microscope according to Utermöhl (1958). At each magnification, at least 100 individuals of the dominant taxon were counted, which corresponds to an error of about 20% (Lund et

Table 1. Outline of the experiments carried out in the years 1998–2001 in Lake Tovel

Year	Period	Structure	No of encl.	Experimental modifications			
				Nutrients enrichment	Date	Light modific.	Date
1998	23/07 – 02/08	Cylinders	3	Phosphorus	B: +5 $\mu g\,l^{-1}$	23/7	
			A,B,C		C: +15 $\mu g\,l^{-1}$	23/7	
1999	09/08 – 09/09	Pools	3	Phosphorus	B: +20 $\mu g\,l^{-1}$	9/8	
			A,B,C		C: +40 $\mu g\,l^{-1}$	9/8	
2000	03/07 – 27/07	Cylinders	3	Phosphorus B, C: +20 $\mu g\,l^{-1}$	3/7 & 13/7	C: shaded	3/7
			A,B,C	Nitrogen B, C: +200 $\mu g\,l^{-1}$	13/7		
2001	17/07 – 24/08	Cylinders	3	Phosphorus B, C: +20 $\mu g\,l^{-1}$	17/7 & 26/7	C: neon lamps	30/7
			A,B,C				

al., 1958). Algal biovolume was calculated from linear dimensions according to Rott (1981).

Densities and biovolumes of four flagellate taxa (*Gymnodinium uberrimum* (Allman) Kofoid & Swezy, *Dinobryon* cf. *sociale* var. *americana* (Brunnthaler) Bachmann, *Campylomonas* sp. and *Tetraselmis* sp.) are considered in this paper. These taxa were selected based both on their importance in the phytoplankton communities of Lake Tovel (in terms of constant presence in the phytoplankton samples and relative abundance) and the information provided by previous ecological investigations on ecologically similar taxa (e.g. Dokulil, 1988; Rott, 1988; Eloranta, 1989; Hill, 1991). Taxonomic identification was based largely on scanning electron microscopy (SEM). Samples fixed in 2.5% glutaraldehyde or Lugol's iodine were subsequently osmicated, dehydrated and dried following existing protocols for cryptophytes and other planktonic nanoflagellates (Novarino, 1991; Novarino et al., 1997). Observations were carried out on material previously sputter-coated with a 20 nm layer of gold-palladium, using a field-emission Philips XL 30 SEM operating at 5 kV.

Densities and biovolumes of the four selected taxa were elaborated statistically by calculating correlation coefficients and one-way analyses of variance to test for any significant difference among the averages densities and biovolumes in the different experimental structures (ANOVA 1; Sokal & Rohlf., 1981). The test results were checked for significance at $p < 0.001$ or $p < 0.05$.

Results

Average, standard deviation and median of physical and chemical variables (Table 2)

The highest temperature values were recorded in summer 1998 (mean = 16.1 °C), the lowest in summer 2000 (mean = 12.7 °C). Solar radiation was similar in summer 1998, 2000 and 2001 (mean = 0.72–0.74 Mj $m^{-2}h^{-1}$), while during summer 1999 weather conditions were poor with low irradiance (mean = 0.55 Mj m^{-2} h^{-1}) (cf. Eccel, 2000). Water temperature in the different years was influenced not only by meteorological conditions but also by the position of the enclosures in the lake. In 1999, mean temperatures around 15.5 °C were measured in very shallow waters close to the shore. In the other years, when the cylinders were placed in the deeper waters of the W shore of the Red Bay average temperatures were between 10.7 and 13.1 °C. The highest values of oxygen saturation were observed in 2000 (mean = 132 in enclosure C, the one that had been shaded), the lowest in 1999 in the pool enclosure A (mean = 102). The moderately alkaline pH values typical for Lake Tovel (around 8.3, Corradini et al., 2000) were found also within the enclosures and increased up to about 9, usually in good accordance with changes in phytoplankton biovolume during the experiments. Total phosphorus concentrations reflected the nutrient additions to the enclosures, while nitrate-nitrogen concentrations indicated that the consumption of this nutrient was highest in the enclosures showing the highest increases of phytoplankton biovolume.

Table 2. Table 2. Mean (m), standard deviation (s) and median of the physical and chemical parameters in 1998-2001 in Lake Tovel. $*$ = in this enclosure light had been artificially attenuated by about 60%. [a]= data obtained from the Internet site *http://relay.ismaa.it:12164/meteo/dati .html* run by the Agrarian Institute of S. Michele all'Adige

	Physical & chemical conditions	Encl. A			Encl. B			Encl. C		
		m	s	median	m	s	median	m	s	median
	Air temperature (°C)[a]	16.1	3.4	15.5	16.1	3.4	15.5	16.1	3.4	15.5
	Solar radiation (Mj m^{-2} h^{-1})[a]	0.72	0.98	0.20	0.72	0.98	0.20	0.72	0.98	0.20
	Water temperature (°)	12.7	0.4	12.6	12.7	0.4	12.6	12.7	0.4	12.6
1998	& O_2 saturation	113	4	113	114	4	113	116	4	113
	pH	8.8	0.08	8.8	8.9	0.08	8.8	9	0.08	8.8
	Total P (μg l^{-1})	10	4.0	11	15	4	11	31	4	11
	–Nitrate-Nitrogen (μ l^{-1})	397	108	350	287	108	350	226	108	350
	Air temperature (°)[a]	13.8	3.1	13.7	13.8	3.1	13.7	13.8	3.1	13.7
	Solar radiation (Mj m^{-2} h^{-1})[a]	0.55	0.83	0.10	0.55	0.83	0.10	0.55	0.83	0.10
	Water temperature (°C)	15.6	1.9	15.3	17.7	2.0	15.4	15.7	1.9	15.5
1999	% O_2 saturation	102	3	102	105	4	105	109	9	104
	pH	8.6	0.2	8.6	8.7	0.3	8.7	8.7	0.3	8.8
	Total P *μ l^{-1})	16	9	20	32	10	37	40	12	41
	–Nitrate-Nitrogen (μg l^{-1})	237	9	20	193	141	170	176	144	150
	Air temperature (°C)[a]	12.7	4.2	12.4	12.7	4.2	12.4	12.7	4.2	12.4
	Solar radiation (Mj m^{-2} h^{-1a}	0.74	1.04	0.10	0.74	1.04	0.10	0.74*	1.04*	0.10*
	Water temperature (°C)	13.1	1.2	12.8	13.0	1.7	13.2	13.0	1.2	12.8
2000	% O_2 saturation	116	4	116	127	9	129	132	13	135
	pH	8.6	0.1	8.6	8.8	0.2	8.9	8.8	0.2	8.9
	Total P (μ l^{-1})	6	2	6	31	11	25	42	21	34
	–Nitrate-Nitrogen (μg l^{-1})	320	13	325	252	84	280	263	69	280
	Air temperature (°C)[a]	15.2	4.2	14.8	15.2	4.2	14.8	15.2	4.2	14.8
	Solar radiation (Mj m^{-2} h^{-1})[a]	0.72	0.99	0.20	0.72	0.99	0.20	0.72	0.99	0.20
	Water temperature (°C)	10.7	0.5	10.8	10.9	0.4	10.8	10.7	0.5	10.8
2001	% O_2 saturation	108	5	109	117	12	109	113	7	109
	pH	8.8	0.1	8.9	9.0	0.3	8.9	8.9	0.2	8.9
	Total P (μ l^{-1})	7	2	6	29	15	6	16	9	6
	Nitrate-Nitrogen (μg l^{-1})	230	38	220	108	123	220	196	50	220

Densities of the selected phytoplankton taxa (Figure 1)

In 1998, *Tetraselmis* sp. was the species showing the highest proliferation (max. density = ca. 2.5 10^6 individuals per litre) in the enclosures, which received the maximum P-enrichment. In 1999 *Campylomonas* sp. reached the highest number of individuals per litre (= 1.5 10^6). *D. sociale* had a marked density increase only at the beginning of each experiment in 1999, 2000 and 2001. *G. uberrimum* did not show changes in density in any of the enclosures.

Biovolumes of the selected phytoplankton taxa (Figure 2)

A marked increase of *G. uberrimum* (characterised by larger cell size compared to the other flagellate algae) was observed in the enclosure with highest enrichment (max = 3 mm^3 l^{-1}) in 1999 and in the shaded enclosure in 2000. The biovolume values confirmed both the increase of *Tetraselmis* sp. and the proliferation of *D. sociale* at the beginning of the experiments. The development of the small *Campylomonas* sp. did not produce a marked biovolume increase.

Table 3. Correlations among algal density and biovolume of four flagellate algae and physical and chemical conditions in 1998–2001 in Lake Tovel.

Gymnodinium — Density

	A98	B98	C98	A99	B99	C99	A00	B00	C00	A01	B01	C01
Water temp. (°C)							−**					
% O_2 saturation					*	**			*		**	
pH									*		**	
Total P ($\mu g\ l^{-1}$)							**		**			−*
NO_3-N ($\mu g\ l^{-1}$)		−*				−**			*		−*	

Gymnodinium — Biovolume

	A98	B98	C98	A99	B99	C99	A00	B00	C00	A01	B01	C01
Water temp. (°C)							−**					
% O_2 saturation					*	**			*		**	
pH									*		**	
Total P ($\mu g\ l^{-1}$)							**		**			−*
NO_3-N ($\mu g\ l^{-1}$)				−*	−*	−**			*		−*	

Dinobryon — Density

	A98	B98	C98	A99	B99	C99	A00	B00	C00	A01	B01	C01
Water temp. (°C)	**								**			
% O_2 saturation												
pH				−*	−**	−**						
Total P ($\mu g\ l^{-1}$)				−*			**	−**			−*	
NO_3-N ($\mu g\ l^{-1}$)	*			**	**	**				**	*	*

Dinobryon — Biovolume

	A98	B98	C98	A99	B99	C99	A00	B00	C00	A01	B01	C01
Water temp. (°C)	**			*		*			*			
O_2 saturation												
pH				−*	−**	−**						
Total P ($\mu g\ l^{-1}$)				−*			**	−**			−*	
NO_3-N ($\mu g\ l^{-1}$)	*			**	**	**				**	*	*

Campylomonas — Density

	A98	B98	C98	A99	B99	C99	A00	B00	C00	A01	B01	C01
Water temp. (°C)							−**					
% O_2 saturation					**	**			**			*
pH						*			*			*
Total P ($\mu g\ l^{-1}$)	−**						**	−**	**	*		
NO_3-N ($\mu g\ l^{-1}$)	−**				−**	−**		−**	−**			

Campylomonas — Biovolume

	A98	B98	C98	A99	B99	C99	A00	B00	C00	A01	B01	C01
Water temp. (°C)							−**					
O_2 saturation					**	**			*	*		
pH						*			*			*
Total P ($\mu g\ l^{-1}$)	−**						**	−**	**	*		
NO_3-N ($\mu g\ l^{-1}$)	−**				−**	−**	−**		−**			

Tetraselmis — Density

	A98	B98	C98	A99	B99	C99	A00	B00	C00	A01	B01	C01
Water temp. (°C)	**								**			
% O_2 saturation						**			−*			
pH	*			−**								
Total P ($\mu g\ l^{-1}$)				−**					−**			
NO_3-N ($\mu g\ l^{-1}$)	**					−**						

Tetraselmis — Biovolume

	A98	B98	C98	A99	B99	C99	A00	B00	C00	A01	B01	C01
Water temp. (°C)	**								*			
O_2 saturation						**						
pH	*			−**								
Total P ($\mu g\ l^{-1}$)				−**					−**			
NO_3-N ($\mu g\ l^{-1}$)	**					−**						

□ = not significant (n.s.).
* = $0.01 < P < 0.05$ significant.
** = $0.001 < P < 0.01$ highly significant.

Correlations between algal density and biovolume and physical and chemical conditions

In 1999, *Gymnodinium uberrimum* densities and biovolumes were positively correlated with oxygen saturation and negatively correlated with NO_3-N concentrations (Table 3). In 2000 the same species correlated positively with several chemical variables in the shaded enclosure.

Dinobryon cf. *sociale* showed a marked negative correlation with pH values and total phosphorus and positive correlation with NO_3-N in all years. *Campylomonas* sp. showed a marked development in summer 1999 and in the shaded enclosure in 2000. In both experiments, the species was positively correlated with the parameters directly related with the algal proliferation (TP, pH and% O_2 saturation). A negative correlation was found with NO_3-N (Table 3).

Tetraselmis sp., whose abundance was relevant in the enriched enclosures only in summer 1998, did not show any significant correlation with any of the physical or chemical parameters.

Relationships within the flagellate taxa during the experiments

The increase in the biovolume of the four taxa observed in the different enclosures and in the different years were compared (see Table 4 for significance levels). The biovolume changes of *Gymnodinium uberrimum* in the different enclosures were significantly correlated to each other, especially in the enclosures of 1999. A similar positive relationship among the different enclosures was found for *Dinobryon* cf. *sociale* in the years 1999 and 2001. In summer 2000, a positive relationship between the two enriched enclosures was found. The abundances of *Campylomonas* sp. showed positive correlations with each other during the summer characterised by the most intense proliferation. *Tetraselmis* sp., which developed only in 1998,

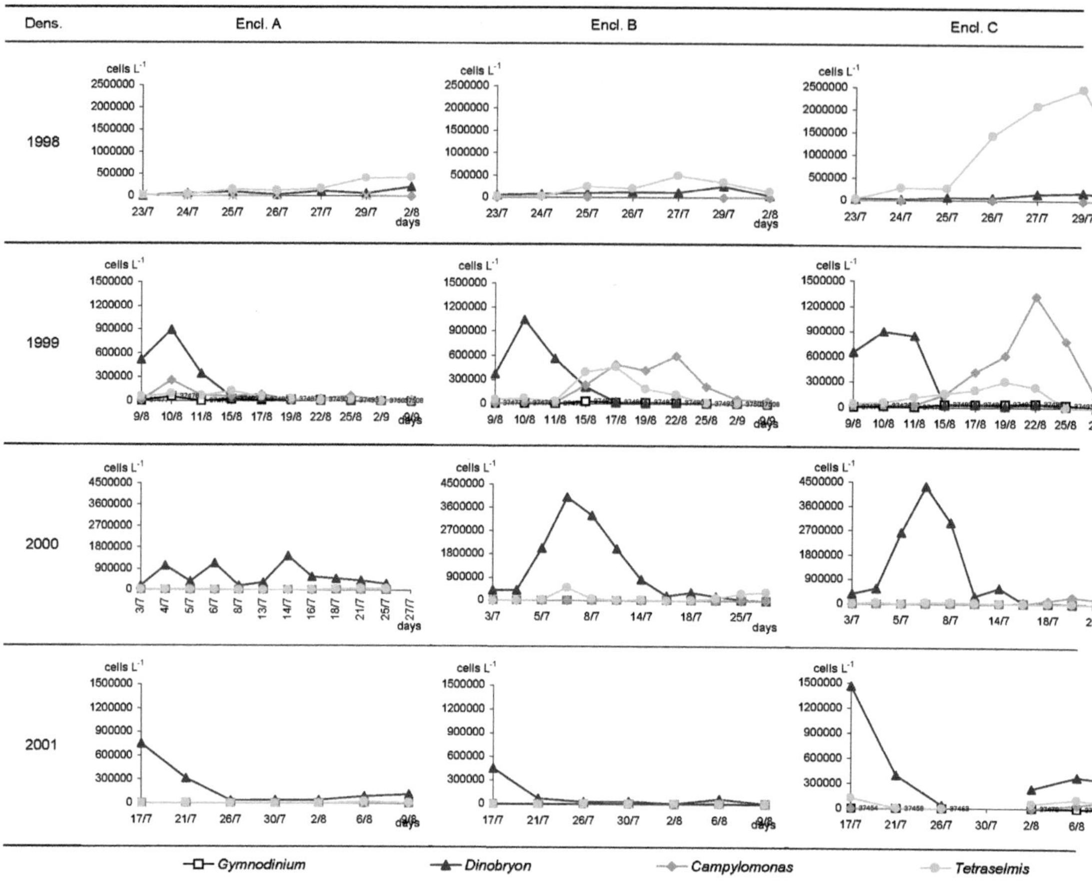

Figure 1. Density (cell l^{-1}) of four flagellated algae taxa in 1998–2001 in Lake Tovel.

did not show any significant correlation in this year (Table 4).

One-Way Analysis of Variance (ANOVA1)

The mean biovolumes of *Gymnodinium uberrimum* in the different enclosures in 1999 and 2000 were significantly different ($p < 0.03$; Table 5). None of the mean biovolume values of *Dinobryon* cf. *sociale* were significantly different. *Campylomonas* sp. showed significant differences in summer 2000 ($p = 0.04$; Table 5), while the differences observed in 1999 were near-significant ($p < 0.09$). Values for *Tetraselmis* were significant in summer 1998 ($p < 0.02$) and 2001 ($p < 0.01$).

Taxonomic observations

The dominant cryptophyte, *Campylomonas* sp. (Fig. 3), was identified based on the presence of a sheet-like (non-plated) periplast, a permanently open ventral furrow and phycoerythrin pigment (as shown by the reddish cell colour visible with the light microscope). These features agree well with those reported in the original taxonomic protologue of the genus *Campylomonas* (Hill, 1991).

Cells of *Campylomonas* sp. were about 15–18 × 5–7 μm in size. The cell shape was distinctly acute at the posterior end and slightly rostrate at the anterior end. The ventral furrow extended posteriorly to about one-third of the cell length. On the left-hand edge of the furrow the discharge site of the contractile vacuole was visible. At the anterior end of the furrow there were two more or less equal flagella, inserted in a slightly displaced position towards the right-hand side of the cell. The length of the flagella was usually about one-half the cell length. The sheet-like periplast often had a papillate appearance produced by the underlying ejectosomes bulging towards the cell surface. These

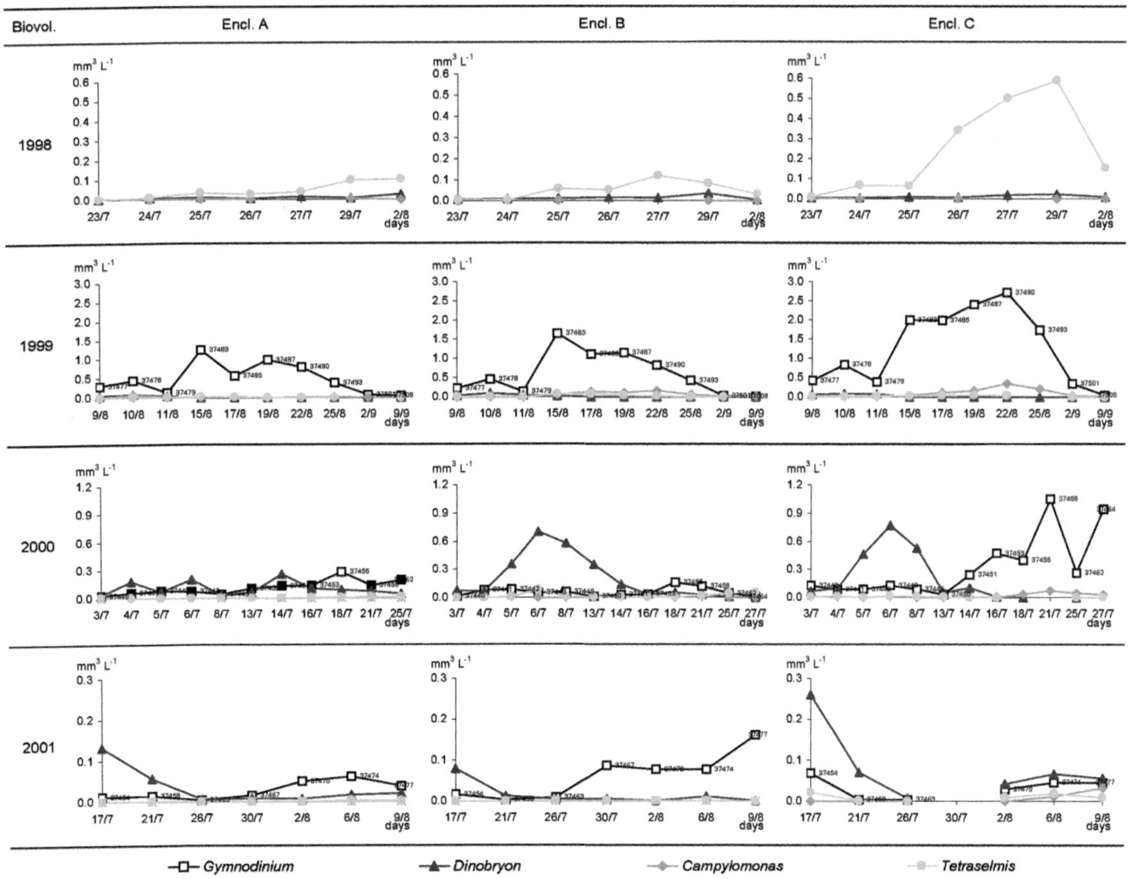

Figure 2. Biovolume (mm^3 l^{-1}) of four flagellated algae taxa in 1998–2001 in Lake Tovel.

ejectosomes were frequently discharged, producing numerous pores on the cell surface.

A conclusive identification at the specific level was not possible. At the light microscopical level this cryptophyte resembled *Cryptomonas marssonii* Skuja (1948), in which SEM has revealed a sheet-like periplast and a ventral furrow (Novarino, 1991), similar to *Campylomonas* sp. observed here. This suggests that *Cryptomonas marssonii* may have to be transferred to the genus *Campylomonas*, but further information is necessary in this respect.

Discussion

The enclosure experiments carried out at lake Tovel in the years 1998 – 2001 confirmed the importance of flagellate algae in the phytoplankton communities of mountain lakes (Pechlaner, 1971; Tilzer, 1972, 1973; Rott, 1988). Several observations on the ecology of the four taxa selected in this study are in accordance with previously available results. The growth of *Gymnodinium uberrimum* is reported to be negatively affected by UV-A plus photosynthetically active radiation (PAR 400–700 nm, Sommaruga, 2001). In Lake Tovel *Gymnodinium uberrimum* increased significantly in the P-enriched enclosures under variable weather and low irradiance conditions in 1999 (Table 2, Fig. 2) and in the P-enriched and shaded enclosure in the year 2000. *G. uberrimum* proliferation is confirmed by the results of the ANOVA test (Table 5) and suggests the ability of the alga to assimilate nutrients and grow even in conditions of low light intensity. The ability to exploit P-availability irrespective of low light conditions could be linked to the supposed mixotrophy of this alga (Porter, 1988; Jansson et al., 1996). The proliferation of *Gymnodinium uberrimum* is known to be affected not only by environmental variables, but also by its opportunistic behaviour and competitive ability (e.g. Rott, 1988;

Table 4. Relationships among the flagellate taxa during the different experiments in 1998–2001 in Lake Tovel

Densitiy Ind l^{-1}	*Gymno*	*Dino*	*Campylo*	*Tetra*	Biovolume $mm^3\ l^{-1}$	*Gymno*	*Dino*	*Campylo*	*Tetra*
A-C 98			*		A-C 98			*	
B-C 98		*	-**	*	B-C 98		*	-**	*
A-B 99		**			A-B 99	**	**		
A-C 99		**			A-C 99	**	**		
B-C 99	**	**	**	*	B-C 99	**	**	**	
A-B 00	*				A-B 00	*		**	
A-C 00				-*	A-C 00				
B-C 00		**			B-C 00		**	*	
A-B 01		**			A-B 01		**		
A-C 01		**			A-C 01		**		
B-C 01		**			B-C 01		**		
A98-A00				-*	A98-A00				-*
A98-A01				*	A98-A01				*
A00-A01			**		A00-A01			**	
C98-B99				*	C98-B99				*
C98-C99				*	C98-C99				*
B99-C00				**	B99-C00				**
B99-C01			*		B99-C01			*	
C99-B00	-*				C99-B00	-*			
C99-B01	**				C99-B01	**			
C99-C01			**		C99-C01			**	
B00-C01	-**				B00-C01	-**			
C00-C01			**		C00-C01			**	

□ = not significant (n.s.).
* = $0.01 < P < 0.05$ significant.
** = $0.001 < P < 0.01$ highly significant.

Table 5. Results of the One-Way Analyses of Variance (ANOVA1) in 1998–2001 in Lake Tovel

		ANOVA1			
Year	Parameters	*Gymnodinium*	*Dinobryon*	*Campylomonas*	*Tetraselmis*
1998	*F*		0.29	2.49	4.87
	P		0.75	0.11	0.02
	sign. level		n.s.	n.s.	*
1999	*F*	3.90	0.096	2.63	1.55
	P	0.03	0.91	0.09	0.23
	sign. level	*	n.s.	n.s.	n.s.
2000	F	4.10	0.88	3.46	2.50
	P	0.03	0.43	0.04	0.10
	sign. level	*	n.s.	*	n.s.
2001	F	0.90	1.77	1.78	6.21
	P	0.43	0.20	0.20	0.01
	sign. level	n.s.	n.s.	n.s.	*

(n.s. = not significant; * = 0.05 – 0.01; ** = 0.01 – 0.001; *** = < 0.001).
F = *F*-statistic.
P = probability.

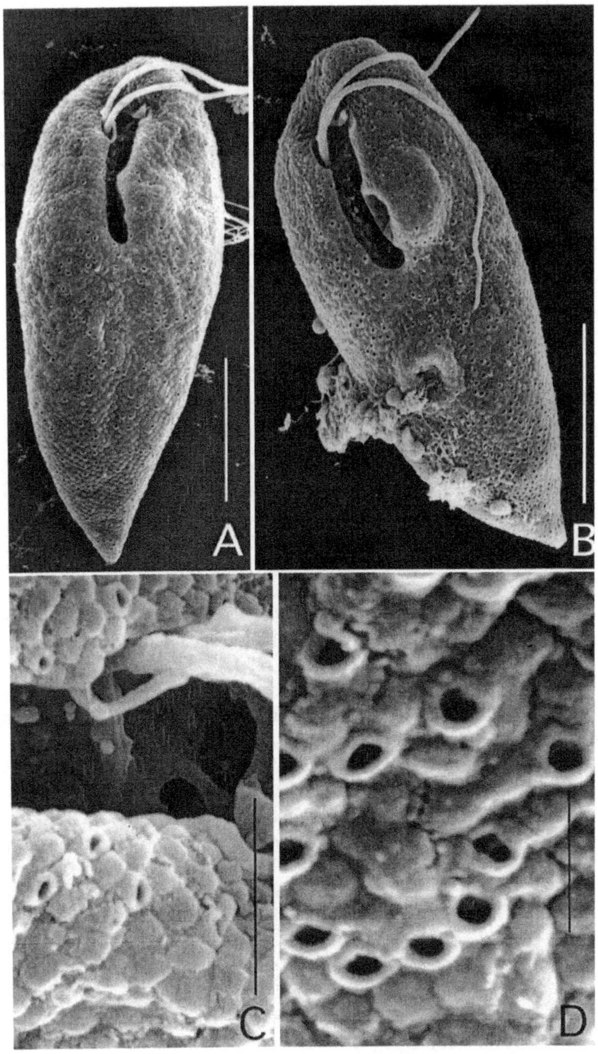

Figure 3. (A–D) *Campylomonas* sp., SEM. (A–B). Two cells, showing the flagella and the ventral furrow with the dischange site of the contractile vacuole. Scale bars = 5 μm. (C). Detail of the furrow region and flagellar insertion. Scale bar = 1 μm. (D) Detail of the sheet-like periplast, showing numerous ejectosome pores. Scale bar = 500 nm.

Reynolds 1997), which allowed the alga to dominate when other algae (e.g. *Glenodinium sanguineum*) were at a minimum.

Dinobryon sociale is reported to be an important component of phytoplankton assemblages inlarge and transparent lakes characterised by mid-range conductivity and pH values (Eloranta, 1989), oligothrophic conditions and high N/P ratio. The observed dominance of this species in oligotrophic lakes (Wetzel, 1983; Van den Hoek et al., 1995) could be promoted by its ability to assimilate even low phosphorus concentrations and its mixotrophic behaviour (Porter, 1988; Jansson et al., 1996; Holen & Boraas, 1996). The abundance of *Dinobryon* cf. *sociale* in the oligotrophic lake Tovel during spring and early summer agrees with these physiological characteristics. The development of this species ceased almost completely in both the control and the nutrient-enriched enclosures by the end of the first week of experiments (Figs 1 and 2), indicating that P-enrichment did not promote a pronounced growth of this taxon. The progressive decline of *D. sociale* observed during the whole sampling month could be due to direct competition with other taxa. The negative relationship with NO_3-N suggests a sensitivity of this taxon to decreased nitrogen concentration. Nitrate concentration is known to be an important environmental factor regulating the distribution and abundance of planktonic Chrysophyceae, as has been shown recently also for high mountain lakes (Tolotti et al., 2003). Recent investigations (Jansson et al.,1996) showed that this mixotrophic species is limited by N, but indifferent to light intensity and high P availability.

The ecological requirements of *Campylomonas* sp. identified in Lake Tovel are comparable to those of some *Cryptomonas*. *C. marssonii* shows a rapid exponential increase at the beginning of the population development period (Dokulil, 1988) in the case of optimal environmental conditions (Stewart & Wetzel, 1986). In Lake Tovel, both the natural (1999) and experimentally imposed light reduction promoted the proliferation of *Campylomonas* sp. in addition to the nutrient enrichments in 2000 (Fig. 1, Table 4). The ability of this alga to grow under low light intensity is confirmed by the negative correlation with this variable (Table 3). The tolerance to low light intensity is a common characteristic of Cryptophyceae (e.g. Reynolds, 1984; Dokulil, 1988), including *C. marsonii* (Pithart, 1995) which is close (or possibly coincident) to the cryptomonad found in Lake Tovel during our investigation. These observations support the hypothesis that Cryptophyceae (Barone & Noselli-Flores, 2003; Lepisto & Holopainen, 2003; Menezes & Novarino, 2003) are also able to improve their nutrient uptake by means of phagotrophic nutrition (e.g. Porter, 1988; Jansson et al., 1996; Holen & Boraas, 1996). During the enclosure experiments in Lake Tovel a negative response of *Campylomonas* sp. to the decrease in NO_3-N concentration was also observed (Table 3).

Since *Tetraselmis* species are commonly used as live food for aquatic animal larvae (Okauchi & Kawamura, 1997), several ecological requirements of this genus are well-known, e.g. an ideal N:P ratio between 5 and 80 and temperatures around 19 – 25 °C (Molina et al., 1991; Abu-Rezq et al., 1999). This genus has a high specific nitrate and ammonium uptake rate, but with a very long (48 hr) lag phase before the start of cell division (Stolte et al., 1994; Stolte & Riegman, 1995). In Lake Tovel *Tetraselmis* sp. showed significant increases in the P-enriched enclosures (Table 5) during the sunny and warm summer 1998 (Figures 1 and 2). During all the other experiments the abundance of the alga remained almost unchanged.

Glenodinium sanguineum can increase and dominate the summer algal community under a combination of phosphorus availability and high light radiation (Tolotti et al., 1999; Tardio et al., 2001; Cantonati et al., 2003). These results point towards an obligate autotrophy of this dinoflagellate. This would be in accordance with the lack of trichocysts reported by Dodge et al. (1987). It should however be remembered that heterotrophy in dinoflagellates can occur by a wide variety of mechanisms (e.g. Larsen & Sournia, 1991).

Among the species investigated, only *Tetraselmis* sp. showed a behaviour similar to that of *Glenodinium sanguineum*, as indicated by the significant increase which occurred in summer 1998. *Gymnodinium uberrimum* and *Campylomonas* sp. replaced *Glenodinium sanguineum* in the enclosures where light radiation was lower. *Dinobryon* cf. *sociale* does not seem to compete with *Glenodinium sanguineum*, as suggested by its intense development in spring and the decrease which follows during summer, even under high nutrient concentrations.

Conclusions

As shown previously for *Glenodinium sanguineum*, *Gymnodinium uberrimum* also showed a prompt and positive response to P-enrichment in all experiments. However, the former species showed a pronounced development only when nutrients were coupled with high light intensities, while the latter appeared to be a better competitor under low light intensities and low water temperature. *Tetraselmis* sp. showed a behaviour similar to that of *G. sanguineum.* On the contrary, the development pattern of *Campylomonas* sp. was similar to that of *Gymnodinium uberrimum. Dinobryon* cf. *sociale* did not show a clear relationship either to nutrient enrichment or weather conditions.

Acknowledgements

The enclosure experiments at Lake Tovel were made possible by financial support from the Adamello-Brenta Natural Park (1998-2000) and the Autonomous Province of Trento (Research Funds awarded to the Project "SALTO / BEST = Blooms and Environment: Science for Tovel", Work Package 6) (2001).

References

Abu-Rezq, T. S., L. Al-Musallam, J. Al-Shimmari & P. Dias, 1999. Optimum production conditions for different high quality marine algae. Hydrobiologia 403: 97–107

Baldi, E., 1941. Ricerche idrobiologiche sul Lago di Tovel – Memorie del Museo di Storia Naturale della Venezia Tridentina 6: 1–279.

Barone, R. & L. Naselli-Flores, 2003. Distribution and seasonal dynamics of cryptomonads in Sicilian water bodies. Hydrobiologia 502 (Dev. Hydrobiol. 172): 325–329.

Cantonati, M., M. Tardio, M. Tolotti & A. Boscaini, 2002. Caratteristiche di mesocosmi utilizzati per lo studio del fitoplancton. Atti XV Convegno Nazionale del Gruppo d'Ecologia di Base "G. Gadio", Trento, 5–7 maggio 2001. Studi Trentini di Scienze Naturali – Acta biologica 78: 167–172.

Cantonati, M., M. Tardio, M. Tolotti & F. Corradini, 2003. Blooms of the dinoflagellate *Glenodinium sanguineum* obtained during enclosure experiments in Lake Tovel. J. Limnol. In press.

Chorst, R. J., & B. Riemann, 1994. Storm-simulated enzymatic decomposition of organic matter in benthonic/pelagic coastal mesocosms. Mar. Ecol. Prog. Ser. 108: 185–192.

Corradini, F., G. Flaim & V. Pinamonti, 2001. Five years of limnological observations on Lake Tovel (1995–1999): some considerations and comparisons with past data. Proc. Ital. Ass. Oceanol. Limnol. 14: 209–218.

Dodge, J. D., P. Mariani, A. Paganelli & R. Trevisan, 1987. Fine structure of the red-bloom dinoflagellate *Glenodinium sanguineum*, from Lake Tovel (N. Italy). Algol. Stud. 47: 125–138.

Dokulil, M., 1988. Seasonal and spatial distribution of chryptophycaen species in the deep, stratifying, alpine lake Mondsee and their role in the food web. Hydrobiologia 161: 185–201

Eloranta, P., 1989. Ecological studies on the ecology of the genus *Dinobryon* in Finnish lakes. Supplement to Nova Hedwigia 95: 99–109.

Eccel, E., 2000. Commento all'annata meteorologica 1999 in Trentino. Natura alpina 51: 19–28.

Halac, S., M. Felip, L. Camarero, S. Sommaruga-Wögrath, R. Psenner, J. Catalan & R. Sommaruga, 1997. An *in situ* experiment to test the solar UVB impact on plankton in a high altitude mountain lake. I. Lack of effect on phytoplankton species composition and growth. J. Plankton Res. 19: 1671–1686.

Hill, D. R. A., 1991. A revised circumscription of *Cryptomonas* (Cryptophyceae) based on examination of Australian strains. Phycologia 30: 170–180.

Holen, D. A. & M. E. Boraas, 1996. Mixotrophy in chrysophytes. In Craig, D., C. D. Sandgren, J. P. Smol & J. Kristiansen (eds.), Chrysophyte Algae. Ecology, Phylogeny and Development. University Press, Cambridge: 119–140.

I. A. S. M. A., 1998. Caratteristiche limnologiche dei laghi del Trentino. Rapporto 1997. Istituto Agrario di San Michele a/A. 176 pp.

I. R. S. A. - C. N. R, 1994. Metodi analitici per le acque. Istituto di Ricerca sulle Acque, Roma. Quaderni.

Jansson, M., P. Blomqvist & A. Johnsson, 1996. Nutrient limitation of bacterioplankton, autotrophic and mixotrophic phytoplankton, and heterotrophic nanoflagellates in Lake Örträstet. Limnol. Oceanog. 41: 1552–1559.

Javornický, P., 2001. Freshwater Rhodomonads (cryptophyceae). Algol. Stud. 102: 93–116.

Javornický, P., 2003. Taxonomic notes on some freshwater planktonic Cryptophyceae based on light microscopy. Hydrobiologia 502 (Dev. Hydrobiol. 172): 271–283.

Larsen, J. & A. Sournia, 1991. The diversity of heterotrophic dinoflagellates. In Patterson, D. J. & J. Larsen (eds), The Biology of Free-Living Heterotrophic Flagellates. Clarendon Press, Oxford: 313–332.

Lepistö, L. & A.-L. Holopainen, 2003. Occurrence of Cryptophyceae and katablepharids in boreal lakes. Hydrobiologia 502 (Dev. Hydrobiol. 172): 307–314.

Lorenzen, C. J., 1967. Determination of chlorophyll and phaeopigments: spectrophotometric equations. Limnol. Oceanogr. 12: 343–346.

Lund, J. W. G., G. Kipling & E. D. LeCren., 1958. The inverted microscope method for estimating algae number and statistical basis of estimating by counting. Hydrobiologia 11: 143–170.

Menezes, M. & G. Novarino, 2003. How diverse are planktonic cryptomonads in Brazil? Advantages and difficulties of a taxonomic-biogeographical approach. Hydrobiologia 502 (Dev. Hydrobiol. 172): 297–306.

Molina, E., E. Martinez, S. Sanchez, F. Garcia & A. Contreras, 1991. The influence of temperature and the initial N:P ratio on the growth of microalgae *Tetraselmis* sp. Proc. Biochem. 26: 183–187

Novarino, G., 1991. Observations on some new and interesting Cryptophyceae. Nordic J. Bot. 11: 599–611.

Novarino, G., I. A. N Lucas. & S. Morrall, 1994. Observation on the genus *Plagioselmis* (Cryptophyceae). Cryptogamie, Algologie 15: 87–107.

Novarino, G., D. K. Mills & F. Hannah, 1997. Pelagic flagellate populations in the southern North Sea, 1988- 89.1. Qualitative observations. J. Plankton Res. 19: 1081–1109.

Novarino, G., 2003. A companion to the identification of cryptomonad flagellates (Cryptophyceae = Cryptomonadea). Hydrobiologia 502 (Dev. Hydrobiol. 172): 225–270.

Okauchi, M. & K. Kawamura, 1997. Optimum medium for large-scale culture of *Tetraselmis tetrathele.* Hydrobiologia 358: 217–222

Paganelli, A., 1992. Lake Tovel (Trentino): limnological and hydrobiological aspects. Mem. Ist. ital. Idrobiol. 50: 225–257.

Paris, G., G. Rossetti, M. Cattadori & G. Giordani, 1995. Interazioni fitoplancton-zooplancton in un lago oligotrofico d'alta quota (Lago Scuro Parmense): risultati di esperimenti con enclosures. Atti della Società Italiana di Ecologia 16: 467–470.

Pechlaner, R., 1971. Factors that control the production rate biomass of phytoplankton in high mountain lakes. Mitt. int. Ver. theor. angewan. Limnol. 19: 125–145.

Pithart, D., 1995. Ecological study of Cryptophyceae from two pools in the Luinice river floodplain. Dissert. Institute of Botany, Czech Academy of Sciences, Trebon. 108 pp.

Porter, K. G., 1988. Phagotrophic phytoflagellates in microbial food webs. Hydrobiologia 159: 89–97.

Premazzi, G., G. Chiaudani, A. Pereira, A. C. Cardoso & E. Rodari, 1999. A restoration programme in Italy: the experience of lake Varese – Proceedings of the 8th International Conference on the Conservation and Management of Lakes, 17-21 May 1999, Copenhagen – Denmark S14A-2.

Ravera, O., 1989. Lake ecosystem degradation and recovery studied by the enclosure method. In: Ravera, O. (ed.), Ecological Assessment of Environmental Degradation Pollution and Recovery, Elsevier, Amsterdam: 217–243.

Ravera, O., 1990. The effects of hypolimnetic oxygenation in the shallow and eutrophic Lake Comabbio (Northern Italy) studied by 'enclosure'. Int. Ver. theor. angewan. Limnol.: Verh. 24: 188–194.

Reynolds, C. S., 1984. The ecology of freshwater phytoplankton. Cambridge University Press, Cambridge. 384 pp.

Reynolds, C. S., 1997. Vegetation processes in the pelagic: a model for ecosystem theory. In: Kinne O. (ed.), Excellence in Limnology 9. 371 pp.

Riemann, B. & M. Sondergaard, 1986. Regulation of bacterial secondary production in two eutrophic lakes and experimental enclosures. J. Plankton Res. 6: 519–536.

Riemann, B., T. G. Nielsen, S. J. Horsted, P. K. Bjornsen & J. Pock-Steen, 1988. Regulation of phytoplankton biomass in estuarine enclosures. Mar. Ecol. Prog. Ser. 69: 171–178.

Rott, E., 1981. Some results from phytoplankton counting intercalibrations. Schweizerische Zeitschrift für Hydrologie 43: 35–62.

Rott, E., 1988. Some aspects of the seasonal distribution of flagellates in mountain lakes. Hydrobiologia 161: 159–170.

Salonen, K. & M. Rosenberg, 2000. Advantages from diel vertical migration can explain the dominance of *Gonyostamum semen* (Raphidophyceae) in a small, steephy-stratifical humic lake, J. Plankton Res. 22: 1841–1853.

Sokal, R. R. & F. J. Rohlf, 1981. Biometry. 2nd edition. Freeman and Co., New York. 859 pp.

Skuja, H., 1948. Taxonomie des Phytoplanktons einiger Seen in Uppland, Schweden. Sym. Bot. Ups. 9: 1–399.

Sommaruga, R., 2001. The role of solar UV radiation in the ecology of alpine lakes. J. Photochem. Photobiol. B: Biology 62: 35–42

Stewart, A. J. & R. G. Wetzel, 1986. Cryptophytes and other microflagellates as couplers in planktonic community dynamics. Archiv für Hydrobiol. 106: 1–19

Stolte, W., T. McCollin, A. A. M. Noordeloos & R. Riegman, 1994. Effect of nitrogen source on the size distribution within marine phytoplankton populations. J. exp. mar. Biol. Ecol. 184: 83–97

Stolte, W. & R. Riegman, 1995. The effect of phytoplankton cell size on transient state nitrate and ammonium uptake kinetics. Microbiology 141: 1221–1229

Tardio, M., M. Cantonati & M. Tolotti, 2001. Condizionamento meteorologico nello sviluppo del fitoplancton del Lago di Tovel (Trentino): indicazioni da ricerche in mesocosmi. XI Congresso Nazionale della Società Italiana di Ecologia, Sabaudia 12–14 settembre 2001. Atti 25: CD-rom.

Tilzer, M., 1972. Dynamik und Produktivitat von Phytoplankton und pelagischen Bakterien in einem Hochgebirgssee (Vorderer Finstertaler See, Osterreich). Archiv für Hydrobiol. 3: 201–273.

Tilzer, M., 1973. Diurnal periodicity in the phytoplankton assemblage of a high mountain lake. Limnol. Oceanogr. 18: 15–30.

Tolotti, M., M. Cantonati & F. Corradini, 1999. Investigating dinophyte blooms with P-enriched enclosures in a mountain lake (lake Tovel, Southern Alps, Italy). Proceedings of the 8th International Conference on the Conservation and Management of Lakes, 17–21 May 1999. Copenaghen, Denmark S16B-5.

Tolotti, M., H. Thies, M. Cantonati, C. M. E. Hansen & B. Thaler, this volume. Flagellate algae (Chrysophyceae, Dinophyceae, Cryptophyceae) in 48 high mountain lakes of the northern and southern slope of the Eastern Alps: biodiversity, taxa distribution and their driving variables.

Tomasi, G., 1989. Dall'immaginario al plausibile. Natura Alpina 40: 1–72.

Tranvik, L. J., K. G. Porter & J. Mc N. Sieburth, 1989. Occurence of bacterivory in *Cryptomonas*, a common freshwater phytoplankter. Oecologia 78: 473–476.

Utermöhl, H., 1958. Zur Vervollkommnung der quantitativen Phytoplankton Methodik. Mitt. int. Ver. theor. angewan. Limnol. 9: 1–38.

Van den Hoek, C., D. G. Mann & H. M. Jahns, 1995. Algae. An introduction to phycology. Cambridge university press. 627 pp.

Vinebrooke, R. D. & P. R. Leavitt, 1999. Differential responses of littoral communities to ultraviolet radiation in an alpine lake. Ecology 80: 223–237.

Wetzel, R. R., 1983. Limnology. Saunders College Publishing. 341 pp.

Hydrobiologia **502:** 297–306, 2003.
L. Naselli-Flores, J. Padisák, M. T. Dokulil (eds),
Phytoplankton and Equilibrium Concept: The Ecology of Steady-State Assemblages.

How diverse are planktonic cryptomonads in Brazil? Advantages and difficulties of a taxonomic-biogeographical approach

Mariângela Menezes[1] & Gianfranco Novarino[2]
[1]*Laboratório de Ficologia, Departamento de Botânica, Museu Nacional-Universidade Federal do Rio de Janeiro, São Cristóvão, 20940-040, Rio de Janeiro, Brazil.*
E-mail: mmenezes@mn.ufrj.br
[2]*Department of Zoology, The Natural History Museum, Cromwell Road, London SW7 5 BD, U.K.*
E-mail: gn@nhm.ac.uk

Key words: Cryptophyceae, biodiversity, taxonomy, biogeography, light microscopy, phytoplankton, tropical waters, Brazil

Abstract

We compare 'temperate data' with information contained in 49 surveys of cryptomonads from Brazilian tropical regions. These surveys contain a total of 351 records relative to a total of 38 taxa at the species level or below. All taxa were identified using traditional morphological characters visible with the light microscope. Two taxa appear to be new species, and amongst the known species *Cryptomonas erosa* Ehrenberg is the most frequently recorded one. To date two species of *Pseudocryptomonas* Bicudo & Tell (all containing multiple photosynthetic endosymbionts/chloroplasts) and the species *Cryptomonas brasiliensis* Castro, C. Bicudo & D. Bicudo have not been found outside Brazil. With respect to their geographical distribution in Brazilian tropical regions, cryptomonads are better represented in reservoirs and coastal lagoons located in the south-eastern part of the country. Some species appear to be restricted to brackish coastal ecosystems (oligo- to mesohaline and meso- to eutrophic waters). There are only four records of marine cryptomonads. A comparative study of three different kinds of shallow aquatic environments in Brazil – north-eastern reservoirs, a flood-plain Amazonian lake, and brackish lagoons on the south-eastern coast – shows that cryptomonads are to be found constantly, and their species number density, and biomass are generally low (usually $<10\%$) in relation to the other phytoplankton. The case of *Plagioselmis* Butcher emend. Novarino, Lucas & Morrall, amongst others, exemplifies the great difficulties encountered while attempting to interpret biogeographical information based on light microscopy within the context of a modern systematic and taxonomic framework based on electron microscopy.

Introduction

The cryptomonads (Cryptophyceae = Cryptomonadea) are a small group of autotrophic and heterotrophic flagellates with some unique morphological features of high evolutionary significance. They constitute a very important component of the plankton in freshwater, coastal and marine ecosystems from polar areas to tropical regions. In many environments they are frequently present throughout the entire year, often with relatively constant population numbers although they can sometimes multiply very rapidly and produce blooms of varying proportions.

Taxonomic identification at the generic and specific levels can be problematic owing to numerous reasons, first and foremost the cells' fragility and sensitivity towards chemical fixatives, coupled with the small number of conspicuous morphological features (Klaveness, 1985, 1989). However, recent systematic and taxonomic studies based on electron microscopy have provided additional morphological information and this has considerably modified many of the traditional identification criteria. Although electron microscopy fully supports the designation of several genera and species as separate taxonomic entities, the interpretation of the species and genus concepts in the cryptomonads is still the object of much discussion (Novarino & Lucas, 1993; Clay et al., 1999).

Recent taxonomic investigations have focused on a relatively small number of cryptomonads grown in laboratory cultures, especially among the marine forms. Because cryptomonad systematics is a rapidly evolving subject which is inevitably subject to change as our knowledge of natural diversity and variability increases and improves, there is still an urgent need for alpha-taxonomic and biogeographical information on natural cryptomonad populations from many different habitats and regions. Such information could be highly beneficial from a systematic point of view because it could enrich the spectrum of morphological and other characters useful for classification purposes, whilst also providing a means for judging on the taxonomic value of existing characters.

The planktonic cryptomonads in tropical regions have been particularly understudied and therefore they deserve urgent attention. Our knowledge of their morphology, nomenclature, taxonomy and geographical distribution is still based largely on traditional light-microscopical concepts. Therefore, accurate comparisons with cryptomonad taxonomic lists from geographical regions outside the tropical areas may be problematic when electron microscopical information is also involved, because such information is almost entirely lacking for the tropical cryptomonads.

The aim of this paper is to present a first taxonomic and biogeographical appraisal of planktonic cryptomonads in Brazilian tropical regions. We also compare the available data with published surveys of cryptomonads from temperate regions. Some of the advantages and difficulties of this approach are highlighted.

Materials and methods

We analyzed 49 floristic and ecological surveys of planktonic cryptomonads from freshwater, brackish and marine environments in Brazilian tropical regions. The surveys were carried out between 1934 and February 2002 and 30 of them have been published in the scientific literature. A total of 1534 collections from 125 sites were analyzed. Additional seven collections were also analyzed using results from phytoplankton sampling programmes carried out by *Laboratório de Ficologia do Museu Nacional*, UFRJ, Rio de Janeiro, Brazil. The localizations of the sites which have been surveyed are shown in Figure 1.

The cryptomonad taxa taken into account for the purposes of the present study are all those specific

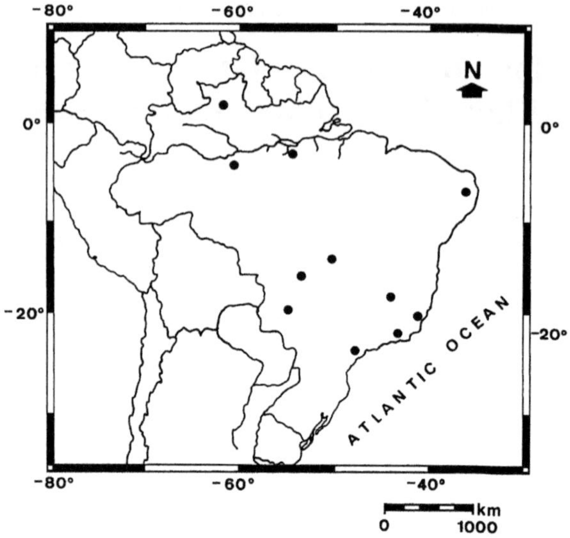

Figure 1. Brazilian tropical regions for which records of cryptomonads are available, based on 49 surveys carried out between 1934 and February 2002.

taxa listed as such in the surveys. Taxa identified at the generic level but not at the specific level have been taken into account only when illustrations (line-drawings and/or micrographs) where provided in the original reports.

In order to better evaluate the contribution of cryptomonads in relation to the other components of the plankton, a comparative study was carried out in three different kinds of shallow aquatic environments in Brazil: (1) Tabocas and Tapacurá freshwater reservoirs, State of Pernambuco, north-eastern Brazil; (2) Batata Lake, a flood-plain Amazonian lake in the northern part of the country; and (3) Imboassica and Rodrigo de Freitas lagoons, two brackish ecosystems in the Rio de Janeiro State, south-eastern Brazilian coast. Some characteristics of the five selected waterbodies are shown in Table 2.

Results

The 49 surveys analyzed here contained a total of 351 records (see Table 1 for details of surveys) relative to a total of 38 cryptomonad taxa at the specific level or below and one at generic level (Table 3). *Cryptomonas* Ehrenberg was the most species-rich genus (21 species), and also the most frequently reported one since it was mentioned in 286 records (Figs 2 and 3). *Cryptomonas erosa* Ehrenberg and *C. marssonii* Skuja were the most frequently recorded species (Fig. 4).

Table 1. Study areas, sites, records, taxa and surveys of cryptomonads in Brazilian tropical regions

Study Area	Sites number	Records number	Taxa number	Surveys number	References
Amazonas[1]	2	7	5	2	Bittencourt-Oliveira (1994)
Espírito Santo[2]	2	2	2	2	Dias Jr. (1991); Huszar et al. (1990)
Goiás[3]	1	12	12	1	Nogueira (1999)
Mato Grosso do Sul[3]	1	6	6	1	Train (1998)
Minas Gerais[2]	1	4	4	1	Giani et al. (1999)
Pará [1]	3	6	6	1	Huszar (1996)
Pernambuco[4]	2	4	3	2	Bressan (2001); Ferreira (2002)
Rondônia[1]	1	1	1	1	Magrin (1993)
São Paulo[2]	94	218	27	8	Bicudo & Bicudo (1967); Bicudo & Tell (1988); Bicudo et al. (1992); Castro (1990); Castro et al. (1991); Castro (1993); Marinho (1996); Silva (1999).
Rio de Janeiro[2]	16	94	18	24	Dias et al. (2001); Menezes & Souza (2001); Melo (2001); Souza (2002)
Distrito Federal[3]	2	6	3	6	Giani & Pinto Coelho (1986); Senna et al. (1998)

Regions of Brazil. [1]North. [2]South-east. [3]Central-west. [4]North-east.

Table 2. Some characteristics of the five selected waterbodies from Brazilian tropical regions

	Tabocas reservoir	Tapacurá reservoir	Batata lake	Imboassica lagoon	Rodrigo de Freitas lagoon
Latitude	08° 06′ 15″ S	08° 06′ 15″ S	1° 28′ S	22° 24′ S	22° 58′ 20″ S
Longitude	36° 13′ 23″ W	36° 13′ 23″ W	56° 14′ W	41° 42′ W	42° 12′ 65″ W
Area (km^2)	1.6	9.8	29.5	3.3	2.4
Depth (m) *	1.1	5.0	2.2	1.0	2.1
Trophic status	Hypertrophic	Hypertrophic	Mesotrophic to eutrophic	Eutrophic to hypertrophic	Eutrophic to hypertrophic

*Average values.

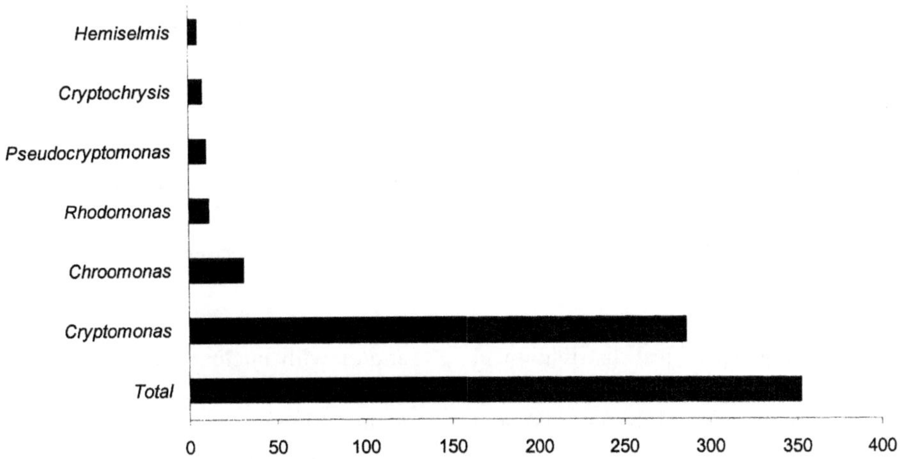

Figure 2. Total number of cryptomonads records by genus.

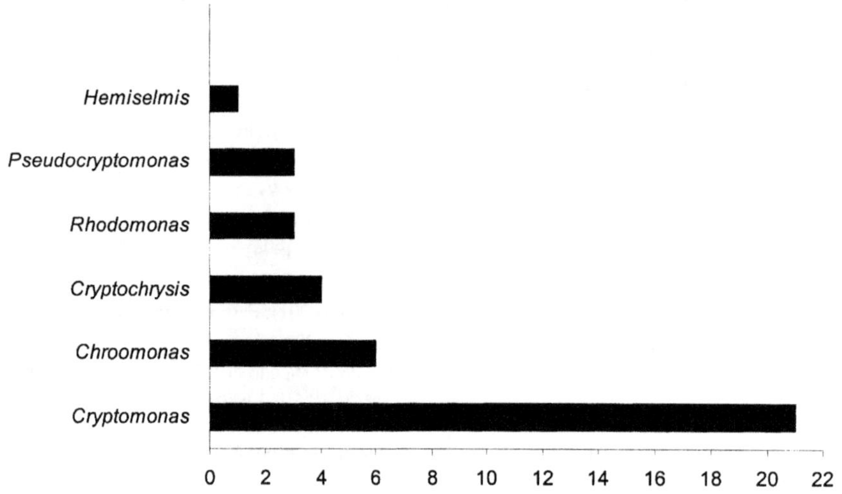

Figure 3. Total number of cryptomonads species by genus.

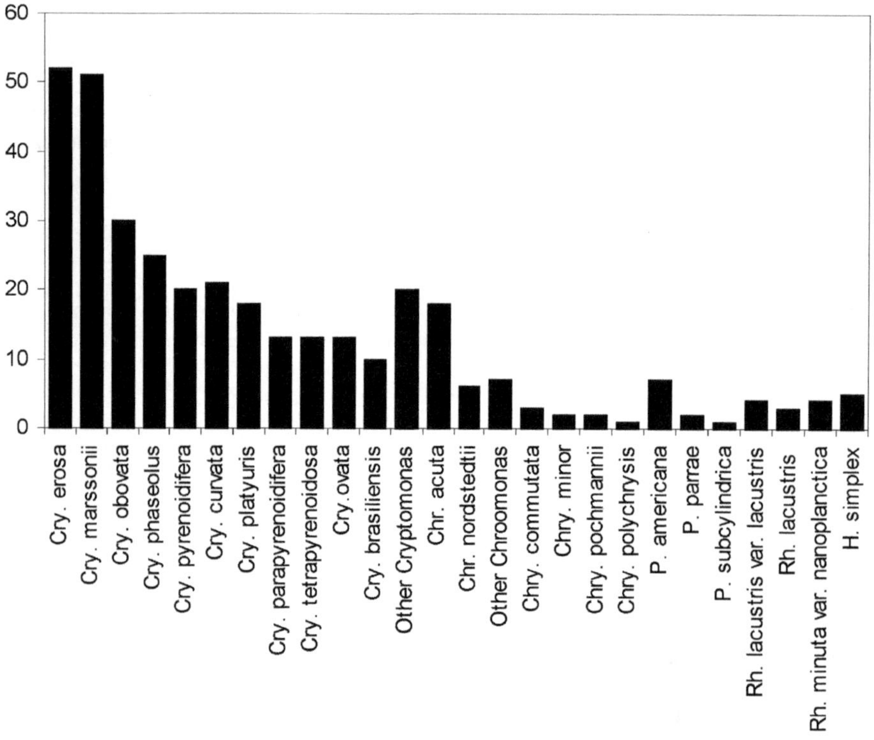

Figure 4. Total number of cryptomonads records by species: Cry. (*Cryptomonas*), Chr. (*Chroomonas*), Chry. (*Chryptocrysis*), P. (*Pseudocryptomonas*), Rh. (*Rhodomonas*), H. (*Hemiselmis*).

With respect to their geographical distribution in Brazilian tropical regions (Fig. 5), cryptomonads are better represented in reservoirs (138 records) and coastal lagoons (97 records) located in the south-eastern part of the country. To date, there have been only four records of marine cryptomonads unidentified and/or with unclear identification: *Cryptomonas* sp., *Hemiselmis* sp., *Rhodomonas* sp., and *Rhodomonas* cf. *minuta* Skuja.

We have also found or noticed in the literature some hitherto undescribed cryptomonads and, as far as is possible to judge on what data we have, there

Table 3. Cryptomonads recorded in Brazilian tropical waters

Chroomonas acuta Utermöhl
Chroomonas coerulea (Geitler) Skuja
Chroomonas nordstedtii Hansgirg
Chroomonas nordstedtii Hansgirg f. *nordstedtii*
Chroomonas nordstedtii Hansgirg f. *minor* Nygaard
Chroomonas salina (Wislouch) Butcher
Cryptochrysis commutata Pascher
Cryptochrysis minor Nygaard
Cryptochrysis pochmannii Huber-Pestalozzi
Cryptochrysis polychrysis Pascher
Cryptomonas anas Javornický
Cryptomonas brasiliensis Castro, C. Bicudo & D. Bicudo
Cryptomonas caudata Schiller
Cryptomonas curvata Ehrenberg emend. Penard
Cryptomonas cylindrica Ehrenberg
Cryptomonas erosa Ehrenberg
Cryptomonas gracilis Skuja
Cryptomonas lobata Kosrsikov
Cryptomonas marssonii Skuja
Cryptomonas obovata Skuja
Cryptomonas ovata Ehrenberg
Cryptomonas ozolini Skuja
Cryptomona phaseolus Skuja
Cryptomonas parapyrenoidifera Skuja
Cryptomonas platyuris Skuja
Cryptomonas pyrenoidifera Geitler
Cryptomonas rostratiformis (Skuja) Huber-Pestalozzi
Cryptomonas stigmatica Wislouch
Cryptomonas tenuis Pascher
Cryptomonas tetrapyrenoidosa Skuja
Cryptomonas sp
Hemiselmis simplex Butcher
Pseudocryptomonas americana Bicudo & Tell
Pseudocryptomonas parrae Bicudo & Tell
Pseudocryptomonas subcylindrica Bicudo & Tell
Rhodomonas lacustris Pascher & Ruttner
Rhodomonas lacustris Pascher & Ruttner var. *lacustris*
Rhodomonas minuta Skuja var. *nanoplanctica* Skuja
aff. *Plagioselmis nanoplanctica* (Skuja) Novarino, Lucas & Morrall
aff. *Teleaulax* Hill

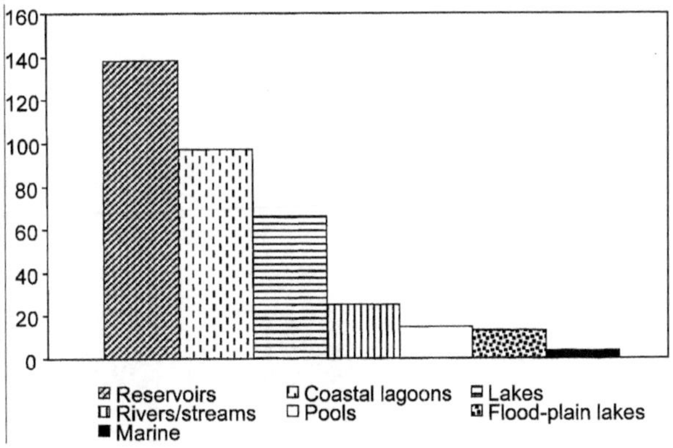

Figure 5. Total number of cryptomonads records by environments.

are at least two species which are new to science. One of these was described from samples collected from lakes, reservoir and rivers located in the State of São Paulo (Castro, 1993, Figs 81–87). This *Cryptomonas* sp. is closely related to *Cryptomonas curvata* Ehrenberg, but differs in the presence of two lateral chloroplasts and two pyrenoids per chloroplast.

The other species has produced a summer bloom (February 2000) in the Rodrigo de Freitas lagoon, a brackish coastal ecosystem in Rio de Janeiro State. It resembles the genus *Teleaulax* but it differs in the cell shape and the presence of at least two pyrenoids (Figs 6–8). Records of this cryptomonad have been found in three more brackish coastal lagoons by Melo (2001), by Menezes & Domingos (1994) and by Souza (2002). All those lagoons vary from oligohaline (salinity 2‰) to eurihaline (salinity >30‰) and from mesotrophic to hypertrophic waters.

A third species (Figs 9 and 10) has been found frequently in almost every brackish coastal lagoon which has been sampled. On the basis of morphological characters (as revealed by the light microscope) and the kind of habitat (mesohaline waters), this cryptomonad may be related to *Plagioselmis nannoplanctica* (Skuja) Novarino, Lucas & Morrall (Novarino et al., 1994), but electron microscopy is necessary in order to confirm the identification at the generic level.

From the five water bodies investigated in Brazilian tropical areas we observed that cryptomonads were generally more or less constantly present in all the ecosystems under study, although in low numbers and biomass (usually <10%) in relation to the other components of the plankton (Fig. 11). This agrees well with other ecosystems in subtropical and temperate areas where cryptomonads like other microflagellates show term growth episodes followed

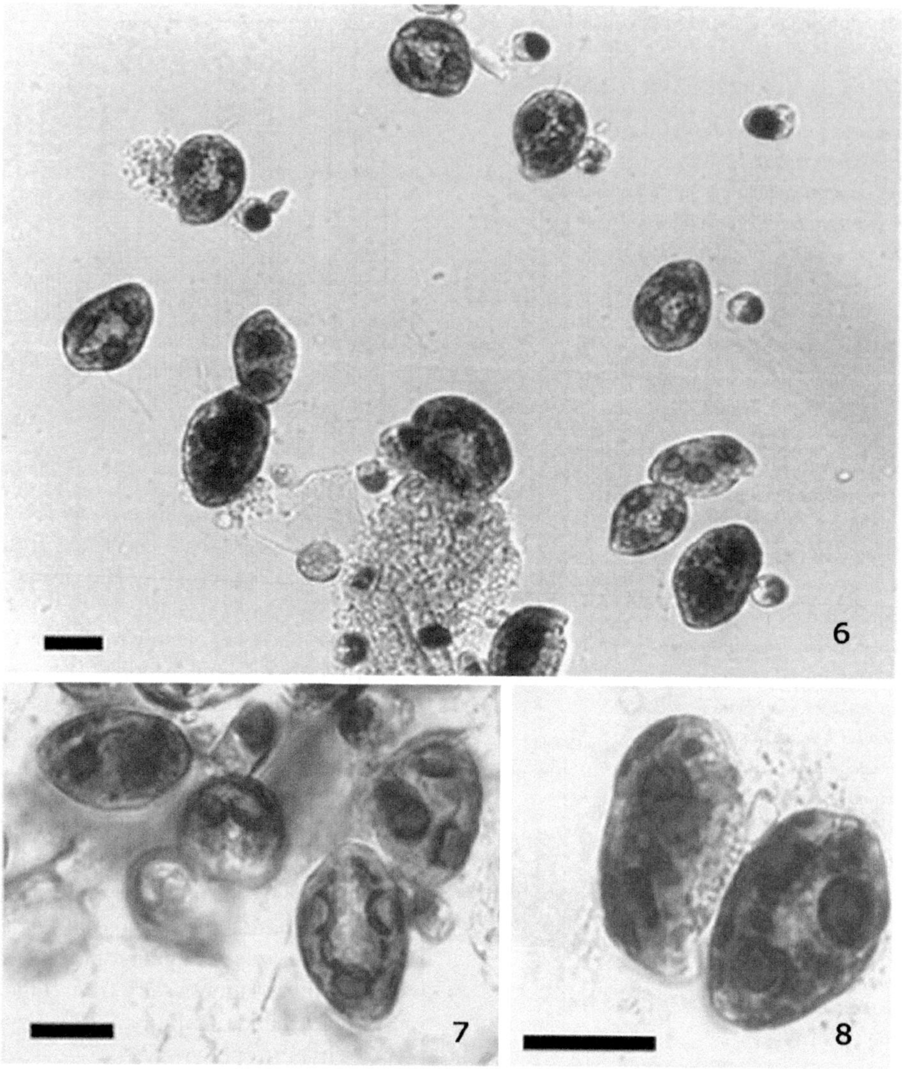

Figure 6–8. aff. *Teleaulax* from Rodrigo de Freitas lagoon, Rio de Janeiro, March 2000. *Figures 6–7.* Mass development of cells. *Figure 8.* Vegetative cells, stained with Lugol, lateral (left) and dorsal view (right). - Scale bar = 10 μm.

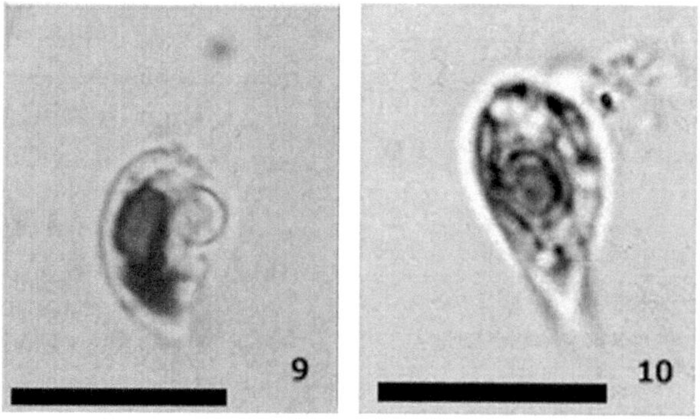

Figure 9–10 aff. *Plagioselmis nannoplanctica. Figure 9.* From Imboassica lagoon, Rio de Janeiro, March 1997, lateral view. *Figure 10.* From Rodrigo de Freitas lagoon, Rio de Janeiro, March 2000, dorsal view. - Scale bar = 10 μm.

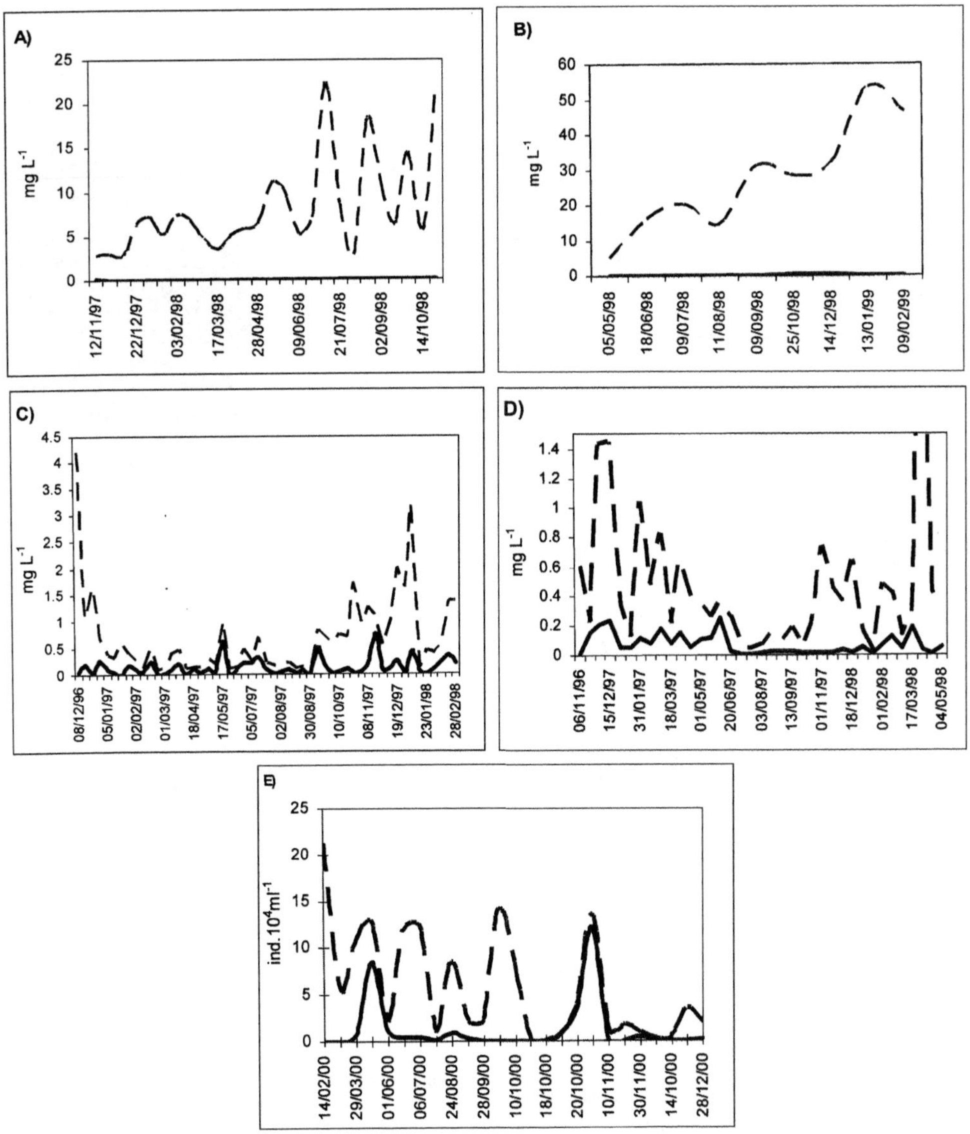

Figure 11. Biomass and density variations of total phytoplankton (– –) and total cryptophytes (—) in five selected waterbodies from Brazilian tropical regions: (A) Tabocas reservoir; (B) Tapacurá reservoir; (C) Batata Lake; (D) Imboassica lagoon; (E) Rodrigo de Freitas lagoon.

by rapid declines in abundance (Willén et al., 1980; Moustaka-Gouni, 1996; Barone & Naselli-Flores, 2003; Tolotti et al., 2003).

Discussion

Because cryptomonad taxonomy in Brazil still follows the traditional criteria revealed by the light microscope, taxa have been identified using the conventional taxonomic accounts and floras of temperate cryptomonads based on light microscopy only. Light microscopy reveals only few characters of taxonomic usefulness, and therefore it makes identification at the specific level uncertain. In spite of this it can still be safely assumed that numerous, morphologically distinct cryptomonad 'species' (morphospecies) occur in Brazilian tropical waters, but some important questions arise:

- Are the tropical cryptomonad floras any different from the temperate ones, and how can the degree of similarity be estimated in the first place?
- How do the light-microscopical cryptomonad morphospecies relate to those taxa which have

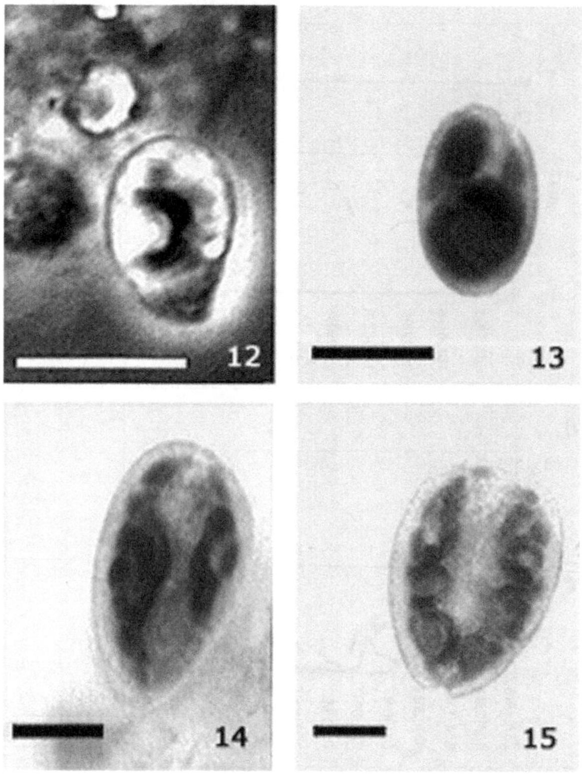

Figure 12–15. Cells of species of *Chroomonas* and *Cryptomonas*, stained with Lugol. *Figure 12. Chroomonas acuta*, lateral view. *Figure 13. Chroomonas nordstedtii*, lateral view. *Figure 14. Cryptomonas pyrenoidifera*, doral view. *Figure 15. Cryptomonas parapyrenoidosa*, dorsal view. - Scale bar = 10 μm.

been described or re-examined using electron microscopy?

Based on what data are available, two contrasting answers emerge to the first question. Firstly, many taxa which have been described from temperate regions, e.g. *Chroomonas acuta* Utermöhl (Fig. 12), *Chroomonas nordstedtii* Hansgirg (Fig. 13), *Cryptomonas platyuris* Skuja, *Cryptomonas pyrenoidifera* Geitler (Fig. 14), *Cryptomonas tetrapyrenoidosa* Skuja (Fig. 15), have been found also in Brazil. Among the 38 taxa recorded from tropical Brazilian waters, 81% are also found in temperate areas. We have also found some similarities between temperate areas and Brazilian and other warm water regions in terms of the occurrence of certain species in relation to the kind of habitat. In general, large species such as *Cryptomonas curvata* Ehrenberg emend. Pennard, *Cryptomonas marssonii* Skuja, possibly allied to *Campylomonas reflexa* (Skuja) Hill, (Fig. 16), and *Cryptomonas erosa* Ehrenberg have been reported from shallow, eutrophic aquatic ecosystems, both in Brazil

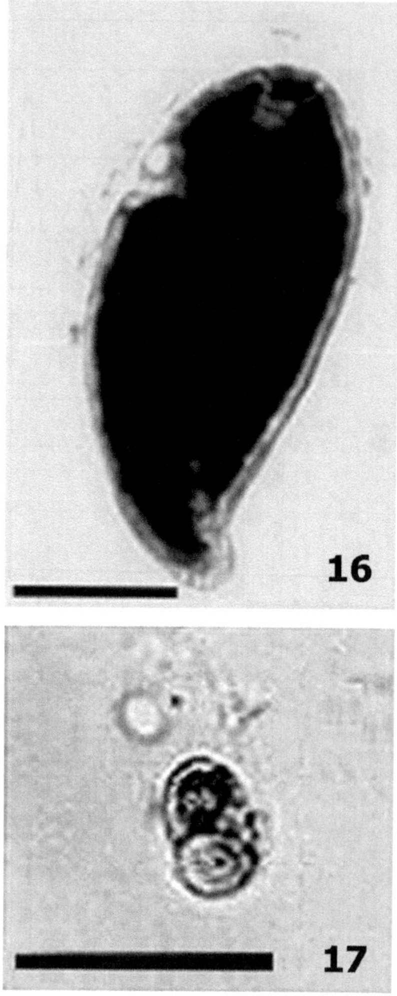

Figure 16–17. Cells of *Cryptomonas marssonii* and *Hemiselmis simplex*. *Figure 16. Cryptomonas marssonii*, lateral view, stained with Lugol. *Figure 17. Hemiselmis simplex*, lateral view, Imboassica lagoon, Rio de Janeiro, March 1997. - Scale bar = 10 μm.

and in temperate regions, and therefore they appear to have a cosmopolitan distribution. *Hemiselmis simplex* Butcher (Fig. 17) is another candidate for cosmopolitan distribution, since it has already been reported from brackish waters in Europe and Japan in addition to Brazil.

Secondly, other cryptomonad species such as *Cryptomonas brasiliensis* described by Castro et al. (1991) (Figs 18–22) and two species of *Pseudocryptomonas* – *P. americana* (Figs 23 and 24) and *P. subcylindrica* (Fig. 25) – described by Bicudo & Tell (1988) from material collected in the state of São Paulo, have not been found outside Brazil. This may suggest that they are geographically restricted

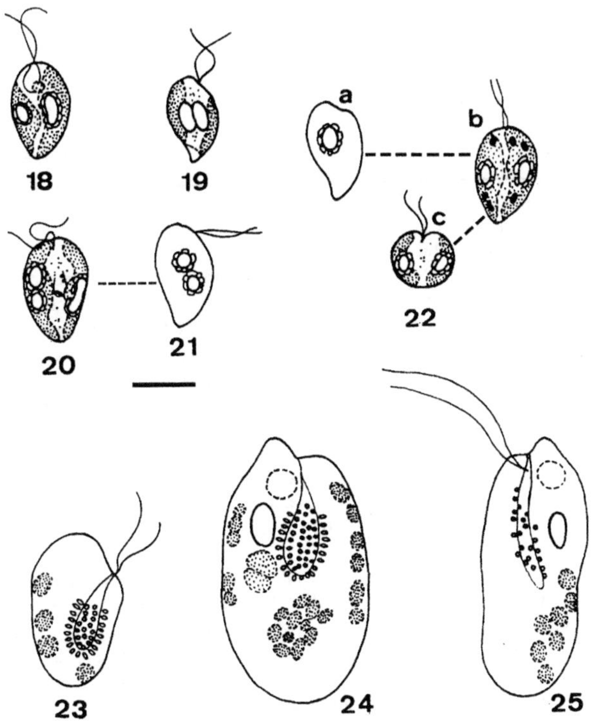

Figure 18–25. 18 – 25. Cells of species of *Cryptomonas* and *Pseudocryptomonas* from Parque Estadual das Fontes do Ipiranga, São Paulo State (from Castro et al., 1991). *Figures 18–22. Cryptomonas brasiliensis. Figures 18, 20.* Ventral view. *Figures 19, 21.* Dorso-lateral view. *Figure 22a.* Lateral view. *Figure 22b.* Dorsal view. *Figure 22c.* Vertical view. *Figures 23, 24. Pseudocryptomonas americana,* lateral view. *Figure 25. Pseudocryptomonas subcylindrica,* lateral view. - Scale bar = 10 μm.

to Brazilian tropical waters, and possibly endemic in some particular ecosystems.

The second question is still unresolved and as such it is one of the greatest difficulties hindering the study of cryptomonad diversity in Brazil. The lack of adequate infrastructure in this country makes it very difficult for the very few Brazilian cryptomonad specialists to adopt efficient and consistent investigation methods. In turn, this makes it very difficult to obtain results which are significant within the context of a modern systematic and taxonomic framework based on electron microscopy (for more details see Javornický, 2003; Novarino, 2003). Therefore, it is likely that the total number of cryptomonad taxa effectively present in Brazil is incorrect and most probably underestimated. In spite of these difficulties, light microscopy is still a valuable tool in the study of cryptomonad diversity because it yields information on taxonomically significant characters such as cell size and shape (Lepistö & Holopainen, 2003), and chloroplast number, size, and colour (the latter reflecting the phycobiliprotein present – see Novarino, 2003). Furthermore, light microscopy is the choice tool for studying the morphological variability of natural cryptomonad populations, which is still largely unknown.

Conclusions

Based on the present review, a number of recommendations emerge for future research. It is overwhelmingly clear that many more baseline investigations are necessary on cryptomonad diversity in Brazil, based on natural and cultured populations studied using light- and electron microscopy. During routine phytoplankton ecological studies based on light microscopy only, it will be highly desirable to include accurate verbal descriptions and quality illustrations (line-drawings and/or photographs) of all the cryptomonads found, even if they are identified only in part, together with a mention of the keys/taxonomic works used for identification. In order to relate the morphology of living cells to that of fixed material, it would be particularly useful to provide observations on living material in sedimentation chambers prior to quantitative analysis using inverted microscopy. Ideally, it would be extremely useful to carry out correlative studies of living material under ordinary light microscopy first, then Lugol-fixed material under the inverted microscope, and finally fixed material using electron microscopy. Finally, we envisage that the development of international collaborative research projects between Brazilian institutions and taxonomic specialists will lead to a better understanding of cryptomonad diversity in Brazil, whilst at the same time improving the training of younger specialists in this country.

Acknowledgements

We thank Dr Vera Lucia de Moraes Huszar and FEEMA for supplying unpublished ecological data on the Rodrigo de Freitas lagoon; Dr Sergio Melo for providing the data on Imboassica lagoon. This study was partly funded by CNPq, FUJB and FAPERJ.

References

Barone, R. & L. Naselli-Flores, 2003. Distribution and seasonal dynamics of cryptomonads in Sicilian water bodies. Hydrobiologia 502 (Dev. Hydrobiol. 172): 325–329.

Bicudo, C. E. M. & R. M. T. Bicudo, 1967. Floating communities of algae in an artificial pond in the Parque do Estado, São Paulo, Brazil. J. Phycol. 3: 233–234.

Bicudo, C. E. M. & G. Tell, 1988. *Pseudocryptomonas*, a new genus of Cryptophyceae from Southern Brazil. Nova Hedwigia 46: 407–411.

Bicudo, C. E. M., D. C. Bicudo, A. A. J. Castro & M. M. Picelli-Vicentin, 1992. Fitoplâncton do trecho a represar do Rio Paranapanema (Usina Hidrelétrica de Rosana), Estado de São Paulo, Brasil. Revista Brasileira de Biologia 52: 293–310.

Bittencourt-Oliveira, M. C., 1994. Ficoflórula do reservatório de Balbina, Estado do Amazonas, IV. Nostocophyceae, Cryptophyceae e Tribophyceae. Hoehnea 21: 163–173.

Bressan, F., 2001. Fatores reguladores da dominância de *Cylindrospermopsis raciborskii* (Woloszynska) Seenayya & Subba-Raju no reservatório Tabocas, Caruaru, PE. Dissertation, Universidade Federal do Rio de Janeiro. 35 pp.

Castro, A. A. J., 1990. Catálogo das Cryptophyceae de águas continentais do Brasil. Revista Brasileira de Biologia 5: 109–112.

Castro, A. A. J., 1993. Cryptophyceae do Estado de São Paulo: inventário taxonômico. Doctoral Thesis, Universidade Estadual Paulista, Rio Claro. 206 pp.

Castro, A. A. J., C. E. M. Bicudo & D. C. Bicudo, 1991. Criptógamos do Parque Estadual das Fontes do Ipiranga, São Paulo, SP. Algas, 2: Cryptophyceae. Hoehnea 18: 87–106.

Clay, B. L., P. Kugrens & R. Lee, 1999. A revised classification of Cryptophyta. Bot. J. linn. Soc. 131: 131–151.

Dias Jr., C., 1991. Ciclo anual do fitoplâncton e algumas variáveis ambientais na Lagoa do Infernão (SP). Dissertation, Universidade Federal de São Carlos. 108 pp.

Dias, I. C. A., M. Menezes, M. G. Sophia, C. A. Souza, A. M. Araújo, B. P. Carmo & C. P. Vianna, 2001. Listagem Taxonômica e Considerações Fitogeográficas. In Costa, A. F. & I. C. A. Dias (eds), Flora do Parque Nacional da Restinga de Jurubatiba e Arredores, Rio de Janeiro: Listagem, Florística e Fitogeografia – Parte II: Biodiversidade das Algas Continentais. Série Livros 8. Museu Nacional. Rio de Janeiro: 171–200.

Ferreira, A. C. S., 2002. Dinâmica do fitoplâncton de um reservatório hipereutrófico (Reservatório Tapacurá, Recife, PE), com ênfase em *Cylindrospermopsis raciborskii* e seus morfotipos. Dissertation, Universidade Federal do Rio de Janeiro. 79 pp.

Giani, A. & R. M. Pinto-Coelho, 1986. Contribuição ao conhecimento das algas fitoplanctônicas do reservatório do Paranoá, Brasília, Brasil: Chlorophyta, Euglenophyta, Pyrrophyta e Schizophyta. Revista Brasileira de Botânica 9: 45–62.

Giani, A., C. C. Figueiredo & P. C. Eterovick, 1999. Algas planctônicas do reservatório da Pampulha (MG); Euglenophyta, Chrysophyta, Pyrrophyta, Cyanobacteria. Revista Brasileira de Botânica 22: 107–116.

Huszar, V. L. M., 1996. Planktonic algae, other than desmids, of three Amazonian systems (Lake Batata, Lake Mussurá and Trombetas River), Pará, Brazil. Amazoniana 14: 37–73.

Huszar, V. L. M., L. H. S. Silva & F. A. Esteves, 1990. Estrutura das comunidades fitoplanctônicas de 18 lagoas da região do Baixo Rio Doce, Linhares, Espírito Santo, Brasil. Revista Brasileira de Biologia 50: 585–598.

Javornický, P., 2003. Taxonomic notes on some freshwater planktonic Cryptophyceae based on light microscopy. Hydrobiologia 502 (Dev. Hydrobiol. 172): 271–283.

Klaveness, D., 1985. Classical and modern criteria for determining species of Cryptophyceae. Bull. Plankton Soc. Japan 31: 111–128.

Klaveness, D., 1989. Biology and ecology of the Cryptophyceae: status and challenges. Biol. Oceanogr. 6: 257–270.

Lepistö, L. & A.-L. Holopainen, 2003. Occurrence of Cryptophyceae and katablepharids in boreal lakes. Hydrobiologia 502 (Dev. Hydrobiol. 172): 307–314.

Magrin, A. G. E., 1993. Aspectos limnológicos, composição e distribuição da comunidade fitoplanctônica em um reservatório da Amazônia brasileira (Reservatório da U.H.E. Samuel, Estado de Rondônia), durante sua fase de enchimento. Dissertation, Universidade Federal de São Carlos. 322 pp.

Marinho, M. M., 1996. Algas, exceto desmídias, de um pequeno reservatório do sudeste brasileiro (Açude do Jacaré, São Paulo, Brasil). Revista Brasileira de Biologia 56: 99–110.

Melo, S., 2001. Fitoplâncton da Lagoa Imboassica (Macaé, RJ-Brasil): flora, estrutura da comunidade e variações espaciais e temporais. Doctoral Thesis, Universidade Federal do Rio de Janeiro. 153 pp.

Menezes, M. & P. Domingos, 1994. La flore planctonique d'une lagune tropicale (Brésil). Rev. d'Hydrobiol. Trop. 27: 273–297.

Menezes, M. & C. A. Souza, 2001. Cryptophyta- Cryptophyceae. In Menezes, M. & I. C. A. Dias (eds), Biodiversidade de Algas de Ambientes Continentais do Estado do Rio de Janeiro. Série Livros 9. Museu Nacional., Rio de Janeiro: 99–103.

Moustaka-Gouni, M., 1996. Some aspects on the morphology and ecology of *Rhodomonas minuta* var. *nannoplanctica* and *R. lens* (Cryptophyceae) in two Greek lakes. Nordic J. Bot. 16: 335–343.

Nogueira, I. S., 1999. Estrutura e dinâmica da comunidade fitoplanctônica da Represa Samambaia, Goiás, Brasil. Doctoral Thesis, Universidade de São Paulo. 340 pp.

Novarino, G., 2003. A companion to the identification of cryptomonad flagellates (Cryptophyceae = Cryptomonadea). Hydrobiologia 502 (Dev. Hydrobiol. 172): 225–270.

Novarino, G. & I. A. N. Lucas, 1993. Some proposal for a new classification system of the Cryptophyceae. Bot. J. linn. Soc. 111: 3–21.

Novarino, G., I. A. N., I. A. N. Lucas & S. Morrall, 1994. Observations on the genus *Plagioselmis* (Cryptophyceae). Cryptogamie, Algologie 15: 87–107.

Senna, P. A. C., M. G. Souza & P. Compère, 1998. A check-list of the algae of the Federal District (Brazil). Scripta Botanica Belgica 16. Natural Botanical Garden of Belgium, M. Verhaegen, Meise. 88 pp.

Silva, L. H. S., 1999. Fitoplâncton de um reservatório eutrófico (Lago Monte Alegre), Ribeirão Preto, São Paulo, Brasil. Revista Brasileira de Biologia 59: 281–303.

Souza, C. A., 2002. Taxonomia e aspectos biogeográficos da flora planctônica de um sistema costeiro salobro- Lagoa Carapebus, Município de Carapebus, RJ. Dissertation, Universidade Federal do Rio de Janeiro, Rio de Janeiro. 201 pp.

Train, S., 1998. Flutuações temporais da comunidade fitoplanctônica do sub-sistema rio-Baía-lagoa do Guaraná, Planície de Inundação do alto Rio Paraná (Baitaporã, Mato Grosso do Sul). Doctoral Thesis, Universidade de São Paulo, São Carlos. 190 pp.

Willén, E., M. Oké & F. González, 1980. *Rhodomonas minuta* and *Rhodomonas lens* (Cryptophyceae) – aspects on form variation and ecology in lakes Mälaren a Vättern Central Sweden. Acta Phytogeographica Suecica 68: 163–172.

Hydrobiologia **502:** 307–314, 2003.
L. Naselli-Flores, J. Padisák, M. T. Dokulil (eds),
Phytoplankton and Equilibrium Concept: The Ecology of Steady-State Assemblages.

Occurrence of Cryptophyceae and katablepharids in boreal lakes

Liisa Lepistö[1] & Anna-Liisa Holopainen[2]
[1] *Finnish Environment Institute, P.O.Box 140, Finland*
E-mail: liisa.lepisto@vyh.fi
[2] *Karelian Research Institute, University of Joensuu, P.O.Box 111 and North Karelian Environment centre, P.O.Box 69, FIN-80101 Joensuu, Finland*

Key words: Cryptophyceae, community structure, assemblages, cell size variations, phytoplankton, quality of waters, boreal lakes, Eastern Finland

Abstract

One of the most important algal groups in Finnish lakes are the Cryptophyceae. Changes in the community structure of Cryptophyceae in a total of 22 lakes belonging to the Vuoksi river basin in eastern Finland were studied. The existence of lakes with water qualities varying from oligotrophic to eutrophic, often loaded by human activities, provides a good opportunity to study the effects of environmental variables on the occurrence and size variation of Cryptophyceae. In the Vuoksi river basin, the main soil type is moraine. Twelve of the lakes were large or moderately large and with clear, i.e. oligo-humic water, and one lake could be described as a small clear water lake. Eight large or moderately large lakes were humic, with a water colour number of 40–70 mg l^{-1} Pt, including three lakes impacted by nutrient loads. One lake was naturally eutrophic, with a high water colour number of 100 mg l^{-1} Pt, and was also impacted by municipal and pulping effluents. CCA-ordination analysis grouped the studied lakes into: (1) clear water lakes, (2) humic lakes and (3) the naturally eutrophic brown water lake. In the CCA-ordination analysis based on cell numbers small Cryptophyceae (Cryptomonadales), *Rhodomonas lacustris* and the katablepharid *Katablepharis ovalis* were grouped into the first axis, which was positively correlated with Secchi depth ($r = 0.58$) and NO_3N - nitrogen ($r = 0.24$) and negatively with Ptot ($r = -0.69$), PO_4P ($r = -0.69$) and water colour number ($r = -0.66$). In humic lakes, medium-sized Cryptophyceae were abundant. The naturally eutrophic lake was grouped into first axis, which is positively correlated with Ptot ($r = 0.69$), PO_4P ($r = 0.69$) and water colour number ($r = 0.66$). The lake formed a distinct group with large Cryptophyceae. Only in this lake was the heterotrophic *Katablepharis ovalis* rather abundant. However, large-sized taxa dominated the biomass of the Cryptophyceae assemblage in all lake types excluding large clear water lakes, where *Rhodomonas lacustris* dominated and large Cryptophyceae co-dominated.

Introduction

The proportion of cryptophycean biomass is approximately equal to that of other algal groups in oligotrophic Finnish clear water lakes (Lepistö, 1999). With increasing eutrophy, Cryptophyceae and Bacillariophyceae become dominant, and ultimately the Cyanophyceae (Cyanobacteria) account for the main part of the biomass in highly eutrophic lakes (Heinonen, 1980; Lepistö, 1999; Willén, 2000). In meso-humic and humic waters Cryptophyceae are abundant throughout the summer (Arvola et al., 1991; Salonen et al., 1992; Lepistö & Saura, 1998; Arvola et al., 1999). Cryptophyceae have a wide environmental spectra, as they are typical in waters rich in nutrients and organic matter (Rosén, 1981) but also tolerate occasional nutrient depletion, especially of nitrogen (Haffner et al., 1980). Furthermore, they tolerate waters with mud and clay particles (Brettum, 1989). Their competitiveness is enhanced by the presence of flagella, which allow the migration vertically between water layers of optimal light conditions and nutrient concentrations (Salonen et al., 1984; Jansson et al., 1996). However, both the seasonal succession and the spatial variation of Cryptophyceae in large Finnish lakes are poorly known (e.g. Karjalainen et al., 1996).

In spring during the rapid development of thermal stratification, the Cryptophyceae replace the Bacillariophyceae (Reynolds, 1984 and references therein, Lepistö et al., 1999).

Cryptophyceae are sensitive to formalin, used as a preservative until the early 1970s in Finnish monitoring studies (Lepistö, 1999). This caused their underestimation during the 1960s and early 1970s. Furthermore, unequivocal identification cannot be accomplished using light microscopy (Javornický, 2001; Javornický 2003; Novarino, 2003), since it is impossible to observe the structure of the depression-furrow-gullet system for correct identification as proposed by Butcher (1967) and Bourrelly (1970).

The aim of this paper was to study the changes in community structure of Cryptophyceae flagellates within the Vuoksi river basin. The existence of lakes with water quality from oligotrophic to eutrophic, some of them with human impact provided a good opportunity to study the effects of environmental variables on Cryptophyceae.

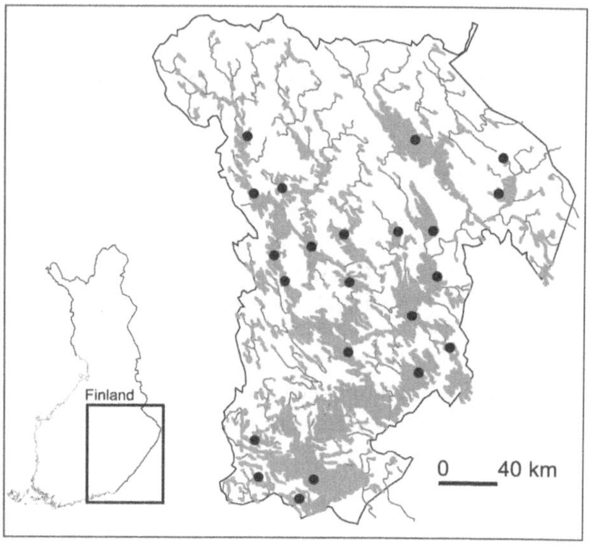

Figure 1. The study lakes and their location in the Vuoksi river basin in Eastern Finland.

Study lakes

The lakes in this study are situated in the Vuoksi river basin in eastern Finland (Fig. 1). The area of the Vuoksi river basin is 61 265 km^2 and lakes therein form more than one third of the total lake surface area of Finland. In the Vuoksi river basin the lake density is 43 lakes per 100 km^2 (Kuusisto, 1978; Raatikainen & Kuusisto, 1988). Lake Saimaa is the largest lake, with an area of 4400 km^2, and it forms a labyrinth of narrow sounds and open waters. In this study 22 lakes were studied. The mean lake depth in the study lakes varied from approximately 4.3 to 17.2 metres, and the retention time from 293 to 4015 days, which is the longest measured retention time of the Finnish lakes.

The studied lakes were grouped according to their surface area, water colour number and main soil type in the catchment area. Different compositions of lake types, altogether eight, are presented as letter combinations (Table 1). For example the combination CAA means that the lakes are large, with mainly moraine soil in the catchment area and the lake water is clear, i.e. oligo-humic. In the Vuoksi river basin, the main soil type is moraine. Twelve of the lakes are large or moderately large and with clear water (CAA, BAA). One of these lakes could be considered as slightly impacted (CAAi). One lake could be described as a small clear water lake (AAA). Eight large or moderately large lakes are humic (CAB, BAB), with water colour number varying from 40 to 70 mg l^{-1} Pt. Three of these are impacted by human activities (CABi). One lake is naturally eutrophic, with a high water colour number of 100 mg l^{-1} Pt, and is also impacted by municipal and pulping wastes (BCCi).

Sampling and analytical methods

Phytoplankton samples were collected by the Regional Environmental Centres as part of the national monitoring of Finnish lakes. Contemporarily with phytoplankton sampling, samples for water chemistry were also taken and analysed according to standard methods (Niemi et al., 2000). Phytoplankton was sampled from a depth of 0–2 m in July in 1982–1999. Samples were preserved using acidic Lugol's solution. Phytoplankton from a constant area of the sedimentation chamber bottom was counted and identified under an inverted microscope using phase-contrast illumination at $\times 200$ and $\times 800$–1200 magnifications by a permanent group of researchers. Cell counts were converted to biovolumes using the cell volumes of the phytoplankton database of the Finnish Environment Institute (for Cryptophyceae, see Table 2). The biomass values are given as fresh weight (mg l^{-1}).

The quantity and composition of the whole phytoplankton assemblage was studied, but particular attention was paid to Cryptophyceae as cell density and biomass and to the occurrence of different size classes in studied lake types. *Rhodomonas lacustris* Pascher et

Table 1. Lake types based on the surface area, the main soil type in the catchment area and water colour number. Note that 'i' after the letter combination means that the lake is impacted by human activities

Type	Number of the lakes	Surface area (km^2)	The soil in the catchment mainly of clay (> 30%)	Water colour (mg l^{-1} Pt)
CAA	8	>100	No	<40
CAAi	1	>100	No	<40
BAA	3	100–10	No	<40
AAA	1	<10	No	<40
CAB	4	>100	No	40–80
CABi	3	>100	No	40–80
BAB	1	100–10	No	40–80
BCCi	1	100–10	Yes	>80

Table 2. Cell-size classes based on cell length. The diameter, length and volumes of observed Cryptophyceae are given

Size (length)	Cell diameter, μm	Cell length, μm	Volume μm^3
Small, ≤16 μm	3	6	24
	5	9	94
	6	12	151
	7	13	262
	8	15	377
	8	16	435
Medium–large, 16–26 μm	8	18	452
	10	20	754
	13	26	1769
Large, > 26 μm	14	33	2418
	16	38	4136

Ruttner and the colourless *Katablepharis ovalis* Skuja were identified at the species level. Other taxa were identified at the order level or genus level and treated as a group (Cryptophyceae). A total of 11 size classes were determined during microscopical investigation and combined to three classes: small, medium-large and large (Table 2).

The correlations between Cryptophyceae cell density and water quality were examined by canonical correspondence analysis (CCA) using the CANOCO computer program (ter Braak, 1987, 1990). Water quality data is presented by Pietiläinen and Räike (1999) and water quality classification is based to Forsberg & Ryding (1980) and OECD (1982). These data are presented below without repeated references.

Results and discussion

CCA-ordination analysis grouped the studied lakes into three different groups: (1) large, moderately large and small clear water lakes, (2) large and moderately large humic lakes and (3) the naturally eutrophic lake.

Clear water lakes

Total phosphorus concentration was low in the large clear water lakes (group CAA), reflecting mainly oligotrophic conditions. The most oligotrophic lake was L. Puruvesi and the highest concentration of total phosphorus was found in L. Saimaa, Riutanselkä (Table 3). One of the lakes, L. Saimaa, Ilkonselkä, could be considered as slightly impacted, due to still ongoing loading from the pulp industry (Laine, 2001). However, neither nutrient concentrations (Ptot 6.8 μg l^{-1}, as median) nor phytoplankton (0.3 mg l^{-1}, as mean) in the epilimnetic water layer (Fig. 2) reflect any signs of human impact. In large clear water lakes the mean phytoplankton biomass was 0.4 mg l^{-1} (Table 3). Cryptophyceae contributed ca. 27% to the total biomass when the proportion of Chrysophyceae was slightly lower, ca. 26% (Fig. 2).

In the moderately large lakes (group BAA), the water colour was somewhat darker than in the large lakes. The total phosphorus concentration was low (Table 3). The most oligotrophic lake was L. Kuolimo and the highest median concentration of phosphorus was observed in L. Louhivesi. The average total phytoplankton biomass was 0.4 mg l^{-1} and Cryptophyceae contributed 35% to the total biomass (Fig. 2).

In the only small lake (group AAA) in this study, phytoplankton biomass was 0.5 mg l^{-1} and Crypto-

Table 3. Some water quality variables in lakes of different lake types in the Vuoksi river basin. The water colour number and total phosphorus concentration (Ptot) are as median values, according to Pietiläinen & Räike (1999). Total biomass and the biomass of Cryptophyceae as mean values are used when estimating the proportion of Cryptophyceae (%). The mean biomass in each lake type is given in brackets

Type (bm mg l^{-1})	Lake	Water colour mg l^{-1} Pt	Ptot $\mu g\ l^{-1}$	Total biom. mg l^{-1}	Crypt. biom. mg l^{-1}	Crypt.%
CAA (0.4)	Höytiäinen	28	7	0.24	0.06	26
	Pyhäjärvi	10	6	0.27	0.07	28
	Iso-Haukivesi	35	10	0.29	0.12	42
	Puruvesi	5	5	0.33	0.04	13
	Juojärvi	28	6	0.34	0.07	22
	Lietvesi	30	8	0.41	0.12	28
	Saimaa, Riutanselkä	20	15	0.66	0.21	31
	Suvasvesi	30	8	0.46	0.11	25
CAAi (0.3)	Saimaa, Ilkonselkä	30	7	0.31	0.07	25
BAA (0.4)	Kermajärvi	20	6	0.17	0.07	42
	Kuolimo	18	4	0.21	0.06	29
	Louhivesi	13	9	0.69	0.25	36
AAA (0.4)	Iso-Hietajärvi	20	5	0.45	0.02	4
CAB (0.5)	Koitere	70	11	0.34	0.14	42
	Pielinen	50	10	0.46	0.08	17
	Paasivesi	40	9	0.44	0.20	47
	Viinijärvi	40	13	0.90	0.11	13
CABi (0.8)	Pyhäselkä	65	11	0.41	0.16	38
	Juurusvesi	60	18	0.62	0.35	56
	P-Kallavesi	65	27	1.32	0.50	38
BAB (1.4)	Koirus	45	16	1.42	0.17	12
BCCi (4.0)	Onkivesi	100	56	3.99	0.82	21

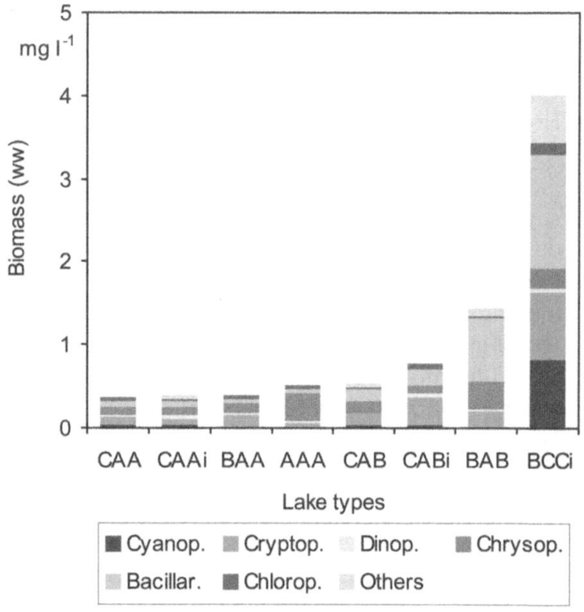

Figure 2. Phytoplankon composition and biomass in different lake types in the Vuoksi river basin in July during the period 1982–1999.

phyceae contributed only 4% to the total biomass when Chrysophyceae dominated (Fig. 2). The water in this lake is characterised by a moderately high colour number but a low total phosphorus concentration (Table 3).

Humic lakes

The median total phosphorus concentration varied from 9 to 13 $\mu g\ l^{-1}$ in the large humic lakes (group CAB) reflecting (oligo-)mesotrophy. The most oligotrophic was L. Paasivesi and the highest phosphorus concentration was observed in L. Viinijärvi (Table 3). The mean total phytoplankton biomass was 0.5 mg l^{-1}. Cryptophyceae contributed ca. 30% and Chrysophyceae ca. 27% to the total biomass (Fig. 2). In the impacted humic lakes (group CABi) Cryptophyceae contributed 44% to the total biomass (Fig. 2). The total phosphorus concentration varied from 11 to 27 $\mu g\ l^{-1}$ and total phytoplankton biomass was 0.8 mg l^{-1} on average. These values were twice as high as in the non-impacted large humic lakes (Table 3).

In the moderately large L. Koirus (group BAB) total phosphorus concentration reflected mesotrophy, and the average phytoplankton biomass was 1.4 mg l^{-1} (Table 3). Bacillariophyceae dominated the total biomass at 53%, whereas Cryptophyceae contributed only 12% to the biomass (Fig. 2).

The only naturally eutrophic lake in this lake material is characterized by very dark water (group BCCi). Furthermore, the lake is loaded by municipal and pulping effluents. The total phosphorus concentration of 56 $\mu g\ l^{-1}$ (Table 3) clearly reflects eutrophy, as does the high phytoplankton biomass, 4.0 mg l^{-1} (Heinonen, 1980). Cryptophyceae contributed 20%, Bacillariophyceae 34% and Cyanophyceae 21% to the total biomass (Fig. 2).

Cryptophyceae asssemblages in lakes

CCA-ordination analysis based on cell numbers grouped Cryptophyceae into the lake types according to cell size. Small Cryptophyceae (Table 2), *Rhodomonas lacustris, Katablepharis* spp. and *Katablepharis ovalis* of the clear water lakes (CAA, BAA, AAA) were grouped into the first axis (Fig. 3). This axis is positively correlated with Secchi depth ($r = 0.58$) and NO_3N - nitrogen ($r = 0.24$) and negatively with Ptot ($r = -0.69$), PO_4P ($r = -0.69$) and water colour number ($r = -0.66$). The environmental variables explained only 15% of the variation. According to Sciandra et al. (2000), the mean cell size reacts significantly to change in growth conditions. However, light and N limitations have opposite effects on cell size. Nitrate deprivation and enrichment induce, respectively an increase and a decrease in cell diameter. Somatic growth is inhibited by low light resulting in a rapid and transient reduction in cell diameter. The results about small cells positively correlated with Secchi depth and also with NO_3N - nitrogen corroborate with the responses of Cryptophyceae to the habitat properties noted also by Reynolds et al. (2002).

The biomass of Cryptophyceae in large clear water lakes was also dominated by *Rhodomonas lacustris* and co-dominated by large *Cryptomonas* spp. In moderately large clear water lakes large *Cryptomonas* sp. dominated the biomass and *Rhodomonas lacustris* co-dominated (Fig. 4). As cell numbers the medium-large Cryptophyceae were grouped into humic lakes (CAB, CABi, BAB). The observations by Tardio et al. (2003) support our results concerning the proliferation

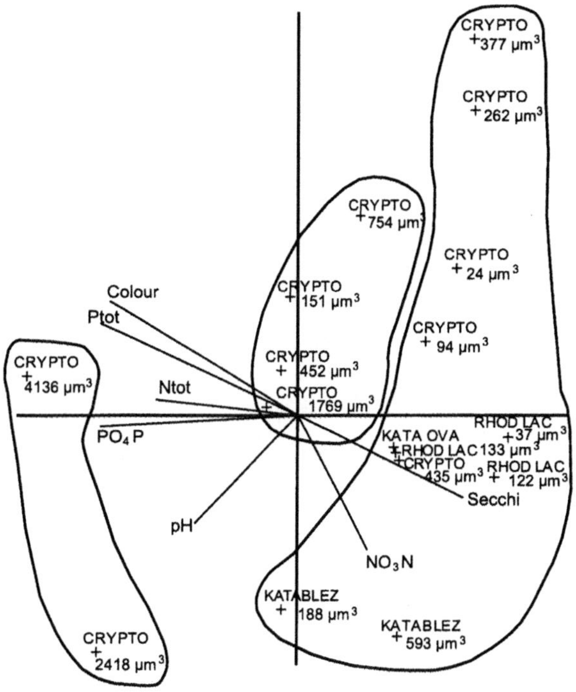

Figure 3. CCA ordination diagram of Cryptophyceae taxa as cell density and environmental variables. Clear water lakes were grouped to the right, humic lakes in the middle and the eutrophic lake to the left.

of these medium-large algae (including *Cryptomonas marsonii*) in humic waters promoted by low light intensity. In this lake group the biomass of Cryptophyceae was dominated by large-sized taxa, particularly in the impacted lakes, and *Rhodomonas lacustris* co-dominated (Fig. 4).

The naturally eutrophic lake formed a distinct group, and CCA-ordination analysis based on cell numbers grouped large *Cryptomonas* spp. (Table 2) into the first axis (Fig. 3). This axis is correlated with Ptot ($r = 0.69$), PO_4P ($r = 0.69$) and water colour number ($r = 0.66$). However, the environmental variables explained only 15% of the variation. In this lake the heterotrophic *Katablepharis ovalis* was rather abundant. The large *Cryptomonas* spp. also dominated the biomass (Fig. 4).

Cryptophyceae are an algal group known to benefit from nutrient-rich waters (Rosén, 1981; Brettum, 1989). When a lake is recovering from loading, not only does water quality improve but the phytoplankton assemblage also shifts towards smaller cell sizes (Wattson & Kalff, 1981; Willén 1992). A good example is Lake Saimaa, Ilkonselkä, a large clear water lake, which was heavily polluted by paper mill effluents especially during the 1960s. The loading has been

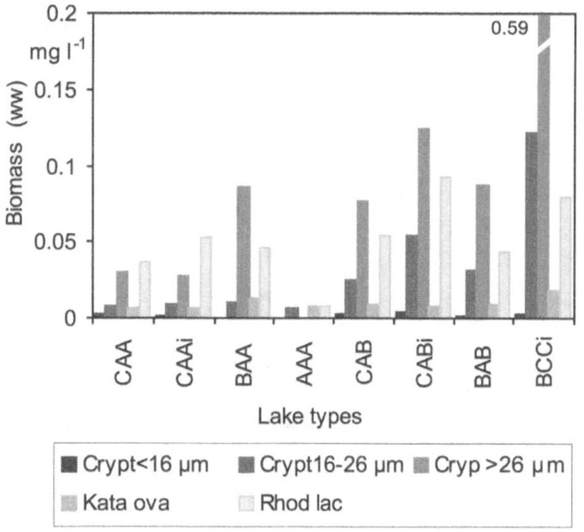

Figure 4. The size fraction of Cryptophyceae taxa as biomass in different lake groups in the Vuoksi river basin in July during the period 1982–1999. Crypt = Cryptophyceae, Kata ova = *Katablepharis ovalis* and Rhod lac = *Rhodomonas lacustris*.

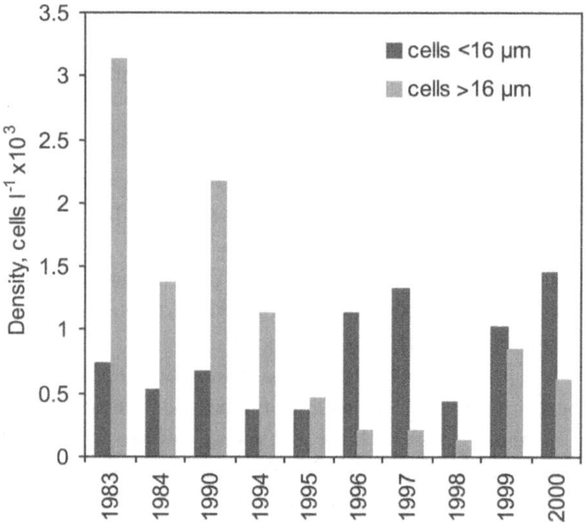

Figure 5. Long-term changes in cell sizes of the Cryptophyceae assemblage in L. Saimaa, Ilkonselkä in July during the period 1983–2000.

reduced almost totally today, although small amounts of waste waters are still discharged (Laine, 2001). The recovery of Saimaa, Ilkonselkä is manifested by the decrease of total phosphorus, phytoplankton biomass and large / moderately large Cryptopyceae-flagellates, and by the increase of small Cryptophyceae in the 1990s (Fig. 5). The increased proportion of nitrogen as a consequence of changes in the N/P ratio is reflected as decrease of the cell size as concluded by Sciandra et al. (op. cit.).

Cryptophyceae tolerate occasional nutrient depletion, especially of nitrogen (Haffner et al., 1980). According to Sciandra et al. (op. cit.), the nitrate deprivation increases the cell diameter. Furthermore, these flagellates have the ability to consume bacteria (Tranvik et al., 1989). Many flagellates are facultatively auto- or phagotrophic and thus able to outcompete purely autothrophic species during nutrient limited conditions (Bergström et al., 2001).

Cryptophyceae are able to migrate vertically to water columns with optimal light conditions and nutrient concentrations (Salonen et al., 1984; Smolander & Arvola, 1988; Jansson et al., 1996; Reynolds et al., 2002), and to avoid the grazing pressure by vertical movements (Salonen & Rosenberg, 2000). The competitive advantage conferred by migration seems to benefit particularly the median-sized taxa abundant in humic waters and large taxa in eutrophic turbid waters. In brown water lakes, the grazing effect of predators is better buffered than in more productive clear water lakes, according to del Giorgio & Casol (1995). Järvinen (2002) concluded that phytoplankton production is limited by nitrogen rather than phosphorus if large herbivorous cladocerans are few in number.

In clear water lakes, small-sized Cryptophyceae dominate as cell density but only in large clear water lakes the small-sized *Rhodomonas lacustris* are abundant enough to dominate the biomass. One reason could be that large cells are easily captured by zooplankton. According to Arvola & Salonen (2001), the cladoceran *Daphnia* affects the abundance of its single-celled prey, and thus the biomass of phytoplankton. Furthermore, cell division is very rapid in the smaller organisms. Here it must be kept in mind that only the upper water layer from the surface to a depth of two metres was sampled. Large taxa might always remain deeper in clear waters (Arvola et al., 1999).

Concluding remarks

The Cryptophyceae comprise a very important group in Finnish humic lakes, but are also very abundant in clear water lakes. Variation of cell size and in abundance appears to reflect different types of waters and also their nutritional status. However, under the light microscope used for monitoring phytoplankton it is impossible to observe details of the cell structure for correct identification. Therefore, during microscopical investigation cell size measurements are very import-

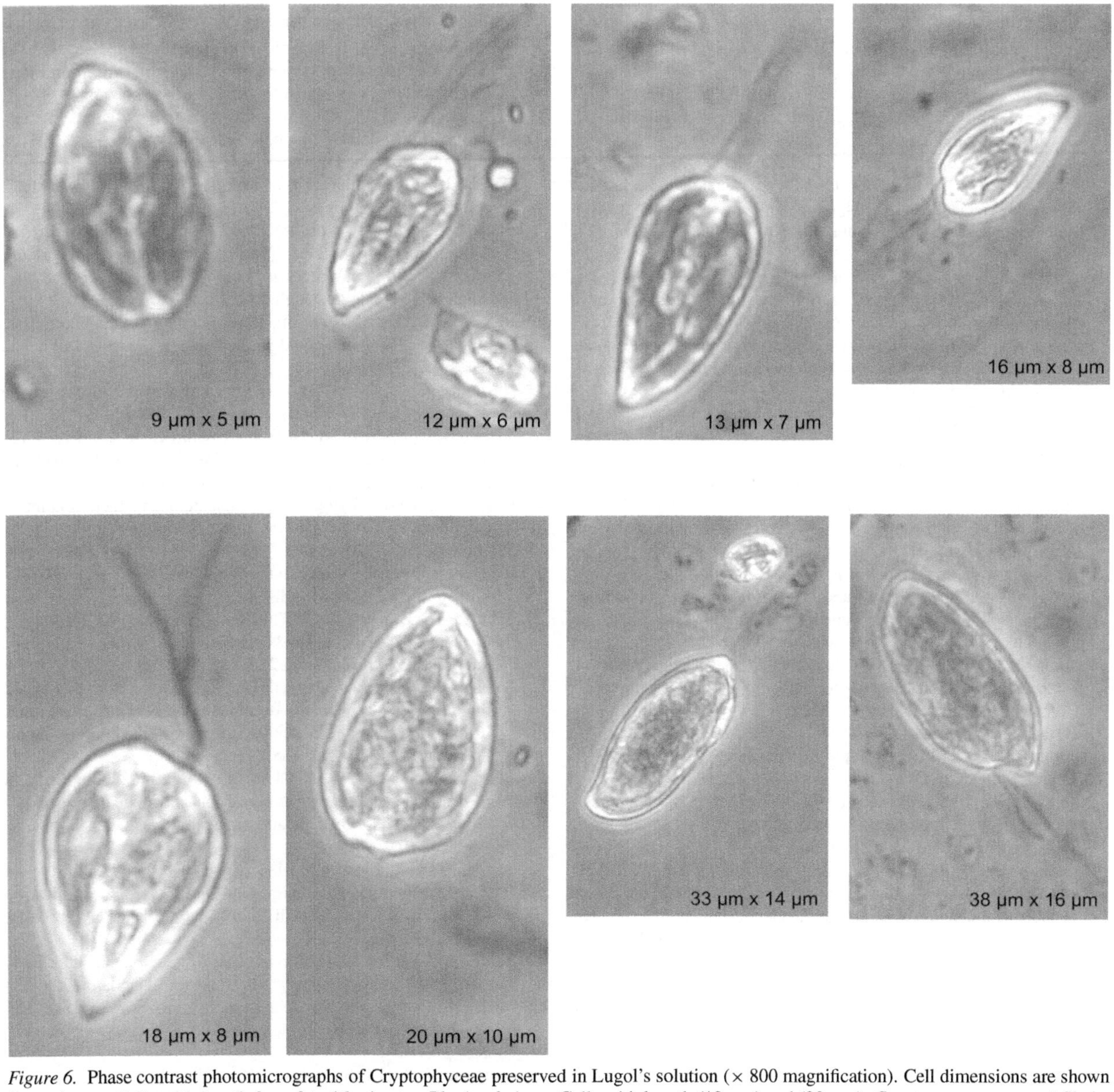

Figure 6. Phase contrast photomicrographs of Cryptophyceae preserved in Lugol's solution ($\times$ 800 magnification). Cell dimensions are shown for each taxa. Cells with length from 9 to 16 microns: *Plagioselmis* sp. Cells with length (18 μm) and (20 μm): *Cryptomonas marssonii* Skuja, with length (33 μm): *Cryptomonas reflexa* (Marsson) Skuja and with length (33 μm): *Cryptomonas curvata* Ehrenb. (Photos Reija Jokipii).

ant. More attention should be paid to species-level identification (Fig. 6), although size variation does appear to provide the information needed for monitoring purposes.

Acknowledgements

We thank Reija Jokipii and Maija Niemelä for analysing the phytoplankton of the study lakes, Sirkka Vuoristo for drawing the figures, and Michael Bailey for improving the English. We also thank Dr Pavel Javornický for valuable comments on the identification of Cryptophyceae.

References

Arvola, L., A. Ojala, F. Barbosa & S. I. Heaney, 1991. Migration behaviour of three Cryptophytes in relation to environmental gradients: an experimental approach. British Phycol. J. 26: 361–373.

Arvola, L., P. Eloranta, M. Järvinen, J. Keskitalo & A.-L. Holopainen. 1999. Phytoplankton. In Keskitalo, J. & P. Eloranta (eds), Limnology of Humic Waters. Bachuys Publishers, Leiden, The Netherlands: 137–171.

Arvola, L. & K. Salonen. 2001. Plankton community of a polyhumic lake with and without *Daphnia longispina* (Cladocera). Hydrobiologia 445: 141–150.

Bergström, A.-K., M. Jansson, P. Blomqvist & S. Drakare. 2001. The influence of water colour and effective light climate on mixotrophic phytoflagellates in three small dystrophic Swedish lakes. Int. Ver. theor. angewand. Limnol: Verh 27: 1861–1865.

Brettum, P., 1989. Alger som indikator på vannkvalitet i norske innsjøer. Planteplankton. [Algae as indicators of water quality in Norwegian lakes. Phytoplankton]. Niva-Rapport 0-86116. 111 pp.

Bourrelly, P., 1970. Les algues d'eau douce 3. Les algues bleues et rouges, les Eugléniens, Péridiniens et Cryptomonadines. Éditions N. Boubée & Cie, Paris. 512 pp.

Butcher, R.W., 1967. An introductory account of the smaller algae of British coastal waters. 4. Cryptophyceae. Fishery Investigations, Ser. 4, London. 54 pp.

Forsberg, C. & S.-O. Ryding, 1980. Eutrophication parameters and trophic state indices in 30 Swedish waste-receiving lakes. Archiv für Hydrobiol. 89: 189–207.

del Giorgio, P. A. & J. M. Gasol, 1995. Biomass distribution in freshwater plankton communities. Am. Nat. 146: 135–152.

Haffner, G. D., G. P. Harris & M. K. Jarais, 1980. Physical variability and phytoplankton communities III. Vertical structure in phytoplankton populations. Archiv für Hydrobiol. 89: 363–381.

Heinonen, P., 1980. Quantity and composition of phytoplankton in Finnish inland waters. Publications of the Water Research Institute 37. 91 pp.

Jansson, M., P. Blomqvist & A. Jonsson, 1996. Nutrient limitation of bacterioplankton, autotrophic and mixotrophic phytoplankton, and heterotrophic nanoflagellates in Lake Örträsket. Limnol. Oceanogr. 41: 1552–1559.

Javornický, P., 2001. Freshwater *Rhodomonas* (Cryptophyceae). Algol. Stud. 102: 93–116.

Javornický, P., 2003. Taxonomic notes on some freshwater planktonic Cryptophyceae based on light microscopy. Hydrobiologia 502 (Dev. Hydrobiol. 172): 271–283.

Järvinen, M., 2002. Control of plankton and nutrient limitation in small boreal brown-water lakes: evidence from small- and large scale manipulation experiments. Academic Dissertation in Hydrobiology. Dep. Ecol. And Systematic. University of Helsinki, Finland. 41 pp.

Karjalainen, J., A.-L. Holopainen & P. Huttunen, 1996. Spatial patterns and relationships between phytoplankton, zooplankton and water quality in the Saimaa lake system, Finland. Hydrobiologia 322: 267–276.

Kuusisto, E., 1978. Suur-Saimaan vesitase ja tulovirtaaman ennustaminen. [Forecasting of the water budget and the inflow in Lake Suur-Saimaa]. Publications of the Water Research Institute 26. 66 pp.

Laine, P., 2001. Kohti vesiensuojelun aikaa; Veden laadun muutokset eteläisellä Saimaalla. Towards the time of water protection; the changes of water quality in the southern part of Lake Saimaa. Acta Universitatis Lappeenrantaensis 111. Dissertation, Lappeenranta University of Technology. 264 pp. English abstract.

Lepistö, L., 1999. Phytoplankton assemblages reflecting the ecological status of lakes in Finland. Monographs of the Boreal Environment Research 16. 43 pp.

Lepistö, L. & M. Saura, 1998. Effects of forest fertilization on phytoplankton in a boreal brown-water lake. Boreal Environ. Res. 3: 33–43.

Lepistö, L., A. Räike & O.-P. Pietiläinen, 1999. Long-term changes of phytoplankton in a eutrophicated boreal lake during the past 100 years (1893–1998). Algol. Stud. 94: 223–244.

Niemi, J., P. Heinonen, S. Mitikka, H. Vuoristo, O.-P. Pietiläinen, M. Puupponen & E. Rönkä, 2000. The Finnish EUROWATERNET: The European agencys monitoring network for Finnish inland waters. The Finnish Environment 445. 62 pp.

Novarino, G., 2003. A Companion to the identification of cryptomonad flagellates (Cryptophyceae = Cryptomonadea). Hydrobiologia 502 (Dev. Hydrobiol. 172): 225–270.

OECD, 1982. Eutrophication of water, monitoring, assessment and control. Paris. 150 pp.

Pietiläinen, O.-P. & A. Räike, 1999. Typpi ja fosfori Suomen sisävesien minimiravinteina. [Nitrogen and phosphorus as algal growth limiting nutrients in Finnish inland waters]. Suomen ympäristö 313. 64 pp. English abstract.

Raatikainen, M. & E. Kuusisto, 1988. Suomen järvien lukumäärä ja pinta-ala. [The number and surface area of the lakes in Finland]. Terra 102: 97–110.

Reynolds, C. S., 1984. The ecology of freshwater phytoplankton. Cambridge University Press, Cambridge. 384 pp.

Reynolds, C. S., V. L. M. Huszar, C. Kruk, L. Naselli-Flores & S. Melo, 2002. Towards a functional classification of the freshwater phytoplankton. J. Plankton Res. 24: 417–428.

Rosén, G., 1981. Phytoplankton indicators and their relations to certain chemical and physical factors. Limnologica 13: 263–290.

Salonen, K., R. I. Jones & L. Arvola, 1984. Hypolimnetic phosphorus retrieval by diel vertical migrations of phytoplankton. Freshwat. Biol. 14: 431–438.

Salonen, K., P. Kankaala, T. Tulonen, T. Hammar, M. James, T.-R. Metsälä & L. Arvola, 1992. Planktonic food chains of a highly humic lake. Hydrobiologia 229: 143–157.

Salonen, K. & M. Rosenberg, 2000. Advantages from diel vertical migration can explain the dominance of Gonyostomum semen (Raphidophyceae) in a small, steeply-stratified humic lake. J. Plankton Res. 22: 1841–1853.

Sciandra, A., L. Lazzara, H. Claustre & M. Babin, 2000. Responses of growth rate, pigment composition and optical properties of *Cryptomonas* sp. to light and nitrogen stress. Mar. Ecol. Prog. Ser. 201: 107–120.

Smolander, U. & L. Arvola, 1988. Seasonal variation in the diel vertical distribution of the migratory alga *Cryptomonas marssonii* (Cryptophyceae) in a small, highly humic lake. Hydrobiologia 161: 89–98.

Tardio, M., M. Tolotti, G. Novarino & M. Cantonati, 2003. Ecological and taxonomic observations on the flagellate algae characterising four years of enclosure experiments in Lake Tovel (Southern Alps). Hydrobiologia 502 (Dev. Hydrobiol. 172): 285–296.

ter Braak, C. J. F. 1987. CANOCO – a FORTRAN program for canonical community ordinations by [partial] [detrended] [canonical] correspondence analysis and redundancy analysis (version 2.1). Agricultural Mathematics Group, Wageningen. 95 pp.

ter Braak, C. J. F., 1990. Update notes: CANOCO version 3.10. Agricultural Mathematics Group, Wageningen, 35 pp.

Tranvik, L. J., K. G. Porter & J. McN. Sieburth, 1989. Occurrence of bacterivory in Cryptomonas, a common freshwater phytoplankton. Oecologia 78: 473–476.

Wattson, S. & J. Kalff, 1981. Relationships between nanoplankton and lake trophic status. Can. J. Fish. aquat. Sci. 38: 960–967.

Willén, E., 1992. Long-term changes in the phytoplankton of large lakes in response to changes in nutrient loading. Nord. J. Bot. 12: 577–587.

Willén, E., 2000. Phytoplankton in water quality assessment – an indicator concept. In Heinonen, P., G. Ziglio & A. Van der Beken (eds), Hydrological and Limnological Aspects of Lake Monitoring. John Wiley & Sons, Ltd, Chichester: 57–80.

Hydrobiologia **502:** 315–324, 2003.
L. Naselli-Flores, J. Padisák, M. T. Dokulil (eds),
Phytoplankton and Equilibrium Concept: The Ecology of Steady-State Assemblages.

Dominance patterns of planktonic algae in Swedish forest lakes

Eva Willén
Swedish University of Agricultural Sciences, Department of Environmental Assessment, P.O. box 7050, SE-750 07 Uppsala, Sweden
E-mail: eva.willen@ma.slu.se

Key words: forest lakes phytoplankton, summer dominance, species richness, characteristic assemblages, *Gonyostomum semen*

Abstract

Dominance patterns of phytoplankton during a late phase (August) in the seasonal sequence of species are presented from 15 Swedish forest lakes with little or no local anthropogenic impact. The main question to elucidate is if predominance of a small number of species (1–3) occurs during a mature phase of the annual succession i.e. if there is an evident competitive exclusion of species in favour of a few dominants. Ten August months per lake are used to illustrate interannual variations caused mainly by differences in weather conditions. In general, 1–3 dominant taxa do not reach more than 60% of the total phytoplankton volume. Only lakes exposed to some stress factors exhibit a more pronounced dominance pattern with 1–3 species occupying >80% of the phytoplankton biomass. Stress factors are harsh climate (arctic lake), light deficiency (very brown water), acidification and occurrence of the invasive raphidophycean species *Gonyostomum semen*. The variation in phytoplankton assemblages in relation to environmental variables and years was tested by classification and ordination methods (TWINSPAN, CCA). The consistency of the species/lake groupings and the set of explanatory environmental variables was checked in a discriminant function analysis. Species associations during investigated years and environmental variables show a very good consistency and 75% of the lakes was classified in the same species group irrespective of year, indicating stable species assemblages from summer to summer.

Introduction

Sweden is among the most forest-rich countries in Europe with 70% of its area covered by mainly coniferous forests (Nilsson, 1996). Consequently, the main lake-type has a large part of its catchment covered by forested areas. The total number of lakes $\geq$0.01 km^2 is 95 700 and additionally there exists a large number of smaller tarns, ponds and pools. The majority of lakes are humic with low concentrations of nutrients and dissolved solids. Median values of pH, conductivity, nutrients, carbon and colour in Swedish lakes, calculated from a synoptic study of 3000 randomly selected lakes >0.04 km^2 in the summer of 1995, were: pH 6.8, conductivity mS m^{-1} 3.9, total nitrogen 500 μg l^{-1}, total phosphorus 10 μg l^{-1}, total organic carbon 7 μg l^{-1} and watercolour 40 mg Pt l^{-1} (Wilander et al., 1998). Ten percent of the lakes are ultra-oligotrophic and at the other end of the trophic spectrum about the same proportion are seriously eutrophicated. Only some 65 lakes in the country have a total phosphorus concentration exceeding 100 μg l^{-1} and the highest recorded value is 2 mg l^{-1} (Persson, 1994). The most variable measure, considering the lakes in general, is water-colour, reflecting the occurrence of very clear lakes to dark brown waters.

Fifteen lakes distributed across the country, more densely spread in the southern parts where the majority of the population lives have been monitored annually (monthly sampling of phytoplankton and environmental variables) since the end of the 1980s (Fig. 1). These lakes are considered as reference lakes to more disturbed ones in comparable ecoregions in the same sense as what is stated in ECs Water Framework Directive, and hence are situated in catchments with only minor disturbance by human activities. Acid deposition is, however, not possible to avoid, and some of the reference lakes are affected by severe acidification which has directed the lakes into a new state which seems to be rather stable. As many as 20 000 lakes in the country are influenced by acidification with a substantial depauperization of the flora and fauna.

Data from these 15 lakes were used to determine if a mature stage (an equilibrium phase) in the annual phytoplankton succession could be detected by finding a period with an 80% dominance in the total biomasss by only 1–3 species as suggested by Sommer et al. (1993). Sommer et al. further suggested a persistence of the dominants for 1–2 weeks in addition to a rather stable level of the total biomass. The sparse sampling at monthly occasions makes it difficult to closely follow if a longer period of biomass and species stability occurs. Ten years of studies of the late successional phases and their dominance patterns could however give an indication of a possibility for competitive exclusions. In oligotrophic waters there is a rather uniform phytoplankton development during the late-summer period with potential to make a reasonably appropriate prognosis of the main components of the phytoplankton assemblage. The late summer phase in phytoplankton development as reflected in August has since long been chosen as the best for extensive studies over the country and for comparisons of lakes from the nemoral to the alpine vegetation zones (Rosén, 1981; Willén et al., 1990; Hörnström, 2002). The late summer period has been chosen for its climatic stability with high water temperatures, stratified water columns and little influence from the catchment, giving prerequisites for internal biological processes to prevail. Although the ice-break in the alpine lakes does not occur until June the quick successional replacements of species makes August a suitable season for comparison of phytoplankton assemblages with southerly areas. Already in September may an autumn situation prevail in many lakes especially north of latitude 61 °N.

The contribution of Swedish lakes to a discussion of successional stages and particularly maturing periods and their prevailing species composition is valuable, because the majority of lakes in most Nordic countries deviates from many others considering climate, catchment characteristics and the nutrient state, which is reflected in the biota. The main question forwarded is if nutrient-poor forest lakes in Sweden exhibit an evident dominance in a late phase of the annual sequence of species. Two other aspects are also treated: the interannual variations and the main species structure in various lake clusters.

Table 1. Mean number of analysed taxa in 15 Swedish forest lakes during a late summer period (August) 1992–2001 and the registered min–max numbers. Counting procedure according to Olrik et al. (1998)

Lake	Mean no of taxa	Max.	Min.
Abiskojaure[1,2]	25	36	14
Brunnsjön[5]	25	36	13
Härsvattnet[3]	16	24	9
Rotehogstjärnen[3,4]	25	44	12
Älgsjön[4]	47	72	25
Ö. Särnamannasjön[1,2,3]	16	20	13
Övre Skärsjön[5]	25	34	14
Allgjuttern[6]	39	51	25
Fiolen[6]	39	60	13
Fräcksjön[6]	42	67	29
Stora Envättern[6]	52	82	36
Stora Skärsjön[6]	51	79	33
Stensjön[2,6]	39	62	25
Remmarsjön[2,6]	39	53	25
Jutsajaure[2,6]	36	46	25

1. Alpine lake with cold water (TP <5 $\mu g\, l^{-1}$).
2. Lakes situated above 61 °N.
3. Very acid lake, extremely depauperized (pH 4.5–5.5).
4. *Gonyostomum*-dominated lake.
5. Acid lake (pH 5.6–5.8).
6. Circumneutral or slightly acid lake (TP 8–12 $\mu g\, l^{-1}$).

Lake characteristics

The selected lakes vary in size from 0.2 to 1.7 km^2 with a median value of 0.3 km^2. The mean depth varies between 2 and 12 m, and the median value is 5 m. All lakes are stratified during the summer season. The retention time of the water in the lakes varies with a minimum of 2.5 months and a maximum of 12 years. Some environmental characters relevant for the phytoplankton development are illustrated in Figure 2, where it appears that a few lakes has a pH below 5.5 and a majority are just slightly acid or circumneutral (pH 6–7, alkalinity <0.1 meq l^{-1}). Most lakes have low nutrient contents and are classified as oligo- or mesotrophic. According to assessments of environmental conditions worked out for Swedish lakes, total phosphorus values <12.5 $\mu g\, l^{-1}$ indicate oligotrophic conditions while concentrations between 12.5 and 25 indicate a mesotrophic state (SEPA 2000). The northernmost lake in this material is situated in the arctic region and the southernmost lake in the boreo-nemoral zone (Påhlsson, 1994).

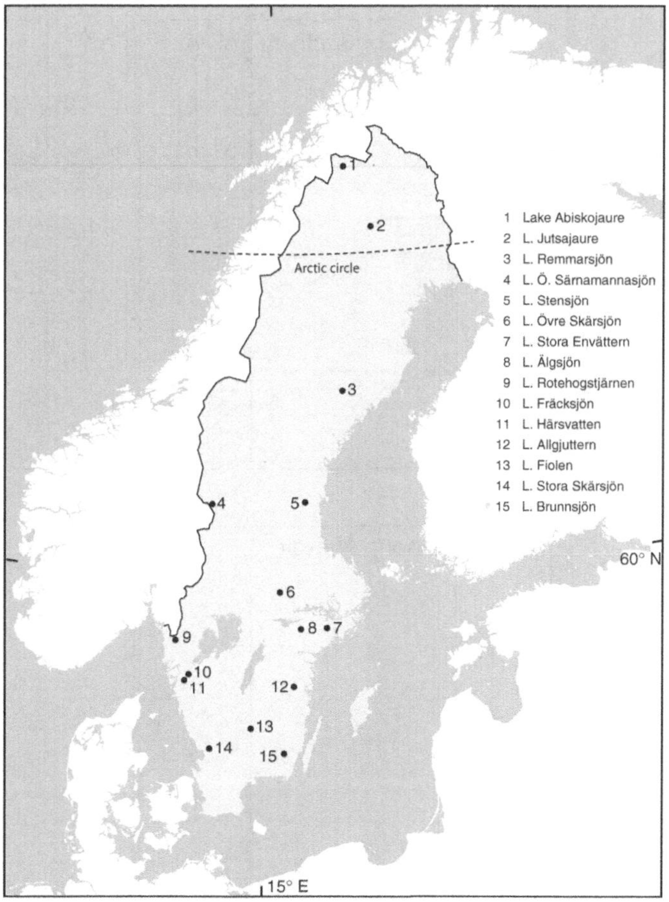

Figure 1. Location of Swedish forest lakes with monthly sampling of plankton and water-chemical variables during the ice-free season.

Sampling and analysis

The 15 lakes are sampled monthly during the growth season, which in the southern half of Sweden means April–October. In other regions, May–October is a common sampling period and in the alpine areas June–September. The sampling started in the late 1980s with investigations of waterchemical variables, phytoplankton, zooplankton and benthic invertebrates. Phytoplankton is sampled with a tube in the epilimnetic zone. In lakes <1 km^2 water from five different sites in a lake is mixed, and a subsample is taken and preserved with JJK-solution supplemented with acetic acid. A complementary net sample (mesh size 20 μm) is collected to facilitate species determinations, when necessary. In larger lakes, just one centrally situated site is used for sampling because the fetch (the free water surface where the wind may act upon the waves) is here considered to produce an adequate wind-mixing. Water chemical variables are analysed according to Swedish standards and phytoplankton according to the Utermöhl technique modified and commonly used in the Nordic countries (Olrik et al., 1998). A detailed description of the counting procedure used for evaluation of species richness is given in Willén (2001). Species nomenclature follows that given in Huber-Pestalozzi (1985), Knudson (1952), Komárek & Marván (1992), Novarino et al. (1994), Preisig (1995) and Süsswasserflora von Mitteleuropa (1985–1999).

The values of water-colour used are recalculated to mg Pt l^{-1} from the assessment of light conditions based on absorbency readings in a photometer from filtered water (0.45 μm membrane filter) at a wavelength of 420 nm (SEPA 2000).

The phytoplankton data presented here represent the period 1992–2001 when a uniform counting method was used. To illustrate the dominance patterns in the lakes the proportion of 1–3 species is shown. The species considered had to reach at least 10% of

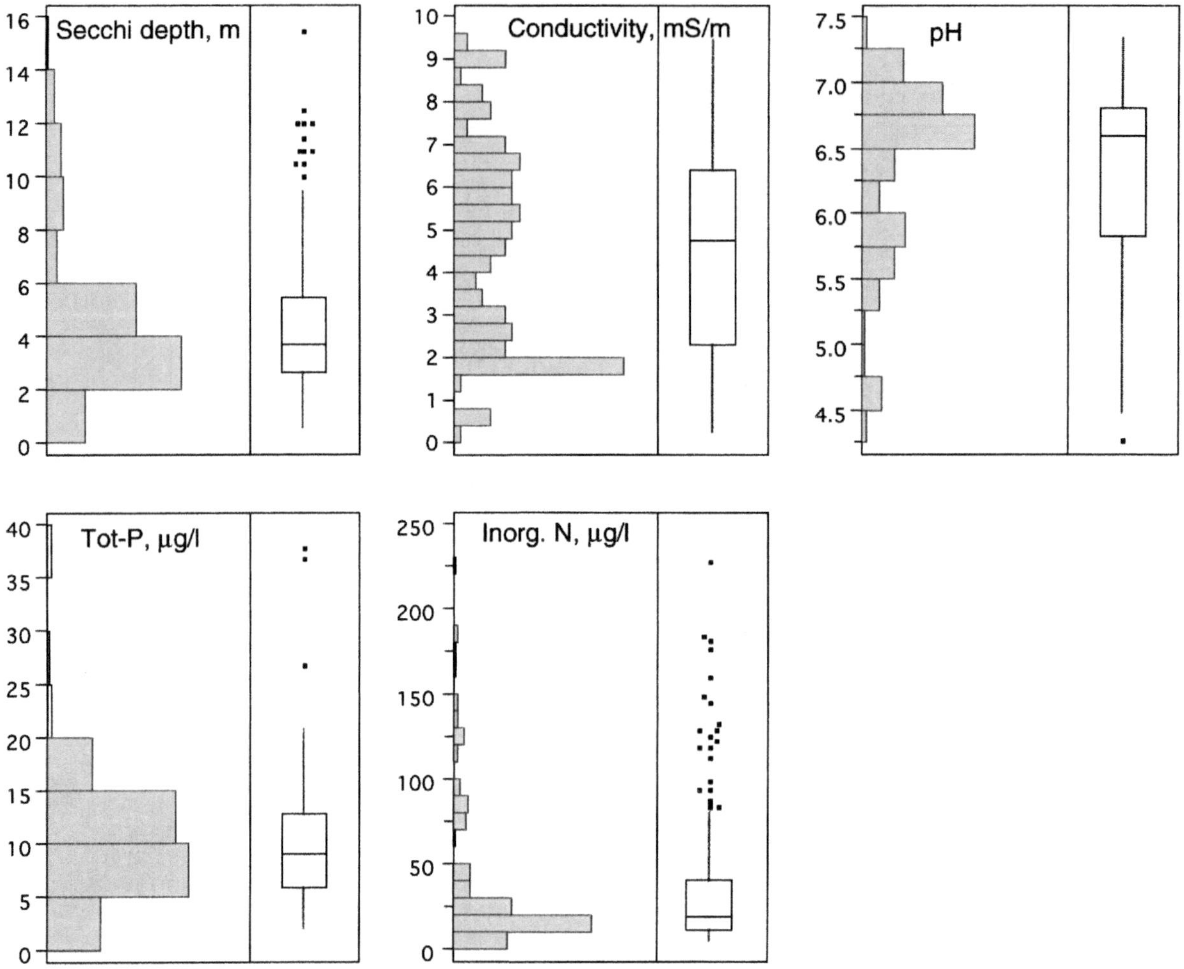

Figure 2. Some water quality variables of importance for phytoplankton development in 15 forest lakes in Sweden. Data represents mean values of the summer seasons 1992–2001. The x-axis give a rough indication of number of lakes and years.

the total phytoplankton volume. Only in some very acidified lakes and in lakes dominated by the invasive species *Gonyosotomum semen* does just one single species reach a main proportion of the total biomass. In most lakes however, three species are included among the dominants. To investigate the variation in phytoplankton assemblages in relation to environmental variables, and separate the lakes, a canonical correspondence analysis (CCA) was used in the programme CANOCO (version 3.15 PPC). The CCA analysis claims a unimodal respons of the species which in this case was checked by a detrended correspondence analysis (DCA). Non-significant environmental variables were identified by a forward selection procedure using a Monte Carlo permutation test (999 permutations) with a p-value ≤ 0.05 in addition to a checking of variance inflation factors (VIF <20) and t-values (ter Braak & Verdonschot, 1995). Lakes and phytoplankton assemblages were classified into biologically similar groups using the TWINSPAN (two-way indicator species analysis) technique based on log-transformed phytoplankton volumes (Hill, 1979). Environmental data were not transformed. In a discriminant function analysis the consistency of TWINSPAN groups of lakes and the set of explanatory environmental variables were checked.

Results and discussion

Dominance patterns

The summarized proportion of the total phytoplankton

volumes achieved by dominant taxa does not reach more than 60% in Swedish forest lakes as a mean value when comparing ten late summer successional years in each lake. Some lakes do however exhibit a more pronounced dominance with a few species occupying >80% of the total biomass (Fig. 3). These lakes, shaded in Figure 3, are all characterized by some stress factor like harsh climate (alpine lakes Abiskojaure, Övre Särnamannasjön), dark brown water (L. Brunnsjön), very acid water (L. Härsvatten, L. Övre Skärsjön) or dominance of the invasive raphidophycean species *Gonyostomum semen* (L. Rotehogstjärnen, L. Älgsjön). Especially large inter-annual variations occur in the *Gonyostomum*-dominated lakes where from time to time this species accumulates in the hypolimninon (Cronberg et al. 1988; Le Cohu et al., 1989).

The proportion of dominant taxa within the growth season is usually high all sampled months among the alpine lakes in this data-set (Fig. 4). Small chrysophycean flagellates of the order Chromulinales and Pedinellales (Preisig, 1995) are common, and during spring and autumn overturn also small *Cyclotella* species. Small-sized cryptophyceans like *Plagioselmis nannoplanctica* (formely *Rhodomonas lacustris v. nannoplanctica*) and naked dinoflagellates are subdominants.

In low-buffered lakes, which were further acidified during the 20th century, the water has a high transparency and the biota is very depauperized (cf L. Härsvattnet, Fig. 4). Here a few species dominate the whole growth season. *Peridinium inconspicuum* usually is a primary dominant but naked flagellates (Ochromonadales), *Dinobryon sociale v. americanum* (spring), *Dinobryon pediforme* (summer), *Istmochloron trispinatum* and *Botryococcus braunii* (autumn) are important subdominants depending on the successional phase of the community. Another lake-type with large dominance of a few species is that invaded by *Gonyostomum* (cf L. Älgsjön, Fig. 4). Here *Gonyostomum semen* is the sole dominant species except in early spring and in the autumn when chrysophyte taxa as *Uroglena*, *Synura* or *Dinobryon* (spring) and diatoms as *Tabellaria flocculosa v. asterionelloides* (autumn) also occur.

Species richness

The species richness in the lakes subjected to some stress factor is significantly lower than the other lakes in this dataset, except for L. Älgsjön where apparently many species co-occur with the invasive one, although in low abundances (Table 1). The nutrient-richer water compared to the other lakes is here a contributive factor (TP $\approx$ 20 μg l^{-1}). In general, however, the *Gonyostomum*-dominated lakes are characterized by a low diversity in terms of higher taxonomic levels as algal classes or functional characters (Cronberg et al., 1988; Willén et al., 1990). Flagellated species usually prevail. In a large number of forest-lakes (n = 113) in Sweden tested in August, the species richness of phytoplankton generally reaches 40–55 taxa, without going into details with flagellates and some other taxa which are hard to separate in an inverted microscope dealing with preserved material. A species richness <20 occurs in very acid lakes with low content of organic matter (i.e. L. Härsvatten). Alpine lakes usually have a species number of ca 30. Large lakes, several tens of km^2 or more are more species rich and especially if the water-mass is unstratified. Then the mean species number may reach >70. The species richness in most lakes is somewhat lower in spring and autumn but varies only little during July–September: i.e. Spp. in July=0.8 x spp in August +7.4; $r^2 = 0.69$ and Spp. in September= 0.8 x Spp. in August + 9.0; $r^2 = 0.66$ (Willén unpublished data).

Annual biomass development

The main algal groups and the mean biomass variation during the growth seasons of 1992–2001 is illustrated for selected lakes in Figure 5. The chosen lakes illustrate those discussed above subjected to some stress-factor, and the lake Fiolen, well-known from many classical limnological works (Naumann, 1917; Thunmark, 1931; Åberg & Rodhe, 1942). The latter lake is chosen as a representative of a typical south Swedish forest lake in a near-natural state (pH 6.5–7, TP 10–15 μg l^{-1}, TN 350–500 μg l^{-1}, colour 15–25 mg Pt l^{-1}). Many forest lakes are, however, more humic than this one.

The biomass usually peaks early in the season except in the *Gonyostomum*-lake, because this alga develops its largest biomass during warm water periods. The biomass in most lakes then successively falls, but a small increase may occur in August. There are fairly even proportions between dominating algal groups after the spring peak. Swedish oligotrophic forest lakes do not show dramatic algal developments during their seasonal cycles. In all three lakes illustrated here, except the one dominated by *Gonyostomum*, chrysophytes and dinoflagellates are important components and diatoms in the alpine lake. Cryptophyceae, which

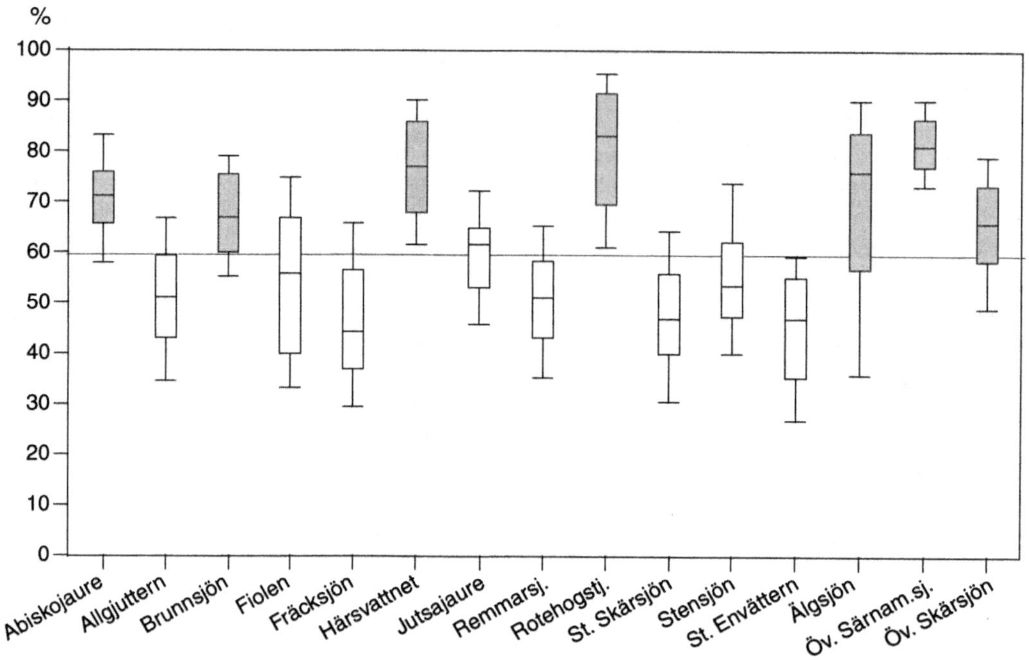

Figure 3. Dominance patterns of 1–3 phytoplankton taxa in 15 Swedish forest lakes during a late summer period (August). Data represents in general 10 late summer seasons per lake, 1992–2001. The boxes show the 25th, 50th and 75th quantiles. The vertical lines end at the 10th and 90th percentiles.

is a common group in many kinds of forest lakes especially if the nutrient state increases, has a subordinate role in the stressed lakes.

Classification of lakes

The summer (August) phytoplankton assemblages in the 15 forest lakes, with some exceptions sampled during 10 years, was separated in five main clusters in the TWINSPAN classification (Fig. 6). A total of 13 environmental variables were tested for significance against the species dataset (cf. Sampling and analysis). The tested variables were: Secchi depth, pH, total nitrogen, total phosphorus, conductivity, lake area, soluble reactive silicon, mean depth, water-colour (measured as absorbency), depth of the surface mixed layer, water temperature and inorganic nitrogen. The first mentioned eight variables were identified as the best to explain the variability. These significant variables ($p = 0.05$) explained 78% of the variance in the dataset. Secchi depth, pH, total nitrogen and total phosphorus were together the most influential ones, explaining 50%. The significance of silicon although explaining only 7% of the variation is due to its binding to humic matter. Lakes with a high content of humic matter usually have a lowered pH which is illustrated in Figure 6 where the arrows for pH and silicon have opposite directions. By the discriminant function analysis the TWINSPAN species groupings and the lake/year scores obtained in the CCA analysis were compared. The correspondence of the two groupings was high from 83 to 100% (Table 2). Interannual variations explain less than 1% in the dataset when analysed as a covariable in the CCA. Years are thus interpreted as having no directed influence on the species assemblages but it is a variable with stochastic directions. Eleven of the 15 lakes are classified in the same TWINSPAN group irrespective of year which indicate rather stable species assemblages from summer to summer. Some lake (Rotehogstjärnen) with irregular occurrence of *Gonyostomum* in the epilimnion is classified two years in group 3 and eight years in group 5. One lake Fräcksjön varies a lot in water colour (absorbancy 0.07–0.14) between separate August periods which causes the species assemblage also to change. In two clear-water lakes in southern Sweden, Fiolen and Allgjuttern, the nutrient content varies at least two-fold from summer to summer causing a change in species compositions. The former lake has 70% of its sampled August periods classified in TWINSPAN

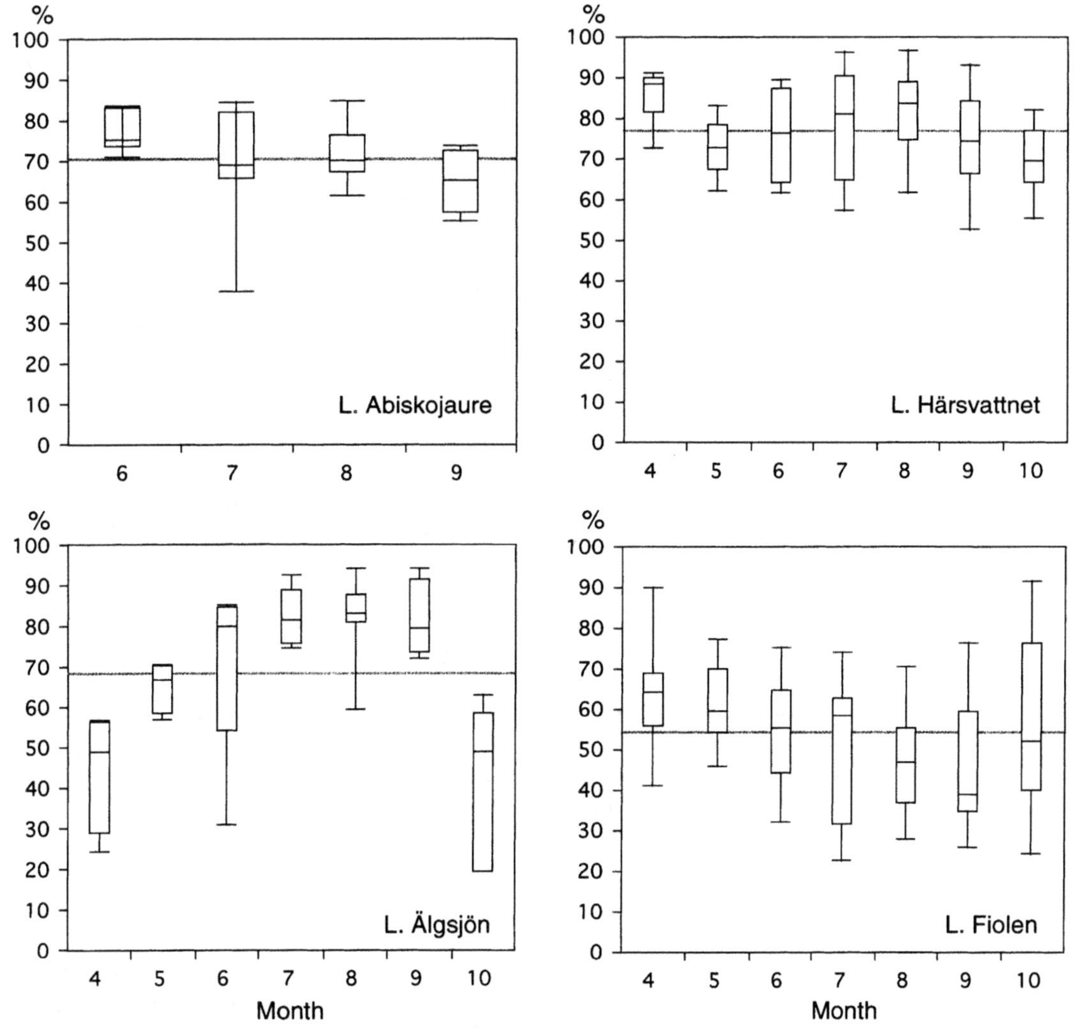

Figure 4. Dominance patterns of 1–3 phytoplankton taxa in an alpine lake (Abiskojaure), a very acid/acidified lake (Härsvattnet), a lake invaded by *Gonyostomum semen* (Älgsjön) and a common forest lake without influence of some extreme stressfactor (L. Fiolen). Data represent growth seasons 1992–2001.

Table 2. Comparisons of TWINSPAN lake-groups and the predictions based on lake/year combined with species scores obtained in the CCA

	TWINSPAN groups					
Groups	1	2	3	4	5	
1	**16 (100%)**	0	0	0	0	16
2	0	**31 (94%)**	2	0	3	36
3	0	0	**47 (84%)**	1	0	48
4	0	1	2	**17 (94%)**	0	20
5	0	1	5	0	**15 (83%)**	21
	16	33	56	18	18	141

Table 3. Groups of lakes and years among the Swedish forest lakes tested in August 1992–2001 and their main species assemblages

Group	Lake character	Characteristic species in the phytoplankton assemblages
1	Very acid (pH <5.5), ultra-oligotrophic clear lakes situated both in northern and southern Sweden. Species poor.	*Dinobryon pediforme, Istmochloron trispinatum, Pseudokephyrion* spp., *Mougeotia* sp., *Chrysolykos skujae*
2	Circumneutral oligo-mesotrophic clear lakes in southern Sweden. Very species rich.	*Chrysochromulina, Stichogloea,* Picocell-sized colonial prokaryotes, *Anabaena lemmermannii, Monoraphidium griffithii, M. dybowskii, Botryococcus terribilis, Cosmarium, Plagioselmis nannoplanctica*
3	Weakly buffered oligotrophic, humic lakes (meso-polyhumic) with variable size, depth and z_{mix}. Low content of inorganic N.	*Merismopedia tenuissima, Mallomonas caudata, Botryococcus braunii, Cryptomonas, unicellular green algal flagellates, Oocystis, Bitrichia chodati*
4	Heterogenous group although well predicted. An alpine lake with small interannual variations and a south-Swedish lake: very transparent water, ion-rich, poor in nitrogen and phosphorus.	*Tetraedron trigonum v. tetralobulatum, Chrysochromulina, Cosmarium* and *Cyclotella small* spp., *Dinobryon cylindricum, Rhizosolenia eriensis, Chrysolykos skujae*
5	A group of south Swedish lakes with humic, comparatively nutrient-rich water (but not really eutrophic) of varying pH state (5.5–7.)	Mass developments and dominance of *Gonyostomum semen.* Subdominant indicators: *Woronichinia naegeliana, Dinobryon divergens, Aulacoseira distans*

class 2 and the latter the same proportion of years in group 4 (Table 3).

Conclusions

The forest lakes discussed here are usually not characterized by strong dominance of a few species, unless exposed to some stress factor. A competitive exclusion of species is apparently a rare feature during the late successional period in Swedish nutrient-poor lakes. Additionally, they also have a rather even interannual development especially reflected in biomasses and algal groups, but usually also in species assemblages, although there may be a variation in the order of dominance between the taxa. Exceptions to this circumstance are connected to lakes with large interannual variations of some important resource.

A comparison of groupings of lakes and species in this material with the functional coda outlined in Reynolds et al. (2002) gives the most evident consistency between codon **Q** characterized by *Gonyostomum* and group 5 presented here. When dealing with other seasons codon **A** is recognizable during the spring mixing in Lake Abiskojaure in Lapland. A group of lakes easy to classify among Swedish forest lakes due to its special flora, is the very acid, ultraoligotrophic, often well stratified clearwater lakes, in the Swedish material separated as group 1. Except the, in many forest lakes commonly occurring dinoflagellate *Peridinium inconspicuum*, which in this acid lake-group has an especially large proportion, also the species given in Table 3 are typical representatives. The relation to the codons mentioned above needs additional penetration, but at present the group 1 is mentioned as a part of $\mathbf{L_o}$. The oligotrophic forest lakes in Sweden and also in Finland and Norway, with apparent dissimilarities in water quality and species structure to many other lakes in Europe need further evaluations. Recently a co-operation project

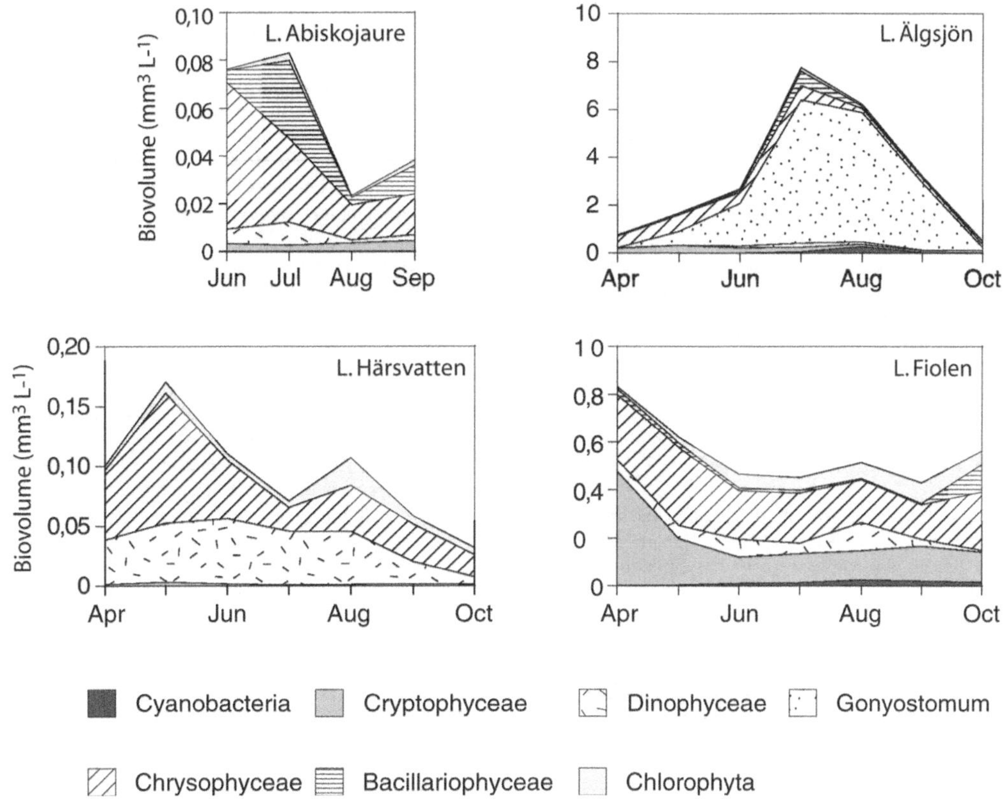

Figure 5. Development of total biovolumes and main algal groups in four Swedish forest lakes sampled monthly during ten years. Mean values 1992–2001. The characters of the lakes is given in Figure 2 and Table 1.

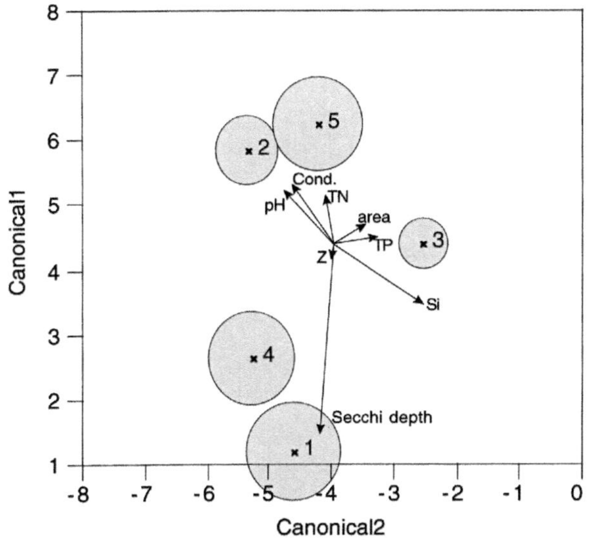

Figure 6. Centroid plot of TWINSPAN groups of lakes and significant environmental variables of 15 Swedish forest lakes sampled in August (10 years). The centroid for each group is marked. The size of the circle denote a 95% confidence limit.

has been initiated between these countries in order to compare species assemblages and their characteristic environments with the object of a detailed and adequate separation.

References

Åberg, B. & W. Rodhe, 1942. Über die Milieufaktoren in einigen südschwedischen Seen. Symb. Bot. Ups. V:3.

Cronberg, G., G. Lindmark & S. Björk, 1988. Mass development of the flagellate *Gonyostomum semen* (Raphidophyta) in Swedish forest lakes – an effect of acidification? Hydrobiologia 161: 217–236.

Hill, M. O., 1979. TWINSPAN – a FORTRAN program for arranging multivariate data in an ordered two-way table by classification of individuals and attributes. Cornell University, Ithaca, N.Y.

Hörnström, E., 2002. Phytoplankton in 63 limed lakes in comparison with the distribution in 500 untreated lakes with varying pH. Hydrobiologia 470: 115–126.

Komárek, J. & P. Marván, 1992. Morphological differences in natural populations of the genus *Botryococcus* (Chlorophyceae). Archiv für Protistenkunde 141: 65–100.

Knudson, B., 1952. The diatom genus *Tabellaria*. 1. Taxonomy and morphology. Ann. Bot. N.S. 16: 421–440.

Le Cohu, R., J. Guitard, N. Comoy & J. Brabet, 1989. Gonyostomum semen (Raphidophycées), Nuisance potentielle des grands réservoirs français? L'exemple du lac de Pareloup. Archiv für Hydrobiol. 117: 225–236.

Huber-Pestalozzi, G. (ed.), 1985. Das Phytoplankton des Süsswassers. Chlorophyceae. Schweizerbartsche Verlagsbuchhandlung. Stuttgart.

Naumann, E., 1917. Undersökningar över fytoplankton och under den pelagiska regionen försiggående gyttje- och dybildningar inom vissa syd- och mellansvenska urbergsvatten. Kungliga Svenska Vetenskapsakademins Handlingar 56. Stockholm.

Nilsson, N-E., (ed.),1996. The Forest. Sveriges Nationalatlas. Bra Böcker, Höganäs (in Swedish).

Novarino, G., I. Lucas & S. Morrall, 1994. Observations of the genus *Plagioselmis* (Cryptophyceae). Cryptogamie, Algologie 15: 87–107.

Olrik, K., P. Blomqvist, P. Brettum, G. Cronberg & P. Eloranta, 1998. Methods for quantitative assessment of phytoplankton in freshwaters, part I. Swedish Environmental Protection Agency, Report 4860.

Persson, G., 1994. Eutrophication of soil, fresh water and the sea. Swedish Environmental Protection Agency, Report 4244.

Preisig, H., 1995. A modern concept of chrysophyte classification. In Sandgren, C., J. Smol & J. Kristiansen (eds), Chrysophyte Algae. Ecology, Phylogeny and Development. Cambridge University Press.

Påhlsson, L., 1994. Vegetation types in the Nordic countries. Nordic Council of Ministers. Tema Nord 1994:665. Copenhagen.

Reynolds, C., V. Huszar, C. Kruk, L. Naselli-Flores & S. Melo, 2002. Towards a functional classification of the freshwater phytoplankton. J. Plankton Res. 24: 417–428.

Rosén, G., 1981. Phytoplankton indicators and their relations to certain chemical and physical factors. Limnologica (Berlin) 13: 263–290.

SEPA, 2000. Environmental quality criteria. Lakes and watercourses. Swedish Environmental Protection Agency, Report 5050.

Sommer, U., J. Padisák, C. S. Reynolds & P. Juhász-Nagy, 1993. Hutchinson's heritage: the diversity-disturbance relationship in phytoplankton. Hydrobiologia 249: 1–7.

Süsswasserflora von Mitteleuropa 1985– 1999. Teil 1, 2, 3, 6, 19. Gustav Fischer Verlag. Jena. ter Braak, C. J. F. & P. F. M. Verdonschot, 1995. Canonical correspondence analysis and related multivariate methods in aquatic ecology. Aquat. Sci. 57: 255–289.

Thunmark, S., 1931. Der See Fiolen und seine Vegetation. Acta Phytogeographica Suecica 2: 1–198.

Wilander, A., R. Johnson, W. Goedkoop & L. Lundin, 1998. The country-wide inventory 1995. A synoptic study of water.chemistry and bottom-fauna in swedish lakes and watercourses. Environmental Protection Agency, Report 4813 (in Swedish with an English summary).

Willén, E., 2001. Phytoplankton and water quality characterization: experiences from the Swedish large lakes Mälaren, Hjälmaren, Vättern and Vänern. Ambio 30: 529–537.

Willén, E., S. Hajdu & Y. Pejler, 1990. Summer phytoplankton in 73 nutrient-ppor Swedish lakes. Classifiction, ordination and choice of long-term monitoring objects. Limnologica (Berlin) 20: 217–227.

Hydrobiologia **502:** 325–329, 2003.
L. Naselli-Flores, J. Padisák, M. T. Dokulil (eds),
Phytoplankton and Equilibrium Concept: The Ecology of Steady-State Assemblages.

Distribution and seasonal dynamics of Cryptomonads in Sicilian water bodies

Rossella Barone & Luigi Naselli-Flores
Dipartimento di Scienze Botaniche, Università di Palermo, Via Archirafi, 38 I-90123 Palermo, Italy
E-mail: rossella.barone@unipa.it

Key words: Cryptomonads, phytoplankton seasonal dynamics, freshwater, Sicily

Abstract

Several species with a relevant ecological importance belong to Cryptophyta. Nevertheless, species-level identification from microscopic observations is problematic, lacking recent taxonomic keys. In this study we report our observations on distribution and seasonal dynamics of Cryptomonads in 33 Sicilian water bodies, as well as the main taxonomical problems we encountered. Species of the genera *Cryptomonas* and *Plagioselmis* are the most common in the examined water bodies. Their biomass seasonal trends usually show a peak in late winter and early spring. In advanced spring, and also in summer, due to the higher grazing pressure, the Cryptomonads biomass reaches its lowest values. Moreover, in a small, eutrophic, temporary pond (Santa Rosalia), without grazing pressure in spring, the Cryptomonads bloomed in summer. From our results, we propose to focus the attention on the sensitiveness of these organisms to filter feeding to explain their seasonal dynamics.

Introduction

The algal group of Cryptomonads belongs to the Phylum Cryptophyta, Order Cryptomonadales. The genera currently recognized in this relatively small, but ecologically and evolutionary important group, are 17 (Clay et al., 1999). Nevertheless, there are several difficulties in species-level identification from observations with optical microscope (Novarino, 2003). The major problem is mainly represented by their small size and brittleness. In particular, the cells are extremely delicate and frequently rupture when fixatives, as formaldehyde, are added. As a consequence, their numbers are generally low in preserved samples and their taxonomical identification is rather difficult in natural phytoplankton assemblages.

However, it is widely accepted that Cryptomonads are important primary producers in both freshwater and marine habitats, and many species are cosmopolitan in their distribution, although they appear to be more common in cooler waters (Burlow & Kugrens, 2002). Moreover, the Cryptomonads are 'high quality' food for herbivores because they are readily ingested and digested, contain relatively high proportion of some essential fatty acids, and lack toxins.

Cryptomonads populations generally reach a maximum following periods of moderate turbulence (Reynolds, 1984) and when grazing is reduced (Burlow & Kugrens, 2002).

In this study, we report some observations on distribution and seasonal dynamics of the more common taxa identified in the Sicilian lakes. In addition, an attempt to track the major factors affecting their population dynamics is offered.

Methods

The Cryptomonads identification was based upon light microscopy observations, on living and Lugol's fixed samples. In particular, we considered five principal characteristics traditionally used in their taxonomy: cell size, cell shape, chloroplast number and starch patterns, presence and direction of a furrow, arrangement of large ejectisomes. Identification mainly followed Huber-Pestalozzi (1968), Hill (1991), Novarino et al. (1994), Clay et al. (1999) and Javornický (2001).

Limnological investigations in Sicily are up to now developed on 32 water bodies (Naselli-Flores, 1999). Our observations on the Cryptomonads distributed in

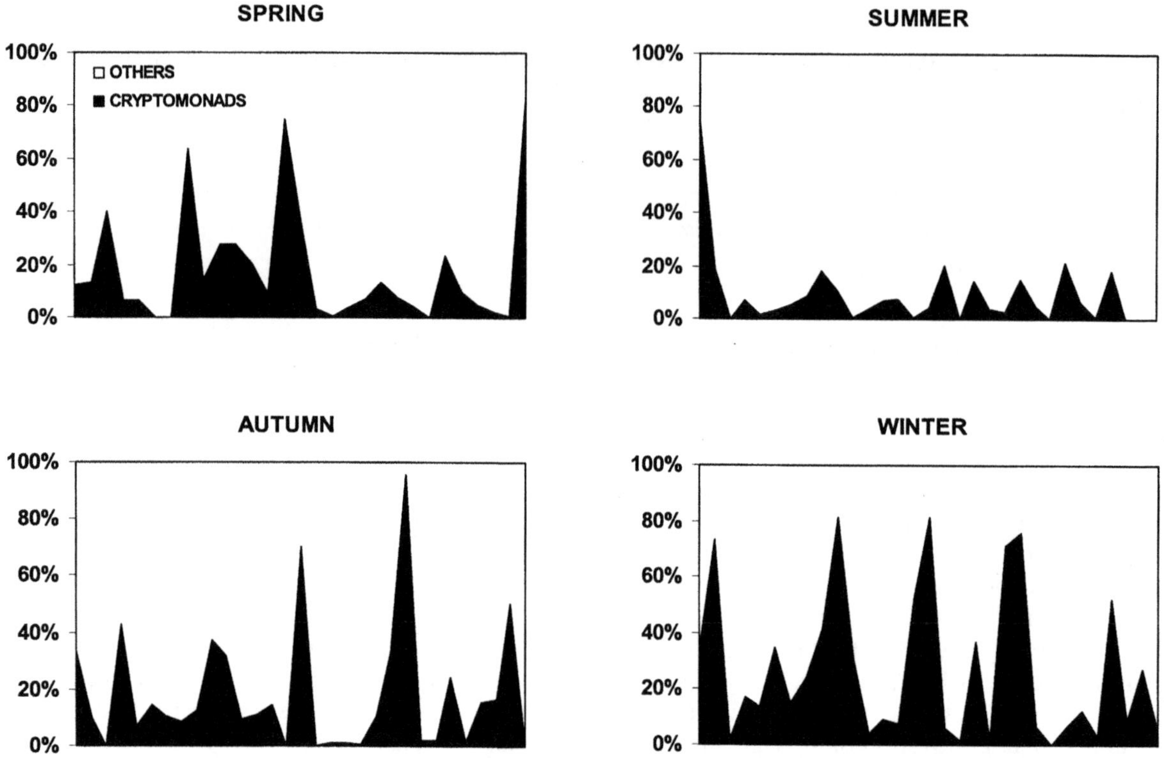

Figure 1. Seasonal distribution of Cryptomonads biomass in 31 Sicilian lakes, ordered along a decreasing trophic gradient.

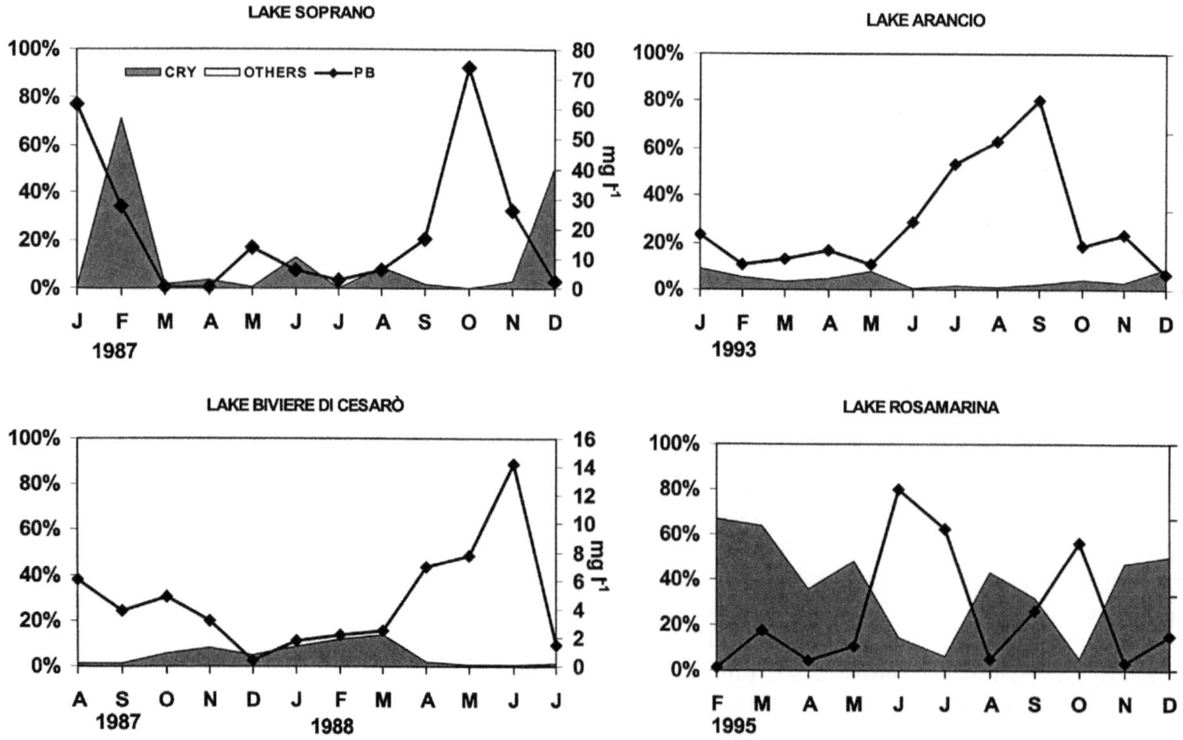

Figure 2. Monthly trends of Cryptomonads biomass in 4 Sicilian lakes. CRY = Cryptophyceae; PB = phytoplankton biomass.

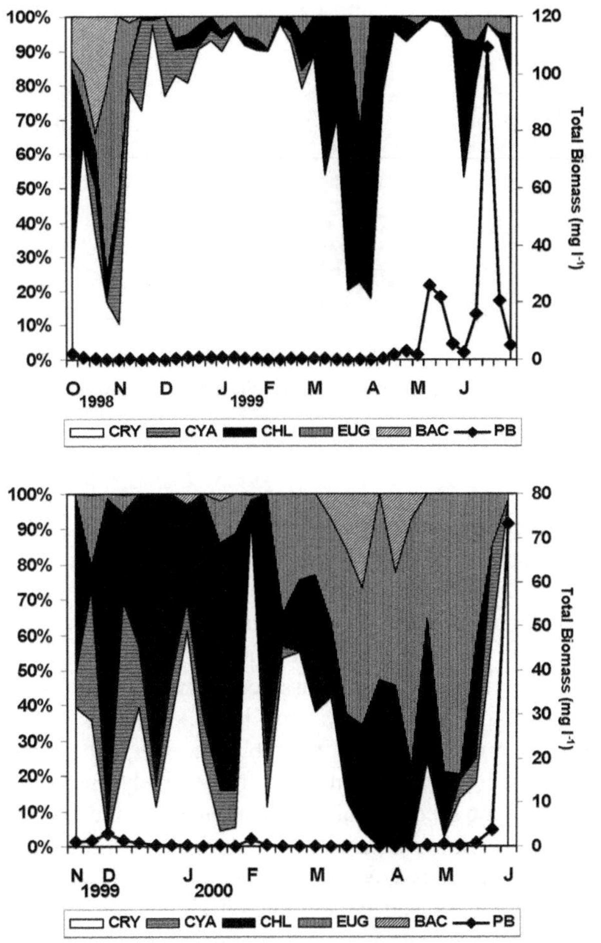

Figure 3. Trend of phytoplankton biomass in Santa Rosalia Pond. CRY = Cryptophyceae; CYA = Cyanophyceae; CHL = Chlorophyceae; EUG = Euglenophyceae; BAC = Bacillariophyceae; PB = phytoplankton biomass.

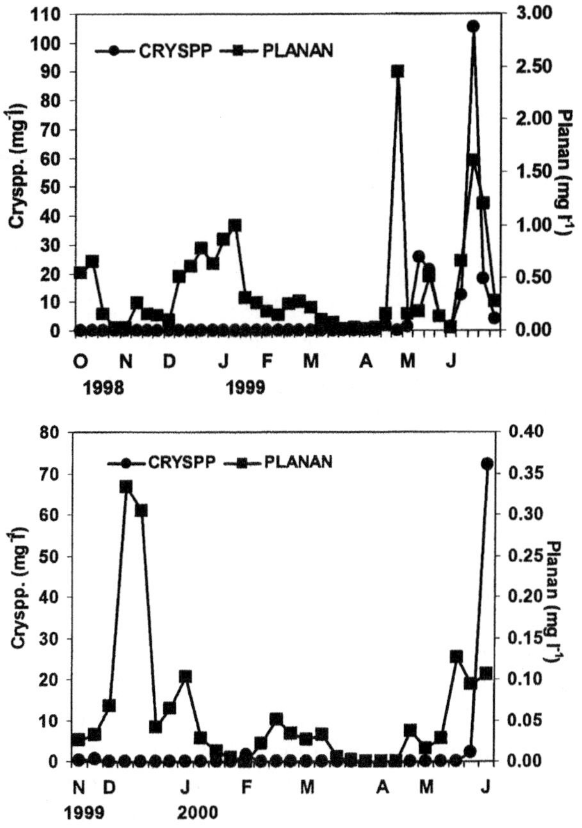

Figure 4. Trend of Cryptomonads biomass in S. Rosalia Pond. CRYSPP: *Cryptomonas* spp; PLANAN: *Plagioselmis nannoplantica.*

these lakes are reported. In addition, we refer on the dynamics of these organisms as resulted from a limnological investigation carried out for two years in a small pond (Naselli-Flores & Barone, 2003) where this group, dominant for most of the year, bloomed in summer.

Results

The Cryptomonads, as a group, are present in all the lakes examined. The identified species are: *Campylomonas reflexa* (Skuja) Hill, *Cryptomonas* cf. *erosa* Ehrenberg, *C.* cf. *ovata* Ehrenberg, *C.* cf. *rostrata* Troitzkaja, *C.* cf. *pyrenoidifera* Skuja, *Plagioselmis nannoplanctica* (Skuja) Novarino, Lucas et Morrall, *Plagioselmis lacustris* (Pascher et Ruttner) Javornický. The more common species are *Cryptomonas* cf. *erosa* and *Plagioselmis nannoplanctica.*

A limnological study carried out on 31 body water (Calvo et al., 1993) allowed to underline the relative contribution of Cryptomonads to total biomass (Fig. 1), along a decreasing trophic gradient (based on annual average values of phytoplankton biomass). Their seasonal dynamics show a major significance of this group in winter, when the lowest values of water temperature, irradiance, z_{eu}/z_{mix}, and grazing pressure were recorded. In particular, the relative importance of their populations showed a seasonal cycle with the highest values in autumn-winter and the lowest in spring-summer. Filter feeders biomass, showed an inverse trend with the highest significance in spring–summer and the lowest in autumn-winter (Calvo et al., 1993). The lake trophy didn't appear as a major

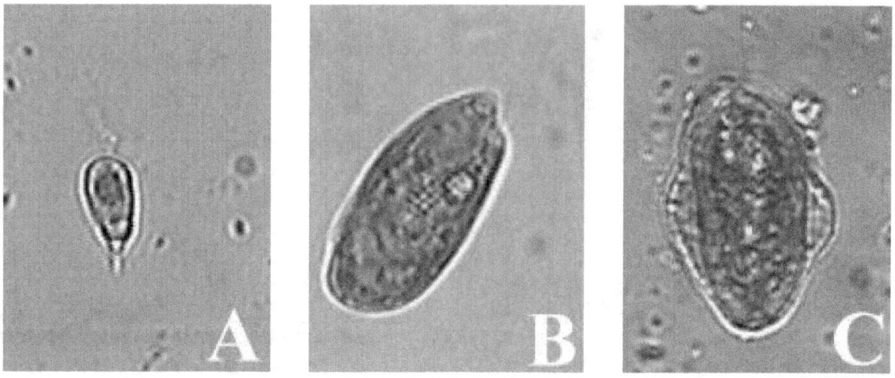

Figure 5. A = *Plagioselmis nannoplanctica* (Skuja) Novarino, Lucas et Morrall; B = *Cryptomonas* cf. *erosa* Ehrenberg; C = swelling cell of *Cryptomonas* sp. (400×).

factor affecting their relative contribution to the total biomass.

Some comparative limnological observations carried out on four natural and artificial lakes (Naselli-Flores & Barone, 2000), showing different dimensions and trophic states, allowed to calculate the contribution of the Cryptomonads to the monthly values of total phytoplankton biomass (Fig. 2). In particular, Lake Soprano was a small, natural, and hypertrophic water body, Lake Arancio a large eutrophic reservoir, Lake Biviere di Cesarò was a small, natural, and meso-eutrophic lake, and Lake Rosamarina a large mesotrophic reservoir. The trend of the group mostly confirmed a major importance in winter and a minor one in summer, according to the common cycle of herbivores biomass, that showed the lowest values in winter and the highest in summer. The Cryptomonads, with a clear waxing and waning of the populations, appeared more significant in the large Lake Rosamarina.

Quite different is the case of Santa Rosalia pond, a small and eutrophic temporary water body at 398 m a.s.l., where a series of limnological observations were carried out for two years. The principal hydrological and biological characteristics are reported in Naselli-Flores & Barone (2002). In the studied years, the emptying phases started in the first week of July 1999 and in the second week of June 2000. The water body was characterized by high nutrient levels, rather low pH values (<6.5) and bicarbonate-alkaline waters. The most important primary producers were metaphytic and rhyzophytic green algae. The study was carried out by weekly sampling from autumn 1998 to the beginning of summer 2000. The first year of study was characterised by a particularly cold winter and the surface of the pond unusually froze at the end of January. Zooplankton biomass showed lowest values in the first year and, in both the studied periods, an abrupt significant decrease in late spring.

Phytoplankton biomass generally showed low values (Fig. 3), mostly <1 mg l^{-1}, except for the cyclic summer Cryptomonads bloom, when the peaks, reached 109 and 73 mg l^{-1}, respectively, in the first and in the second year of investigation. In the first year phytoplankton assemblage was mostly dominated by Cryptomonads species, while in the second year they frequently pulsed and dominated, along with small flagellate (e.g. *Spermatozopsis*, *Chlamydomonas*), coccal green algae (e.g. *Ankyra*) or small euglenophytes (e.g. *Trachelomonas*). However, in both the studied years a major significance of Cryptomonads was recorded in winter and a Cryptophytes bloom occurred in late spring-early summer, abruptly truncated by the drying of the pond.

The biomass trends of the identified taxa are reported in Figure 4. *Plagioselmis nannoplanctica* (Fig. 5A) population didn't show any particular seasonality, while *Cryptomonas* spp., with prevailing *Cryptomonas* cf. *erosa* (Fig. 5B), appeared more significant in summer. When the Cryptomonads bloom occurred in early June, just before the drying up of the pond, many individuals of the genus *Cryptomonas* were showing a particular morphology, characterised by swelling and twisting of the cells (Fig. 5C), probably relative to forthcoming cysts production.

Concluding remarks

From our results, it is possible to infer some remarks:

- Cryptomonads are present in all the water bodies examined, according to their ubiquitous distribution (Klaveness, 1988).
- Their importance and populations dynamics appear mostly trophic state independent, according to Ojala (1993).
- The group shows a seasonal cycle, with a major significance in winter, when lower values of temperature, a decreased grazing pressure and a reduced light climate occur.

The summer Cryptomonads bloom in Santa Rosalia Pond shows that the species of this 'winter-group' are not directly affected by the temperature, which, on the contrary, more directly affects the herbivores. Moreover, the high turbidity, typical of a shallow, drying up pond, highlights a tolerance to low irradiances values.

Some physiological properties common to the group, as photoadaptation strategies and resting spore formation, in spite of their sensitiveness to grazing pressure, clearly explain the ecological success of the Cryptomonads in this small water body.

According to the proposal of a functional classification of the freshwater phytoplankton by Reynolds et al. (2002), *Cryptomonas* spp. and *Plagioselmis* spp. are respectively distributed in the **Y** and $\mathbf{X_2}$ coda. In particular, the **Y** functional group shows *Cryptomonas* spp. as typical representatives, well-adapted to live in a wide range of habitats, mostly in small, enriched lakes; nevertheless, their dynamics are vulnerable to the feeding activity of herbivores. The $\mathbf{X_2}$ is represented by small organisms, generally inhabiting meso-eutrophic lakes; also this association is vulnerable to filter feeding.

Our results confirm this ecological requirements and detect filter feeding as a major factor affecting the populations dynamics of these small edible phytoflagellates in the Sicilian lakes.

References

Barlow, S. B. & P. Kugrens, 2002. Cryptomondads from the Salton Sea, California. Hydrobiologia 473: 129–137.

Calvo, S., R. Barone, L. Naselli Flores, C. Fradà Orestano, G. Dongarrà, A. Lugaro & G. Genchi, 1993. Limnological studies on lakes and reservoirs of Sicily. Naturalista siciliano, Palermo, 17 (suppl.): 1–292.

Clay, B. L., P. Kugrens & R. E. Lee, 1999. A revised classification of Cryptophyta. Bot. J. linn. Soc. 131: 131–151.

Hill, D. R. A., 1991. A revised circumscription of *Cryptomonas* (Cryptophyceae) based on examination of australian strains. Phycologia 30: 170–188.

Huber-Pestalozzi, G., 1968. Das Phytoplankton des Süßwassers. 2 Aufl., ed. B. Fott. Die Binnengewässer. E. Scheizerbart'sche Verlagsbuchhandlung, Stuttgart. 322 pp.

Javornický, P., 2001. Freshwater Rhodomonas (Cryptophyceae). Archiv für Hydrobiol./Algol. Stud. 102: 93–116.

Klaveness, D., 1988. Ecology of the Cryptomonadida: a first review. In Sandgren, C. D. (ed.), Growth and Reproductive Strategies of Freshwater Phytoplankton. Cambridge University Press, Cambridge: 105–133.

Naselli Flores, L., 1999. Limnological studies on Sicilian reservoirs: an ecosystemic, comparative approach. In Tundisi, J. G. & M. Straskraba (eds), Theoretical Ecology of Reservoirs and its Applications. Backhuys Publishers, Leiden: 283–311.

Naselli Flores, L. & R. Barone, 2000. Phytoplankton dynamics and structure: a comparative analysis in natural and man-made water bodies of different tropic state. Hydrobiologia 438: 65–74.

Naselli Flores, L. & R. Barone, 2003. Limnology of a small, temporary water body: the Pond of Santa Rosalia (Sicily, Italy). Int. Ver. theor. angewand. Limnol: Verh. 28: 1673–1677.

Novarino, G., 2003. A companion to the identification of cryptomonad flagellates (cryptophyceae = cryptomonadea), Hydrobiologia 502 (Dev. Hydrobiol. 172): 225–270.

Novarino, G. & I. A. N. Lucas, 1993. Some proposals for a new classification system of the Cryptophyceae. Bot. J. linn. Soc. 111: 3–21.

Novarino, G., I. A. N. Lucas & S. Morrall, 1994. Observations on the genus *Plagioselmis* (Cryptophyceae). Cryptogamie, Algologie 15: 87–107.

Ojala, A., 1993. Effects of temperature and irradiance on the growth of two freshwater photosynthetic cryptophytes. J. Phycol. 29: 278–284.

Reynolds, C. S., 1984. The Ecology of Freshwater Phytoplankton. Cambridge University Press, Cambridge. 384 pp.

Reynolds, C. S., V. L. Huszar, C. Kruk, L. Naselli-Flores & S. Melo, 2002. Towards a functional classification of the freshwater phytoplankton. J. Plankton Research 24: 417–428.

Hydrobiologia **502:** 331–348, 2003.
L. Naselli-Flores, J. Padisák, M. T. Dokulil (eds),
Phytoplankton and Equilibrium Concept: The Ecology of Steady-State Assemblages.

Flagellate algae (Chrysophyceae, Dinophyceae, Cryptophyceae) in 48 high mountain lakes of the Northern and Southern slope of the Eastern Alps: biodiversity, taxa distribution and their driving variables

Monica Tolotti[1,2], Hansjörg Thies[1], Marco Cantonati[2], Claude M. E. Hansen[1] & Bertha Thaler[3]
[1] *Institute of Zoology & Limnology, University of Innsbruck, Technikerstrasse 25, A – 6020 Innsbruck.*
E-mail: Monica.Tolotti@uibk.ac.at
[2] *Museo Tridentino di Scienze Naturali, Limnology and Phycology Section, Via Calepina 14, I – 38100 Trento*
[3] *Laboratorio Biologico Provinciale, APPA, Unterbergstrasse 2, I – 39055 Laives*

Key words: high mountain lakes, flagellate algae, biodiversity, functional classification, multivariate analysis

Abstract

Survey samplings on late summer phytoplankton communities were carried out on 48 high mountain lakes located on the Austrian and Italian slopes of the Eastern Alps. The lakes of North Tyrol (A) and South Tyrol (I) were sampled in 2000 as part of the EU project EMERGE (EVK1-CT-1999- 00032). The lakes of Trentino (I) were investigated in other research projects during 1996 and 1997 (Adamello mountain range) and 2000 (catchment of the River Avisio), respectively. The objectives of this paper are: (1) to study taxonomy and biodiversity of Chrysophyceae, Dinophyceae and Cryptophyceae in high altitude lakes of the Eastern Alps; (2) to identify functional flagellate groups characterising lakes with similar habitat properties, (3) to identify the environmental variables driving abundance and distribution of the three selected algal groups, thus contributing to the selection of sensitive bioindicator taxa. The lakes investigated show rather wide morphological, chemical and trophic state gradients. Flagellate algae account for a median relative abundance (R.A.) of 68%. Chrysophyceae are the most important group in terms of biodiversity and R.A.. Special flagellate associations could be related to lake features, like catchment geology, mineralization level and nutrient concentrations. However, the distribution of flagellate algae did not allow a complete geographical separation of the lakes studied in the different districts. Multivariate canonical analyses indicate that the distribution of Chrysophyceae is mainly driven by NO_3-N concentration and thermal conditions, while Dinophyceae are driven by a combination of alkalinity, altitude, thermal condition and, less importantly, nutrient concentration. Physical properties of the lakes, such as thermal condition and lake depth, represent the principal driving variables for Cryptophyceae. The responses to the different environmental variables suggest that the three flagellate groups analysed might be used as indicators for environmental changes in high mountain lakes of the Eastern Alps.

Introduction

Chrysophyceae, Dinophyceae and Cryptophyceae are among the most common and abundant phytoplankton groups in temperate lakes of the Northern hemisphere (Dokulil, 1988; Sandgren, 1988; Eloranta, 1989; Reynolds, 1997; Barone & Naselli-Flores, 2003; Padisák et al., 2003; Willén, 2003). Due to their usually small cell size, motility and the ability of several species to supplement their nutrient uptake by phagotrophy (e.g. Holen & Borass, 1995; Gervais, 1997; Reynolds et al., 2002), these algae are well adapted to the extreme temperature, light and nutrient concentrations characterising high altitude lakes (e.g. Pechlaner, 1971; Tilzer & Schwarz, 1976). Therefore, flagellates often dominate the summer phytoplankton of high mountain lakes and are of major importance for the survival of the phytoplankton community under ice and snow cover in winter (e.g. Nauwerck, 1966; Schneider, 1981; Rott, 1988; Psenner, 1998).

Because of their physiological characteristics and special ecological preferences flagellate algae are regarded as useful indicators of environmental changes occurring in temperate lakes. Chrysophyceae are particularly sensitive to nutrient availability, as highlighted by several studies in Scandinavian lakes (e.g. Rosèn, 1981; Brettum, 1989; Eloranta, 1989). Some dinoflagellates are known to be indicators of acidification processes (e.g. Schindler et al., 1985; Willén et al., 1990; Nauwerck, 1994; Pugnetti & Bettinetti, 1995; Willén, 2003), oligotrophy and arctic-like environmental conditions (e.g. Rott, *loc. cit.*). Cryptophyceae mainly include cosmopolitan taxa typically present in moderate numbers throughout the year in lakes of varying morphology and trophic level (Dokulil, 1988; Reynolds, 1997; Lepistö & Holopainen, 2003). Because of their tolerance to low light intensity and temperature (e.g. Cloern, 1977; Gervais, 1997; Tardio et al., 2003) Cryptophyceae are normally more frequent and abundant in deep and stratified lakes (e.g. Dokulil, *loc. cit.*).

The use of flagellate algae as bio-indicators appears to be particularly important for remote high altitude lakes considering their high sensitivity to environmental changes, which is mainly due to the extreme climate conditions and low buffering capacity characterising these ecosystems. Additionally, high altitude lakes of several Alpine regions are being increasingly impacted by eutrophication (e.g. alpine pasture), tourism and airborne contaminants (e.g. organic pollutants, heavy metals, Carrera et al., 2002), whose effects are often difficult to quantify.

However, several practical difficulties affect the investigation of flagellate algae. The identification of taxa without a hard cell wall is hindered by modifications in cell shape and size (shrinking) due to the application of common fixatives (Rott, 1981). The identification keys currently used in the majority of laboratories are rather old, while new taxonomical achievements and revisions are scattered across a large number of scientific journals and papers. This situation is particularly critical for the Cryptophyceae, whose taxonomy is undergoing a major revision (e.g. Hill, 1991; Novarino et al., 1994; Javorniký, 2001). Finally, information about ecology and geographical distribution of phytoflagellates is often scarce for high mountain lakes, due mainly to the logistic difficulties of investigating such extreme environments.

The objectives of this paper are: (1) to study the taxonomy, biodiversity and geographical distribution of Chrysophyceae, Dinophyceae and Cryptophyceae in a large group of high-altitude lakes located in four districts on both the Northern and Southern slope of the Eastern Alps; (2) to identify functional groups of flagellate algae characterising lakes with similar habitat properties; (3) to identify the most important environmental variables driving abundance and distribution of the three selected algal groups, thus contributing to the understanding of the ecology of flagellate algae in high mountain lakes and to the selection of sensitive bioindicator taxa.

Description of the sites studied

The data used in this study were provided by three different research projects focussing on biodiversity and ecology of high mountain lakes. The lakes of North Tyrol (NTY) in Austria and of South Tyrol (STY) in Italy (Fig. 1) were surveyed in the year 2000 under the EU project EMERGE (European Mountain lake Ecosystem: Regionalisation, diaGnostics and socio-economic Evaluation, EVK1-CT-1999-00032). The lakes of Trentino (Fig. 1) located in the Adamello-Brenta Natural Park (ADA, cf. Cantonati et al., 2002b) and in the basin of the River Avisio (AVI), were investigated in local research projects in late summer 1996–1997 and 2000, respectively. All the lakes studied are located above the timberline (ca. 1800 m a.s.l.) and have a lake area larger than ca. 0.5 ha (Table 1). The lakes studied within the EMERGE project (NTY and STY) are located on different geological substrates, ranging from calcareous to acid gneiss bedrock (Table 1). The lakes of the ADA district are located on granite/tonalite (Cantonati et al., 2002b), the majority of the AVI lakes on porphyric bedrock (Table 1). The AVI lakes show a narrower altitudinal range with a lower median value, in comparison to the other lakes (Table 2). The largest catchment areas were found in the ADA and STY districts, as well as the highest median values of maximum lake depths (Tables 1 and 2). The majority of the lake catchments present low vegetation-cover percentage (sparse shrubs, small trees, alpine meadows) even though several lower altitude lakes show a vegetation cover of up to 98% (Table 1). The majority of the lakes with higher vegetation cover are located in the AVI district (Table 2).

Lake altitude, maximum depth and vegetation cover were measured in the field, while geology, lake and catchment area and other related parameters not presented here were determined using orthorectified digital maps and GIS (Geographic Information System) techniques (Cantonati et al., 2002a; Hansen et

Table 1. Morphometrical characteristics of the lakes studied; sup. = superior; inf. = inferior; med. = medium; S. = South; N = North; S.D. = sampling date; * = Secchi transparency down to the bottom of the lake; eutrophic. = eutrophication processes

N.	Code	Name	Altitude m a.s.l.	Area ha	Depth m	Geology	Veget. cover %	Indication of human impact condition at s. d.	Thermal m	Secchi m	S.D.
North Tyrol (A)											
1	TY0047	Seewisee – inf.	2229	1.8	1.8	carbonatic	95	(pasture)	mixed	1.8 *	1
2	TY0048	Seewisee - sup.	2469	1.4	13.5	carbonatic	5	not detectable	stratified	13.5	11
3	TY0049	Seewisee – med.	2425	0.5	4.5	carbonatic	5	not detectable	mixed	4.5 *	0.5
4	TY0077	Steinsee	2222	1.9	7.2	carbonatic	25	(pasture)	mixed	7.2*	0.5
5	TY0166	Drachensee	1874	5.0	24.0	carbonatic	30	fish stocking (pasture)	stratified	8.0	20
6	TY0170	Schwarzsee ob Sölden	2796	3.7	18.0	gneiss	5	winter sports	stratified	6.0	15
7	TY0188	Gossenköllesee	2413	1.6	9.9	gneiss	30	(pasture)	stratified	9.0	8
8	TY0189	Rotfelssee	2485	1.0	5.8	gneiss	10	not detectable	stratified	5.8 *	5
9	TY0194	Plenderlesee -sup.	2344	2.2	7.5	gneiss	25	tourism (nutrient load)	stratified	6.0	6
10	TY0195	Plenderlesee - inf.	2317	1.5	5.7	gneiss	25	not detectable	gradient	5.7 *	0.5
11	TY0229	Mutterbergersee	2479	3.1	8.1	gneiss	30	pasture (acid soil)	mixed	8.1 *	0.5
12	TY0303	Friesenbergersee	2450	1.9	4.5	gneiss	10	hut, eutrophic.	gradient	4.5 *	0.5
13	TY0305	Wesendlkarsee - sup.	2660	1.7	16.5	gneiss	5	not detectable	stratified	8.0	14
14	TY0306	Wesendlkarsee - inf.	2368	2.5	21.5	gneiss	40	(pasture) fish stocking	stratified	7.5	20
15	TY0338	Langer See	2232	3.7	15.4	mica scists	80	pasture, fish stocking	stratified	4.5	14
16	TY0339	Scheibensee	2290	1.5	4.0	mica scists	75	pasture, fish stocking	gradient	4 *	0.5
17	TY0427	Ascher See	2532	1.4	7.0	gneiss	80	(pasture)	mixed	5.0	6.5
18	TY0428	Anraser See	2538	2.1	6.5	gneiss	50	(pasture)	gradient	2.5	0.5
19	TY0444	Wildsee	1840	3.2	14.5	carbonatic	50	hut, eutrophic., fish stocking	stratified	8.0	12
20	TY0459	Alkuser See	2432	7.2	49.0	gneiss	35	not detectable	stratified	15.0	47
21	TY0463	Neualplsee - S	2438	2.0	13.0	gneiss	30	not detectable	gradient	3.0	12.5
22	TY0464	Neualplsee - N	2440	2.1	7.0	gneiss	50	not detectable	gradient	3.0	0.5
South Tyrol (I)											
23	TY0043	Rasasser See	2682	1.5	9.3	gneiss	39	not detectable	mixed	8.05	
24	TY0152	Alplanersee	2387	3.0	20.0	gneiss	68	pasture, fish stocking	mixed	6.0	4.5
25	TY0168	Kofelraster Süd	2405	3.1	9.0	mica scists	58	(fish stocking) pasture	mixed	3.5	5
26	TY0237	Timmelsschwarzsee	2514	6.4	27.0	gneiss	56	fish stocking	gradient	6.0	9
27	TY0270	Kratzbergersee	2119	2.4	12.0	gneiss	78	(fish stocking) pasture	gradient	2.3	4
28	TY0352	Grünsee	2043	1.1	2.7	carbonatic	45	pasture	gradient	2.7 *	0.5
Adamello group (I)											
29	GARZ	Garzonè	1941	10.7	27.0	granite	67	dam	stratified	15	20
30	GHIA	Ghiacciato	2571	0.3	8.3	tonalite	44	not detectable	mixed	8.0	5
31	MAND	Mandrone	2399	2.6	7.3	tonalite	47	hut, eutrophic.	gradient	7.3 *	7
32	NERO	Nero	2233	2.8	33.7	tonalite	72	fish stocking	stratified	8.5	15
33	ROTO	Rotondo	2523	1.2	16.3	tonalite	56	not detectable	gradient	16.3*	15
34	SERO	Serodoli	2371	6.5	33.2	granite	29	damstratified	18.3	25	
35	SGIU	San Giuliano	1936	4.3	5.7	granite	70	hut, eutrophyc., pasture, dam	mixed	5.7 *	4.5
36	SMAN	Scuro del Mandrone	2663	7.9	30.0	tonaliti	12	dam	stratified	26.5	25
37	SERM	Serodoli Medio	2379	0.3	3.0	granite	34	not detectable	mixed	3.0 *	2.5
38	TLPR	Tre Laghi Principale	2257	1.7	16.1	tonalite	42	fish stocking	stratified	10.5	14
River Avisio basin (I)											
39	ANTE	Antermoia	2496	2.6	4.4	dolomite	0	not detectable	mixed	4.4 *	1
40	BRUT	Brutto	2210	1.7	20	porphyry	75	not detectable	stratified	11.5	16
41	BUSE	Buse	2059	2.3	2.7	porphyry	95	not detectable	mixed	1.7	1
42	CAVA	Cavallazza	2138	1.0	7.1	porphyry	85	not detectable	stratified	3	5

Table 1. Continued

N.	Code	Name	Altitude m a.s.l.	Area ha	Depth m	Geology	Veget. cover %	Indication of human impact at s. d.	Thermal condition	Secchi m	S.D.
43	LAGU	Lagusel	2102	0.7	1.7	marl	98	pasture	mixed	1.7 *	1
44	LUS1	Lusia I	2332	0.7	5.2	porphyry	85	not detectable	mixed	4.8	1
45	LUS2	Lusia II	2352	0.4	2.5	porphyry	80	not detectable	mixed	2	1
46	MORE	Moregna	2058	1.8	6.9	porphyry	80	pasture eutrophication	stratified	4.5	3
47	STEL	Stellune	2099	2.8	18.5	porphyry	70	pasture	stratified	9.5	6
48	TROT	Trote	2101	1.0	5.3	porphyry	60	not detectable	stratified	5.3 *	4

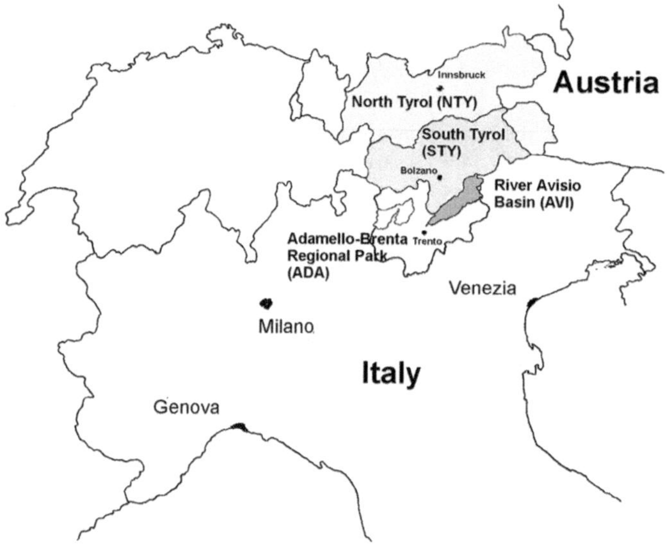

Figure 1. Geographical location of the lakes studied in the four districts. Lake codes as in Table 1.

al., 2002). Human impact was partly estimated in the field and partly gathered from the literature. Although no direct impact was often detectable, several lakes within each district appeared to be potentially affected by alpine pasture, fish stocking or tourism activities (Table 1). Several deep and large lakes of the ADA district were dammed at the beginning of the 1950's for hydroelectric power purposes (Cantonati et al., 2002b).

The knowledge of general limnology of the lakes studied was rather poor and scattered previous to the research projects considered in the present contribution. An exception is represented by some lakes in NTY, which were intensively studied since the 1960's (e.g. Nauwerck, 1966; Pechlaner, 1971; Rott, 1988), during the 1980's and 1990's under national projects concerned with acidification processes (e.g. Psenner, 1998) and recently in the European MOLAR project (e.g. Thies et al., 2000; 2002). Few specific studies were carried out on the lakes located in the Italian districts (e.g. Stella, 1931; Baldi, 1932, 1935; Poda, 1935; Zucchelli, 1952) and most of the previous assessments were part of general limnological investigations devoted to broader areas of the Italian Alps (e.g. Tomasi, 1962; Giussani et al., 1986).

Materials and methods

Sampling

All lakes considered in this contribution were sampled once in late summer (end of July–end of September), which usually coincides with the complete development of the planktonic community and with the highest phytoplankton biovolume in high mountain

Table 2. Important morphological, chemical and biological variables of the 48 lakes studied. R.A. = relative abundance; U = ultra-oligotrophic; O = oligotrophic; OM = oligo-mesotrophic; M = mesotrophic; E = eutrophic

	NTY (A)			STY (I)			ADA (I)			AVI (I)			ALL LAKES		
	Min	Max	Median	Min	Max	Median	Min	Max	Median	Min	Max	Median	Min	Max	Median
Altitude [m a.s.l.]	1840	2796	2429	2043	2682	2396	1936	2663	2375	2058	2496	2120	1840	2796	2370
Lake area [ha]	0.5	7.2	1.9	1.1	6.4	2.7	0.3	10.7	2.7	0.4	2.8	1.4	0.3	10.7	1.9
Catchment area [ha]	14	204	37	24	210	62	5	275	94	5	131	50	5	275	51
Catchment/ lake area	4.4	76.6	19.3	10.1	185.8	16.1	4.0	138.0	35.0	10.4	117.9	31.7	4.0	185.8	22.2
Max. Depth [m]	1.8	49.0	7.8	2.7	27.0	10.7	3.0	33.7	16.2	1.7	20.0	5.3	1.7	49.0	8.2
Vegetation cover [%]	5	95	30	39	78	57	12	72	46	0	98	80	0	98	49
Surface temperature [°C]	7.1	17.7	11.5	6.5	9.3	9.3	9.1	16.4	12.1	10.4	17.5	14.5	6.5	17.7	12.1
Sampling Temperature[°C]	3.9	17.2	8.0	6.0	9.3	7.0	4.3	15.9	7.2	6.3	16.0	11.7	3.9	17.2	7.8
Conductivity [μS cm^{-1}]	6	182	36	24	254	55	8	22	10	14	223	28	6	254	26
Alkalinity [μeq l^{-1}]	4	1586	196	20	1949	194	0	102	65	100	2020	260	0	2020	159
pH	5.7	8.8	7.2	5.7	9.1	7.5	5.7	6.7	6.2	6.4	8.9	6.6	5.7	9.1	6.9
DP [μg l^{-1}]	1	5	2	0	2	1	1	8	2	1	13	2	0	13	2
TP [μg l^{-1}]	2	19	5	1	6	3	1	16	6	3	24	15	1	24	5
NO_3-N [μg l^{-1}]	0	236	102	30	360	155	56	240	151	15	110	15	0	360	102
NH_4-N [μg l^{-1}]	1	226	5	4	20	9	9	50	17	10	78	10	1	226	10
Si [mg l^{-1}]	0.1	1.5	0.6	0.2	1.9	1.2	0.8	3.3	1.7	1.3	4.6	2.4	0.1	4.6	1.1
$SO^{2-}{}_4$ [mg l^{-1}]	0.5	15.2	2.7	1.7	32.4	7.0	0.7	2.7	1.5	0.8	27.0	1.2	0.5	32.4	1.7
Mg^{2+} [mg l^{-1}]	0.1	9.7	0.4	0.3	16.3	1.4	0.1	0.2	0.1	0.1	11.2	0.2	0.1	16.3	0.3
Phytoplankton taxa [N]	10	48	19	7	25	22	6	43	19	12	37	27	6	48	20
Flagellated taxa [N]	4	20	11	3	17	11	2	13	6	5	19	12	2	20	11
Chlorophyll *a* [μg l^{-1}]	0.2	18.6	0.9	0.1	29.0	3.0	0.2	4.8	1.4	–	–	–	0.1	29.0	1.3
Phytoplankton Biovol. [mm^{3}m^{-3}]	83	3342	384	7	4626	289	14	424	57	25	6343	1504	7	6343	334
Flagellates [R.A.]	4	99	82	35	100	59	1	100	74	2	89	34	1	100	68
Chrysophyceae [R.A.]	1	88	30	2	43	38	1	72	12	1	42	6	1	88	20
Dinophyceae [R.A.]	0	86	12	0	98	8	0	94	12	0	84	4	0	98	10
Cryptophyceae [R.A.]	0	70	1	0	51	4	0	58	4	0	62	2	0	70	2
Shannon Index [S.I.]	1.0	3.8	2.3	0.8	3.2	2.8	0.1	2.7	1.7	0.3	2.9	2.1	0.1	3.8	2.2
S.I. Chrysophyceae	0.3	2.7	1.8	1.1	2.0	1.7	0.0	2.5	0.8	0.8	2.7	2.2	0.0	2.7	1.6
S.I. Dinophyceae	0.0	1.8	1.0	0.1	0.9	0.7	0.0	1.0	0.1	0.0	1.7	0.2	0.0	1.8	0.6
S.I. Cryptophyceae	0.0	1.8	0.6	0.4	1.9	1.7	0.0	1.5	0.0	0.0	1.0	0.0	0.0	1.9	0.5
Trophic state [OECD-TP]	U	M	O	U	O	U	U	O	O	U	M	M	U	M	O
Trophic state [Heinonen, 1980]	U	E	O	U	E	O	U	O	U	U	E	M	U	E	O

lakes (Pechlaner, 1971; Rott, 1988). Temperature profiles were measured at one-meter intervals at the deepest point of the lakes, using a multi-parametric probe. Water samples for chemical (1 l), chlorophyll (2 l) and phytoplankton analyses (0.1–1 l) were taken with a Patalas-Schindler sampler. For the analyses presented in this contribution, a singular sampling depth per lake was used, according to the EMERGE protocol No 07 (Straþkrabová & Stuchlík, 2000). The sampling depth (Table 1) corresponds to the expected depth of the peak of chlorophyll *a* concentration in each lake, which was estimated as follows: (a) 0.5–1 m in lakes with a maximum depth of 5 m or less as well as in thermally mixed lakes; (b) ca. 1 m above the lake bottom in thermally stratified lakes with Secchi disc transparency – measured with 30 cm Secchi disk – reaching the bottom; (c) 1.5 × Secchi depth in thermally stratified lakes with Secchi disk transparency not reaching the bottom.

Laboratory and data analyses

Chemical analyses, including pH, alkalinity, conductivity, major ions and nutrient concentrations, were carried out at the Istituto Agrario di S. Michele all'Adige (ISMAA) for the lakes located in Trentino, at the Laboratorio Biologico Provinciale of Laives (APPA) for the lakes located in STY and at the Institute of Zoology and Limnology of the University of Innsbruck (UIBK-IZL) for the lakes located in NTY. The three laboratories participate in the intercalibration exercises periodically organised by the CNR - Istituto per lo Studio degli Ecosistemi, formerly Istituto Italiano di Idrobiologia, in Pallanza (Italy). Alkalinity was determined using a potentiometric titration (double flex-point method) and a modified Gran titration (Psenner, 1988). Ion and nitrate nitrogen analyses were performed by ion-exchange chromatography. Ammonia, dissolved and total phosphorus (DP and TP, respectively) and soluble reactive silica were analysed by spectrophotometry (A.P.H.A., A.W.W.A., W.C.P.F, 1992, see also EMERGE protocol N° 04, Mosello & Wathne, 2000). Chlorophyll *a* was extracted with acetone 90% and calculated by the equations of Lorenzen (1967) and Jeffrey & Humphrey (1977). No chlorophyll analyses were performed for the lakes in the AVI district.

The phytoplankton counts for the lakes in NTY, ADA and AVI districts were performed by one person (M.T.) according to Utermöhl (1958) and Lund (1958). The biovolume of each taxon was calculated by approximating the cell shape to simple geometrical solids (Rott, 1981). The counts for the lakes located in STY were carried out by ARGE Limnologie (Innsbruck), following the same methods and keys for the taxa determination (Das Phytoplankton des Süßwasser established by G. Huber-Pestalozzi and Süßwasserflora von Mitteleuropa, established by A. Pascher). According to the EMERGE protocols each taxon was identified with an 8-character code, including letters, numbers and, in the case of unidentified taxa, district abbreviations (TY = Tyrol, AD = Adamello, AV = Avisio).

The trophic level of the lakes studied (Table 3) was estimated according to the OECD method (Vollenweider & Kerekes, 1982) based on TP concentrations and to the Heinonen method (1980), based on midsummer phytoplankton biovolume, which appeared to be satisfactory for the trophic classification of high mountain lakes in the Alps sampled once during the vegetation period (Tolotti, 2001). Species diversity was assessed using the Shannon-Wiener function (S.I., 2-based logarithms, Shannon & Weaver, 1949) calculated from the number of individuals for each taxon (Sommer et al., 1993).

The TWINSPAN program (Hill et al., 1975) was used for the divisive hierarchical clustering of the lakes based on the relative abundance (R.A.) of the flagellate taxa. TWINSPAN analysis was restricted to taxa occurring in at least 5 lakes (ca. 10% of the lakes studied), while no restriction of the R.A. was applied. Canonical Correspondence Analysis (CCA, CANOCO 4.0, ter Braak & Smilauer, 1998) was performed for a combined evaluation of abundance and distribution of flagellate algae and environmental variables. Canonical analyses of Chrysophyceae and Dinophyceae were restricted to taxa occurring in at least two lakes. The ordinations were improved by excluding highly-correlated variables (e.g. conductivity, pH, Mg, DP and NH_4–N) and performing a stepwise selection of the variables based on a Monte Carlo permutation test (199 permutations, accepted $p <$ ca. 0.20). The significance of the CCA eigenvalues was tested by a Monte Carlo permutation test carried out on the first canonical axis and on the sum of all canonical axes, respectively.

Results

Environmental variables

The majority of the lakes deeper than ca. 10 m and some shallow lakes (5–7 m depth) located at lower altitude or sampled at the end of August 2000 showed a well-developed thermocline at the sampling date (Table 1). Several of the deepest lakes showed low hypolimnetic oxygen saturation (Cantonati et al., 2002b). The higher median surface and sampling temperature recorded in the AVI district are related to the lower median altitude and smaller depths of these lakes (Tables 1 and 2).

The description of the chemical variables of the study lakes given in this paper refers to the values measured at the sampling depths. Due to the different catchment lithology (Table 1) the water chemistry of the lakes varied considerably (Table 2). The lowest conductivity and pH values were recorded in the lakes of the ADA district, located on granite, and in those of the AVI district, located on porphyry. The low pH (<6.5) and alkalinity values (<100 μeq l^{-1}) measured in several lakes of ADA, NTY and STY districts (Table 2) indicate the acid sensitivity of these lakes (Psenner, 1998; Cantonati et al., 2002b). The higher alkalinity values measured in the AVI district (median = 260 μeq l^{-1}, Table 2) may be related to the lower lake altitude, the higher percent of vegetation cover in the catchment and to internal alkalinity generation processes generated by intense phytoplankton activity (Thies, 2000). The last hypothesis seems to be confirmed by the higher TP and particularly low NO_3–N concentrations measured in the majority of these lakes, in comparison to the other districts (Table 2).

The distribution of TP, NO_3–N and Si in the lakes of the four lake districts is shown in Figure 2. TP concentrations follow a rather regular Gaussian dis-

Table 3. Essential characterisation of phytoplankton communities and trophic status of the lakes studied. BCL = Bacillariophyceae; CHL = Chlorophyceae sensu lato; CHR = Chrysophyceae; CON = Conjugatophyceae; CVO = Volvocales; CYA = Cyanophyceae; DIN = Dinophyceae; S.I. = Shannon-Wiener Index; TL = trophic level; He. = Heinonen, 1980; Lake codes and trophic status abbreviations as in Tables 1 and 2, respectively

N.	Code	TP	Chl. *a*	Phytopl. Biovolume	Dominant group	Dominance	S. I.	TL TP	TL Phy.-Vol.
		$\mu g\, l^{-1}$	$\mu g\, l^{-1}$	$mm^3\, m^{-3}$		%	OECD	He.	
North Tyrol (A)									
1	TY0047	6.5	0.5	222	CHR	64	2.6	O	O
2	TY0048	5.0	0.5	220	DIN	70	1.2	O	O
3	TY0049	2.1	0.2	87	CHR	37	2.4	U	U
4	TY0077	1.9	0.7	355	CHL	87	2.9	U	O
5	TY0166	4.7	2.1	141	CHR	48	2.8	O	U
6	TY0170	3.5	1.6	816	DIN	69	1.1	U	OM
7	TY0188	4.7	2.9	611	BCL	58	2.2	O	OM
8	TY0189	2.6	0.5	165	CHR	66	3.0	U	U
9	TY0194	1.5	0.9	155	DIN	60	2.2	U	U
10	TY0195	3.5	1.3	344	DIN	53	2.7	U	O
11	TY0229	2.4	0.4	466	DIN	86	3.2	U	O
12	TY0303	3.2	0.2	83	CVO	54	2.6	U	U
13	TY0305	5.9	2.0	925	DIN	53	1.7	O	OM
14	TY0306	3.8	1.3	584	CHR	88	2.0	U	OM
15	TY0338	18.8	7.5	2186	CHR	73	3.3	M	M
16	TY0339	6.5	1.5	412	CHR	41	3.8	O	O
17	TY0427	7.2	0.5	161	CHL	57	1.5	O	U
18	TY0428	6.0	1.0	1064	BCL	55	1.1	O	M
19	TY0444	9.4	0.8	187	CRY	70	1.8	O	U
20	TY0459	4.0	2.7	1942	BCL	74	2.1	O	M
21	TY0463	14.3	18.6	3342	CON	70	1.0	M	E
22	TY0464	13.1	1.6	1119	DIN	85	2.5	M	M
South Tyrol (I)									
23	TY0043	0.9	0.1	10	DIN	98	2.7	U	U
24	TY0152	3.0	1.5	334	CRY	49	2.9	U	O
25	TY0168	6.3	6.8	1773	CHR	42	3.2	O	M
26	TY0237	3.0	4.4	244	CHR	44	0.8	U	O
27	TY0270	4.0	29.0	4626	CON	70	2.3	O	E
28	TY0352	1.8	0.1	7	CHL	39	3.0	–	U
Adamello group (I)									
29	GARZ	16.0	2.6	308	CRY	58	2.7	O	O
30	GHIA	2.0	0.7	68	CHL	99	0.1	U	U
31	MAND	1.0	–	17	CHR	64	1.7	U	U
32	NERO	4.5	4.8	334	CHR	52	1.4	O	O
33	ROTO	6.0	0.3	14	CHR	72	1.8	O	U
34	SERO	8.0	1.7	32	CHL	52	2.6	O	U
35	SGIU	9.0	1.4	424	CHL	82	1.7	O	O
36	SMAN	2.0	0.2	46	DIN	91	1.7	U	U
37	SERM	8.7	0.4	19	CRY	58	2.3	O	U
38	TLPR	5.0	4.6	240	CON	58	1.3	O	O

Table 3. Continued

N.	Code	TP $\mu g\ l^{-1}$	Chl. *a* $\mu g\ l^{-1}$	Phytopl. Biovolume $mm^3\ m^{-3}$	Dominant group	Dominance %	S. I. OECD	TL TP He.	TL Phy.-Vol.
River Avisio basin (I)									
39	ANTE	3.0	–	40	CHR	42	1.9	U	U
40	BRUT	12.0	–	1284	DIN	84	2.1	M	M
41	BUSE	14.0	–	5299	CHL	98	0.3	M	E
42	CAVA	15.0	–	1859	CHL	41	2.7	M	M
43	LAGU	16.0	–	96	CRY	62	2.9	M	U
44	LUS1	10.0	–	620	CON	54	2.2	M	OM
45	LUS2	24.0	–	6343	CYA	49	1.7	M	E
46	MORE	22.0	–	1724	DIN	40	2.1	M	M
47	STEL	11.0	–	2788	BCL	64	2.6	M	E
48	TROT	23.0	–	25	CON	45	1.1	M	U

tribution, with the majority of the lakes (N = 18) ranked in the central class (4–9.9 $\mu g\ l^{-1}$). The NYT and ADA lakes are uniformly distributed among the different classes, while the AVI lakes are almost completely concentrated in the classes corresponding to the highest TP concentrations. The distributions of NO_3-N and Si are rather irregular because of the accumulation of the majority of the AVI lakes in the classes corresponding to the lowest NO_3-N (<50 $\mu g\ l^{-1}$) and to the highest Si concentrations (>2 mg l^{-1}), respectively.

The trophic classification of the lakes based on TP concentrations (Vollenweider & Kerekes, 1982) ranked 73% the lakes as oligotrophic. Nine of the remaining mesotrophic lakes are located in the AVI district (Tables 2 and 3).

Phytoplankton community

The number of phytoplankton taxa identified in each lake showed a wide range (Table 2), while the number of flagellate taxa per lake was comparable in the four districts, with slightly lower values recorded only in the ADA district. The S.I. values relative to the whole phytoplankton community were relatively high and showed comparable ranges in the four districts (Tables 2 and 3). Median chlorophyll *a* concentration and total phytoplankton biovolume were in general low (1.3 $\mu g\ l^{-1}$ and 334 $mm^3\ m^{-3}$ respectively, Table 2), although the values of both variables were distributed along relatively wide ranges (Tables 2 and 3). The trophic classification based on total phytoplankton biovolume (Heinonen, 1980) indicated 63% of the lakes studied as oligotrophic (Table 3), while 5 lakes (10.4%) with phytoplankton biovolume higher than 2500 $mm^3\ m^{-3}$ resulted as eutrophic. The majority of the mesotrophic and eutrophic lakes were located in the AVI district (Tables 2 and 3), while all the ADA lakes showed pronounced oligotrophic conditions.

Non-flagellate algae were dominant in terms of R.A. in 18 (37.5%) lakes and in general were more abundant in lakes of high trophic level (Table 3). The dominance of Conjugatophyceae was mainly due to the small *Cosmarium asphaerosphorum* var. *strigosum* Nordstedt, while small centric diatoms of the group *Cyclotella comensis* Grunow dominated in four mesotrophic or eutrophic lakes (Table 3). Filamentous Cyanophyceae (*Anabaena* sp. Bory) prevailed in a sole shallow, eutrophic lake of the AVI district (Tables 1 and 3), while in all other lakes they were almost completely absent.

Flagellate algae contributed to the total phytoplankton biovolume of the lakes studied with a median R.A. of 68%, even though the percentage varied in the different districts (Table 2). However, the maximum R.A reached by flagellates was higher than ca. 90% in all the four districts (Table 2). As flagellate Chlorophyceae (*Carteria cordiformis* Carter) were dominant in only one lake (Table 3), they will not be further discussed in this contribution. Among the flagellate algae considered in this paper, Chrysophyceae, Dinophyceae and Cryptophyceae were dominant in 13 (27%), 11 (23%) and five (10%) lakes, respectively (Table 3).

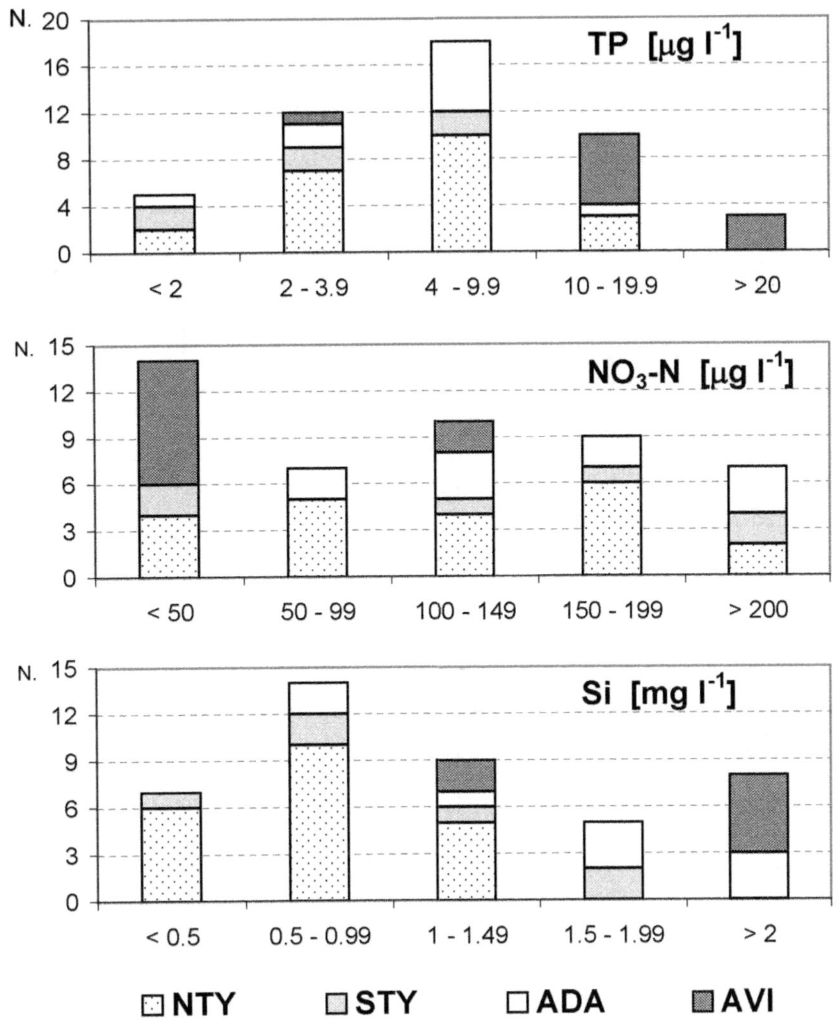

Figure 2. Frequency distribution of the principal algal nutrient concentrations (TP, NO_3-N and Si) in the lakes studied.

Chrysophyceae were the most common group, being present in all the lakes studied. The S.I. values calculated for this group (median = 1.6, Table 2) indicated a general high Chrysophyceae diversity, with somewhat lower S.I only in the ADA district (Table 2). Many of the taxa reported in Table 4 (occurring in at least two lakes) were found in different districts, although often in a small proportion of lakes. However, the majority of the taxa found in STY, are exclusively found in this district (Table 4). The most common taxa were represented by very small flagellates, including *Chrysochromulina parva*, a Haptophyceae grouped together with the Chrysophyceae in this contribution (Table 4). Chrysophyceae were the most important flagellate group also in terms of R.A. (median for all the lakes = 20% of the total phytoplankton biovolume), although their role was rather different in the four districts (Table 2).

The Dinophyceae were present in 44 lakes with 19 taxa. Although they showed a lower median R.A. (10%) in comparison to Chrysophyceae, the Dinophyceae reached particularly high R.A. (up to 98%) in several lakes of all the districts (Tables 2 and 3). Both the lower number of taxa and the higher dominance within the phytoplankton community (compared with Chrysophyceae) are reflected by the lower S.I. calculated for this group (median = 0.6, Table 2). About 50% of the Dinophyceae found in this study, including the majority of the small taxa without a rigid cell wall, could only be identified at genus level (Table 4). *Gymnodinium uberrimum*, which reached extremely high R.A. in some lakes (up to 94%, Table 4), was the most common taxon, but several Dinophyceae were found in different districts and in more than 10 lakes.

Cryptophyceae were less important both in terms of number of taxa identified (Table 4) and R.A. (median = 2%, Table 2). Moreover, they were irregularly distributed, being completely absent in 25% of the lakes and dominant in others (Tables 2 and 3). This situation is reflected by the low S.I. values calculated for this group (median = 0.5) in all the districts except in STY (Table 2).

Due to the peculiar status of the taxonomical knowledge of Cryptophyceae only few taxa could be certainly identified at species level (Novarino, 2003). For instance, one of the most common taxa was a large form (ca. 27 μm length), which we called *Cryptomonas erosa/reflexa*, because of its similarity to the still un-clearly separated *C. erosa* Ehrenberg and *C. reflexa*. This taxon could be certainly ascribed to *C. reflexa* only in two lakes of the NTY district (Table 4). The most common taxon was *Rhodomonas minuta* var. *nannoplanctica*, which is now called *Plagioselmis nannoplanctica* (Skuja) Novarino, Lucas & Morrall according to most recent accepted nomenclature (Novarino et al., 1994; Novarino, 2003; Javornický, 2003). This was the sole Cryptophycea found in all the districts. *Katablepharis ovalis* (an eterotrophic Cryptophycea), *C. obovata* and *C. marssonii*, now ascribed to the new Genus *Campylomonas* Hill (1991), were the only other taxa found in more than three lakes (Table 4).

Sample classification and species ordination

According to the flagellate species composition of the lakes studied, the first TWINSPAN division level significantly (eigenvalue = 0.66) separated the six lakes of STY (group A in Figure 3) from the lakes located in the other districts. This neat separation is due to the abundance in the STY lakes of several exclusive taxa identified at genus level (Table 4). However, apart from the general oligotrophic conditions, these lakes did not show other common environmental features.

The next three TWINSPAN division levels ranked the remaining 42 lakes into five major groups (Groups B-F in Figure 3). Group B1 includes five lakes of NTY, whose phytoplankton was mainly dominated by Dinophyceae (Table 3) and with high R.A. of *Peridinium inconspicuum*, *Gymnodinium* cf. *impatiens* and *Katodinium* sp.. The lakes ranked into this group are located at high altitude (median = 2450 m a.s.l.), on catchments composed of acid rocks (gneiss) and with

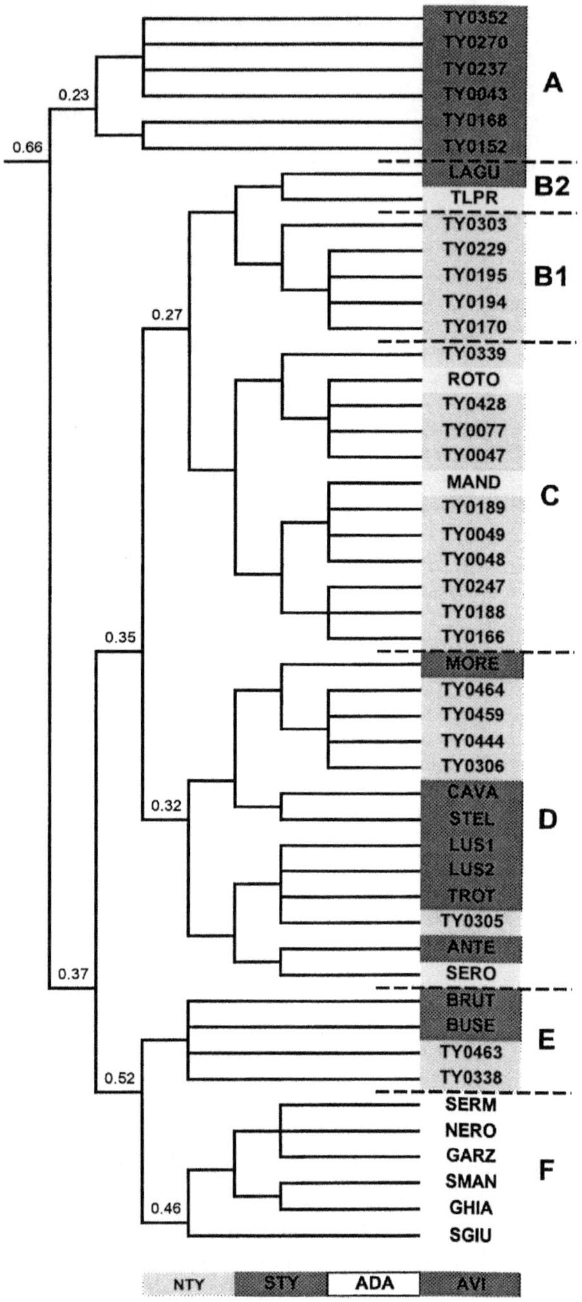

Figure 3. TWINSPAN classification of the lakes studied according to the distribution of flagellate taxa present at least in 5 lakes (*N.* = 37). Eigenvalues of the first four divisions are reported. Lake codes as in Table 1.

low vegetation cover (max. = 30%, Table 1). All show acid pH (min. = 5.7), low alkalinity values (median = 25 μeq l^{-1}) and clear oligotrophic conditions. The lakes LAGU and TLPR, located on calcareous and granitic rocks respectively, were ranked in a partially separated group (B2 in Figure 3) according to the high R.A. of *Cryptomonas obovata*, *Gymnodinium* sp. (GYMN07TY) and of small Chrysophceae (e.g. CHRF03TY, Table 4). In comparison with the lakes of Group B1, these two oligo-mesotrophic lakes are located at lower altitude into catchments with higher vegetation cover (Tables 1 and 3).

Group C includes 12 oligotrophic to mesotrophic lakes mainly located in the NTY district (Table 3) and characterised by high R.A. of different *Chromulina* species (Chrysophyceae). Even though five of the seven calcareous lakes studied were ranked into Group C (Table 1), the catchment lithology of these lakes is rather variable (Table 1), thus preventing the indication of common chemical characteristics. Group D ranks 13 lakes with prevalent location on porphyry (AVI) and gneiss (NTY) bedrock and with high R.A. of *Gymnodinium uberrimum* and small Chrysophyceae (like *Chrysococcus rufescens* and *Monochrysis* cf. *parva*). The majority of these lakes showed mid vegetation cover in their catchments (median = 50%), very low NO_3–N and high Si concentrations (median = 30 μg l^{-1} and 1300 μg l^{-1}, respectively) and were mainly of mesotrophic status (Table 3).

Group E includes four mesotrophic/eutrophic lakes (Table 3) located in catchments with high vegetation cover (up to 95%) of the NTY and AVI districts, characterised by high R.A. of *Cryptomonas erosa/reflexa* and *Mallomonas acaroides* (Table 4). The discriminating chemical features of this group of lakes are low conductivity (median = 23 μS cm^{-1}), pH (median = 6.5) and alkalinity values (median = 150 μeq l^{-1}) and low NO_3–N and Si concentrations (median = 65 μg $l^{-1}$and 791 μg l^{-1}, respectively). Finally group F, including relatively large and deep lakes (median = 3.5 ha and 18 m, respectively), located in catchments with scarce to medium vegetation cover of the ADA district (Table 1), was significantly separated (eigenvalue = 0.52) from group E on the basis of high R.A. of *Gymnodinium uberrimum, Peridinium inconspicuum* and *Dinobryon cylindricum* var. *alpina*. In comparison to Group E, the lakes of Group F present even lower water mineralization levels (median conductivity, pH and alkalinity = 9 μS cm^{-1}, 6.2 and 51 μeq l^{-1}, respectively), higher NO_3–N concentrations (median = 140 μg l^{-1}) and pronounced oligotrophy.

The CCA ordinations of the three groups of flagellate algae in relation to the most significant environmental variables measured at the 48 lakes studied are plotted in Figures 4, 5 and 6, where the posi-

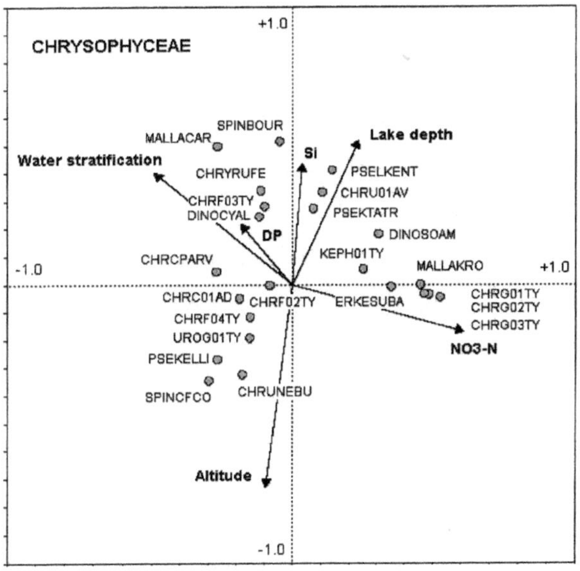

Figure 4. CCA ordination of the 38 Chrysophyceae taxa occurring at least in two lakes in relation to the six most significant environmental variables. (λ_1 = 0.486; λ_2 = 0.296). The 23 taxa with minimum fit = 4% are plotted. Taxon codes as in Table 4.

tion of each taxon reflects its relationships with the different driving variables. Figure 4 shows the distribution of the 23 Chrysophyceae taxa with minimum fit = 5%. The sum of the four canonical axes accounts for the 84% of the total variance of species

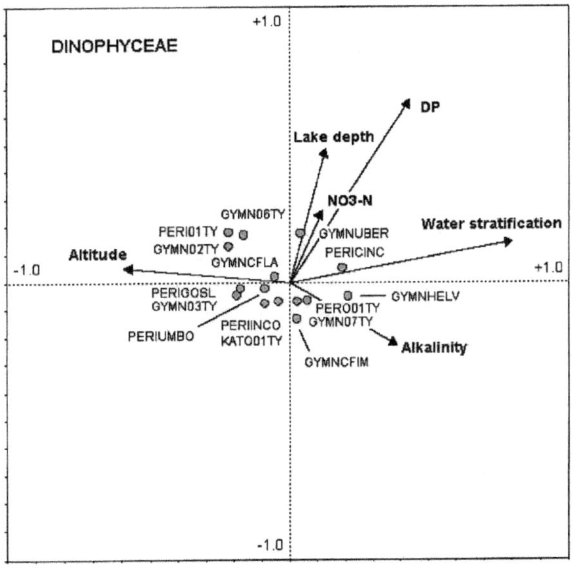

Figure 5. CCA ordination of the 15 Dinophyceae taxa occurring at least in two lakes in relation to the six most significant variables. (λ_1 = 0.506; λ_2 = 0.474). Taxon codes as in Table 4.

Table 4. Flagellate taxa of the 48 lakes studied used for TWINSPAN (Occurring in at least in 5 lakes) and CANOCO analyses. N = number of occurrences; Max. = maximum R.A. reached by each taxon

CODE	NAME	AUTHOR	NTY		STY		ADA		AVI		ALL	
			N	Max	N	Max	N	Max	N	Max	N	Max
Chrysophyceae												
BITRCHOD	*Bitrichia chodatii*	(REV.) HOLLANDE	0	–	1	22.6	2	0.3	3	0.2	6	22.6
BITRDANU	*Bitrichia danubiensis*	JURIŠ	1	0.1	0	–	0	–	2	0.1	3	0.1
CHRUNEBU	*Chromulina nebulosa*	CIENKOWSKI	9	7.3	0	–	0	–	4	0.9	13	7.3
CHRU01AV	*Chromulina* sp. 01 – 3.5–4 μm	CIENKOWSKI	0	–	0	–	1	3.8	1	0.3	2	3.8
CHRU01TY	*Chromulina* sp. 01 – 5 μm	CIENKOWSKI	9	2.8	0	–	0	–	4	5.7	13	5.7
CHRU02AV	*Chromulina* sp. 02 – 6 μm	CIENKOWSKI	0	–	0	–	2	3.1	1	2.0	3	3.1
CHRU02TY	*Chromulina* sp. 02 – 8 μm	CIENKOWSKI	16	83.0	0	–	5	26.9	0	–	21	83.0
CHRCPARV	*Chrysochromulina parva*	LACKEY	11	5.8	0	–	0	–	7	3.2	18	5.8
CHRYRUFE	*Chrysococcus rufescens*	KLEBS	4	0.3	0	–	1	7.2	5	1.7	10	7.2
CHRY02TY	*Chrysococcus* sp. 02 – 5 μm	KLEBS	3	0.2	0	–	3	6.3	3	17.0	9	17.0
CHRLSKUJ	*Chrysolykos skujai*	(RAMB.) BOURRELLY	1	–	1	0.2	0	–	0	–	2	0.2
CHRSBREV	*Chrysosphaerella brevispina*	KORŠIKOV	1	1.4	0	–	0	–	1	2.3	2	2.3
DINOCREN	*Dinobryon crenulatum*	W. & G.S. WEST	0	–	0	–	0	–	2	0.1	2	0.1
DINOCYAL	*Dinobryon cylindricum* var. *alpina*	IMHOF	4	3.8	0	–	4	4.9	0	–	8	4.9
DINOSOAM	*Dinobryon sociale* var. *americana*	(BRUN.) BACHMANN	1	25.3	0	–	2	3.7	1	–	4	25.3
ERKESUBA	*Erkenia subaequaliciliata*	SKUJA	0	–	5	10.5	0	–	0	–	5	10.5
KEPHMAST	*Kephyrion mastigophorum*	SCHMID	3	0.6	0	–	0	–	0	–	3	0.6
KEPH01TY	*Kephyrion* sp. 01	PASCHER	0	–	3	0.2	0	–	0	–	3	0.2
MALLACAR	*Mallomonas acaroides*	PERTY	2	11.7	0	–	0	–	3	21.1	5	21.1
MALLAKRO	*Mallomonas akrokomos*	RUTTNER	0	–	1	0.6	0	–	2	0.1	3	0.6
MALLPEDI	*Mallomonas pediculus*	TEILING	1	10.7	0	–	0	–	1	1.0	2	10.7
MONOPARV	cf. *Monochrisis parva*	SKUJA	5	1.3	0	–	0	–	4	1.2	9	1.3
PSEKELLI	*Pseudokephyrion ellipsoideum*	(PASCHER) SCHMID	2	19.9	0	–	0	–	0	–	2	19.9
PSEKENTZ	*Pseudokephyrion entzii*	CONRAD	3	5.1	0	–	0	–	3	0.4	6	5.1
PSEKTATR	*Pseudokephyrion tatricum*	(JURIŠ) STARMACH	2	0.3	0	–	2	1.5	1	–	5	1.5
PSEPERKE	*Pseudopedinella erkensis*	SKUJA	3	0.5	0	–	4	28.0	1	–	8	28.0
SPINBOUR	*Spiniferomonas* cf. *bourrellyi*	TAKAHASHI	0	–	0	–	1	0.7	1	0.1	2	0.7
SPINCFCO	*Spiniferomonas* cf. *cornuta*	BALANOV	2	5.5	0	–	0	–	0	–	2	5.5
STICDOER	*Stichogloea doerderleinii*	(SCHMIDLE) WILLE	1	4.8	0	–	1	1.5	2	0.6	4	4.8
UROG01TY	*Uroglena* sp. 01	EHRENBERG	13	53.5	3	1.4	0	–	0	–	16	53.5
CHRG01TY	Chrysophycea Genus 01 – 3 μm		0	–	6	1.1	0	–	0	–	6	1.1
CHRG02TY	Chrysophycea Genus 02 – 5 μm		0	–	6	14.2	0	–	0	–	6	14.2
CHRG03TY	Chrysophycea Genus 03 – 10×7 μm		0	–	5	33.9	0	–	0	–	5	33.9
CHRF02TY	Flagellate type 02 – 3 μm		20	3.0	0	–	3	1.6	8	1.1	31	3.0
CHRF03TY	Flagellate type 03 – 5μm		7	1.7	0	–	4	43.6	7	2.2	18	43.6
CHRF04TY	Flagellate type 04 – heart shaped		13	22.4	0	–	0	–	2	2.9	15	22.4
CHRF05TY	Flagellate type 05 – 8 μm		1	7.7	0	–	2	10.2	1	1.1	4	10.2
CHRC01AD	Coccal type 01 – 4.5 μm		0	–	0	–	1	2.7	7	21.4	8	21.4
Dinophyceae												
GYMNCFIM	*Gymnodinium* cf.*impatiens*	SKUJA	8	17.7	0	–	0	–	3	9.0	11	17.7
GYMNCFLA	*Gymnodinium* cf. *lantzschii*	UTERMÖHL	8	5.9	0	–	3	2.4	1	0.0	12	5.9
GYMNHELV	*Gymnodinium helveticum*	PENARD	5	70.0	0	–	0	–	0	–	5	70.0
GYMNUBER	*Gymnodinium uberrimum*	(ALL.) KOF. & SWEZY	7	81.1	0	–	6	93.9	4	0.6	17	93.9
GYMN02TY	*Gymnodinium* sp. 02 – 10×8 μm	STEIN	0	–	2	3.8	0	–	0	–	2	3.8
GYMN03TY	*Gymnodinium* sp. 03 – 15×12 μm	STEIN	0	–	3	6.1	0	–	0	–	3	6.1

abundances as constrained by the six most significant environmental variables. The ordination is highly significant ($p < 0.005$) according to the Monte Carlo permutation test performed both on the first axis and on the sum of the canonical axes. The distribution of the Chrysophyceae appears to be mainly driven by NO_3–N concentration and thermal stratification of the water column ($p < 0.01$) and secondly by altitude ($p < 0.06$). Lake depth, Si and DP appear to be less important in driving the distribution of Chrysophyceae. The taxa plotted in Figure 4 can be divided in three groups according to their position within the ordina-

Table 4. Continued

CODE	NAME	AUTHOR	NTY N	NTY Max	STY N	STY Max	ADA N	ADA Max	AVI N	AVI Max	ALL N	ALL Max
GYMN06TY	*Gymnodinium* sp. 06 – small	STEIN	0	–	0	–	1	6.6	1	0.5	2	6.6
GYMN07TY	*Gymnodinium* sp. 07 – no plastids	STEIN	2	4.7	0	–	1	3.1	2	1.9	5	4.7
KATO01TY	*Katodinium* sp. 01	SKUJA	7	4.5	2	11.4	0	–	2	2.0	11	11.4
PERICINC	*Peridinium cinctum*	PENARD	2	0.9	0	–	0	–	4	39.7	6	39.7
PERIINCO	*Peridinium inconspicuum*	LEMMERMANN	6	59.0	0	–	3	20.7	2	0.5	11	59.0
PERIGOSL	*Peridinium goslaviense*	WOLOSZYNSKA	0	–	1	93.9	0	–	1	7.3	2	93.9
PERIUMBO	*Peridinium umbonatum*	STEIN	5	64.6	2	1.9	2	16.6	1	0.4	10	64.6
PERI01TY	*Peridinium* sp. 01 – 20×18 μm	EHRENBERG	0	–	3	5.3	0	–	0	–	3	5.3
PERO01TY	*Peridiniopsis* sp. 01	LEMMERMANN	9	25.3	0	–	1	10.3	3	0.5	13	25.3
Cryptophyceae												
CHRMCFNO	*Chroomonas* cf. *nordstedtii*	HANSGIRG	3	0.8	0	–	0	–	0	–	3	0.8
CRYPERRE	*Cryptomonas erosa/reflexa*		5	20.5	0	–	4	57.9	4	56.0	13	57.9
CRYPMARS	*Cryptomonas marssonii*	SKUJA	3	14.7	0	–	2	4.4	0	–	5	14.7
CRYPCFOB	*Cryptomonas* cf. *obovata*	SKUJA	5	2.9	0	–	1	0.6	3	5.6	9	5.6
CRYPPYRE	*Cryptomonas pyrenoidifera*	GEITLER	0	–	0	–	1	0.5	2	2.2	3	2.2
CRYPREFL	*Cryptomonas reflexa*	SKUJA	2	3.2	0	–	0	–	0	–	2	3.2
CRYP04TY	*Cryptomonas* sp. 04 – 20×10 μm	EHRENBERG	0	–	3	1.8	0	–	0	–	3	1.8
CRYP05TY	*Cryptomonas* sp 05 – 25×12 μm	EHRENBERG	0	–	3	9.7	0	–	0	–	3	9.7
CRYP06TY	*Cryptomonas* sp 06 – 30×15 μm	EHRENBERG	0	–	3	19.7	0	–	0	–	3	19.7
KATACFOV	*Katablepharis* cf.*ovalis*	SKUJA	8	4.1	2	2.2	0	–	3	0.5	13	4.1
RHODMINA	*Rhodomonas minuta* var. *nannoplanctica*	SKUJA	13	44.5	3	8.8	3	2.0	4	40.9	23	44.5
RHOD01AD	*Rhodomonas* sp. - dark brown	KARSTEN	0	–	0	–	2	20.4	0	–	2	20.4
CRYG01AD	Cryptophycea Genus 01 – 17×6 μm		0	–	0	–	2	10.1	0	–	2	10.1
CRYG01TY	Cryptophycea Genus 01 – 16×7 μm		1	0.1	0	–	0	–	0	–	1	0.1

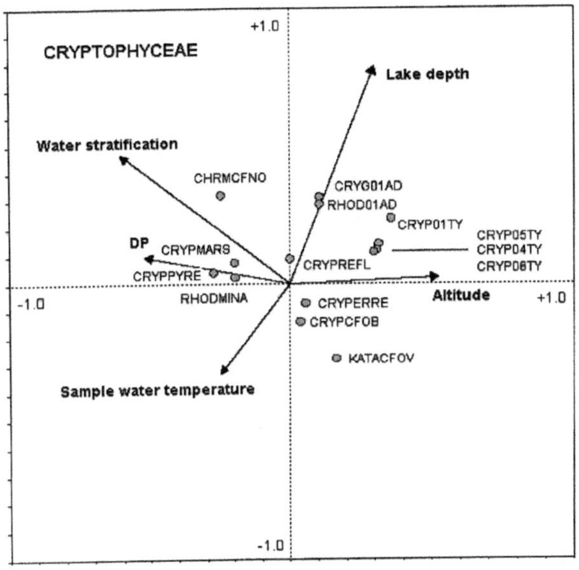

Figure 6. CCA ordination of the 15 Cryptophyceae taxa identified in the lakes studied in relation to the five most significant variables. ($\lambda_1 = 0.651$; $\lambda_2 = 0.368$). Taxon codes as in Table 4.

tion space. *Dinobryon sociale* var. *americana*, *Erkenia subaequaliciliata*, *Mallomonas akrokomos* and several taxa identified at genus level (see Table 4 for the taxon codes) are located in a region characterised by high NO_3–N concentrations, mixed thermal conditions and mid altitude. *Mallomonas acaroides*, *Spiniferomonas bourrelly*, *Chrysococcus rufescens*, *Dinobryon cylindricum* var. *alpina*, *Pseudokephyrion entzii* and *P. tatricum* show preference for stratified lakes located at lower altitude, with lower NO_3–N and higher Si concentrations. The third group, mainly including small flagellates, such as *Spiniferomonas* cf. *cornuta*, *Chromulina nebulosa*, *Pseudokephyrion ellipsoideum*, *Uroglena* sp. and *Chrysochromulina parva*, seen to be associated with mid-high altitude, mid NO_3–N and low Si concentration.

The four canonical axes of the CCA ordination relative to the Dinophyceae (Fig. 5) account for ca. 89% of the total variance in species abundances as constrained by the six most significant environmental variables. The ordination is highly significant according to the Monte Carlo permutation test performed on the first axis ($p < 0.04$) and on the sum of the canonical axes ($p < 0.005$). Dinophyceae appear to be driven by a different combination of environmental variables, including thermal stratification ($p < 0.005$), altitude ($p < 0.11$), alkalinity ($p < 0.08$) and DP concentrations ($p < 0.01$). Lake depth and NO_3–N show a less significant role in affecting the distribution of this algal group (Fig. 5). The majority of the identified Dinophyceae taxa are located in the central region of the ordination space, corresponding to non-stratified conditions, mid altitude and alkalinity and low DP concentration. *G. uberrimum* and *Peridinium cinctum* are the only two species showing a preference for deep and stratified water with higher nutrient concentrations (Fig. 5), while the heterotrophic *G. helveticum*, exclusively found in calcareous lakes, appears to be associated with stratified lakes with higher alkalinity values.

The CCA ordination obtained for the Cryptophyceae (Fig. 6) is significant ($p < 0.04$) only according to the Monte Carlo permutation test performed on the sum of all canonical axes, which explains ca. 98% of the total variance of the species as constrained by the five most significant environmental variables. Only two physical variables, i.e. thermal stratification and lake depth, appear to be highly significant ($p < 0.015$ and < 0.05, respectively) for the explanation of the Cryptophyceae distribution in the lakes studied. Altitude, DP concentration and water temperature at the sampling depth show p values ranging from 0.14 to 0.26. The distribution of several common taxa (Table 4) along a gradient of thermal condition and altitude, confirms the importance of these two variables as Cryptophyceae driving factors. *Cryptomonas marssonii*, *C. pyrenoidifera* and *Rhodomonas minuta* var. *nannoplanctica* are the sole taxa related to lakes with lower altitude and higher DP concentrations (Fig. 6).

Discussion

The environmental characteristics of the lakes studied indicates some differences among the four districts, especially regarding catchment geology and vegetation cover, lake altitude and lake size. These variables are of crucial importance for high mountain lakes, not only because of their direct and indirect influence on physical, chemical and biological water properties (Reynolds, 1997), but also because of their ability to modulate the response of such sensitive ecosystems to the various anthropogenic impacts, which are increasingly affecting mountain ranges in Europe (e.g. Psenner & Catalan, 1994). In spite of the wide environmental and biological ranges, the median values relative to the whole group of lakes investigated correspond to a very typical 'high mountain condition', characterised by small lake size, low water mineralization, neutral or slightly acid pH and low nutrient concentration (Table 2). Median chlorophyll *a* concentration and phytoplankton biovolume were comparable with those reported in previous surveys of high mountain lakes in the Alps (e.g. Schneider, 1981; Pugnetti & Bettinetti, 1995; Psenner, 1998) and in Northern Europe (e.g. Willén et al., 1990; Nauwerck, 1994) and confirmed the oligotrophic status of the majority of the lakes studied.

The phytoplankton analyses confirmed the substantial role of flagellate algae in determining composition, quantity and functionality of the phytoplankton communities of high mountain lakes. In accordance with previous assessments from high altitude lakes of the Alps (e.g. Rott, 1988; Psenner, 1998; Tolotti, 2001), Chrysophyceae showed the highest biodiversity, while Dinophyceae played a major role in determining the total biomass and the functional structure of the phytoplankton community. Cryptophyceae were a less important group both in terms of biodiversity and biovolume, thus confirming previous studies (Dokulil, 1988; Psenner, *loc. cit.*; Rott, *loc. cit.*; Tolotti, *loc. cit.*). The irregular distribution and abundance of Cryptophyceae in the lakes studied reflects their opportunistic behaviour, i.e. their ability to rapidly grow when other algal groups are at a minimum (e.g. Dokulil, *loc. cit.*; Reynolds, 1997; Rott, *loc. cit.*).

The hierarchical clustering (TWINSPAN) of the lakes on the basis of their flagellate communities allowed the identification of several lake groups characterised by particular flagellate associations and environmental features (Fig. 3). The flagellate algae appear to be essentially distributed along catchment geology and lake altitude gradients, which, as already mentioned before, can affect other important variables, such as mineralization level and nutrient concentration of the lake water. Starting from one extreme of the geological-altitudinal gradient, oligotrophic high altitude lakes (around 2400 m a.s.l.) with very low mineralization level were characterised by a high proportion of different Dinophyceae taxa (Groups B1 and F in Figure 3). This distribution agrees with the tolerance of several Dinophyceae towards poorly buffered and oligotrophic conditions, as stated by previous studies (e.g. Rosén, 1981; Brettum, 1989; Willén et al., 1990; Nauwerck, 1994; Pugnetti & Bettinetti,

1995; Willén 2003). The role of *Gymnodinium uberrimum* as discriminating taxon for two very different groups (C and F) confirms its pronounced tolerance to trophic conditions (Brettum, 1989) and its opportunistic behaviour (e.g. Rott, 1988; Tardio et al., 2003), thus suggesting a poor value as a bio-indicator. Lakes located at lower altitude (2100–2200 m a.s.l.), characterised by low water mineralization, but with higher nutrient concentrations (Groups B2 and E in Fig. 3) were separated on the basis of abundance of large and tolerant Cryptophyceae (i.e. *Cryptomonas erosa/reflexa* and *C. obovata*), in association with small Chrysophyceae. The preference of large *Cryptomonas* species for more productive lakes has been assessed by previous ecological studies (e.g. Rott, 1988; Brettum, 1989; Reynolds, 1997). As shown by studies on Scandinavian lakes (e.g. Brettum, 1989; Eloranta, 1989), the distribution of Chrysophyceae appeared to be strongly related to the nutrient concentration of the lakes, as indicated by their preference either for oligotrophic lakes with relatively high NO_3-N concentrations (Group F), or for meso-eutrophic lakes showing low NO_3-N concentrations (groups D and E).

The TWINSPAN classification of the lakes did not lead to a complete geographical separation of the lakes studied even though the ADA and AVI districts showed a more pronounced tendency to group according to their homogeneous catchment lithology. The separation of the six lakes studied in STY was not supported by the identification of common environmental variables and it was probably related to the fact that a different person counted the phytoplankton samples from STY. This aspect highlights the importance of inter-calibration tests for wide regional surveys involving more than one working group. The poor geographical separation of the lakes indicates that the algal community of each lake has to be regarded as the result of a unique combination of morphological, chemical and biological features, including factors such as light conditions, inter-specific relationships with other algal groups, grazing effects, presence of stable fish communities in the lakes, and human impact, which could not be considered in this study.

The canonical correspondence analyses confirmed several aspects already highlighted by the TWINSPAN classification of the lakes, adding information on the ecological preferences of individual species. An interesting aspect not highlighted by the TWINSPAN analysis is the importance of the water thermal conditions as driving variable for the distribution of the three flagellate groups, in particular Cryptophyceae. This may be related to the fact that the vertical water movements strongly affect the motility of flagellate algae and their ability to maintain an optimal position within the water column (Reynolds, 1997). All three flagellate groups include taxa preferring either mixed or stratified conditions.

The distribution of the Chrysophyceae according to a combination of nutrient and altitude confirms the TWINSPAN output and suggests a possible use of several taxa as indicators of nutrient enrichment, caused for instance by airborne pollutants (Schmidt & Psenner, 1992; Psenner & Catalan, 1994). Dinophyceae are known to be a more heterogeneous group, showing a wide variety of ecological and environmental preferences (Rott, 1988; Brettum, 1989; Reynolds, 1997; Tardio et al., 2003). However, the majority of the taxa identified in this study showed preference for mid-high altitude lakes with low mineralization levels and oligotrophic conditions, thus indicating their possible use as sensitive indicators of acidification processes and trophic changes.

The canonical ordination of Dinophyceae allowed also the identification of *Peridinium cinctum* and *Gymnodinium uberrimum* as indicators of higher nutrient concentrations (Fig. 5). This aspect, while only partially supported by the TWINSPAN classification, is reported by several ecological studies (e.g. Brettum, *loc. cit.*; Reynolds, *loc. cit.*; Rosén, 1981; Willén et al., 1990). However, the contrasting results given by TWINSPAN and CANOCO for *Gymnodinium uberrimum*, suggest a low value of this taxon as a bio-indicator.

The ordination of Cryptophyceae reflects their pronounced tolerance to the majority of the chemical variables and their sensitivity to physical properties such as thermal conditions and lake depth, the last one related with underwater light conditions (cf. Dokulil, 1988; Rott, 1988; Reynolds, 1997; Rosén, *loc. cit.*). Some discrepancies were evident between the output of CCA ordination of Cryptophyceae and TWINSPAN classification of the lakes studied. *Rhodomonas minuta* var. *nannoplanctica*, *Cryptomonas marssonii* and *C. pyrenoidifera*, which were not indicated as discriminating species by TWINSPAN classification, were clearly related to the trophic condition of the lakes in the CCA analysis. *Cryptomonas erosa/reflexa* and *C. obovata*, which seemed to prefer more productive lakes according to TWINSPAN classification, seem to respond almost exclusively to physical water conditions in CCA analysis (Fig. 6). These contrasting results indicate the Cryptophyceae as the most problematic group to be used as bio-indicators.

On the whole, the multivariate analyses were useful to validate and refine the three major functional groups already identified by TWINSPAN classification, which can be tentatively summarised as follows. Group I includes taxa belonging to the section *Peridinium umbonatum* and small *Gymnodinium* species, which are prominent in high altitude (> 2400 m a.s.l.), scarcely buffered and clearly oligotrophic lakes. Group II includes several small Chrysophyceae of different genera, which are characteristic of oligotrophic high altitude lakes, neutral or slightly acid. This group could be further divided into three subgroups according to the preference of the taxa for NO_3-N and TP concentration and altitude, as indicated in Figure 4. The combination of *Peridinium cinctum* and eutraphentic Cryptophyceae is representative of Group III, which characterises lakes at lower altitude (<2200 m a.s.l.) with higher mineralization levels and higher trophic conditions. A tentative comparison of these functional groups with those commonly used in phytoplankton ecology (Reynolds, 1997; Reynolds et al., 2002) proved inconclusive, although some correspondence could be found with the Reynolds' Groups X2, X3 and especially E, all characterised by Chrysophyceae. This poor correspondence might be explained by the fact that the functional groups identified here include only part of the phytoplankton community, thus excluding eventual important interspecific relationships. Additionally, the majority of the Reynolds groups have been proposed for temperate lakes at lower altitude (cf. Reynolds, 2002), while a special calibration might be necessary for high mountain lakes.

Conclusions

The present investigation contributes to the knowledge of algal taxonomy and diversity in high mountain lakes of the Eastern Alps, showing the important role of flagellates in determining the high algal biodiversity, which often characterises these extreme environments. The analyses of abundance and distribution of the three flagellate groups has improved the understanding of their ecological preferences and has allowed the creation of some functional groups in relation with particular morphological, chemical and biological lake features. However, the poor correspondence between the functional groups identified during this study and those currently used in ecological studies (Reynolds et al., 2002) suggests the opportunity to improve the knowledge of flagellate algae in mountain lakes and to extend these studies to other regions of the Alps.

The multivariate analyses allowed the identification of several taxa and taxa associations for use as bio-indicators of environmental change in high mountain lakes of the Eastern Alps. This aspect might become increasingly important in the future, considering the environmental changes expected and forecasted by the scientific community, as a consequence of the increasing human impacts (e.g. eutrophication, airborne pollution, local and global climatic changes) affecting mountain regions all around the world.

Acknowledgements

This study was financially supported by the European Commission (Project EMERGE, EVK1-CT-1999-00032), the Austrian Ministry of Science (BM:BWK), the Adamello-Brenta Natural Park and the "Consorzio dei Comuni del Bacino Imbrifero Montano dell'Adige" (Authority of the Avisio River Basin). The authors wish to thank all colleagues of the Museo Tridentino di Scienze Naturali, of the Institute of Zoology and Limnology and of the Institute of Botany of the University Innsbruck, who supported the study in the field and the laboratory. Thanks also to Michael Huges (Environmental Change Research Centre, University College London) for the linguistic revision of the manuscript.

References

A.P.H.A., A.W.W.A., W.P.C.F., 1992. Standard Methods for the examination of water and wastewater. American Public Health Association, Washington DC.

Baldi, E., 1932b. Note zoologiche sopra alcuni laghi della Presanella-II. Il Lago Serodoli. Memorie del Museo di Storia Naturale della Venezia Tridentina 1: 167–192.

Baldi, E., 1935. I Planctonti del Lagorai. Bollettino della Pesca, Piscicoltura e Idrobiologia, Supplemento 10: 1–47.

Barone, R. & L. Naselli-Flores, 2003. Distribution and seasonal dynamics of cryptomonads in Sicilian water bodies. Hydrobiologia 502 (Dev. Hydrobiol. 172): 325–329.

Brettum, P., 1989. Alger som indikatorer på vannkvalitet i norske innsjøer. Planteplankton. NIVA Report. 111 pp.

Cantonati, M., A. Boscaini, F. Corradini & M. Lazzara, 2002a. Morfometria e idrochimica di laghi d'alta quota del bacino del fiume Avisio (Trentino orientale). Studi Trentini di Scienze Naturali, Acta Biologica 78: 101–116.

Cantonati, M., M. Tolotti & M. Lazzara, 2002b. I laghi del Parco Naturale Adamello-Brenta. Documenti del Parco 14, Parco Naturale Adamello-Brenta & Museo Tridentino di Scienze Naturali, Strembo, Italy. 285 pp.

Carrera, G., P. Fernandez, J. O. Grimalt, M. Ventura, L. Camarero, J. Catalan, U. Nickus, H. Thies & R. Psenner, 2002. Atmospheric deposition of organochlorine compounds to remote high mountain lakes of Europe. Environ. Sci. Technol. 36: 2581–2588.

Cloern, J. E., 1977. Effects of light intensity and temperature on *Cryptomonas ovata* (Cryptophyceae) growth and nutrient uptake rates. J. Phycol. 13: 389–395.

Dokulil, M., 1988. Seasonal and spatial distribution of cryptophycean species in the deep, stratifying, alpine lake Mondsee and their role in the food web. Hydrobiologia 161: 185–201.

Eloranta, P., 1989. Ecological studies on the ecology of the genus *Dinobryon* in Finnish lakes. Supplement to Nova Hedwigia 95: 99–109.

Gervais, F., 1997. Light-dependent growth, dark survival and glucose uptake by cryptophytes isolated from a freshwater chemocline. J. Phycol. 33: 18–25.

Giussani, G., R. De Bernardi, R. Mosello, I. Origgi & T. Ruffoni, 1986. Indagine limnologica sui laghi alpini d'alta quota. Documenta Ist. ital. Idrobiol. 9: 1–415.

Hansen, C. M. E, M. Tolotti, R. Ettinger, H. Thies, D. Tait & R. Psenner, 2002. The application of Geographic Information Science (GIS) for data synthesis and evaluation of remote high mountain lakes in the lake district Tyrol (Austria & Italy). In Pillmann, W. & K. Tochtermann (eds), Environmental Communication in the Information Society, Part 1. ISEP, Vienna. 678 pp.

Heinonen, P., 1980. Quantity and composition of phytoplankton in Finnish inland waters. Publications of the Water Research Institute 37, Vesihallitus-National Board of Waters, Finland.

Hill, D. R. A., 1991. A revised circumscription of *Cryptomonas* (Cryptophyceae) based on examination of Australian strains. Phycologia 30: 170–180.

Hill, M. O., R. G. H. Bunce & M. W. Shaw, 1975. Indicator species analysis, a divisive polythetic method of classification and its application to a survey of native pinewood in Scotland. J. Ecol. 63: 597–613.

Holen, D. A. & M. E. Boraas, 1996. The mixotrophy in chrysophytes. In: Craig, D., C. D. Sangren, J. P. Smol & J. Kristiansen, (eds), Chrysophyte Algae. Ecology, Phylogeny and Development. University Press, Cambridge: 119–140.

Javorniký P., 2001. Freshwater Rhodomonads (Cryptophyceae). Algol. Stud. 102: 93–116.

Javornický, P., 2003. Taxonomic notes on some freshwater planktonic Cryptophyceae based on light microscopy. Hydrobiologia 502 (Dev. Hydrobiol. 172): 271–283.

Jeffrey, S. W. & G. F. Humphrey, 1975. New spectrophotometric equations for determining chlorophyll *a*, *b*, c_1 and c_2 in higher plants, algae and natural phytoplankton. Biochemie und Physiologie der Pflanzen 167: 191–194.

Lepistö, L. & A.-L. Holopainen, 2003. Occurrence of Cryptophyceae and katablepharids in boreal lakes. Hydrobiologia 502 (Dev. Hydrobiol. 172): 307–314.

Lorenzen, C. J., 1967. Determination of chlorophyll and phaeopigments: spectrophotometric equations. Limnol. Oceanogr. 12: 543–556.

Lund, J. W. G., G. Kipling & E. D. Le Cren, 1958. The inverted microscope method for estimating algae number and statistical basis of estimating by counting. Hydrobiologia 11: 143–170.

Mosello, R. & B. M. Wathne, 2000. EMERGE Protocol N. 4 – Chemical analyses of major ions and nutrients. http:/www.mountain-lakes.com.

Nauwerck, A., 1966. Beobachtungen über das Phytoplankton klarer Hochgebirgsseen. Schweizerische Zeitschrift für Hydrologie 28: 4–28.

Nauwerck, A., 1994. A survey on water chemistry and plankton in high mountain lakes in northern Swedish Lapland. Hydrobiologia 274: 91–100.

Novarino, G., 2003. A companion to the identification of cryptomonad flagellates (Cryptophyceae = Cryptomonadea), Hydrobiologia 502 (Dev. Hydrobiol. 172): 225–270.

Novarino, G., I. A. N. Lucas & S. Morrall, 1994. Observaion on the genus *Plagioselmis* (Chryptophyceae). Cryptogamie, Algologie 15: 87–107.

Padisák, J., G. Borics, G. Fehér, I. Grigorszky, I. Oldal, A. Schmidt & Z. Zámbóné-Doma, 2003. Dominant species, functional assemblages and frequency of equilibrium phases in late summer phytoplankton assemblages in Hungarian small shallow lakes. Hydrobiologia 502 (Dev. Hydrobiol. 172): 157–168.

Pechlaner, R., 1971. Factors that control the production rate biomass of phytoplankton in high mountain lakes. Mitt. int. Ver. theor. angewan. Limnol. 19: 125–145.

Poda, E., 1935. Composizione chimica e proprietà fisico-chimiche dell'acqua dei laghi alpini del Lagorai. Bollettino della Pesca, Piscicoltura e Idrobiologia, Supplemento, 10: 447–457.

Psenner, R., 1988. Alkalinity generation in a soft-water lake: watershed and in-lake processes. Limnol. Oceanogr. 33: 1463–1475.

Psenner, R., 1998. Versauerung von Hochgebirgsseen 1985 - 1995. Bundesministerium für Land- und Forstwirtschaft, Wien. 134 pp.

Psenner, R. & J. Catalan, 1994. Chemical composition of lakes in crystalline basins: a combination of atmospheric deposition, geologic background, biological activity and human action. In: Margalef, R. (ed.), Limnology Now: A Paradigm of Planetary Problems, Elsevier Science, Amsterdam: 255–314.

Pugnetti, A. & R. Bettinetti, 1995. The phytoplankton communities of two acid sensitive alpine lakes (Lakes Paione, Central Alps, Italy). Mem. Ist. ital. Idrobiol. 53: 39–52.

Reynolds. C. S., 1997. Vegetation processes in the pelagic: a model for ecosystem theory. In: Kinne, O. (ed.), Excellence in Limnology 9: 371 pp.

Reynolds. C. S., V. Huszar, C. Kruk, L. Naselli-Flores & S. Melo, 2002. Toward a functional classification of the freshwater phytoplankton. J. Plankton Res. 24: 417–428.

Rosén, G., 1981. Phytoplankton indicators and their relations to certain chemical and physical factors. Limnologica 13: 263–290.

Rott, E., 1981. Some results from phytoplankton counting intercalibrations. Schweizerische Zeitschrift für Hydrologie 43: 34–62.

Rott, E., 1988. Some aspects of the seasonal distribution of flagellates in mountain lakes. Hydrobiologia 161: 159–170.

Sandgren, C. D., 1988. The ecology of chrysophyte flagellates: their growth and perennation strategies as freshwater phytoplankton. In Sandgren, C. D. (ed.), Growth and Reproductive Strategies of Freshwater Phytoplankton. Cambridge University press: 9–104.

Schindler, D. W., K. H. Mills, D. F. Malley, D. L. Findlay, J. A. Shearer, I. J. Davies, M. A. Turner, G. A. Linsey & D. R. Cruikshank, 1985. Long-term ecosystem stress: the effect of years of experimental acidification on a small lake. Science 228: 1395–1401.

Schmidt, R. & R. Psenner, 1992. Climate changes and anthropogenic impacts as causes for pH fluctuation in remote alpine lakes. Documenta Ist. ital. Idrobiol. 32: 31–57.

Schneider, U., 1981. Phytoplankton und Primärproduktion in einigen Hochgebirgsseen des Kantons Tessin. Ph.D. Thesis, University Bern. 152 pp.

Shannon, C. E. & W. Weaver, 1949. The mathematical theory of communication. University of Illinois Press, Urbana. 117 pp.

Sommer, U., J. Padisák, C. S. Reynolds & P. Jusász-nagy, 1993. Hutchinson's heritage: the diversity-disturbance relationship in phytoplankton. Hydrobiologia 249: 1–7.

Stella, E., 1931. Intorno ad alcuni laghi alpini del Trentino, dell'Ampezzano e dell'Alto Adige. Memorie del Museo di Storia Naturale della Venezia Tridentina 1: 45–46.

Straškrabová, V. & E. Stuchlík, 2000. EMERGE Protocol N. 7 – Sampling of the plankton food web. http:/www.mountain-lakes.com.

Tardio, M., M. Tolotti, G. Novarino & M. Cantonati, 2003. Ecological and taxonomic observations on the flagellate algae characterising four years of enclosure experiments in Lake Tovel (Southern Alps). Hydrobiologia 502 (Dev. Hydrobiol. 172): 285–296.

ter Braak, C. J. F. & P. Smilauer, 1998. CANOCO reference manual and user's guide to Canoco for Windows: software for canonical community ordination (version 4). Microcomputer Power, Ithaca, New York, U.S.A. 195 pp.

Thies, H., 2000. The importance of internal alkalinity generation for an acidified mountain lake as revealed by mass budgets. Silva Gabreta 4: 97–104.

Thies, H., U. Nickus, C. Arnold & R. Psenner, 2002. A hydrological tracer experiment with LiCl in a high mountain lake. Hydrol. Proc. 16: 2329–2337.

Thies, H., U. Nickus, C. Arnold, R. Schnegg, A. Wille & R. Psenner, 2000. Biogeochemistry of a high mountain lake in the Austrian Alps. Int. Ver. theor. angewan. Limnol.: Verh. 27: 517–520.

Tilzer, M. & K. Schwarz, 1976. Seasonal and vertical patterns of phytoplankton light adaptation in a high mountain lake. Archiv für Hydrobiol. 44: 488–504.

Tolotti, M., 2001. Phytoplankton and littoral epilithic diatoms in high mountain lakes of the Adamello-Brenta Regional Park (Trentino, Italy) and their relation to trophic status and acidification risk. J. Limnol. 60: 171–188.

Tomasi, G., 1962. Origine, distribuzione, catasto e bibliografia dei laghi del Trentino. Studi Trentini di Scienze Naturali 1-2: 1–333.

Utermöhl, H., 1958. Zur Vervollkommung der quantitative Phytoplankton-Methodik. Mitt. int. Ver. theor. angewan. Limnol. 9: 1–38.

Vollenweider, R. A. & J. Kerekes, 1982. Eutrophication of Waters, Monitoring, Assessment and Control. OECD, Paris. 154 pp.

Willén, E., 2003. Dominance patterns of planktonic algae in Swedish forest lakes. Hydrobiologia 502 (Dev. Hydrobiol. 172): 315–324.

Willén, E., S. Hajdu & Y. Pejler, 1990. Summer phytoplankton in 73 nutrient-poor Swedish lakes – classification, ordination and choice of long-term monitoring objects. Limnologica 20: 217–227.

Zucchelli, M., 1952. Osservazioni morfologiche e fisiche su alcuni laghi del Gruppo Adamello-Presanella. Studi Trentini di Scienze Naturali 29: 142–170.

Hydrobiologia **502:** 349–355, 2003.
L. Naselli-Flores, J. Padisák, M. T. Dokulil (eds),
Phytoplankton and Equilibrium Concept: The Ecology of Steady-State Assemblages.

Redefinition of *Peridinium lomnickii* Wołoszynska (Dinophyta) by scanning electronmicroscopical survey

István Grigorszky[1], Lothar Krienitz[2], Judit Padisák[3], Gábor Borics[4] & Gábor Vasas[1]
[1]*Botanical Department, Debrecen University, H-4010 Debrecen, P. O. Box 14, Hungary*
E-mail: gege@tigris.klte.hu
[2]*Institut für Gewässerökologie und Binnenfischerei im Forschungsverbund Berlin e. V., Abteilung Limnologie Geschichteter Seen, Alte Fischerhütte 2, D-16775 Neuglobsow, Germany*
[3]*Limnological Department, University of Veszprém, H-8201 Veszprém P.O.Box. 158., Hungary*
[4]*Transtisanian Region, Environmental Protection Inspectorate, H- 4025 Debrecen, Piac u. 9/b, Hungary*

Key words: Peridinium lomnickii, Peridinium wierzejskii, Peridinium lomnickii var. *splendida*, life cycle, gymnodinoid-, glenodinoid- and peridinoid lifecycle forms

Abstract

Peridinium lomnickii Wołoszynska was investigated by scanning electron microscopy with special attention to the importance of development of thecal plate during the life cycle. Different life cycle stages (gymnodinoid-, glenodinoid-, peridinioid) are described on the basis of development of cell wall, presence and development of sutures, appearance of pore and the change of the cell shape. Differences and possible relationships between the three existing varieties of the species are discussed. We suggest that the three varieties of *P. lomnickii, P. lomnickii* var. *lomnickii, P. lomnickii* var. *wierzejskii* and *P. lomnickii* var. *splendida,* represent the different life cycle stages of the species. These results and the known ontogenic cycle of dinophyta taxa should be taken into consideration, when a phylogenic tree of the dinophytes is constructed.

Introduction

Distinguishing the different life cycle forms of dinoflagellates is of basic importance both from taxonomic and ecological point of view. Detailed life cycles are known only for some freshwater dinoflagellate monads, *Ceratium cornutum* (Stosch, 1965), *Ceratium furcoides* (Hickel, 1988a), *Ceratium hirundinella* (Zederbauer, 1904; Huber & Nipkow, 1922, 1923; Entz, 1931), *Katodinium fungiforme* (Spero & Morée, 1981), *Peridiniopsis lubiensiforme* (Diwald, 1938), *Peridinium cinctum* (Pfiester, 1975; Pfiester et al., 1980). Life histories of *Ceratium monoceras* (Temponeras et al., 2000), *Ceratium rhomvoides* (Hickel, 1988b), *Hemidinium nasutum* (Popovskỳ, 1971), *P. gatunense* (Pfiester, 1977*), P. inconspicuum* (Pfiester et al., 1984), *P. limbatum* (Pfiester & Skvarla, 1980), *P. volzii* (Pfiester & Skvarla, 1979), *P. willei* (Pfiester, 1976), and *Wołoszynskia pseudopalustris* (v. Stoch, 1973) are partially described. Extremely complex life cycles have been described for some freshwater dinococcalians (Popovskỳ, 1961; Pfiester & Popovskỳ, 1979; Popovskỳ & Pfiester, 1982) and various marine dinoflagellates (Montresor 1995a,b; Montresor & Zignone, 1998). The basic part of the vegetative life cycle of the above mentioned species consists of gymnodinoid-, glenodinoid-, peridinoid-, life cycle forms and cyst (Fig. 1).

The different life cycle stages are already described as gymnodinoid-, glenodinoid- and peridinoid stages (Popovskỳ & Pfiester, 1990). These names come from the resemblance of the different life cycle forms to the *Gymnodinium, Glenodinium* and *Peridinium* genera. Actually, the most recent review in systematics (Popovskỳ & Pfiester, 1990) dropped the use of genus *Glenodinium* and claimed that the representatives of this genus can be considered as different life cycle stages of certain *Peridiniaceae* taxa. The above cited work provides hypothetic suggestions how to relate some species formerly described as *Glenodinium*

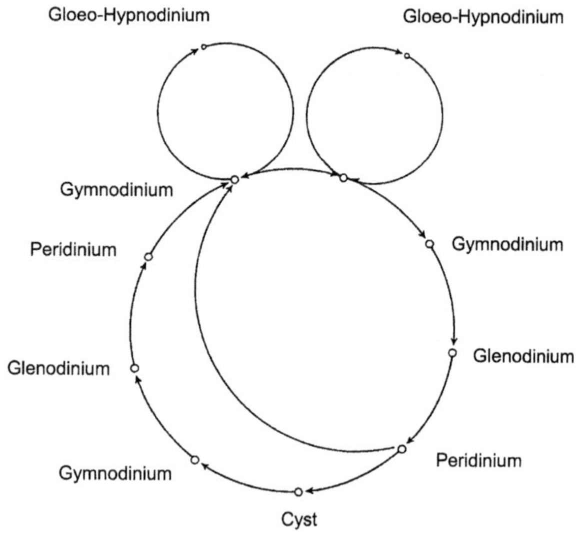

Figure 1. The diagrammatical life history of a idealised dinophyte taxa with armoured and unarmored, motile and nonmotile stages (Entz, 1926).

species to an adequate *Peridinium* and *Peridiniopsis* species and underlined the taxonomic importance of the differentiation of the life cycle stages. Apart from its importance in identification of species formerly belonging to *Glenodinium*, such differentiations are of general importance in the taxonomy of dinoflagellates, because otherwise many taxonomic problems remain unsolved especially concerning younger developmental stages.

The present study describes fine structures – cell shape, sutures and apical pore – during the thecal development of plates. The primary aim is to characterize the different life cycle stages (gymnodinoid-, glenodinoid-, peridiniod stages) based on morphological criteria, such as the development of theca, presence and development of sutures, the appearance of the pore and the change of the cell shape in *P. lomnickii*. Typical plate arrangement (tabulation) of *P. lomnickii* is given in Popovský & Pfiester (1990). Sutures are transversely striated intercalary bands between the plates. They, probably, allow the cell to grow somewhat when plates are already developed.

Our basic preconception is that the different life cycle stages can be distinguished on morphological bases, and moreover that developmental forms can help us to clarify taxonomic relevance of different varieties of a given species, or taxonomy of closely related species. In this paper, the gymnodinoid, glenodinoid and peridinoid life cycle forms of *P. lomnickii* populations isolated from Baas See, Northern Germany and Peresi-Holt-Körös South-East Hungary and the Hungarian culture collection are used to demonstrate the applicability of this concept.

Materials and methods

Plankton samples were collected in March 1995 from a small bog lake named the Kleiner Barschsee, which is close to the Lake Stechlin Area (Brandenburg, Germany). The another samples were collected in March 2002 from a small oxbow lake named Peresi-Holt Körös, which is situated South-East Hungary. The samples were collected with 10 μm mesh size plankton nets and investigated alive.

A part of the collected material was examined after collection without fixation. Another part was fixed with Lugol's iodine or formaldehyde.

The culture of *P. lomnickii* was established from a culture collection of Peresi-Holt Körös in Hungary. The taxa were cultured at 12 °C with a 12:12 h light:dark regime. Light intensity was 8 μmol m^{-2} s^{-1}. The Carefoot's medium (Carefoot, 1968) had been used successfully to culture the *P. lomnicki*. Four hundred cells were examined in each sample.

The system of Kofoid (1909) for the plate designations of the *P. lomnickii* has been used. The algae were studied and identified both under the light microscope (LM LEICA Diaplan and Olympus BX-50), and the scanning electron microscope (SEM HITACHI S-4500). The samples were rinsed in distilled water, then, a drop of the cell suspension was placed on a nucleopore filter (COSTAR Corp.). The algal material was mounted on the filter surface by quick-freezing in nitrogen slush to remove excess water. The algae were frozen for some minutes at −95 °C in a cryo-transfer system CT 1500 HF (OXFORD Instruments), sputtered with gold-palladium, transferred into the cooled field emission SEM and investigated at 5 kV, 10 μA.

Results

The *P. lomnickii* populations consisted of individuals varying markedly in size and shape. Most cells were spherical to ellipsoidal, with rounded poles, and they were somewhat apiculated at the apex (Figs 2–5). There were three main cell types in each sample

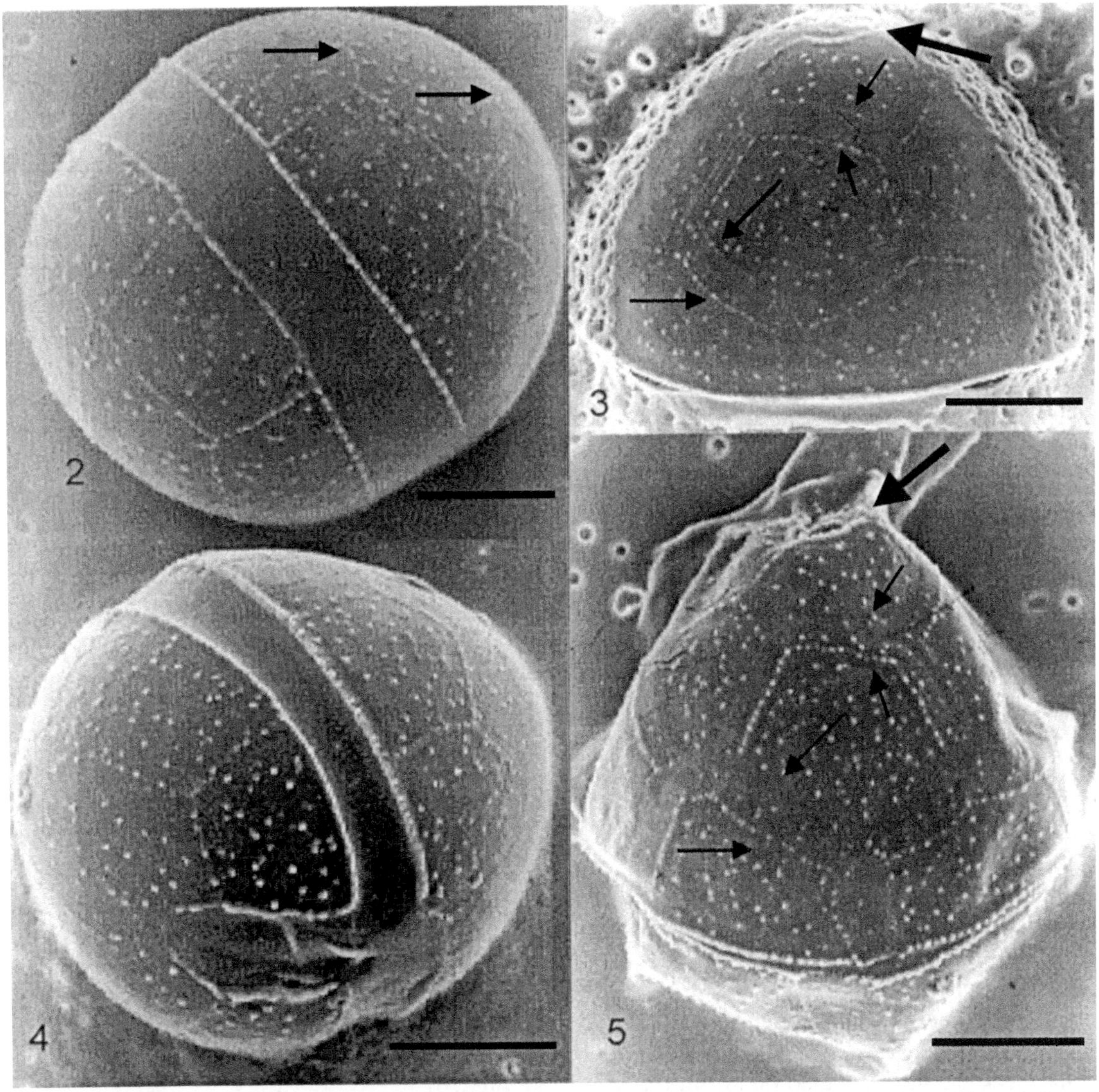

Figure 2–5. SEM photographs of different life cycle forms of *P. lomnickii*. Scale beside figures correspond to 10 mm. *Figure 2*. Gymnodinoid form. The thin arrows show the margins of precingular and intercalary plates. *Figure 3*. Glenodinoid form. The thin arrows show the margins of precingular and intercalary apical plates (sutures). The thick arrow shows the pore. *Figure 4*. Gymnodinoid form. *Figure 5*. Peridinoid form. The thin arrows show the sutures. The thick arrow shows the developed pore complex.

and culture material corresponding: gymnodinoid-, glenodinoid- and peridinoid stage.

Gymnodinoid stage (Figs 2 and 4)

The cells are spherical to slightly egg-shaped and they were not flattened dorso-ventrally. The future epicone and hypocone are of similar size. In this life cycle stage cells were covered with only an elastic periplast. Real sutures have not been developed yet. At the margins of the plates 'dotted lines' were seen, which showed the plate tabulation of *P. lomnicki*. The pore, the cingular and the sulcal plate structure was not visible. The cell sizes are 26–43 × 25–41 μm.

Glenodinoid stage (Fig. 3)

The theca is thicker and the sutures are well developed. The pore is visible. The cell shape in this phase is close to spherical, yet not typical for the developed form of the holotype of *P.lomnickii*. The cell sizes are 28–44 × 34–47 μm.

Peridinoid stage (Fig. 5)

The cell shape is elongated and somewhat conical, typical *Peridinium*, and it is characterized by plate tabulation of 4′ 3a, 7″, 5‴, 2⁗. Sutures were wide, the cingulum and the sulcus were deep. The 1a and 2a intercalary plates have typical pentagonal shape and both touched the 3″ precingular plate. The 3a intercalary plate is hexagonal and together with the 2a intercalary plate touched the 4″ precingular plate. The 3a intercalary plate was bigger than either 1a or 2a intercalary plates. The 1a and 2a intercalary plates were mostly equal in size but sometimes one of them is smaller causing some asymmetry of cells. The cells sizes are 39–53 × 36–47 μm.

Discussion

The fine structure of freshwater Dinophyta species has of great interest during the last 20 years (Dodge & Crawford, 1970; Pfiester, 1975; Boltovskoy, 1976; Bold & Wynne, 1978; Pfiester et al., 1980; Spector et al., 1981; Carty & Cox, 1986; Yamagishi & Akiyama, 1987; Hickel, 1988a,b, Toriumi & Dodge, 1993). Although contemporal investigators have added significant information to our knowledge concerning the plate arrangement, the apical pore forms etc., the majority of freshwater dinoflagellates have not been studied sufficiently yet.

Some SEM photographs of *P. lomnickii* were published by earlier investigators. In Dodge (1985) only the gymnodinoid form is shown. Ling et al.'s (1989) paper has been the most detailed one from the taxonomic point of view: it contains the SEM documentation, four drawings and a short description of the species. These show that the investigated *P. lomnickii* cells were in early gymnodinoid phase since the cells are spherical and the sutures are narrow, somewhat excavated canals. Neither of the above papers mentions which stage can be seen on the figures. In some cases, asymmetrical forms of *P. lomnickii* were observed (Wołoszynska, 1916; Ling et al., 1989); this asymmetry appears if the 1a plate is very small.

Our SEM studies made it clear that sutures appear in gymnodinoid stage and they are seen as thin dotted lines, then they become more and more apparent during development of *P. lomnickii*. In the glenodinoid stage, cells are spherical and real sutures are visible with SEM. Peridinoid stage can be identified either by LM and SEM, by the typical cell shape and presence of pore and sutures.

Based on the above observations the previously described taxa and forms of *P. lomnickii* can be discussed as follows.

The species that was described as *P. wierzejskii* by Wołoszynska (1916) was moved to *P. lomnickii* (Popovský & Pfiester, 1990) as *P. lomnickii* var. *wierzejskii*. The *P. wierzejskii* (Wołoszynska, 1916; Figs 8–9), *P. wierzejskii var. minor* (Wołoszynska, 1916; Fig. 10), and *P. lomnickii* var. *wierzejskii* (Popovský & Pfiester, 1990), can be accepted as the gymnodinoid forms of *P. lomnickii*. A gymnodinoid life cycle stage was previously described as *P. lomnickii* var. *punctulatum* by Lindemann (1924; Figs 6 and 7). In the gymnodinoid stage the presence of holes, places of trichocysts, are visible (Figs 2–4). The name 'punctulatum' refers to the several pores through which the trichocysts are discharged (Pennick & Clarke, 1977; Dürr, 1979; Morrill & Loeblich III, 1981). The trichocyst pores looks like dots sometimes they are visible in LM, therefore, this might be the origin of the name of the epithet 'punctulatum'.

Two drawings from the original description of *P. wierzejkii* show that this taxon can be regarded as the glenodinoid stage of *P. lomnickii* (Figs 12–13).

The peridinoid life cycle stage was described by different authors as different species and different varietas of *P. lomnickii*:

P. lomnickii by Wołoszynska (1916; Figs 18–20) and Lefèvre (1932; Fig. 15)

P. lomnickii var. *splendida* by Wołoszynska (1916; Figs 21–23)

Chalubinskia tatrica by Wołoszynska (1916; Fig. 16)

Peridinium tatrae by Schiller (1937; Fig. 17)

The separation between gymnodinoid life cycle stages of *Peridiniaceae* and *Gymnodinium* species is extremely difficult, many times impossible. Field samples may contain *Gymnodinium* species and gymnodinoid life cycle forms of peridinean. It is proved in several marine *Gymnodinium* species that any plate structure might be completely absent (Spero & More, 1981; Wedemayer et al., 1982; Larsen, 1994). Our investigations with cryotransfer combined SEM has shown that gymnodinoid *Peridiniaceae* forms do posses the characteristic tabulation in *P. lomnickii*. In very careful LM observations, these structures may be recognized as it is documented in drawings of gymnodinoid stages by Wołoszynska (1916) and Lindemann's (1924).

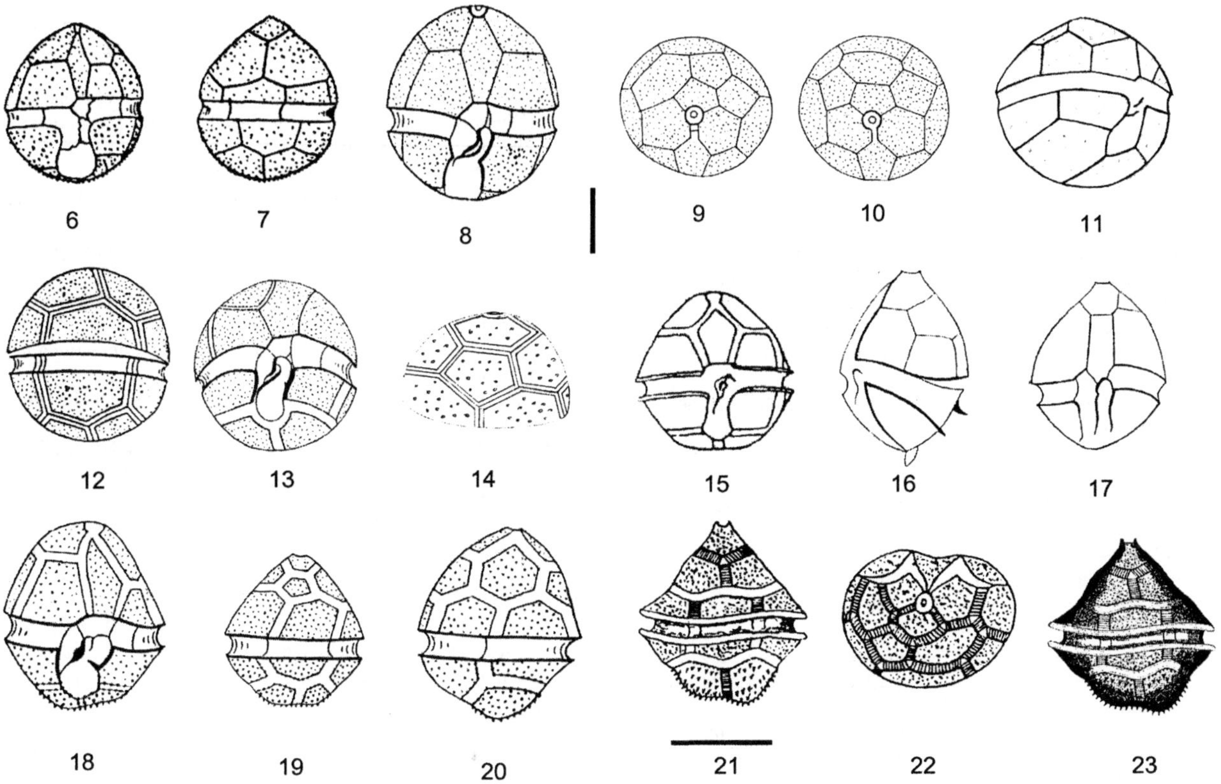

Figure 6–23. Different life cycle forms of *P.lomnickii.* 6–7: gymnodinoid forms after Lindemann(1924); 8–10: gymnodinoid forms after Wołoszynska (1916); 11: gymnodinoid form after Grigorszky et al. (this study); 12, 13: glenodinoid form after Woloszynska (1916); 14: glenodinoid form after Grigorszky et al. (this study); 15: peridinoid after Lefèvre (1932); 16: peridinoid form after Wołoszynska (1916); 17: peridinoid form after Schiller (1937); 18–23. peridinoid form after Wołoszynska (1916). Vertical bar corresponds to 10 μm.

A wide variety of morphological forms has been described, from various Dinophyta taxa. In general, taxonomic conclusions are based on morphological evidence, separately from their genetic background, life cycle, and sexual reproduction. Currently phylogenetic trees based on genetic data include information on pigments as well as various other cellular components as taxonomic markers. Genetic studies should allow more fundamental understanding of morphological and developmental characters widely used in traditional taxonomy. Application of advanced techniques that include morphological, genetic, as well as other taxonomic criteria to systematics has been a slow process. The new information and new techniques have enabled us to ask new questions about the taxonomic problems.

The Dinophyta division is probably not the very first group of biota where the phylogenetic 'tree' has been built, based on morphological observations (Taylor, 1987) and is contrary to their ontogeny, especially when these morphological features are distinct and conservative for each taxon. The more characteristics available to the taxonomist when classifying an organism the more valid that classification will be. Thus, the knowledge of morphology throughout ontogeny (life cycle) gives the taxonomist a vast array of new additional morphological information with which to construct a more accurate phylogenetic relationship.

Acknowledgements

Sampling and cryotransfer-combined SEM was performed in the Institut für Gewässerökologie und Binnenfischerei im Forschungsverbund Berlin e. V., Abteilung Limnologie Geschichteter Seen, Neuglobsow. Additional support was provided by the Hungarian National Science Foundation (OTKA No. T29636, F31802), Hungarian Higher Educational Found (FKFP No. 0195) and Békessy György Science Found.

References

Bold, H. C. & M. J. Wynne, 1978. Introduction to the algae. Structure and Reproduction, Prentice-Hall, New Jersey. 706 pp.,

Boltovskoy, A., 1976. Estructura y estereoultraestructura tecal de dinoflagelados. III. *Peridinium bipes* Stein, forma *apodata* n.f. Physis Section B, 35: 147–155.

Bourrelly, P., 1968. Notes sur les Péridiniens d'eau douce. Protistologica 4: 5–16.

Carefoot, J. R., 1968. Culture and heterotrophy of the freshwater dinoflagellates *Peridinium cinctum* f. *ovoplanum* Lindemann. J. Phycol. 4: 129–131.

Carty, S. & E. R. Cox 1986. *Kansodinium* gen.nov. and *Durinskia* gen.nov.: two genera of freshwater dinoflagellates (Pyrrophyta). Phycologia 25: 197–204.

Diwald, K., 1938. Die ungeschlechtliche und geschlechtliche Fortplanzung von Glenodinium lubieniensiforme. Flora 32: 174–192.

Dodge, J. D., 1985. Atlas of Dinoflagellates. Scanning Electron Microscope Survey. Farrand Press, London.

Dodge, J. D. & R. W. Crawford, 1970. The morphology and fine structure of *Ceratium hirundinella* (Dinophyceae). J. Phycol. 6: 137–149.

Dürr, G., 1979. Elektronenmikroskopische Utersuchungen am Panzer von Dinoflagellaten II. *Peridinium cinctum.* Archiv für Protistenkunde 122: 88–120.

Entz, G., 1926. Beiträge zur Kenntnis der Peridineen. I. Zur Morphologie und Biologie von *Peridinium borgei* Lemmermann. Archiv für Protistenkunde 56: 397–446.

Entz, G., 1931. Analyse des Wachtums und der Teilung einer Population sowie eines Individuums der Protisten *Ceratium hirundinella* unter den natürlichen Verhaltnissen. Archiv für Protistenkunde 74: 311–361.

Hickel, B., 1988a. Sexual reproduction and life cycle of *Ceratium furcoides* (Dinophyceae) in situ in the lake Plu*β*see (F.R.G.). Hydrobiologia 161: 41–48.

Hickel, B., 1988b. Morphology and life cycle of *Ceratium rhomvoides* nov. sp. (Dinophyceae) from the lake Plu*β*see (F.R.G.). Hydrobiologia 161: 49–54.

Huber, G. & F. Nipkow, 1922. Experimentelle Untersuchungen über die Entwicklung und form bildung von *Ceratium hirundinella* O. F. Müller. Zeitschrift für Botanik 14: 337–371.

Huber, G. & F. Nipkow, 1923. Experimentelle Untersuchungen über die Entwicklung und form bildung von *Ceratium hirundinella* O. F. Müller. Flora N. F. 116: 114–215.

Kofoid, C. A., 1909. On *Peridinium steinii* Jörgensen, with a note on the nomenclature of the skeleton of the Peridinidae. Archiv für Protistenkunde 16: 25–47.

Larsen, J., 1994. Unarmoured dinoflagellates from Australian waters. I. The genus *Gymnodinium* (Gymnodiniales, Dinophyceae). Phycologia 33: 24–33.

Lefévre, M., 1932. Monographie des espèces d'eau douce du genre *Peridinium.* Archives de Botanique 2: 1–208.

Lindemann, E., 1924. Über finnische Peridineen. Archiv für Hydrobiol. 15: 1–4

Ling, H. U., R. L. Croome & P. A.Tyler, 1989. Freshwater dinoflagellates of Tasmania, a survay of Taxonomy and distribution. British Phycol. J. 24: 111–129.

Morrill, C. W. & A. R. Loeblich III, 1981. The dinoflagellate pellicular wall layer and its occurrence in the division Pyrrophyta. J. Phycol. 17: 315–323.

Montresor, M., 1995a. The life history of *Alexandrium pseudogonyaulax* (Gonyaulacales, Dinophyceae). Phycologia 34: 444–448.

Montresor, M., 1995b. *Scrippsiella ramonii* sp. nov. (Peridiniales, Dinophyceae), a marine dinoflagellate producing a calcareous resting cyst. Phycologia 34: 87–91.

Montresor, M. & A. Zingone, 1998. *Scrippsiella-precarina* sp-nov (Dinophyceae), a marine dinoflagellate from the gulf naples. Phycologia 27: 387–394.

Pennick, N. C. & K. J. Clarke, 1977. The occurrence of scales in the peridinean dinoflagellate *Heterocapsa triquetra* (Ehrenberg). British Phycol. J. 12: 63–66.

Pfiester, L. A., 1975. Sexual reproduction of *Peridinium cinctum* f. *ovoplanum* (Dinophyceae). J. Phycol. 11: 259–265.

Pfiester, L. A., 1976. Sexual reproduction of *Peridinium willei* (Dinophyceae). J. Phycol. 12: 234–238.

Pfiester, L. A., 1977. Sexual reproduction of *Peridinium gatunense* (Dinophyceae). J. Phycol. 13: 92–95.

Pfiester, L. A. & J. Popovskỳ, 1979. Parasitic, amoeboid Dinoflagellates. Nature 279: 421–424.

Pfiester, L. A. & J. J. Skvarla, 1979. Heterothallism and thecal development in the the sexual life history of *Peridinium volzii* (Dinophyceae). Phycologia 18: 13–18.

Pfiester, L. A., R. A. Lynch & J. J. Skvarla, 1980. Comparative ultrastructure of vegetative and sexual thecae of *Peridinium limbatum* and *Peridinium cinctum* (Dinophyceae). Am. J. Bot. 67: 955–958.

Pfiester, L. A., P. Timpano, J. J. Skvarla & J. R. Holt, 1984. Sexual reproduction and meiosis in *Peridinium inconspicuum* Lemmermann (Dinophyceae). Am. J. Bot. 71: 1121–1127.

Popovskỳ, J., 1961. *Cystodinedria maxima* sp. nova. Preslia 33: 291–296.

Popovskỳ, J., 1971. Some remarks to the life cycle of *Gloeodinium montanum* Klebs and *Hemidinium nasutum* Stein (Dinophyceae) Archiv für Protistenkunde 113: 131–136.

Popovskỳ, J. & L. A. Pfiester, 1982. The life history of *Stylodinium sphaera* Pascher and *Cystodinedria inermis* (Geitler) Pascher (Dinophyceae), two freshwater facultative predator-autotrophs. Archiv für Protistenkunde 125: 115–127.

Popovskỳ, J. & L. A. Pfiester, 1990. Süsswasserflora von Mitteleuropa Band 6: Dinophyceae (Dinoflagellida). Gustav Fischer Verlag, Stuttgart. 272 pp.

Schiller, J., 1937. Dinoflagellatae (Peridineae). In Kolkwitz, R. (ed.), Rabenhorst's Kryptogamenflora von Deutschland, Österreich und Schweiz. 2. Aufl. 10/3 2. Akadem. Verlagsges. 589 pp.

Spector, D. L. (ed.), 1984. Dinoflagellates. Academic Press, New York, NY. 545 pp.

Spector, D. L., L. A. Pfiester & R. E. Triemer, 1981. Ultrastructure of dinoflagellate *Peridinium cinctum* f. *ovoplanum.* II. Light and electron microscopic observation on fertilization. Am. J. Bot. 68: 34–43.

Spero, H. J., Morée 1981. Phagotrophic feeding and its importance to the life cycle of the holozoic dinoflagellate, *Gymnodinium fungiforme.* J. Phycol. 17: 43–51.

Starmach, K., 1974. Flora słodkowodna Polski 4. Polska Akademia Nauk. Warszawa. 520 pp.

Stosch, H. A., 1965. Sexualitat bei *Ceratium cornutum* (Dinophyta). Naturwissenschaffen 52: 112–113.

Stosch, H. A., 1973. Observations on vegetative reproduction and sexual life cycle of two freshwater dinoflagellates, *Gymnodinium pseudopalustre* Schiller and *Wołoszynska apiculata* sp. nov. British Phycol. J. 8: 105–134.

Taylor, F. J. R., 1987. The Biology of Dinoflagellates. Botanical Monographs Vol. 21., Blackwell Scientific Publications, London. 785 pp.

Temponeras, M., J. Kristiansen & M. Moustaka-Gouni, 2000. A new Ceratium species (Dinophyceae) from Lake Doriani, Macedonia, Greece. Hydrobiologia 424: 101–108.

Wedemayer, G. J., L. W. Wilcox & L. E. Graham, 1982. *Amphidinium cryophilum* sp. nov. (Dinophyceae) a new freshwater Dinoflagellate. I. Species description using light and scanning electron microscopy. J. Phycol. 18: 13–17.

Wołoszynska, J., 1916. Polnische Süsswasser-Peridineen. Bulletin de L'Académie des Sciences. Cracovie (B) Classe mathématiques naturelles. Série B: Sciences naturelles. (Okt–Nov–Dec 1915): 260–285.

Yamagishi, T. & M. Akiyama, 1987. Photomicrographs of the Freshwater Algae. Vol. 15. 1–100.

Zederbauer, E., 1904. Geschlechtliche und ungeschlechtliche Fortpflanzung von *Ceratium hirundinella*. Berliner Deutsche Botanische Gesellschaft 22: 1–8.

Hydrobiologia **502:** 357–366, 2003.
L. Naselli-Flores, J. Padisák, M. T. Dokulil (eds),
Phytoplankton and Equilibrium Concept: The Ecology of Steady-State Assemblages.

Long-term trends in species composition and diurnal migration of dinoflagellates in Lake Tovel (Trentino, Italy)

Giovanna Flaim[1], Eugen Rott[2], Flavio Corradini[1], Giambattista Toller[1] & Basilio Borghi[1]
[1]*Istituto Agrario – 38028 S. Michele all'Adige TN Italy*
E-mail: flaim@mail.ismaa.it
[2]*University of Innsbruck, Institute of Botany, Austria*

Key words: Lake Tovel, dinoflagellates, bloom, vertical migration, phototaxis, temperature preference, wind pattern

Abstract

Lake Tovel, an oligotrophic mountain lake (Trentino, N. Italy) is famous for its past bright red summer blooms of the dinoflagellate *Glenodinium sanguineum* Marchesoni. In 1938 and 1941 Baldi studied the lake and bloom phenomenon in detail. In particular, he found that the dinoflagellate showed high morphological plasticity and that it was present in the whole lake, but concentrated in the Red Bay, where it exhibited an active vertical migration. In 2001 the presence of *G. sanguineum* in the whole lake, with higher densities in the Red Bay could be confirmed. However the cell densities found in the Red Bay are up to 10^3 lower than those reported by Baldi. During two diurnal studies in August 2001 almost identical diurnal depth variations of maximum density of *G. sanguineum* were found, likely to be related to positive phototaxis. Although in the Red Bay the patterns of vertical variations were less clear, potential influential factors (wind, sun, temperature) are analysed and discussed. Dinoflagellate species composition from this and earlier studies show an increase in the number of taxa recorded from the lake in recent years and a shift in species composition possibly related to changes in the trophic status of the lake.

Introduction

Lake Tovel, a small, oligotrophic mountain lake, situated in the heart of the Brenta Adamello Natural Park (Trentino, N. Italy) is one of the few Italian habitats protected under the 1980 Ramsar Convention. Apart from its pristine beauty, Tovel is famous for its blood red summer blooms in the SW Red Bay. The English traveller Freshfield (1875) was the first to briefly mention the phenomenon, which he encountered during his trip to the Italian Alps during 1864. A more thorough description was given a few years later by Loss (1872) who credited the phenomenon to infusoria. Largaiolli (1907a) first attributed the bloom formation to the presence of a dinoflagellate, which he classified as *Glenodinium pulvisculus* Stein var. *oculatum* Largaiolli. Detailed LM studies permitted Marchesoni (1941) to reclassify it as *Glenodinium sanguineum* March. but also this name needs to be discussed for priority reasons. Recently this dinoflagellate has been included in the *Woloszynskia coronata* group (Popovský & Pfiester, 1990), although the type species of the taxon is unknown. Since we can not go into the details of botanical nomenclature in this context we prefer to continue using Marchesoni's classification. Interest in the lake and the catchment has generated almost 400 publications during the last 150 years; the most recent studies have focused on the possible causes of bloom inhibition, blooms which have essentially disappeared since 1964 except for sporadic and chromatically minor events (Arrighetti & Siligardi, 1979; Miola & Trevisan, 1982; Dodge et al., 1987). Of particular merit for an exhaustive description of the bloom phenomenon are Baldi's studies (1938, 1941).

The spectacular reddening of oligotrophic lake Tovel, which in the past occurred almost every summer, represented a unique event with relevant positive implications for the Trentino province. While definitely not excluding *a priori* other causes and interactions, the most promising single reason for the lack of bloom formation to date seems to be one first proposed by Dodge (1970) who suggested that changes in land use (a decrease in the number of dairy herds that summer in the alpine pastures in the lake's catchment) were limiting the amount of nutrients entering the lake (Dokulil & Teubner, 2003; Salmaso, 2003).

In their review, Corradini et al. (2001) underlined the lack of nutrient data prior to 1964. The first reliable data for phosphorus, nitrogen and silicon are from the late 1970s and do not indicate any significant changes since then. Nutrient analysis indicates oligotrophic conditions with P as the limiting factor. The same authors also compared the few temperature profiles available from the pre-1964 period with the numerous profiles available from recent years, finding no substantial differences in thermal stratification patterns or surface water temperatures at least for the pelagic zone.

The literature on Lake Tovel and the Institute's archives were searched looking for clues, other than nutrient indicators, which would suggest a change in trophic status, including a general comparison of dinoflagellate species present in current and past surveys. Baldi's (1941) migration studies were repeated in summer 2001 and extended to the centre of the lake as a good starting point for understanding the present behaviour of *Glenodinium* in Lake Tovel.

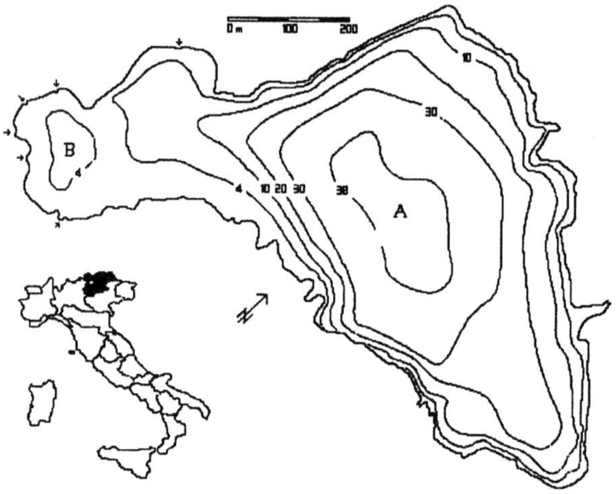

Figure 1. Bathymetric map of Lake Tovel: A = central lake station, B= Red Bay station.

Study site

Lake Tovel (A = 0.38 km^2, V = 7.4 10^6 m^3, z_{max} = 39 m; z_{mean} = 19 m) has two basins, a large deep (z_{max} = 39 m) north east one and a much smaller, shallower (z_{max} = 5 m) south west basin (Fig. 1) which is usually dry in winter. Geological substrate of the catchment area (39.9 km^2) is dolomite and limestone. The lake is ice-covered from December to April. Detailed descriptions of geomorphological and environmental aspects of the lake have already been given (Paganelli, 1992; Fuganti & Morteani, 1999, Corradini et al., 2001). The high catchment to surface area ratio (app. 100) is responsible for the low water retention time in the lake; Paganelli (1992) calculated 33 days renewal time for summer. High elevation of the catchment causes fast water renewal in spring and early summer concomitant with snow melt.

Methods

Migration trials

Diurnal migration studies on 7 and 28 August 2001 followed in principle Baldi (1941), but with additional sampling depths and a central lake station added. Both sampling days were clear and sunny; meteorological data are from the Tovel Meteo Station located at the SW corner of the lake (Fig. 2). Water was sampled using a Ruttner bottle from 0, 0.5, 1, 2, 5, 10, 15, 20 and 25 m for station A (lake centre) and 0, 0.5, 1, 2 and 3.5 m for station B (Red Bay) and immediately fixed in acidified Lugol's solution. Sampling was done every 2 hours starting at 5:00 till 21:00 (Solar Time). Subsamples of 25 ml (Station B) and 50 ml, respectively (Station A) were counted using an inverted Leica microscope with gridded eyepiece, considering only dinoflagellates between 15 and 31 μm as *G. sanguineum.* Since fixation did not permit distinguishing between red and green forms, on 7 August (station B at 5-7-9-11-13-15 and 17 hours) samples were collected in double and examined live. All *Glenodinium sanguineum* appeared green with one or more red spots, but the temperature sensitive cells quickly formed a round 'temporary cyst' when placed on a slide. Results are presented as % of column total for the hour of sampling in order to reduce density variations by cell patchiness between samplings series (Dagg et al., 1997). Temperature profiles at 1 m intervals in station A and 0.5 m intervals for station B were monitored with a Hydrolab multiparametric probe at 8, 12 and 18 h (Fig. 3). Lake water was analysed for basic nutrients on 28 August 2001. Nutrient analyses were performed using standard analytical techniques (IRSA-CNR, 1994) (Table 1).

Table 1. Nutrient analyses from the central station (A) and in the Red Bay (B) at various depths in L. Tovel. All analyses are from 28 Aug 2001

Parameter	Station / Depth (m)	A 0	A 2.5	A 5	A 12	B 0	B 2.5
pH 20 °C		8.4	8.5	8.6	8.6	8.4	8.4
Cond. 20 °C	$\mu S\ cm^{-1}$	160	161	162	169	162	160
Alkalinity	$\mu eq\ l^{-1}$	1.84	1.84	1.86	1.96	1.84	1.82
N-NO_3	$\mu g\ l^{-1}$	260	260	250	230	250	260
N-NH_3	$\mu g\ l^{-1}$	<30	<30	<30	<30	<30	<30
P-PO_4^{3-}	$\mu g\ l^{-1}$	2	2	2	2	1	2
Total P	$\mu g\ l^{-1}$	3	3	3	3	3	3
Reactive SiO_2	$mg\ l^{-1}$	0.21	0.21	0.16	0.14	0.24	0.28

Results and discussion

Migration studies

It is well known that most freshwater dinoflagellates tend to migrate up and down the water column within a circadian rhythm (Levandowsky & Kaneta, 1987), and show very frequently surface aggregations during the day and subsurface dispersal at night (Kamykowski et al., 1998). In the lake centre Baldi (1941) found *Glenodinium sanguineum* at all depths (0, 5, 10 and 32 m) with a strong concentration in the surface layers. In 2001 the alga was found to be present in the whole water column down to 25 m with a strong concentration in the upper layers (Fig. 4). Precision of migration is strongest when each individual cell in a population exhibits the same response to a stimulus (Kamykowski et al., 1998). In our case precision of migration was not strong, but a clear shift in the depth of maximum cell density is evident for both sampling dates in the central lake station (Fig. 4a,b) and likely to be related to positive phototaxis since the shift of maximum density apparently moved upward in the morning (5–13 h) and down in the afternoon. However variations in nutrient supply in different depths, water movements and multiple other reasons can also modulate the pattern of diel vertical migration (Heaney & Eppley, 1981; Karp-Boss et al., 2000). Wind shear during the central hours of the day should move the algae from the central parts of the lake to the shores and might act against migration, so it is likely that what we see in Figures 4a,b is a true upward migration towards light and warm surface waters. For both days, wind velocity never exceeded 3 m s^{-1} and can therefore be considered a light wind (Frempong, 1984). The wind increases to over 1 m s^{-1} in the central hours of the day when the dinoflagellate is near the surface and more prone to wind shear (Fig. 2). Meteorological conditions for the two days are similar (Fig. 2) but an earlier and longer eastern upwing circulation and a longer sunshine duration are observed on 7 August. Ionic composition or nutrient concentrations do not seem to indicate the presence of a chemical gradient (Table 1). The main difference between the two sampling days seems to be related to water temperature (Fig. 3). If we look at the temperature profiles for the 2 days we see that maximum cell density never occurs below 10–12 °C. This happens at 1 m on 7 August and at 2 m on 28 August.

Glenodinium sanguineum tends to concentrate in the warmer surface waters. The more defined vertical gradient = greater precision seen on 7 August (migration limited to 1 m) with respect to 28 August (migration up to 2 m) seems to confirm the importance of temperature.

Paganelli (1992) and Cavalca et al. (2001) have previously described changes in *Glenodinium* density in the Red Bay. During both sampling dates, *Glenodinium sanguineum* density in the Red Bay was about 10^2–10^3 times that found in the central basin surface waters. Historically, this density difference has been attributed to a morning breeze, which transports the cells in a southern direction towards the Red Bay. Compared to Baldi (1941) cell density during both our sampling days is similar to his green morphs but about 10^1–10^2 times less compared to his red series and up to 10^3 times less when all morphs are considered combined (Table 2).

When vertical time series for the two sampling dates and Baldi's (1941) red and green series are com-

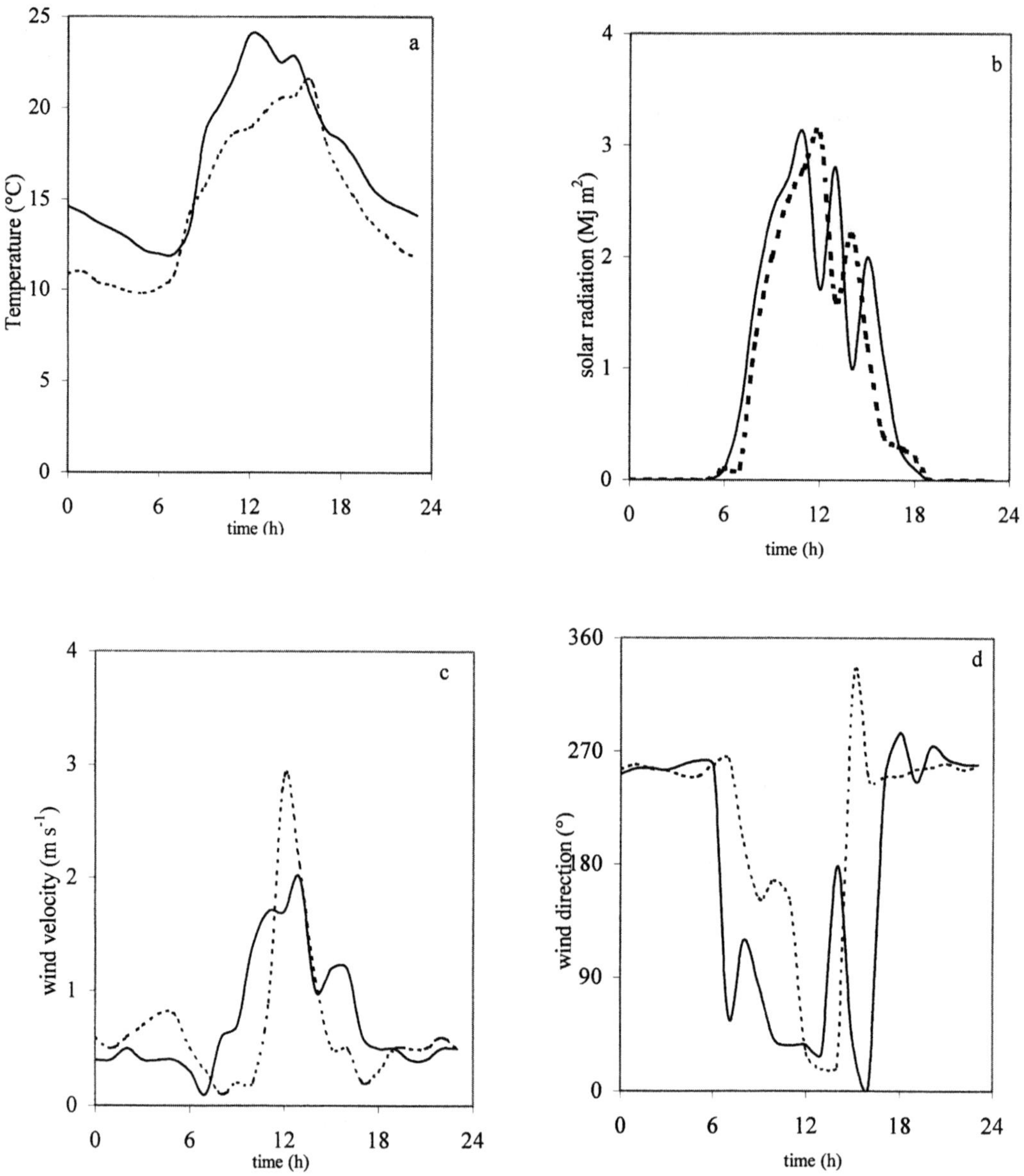

Figure 2. Air temperature (a), solar radiation (b), wind velocity (c) and wind direction (d) for the two sampling dates: 7 Aug (solid line) and 28 Aug 2001 (dotted line).

pared (Table 2), the data do not seem to indicate a readily discernible diurnal migration pattern in this area of the lake. However, *Glenodinium sanguineum* does seem to concentrate in the first meter of water and avoids the bottom waters. Here again, if we look at the temperature profiles (Fig. 2) we can see that for both days waters 2 m and above are >11–12 °C while under this depth, waters are quite cold (5–7 °C) due to underground inflow. Figure 5 shows the correlation between temperature and cell density, especially strong in the Red Bay. Baldi's (1941) green morphs seem to follow the same pattern of avoidance of the colder deeper waters. His red series however seem to migrate down during the late afternoon. Preliminary studies on laboratory cultures indicate that at least in laboratory conditions the alga prefers cold water (<10 °C), but this may be tied to the physiological condition of the cells (Heaney & Furnass, 1980; Heaney & Eppley, 1981).

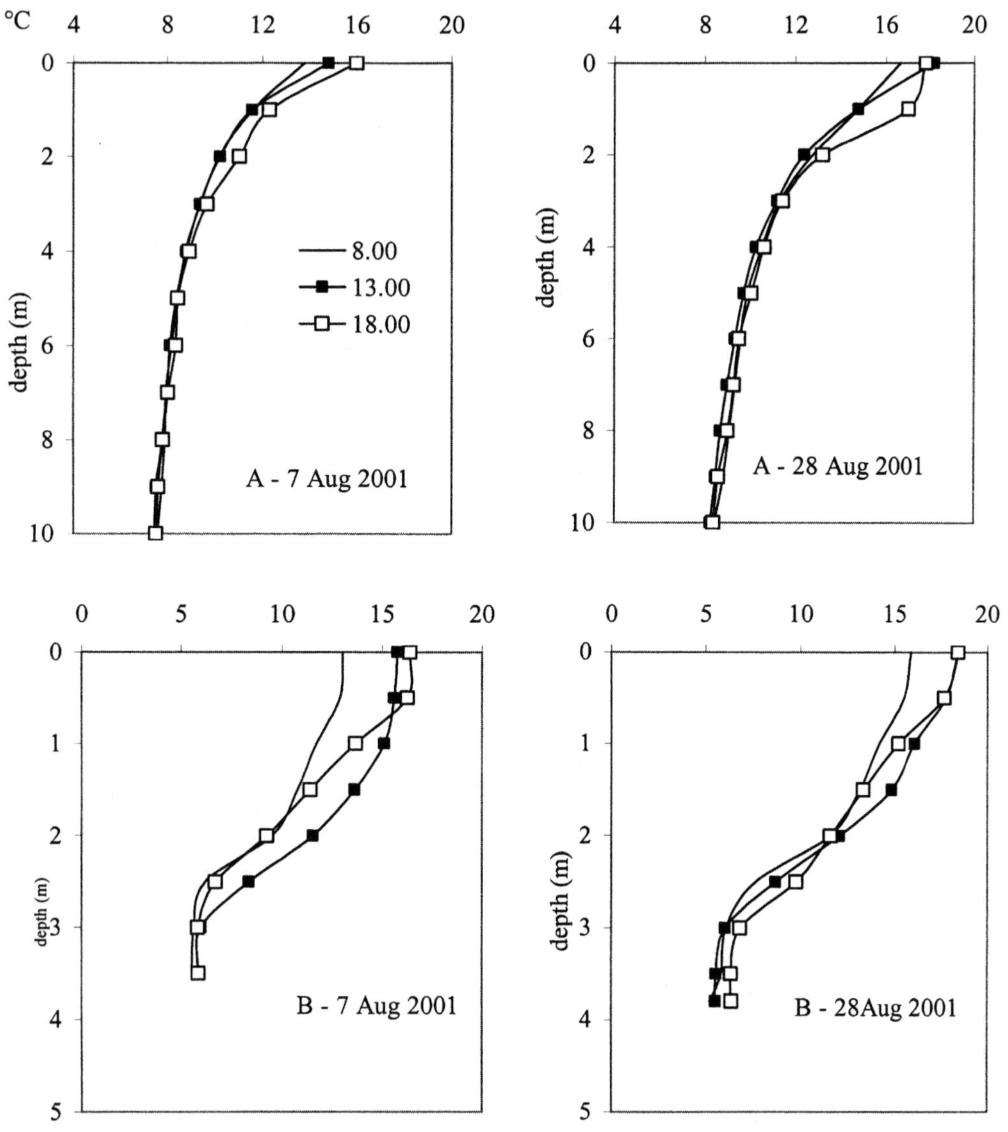

Figure 3. Temperature profiles for stations A and B in August 2001. Each of the two sampling dates has three profiles taken at 8:00, 13:00 and 18:00 hours.

Long-term trend based on data from literature and own observations

The constant presence of *Glenodinium sanguineum* from its first scientific description in Largaiolli (1907) up to this study is evident (Table 3), but most studies have concentrated attention on *Glenodinium sanguineum* while ignoring other dinoflagellates. The examination of the literature shows that a long term variation in the dinoflagellates species composition is likely, although there is a constant uncertainty in the identification / delimitation of the taxa. Difficulties in dinoflagellate taxonomy have been outlined by Grigorszky et al. (2003). As a first step to improve the situation, the size data of dinoflagellates found in the lake are listed and supplemented by our own observations (Table 4). *Ceratium hirudinella*, one of the taxa hard to confuse with others, is the only other dinoflagellate besides *Glenodinium sanguineum* mentioned by Largaiolli (1907), Baldi (1941) and Marchesoni (1959) and was found in the pelagic during spring–summer samplings. Arrighetti & Siligardi (1979) re-

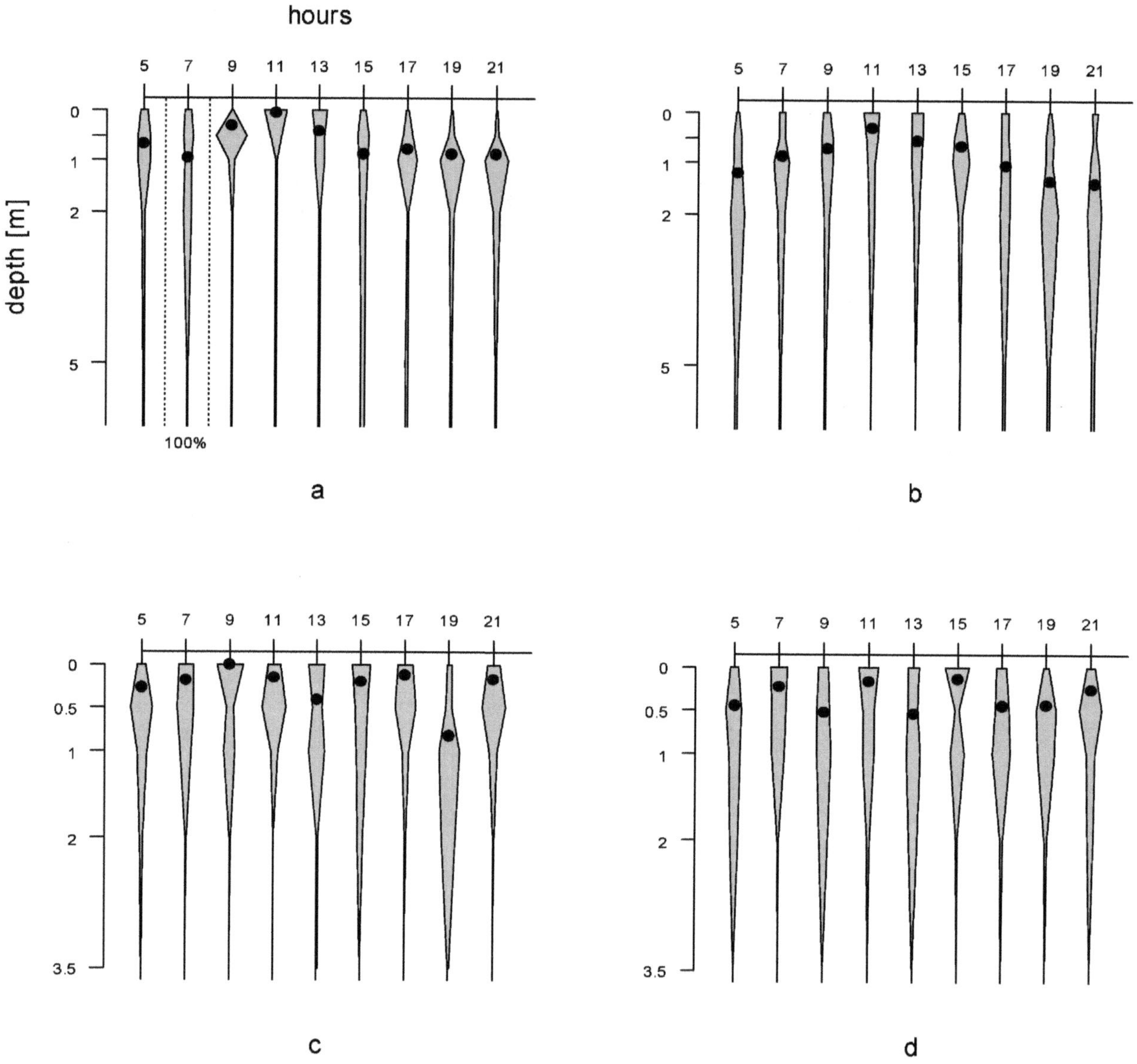

Figure 4. Profiles of% distribution of *Glenodinium sanguineum* on two sampling dates for station A (a-7 Aug and b-28 Aug 2001) and B (c-7 Aug and d-28 Aug 2001) sampled at various depths at two hour intervals. Solid dots indicate depth of median cell density (59%) for that series of samples calculated on whole column data.

ported it from their 1977–1978 campaign; Paganelli et al. (1981) and Paganelli (1992) do not mention it from their surveys. We found *Ceratium* in low densities both in 1995 (2000 cells l^{-1}) and in 1996 (500 cells l^{-1}) but it is virtually absent since then (in the over 300 phytoplankton samples examined in 2001, only one single *Ceratium* cell was found). Small dinoflagellates, presumably species of *Gymnodinium* have been present at least since the late 1970s (Miola & Trevisan, 1982). Recent surveys have found *Gymnodinium helveticum* and *Gymnodinium uberrimum*. *G. helveticum*, a heterotrophic dinoflagellate, is a common euplanktonic species in Lake Tovel, which we can find in our samples all year round. Its distinct shape would make it rather difficult to overlook when it is present. *G. uberrimum* is well represented both in the lake centre and in the Red Bay. In the central basin its distribution seems to be similar to that found for *G. uberrimum* in a high mountain lake (Tilzer, 1973) with maximum densities at 10–20 m depth. Un-

Table 2. Percent distribution of red and green morphs of *Glenodinium sanguineum* from Baldi (1941) and from two sampling days in 2001 for each time series considered at various depths in the Red Bay of L. Tovel. Maximum concentration for each time series is highlighted. The last line is the total column density given as 10^6cells l^{-1}

	hour	1	3	5	7	9	11	13	15	17	19	21	23
Baldi (1941) Red morphs													
depth (m)	0	**57**	**73**	**44**	**79**	**59**	**78**	**69**	**67**	**36**	36	14	**84**
	2	21	10	30	13	15	11	7	6	32	**42**	33	8
	4	21	17	26	8	26	11	24	27	32	22	**53**	7
10^6 cell l^{-1}		1.21	0.78	0.45	1.28	2.24	3.62	2.77	3.10	1.01	1.82	0.86	1.70
Baldi (1941) Green morphs													
	0	**68**	**77**	36	**74**	**66**	**77**	**87**	**68**	**92**	**87**	**49**	**83**
	2	23	14	**50**	22	20	23	12	30	8	10	42	13
	4	9	9	14	4	13	0	1	1	0	3	8	4
		0.12	0.06	0.04	0.14	0.13	0.06	0.33	0.40	0.33	0.19	0.09	0.19
7 Aug 2001													
	0			19	35	**59**	28	37	**39**	34	9	27	
	0.5			**51**	**37**	14	**55**	20	27	**48**	12	**52**	
	1			23	25	25	16	**38**	21	16	**46**	19	
	2			7	3	2	1	5	13	2	31	2	
	3.5			0	0	0	0	1	0	0	2	0	
				0.20	0.17	0.16	0.21	0.07	0.28	0.13	0.11	0.06	
28 Aug 2001													
	0			19	**36**	21	**43**	25	**53**	21	13	20	
	0.5			**34**	32	27	25	22	10	33	**43**	**49**	
	1			26	28	**32**	25	**32**	33	**40**	37	18	
	2			20	4	19	5	19	4	6	7	11	
	3.5			1	0	0	1	2	0	1	0	1	
				0.05	0.08	0.20	0.09	0.13	0.21	0.14	0.11	0.08	

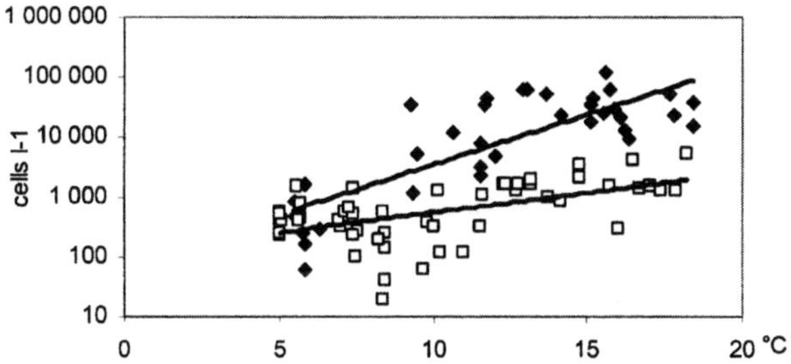

Figure 5. Cell density of *Glenodinium sanguineum* in relation to temperature in the lake centre (A – open squares) ($y = 119.51e^{0.15x}$, $r = 0.54$, $P < 0.01$, $n = 54$) and in the Red Bay (B – filled squares) ($y = 75.80^{e0.38x}$, $r = 0.80$, $n = 35$, $P < 0.01$).

fortunately its size range (24–90 μm as reported in Popovskỳ & Pfiester, 1990) partially overlaps the size range reported for *Glenodinium sanguineum* (Table 4). Especially in fixed samples, small *G. uberrimum* cells might be easily confused with large *Glenodinium sanguineum* cells; we considered cells over 32 μm to be *G. uberrimum.* Two winter species *(Peridinium aciculiferum* and *P. lomnickii)* have only been found recently; that in part may be due to logistic difficulties in sampling Lake Tovel during the winter months, but Baldi (1938, 1941), who sampled several times under ice never mentioned the presence of *Peridinium.* The combined density of these species in a January 2001 sample under ice was over 10^5 cells l^{-1}. Dodge (1970) mentioned the presence of *Peridinium cinctum* and *Gonyaulax* sp.

Our investigations indicate a long-term shift of dinoflagellate taxa from the large mesotraphentic species (*Ceratium*) to smaller sized oligotraphentic spe-

Table 3. Species list and chronology of dinoflagellate taxa in Lake Tovel arranged by first year of record. References: (1) Largaiolli, 1907a; (2) Largaiolli, 1907b; (3) Baldi, 1941; (4) Marchesoni, 1959; (5) Dodge, 1970; (6) Arrighetti & Siligardi, 1979; (7) Paganelli et al 1981; (8) Miola & Trevisan, 1982; (9) Candioli & Tamanini, 1998; (10) present study

Dinoflagellate taxon	Year of 1st record	Reference	Further citations	Comments
Ceratium hirundinella O.F. Müller	1907	2	2,3,5	common summer form in past, now rare
Glenodinium sanguineum March.	1907	1	2–10	always present - dominant Red Bay dinoflagellate during warm summers
Gonyaulax sp.	1970	5		only this record
Gymnodinium cf. *simplex* (Lohmann) Kofoid & Swezy	1970	5		only this record
Peridinium cinctum Müller Ehrenberg	1970	5		only this record
Gymnodinium ordinatum Skuja	1982	7		probably equivalent to small dinoflagellates
Gymnodinium helveticum Penard	1998	9	10	found only in the pelagic zone
small dinoflagellates	1998	9	10	abundant in the Red Bay
Gymnodinium uberrimum (Allman) Kofoid & Swezy	1998	9	10	always present - dominant Red Bay dinoflagellate during cold summers
Peridinium aciculiferum Lemmermann	2002	10		winter species
Peridinium lomnickii Woloszynska	2002	10		winter species

Table 4. Cell lengths in μm for some Lake Tovel dinoflagellates (A *Glenodinium sanguineum* March.; B *Gymnodinium ordinatum* Skuja; C small dinoflagellates; D *Gymnodinium uberrimum*) as reported in the literature. References: (1) Largaiolli, 1907a; (2) Baldi, 1941; (3) Marchesoni, 1941; (4) Arrighetti & Siligardi, 1979; (5) Miola & Trevisan, 1982 (6) Dodge et al., 1987; (7) Popovsky & Pfiester, 1990; (8) Candioli & Tamanini, 1998; (9) present study

Taxon	Reference	Min	Max	Mean	No.	Comments
A	1	12	33	22		red morph
A	2			22		red morph as cited in (1)
A	2			>>22		other morphs
A	3			23		red morph
A	4	25	40			
A	5	15	20		130	mobile cells, 80% within range
A	5	18	25		40	immobile cells, 80% within range
A	6			18.7	>100	SD = −+1.7
A	8			16.9	20	2 length groups
A	8			25.8	20	
A	9	15	31			all morphs
B	5	10	16			length as reported in (7)
C	8		<15			
D	8			38		
D	9	>32				

cies. *Ceratium hirudinella* is frequently recorded to favour mesotrophic waters (Popovskỳ & Pfiester, 1990; Reynolds, 1998; Reynolds et al., 2002). *Gymnodinium helveticum* is considered to be an oligotraphentic species (Popovskỳ & Pfiester, 1990) as well as *P. lomnickii* (Ling et al., 1989) and *G. uberrimum* an oligo-mesotraphentic (Rott, 1988) or oligotraphentic taxon (Popovskỳ & Pfiester, 1990).

What seems to emerge from this notably fragmentary picture is (*i*) an inherent difficulty in deciding where, at least in terms of dimensions, *Glenodinium sanguineum* begins and ends and (*ii*) a tendency towards the presence of more oligotraphentic species such as *Gymnodinium helveticum* and *Peridinium lomnickii* accompanied by the reduction of more nutrient tolerant species such as *Ceratium hirudinella* and possibly *Glenodinium sanguineum.*

Conclusions

The alga *Glenodinium sanguineum*, or at least what we think corresponds to the taxon studied by Baldi (1941), is still present in the whole lake. From this first series of trials we suspect that the cell plasticity reported by Baldi (1938; 1941), especially his 'conspicuously larger' forms, might actually have been related to the presence of more than one dominant species during the period of the blooms. Similar taxa can not be differentiated with LM, especially when working with live samples notorious for their lack of temperature tolerance. In the centre of the lake vertical migration is limited to the first few meters and seems to be defined by a combination of positive phototaxis and avoidance of wind drift and cold temperatures. In the Red Bay the smaller vertical shifts compared to the lake centre are likely to be related to different effects of wind pattern and sunshine and a stronger temperature gradient within the bay. Further work will require additional temperature profiling and sampling also on cloudy days to separate the effects of sunshine and temperature on migration. Sampling is underway to quantify precisely the wind driven passive transport from the centre towards the Red Bay. Differences in the published dinoflagellate community tend to indicate a more oligotrophic state in recent years compared to Baldi (1941) and Marchesoni (1959), giving weight to the theory first proposed by Dodge (1970) that the lake is receiving less nutrients. Most probably changes in the dinoflagellate species during and after the bloom period may best be seen from cysts in the sediments. Since dinoflagellate cysts are no easier to identify than vegetative cells, probably the best approach would be to try to develop a single cell/ cyst based molecular inventory.

Acknowledgements

This study is part of Workpackage III of SALTO, a research project funded by the Province of Trento. The authors are indebted to Øjvind Moestrup and Gert Hansen for identification of *P. lomnickii* and to Vigilio Pinamonti for all his help.

References

Arrighetti, A. & M. Siligardi, 1979. Analisi idrobiologiche al Lago di Tovel (Trento). Esperienze e Ricerche 5: 5–69.

Baldi, E., 1938. Relazione preliminare sulle ricerche al lago di Tovel. Studi Trentini di Scienze Naturali 19: 247–339.

Baldi, E., 1941. Ricerche idrobiologiche sul lago di Tovel. Memorie del Museo di Storia Naturale della Venezia Tridentina 6: 1–297.

Candioli, E. & S. Tamanini, 1998. Phytoplankton species composition and biomass in Lake Tovel and four enclosures. In Rott, E. & B. Maiolini (eds), Limnology of a Subalpine Lake: Lago di Tovel. An Advanced Student's Course. Institute of Botany, Univ. Innsbruck and Natural History Museum Trento. 82 pp.

Cavalca, L., P. Ferrari & V. Andreoni, 2001. *Glenodinium sanguineum* March. and the reddening phenomenon of Lake Tovel: biological and environmental aspects. Ann. Microbiol. 51: 159–177.

Corradini, F., G. Flaim & V. Pinamonti, 2001. Five years of limnological observations on lake Tovel (1995–1999): some considerations and comparisons with past data. Atti Ass. ital. Oceanol. Limnol. 14: 209–218.

Dagg, M. J., B. W. Frost & J. A. Newton, 1997. Vertical migration and feeding behavior of *Calanus pacificus* females during a phytoplankton bloom in Dabob Bay, U.S. Limnol. Oceanogr. 42: 974–980.

Dodge, J. D., 1970. Report of limnological investigation of Lake Tovel (Trentino – N. Italy). Studi Trentini di Scienze Naturali 47: 91–94.

Dodge, J. D., P. Mariani, A. Paganelli & R. Trevisan, 1987. Fine structure of the red bloom dinoflagellate *Glenodinium sanguineum* from lake Tovel (N. Italy). Archiv für Hydrobiologie (Suppl.) 78: 125–138.

Dokulil, M. T. & K. Teubner, 2003. Steady state phytoplankton assemblages during thermal stratification in deep alpine lakes. Do they occur? Hydrobiologia 502 (Dev. Hydrobiol. 172): 65–72.

Frempong, E., 1984. A seasonal sequence of diel distribuiton patterns for the planktonic dinoflagellate *Ceratium hirundinella* in a eutrophic lake. Freshwat. Biol. 14: 401–421.

Freshfield, D. W., 1875. Italian Alps. G. Srobele, London.

Fuganti, A. & G. Morteani, 1999. La storia del Lago di Tovel in base ai sedimenti della Baia Rossa (Trentino). Geologia Tecnica & Ambientale 3: 21–32.

Grigorszky, I., L. Krienitz, J. Padisák, G. Borics & G. Vasas, 2003. Redefinition of *Peridinium lomnickii* Wołoszynska (Dinophyta) by scanning electronmicroscopical survey. Hydrobiologia 502 (Dev. Hydrobiol. 172): 349–355.

Heaney, S. I. & R. W. Eppley, 1981. Light, temperature and nitrogen as interacting factors affecting diel vertical migrations of dinoflagellates in culture. J. Plankton Res. 3: 331–344.

Heaney, S. I. & T. I. Furnass, 1980. Laboratory models of diel vertical migration in the dinoflagellate *Ceratium hirudinella.* Freshwat. Biol. 10: 163–170.

IRSA – CNR, 1994. Metodi analitici per le acque. Istituto di Ricerca sulle Acque, Rome. 342 pp.

Kamykowski, D., E. J. Milligan & R. E. Reed, 1998. Relationships between geotaxis/phototaxis and diel vertical migration in autotrophic dinoflagellates. J. Plankton Res. 20: 1781–1796.

Karp-Boss, L., E. Boss & P. A. Jumars, 2000. Motion of dinoflagellates in a simple shear flow. Limnol. Oceanogr. 45: 1594–1602.

Largaiolli, V., 1907a. Ricerche biolimnologiche sui laghi trentini. Ac. sc. Ven. Tren. Istriano. Padova A4: 1–7.

Largaiolli, V., 1907b. La varietà oculata del *Glenodinium pulvisculus* (Ehr.) Stein. Nuova Notarisia 18: 1–5.

Levandowsky, M. & P. J. Kaneta, 1987. Behaviour of dinoflagellates. In Taylor, F. J. R. (ed.), The Dinoflagellates. Blackwell Scientific Publications, Oxford: 360–397.

Ling, H. U., R. L. Croome & P. A. Tyler, 1989. Freshwater dinoflagellates of Tasmania, a survey of taxonomy and distribution. British Phycol. J. 24: 111–129.

Loss, G., 1872. La Valle di Non – Saggio d'illustrazione delle Alpi Trentine. Seiser Typography. Trento.

Marchesoni, V., 1941. Sulla posizione sistematica del *Glenodinium* determinante l'arrossamento del lago di Tovel. Studi Trentini di Scienze Naturali 22: 11–18.

Marchesoni, V., 1959. La val di Tovel e il "Lago Rosso". Natura Alpina 10: 37–76.

Miola, A. & R. Trevisan, 1982. Un contributo allo studio su ciclo biologico e contenuto in clorofilla di *Glenodinium sanguineum* March. del Lago di Tovel (Trento). Studi Trentini di Scienze Naturali 59: 23–29.

Paganelli, A., 1992. Lake Tovel (Trentino): limnological and hydrobiological aspects. Mem. Ist. ital. Idrobiol. 50: 225–257.

Paganelli, A., R. Trevisan, P. Cordella, A. Miola & M. Cappelletti, 1981. Ricerche limnologiche sul Lago di Tovel (Trentino) ed ipotesi sul mancato arrossamento. Studi Trentini di Scienze Naturali 58: 393–424.

Popovský, J. & L. A. Pfiester, 1990. Dinophyceae (Dinoflagellida). In Ettl, H., J. Gerloff, H. Heynig & D. Mollenhauer (eds), Süßwasserflora Mitteleuropas 6. G. Fischer Verl. Stuttgart. 273 pp.

Reynolds, C. S., 1998. What factors influence the species composition of phytoplankton in lakes of different trophic status? Hydrobiologia 369/370: 11–26.

Reynolds, C. S., V. Huszar, C. Kruk, L. Naselli-Flores & S. Melo, 2002. Towards a functional classification of the freshwater phytoplankton. J. Plankton Res. 24: 417–428.

Rott, E., 1988. Some aspects of the seasonal distribution of flagellates in mountain lakes. Hydrobiologia 161: 159–170.

Salmaso, N., 2003. Life strategies, dominance patterns and mechanisms promoting species coexistence in phytoplankton communities along complex environmental gradients. Hydrobiologia 502 (Dev. Hydrobiol. 172): 13–36.

Tilzer, M. M., 1973. Diurnal periodicity in the phytoplankton assemblage of a high mountain lake. Limnol. Oceanogr. 18: 15–30.

Hydrobiologia **502:** 367–382, 2003.
L. Naselli-Flores, J. Padisák, M. T. Dokulil (eds),
Phytoplankton and Equilibrium Concept: The Ecology of Steady-State Assemblages.

Planktic oscillatorialean cyanoprokaryotes (short review according to combined phenotype and molecular aspects)

Jiří Komárek
Faculty of Biology, University of South Bohemia, České Budějovice, and Institute of Botany, Czech Academy of Sciences, Třeboň, Czech Republic
E-mail: komarek@butbn.cas.cz

Key words: cyanoprokaryotes, cyanobacteria, Oscillatoriales, taxonomy, plankton, identification keys

Abstract

This review presents the main genera and species of the oscillatorialean cyanobacteria, that can be observed in freshwater planktic biotopes. The taxa are classified according to combined modern, ultrastructural and molecular criteria. The review is organised in form of subsequent tables.

Introduction

The developments of the knowledge concerning the cyanobacteria have drastically changed the criteria that are used for their taxonomic classification (Rippka et al., 1979; Anagnostidis & Komárek, 1985; Castenholz & Waterbury, 1989). The application of electron microscopy as well as new biochemical and molecular methods explained a lot of enigmatic characters, and have shed a new light on the characteristics of this group (Castenholz, 2001). Recent studies that inferred the existence of lateral gene transfer within cyanobacterial populations (Rudi et al., 1998; Barker et al., 2000) and recognised the role of acclimation and adaptation processes in their prodigious evolutionary success since early Precambrian (Hagemann, 2002; Komárek & Kaštovský, 2003a), explained more or less the strategy of their 'static' evolution (Schopf, 1974, 1996; Schopf in Pennisi, 1994). It was proved, that the genotype diversity in nature (Garcia-Pichel, 1998; Taton et al., 2002) is distinctly wider than it is recognisable according to phenotype analyses, and only 5–10% of the real diversity particularly in tropical and extreme biotopes was acknowledged according to serious estimations (DiCastri & Younéz, 1994; Watanabe, 1999). All these novel insights influence the taxonomic classification, which must, of course, reflect the whole modern knowledge as much as possible.

Table 1. Main characters of the families within the order Oscillatoriales

CHROOCOCCALES (coccoid and colonial; binary fission of cells)
PLEUROCAPSALES (coccoid and pseudofilamentous; combined cell division by binary and multiple fission)

OSCILLATORIALES (filamentous cyanobacteria, absence of heterocytes and akinetes); combined genetic and phenotype criteria:

- Pseudanabaenaceae (trichomes <4 μm wide, parietal thylakoids, solit. trichomes) * Table 2
- Schizotrichaceae (trichomes < 5 μm wide, parietal thylakoids, fascic. trichomes)

- Borziaceae (trichomes 3-12 μm wide, radial thylakoids, necrids -)
- Phormidiaceae (trichomes 3–5 μm wide, radial thylakoids, necrids +) * Table 8
- Gomontiellaceae (trichomes 6–20 μm wide, irreg. thylakoids, flattened trichomes)
- Oscillatoriaceae (trichomes 6-60 μm wide, irregular thylakoids, cylindr. trichomes)* Table 15

NOSTOCALES (filamentous, heterocytes + akinetes, false branching
STIGONEMATALES (filamentous, facultatively heterocytes + akinetes, true branching)

Even if it is difficult and laborious to select the relevant information in the vast modern literature, this is a necessary task to obtain a taxonomic system that gives a good appreciation of the cyanobacterial

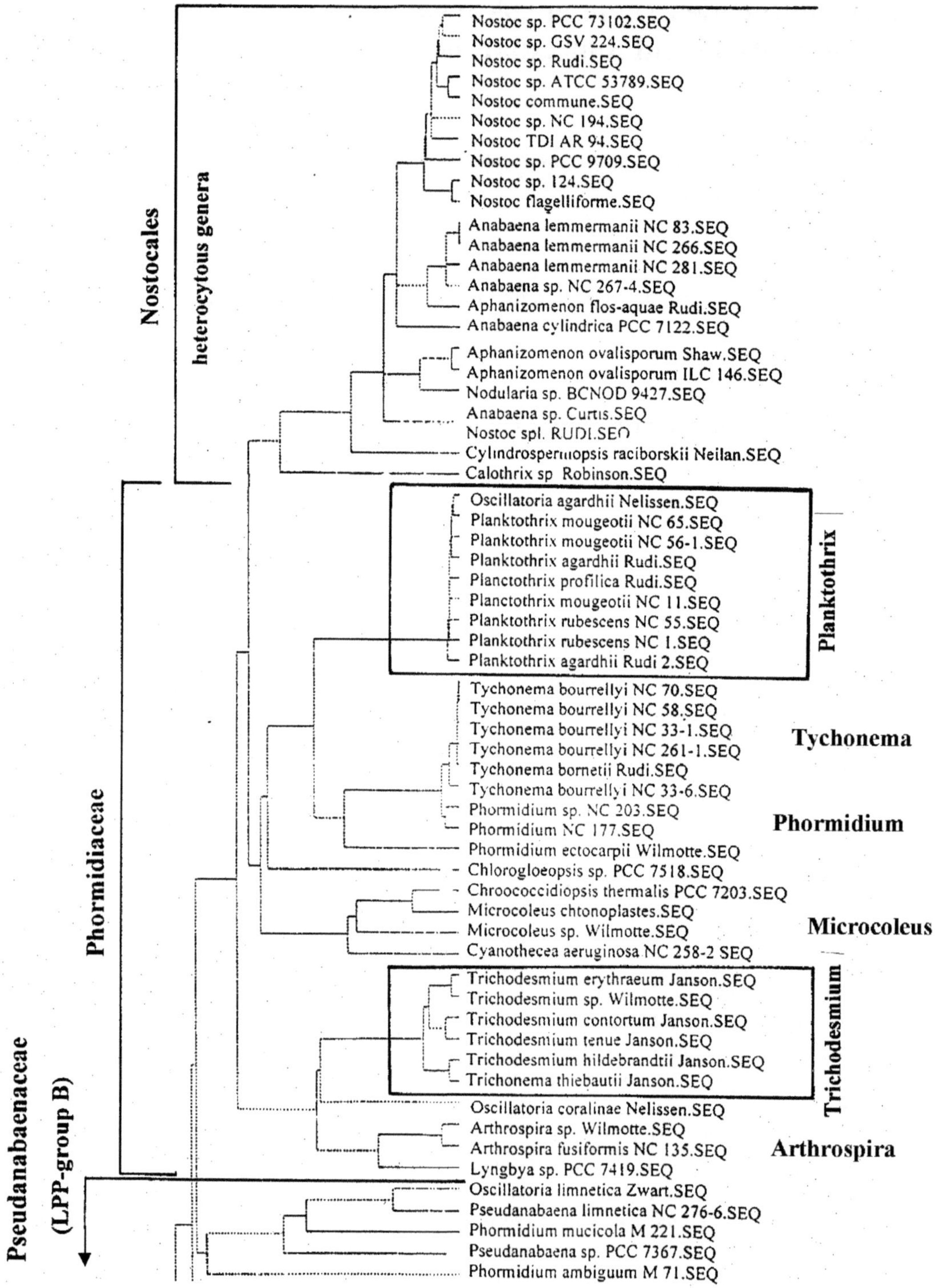

Figure 1. Part of a phylogenetic tree of cyanobacteria constructed by Megalign (DNAStar Inc.) containing filamentous genera (sequences by GenBank of NCBI, February 2002).

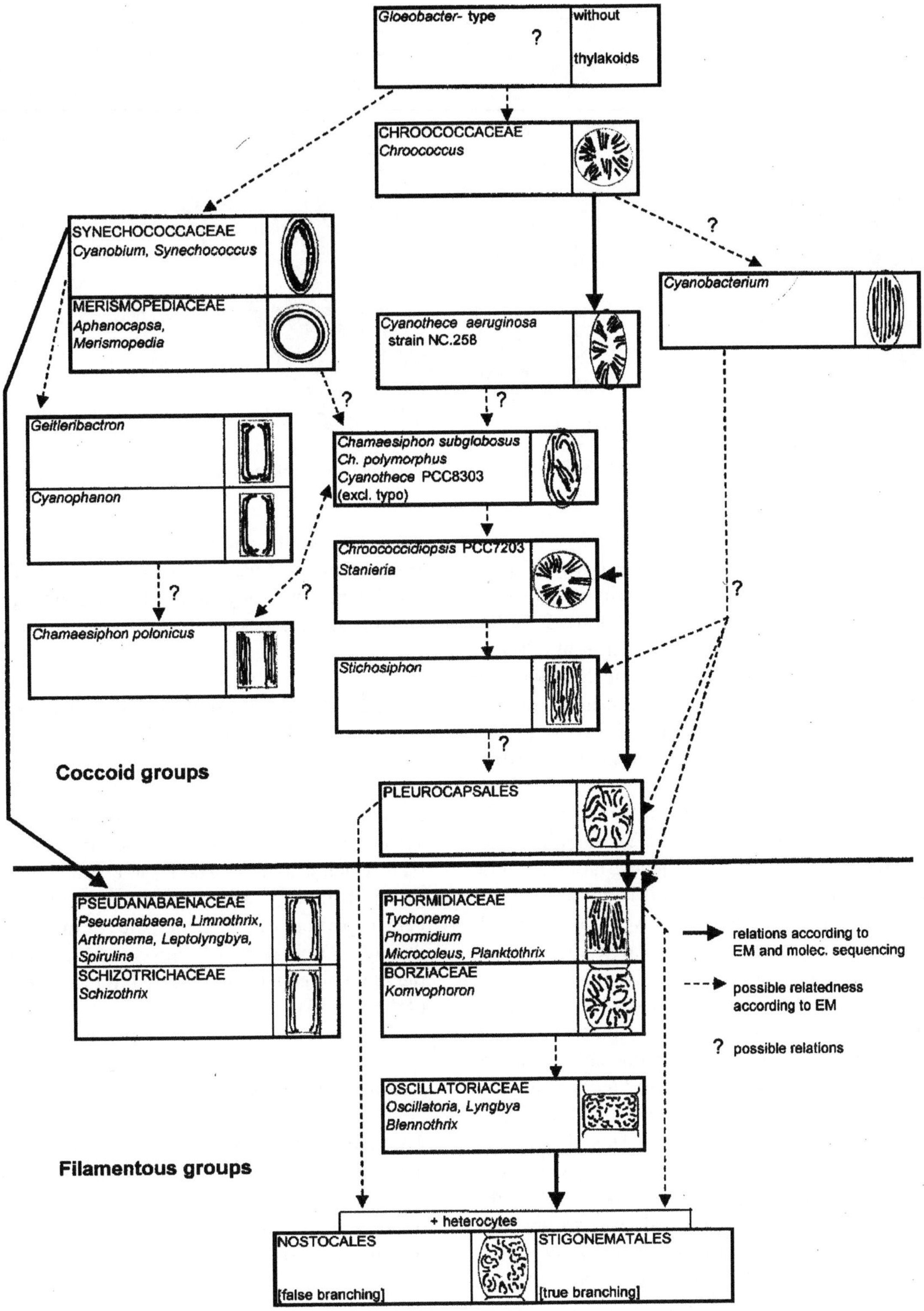

Figure 2. Relations between coccoid and filamentous cyanobacteria according to molecular sequencing and EM characters.

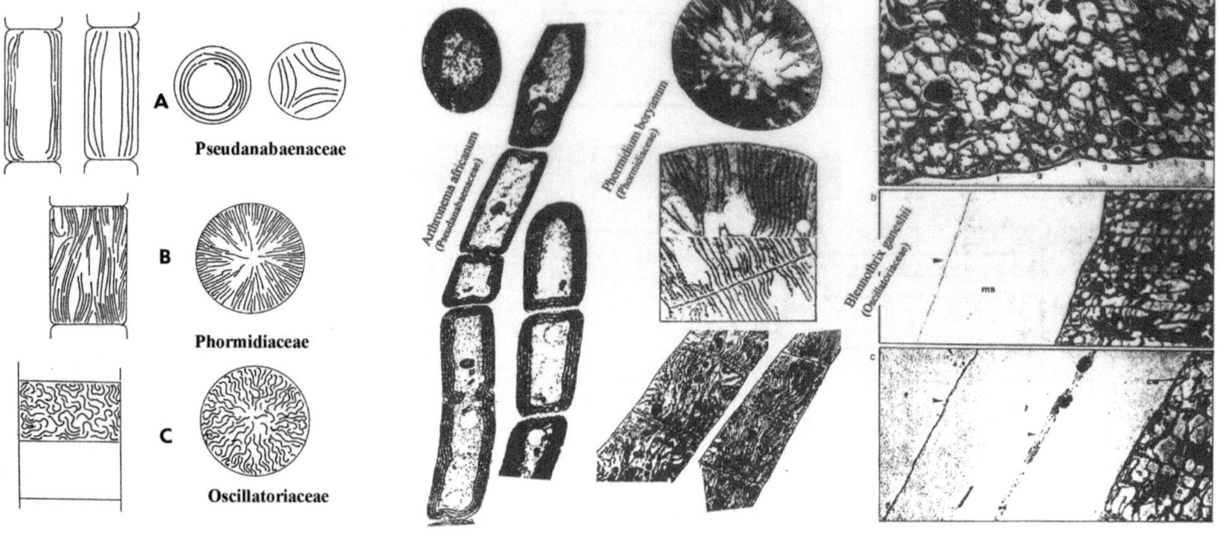

Figure 3. Position of thylakoids in main families of Oscillatoriales.

diversity and that is a good basis for experimental and ecological work. The only correct way to perform systematic work is to use a polyphasic approach (Anagnostidis & Komárek, 1985; Wilmotte & Golubiæ, 1991; Castenholz, 2001; Flechtner et al., 2002), combining all morphological, cytological, as well as ecological, biochemical and molecular criteria. In particular, the 16S rDNA sequence analysis can be a good basis for the evaluation of cyanobacterial genera (Fig. 1), and it can be in good agreement with generic phenotype and ultrastructural characters (Komárek, 2002; Komárek & Kaštovský, 2003b; Fig. 2). In spite of the fact that the revision of the cyanobacterial taxonomic system is still in a starting phase, it is possible to revise and define already a number of well-characterised and clearly distinct cyanobacterial genera (Castenholz, 2001; Komárek et al., 2003).

The infrageneric criteria are less clear (Kohl & Nicklisch, 1981; Kato et al., 1991; Komárek, 2002) and the species concept is not yet defined according to molecular criteria. However, the only method for future taxonomic work is to examine how the traditional species are fitting in revised genera. This also asks a continuous reassessment, on the basis of new information. In this paper, the main planktic species of oscillatorialean cyanobacteria are revised, in accordance to the present state of knowledge of their diversity. For practical reasons, the following conventional species definition is used: Group of populations (+ strains; clusters) which belong to one and the same genotype (genus). They are characterised by stabilised phenotype with defined and recognisable features (that have distinct limits of variation), and by the similar ecological demands. They should be constantly present in the special biotopes to which they are adapted.

Results (Review of oscillatorialean genera and species; Figs 1–3, Tables 1–15)

The present review of oscillatorialean cyanobacteria is a compromise between the traditional morphological system (particularly on the species level), and the new data, when available. The basic taxonomic scheme is given in Table 1. The position of thylakoids within cells is particularly characteristic for main families of the order Oscillatoriales (Figs 2 and 3). This review of the planktic genera is presented in the series of tables (Tables 2–15).

Acknowledgements

The author thanks Dr Jan Kaštovský for help with documentation, Mrs Dana Švehlová for technical preparation of tables, and both anonymous reviewers for improving the English text and numerous valuable recommendations.

Table 2. Diagnostical features and the key of the family Pseudanabaenaceae

Main features of the family Pseudanabaenaceae:

- **simple filaments, mainly less than 3 (5) µm wide**
- **parietal thylakoids (th)**
- **facultative sheaths (sh)**
- **facultative motility**
- **disintegration of trichomes without or with irregular necridic cells (C)**
- **one pore in cross walls (F)**

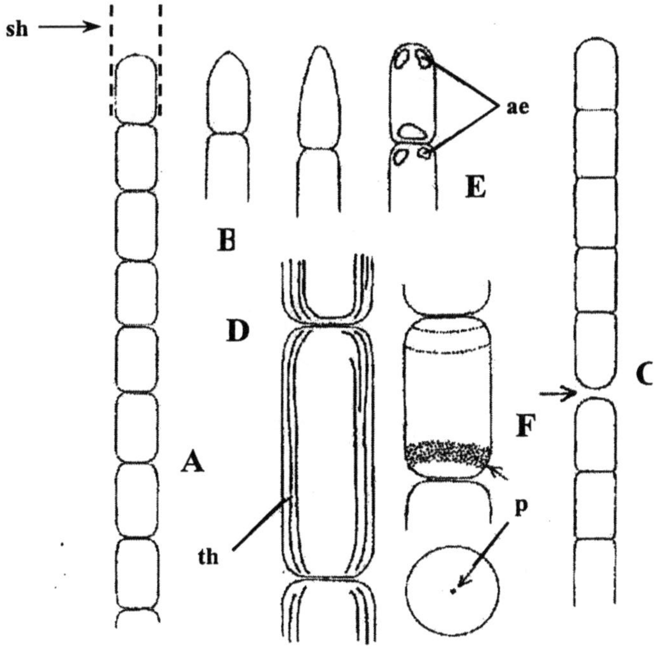

From Anagnostidis & Komárek 1988

Key to the genera containing planktic species:

1a Trichomes always without firm sheaths, but sometimes with mucilaginous envelopes 2

2a Trichomes mainly short, curved or irregularly coiled, sometimes only few-celled, disintegrating, with indistinct mucilaginous envelopes, with neighbouring cells occasionally disorganized .... **Romeria (Tab. 3-4)**

2b Trichomes more or less cylindrical, straight or slightly coiled, multicelled 3

3a Trichomes mostly constricted at cross walls, solitary or in clusters, usually without polar aerotopes .... **Pseudanabaena (Tab. 5)**

3b Trichomes unconstricted or very slightly constricted at cross walls, always solitary, with polar and/or central aerotopes .... **Limnothrix (Tab. 6)**

1b Trichomes always with thin, firm, colourless sheaths; solitary filaments in plankton .... **Planktolyngbya (Tab. 7)**

Table 3. Generic features of the genus *Romeria*

- **trichomes (pseudofilaments?)**

- short (1-12 [23 - > 60] cells)
- shift of neighbouring cells
[- constrictions at cross-walls]
[- mucilaginous envelopes]
- sheaths are lacking

- **cells**

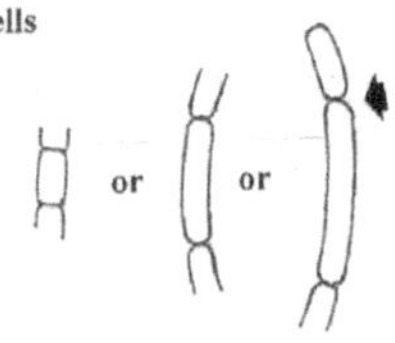

- always longer than wide (1)2-10.5(28):1
- cylindrical
- facultative asymmetrical cell division

- **ultrastructure**

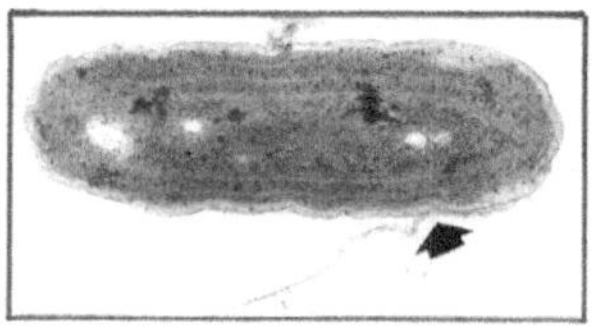

- parietal thylakoids
- cell connection

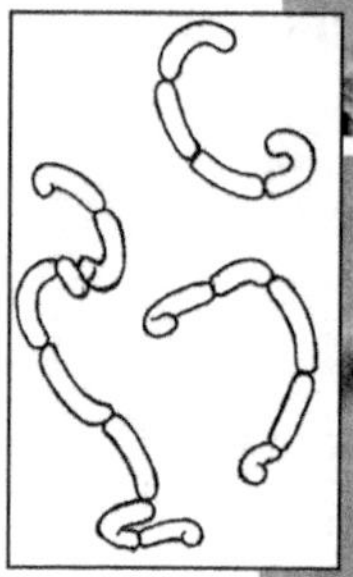

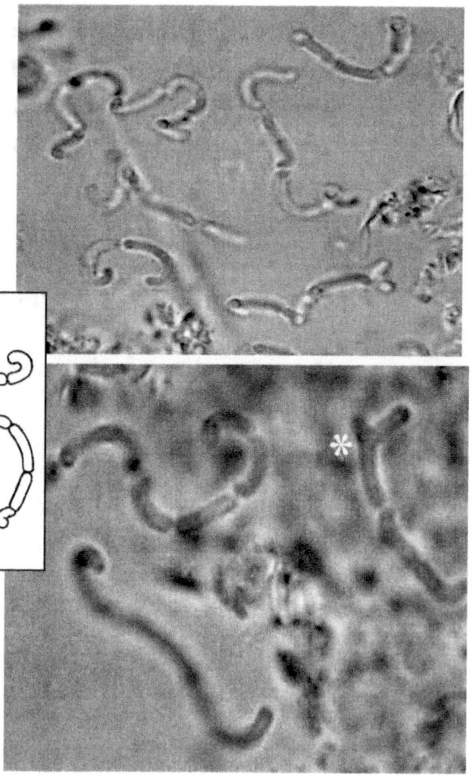

- **ecology**
  - * freshwater plankton and metaphyton (12)
    – mass developments (2 cases: S. Bohemia, Brazil – Caruaru)
  - * marine plankton (1)
  - * cryoseston (2)
  - * benthic – epipelic (2)
  - * benthic → metaphytic – mountain streams (1)
  - * aerophytic (on mosses) (1)

(number of species in parentheses)

From Komárek et al. in prep.

Table 4. Review of planktic species of the genus *Romeria*

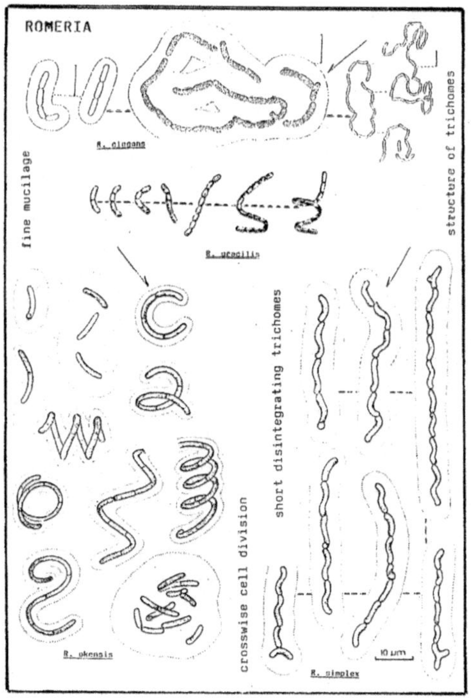

References: Geitler (1932), Komárek (1958), Hindák (1975)

Cells distinctly narrower than 1 µm (max. to 1 µm wide):					
	3.7-8.3 x 0.6-0.8		alascense	planktic in oxidation ponds	Alaska
	1-12.8 x 0.7-1		caruaru	planktic in tropical reserv.	Brazil (Pernambuco)
Cells ± 1 µm wide (0.8-1.3 µm):					
	1-1.5 x ± 1		cylindrocellulare	planktic in eutroph. ponds	Slovakia
	3-6 x 0.8-1.2		leopoliensis	planktic in ponds & lakes	temperate zones
	4.5-9.2 x 1-1.3		hieroglyphica	planktic, small ponds & lakes	Brazil (São Paulo)
Cells 1.2-1.6 µm wide:					
	4-12 x ±1.2-1.6		simplex	planktic, eutr. village ponds	Slovakia
	3.5-4.6 x 1.2-1.5		gracile	planktic in ponds & lakes	cosmopolit.
	4.5-13 x 1.5	± straight	victoriae	planktic in reserv. & lakes	trop. E & SE Africa
Cells to 2 µm wide (± 1.6 – 2.2 µm):					
	3-9 x 1.3-2	curved	elegans	planktic in ponds & rivers	temperate Eurasia
Cells > 2 µm wide (2-3 µm):					
	5.2-12 x 2-3		okensis	planktic, eutr. ponds & rivers	CZ,Russia, Hungary
	3-7 x 2-2.8		crassa	planktic in fishponds	Slovakia
	2.6-8.2 x 1.4-2.8		heterocellularis	planktic in large resevoirs	tropical; Brazil

From Komárek 2001

Table 5. Main planktic species of the genus *Pseudanabaena*

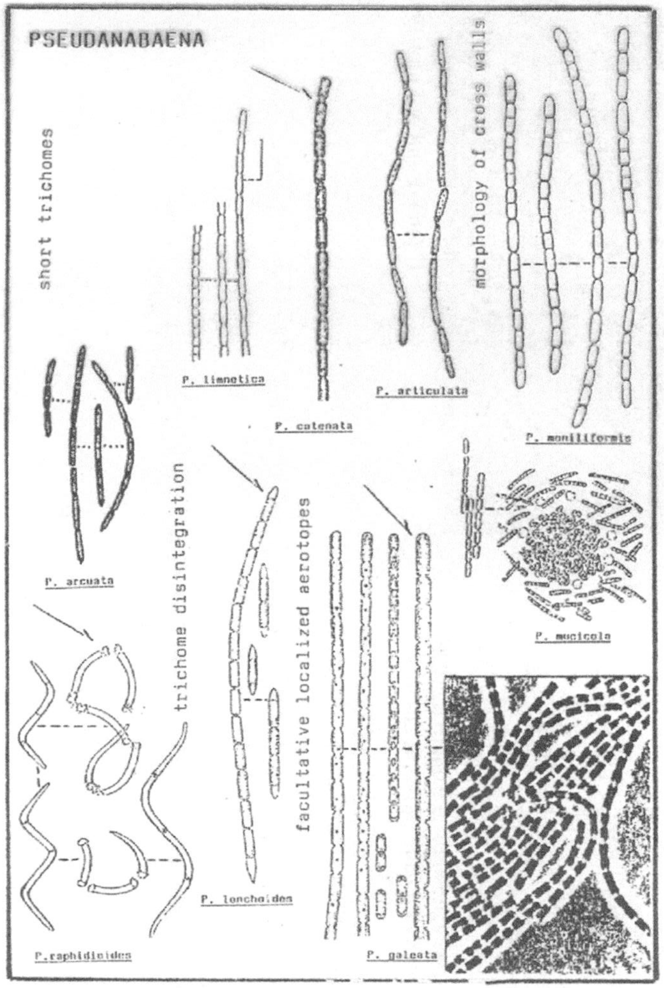

Review of planktic species:

A. **Cells without aerotopes, rounded apical cells, solitary in plankton.**
   P. catenata *
   P. limnetica *
   P. articulata *
   P. moniliformis *

B. **Cells without aerotopes, conical apical cells, solitary in plankton.**
   P. raphidioides *
   P. acicularis
   P. dictyothalla

C. **Endogloeic in planktic cyanobacteria, usually in short trichomes.**
   P. arcuata *
   P. franquetii
   P. woronichinii
   P. endophytica
   P. westiana
   P. mucicola
   P. rosea

D. **Cells with aerotopes, facultatively in plankton.**
   P. galeata *

* = see figures

References: Skuja (1948), Geitler (1932), Starmach (1966), Huber-Pestalozzi (1938), Komárek (1958), Geitler (1959), Anagnostidis (1961), Komárek & Kling (1991).

Table 6. Key of planktic species of the genus *Limnothrix*

Main generic features:
- **pseudanabaenacean structure of cells**
- **planktic type of life**
- **solitary trichomes**
- **cells longer than wide**
- **trichomes not constricted or very slightly constricted at cross-walls**
- **localized aerotopes in cells (polar or central)**
- **absence of sheaths**

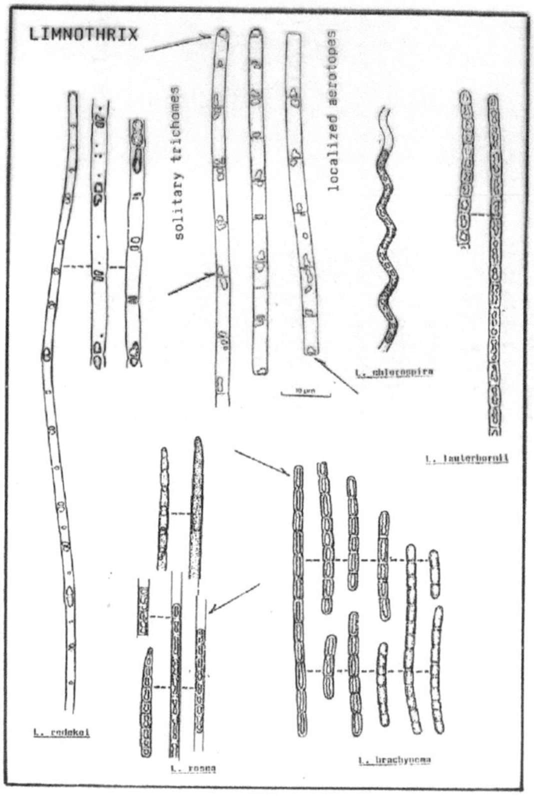

References: Anagnostidis & Komárek (1988; selected from different authors)

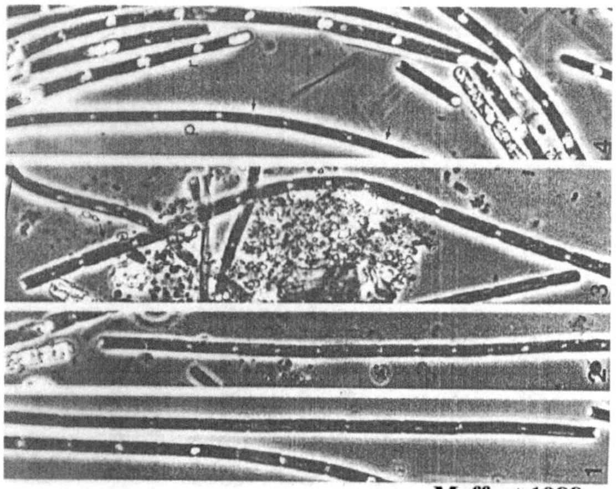

Meffert 1988

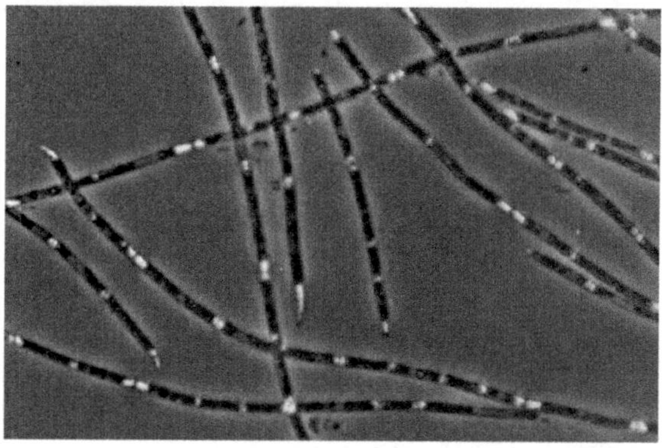

Canter-Lund & Lund 1995

List of main species:

***L. planctonica* ***	***L. brachynema* ***
L. meffertae	***L. hypolimnica***
***L. redekei* ***	***L. chlorospira* ***
L. obliqueacuminata	***L. pseudospirulina***
L. vacuolifera	

* = see figures

Table 7. Generic features and the key of the genus *Planktolyngbya*

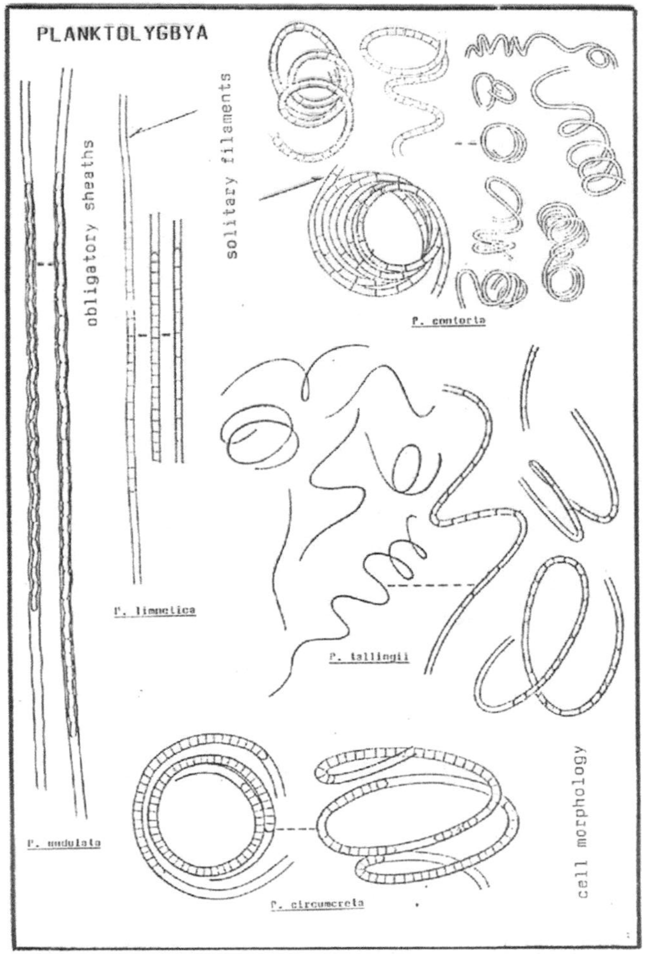

References: Geitler (1932), Kondrateva (1968), Hindák & Mustaka (1988), Komárek & Kling (1991)

Generic features:

- Pseudanabaenacean structure of cells and filaments
- Planktic type of life
- Solitary filaments
- Maximal width of filaments 2.8(4) µm
- Obligatory presence of sheaths

Diacritical interspecific characters:

- Type of coiling
- Shape and size of cells
- Ecology and distribution

Key to the main planktic species:

1. Straight filaments
   P. crassa (northern) – tr. 2.4-3.8 µm w.; cells 10-16 µm lg.
   P. brevicellularis (northern) – tr. ± 1.5 µm w.; cells 1-1.5 µm lg.
   P. limnetica - tr. 0.5–1.8 µm w.; cells (1)2-5 µm lg. *
   P. bipunctata – tr. 1-1.5 µm w.; cells 3.5-5.5 µm lg. (granules)
   P. minor (tropical) – tr. 0.6-0.8 µm w.; cells 2.5-5 µm lg.
   P. undulata (tropical) – tr. 0.8-1.2 µm w.; cells 1.6-3 µm lg. *

2. Irregularly coiled filaments
   P. circumcreta (tropical) – tr. 1.8-2.1 µm w.; cells 1-2.5 µm lg. *
   P. contorta – tr. 1-2 µm w.; cells 2.5-5 µm lg. *
   P. capillaris (northern) – tr. 0.5-0.9 µm w.; cells 2-6 µm lg.
   P. tallingii (tropical) – tr. 0.8-1.2 µm w.; cells 3.5-9.8 µm lg. *

3. ± regularly screw-like filaments
   P. microspira (tropical) – tr. 0.7-1 µm w.; cells 1.8-4.5 µm lg.
   P. regularis (tropical) – tr. 1.6-2.2 µm w.; cells 2.5-6 µm lg.

* = see figures

Table 8. Diagnostic features and key to planktic genera of Phormidiaceae (comp. Fig. 1)

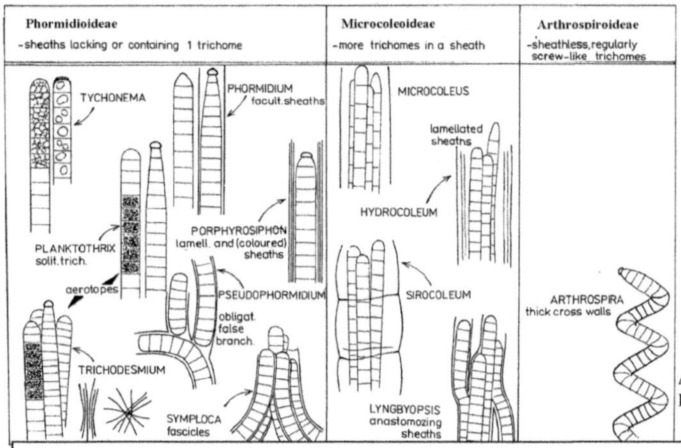

Thylakoid arragement:

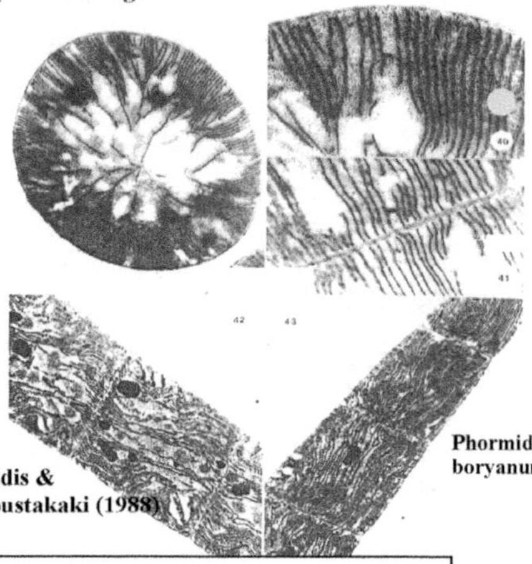

Key to the planktic genera:

1a Filaments ± straight or irregularly coiled **2**
 2a Trichomes solitary **3**
  3a Trichomes pale, always without gas vesicles (aerotopes), but with slightly net-like protoplasts ...... **Tychonema** (Tab. 9)
  3b Cells in trichomes at least facultatively with aerotopes (protoplasts not reticulate) **4**
   4a Trichomes without sheaths, but with mucilaginous envelopes .... **Katagnymene**
   4b Trichomes without or facultatively with sheaths, without envelopes .... **Planktothrix** (Tab. 10-12)
   Planktothricoides (Tab. 11)
  (**Note:** The genera Planktothrix and Planktothricoides were distinguished according molecular sequencing.)
 2b Trichomes joined to clusters or fascicles .... **Trichodesmium** (Tab. 13)

1b Filaments (trichomes) screw-like coiled, always without sheaths .... **Arthrospira** (Tab. 14)

Table 9. The genus *Tychonema* (planktic *T. bourrellyi*)

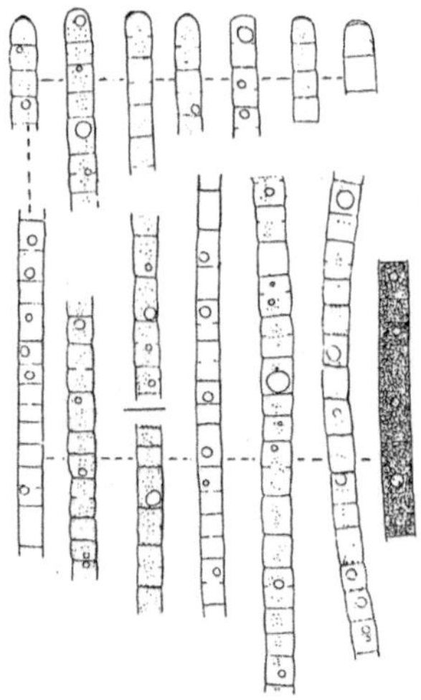

T. bourrellyi

planktic in mesotrophic northern lakes

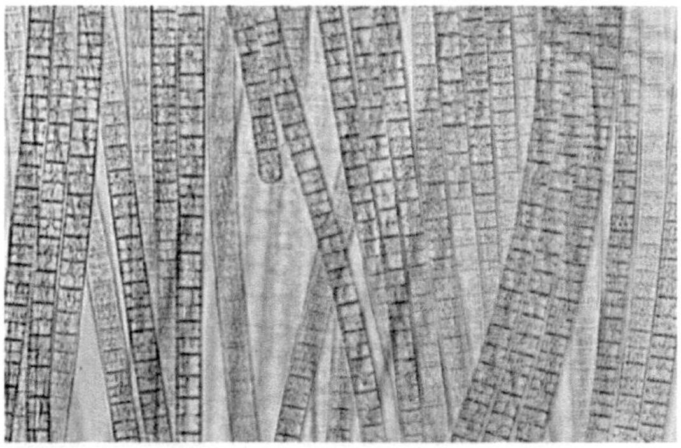

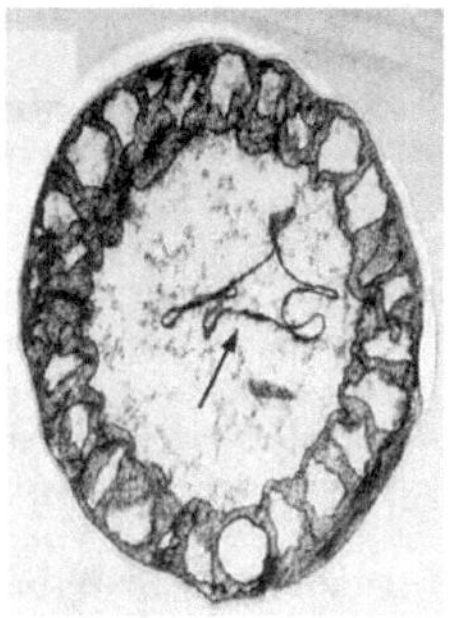

widened thylakoids (keritomy)

From Lund 1955, Canter-Lund & Lund 1995, Komárek & Albertano 1994

Table 10. The genus *Planktothrix*

Generic features:

- **separation according to molecular sequencing**
- **planktic type of life**
- **solitary trichomes**
- **facultatively (rarely) sheaths (usually only under stress conditions)**
- **obligatory presence of gas vesicles (and aerotopes)**
- **diazocytes**

Key to the main planktic species:

1a Filaments vavy and coiled, 7.5-11.5 µm wide, secondary in plankton .... **P. planctonica**
1b Filament straight or bent, free-floating **2**
2a Trichomes cylindrical, not narrowed at the ends **3**
3a Trichomes not constricted at cross-walls, (5)6-10 µm wide .... **P. mougeotii**
3b Trichomes constricted at cross-walls, 5-8 µm wide .... **P. clathrata**
2b Well developed trichomes slightly narrowed to the ends **4**
4a Trichomes red **5**
4b Trichomes more or less green **6**
5a Trichomes 2.2-5.8 µm wide .... **P. prolifica**
5b Trichomes 4-9 µm wide .... **P. rubescens**
6a Trichomes 2-4 µm wide, yellowish green .... **P. suspensa**
6b Trichomes mainly 3.5-6 µm wide .... **P. agardhii**

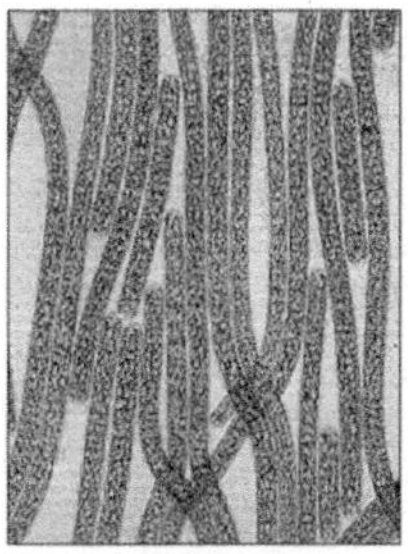

P. mougeotii

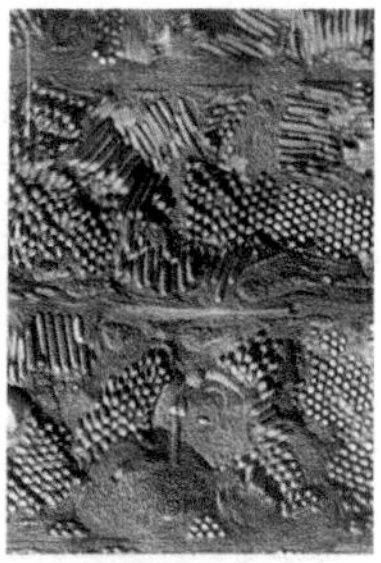

gas vesicles in cells

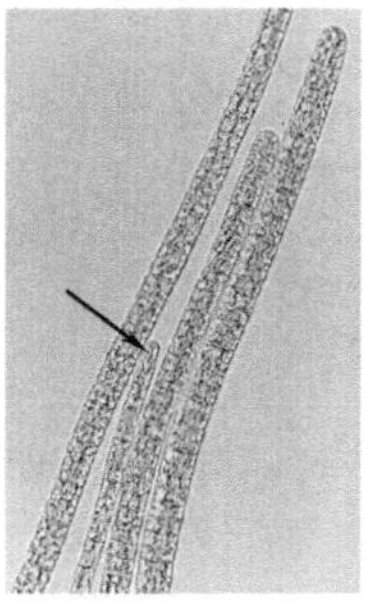

P. agardhii

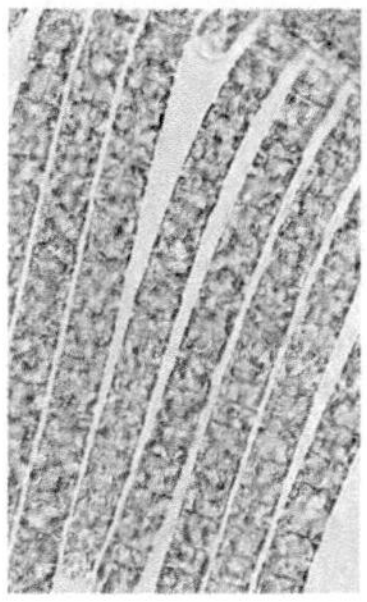

P. rubescens

Microphotos after Canter –Lund & Lund 1995 and Hindák 2001

Table 11. Evaluation of *Planktothrix* species and related planktic phormidiacean genera according to genetic, biochemical and ecophysiological charaters

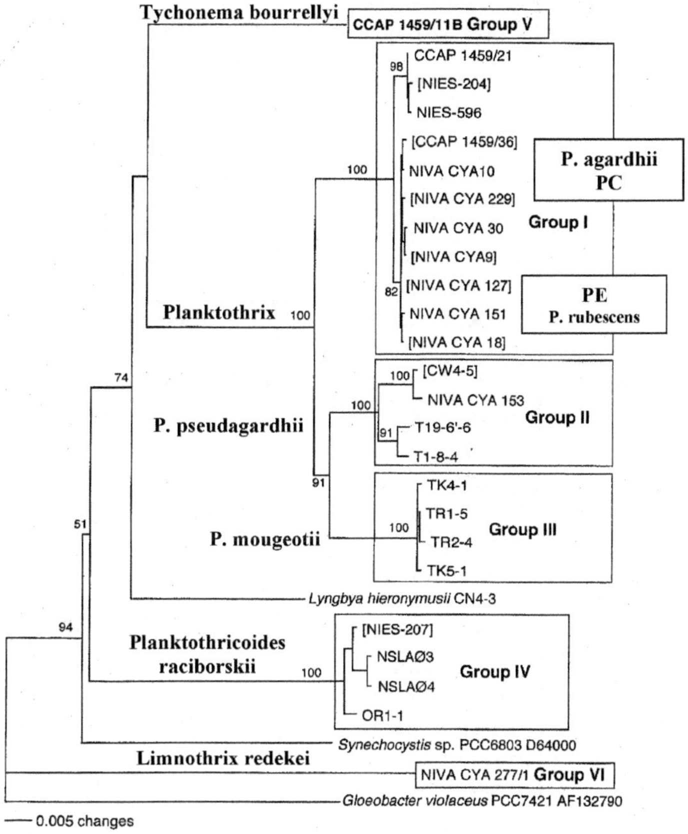

After Suda et al. 2002

Table 12. Separation of *Planktothrix agardhii* and *P. rubescens*

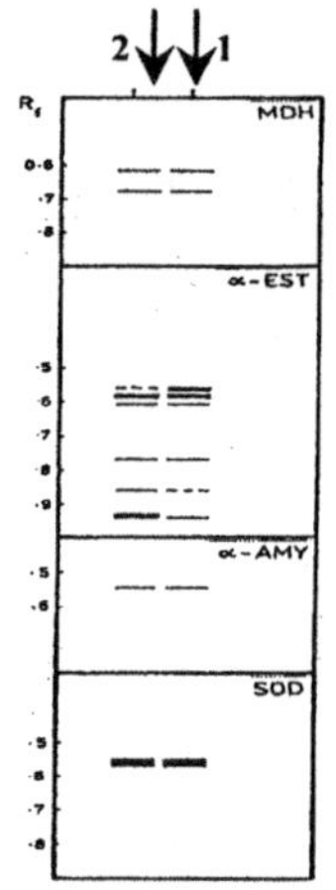

isozyme analysis

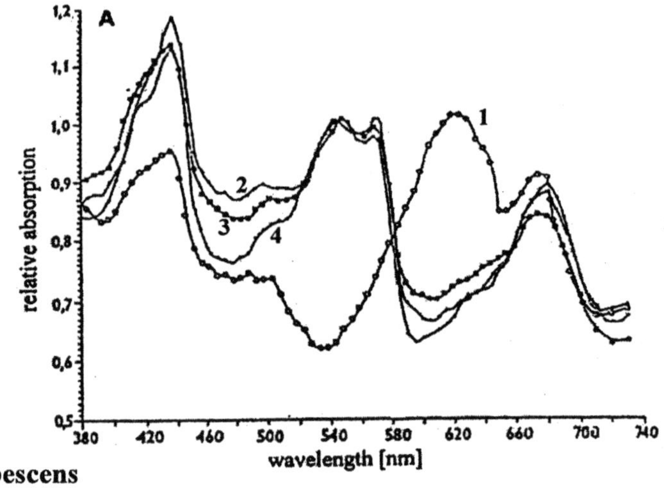

1 - P. rubescens
2-4 - P. agardhii

List of *Planktothrix* species:

	Species	Distribution	Trichome width [µm]	Note
	zahidii	Pakistan	2.5-3.5	straight trichomes
	suspensa	temperate zones	2.0-4.0	calyptrate
	prolifica	northern Europe, central North America	2.0-5.8	red ("red agardhii"), calyptrate
	clathrata	northern parts of temperate zone	5.0-6.0	
→	***agardhii***	temperate and tropical zones	(2.8?)4.0-6.0(9.5?)	narrowed ends, calyptrate
	cryptovaginata	eastern Europe	(4.0)5.0-8.5	facultative sheaths
→	***rubescens***	central and north Europe, central North America	(4.0)6.0-8.2(9.0)	red, calyptrate
	mougeotii	cosmopolitan (excl. polar regions)	(5.0)5.5-9.7(10?)	cylindrical ends
	planctonica	eastern Europe	8.0-10.0	slightly coiled, cylindrical trichomes
	geitleri	scarcely in tropical and southern temperate zone	(6.0)7.0-10.0(11.0)	narrowed straight ends
*	***raciborskii***	eastern Asia, tropical and subtropical Asia, Africa	6.0-9.0(11.0)	narrowed and curved ends

* = *Planktothricoides*

Table 13. The genus *Trichodesmium*

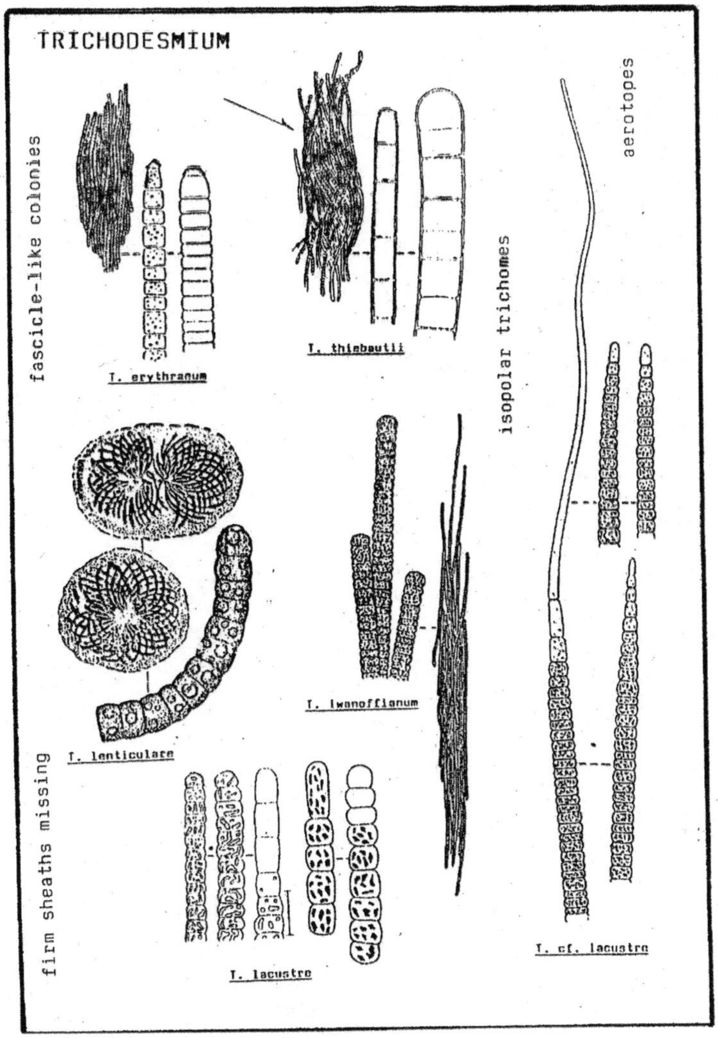

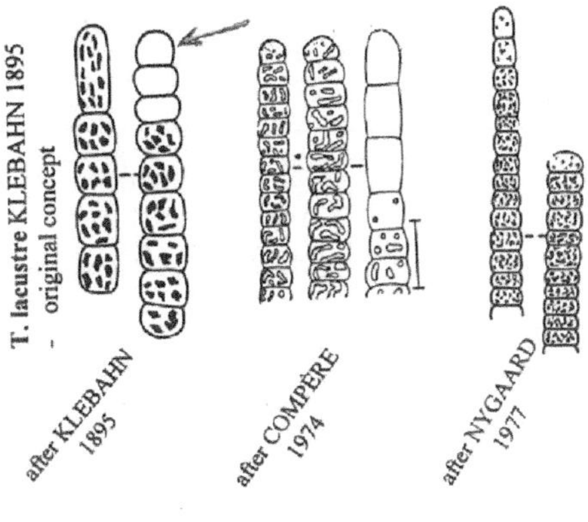

References: Kosinskaja (1948), Kondrateva (1968), Nygaard (1977), Lundgren & al. (2001), Orcutt & al. (2002).

Key to the freshwater species:

1a Terminal cells in well developed trichomes hyaline, sometimes slightly elongated .... **T. lacustre**

1b Terminal cells similar to other vegetative cells **2**

2a Cells always shorter than wide, mainly tropical species .... **T. iwanoffianum**

2b Cells ± isodiametrical .... **T. lacustre sensu G.M.Smith 1920**

Note: Marine species are not included in the key.

Table 14. Planktic species of the genus *Arthrospira*

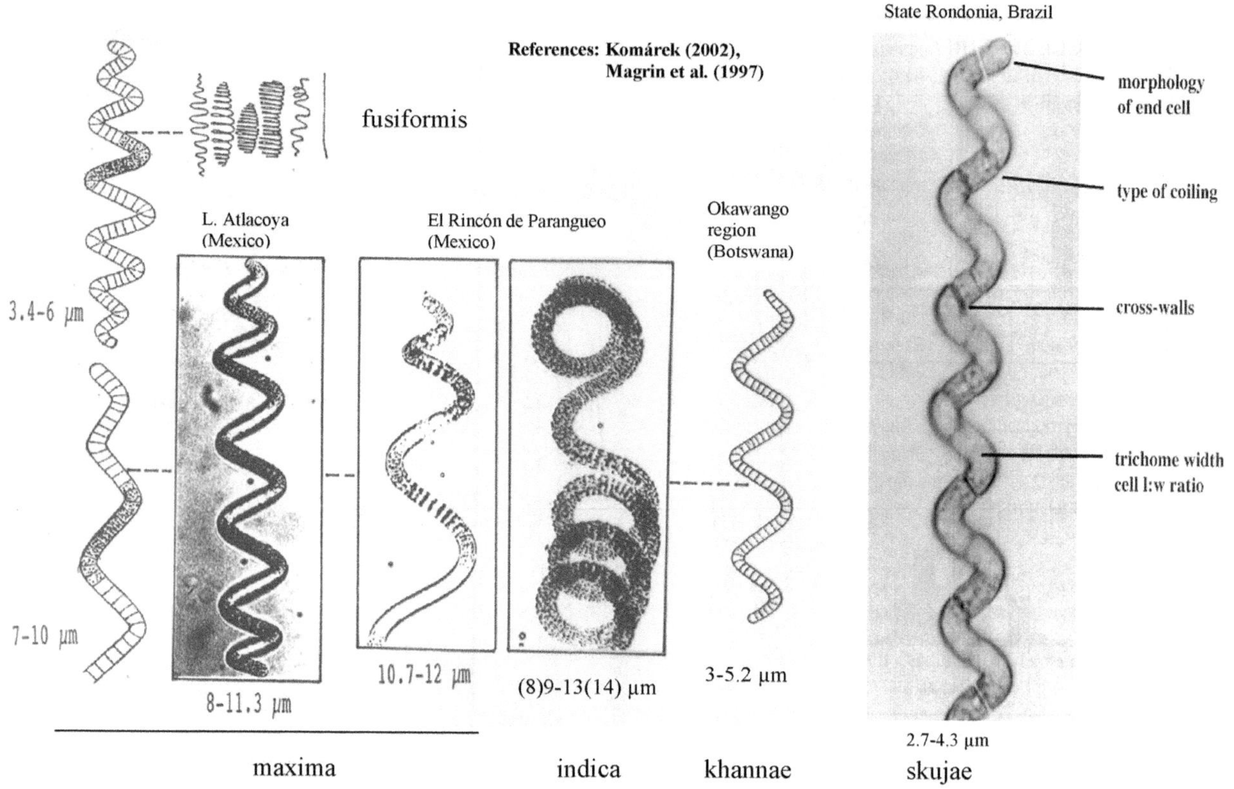

Table 15. Planktic *Lyngbya* species (Oscillatoriaceae)

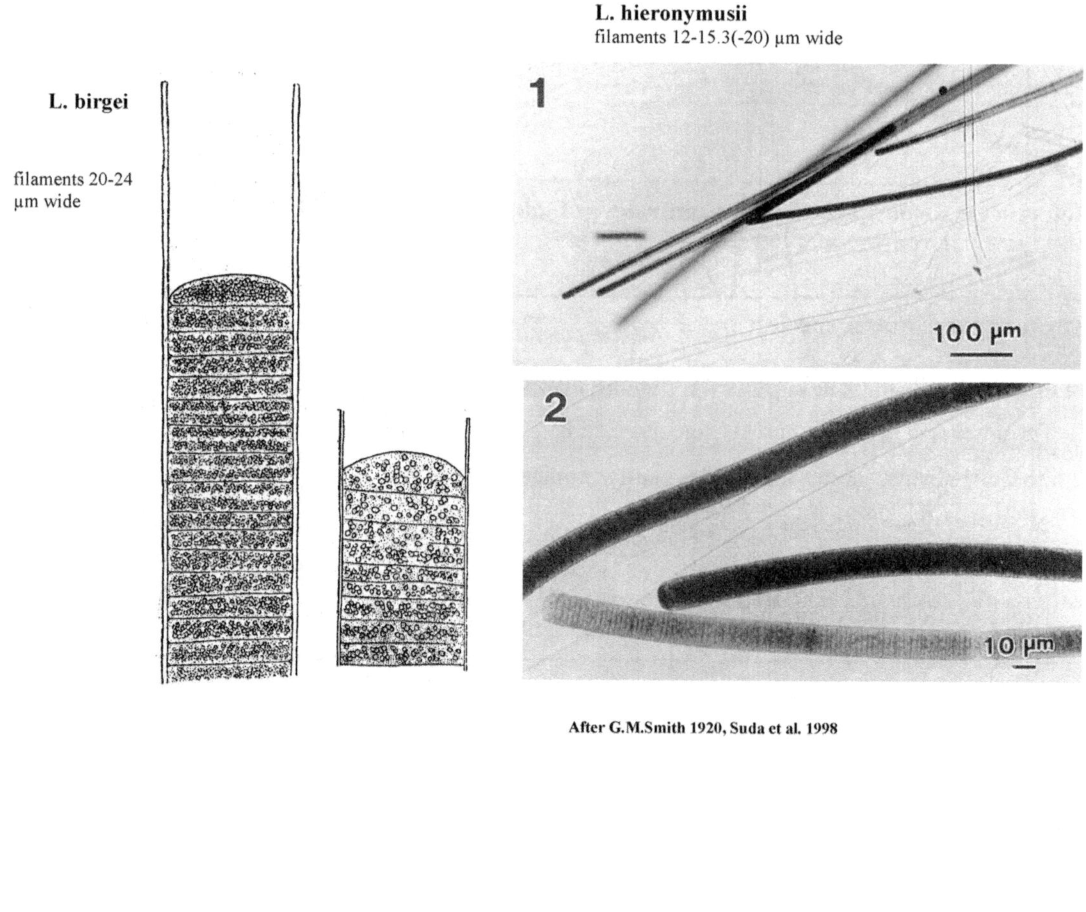

After G.M.Smith 1920, Suda et al. 1998

References

Anagnostidis, K. & J. Komárek, 1985. Modern approach to the classification system of cyanophytes. 1. Introduction. Archiv für Hydrobiol./Algol. Stud. 38/39: 291–302.

Anagnostidis, K. & J. Komárek, 1988. Modern approach to the classification system of cyanophytes. 4. Nostocales. Archiv für Hydrobiol./Algol. Stud. 56: 517–526.

Barker, G. L. A., B. A. Handley, P. Vacharapiyasophon, J. R. Stevens & P. K. Hayes, 2000. Allele-specific PCR showes that genetic exchange occurs among genetically diverse *Nodularia* (Cyanobacteria) filaments in the Baltic Sea. Microbiology 146: 2865–2875.

Canter-Lund, H. & J. W. G. Lund, 1995. Freshwater algae: Their microscopic world explored. Biopress, Bristol, U.K.

Castenholz, R.W., 2001. Phylum BX. Cyanobacteria. Oxygenic Photosynthetic Bacteria. In Boone, D. R. & R. W. Castenholz (eds), Bergey's Manual of Systematic Bacteriology, 2nd Edition, Springer. 473–599.

Castenholz, R. W. & J. B. Waterbury, 1989. Group I. Cyanobacteria. In Staley, J. T. et al. (eds), Bergey's Manual of Systematic Bacteriology, Vol.3, Section 19. 1710–1727.

DiCastri, F. & T. Younèz, 1994. Diversitas: Yesterday, today and a path towards the future. Biol. Int. 29: 3–23.

Flechtner, V. R., S. L. Boyer, J. R. Johansen & M. L. DeNoble, 2002. *Spirirestis rafaelensis* gen. et sp. nov. (Cyanophyceae), a new cyanobacterial genus from arid soils. Nova Hedwigia 74: 1–24.

Garcia-Pichel, F., U. Nübel & G. Muyzer, 1998. The phylogeny of unicellular, extremely halotolerant cyanobacteria. Arch. Microbiol. 169: 469–482.

Geitler, L., 1959. *Spirulina raphidioides* n. sp. und Bemerkungen über ähnliche Planktonalgen. Österreich. botanische Zeitschrift 106: 133–137.

Hagemann, M., 2002. Environmental stress, signalling and basic acclimation reactions. In Solheim, B. et al. (eds), Cyanobacteria and Nitrogen Fixation in Extreme Environments. European Science Foundation CYANOFIX. p. 24.

Hindák, F., 1975. Einige neue und interessante Planktonblaualgen aus der Westslowakei. Archiv für Hydrobiol./Algol. Stud. 15: 330–353.

Hindák, F., 2001. Fotografický atlas mikroskopických siníc. [Microscopic cyanobacteria – microphotos.] Veda Bratislava. 127 pp.

Hindák, F. & M. T. Moustaka, 1988. Planktic cyanophytes of Lake Volvi, Greece. Archiv für Hydrobiol./Algol. Stud. 50/53: 497–528.

Kato, T., M. F. Watanabe & M. Watanabe 1991. Allozyme divergence in *Microcystis* (Cyanophyceae) and its taxonomic interference. Archiv für Hydrobiol./Algol. Stud. 64: 129–140.

Kohl, J. G. & A. Nicklisch, 1981. Chromatic adaptation of the planktonic blue-green algae *Oscillatoria redekei* Van Goor and its ecological significance. Int. Rev. ges. Hydrobiol. 66: 83–94.

Komárek, J., 1958. Die taxonomische Revision der planktischen Blaualgen der Tschechoslowakei. In Komárek, J. & H. Ettl (eds), Algologische Studien. Academia, Praha: 10–206.

Komárek, J., 2002. Problems in cyanobacterial taxonomy; implication for most common toxin producing species. In Melchiorre, S., E. Viaggiu & M. Bruno (eds), Rapporti ISTISAN (Istituto Superiore di Sanitá). Roma: 6–43.

Komárek, J. & P. Albertano, 1994. Cell structure of a planktic cyanoprokaryote, *Tychonema bourrellyi.* Archiv für Hydrobiol./Algol. Stud. 75: 157–166.

Komárek, J. & J. Kaštovský, 2003a. Adaptability in diversification processes of cyanobacteria; the example of *Synechococcus bigranulatus*. Archiv für Hydrobiol./Algol. Stud. 109: 299–304 (Papers of Cyanobacterial Research 4).

Komárek, J. & J. Kaštovský, 2003b. Coincidences of structural and molecular characters in evolutionary lines in cyanobacteria. Archiv für Hydrobiol./Algol. Stud. 109: 305–326 (Papers Cyanobact. Research 4).

Komárek, J., H. J. Kling & J. Komárková, 2003. Filamentous cyanobacteria. In Wehr, J. D. & R. G. Sheath (eds), Freshwater Algae of North America: Ecology and Classification. Academic Press, Massachusetts. 117–196.

Kosinskaja, E. K., 1948. Opredelitel' morskich sinezelenych vodoroslej. [Determination key for marine blue-green algae.], Izd. AN SSSR, Moskva-Leningrad. 278 pp.

Lund, J. W. G., 1955. Contribution to our knowledge of British algae. XIV. Three new species from the English Lake District. Hydrobiologia 7: 219–229.

Lundgren, P., E. Söderbäck, A. Singer, E. J. Carpenter & B. Bergman, 2001. *Katagnymene:* characterization of a novel marine diazotroph. J. Phycology 37: 1052–1062.

Magrin, A. G. E., P. A. C. Senna & J. Komárek, 1997. *Arthrospira skujae,* a new planktic tropical cyanoprokaryote. Archiv für Protistenkunde 148: 479–489.

Meffert, M. E., 1988. *Limnothrix* Meffert nov. gen. The unsheathed planktic cyanophycean filaments with polar and central gas vacuoles. Archiv für Hydrobiol./Algol. Stud. 50/53: 269–276.

Nygaard, G., 1977. New or interesting plankton algae. Kongelige danske videnskaberne Selskab, Biol. Skifter 21: 1–107.

Orcutt, K. M., U. Rasmussen, E. A. Webb, J. B. Waterbury, K. Gundersen & B. Bergman 2002. Characterization of *Trichodesmium* spp. by Genetic Techniques. Appl. Environ. Microbiol. 68: 2236–2245.

Pennisi, E., 1994. Static evolution. Science News, Washington 145: 168–169.

Rippka, R., J. B. Deruelles, J. B. Waterbury, M. Herdman & R. Y. Stanier, 1979. Generic assignments, strain histories and properties of pure cultures of cyanobacteria. J. gen. Microbiol. 111:1–61.

Rudi, K., O. M. Skulberg & K. S. Jakobsen, 1998. Evolution of cyanobacteria by exchange of genetic material among phyletically related strains. J. Bacteriol. 180: 3453–3461.

Schopf, J. W., 1974. The development and diversification of Precambrian life. Origins of Life 5: 119–135.

Schopf, J. W., 1996. Cyanobacteria: pioneers of the early Earth. Nova Hedwigia 112: 13–32.

Skuja, H., 1948. Taxonomie des Phytoplanktons einiger Seen in Uppland, Sweden. Symb. Bot. Ups. 9: 1–399.

Smith, G. M., 1920. Phytoplankton of the Inland Lakes of Wisconsin I. Wisconsin Geology and Natural History Surveys 57, Series on science 12: 1–243.

Suda, S., Y. Liu, J. He, Z. Hu, M. Hiroki & M. M. Watanabe, 1998. Morphological, biochemical and physiological characteristics of *Lyngbya hieronymusii* var. *hieronymusii* (Oscillatoriales, Cyanobacteria). Phycol. Res. 46 (Suppl): 51–55.

Suda, S., M. M. Watanabe, S. Otsuka, A. Mahakahant, W. Yongmanitchai, N. Nopartnaraporn, Y. Liu & J. G. Day, 2002. Taxonomic revision of water bloom-forming species of oscillatorioid cyanobacteria. Int. J. Syst. Evol. Microbiol. 52: 1577–1595.

Taton, A., S. Grubisic & A. Wilmotte, 2002. Cyanobacterial diversity in Antarctic microbial mats. In Solheim, B. et al. (eds), Cyanobacteria and Nitrogen Fixation in Extreme Environments. Eur. Sci. Found. CYANOFIX, Longyearbyen. p. 22.

Watanabe, M. M., 1999. Network approach to make biodiversity information accesible worldwide. As an example of microorganisms. In Species 2000, Abstracts of the International Workshop for Studies on Biodiversity, Tsukuba. p. 13.

Wilmotte, A. & S. Golubić, 1991. Morphological and genetic criteria in the taxonomy of Cyanophyta/Cyanobacteria. Archiv für Hydrobiol./Algol. Stud. 64: 1–24.

Hydrobiologia **502:** 383–388, 2003.
L. Naselli-Flores, J. Padisák, M. T. Dokulil (eds),
Phytoplankton and Equilibrium Concept: The Ecology of Steady-State Assemblages.

A new planktic species of *Pseudanabaena* (Cyanoprokaryota, Oscillatoriales) from North American large lakes

Hedy J. Kling[1] & Susan Watson[2]
[1]*Algal Taxonomy and Ecology Inc, 31 Laval Dr. Winnipeg MB R3T 2X8. E-mail: hkling@mts.net*
[2]*National Water Research Institute, 867 Lakeshore Road, Burlington, ON L7R 4A6*

Key words: Cyanoprokaryota, Oscillatoriales, *Pseudanabaena*, planktic, large lakes, temperate

Abstract

A new species of *Pseudanabaena* (Cyanoprokaryota, Oscillatoriales) from the plankton of North American large lakes is documented, with ecological and distributional data. This taxon differs from previously described *Pseudanabaena* species with respect to a characteristic undulating and coiling filament structure. *Pseudanabaena contorta* spec. nova appears to be specific to large North American Great Lakes with a drainage basin that is not totally Precambrian shield. Recently, it has been found as a consistent but minor component of the planktic cyanoprokaryote populations during the spring and summer months in Lake Superior and Lake Ontario. Comparisons are made with other common species of this genus as well as another undulating morphotype of the genus *Glaucospira,* that could possibly result in confusion, found in other large Alberta and northern Saskatchewan lakes.

Introduction

The genus *Pseudanabaena* first was established in 1915 by Lauterborn, and has since undergone several revisions. Initially the genus contained only two species (*P. catenata* Lauterb. and *P. constricta* (Szafer) Lauterb.), but following detailed characterization by Chang et al. (1985) and others, encompassed >33 species and forms. More recently, Anagnostidis & Komárek (1988) divided the species described to date among two genera and five subgenera. *Pseudanabaena* species are morphologically simple members of the Oscillatoriales commonly present in low numbers in the plankton of many lakes. The taxonomy of this genus is still unresolved. Although its members occur in many different habitats, their morphological variability and range of ecological requirements are relatively unknown. Similar morphotypes may occur under very dissimilar ecological conditions, while different morphotypes sometimes occur under the same conditions, leading to taxonomic confusion. Individual species can demonstrate considerable morphological variability. Romo et al. (1993) observed a range of morphotypes in clonal cultures of *Pseudanabaena galeata* Böcher (subgenus *Ilyonema*), especially in end cell morphology, length to width ratio and sheath production. In addition to this ambiguity within the genus *Pseudanabaena*, there is confusion between this species and such morphologically similar genera as *Limnothrix* (Meffert 1987). Here, we describe a new, coiled morphological species of *Pseudanabaena* from North American Great Lakes and comment on the genus *Glaucospira* another undulating morphotype of filamentous cyanoprokaryote found in large western prairie lakes.

Methods

Samples were collected during routine sampling for phytoplankton and water chemistry by the Dept. of Fisheries and Oceans (DFO), Large Lakes Observatory (LLO; Duluth, MN) and Environment Canada (EC). Subsamples were preserved immediately with Lugol's Iodine for subsequent identification and enumeration.

Analyses were undertaken on an M-40 Wild Inverted Microscope, using Utermöhl chambers and standard methods (Findlay & Kling, 1996). The coiled morphotype of *Pseudanabaena* first was observed in

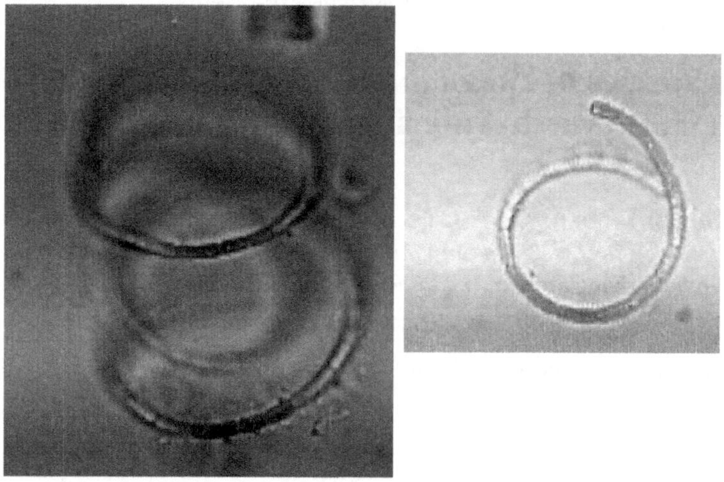

Figures 1 and 2. Coiling morphotypes. Specimens from Lake Ontario. Short and long cells in filaments; granules (gas vacuoles?) apparent in most cells.

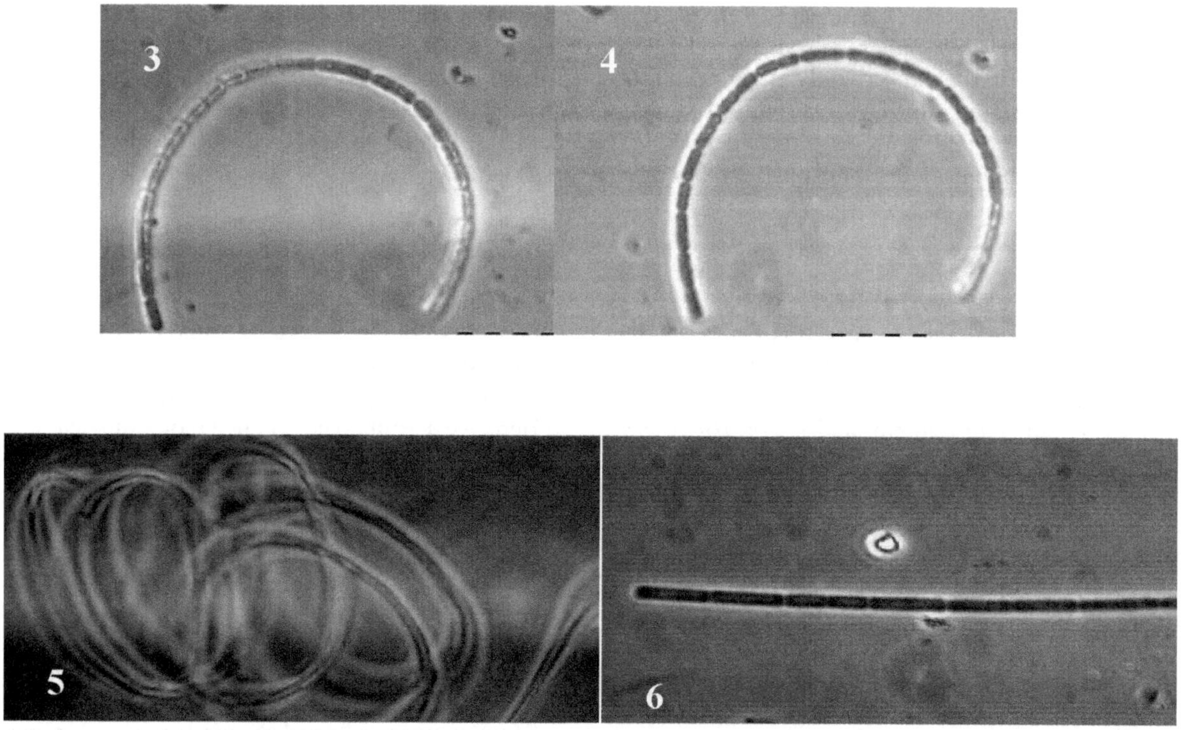

Figures 3–6. Specimens from Lake Superior. 3 and 4: *Coiling morphotype* (2 foci for the same filament), granules (gas vacuoles) present in most cells; 5 and 6: Coiling morphotype of new taxon compared to straight morphotype of *Pseudanabaena limnetica.*

samples from Lake Superior offshore of Thunder Bay in 1990 – 91, during a DFO study of large lakes (Kling, unpub. data). In 1999 and 2000, this coiled filamentous cyanobacterium was observed in samples from the Western tip of Lake Superior (Duluth; Fig. 9), while in 2001, a similar morphotype was observed in samples from Western Lake Ontario (Ontario). Many previous records of phytoplankton in the Great Lakes (1970 – present) have recorded an *Oscillatoria* sp., that may have included this coiled species under its umbrella. The past data used to show depth distribution in Lake Ontario (Fig. 10) was made available from the Environment Canada collection (S. Watson unpub. data).

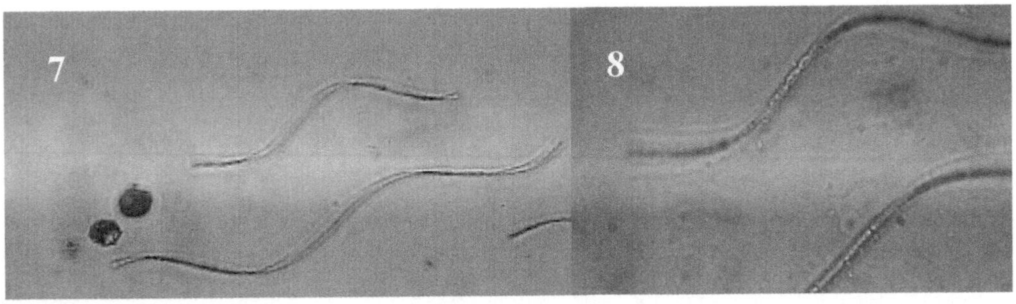

Figures 7 and 8. Undulating morphotype. Specimens from experimental enclosures, Calgary (S. Watson unpubl. data) with morphology similar to the Lac La Ronge morphotypes. Straight to low undulations with some granules (gas vacuoles?) near cell ends.

Another undulating morphotype (included here for comparison) belonging possibly to the genus *Glaucospira* Lagerheim (Komarek pers. comm.) but originally thought to belong to this *Pseudanabaena* genus (Figs 7 & 8) was found in Reindeer Lake, Lac LaRonge (Saskatchewan), Lake Athabasca and enclosure experiments near Glenmore Reservoir Calgary, Alberta (S. Watson). These habitats are very different ecologically from Eastern Great Lake where the *Pseudanabaena* contorta was found.

Results and discussion

We observed a new filamentous cyanoprokaryote species belonging to the Oscillatoriales in recent plankton samples from the Great Lakes. This species has been included under the genus *Pseudanabaena* subgenus *Pseudanabaena*, following the revised classification (Anagnostidis & Komárek, 1988).

Along with the common planktic species *Pseudanabaena limnetica* (Lemm) Kom. (straight trichomes) (Fig. 6), unidentified coiled and undulating morphotypes occurred in the plankton of Lake Superior and other Great Lakes such as Lake Ontario.

These coiled morphotypes do not fit the description of any *Pseudanabaena* species previously described from this group (Table 1; adapted from Komárek & Kling, 1991). The morphology of coiled populations seems to be relatively stable within lakes, and among the large lakes ie lakes and warrants being described as a new species. Lake Superior and Lake Ontario morphotypes were very similar in coiling morphology and cell size (Figs 1–5).

Species description

Habitat: plankton of North American Great Lakes.
Description: Trichomes solitary, irregularly undulating to spirally coiled 50– >200 μm long and up to 2 (2.2) μm wide, without mucilage or sheath, deeply constricted at cross walls. Cells elongated, cylindrical, with ends rounded or bluntly pointed, pale homogenous, blue-green contents, sometimes with a granule (aerotopes?) at the cross-walls, 2.5–7 (9) μm long × 1.6 (2.2) μm wide. Trichome coils 28–30 μm in diameter. Cell division perpendicular, trichomes separate by cell disintegration or fragmentation

This new species has similar cell dimensions and coiling pattern to the coiled species in the genus *Planktolyngbya* (*P. circumcreta* (G.S. West) Anagn. & Kom., *P. contorta* (Lemm.) Anagn. & Kom., *P. capillaris* (Hind.) Anagn. & Kom., and *P. tallingii* Kom. & Kling). The taxa that are most similar, especially in trichome morphology, are *Planktolyngbya contorta* and *P. tallingii*. Our species has deep constrictions at the cross walls and has not been found with a sheath.

Diagnosis: Trichomata solitaria, libere natantia, irregulariter undulata vel spiraliter contorta, 50- >200 um longa, ad 2(2.2) um lata, distincte constricta ad dissepimenta; mucus et vagina carentes. Cellulae elongatae, cylindricae, cum apicalibus rotundatis vel rotundate-obtuse acutis, 2.5–7(9) um longae, 1.6–2.2 um latae, contentu homogeneo, aeruginoso, interdum cum granulis facultativis ad dissepimentibus. Spirae trichomatibus 28–30 um diametro. Divisio cellularum perpendiculariter, reproductio trichomatibus disintegratione vel fragmentatione. Heterocytae akinetesque carentes. - Habitatio: Planktice in lacubus grandis; locus classicus: lacus Lake Superior prope oppido Thunder Bay, Canada centralis; VII. 1990 (legit H. Kling et S. Watson). - **Iconotypus: figura nostra 3, 4.**

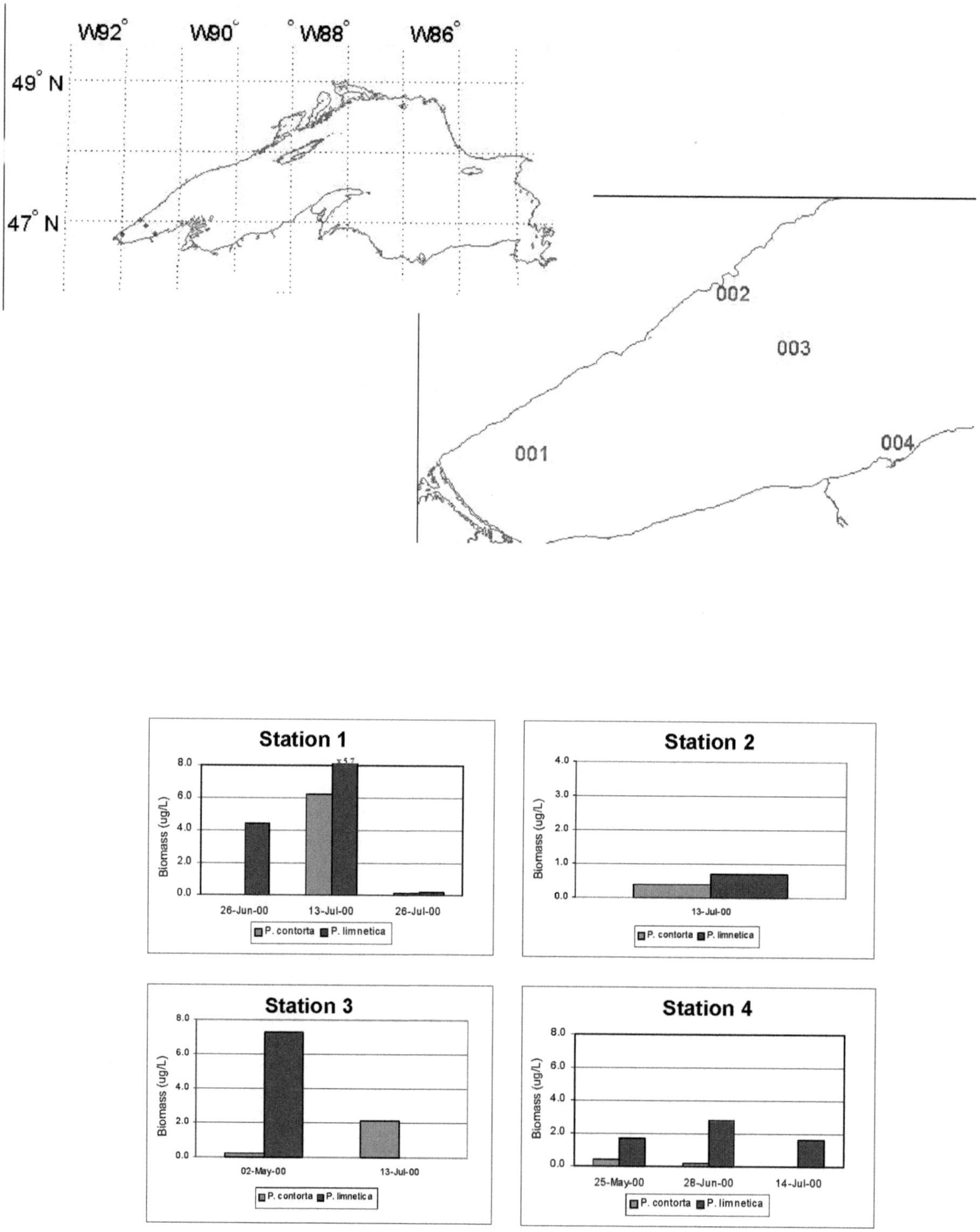

Figure 9. Sampling sites (top) and distribution of *Pseudanabaena. contorta* spec nova and *P. limnetica* (bottom) in Lake Superior, July 2000.

Table 1. Comparison of *Pseudanabaena* species (subg. *Pseudanabaena*; modified from Komárek & Kling 1991)

	Ps. catenata	*Ps. articulata*	*Ps. limnetica*	*Ps. tenuis*	*Ps. moniliformis*	*Ps.* sp.1	*Ps.* sp.2	*Ps. contorta*
Filament length (μm)	–200	40–120(300)	–120(–300)					>28–>100
Cell length (μm)	2–4(7)	5–11	(1.2)4–12	6–8	2.5–5.6	1.2–2.8(7)	–2-3(4)	2.5–7(9)
Cell width (μm)	1.3–1.5(2)	1–1.3	1–1.5(2)	1–1.5	+/–2	0.9–1.2	0.8–1(1.4)	1.4–1.6(2)
Number of cells in trichome		4–40			–18(–26)	–>50	–6	>50
Cell contents	Pale blue-green	Pale grey-blue	Pale blue-green	Pale blue-green	Pale blue-green or olive green	Pale blue-green	Pale yellow-green with several prominent granules	Pale grey-blue-green with or with out granules
Ecology	Benthic or tychoplanktic slightly eutrophic water bodies	Planktic in lakes	Planktic, slightly eutrophic water, lakes reservoirs	Benthic in lakes	Planktic in large lakes	Planktic in large lakes	Planktic in large lakes	Planktic in large Deep temperate lakes
Distribution	Temperate zone	Northern Europe	Temperate zone, India?	Northern Europe	East African Lakes	East African Lakes	East Africa Lakes	North American Large Lakes

Isotypi: preserved sample Cana #75387 Canadian National Museum, Ottawa Ontario and 011Kling, Freshwater Institute, Winnipeg Manitoba.

Slightly thinner undulating morphotypes most likely belonging to the genus *Glaucospira* Lagerheim were also identified from large more nutrient rich, light limited shallow lakes such as Lake Athabasca, Reindeer Lake and Lac La Ronge, and some experimental enclosures (Calgary, S. Watson, unpub. data). These *Glaucospira* morphotypes from Lac La Ronge and the experimental enclosures (Calgary, Alberta) with undulating trichomes (Figs 7 and 8), had an average length of ca. 70 μm (cell sizes 4.2 – 5.6 × 1.4 (2) μm). Reindeer Lake trichomes were similar, with large undulations or waves, ca. 60–70 μm long, cells 4.2 – 5.6 × 1.4 (2) μm. Lake Athabasca morphotypes were broken up into mainly short curved trichomes with average length of ca. 28 μm; 1.4 μm wide. These will be dealt with further in a later publication.

Distribution and seasonality

In Lake Superior coiling *Pseudanabaena contorta* (Figs 3–5) and *Pseudanabaena limnetica* (Fig. 6) were most prevalent at the western basin station near Duluth during late May through to August with a maximum biomass record in mid July 2000 (Fig. 9). *Pseudanabaena contorta* was present in low biomass

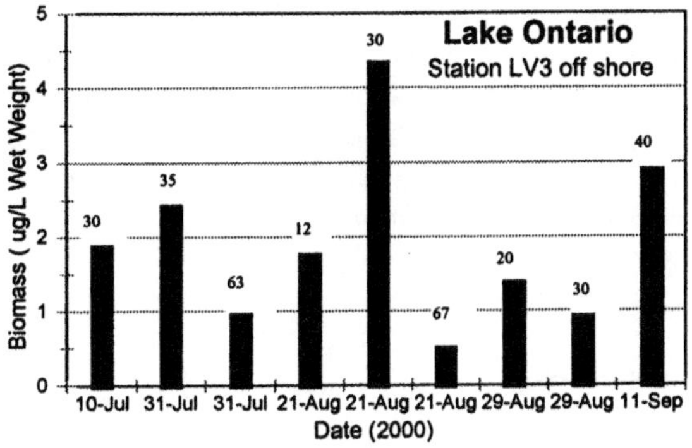

Figure 10. The distribution of P. contorta n. sp. showing the depth (labels above bars) of maximum biomass at Station LV3 off shore Lake Ontario during 2000.

(<6 μg/l) relative to the biomass (maximum 45 μg/l) of *P. limnetica*. In Lake Ontario, these coiling morphotypes of *P. contorta* also were most abundant in late summer and reached their highest density deeper in the water column below 5 m depth (Fig. 10).

The geographic distribution seems to be limited, but is relatively unknown at present. We have recorded it only in low numbers in samples from Lake Superior, L. Ontario and it may be endemic to these types of water bodies. The undulating & coiling morphotypes typical of this taxon seem to prefer the large lake habitat, as they were not recorded from smaller lakes sampled at the same time in the vicinities of these large lakes.

Conclusions

The most common species in these lakes is *Pseudanabaena limnetica*, with a straight trichome which is relatively constant on morphology within a lake but somewhat variable among lakes. This coiling species *Pseudanabaena contorta* sp. nova is present at low abundance in the North American Great lakes. *Pseudanabaena contorta* seems to be limited in distribution, preferring the pelagic regions of the Great Lakes that may be either light- or nutrient- limited and a neutral to high pH. At present, we do not have undulating or coiling morphotypes in culture and do not know the range of morphological variability that can be expected. We also do not know whether or not it has the potential for taste/odour (T/O) production with is a common occurrence among several members of this genus. Cultures of some more common, more eutrophic species of the genus (particularly in the subgenus *Ilyonema*) such as *P. galeata* can produce potent odour compounds such as 2-methylisoborneol (MIB) and geosmin (S. Watson & G. Izaguirre, pers. comm). It maybe possible to answer some of these questions through more *in situ* observations and culture experiments.

Acknowledgements

We are especially indebted to Dr Jiří Komárek for critical comments and translating the species description into Latin. We would also like to extend our thanks Dr Tom Johnson leader of the Large Lakes Observatory, Duluth, Minnesota for supporting the phytoplankton work in Lake Superior Western Basin, Drs Derek Muir and Stephanie Guildford, and Environment Canada (EC) for including phytoplankton as part of their research on bioaccumulation of organic pollutants for the Saskatchewan and Alberta lakes and Murray Charlton and EC for support and collection of Lake Ontario samples.

References

Anagnostidis, K. & J. Komárek, 1988. Modern approach to the classification of cyanophytes 3. Oscillatoriales. Arch. Hydrobiol./Algol. Stud. 50/53:327–472.

Geitler, L., 1932. Cyanophyceae. In Rabenhorst's Kryptogamrn flora von Deutschland, Ostereich und der Schweiz 14. Reprint 1985 Koeltz Scientific Books Koenigstein Germany: 1–1196.

Findlay, D. L. & H. J. Kling, 1996. Methods in Phytoplankton Analysis. EMAN.

Komárek, J. & H. J. Kling, 1991. Variation in six planktic cyanophyte genera in Lake Victoria (East Africa). Arch. Hydrobiol./Algol. Stud. 61: 21–45

Meffert, M. E., 1987. Planktic unsheathed filaments (Cyanophyceae) with polar and central gas vacuoles 1. Their morphology and taxonomy. Arch. Hydrobiol./Suppl. 76, Monographische Beiträge 4: 315–346.

Romo, S., M. R. Miracle & M. Hernandez-Marine 1993. Remarks on the morphology of *Pseudanabaena galeata* Böcher. Arch. Hydrobiol./Algol. Stud. 68: 39–49.

Hydrobiologia **502:** 389–394, 2003.
L. Naselli-Flores, J. Padisák, M. T. Dokulil (eds),
Phytoplankton and Equilibrium Concept: The Ecology of Steady-State Assemblages.

Estimation of minimum sedimentary inoculum (akinete) pool of *Cylindrospermopsis raciborskii*: a morphology and life-cycle based method

Judit Padisák
University of Veszprém, Department of Limnology, H-8200 Veszprém, Egyetem u. 10, P. O. Box 158, Hungary
E-mail: padisak@tres.blki.hu

Key words: Cylindrospermopsis raciborskii, germination, akinete, life cycle, inoculum, Balaton

Abstract

Cylindrospermopsis raciborskii undergoes characteristic morphological changes during its annual population development. Primary filaments (those that derive directly from the akinetes) possess the following morphological features: the trichomes have one or two acuminate ends; they are narrower (2.4–2.6 μm) than older filaments, their lengths vary in a wide range (from 40 to 300 μm); some of them are straight but others are slightly coiled even in populations that later consist exclusively of straight filaments; cell walls between the cells are not or only hardly visible; the filaments have a fine granulation but no gas-vacuoles, polyphosphate- or any other contrasting bodies are seen. The secondary filaments deriving from the first division of primary filaments are similar but with only one acuminate end. This way, the number of germinating akinetes can be estimated as the number of primary filaments + 1/2 number of secondary filaments. The described morphological method to estimate inoculum size of *C. raciborskii* has limitations. (1) The number of akinetes that we get in this way is only a minimum number since germination is not synchronous under field conditions and probably not each akinete germinates that is present in the sediments. (2) The method is applicable only if germination occurs suddenly (relatively synchronously). This is often the case in a temperate region where germination is triggered by temperature. (3) The method cannot be used in tropical or subtropical lakes where water temperature is permanently rather high and akinetes do not develop or develop only very rarely. (4) Being an *a posteriori* method it cannot be used as predictive tool. Estimations for Lake Balaton are in a good agreement with results of studies with other methods (germination in laboratory cultures and direct counting) and have the advantage that they do not need time-consuming experimenting or difficult direct countings.

Introduction

After its description almost a century ago (Woloszynska, 1912), *Cylindrospermopsis raciborskii* (Woloszynska) Seenaya & Subba Raju had been known as pantropical species. Despite accumulating data about its occurrence at higher latitudes it was still specified as 'pantropic' in a relatively recent phycogeographical review (Hoffmann, 1996). It has been shown (Padisák, 1997) that *C. raciborskii* has become an invading species in the 20th century reaching the northernmost margin of its distribution in Germany at latitudes of 53–54° N (Krienitz & Hegewald, 1996) and until now it was recorded from each continent except the Antarctica. Since a detailed review (Padisák, 1997) about its worldwide distribution and expansion pattern was published data are accumulating at an exponential rate (e.g. Nixdorf et al., 2003; Mischke & Nixdorf, 2003; Padisák et al. 2003; Stoyneva, 2003).

C. raciborskii attracted the attention of scientists and managers for two basic reasons. One is its actual (Bourke et al., 1983) or potential toxicity jeopardizing wildlife or even human health. The other is its tendency to form dense, however suspended, blooms in eutrophic and hypertrophic reservoirs interfering with water use (Padisák, 1997).

In spite of many investigations (e.g. Dokulil & Mayer, 1996; Présing et al., 1996; Istvánovics et al., 2000; Shafik et al., 2001; Spröber et al., 2003) on the physiological ecology of the species there are several details that have remained unexplored; among them, the size of the sedimentary akinete pool (number of akinetes) and the role of akinetes in P-metabolism of the species.

C. raciborskii is a perennial species in tropical and subtropical waters where it either does not produce akinetes at all or produces them only very rarely. As it was shown in Gorzó's (1987) rather early paper, temperature optimum of germination of *C. raciborskii* akinetes is different from that of other N-fixing nostocalean species: in his laboratory experiments with akinetes from Lake Balaton sediments, akinetes of *C. raciborskii* showed a high percent of germination only in the temperature range of 22–23.5 °C, while this range for three other heterocytic blue-green algal species was 18–25 °C. This high temperature demand in itself could explain the stochasticity of blooms in Lake Balaton (Padisák, 1998). During the last bloom (1994) of the species in Lake Balaton, the population produced a much higher amount of akinetes per unit biomass than other nostocaleans, which suggested the hypothesis (Padisák & Istvánovics, 1997) that akinetes of this species have a double role: they are inocula for the next year growth and they may assimilate P during their sedimentary stage which can be used as P-pool of the initial P growth (Istvánovics et al., 1993). Concerning trigger of akinete production, Moore et al. (2003) suggested that it is triggered by initial temperature shock (from 25 to 15 °C), and phosphorus is necessary for further growth and full development of akinetes.

In case of any water quality management predictions concerning the expectable result of restoration measures are essential. Dynamic ecological models, especially if parametric based on a robust knowledge of the system variables and transfer functions, can be used as powerful predictive tools. However, each model needs a 'starting point' that could be initial biomass of the target biota (ecosystem) or its inoculum size. It was sufficiently tested in a model (Padisák & Koncsos, 2002) that *C. raciborskii* may still bloom in Lake Balaton, but only if an infinite inoculum size was supposed. Presence and size of inocula, thus, has been a fundamental issue in phytoplankton ecology and in water quality management as well.

In this paper, a simple morphological method is described to estimate the minimum sedimentary inoculum size of C. *raciborskii.*

Results and discussion

Numerous morphological descriptions and drawings are available in the literature (see a list in Padisák, 1997; Shafik et al., 2003) on *C. raciborskii.* Among them, Singh's (1962) early publication is still an excellent one. He observed that during an annual population cycle *C. raciborskii* undergoes characteristic morphological changes (Figs 1–16). He noted the different morphology of the young filaments, namely that they have two acuminate ends and do not have heterocytes.

Population dynamics, including the number of akinetes, of *C. raciborskii* has been regularly monitored in Lake Balaton. During the early 1990s when the species bloomed regularly in the Keszthely basin maximum frequency of akinetes were recorded as 3 534 292 akinetes l^{-1}on 29 August 1994 and approx. half as much akinetes were found in the water samples in 1996 (1 741 811 akinetes l^{-1} on 19 September 1996). These numbers (supposing that each akinete will be buried in surface sediments and the water depth is 2.5 m) correspond to production of 400 000–900 000 akinetes cm^{-2} sediment surface $year^{-1}$. These numbers correspond to kind of maximum inoculum size. In the sediments akinetes may degrade or be consumed by bottom dwelling animals etc.; therefore, the maximum number of akinetes provides a good measure of survival capacity of a population but says little about the number of inoculum for the following year(s).

Although *C. raciborskii* did not bloom since 1994 in the Keszthely basin of Lake Balaton, moderate peaks regularly occurred in late summers (Fig. 17). In 2000, a sudden population increase (from approx. 1 mg $l^{-1}$to 4 mg l^{-1}) was obseved from 22 August to 27 August. On 27 August a large number of young filaments were found with the following features: the trichomes had one or two acuminate ends; they were narrower (2.4–2.6 μm) than older filaments, their lengths varied in a wide range (40 – 300 μm); some of them were slightly coiled while others were straight (normally the Balaton population consists exclusively of straight filaments – for this reason Singh's [1962] coiled filament on Fig. 16 is also a young filament); cell walls between the cells were not or only hardly visible, the filaments had a fine granulation but no aerotopes, polyphosphate- or any other contrasting bodies were seen.

In one case (in 1994) germination of a filament from a fresh akinete was observed. The old trichome was visible like a pale ghost on the akinete, which (unlike the others which had some orange-brown droplets

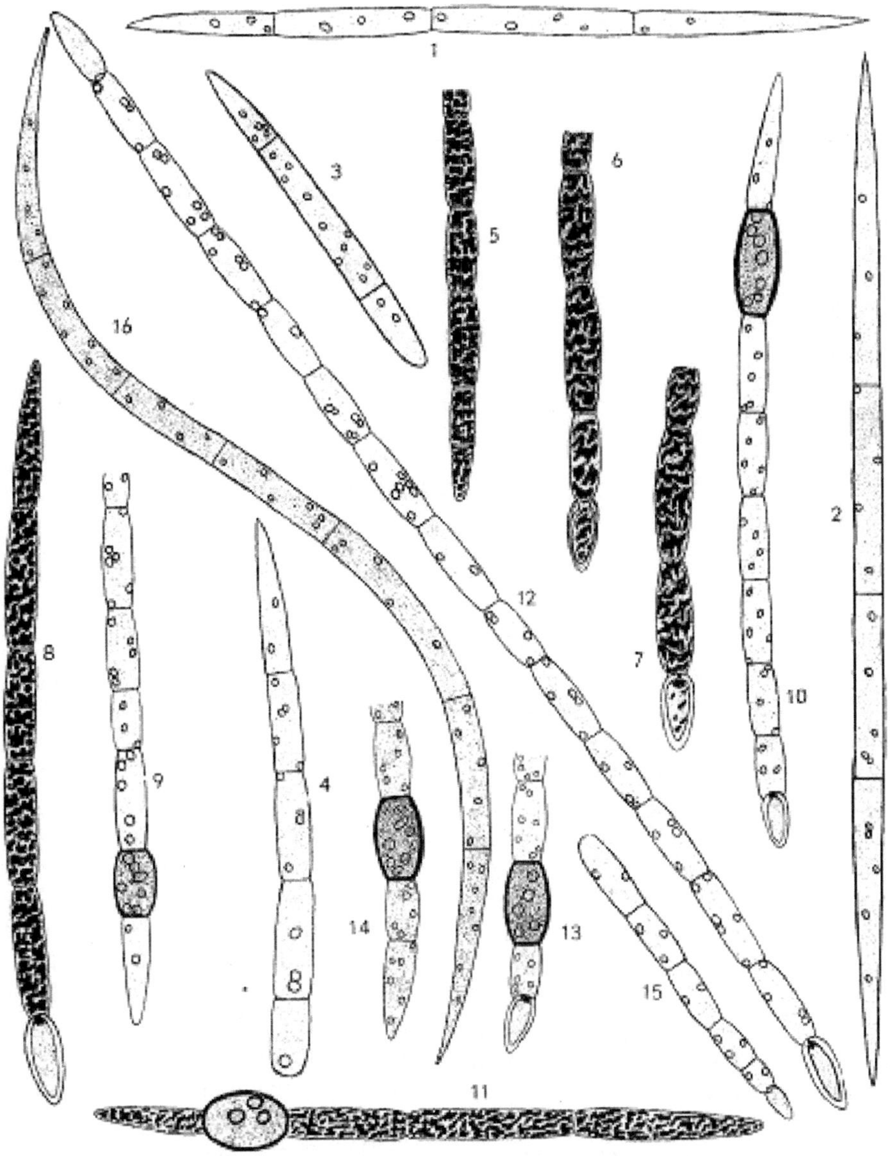

Figures 1–16. Cylindrospermopsis raciborskii (reproduced from Figs 1–16 in Singh, 1962). 1–2, young- trichomes with acuminate ends; 3–4, short fragmented trichomes; 5–8, showing developmental stages of heterocytes; 9–11, 13–14, showing developmental stages of akinetes; 12, long trichome with heterocytes at both ends; 15, short fragmented trichome with heterocytes at one ends; 16, coiled trichome.

[oil droplets?]) was internally unstructured. The young trichome was about 40 μm long, 2.4 μm wide, and it was acuminate at both ends. Cross-walls were not visible. The longitudinal axis of the young filament deviated by approx. 10–15° from the longitudinal axis of the akinete, This observation proves that filaments with two acuminate ends derive directly from the akinete and that *C. raciborskii* akinetes do not have an obligate dormancy period.

The distinct morphological features of primary filaments (those which germinate directly from the akinete and have two acuminate ends) and secondary filaments (that derive from the first division of the primary filaments: they have only one acuminate end) allow us to estimate the minimum number of germinating akinetes. This way, the number of primary filaments + half of the number of secondary filaments provides us the minimum number of akinetes. Calculating as described above, the inoculum size for the

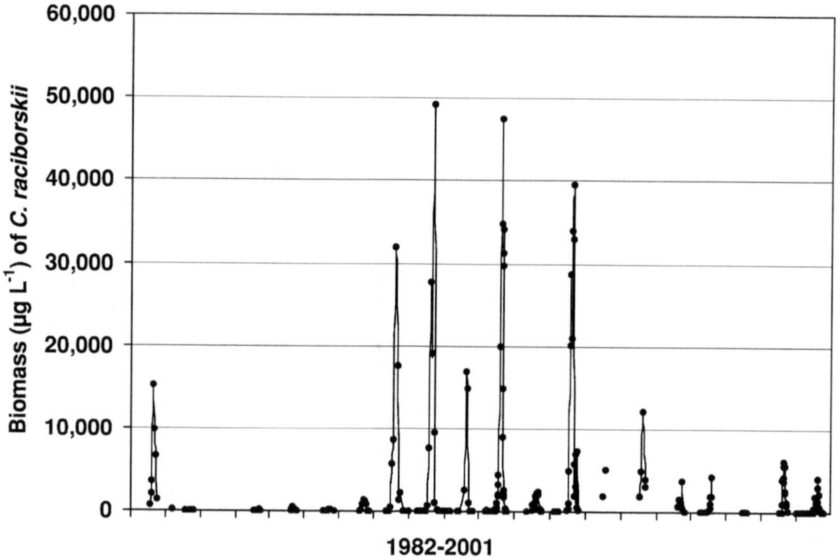

Figure 17. Biomass (μg l^{-1}) of *C. raciborskii* in the Keszthely basin of Lake Balaton between January 1982 and December 2001 (marks on the x axis indicate years).

year 2000 was estimated at 129 000 akinetes cm^{-2} sediment surface. Supposing a 500 μm^3 volume for individual filaments and 2.5 m water depth, this amount of primary filaments corresponds to 0.258 mg l^{-1} *C. raciborskii* biomass in the water column. In field observations (Padisák, 1994; Fabbro & Duivenvoorden, 1996), 3 days doubling time was estimated for exponentially growing *C. raciborskii*. Thus, the 0.258 mg l^{-1} initial filament biomass may increase to usually observed peak biomasses (30–60 mg l^{-1}) within 21–24 days. This calculation is in good agreement with field observations. As it follows, mass germination of *C. raciborskii* enables the species to exhibit explosion-like blooms in freshwaters.

The described morphological method to estimate inoculum size of *C. raciborskii* has several limitations. Namely:

- The number of akinetes that we get in this way is only a minimum number since germination is not synchronous under field conditions and probably not each akinete present in the sediment does germinate.
- The method is applicable only if germination occurs suddenly (relatively synchronously). This is often the case in temperate regions where germination is triggered by temperature.
- The method cannot be used in tropical or subtropical lakes where water temperature is permanently rather high and akinetes do not develop or develop only very rarely.
- Being an *a posteriori* method it cannot be used as a predictive tool.

There are two alternative methods to estimate sedimentary inoculum size of *C. raciborskii.* Gorzó (1985, 1986) collected sediment samples from the upper 5 cm of the sediments of Lake Balaton and germinated them in laboratory cultures. According to his estimations, the number of akinetes increased from 9000 to 20 000 akinete cm^{-3} sediment between 1984 and 1986. In these experiments contribution of *C. raciborskii* akinetes to total akinete number in the Keszthely basin was 80%, which results in an estimate of 7200 to 16 000 akinetes cm^{-3} sediments. Kovács & Koncz (2002) & Kovács et al. (2003) have recently reported about akinete numbers of *C. raciborskii* in sediments of Lake Balaton based on direct counting (combined method with epifluorescene and normal light microscope) method. With this method 4500–6000 akinetes cm^{-3} uppermost sediment was estimated for samples deriving from Lake Balaton sediments and they mentioned that the highest recorded *C. raciborskii* akinete number was 100 000 akinetes cm^{-3} sediment.

Despite differences falling in the range of 2 orders (from thousands to hundred thousands) of magnitude, the results of these three different approaches are in rather good agreement. The morphological approach described in this study is able to provide inoculum size referring to cm^2 of sediment surface while the other studies used the uppermost sediment layers (the uppermost 1 cm for direct counting and the uppermost 5 cm in cultivating). It was shown by both Gorzó (1985) and G.-Tóth et al. (1994) that viable phytoplankton can be found down to 30 cm below sediment surface which might result in underestimation if only upper sediments are considered. Moreover, the distribution of akinetes in the sediments can be quite patchy since they are probably accumulating in deposition zones. However, by accounting the deficiencies of each of the three methods mentioned above we might conclude that sedimentary inoculum pool of *C. raciborskii* in Lake Balaton varies in the orders of 10^3–10^5 akinetes cm^{-2} sediment surface and this amount of akinetes is sufficient for bloom initiation.

Acknowledgements

This work was supported by the "Phytoplankton-online" (EVK1-CT-1999-0037) EC project. I thank Dr. Delio Ruggiu for his careful corrections on the manuscript of this paper.

References

Bourke, A. T. C., R. B. Hawes, A. Neilson & N. D. Stallmann, 1983. An outbreak of hepato-enteritis (the Palm Island mystery disease) possibly caused by algal intoxication. Toxicon 3 (Supplement): 45–48.

Dokulil, M. T. & J. Mayer, 1996. Population dynamics and photosynthetic rates of a *Cylindrospermopsis - Limnothrix* association in a highly eutrophic urban lake, Alte Donau, Vienna, Austria. Algological Studies 83: 179–195.

Hoffmann, L., 1996. Geographic distribution of freshwater blue-green algae. Hydrobiologia 336: 33–40.

Fabbro, L. D., H. Kling & L. J. Duivenvoorden, 1996. Morphological variation of *Cylindrospermopsis* in natural populations. In *Cylindrospermopsis* – A New Toxic Algal Bloom Challenge for Australia, Brisbane City Hall, 24 Oct. 1996, Symposium Abstracts.

Gorzó, Gy., 1985. A Balaton üledékéből kitenyészthető planktonikus, heterocisztás cianobaktériumok [Planktonic heterocystic cyanobacteria cultivated in the sediment of Lake Balaton]. Hidrológiai Közlöny 65: 357–360. (in Hungarian with English summary).

Gorzó, Gy., 1996. A Balaton üledékében előforduló heterocisztás cianobaktérium spórák csírázásának hőfokfüggése [Temperature dependence of germination of heterocytic cyanobacteria spores occurring in the sediments of Lake Balaton]. MHT VI. Országos Vándorgyüles: 499–507 (in Hungarian).

Gorzó, Gy., 1987. Fizikai és kémiai faktorok hatása a Balatonban előforduló heterocisztás cianobaktériumok spóráinak csírázására [The influence of physical and chemical factors on the germination of spores of heterocystic cyanobacteria in Lake Balaton]. Hidrológiai Közlöny 67: 127–133 (in Hungarian with English summary).

G.-Tóth, L., Zs. Langó, J. Padisák & E. Varga, 1994: Terminal electron transport system (ETS)-activity in the sediment of Lake Balaton, Hungary. Hydrobiologia 281: 129–139.

Hoffmann, L., 1996. Geographic distribution of freshwater blue-green algae. Hydrobiologia 336: 33–40.

Istvánovics V., H. M. Shafik, M. Présing & Sz. Juhos, 2000. Growth and phosphate uptake kinetics of the cyanobacterium, *Cylindrospermopsis raciborskii* (Cyanophyceae) in throughflow cultures. Freshwat. Biol. 43: 257–275.

Istvánovics, V., K. Pettersson, M. A. Rodrgio, D. Pierson, J. Padisák & W. Colom, 1993. *Gloeotrichia echinulata*, a colonial cyanobacterium with a unique phosphorus uptake and life strategy. J. Plankton Res. 15: 531–552.

Kovács, A. W. & E. Koncz, 2002. A N_2-kötő cianobaktériumok akinétaszám változása a Balatonban [Changes of akinete number of N_2-fixing cyanobacteria in Lake Balaton, Hungary]. Hidrológiai Közlöny 82: 64–67 (in Hungarian with English summary).

Kovács, A. W., E. Koncz & L. Vörös, 2003. Akinete abundance of N_2-fixing cyanobacteria in sediment of Lake Balaton (Hungary) Hydrobiologia (in press)

Krienitz, L. & E. Hegewald, 1996. Über das Vorkommen von wärmliebenden Blaualgenarten in einem norddeutschen See. Lauterbornia 26: 55–64.

Mischke, U. & B. Nixdorf, 2003. Equilibrium phase conditions in shallow German Lakes: How Cyanoprokaryota species establish a steady state phase in late summer. Hydrobiologia 502 (Dev. Hydrobiol. 172): 123–132.

Moore, D., O'Donohue, G. Shaw & C. Critchley, 2003. Potential triggers for akinete differentiation in an Australian strain of the cyanobacterium *Cylindrospermopsis raciborskii* (AWT 205/1). Hydrobiologia, in press.

Nixdorf, B., U. Mischke & J. Rücker, 2003. Phytoplankton assemblages and steady state in deep and shallow eutrophic lakes – an approach to differentiate the habitat properties of Oscillatoriales. Hydrobiologia 502 (Dev. Hydrobiol. 172): 111–121.

Padisák, J., 1994. Relationships between short-term and long-term responses of phytoplankton to eutrophication of the largest shallow lake in Central Europe (Balaton, Hungary). In Sund, H., H.-H. Geller, W. Xiaogan, Y. Kechang & S. Fengnind (eds), Environmental Protection and Lake Ecosystem. Science and Technology Press, Beijing: 419–437.

Padisák, J., 1997. *Cylindrospermopsis raciborskii* (Woloszynska) Seenayya et Subba Raju, an expanding, highly adaptive cyanobacterium: worldwide distribution and review of its ecology. Arch. Hydrobiol./Supplement 107 (Monographic Studies): 563–593.

Padisák, J., 1998. Sudden and gradual responses of phytoplankton to global climate change: case studies from two large, shallow lakes (Balaton, Hungary and the Neusiedlersee Austria/Hungary). In George, D. G., J. G. Jones, P. Puncochar, C. S. Reynolds & D. W. Sutcliffe (eds), Management of Lakes and Reservoirs During Global Change: Kluwer Academic Publishers, Dordrecht, Boston, London: 111–125.

Padisák, J., G. Borics, G. Fehér, I. Grigorszky, I. Oldal, A. Schmidt & Z. Zámbóné-Doma, 2003. Dominant species, functional assemblages and frequency of equilibrium phases in late summer phytoplankton assemblages in Hungarian small shallow lakes. Hydrobiologia 502 (Dev. Hydrobiol. 172): 157–168.

Padisák, J. & V. Istvánovics, 1997. Differential response of blue-green algal groups to phosphorus load reduction in a large shallow lake: Balaton, Hungary. Int. Ver. theor. angewan. Limnol.: Verh. 26: 574–580.

Padisák, J. & L. Koncsos, 2002. Trend and noise: long-term changes of phytoplankton in the Keszthely-Basin of Lake Balaton, Hungary. Verh. Int. Ver. theor. angewan. Limnol. 28: 194–203.

Présing, M., S. Herodek, L. Vörös & I. Kóbor, 1996. Nitrogen fixation, ammonium and nitrate uptake during a bloom of *Cylindrospermopsis raciborskii* in Lake Balaton. Arch. Hydrobiol. 136: 553–562.

Shafik H. M., S. Herodek, M. Présing & L. Vörös, 2001. Factors effecting growth and cell composition of cyanoprocaryote *Cylindrospermopsis raciborskii* (Woloszynska) Seenayya et Subba Raju. Algological Studies 103: 75–93.

Shafik H. M., L. Vörös, P. Sprőber, M. Présing & A. W. Kovács, 2003. Some special morphological features of *Cylindrospermopsis raciborskii* (isolated from Lake Balaton (Hungary). Hydrobiologia 506–509 (in press)

Singh, R. N., 1962. Seasonal variants of *Anabaenopsis raciborskii* Wolosz. Hydrobiologia 20: 87–91.

Stoyneva, M. P., 2003. Steady-state phytoplankton assemblages in shallow Bulgarian wetlands. Hydrobiologia 502 (Dev. Hydrobiol. 172): 169–176.

Spröber, P., H. M. Shafik, M. Présing, A. Kovács & S. Herodek, 2003. Nitrogen uptake and fixation in the cyanobacterium *Cylindrospermopsis raciborskii* under different nitrogen conditions. Hydrobiologia 506–509 (in press)

Woloszynska, J., 1912. Das Phytoplankton einiger Javanian Seen mit berücksichtigung des Sawa-Planktons. Bulletin International Academiae Scientarium Cracoviae, Series. B. 6: 649–709.

Hydrobiologia **502:** 395–403, 2003.
L. Naselli-Flores, J. Padisák, M. T. Dokulil (eds),
Phytoplankton and Equilibrium Concept: The Ecology of Steady-State Assemblages.

Equilibrium/steady-state concept in phytoplankton ecology

Luigi Naselli-Flores[1], Judit Padisák[2], Martin T. Dokulil[3] & Ingrid Chorus[4]
[1] *University of Palermo, Dipartimento di Scienze Botaniche, Via Archirafi, 38 I-90123 Palermo, Italy*
E-mail: luigi.naselli@unipa.it
[2] *University of Veszprém, Department of Limnology, H-8200 Veszprém, P. O. Box 158, Hungary*
[3] *Institute of Limnology, Austrian Academy of Sciences, Mondseestrasse 9, A-5310 Mondsee, Austria*
[4] *Federal Environmental Agency, P.O. Box 33002, D-14191 Berlin, Germany*

Key words: equilibrium, steady-state, functional groups, phytoplankton associations

Abstract

This paper summarises the outcomes of the 13th Workshop of the International Association of Phytoplankton Taxonomy and Ecology (IAP). The authors mostly addressed their contributions on the following topics: the effect of trophic state on the attainment of a steady-state; the establishment of equilibria in deep and shallow lakes; the role of spatial heterogeneity, disturbance, and stress in the establishment of equilibrium assemblages; the mechanisms leading to the steady state; the frequency and longevity of equilibrium phases, and the role of morphological and physiological plasticity of phytoplankton in maintaining the (apparently) same populations under different environmental conditions. The composition of steady-state assemblages is compared to that of phytoplankton functional groups (coda). Those functional associations recognised as steady-state assemblages appear to be strongly K-determined in many instances.

Introductory remarks

Quite a number of ecological concepts have been using terms such as ecological equilibrium, stability, steady-state, climax, stable state, etc. Apart of the still existing, strong, religion-based belief of the medieval Europe that World, and especially Nature, is balanced, probably the climax concept (Clements, 1916) was the first scientific approach that recognised that temporal development of a community reaches a stage that is rather persistent, self sustaining, and further "development" is limited if at all possible. The term "steady-state" originates in culturing experiments with microrganisms, including algae, and insofar is widely used when ecological concepts are tested in chemostats (e.g. Gaedeke & Sommer, 1986). Stability is well defined in thermodynamics of closed systems. Ecological concepts have often been "translations" of scientific theories developed in physics or chemistry but "translation" almost always lacks scientific corroboration (Krebs, 2001). The problem is that often these concepts remain vague and they are not formally defined (Peters, 1991). In addition, the term stability carries a meaning of "motionless" that certainly does not apply to community ecology. Thus, the term equilibrium was supposed to better depict the dynamic nature of a community when it is, or seems to be, persistent. Actually the persistancy is only related to the temporal scale of the observer. The term stable state was recently proposed by Scheffer (1998) to describe the alternative stages in shallow lake ecosystems. The need for clear definitions and the necessity for clarifying terms are openly addressed in Rojo & Álvarez-Cobelas (2003). Whatever term is selected from the diverse terminology, the ultimate goal is to describe a situation of little variability of species dominance and total biomass across time.

The equilibrium/steady-state concept emerged in phytoplankton ecology when discussing the diversity-disturbance relationship (Sommer et al., 1993). In this context, scientists often refer to equilibrium and non-equilibrium hypotheses to try to describe specific communities.

The non-equilibrium hypothesis posits that equilibrium forces are weak and that competitive interactions are minimal. A non-equilibrium model requires that environmental disturbances occur with sufficient frequency to disrupt the course of competitive exclusion (Harris, 1986: 29–30). This view highlights the influence of disturbance on biological diversity. In particular, higher diversity values are attained when environmental and internal disturbances can prevent the establishment of equilibrium conditions (Sommer et al., 1993). In non-equilibrium models, species diversity was supposed to decrease through random local extinction in a stochastic dynamic system. Species diversity however, can be maintained when the rates of local extinction are low and balanced by local immigration (inocula).

If systems operate hierarchically, spatial and temporal heterogeneity plays a central role in understanding population, community and ecosystem processes (Whittaker & Levine, 1977). Hierarchy theory is a formal approach to the complex influences of scale, and suggests that different phenomena (e.g. long term processes and disturbances) influence systems on different scales (Allen & Starr, 1982). In addition, the different scales (phenomena) are nested and interact. Climate changes alter vegetation over long periods of time. Within a specific climate such as our current one, disturbance can alter the vegetation at a much smaller temporal scale. While the vegetation reacts to the disturbance, it is still constrained by the current climate in how it can react (i.e., what species are available for colonization) thus, disturbance is nested within the climatic regime. Disturbance is recognized as an important factor that increases heterogeneity at all scales. The frequency of disturbance events has been a focus of investigations in phytoplankton dynamics in Baja (Hungary), at the 1991 IAP meeting (Padisák et al., 1993). During the last decade non-equlibrium concepts became increasingly accepted not only in phytoplankton ecology but in ecology in general (Krebs, 2001). However, the equilibrium/steady-state concept nonetheless remains useful to explain long-lasting successional phases dominated by one or several species, especially because such species are often a nuisance and can interfere with water use.

According to Whittaker (1975), each species in an equilibrium community occupies a different niche that results from and reduces direct competition and thus species composition is more or less stable according to environmental predictability. In this respect both temporal and spatial niche segregation is essential to maintain communities in equilibrium. In aquatic environments, this segregation could be reached in the absence of disturbance and would result in an eternal steady state where diversity is reduced to minimal levels by competitive exclusion. In reality, absence of disturbance may occur only for limited time spans, and the question is which time spans are sufficient under which conditions in order to allow equilibrium to be attained.

The main objective of the 13th IAP was therefore to test the applicability of non-equilibrium and equilibrium models to phytoplankton dynamics. This included the attempt to define the occurrence, nature and properties of "equilibrial species" which dominate the assemblages in equilibrium conditions from an ecological point of view. Several questions were addressed at the workshop:

(i) What are the ecological conditions which lead to a steady state assemblage and when do they occur?
(ii) Is the repetitive nature of phytoplankton assemblages in deep lakes (as compared to the "unpredictable" assemblages of shallow ones) only connected to a more resilient environment or is it in some way linked to the pool of species that can live in that environment?
(iii) Do those species which are commonly found in both deep and shallow environments show any physiological and/or morphological plasticity?

A hypothetical situation where a maximum of three species dominate the assemblage for, at least, three weeks without considerable change in total biomass (Sommer et al., 1993) represented the point of departure for discussions.

The equilibrium/steady-state debate in phytoplankton. What has been achieved?

Points which were strongly debated during the 13th IAP meeting were the relation of equilibrium or steady state phases to trophic state, mixing type, environmental stress, lake morphology, biotic interactions other than competition and especially to the functional groups of phytoplankton.

The effect of trophic state on the attainment of a steady-state

Lakes that were tested for equilibria included almost the whole range of the trophic spectrum: from oligotrophy to hypertrophy. High physical stability and resilience characterize large and deep lakes with low water renewal times and moderate trophic states that allows occurrence of an equilibrium (Salmaso, 2003). However, as pointed out by Dokulil & Teubner (2003) steady-state conditions seem to be highly unlikely at the lower end of the trophic spectrum. In oligotrophic, shallow forest lakes equilibrated assemblages were detected (Willén, 2003) but only in lakes with large interannual variations in some resource, or only under stress conditions which will be discussed later. In a mesotrophic reservoir, Leitão et al. (2003) noted the establishment of a sort of equilibrium where only one species contributed most to total biomass, however, they considered it as a "no-change" situation rather than an equilibrium stage within a competitive system. Contributions from eutrophic or hypertrophic systems were unequivocal claimed that steady state assemblages readily occur in such lakes. Most often cyanoprokaryotes occurred as protagonists (Nixdorf et al., 2003; Mischke & Nixdorf, 2003; Naselli-Flores & Barone, 2003; Stoyneva, 2003) but other groups of algae contributed most to equilibria in wetlands (Ortega et al., 2003; Padisák et al., 2003).

Establishment of equilibria in deep and shallow lakes

There seems to be an interesting difference between stratifying and non-stratifying lakes concerning conditions for the establishment of steady-states. Long-lasting stratification allowed phytoplankton assemblage to complete the process of competitive exclusion and attain an equilibrium stage in the deep Lake Maggiore in Italy (Morabito et al., 2003). A quite different stratified lake, the almost always ice-covered Lake Boeckella (Antarctica), showed analogous features (Allende & Izaguirre, 2003). Non-stratifying lakes are traditionally considered as fast-changing ecosystems characterized by quick and unpredictable phytoplankton changes. Interestingly, Nixdorf & Miscke (2003) concluded that steady-state conditions occur more regularly in shallow lakes and they are more predictable than those in deep lakes. Actually, this statement is corroborated by many case studies in this volume. This is probably due to the tendency of shallow lakes towards higher trophic states and overlaps with findings discussed above. In addition, as shown by Chorus & Schlag (1993), for those systems which are always mixed, this represents a sufficiently stable condition to allow succession to proceed towards "late summer climax" (and "intermediate quiescence" would be the actual disturbance).

The role of spatial heterogeneity

Phytoplankton has been traditionally considered a homogenous assemblage inhabiting the pelagic compartment of an aquatic ecosystem. While this may be true for the (few) deep lakes in the world, it became clear during discussions about equilibria that this paradigm cannot be maintained for the most part of aquatic ecosystems. In wetland ecosystems and very shallow waters tychoplanktonic, benthic, periphytic, metaphytic, epilithic, etc. algae contribute to planktonic biomass and thus mask equilibrial communities (Borics et al., 2003, Allende & Izaguirre, 2003; O'Farrell et al., 2003; Ortega et al., 2003; Padisák et al., 2003).

In contrast, it was shown that assemblages without this influence, i.e. in euphotic metalimnia or upper hypolimnia often can reach long-lasting steady-state phases (Albay & Alkcaalan, 2003; Morabito et al., 2003; Teubner et al., 2003).

From a theoretical point of view, habitat heterogeneity within aquatic environments is likely to be reached only in the absence of disturbance: it requires an extended time span of steady state where diversity is reduced to minimal levels by competitive exclusion. Disturbance increases connectivity among the different layers of the same water body or among different environments (e.g. floodplain lakes, cascade reservoirs, wetland systems) and re-opens the competition games to different species. In brief, the homogenisation should decrease the β-diversity and increase the α-diversity.

Thus, it may be very important to evaluate the degree of connectivity and segregation within a single ecosystem or among temporarily inter-connected environments. In the first case, the thermal structure of a given water body as well as its trophic state and morphological traits have to be investigated, in an appropriate time-scale and spatial resolution. In fact, the strength of the gradients due to e.g. the relative density of the different layers or the depth of light penetration may represent an effective mechanism of niche segregation. In the second case the steady-state assemblages may appear in the periods of increased connectivity and may lead to a flow-maintained r-dominance during flooding.

Deductively, steady-state assemblages should be likely to occur only in highly resilient, not stressed, and strongly segregating environments such as deep lakes in temperate climates or in inter-connected environments during the periods of higher connectivity.

These observations may suggest that when dealing with community ecology in vertically heterogenous deep lakes or horizontally/vertically segregated wetlands, extreme care should be taken in sampling methodology, in other words, integrated sampling is not recommended.

The role of disturbance

The role of disturbance in phytoplankton ecology has been thoroughly discussed at a previous meeting (Padisák et al., 1993). As the discussion about disturbance cannot be fruitful without defining equilibria, it is recurrent when trying to understand steady-state conditions. In experimental conditions, steady-state episodes were rare and connected to the absence of sediment resuspension and related nutrient pulses together with changes in light conditions (Ortega et al., 2003). Even in natural conditions, equilibrium did not occur under very rapidly fluctuating conditions (O'Farrell et al., 2003) or at intermittent mixing (Allende & Izaguirre, 2003). As stated by Ortega et al. (2003), autogenic succession leading to a climax state seemed to need an unperturbed environment. However, as pointed out by Rojo & Álvarez-Cobelas (2003) there are steady-state assemblages in both stable and fluctuating environments, and, as found at the above-mentioned previous meeting, continuous perturbation can indeed be interpreted as a steady condition leading to steady-state assemblages (Chorus & Schlag, 1993).

The effect of stress

Stress and disturbance are usually difficult to separate under field conditions. When talking about stress in context of dynamic equilibria we basically include two scenarios:

(i) when one or more environmental factors are at the extreme of the range in which organisms can grow and
(ii) when the change in question is either too abrupt or too ultimate and exceeds the homeostatic plateau of the assemblage.

As observed by Stoyneva (2003), steady state occurred after drastic changes in hydrological regime such as rapid change in water level or seawater ingression combined with extremely high temperatures. Analogous patterns were found in a Sicilian reservoir after a sudden and consistent water addition (Naselli-Flores & Barone, 2003).

Boreal forest lakes with more than 80% dominance of only few species were characterized by one or more stress factors: harsh climate, dark-brown water or extreme acidity (Willén, 2003). Highly saline lakes in Hungary more often accommodated steady-state assemblages than others falling into the normal freshwater range (Padisák et al., 2003). Stress and "regular" disturbance in itself can be seen as factors allowing equilibrium (Komárková & Tavera, 2003). On the whole, extreme environments select for individual adaptations of one or a few species and therefore such environments are suitable to host steady-state assemblages.

Mechanisms leading to steady state

The original description of identifying an equilibrium (Sommer at al., 1993) concentrated on dynamic features of temporal changes in phytoplankton and was based on resource partitioning concepts, more specifically on competition. However, competition is not the only type of biotic interaction that can result in equilibrial assemblages, as is thoroughly discussed by Rojo & Álvarez-Cobelas (2003). There are many different processes which result in steady-state assemblages and therefore there are many different types of steady-state. As stated above, we can distinguish between equilibrium resulting from competitive interactions and equilibrium arising from species specific abilities such as mixotrophy (O'Farrell et al., 2003; Padisák et al., 2003), buoyancy regulation or motility (Albay & Alkcaalan, 2003; Flaim et al., 2003; O'Farrell et al., 2003), shade adaptation (Mischke et al., 2003; Morabito et al., 2003) growth rates (Ortega et al., 2003). Trophic relationships, especially selective grazing may prevent the formation of epilimnetic steady-state assemblages (Teubner et al., 2003). In some extreme environments such as the lakes in Antarctica, heavy grazing pressure may select for the fastest growing phytoplankton species. Although only superficially touched in the contributions to this volume, other mechanisms like allelopathy, invasions (Naselli-Flores & Barone, 2003), sudden outbursts due to sedimentary inocula (Padisák, 2003) and parasitism may contribute to development or disruption of steady-states.

Frequency and longevity of equilibrium phases

Steady-state stages are not frequently attained in phytoplankton successions (Willén, 2003; Padisák et al., 2003). In any case, once reached, the equilibrium phase generally covers a short period not longer than a few weeks, as shown in almost all the contributions. Long-lasting equilibrium phases where recorded in the tropical lake Catemaco (Komárková & Tavera, 2003) and in the Mediterranean reservoir Lake Arancio (Naselli-Flores & Barone, 2003), and both were due to cyanoprokaryotes. The rarity and the ephemeral nature of these events might be attributed to the frequently changing aquatic environment, to the fast response time of phytoplankton, and to the different scenarios upon which phytoplankton play. It seems important to underline that the majority of available studies still comes from temperate climates, where seasonal changes of physical conditions terminate the equilibrium phase. Probably, in tropical or subtropical systems these changes are smoother or quite absent allowing a longer persistence of equilibrium phases. The criteria set to identify an equilibrium phase, a maximum of three species assessing the 80% of total biomass for, at least, three weeks without considerable variation in total biomass itself (Sommer et al., 1993), are probably too strict and any modification to them is welcome. Mischke & Nixdorf (2003) allowed $\pm$ 15% biomass variation for judging constancy of biomass. An interesting observation was referred to by Rojo & Álvarez-Cobelas (2003) who found that the increase of biomass values was due to an increase in average dimension of the organisms and not due to an increase in cell numbers. In addition, some authors (Morabito et al., 2003; Nixdorf et al., 2003) found it necessary to increase the number of species at equilibrium to 4–5. Table 1 summarizes the equilibrated assemblages recognized by the authors of the contributions collected in this volume.

Morphological and physiological plasticity of phytoplankton

According to Salmaso (2003), the existence of different ecotypes within a species may explain the contrasting behaviour of some species in different types of lakes. Even in the same environment, the group recognised as a "species", using morphological criteria, can actually be formed by a consortium of different genotypes with potentially slightly different biochemical and/or physiological properties, or it may require different environmental conditions throughout its life-cycle (Rojo & Álvarez-Cobelas, 2003). Examples of this plasticity were shown by Dokulil & Teubner (2003) and by Naselli-Flores & Barone (2003).

Steady-state assemblages vs phytoplankton functional groups (coda)

Reynolds et al. (2002) attempted to bracket together species with similar morphological and physiological traits and with similar ecologies as functional groups. Some of these represent equilibrated assemblages. Therefore we searched for the functional groups recognized and cited by the contributors. Of the 31 functional groups described until now, 24 were cited (Table 2).

Moreover, only three clear successional sequences were reported: $\mathbf{B \rightarrow R \rightarrow C \rightarrow L_M \rightarrow L_o}$ (Dokulil & Teubner, 2003); **(Y-X2)** $\rightarrow$ **(P-X2)** $\rightarrow$ **(F-P)** $\rightarrow$ **(Lo-T)** $\rightarrow$**Y** (Leitão et al., 2003); $\mathbf{L_M}$ $\rightarrow$ **M(F)** $\rightarrow$ $\mathbf{L_M}$ $\rightarrow$ **P** (Naselli-Flores & Barone, 2003).

Based on habitat properties, Mischke & Nixdorf (2003) suggested to include *Aphanizomenon gracile* into the functional group $\mathbf{S_N}$. Allende & Izaguirre (2003) enriched the group **R** with the following species: *Isocystis pallida, Leptolyngbya antarctica, L. tenue* and *Pseudanabaena catenata.* Leitão et al. (2003) included *Planctonema lauterbornii* into group **T**. Establishment of two new groups were suggested: $\mathbf{W_s}$ for *Synura* dominated assemblages and $\mathbf{Y_{Ph}}$ for those in which *Phacotus* is dominant (Padisák et al., 2003).

It is interesting to underline that among the species listed in Table 1, cyanoprokaryotes, belonging to the functional groups **S1,** $\mathbf{S_N}$**, M, H1** and **R**, are in overwhelming dominance. Other species belong to groups comprising non-edible large sized phytoplankton (**Lo, C, P, T, F, LM** and **Q**). Phytoplankton species of grazable size were found only twice in equilibrated communities. These were *Cryptomonas erosa* (group **Y**) in the warm monomictic hypertrophic lake El Porcal and *Lagherheimia genevensis* (edibility may depend on extension of spines) in the warm monomictic mesotrophic Lake Las Madres (Rojo & Álvarez-Cobelas, 2003).

Conclusions

The outcome of the discussions on equilibrium or steady state phases during the 13th IAP meeting in Sicily can be summarized by the following statements:

- In phytoplankton assemblages, the concept of equilibrium is not intended to imply an energetic equilibrium since the phytoplankton equilibria are most likely far from thermodynamic equilibrium.

Table 1. Species found in equlibrum stages by contributors of the proceedings of the 13th IAP Workshop. Each reference is found in Hydrobiologia 502. Functional group qualification was established by Naselli-Flores and Padisák according to Reynolds et al. (2002)

	Equilibrium phases	Functional group	Lake/country	Type of the lake	Reference
1.	*Planktothrix agardhii*, *Aphanizomenon gracile*, *Pseudanabaena limnetica*	**S1** $\mathbf{S_N}$ $\mathbf{S_N}$	Langersee, Melangsee (Germany)	both shallow, eutrophic	Mischke & Nixdorf, Nixdorf et al.
2.	*Limnothrix amphigranulata*, *Pseudanabaena limnetica*, *Planktothrix agardhii*	**S1** **S1** **S1**	Melangsee (Germany)	shallow, eutrophic	Nixdorf et al.
3.	*Pseudanabaena limnetica* *Aphanizomenon gracile*, *Cylindrospermopsis raciborskii*	**S1** $\mathbf{S_N}$ $\mathbf{S_N}$	Scharmützersee (Germany)	deep, dimictic, eutrophic	Nixdorf et al.
4.	*Pseudanabaena limnetica*	**S1**	Scharmützersee (Germany)	deep, dimictic, eutrophic	Nixdorf et al.
5.	*Microcystis aeruginosa*	**M**	Srebarna (Bulgaria)	shallow, hypertrophic	Stoyneva
6.	*M. wesenbergii*	**M**	Srebarna (Bulgaria)	shallow, hypertrophic	Stoyneva
7.	*M. wesenbergii*, *Anabaenopsis arnoldii*	**M** **H1**	Shabla (Bulgaria)	shallow, eu/hypertrophic	Stoyneva
8.	*Cylindrospermopsis raciborskii*	$\mathbf{S_N}$	Srebarna (Bulgaria)	shallow, eu/hypertrophic	Stoyneva
9.	*C. raciborskii*, *Anabaenposis arnoldii*	$\mathbf{S_N}$ $\mathbf{S_N}$	Srebarna (Bulgaria)	shallow, eu/hypertrophic	Stoyneva
10.	*Peridinium* cf. *umbonatum*	$\mathbf{L_0}$	Srebarna (Bulgaria)	shallow, eu/hypertrophic, under pool condition	Stoyneva
11.	*Microcystis wesenbergii*, *Aphanzomenon flos-aquae*, *Anabaena spiroides*	**M** **H1** **H1**	Vaya (Bulgaria)	shallow, eu/hypertrophic	Stoyneva
12.	*Microcystis wesenbergii* *Aphanizomenon gracile*	**M** $\mathbf{S_N}$	Vaya (Bulgaria)	shallow, eu/hypertrophic	Stoyneva
13.	*Cylindrospermopsis catemaco*	$\mathbf{S_N}$	Catemaco (Mexico)	shallow. hypertrophic, tropical	Komárková & Tavera
14.	*Cylindrospermopsis philippinensis*	$\mathbf{S_N}$	Catemaco (Mexico)	shallow. hypertrophic, tropical	Komárková & Tavera
15.	*Microcystis aeruginosa*	**M**	Ömerly (Turkey)	deep, hypertrophic	Albay & Alkcaalan
16.	*Asterionella formosa*, *Fragilaria crotonensis*	**C** **P**	Ömerly (Turkey)	deep, hypertrophic	Albay & Alkcaalan
17.	*Planktothrix rubescens*	**R**	Ammersee, (Austria)	deep mesotrophic	Teubner et al.
18.	*Planctonema lauterbornii*	**T**	Vouglans (France)	deep, meso-eutroophic, reservoir	Leitão et al.
19.	*Peridiniopsis durandii*	$\mathbf{L_0}$	Vouglans (France)	deep, meso-eutroophic, reservoir	Leitão et al.
20.	*Cryptomonas erosa*, *Limnothrix redekei*	**Y** **S1**	El Porcal (Spain)	warm monomictic, hypertrophic	Rojo & Álvarez-Cobelas

Table 1. contd.

21.	*Plsnktothrix agardhii,* *Cyclotella meneghiniana* *Cryptomonas erosa* *Dictyosphaerium* *Limnothrix redekei*	**S1** **C** **Y** **F** **S1**	El Porcal (Spain)	warm monomictic, hypertrophic	Rojo & Álvarez-Cobelas
22.	*Peridiunium willei* *Peridiniopsis borgei* *Lagerheimia genevensis*	**L**$_O$ **L**$_O$ **X1**	Las Madres (Spain)	shallow, hypertrophic	Rojo & Álvarez-Cobelas
23.	*Microcystis* spp.	**M**	Arancio (Italy)	shallow/deep reservoir	Naselli-Flores & Barone
24.	*Ceratium hirundinella*	**L**$_O$	Arancio (Italy)	shallow/deep reservoir	Naselli-Flores & Barone
25.	*Goniostomum semen*	**Q**	Rotehogstjärnen (Sweden), Älgsjön (Sweden)	acid lakes	Willén
26.	*Botryococcus braunii, Willea wilhelmi, Eutetramorus planctonicus*	**F**	Mohonk (U.S.A.)	stratified, oligotrophic	Huszar et al.
27.	Unidentified dinoflagellate *Nitzschia acicularis*	**Lm** **C**	Wononscopomuc (U.S.A.)	stratified mesotrophic	Huszar et al.
28	*Fragilaria crotonensis* *Woronichinia* sp. *Botryococcus braunii*	**P** **L**$_O$ **F**	Stissing (U.S.A.)	Stratified, meso-eutrophic	Huszar et al.
29.	*Fragilaria crotonensis,* *Asterionella formosa,* *Diatoma tenuis,* *Planktothrix rubescens,* *Pseudanabaena* sp.	**P** **C** **P** **R** **R**	Maggiore (Italy)	deep, mesotrophic	Morabito et al.
30.	*Planktothrix agardhii* *Pseudanabaena catenata*	**S1 (R)** **R**	Maggiore (Italy)	deep, mesotrophic	Morabito et al.

We suggest to call them steady states assemblages. Some are successional near-climaces, others are maintained short of this by allochthonous factors (they are plagioclimaces in fact). All are "steady-state outcomes" of the blend of environmental factors recently attained.

- This may be considered as the opposite of a state of flux, depending upon critical exceedence of the environmental conditions, or what we call forcing and disturbance.
- Since most of water use problems (could it be whatever kind) are almost invariably connected to "overdominance" (steady-state dominance) of some phytoplankton species it is a rather important task for basic science to explore ecological nature of such scenarios.
- Communities are structured by a tension between two forces, abiotic external and biotic internal factors. This point is very simplistic indeed: written in first-level textbooks of ecology. However, there are quite few studies spanning over geographic latitudes and altitudes (including lake depths) that explored it in such detail as given here.
- The external factors constrain traits within limits, the internal forces separate coexisting species. Stressed conditions, including a very high trophic state, may set to zero competitive exclusion allowing the persistence of that stress-

Table 2. Functional groups that were identified by contributors of the proceedings of the 13th IAP Workshop. Each reference is found in Hydrobiologia 502. Characteristic species are cited according to Reynolds et al. (2002)

Functional group	Characteristic species	Reference
A	*Urosolenia*	Huszar et al.
D	*Synedra acus, Nitzschia* spp., *Stephanodiscus hantzschii*	Albay & Akcaalan
E	*Dinobryon, Mallomonas, Synura*	Tolotti et al.
G	*Eudorina, Volvox*	Huszar et al.
J	*Pediastrum, Coelastrum, Scenedesmus*	Albay & Akcaalan
N	*Tabellaria, Cosmarium, Staurodesmus*	Huszar et al.
Q	*Gonyostomum semen*	Willén
S2	*Spirulina, Arthrospira, Raphidiopsis*	O'Farrell et al.
H2	*Anabaena lemmermannii, Gloeotrichia echinulata*	Huszar et al.
Z	*Synechococcus,* prokaryotic picoplankton	O'Farrell et al.
M	*Microcystis, Sphaerocavum*	Albay & Akcaalan; Naselli-Flores & Barone
X1	*Chlorella, Ankyra, Monoraphidium*	O'Farrell et al.; Allende & Izaguirre,
X3	*Koliella, Chrysococcus,* eukaryotic picoplankton;	Allende & Izaguirre; Tolotti et al.
T	*Mougeotia, Tribonema, Geminella*	Morabito et al.; Leitão et al.
W1	euglenoids, *Synura, Gonium*	O'Farrell et al., Huszar et al.
B	*Aulacoseira subartica, A. islandica*	Dokulil & Teubner; Albay & Akcaalan,
H1	*Anabaena flos-aquae, Aphanizomenon*	Mischke & Nixdorf; Stoyneva; Huszar et al.
K	*Aphanotece, Aphanocapsa*	O'Farrell et al.; Allende & Izaguirre; Huszar et al.
R	*Planktothrix rubescens*	Morabito et al.; Dokulil & Teubner; Allende & Izaguirre
SN	*Cylindrospermopsis, Anabaena minutissima*	Mischke & Nixdorf; Stoyneva; Komárková & Tavera
C	*Asterionella formosa, Aulacoseira ambigua*	Dokulil & Teubner; Albay & Akcaalan; Morabito et al.; Huszar et al.
Lm	*Ceratium, Microcystis*	Dokulil & Teubner; Stoyneva; Naselli-Flores & Barone; Huszar et al.
X2	*Plagioselmis, Chrysochromulina*	Leitão et al.; Allende & Izaguirre; Barone & Naselli-Flores; Tolotti et al.
F	colonial Chlorococcales	Leitão et al.; O'Farrell et al.; Albay & Akcaalan; Naselli-Flores & Barone; Huszar et al.
Lo	*Peridinium, Woronichinia*	Dokulil & Teubner; Leitão et al.; O'Farrell et al.; Willén; Huszar et al.
P	*Aulacoseira granulata, Fragilaria, Closterium*	Leitão et al.; Morabito et al.; Albay & Akcaalan; Naselli-Flores & Barone; Huszar et al.
Y	*Cryptomonas*	Leitão et al.; O'Farrell et al.; Barone & Naselli-Flores; Albay & Akcaalan; Huszar et al.
S1	*Planktothrix agardhii, Limnothrix redekei, Pseudoanabaena*	Nixdorf et al.; Mischke & Nixdorf; O'Farrell et al.; Morabito et al.; Naselli-Flores & Barone

tolerant species characterised by special adaptations

- "Steady-state" assemblages in natural conditions appear to be strongly determined by K-strategists in many instances.
- Guilds, functional types and ecological groups can be useful tools to describe and understand rules of community assembly.

References

Albay, M. & R. Akcaalan, 2003. Factors determining the phytoplankton steady state assemblages in a drinking-water reservoir (Ömerli reservoir, Istanbul). Hydrobiologia 502 (Dev. Hydrobiol. 172): 85–95.

Allen, T. F. H. & Starr, T. B. 1982: Hierarchy: perspectives for ecological complexity. University of Chicago Press, Chicago.

Allende, L. & I. Izaguirre, 2003. The role of physical stability on the establishment of steady state in the phytoplankton community of two Maritime Antarctic lakes. Hydrobiologia 502 (Dev. Hydrobiol. 172): 211–224.

Barone, R. & L. Naselli-Flores, 2003. Distribution and seasonal dynamics of cryptomonads in Sicilian water bodies. Hydrobiologia 502 (Dev. Hydrobiol. 172): 325–329.

Borics G., B. Tóthmérész, I. Grigorszky, J. Padisák & G. Várbíró, 2003. Algal assemblage types of bog-lakes in Hungary and their relation to water chemistry, hydrological conditions and habitat diversity. Hydrobiologia 502 (Dev. Hydrobiol. 172): 145–155.

Chorus, I. & G. Schlag, 1993. Importance of intermediate disturbances for the species composition and diversity of phytoplankton in two very different Berlin lakes. Hydrobiologia 249: 67–92.

Clements, F. E. 1916. Plant Succession: an Analysis of the Development of Vegetation. Carnegie Inst. Washington Publ. 242: 1–517.

Dokulil M. & K. Teubner, 2003. Steady state phytoplankton assemblages during thermal stratification in deep alpine lakes. Do they occur? Hydrobiologia 502 (Dev. Hydrobiol. 172): 65–72.

Flaim, G., E. Rott, F. Corradini & G. Toller, 2003. Long-term trends in species composition and diurnal migration of dinoflagellates in Lake Tovel (Trentino, Italy). Hydrobiologia 502 (Dev. Hydrobiol. 172): 357–366.

Gaedeke, A. & U. Sommer, 1986. The influence of periodic disturbances on the maintenance of phytoplankton diversity. Oecologia 71: 98–102.

Harris, G. P. 1986: Phytoplankton Ecology. Strucure, function and fluctuation. Chapman and Hall, London.

Huszar, V., C. Kruk & N. Caraco, 2003. Steady-state assemblages of phytoplankton in four temperate lakes (NE USA). Hydrobiologia 502 (Dev. Hydrobiol. 172): 97–109.

Komárková, J. & R. Tavera, 2003. Steady state of phytoplankton assemblage in tropical Lake Catemaco (Mexico). Hydrobiologia 502 (Dev. Hydrobiol. 172): 187–196.

Krebs, C. H., 2001. Ecology. Experimental Analysis of Distribution and Abundance. Addison Wesley, San Francisco, pp. 695.

Leitão, M., S. Morata, S. Rodriguez & J. P. Vergon, 2003. The effect of perturbations on phytoplankton assemblages in a deep reservoir (Vouglans, France). Hydrobiologia 502 (Dev. Hydrobiol. 172): 73–83.

Mischke U. & B. Nixdorf, 2003. Equilibrium phase conditions in shallow German lakes: How cyanoprokaryota species establish a steady state phase in late summer. Hydrobiologia 502 (Dev. Hydrobiol. 172): 123–132.

Morabito G., A. Oggioni & P. Panzani, 2003. Phytoplankton assemblage at equilibrium in large and deep subalpine lakes: a case study from Lago Maggiore (N. Italy). Hydrobiologia 502 (Dev. Hydrobiol. 172): 37–48.

Naselli-Flores, L. & R. Barone, 2003. Steady-state assemblages in a Mediterranean hypertrophic reservoir. The role of *Microcystis* eco-morphological variability in maintaining an apparent equilibrium. Hydrobiologia 502 (Dev. Hydrobiol. 172): 133–143.

Nixdorf, B., U. Mischke & J. Rucker, 2003. Phytoplankton assemblages and steady state in deep and shallow eutrophic lakes – an approach to differentiate the habitat properties of Oscillatoriales. Hydrobiologia 502 (Dev. Hydrobiol. 172): 111–121.

O'Farrell, I., R. Sinistro, I. Izaguirre & F. Unrein, 2003. Do steady-state assemblages occur in shallow lentic environments from wetlands? Hydrobiologia 502 (Dev. Hydrobiol. 172): 197–209.

Ortega-Mayagoitia, E., C. Rojo & M. A. Rodrigo, 2003. Controlling factors of phytoplankton assemblages in wetlands: an experimental approach. Hydrobiologia 502 (Dev. Hydrobiol. 172): 177–189.

Padisák J., 2003. Estimation of minimum sedimentary inoculum (akinete) pool of *Cylindrospermopsis raciborskii*: a morphology and life-cycle based method. Hydrobiologia 502 (Dev. Hydrobiol. 172): 389–394.

Padisák, J., G. Borics, G. Fehér, I. Grigorszky, I. Oldal, A. Schmidt & Z. Zámbóné-Doma, 2003. Dominant species, functional assemblages and frequency of equilibrium phases in late summer phytoplankton assemblages in Hungarian small shallow lakes. Hydrobiologia 502 (Dev. Hydrobiol. 172): 157–168.

Peters, R. H., 1991. A Critique for Ecology. Cambridge University Press, Cambridge, pp. 366.

Reynolds, C. S. 1997. Vegetation Processes in the Pelagic: A Model for Ecosystem Theory. Ecology Institute, Oldendorf/Luhe.

Rojo C. & M. Álvarez-Cobelas, 2003. Are there steady state phytoplankton assemblages in the field? Hydrobiologia 502 (Dev. Hydrobiol. 172): 3–12.

Salmaso N., 2003. Life strategies, dominance patterns and mechanisms promoting species coexistence in phytoplankton communities along complex environmental gradients. Hydrobiologia 502 (Dev. Hydrobiol. 172): 13–36.

Scheffer, M., 1998. Ecology of Shallow Lakes. Chapman and Hall, London, pp. 257.

Sommer, U., J. Padisák, C. S. Reynolds & P. Juhász-Nagy 1993: Hutchinson's heritage: the diversity-disturbance relationship in phytoplankton. Hydrobiologia 249: 1–7.

Stoyneva, M., 2003. Steady state phytoplankton assemblages in shallow Bulgarian wetlands. Hydrobiologia 502 (Dev. Hydrobiol. 172): 169–176.

Teubner, K., M. Tolotti, S. Greisberger, He. Morscheid, M. T. Dokulil & H. Morscheid, 2003. Steady state phytoplankton in a deep pre-alpine lake: Species and pigments of epilimnetic versus metalimnetic assemblages. Hydrobiologia 502 (Dev. Hydrobiol. 172): 49–64.

Tolotti, M., H. Thies, M. Cantonati, C. M. E. Hansen & B. Thaler, 2003. Flagellate algae (Chrysophyceae, Dinophyceae, Cryptophyceae) in 48 high mountain lakes of the northern and southern slope of the eastern Alps: Biodiversity, distribution of taxa, and their driving variables. Hydrobiologia 502 (Dev. Hydrobiol. 172): 331–348.

Whittaker 1975: Communities and Ecosystems. Macmillan, New York.

Whittaker, R. H. & Levin, S. A. 1977: The role of mosaic phenomena in natural communities. Theor. Popul. Biol. 12: 117–139.

Willén, E., 2003. Dominance patterns of planktonic algae in Swedish forest lakes. Hydrobiologia 502 (Dev. Hydrobiol. 172): 315–324.

GPSR Compliance

The European Union's (EU) General Product Safety Regulation (GPSR) is a set of rules that requires consumer products to be safe and our obligations to ensure this.

If you have any concerns about our products, you can contact us on ProductSafety@springernature.com

In case Publisher is established outside the EU, the EU authorized representative is:

Springer Nature Customer Service Center GmbH
Europaplatz 3
69115 Heidelberg, Germany

Batch number: 10800927

Printed by Printforce, the Netherlands